Selected Scientific Papers
of E.U. Condon

Edward Uhler Condon (1902-1974)

Asim O. Barut   Halis Odabasi
Alwyn van der Merwe

Editors

Selected Scientific Papers of E.U. Condon

Springer-Verlag
New York   Berlin   Heidelberg   London
Paris   Tokyo   Hong Kong   Barcelona

Asim O. Barut
Department of Physics
University of Colorado
Boulder, CO 80309-0390
USA

Halis Odabasi
Department of Physics
University of Colorado
Boulder, CO 80309-0390
USA

Alwyn van der Merwe
Department of Physics
University of Denver
Denver, CO 80208
USA

Library of Congress Cataloging-in-Publication Data
Condon, Edward Uhler, 1902–1974.
  [Selections. 1990]
  E.U. Condon: selected scientific papers / Asim O. Barut, Halis
Odabasi, Alwyn van der Merwe, editors.
    p. cm.
  Includes bibliographical references.
  ISBN-13: 978-1-4613-9085-5   e-ISBN-13: 978-1-4613-9083-1
  DOI: 10.1007/978-1-4613-9083-1

  1. Physics. I. Barut, A.O. (Asim Orhan), 1926– .
II. Odabasi, Halis. III. Van der Merwe, Alwyn. IV. Title.
QC71.C6482   1990
530—dc20   90-45451

Printed on acid-free paper.

Typeset by Asco Trade Typesetting Ltd, Hong Kong.

9 8 7 6 5 4 3 2 1

Preface

"The middle third of the twentieth century was the era of hegemony of physics in American science. During that whole period Edward Uhler Condon was a leader in physics, in research of his own, in stimulating research in others, in applying physics, and in calling attention to the effect on all of us of its indiscriminate and irrational application. When he made his first contribution to theoretical physics in 1926, the word physics was not in the vocabulary of most Americans and the evolutionary concepts of quantum mechanics and relativity were just being worked out in Europe; by 1960 the applications of electronics and solid state physics had begun to change our lives irreversibly, and the implications of nuclear physics were manifest to everyone. Ed Condon contributed to each part of this explosive evolution."—Philip M. Morse.[1]

We are happy to present the Selected Papers of Edward Uhler Condon, one of the most influential American physicists of this century, whose life has touched many fields and many people including our own. The papers are divided into two volumes: this volume of scientific papers, and a companion volume of popular writings. The papers in this volume are arranged chronologically.

Since his death, several obituaries of Condon appeared,[1-3] and the Fourth International Conference on Atomic Physics was dedicated to his memory. The Proceedings of this conference[4] contains contributions by L. M. Branscomb. I. I. Rabi, and B. R. Judd about Condon's accomplishments as a scientist and as a very concerned humanist. Another book,[5] edited as a tribute to Condon on the occasion of his becoming Professor Emeritus in the Department of Physics and Astrophysics in the University of Colorado in the summer of 1970, also has pertinent and detailed information in its preface and in its foreword written by Frederick Seitz. Therefore, rather than repeating the details, we refer the interested reader to these sources, which are reprinted in the companion volume of popular writings.

As listed here, Condon published over one hundred and ten papers; most of them are scientific; a few deal with the social issues related to the applications of physics. He co-authored three books, one of which was published six years after his death, and co-edited a handbook. Perhaps because the book *The Theory of Atomic Spectra* by E. U. Condon and G. H. Shortley has dominated atomic spectroscopy over the last forty-six years and has probably attracted more readers than any of his other writings, Condon is best known for his contributions to atomic spectroscopy. But he actually wrote, mostly with his students, only eleven papers on this subject. As the papers contained in this book attest, he contributed significantly to many other fields of modern physics. Among his most important contributions we list:

(1) His Ph.D. thesis, written in six days and based on the extension of some work done by James Franck, laid the foundation of the pre-quantum mechanical version of what later became widely known in physics and chemistry as the *Franck-Condon principle* (paper 7). The wave-mechanical basis of this principle was given in later papers.

(2) His approach in calculating the binding energy of the ground state of hydrogen molecule by means of molecular orbitals was first of its kind[6] (papers 12 and 13).

(3) With H. D. Smyth, he gave an interpretation of the continuous spectrum of molecular hydrogen (paper 22).

(4) With Ronald Gurney, he proposed the α-particle tunnelling from the nucleus—the first application of quantum mechanics to the details of inner structure of atomic nuclei (papers 19 and 23) (independently discovered by George Gamow).

(5) A suggestion of Condon led to the discovery of a mode of dissociation of molecules under electron impact in which the fragments acquire a considerable amount of kinetic energy[7] (paper 26).

(6) Two papers, one with Benedict Cassen and the other with Gregory Breit and R. D. Present, were the first to draw attention to the extra significance of isotopic spin because of the charge independence of nuclear interactions (papers 50, 51).

(7) He contributed to the theory of mass spectroscopy and infra-red spectroscopy (the latter with R. Bowling Barns) and to the theory of the optical rotatory power of molecules (with Henry Eyring and William Altar) (papers 52 and 58).

Besides being an outstanding scientist, Condon was also an extraordinary person. He was warm, considerate, friendly, generous, very concerned, and very humorous. He was an admired, dear friend to one of us (A.O.B.) and like a father to the other (H.O.). We cherish his memory dearly.

We would like to take this opportunity to acknowledge a grant from the Corning Glass Foundation towards the publication of this book. Thanks are due to Mr. Stephen Catlett, manuscript librarian at the American Philosophical Society Library, where the papers of Professor E. U. Condon are preserved, for his friendly help. We would also like to thank the Condon family and others for permission to edit the papers included in this book.

References

1. P. M. Morse, *Rev. Mod. Phys.* 47, 1 (1975).
2. L. M. Branscomb, *Phys. Today* 27, 68 (1974).
3. C. Eisenhart, *Dimensions* (the Technical News Bulletin of the National Bureau of Standards) 58, 151 (1974).
4. *Atomic Physics*, Vol. 4, G. Z. Putlitz, E. W. Weber, and A. Winnacker, eds. (Plenum, New York, 1975).
5. *Topics in Modern Physics—A Tribute to Edward U. Condon*, W. E. Brittin and H. Odabasi, eds. (Colorado Associated University Press, Boulder, 1971).
6. Linus Pauling, *The Nature of the Chemical Bond*, 3rd edn. (Cornell University Press, Ithaca, 1960), p. 23.
7. W. Bleakney, *Phys. Rev.* 35, 1180 (1930).

Asim O. Barut
Halis Odabasi
Alwyn van der Merwe

Scientific Publications of Edward U. Condon

1. "An erroneous experiment in gaseous diffusion," *School Science and Mathematics 23*, 415–6 (1923).
2. "A possible manifestation of directional hysteresis in iron (abstract)," *Phys. Rev. 23*, 665 (1924).
3. "The age of the stars," *Proc. Nat. Acad. Sci. 11*, 125–30 (1925); also letter in *Nature*, March 21, 1925.
4. "The theory of the radiometer" (with H.E. March and L.B. Loeb), *J. Opt. Soc. Am. 11*, 257–62 (1925).
5. "Curiosities of Mathematics," *Little Blue Book No. 876* (Haldeman-Julius, Girard, Kansas, 1925).
6. "Theory of the range of alpha-particles" (with L.B. Loeb), *J. Franklin Inst. 200*, 595–607 (1925).
7. "A theory of intensity distribution in band systems" (Ph.D. thesis), *Phys. Rev. 28*, 1182–1201 (1926).
8. "Mean free paths in a gas whose molecules are attracting rigid elastic spheres" (with E.V. Van Amringe), *Phil. Mag. 3* (7) 604–14 (1927).
9. "Remarks on penetrating radiation," *Proc. Nat. Acad. Sci. 12*, 323–6 (1926).
10. "The rapid fitting of a certain class of empirical formulae by the method of least squares," *Univ. Calif. Publ. Math. 2*, 55–66 (1927).
11. "Coupling of electronic and nuclear motions in diatomic molecules," *Proc. Nat. Acad. Sci. 13*, 462–6 (1927).
12. "Wave mechanics and the normal state of the hydrogen molecule," *Proc. Nat. Acad. Sci. 13*, 466–70 (1927).
13. "Über den Grundzustand des Wasserstoffmoleküls nach der Wellenmechanik," *Verh. Deut. Phys. Ges. 8*, 19–21 (1927).
14. "The Zeeman effect of the symmetrical top according to wave mechanics," *Phys. Rev. 30*, 781–4 (1927).
15. "Statistics of vocabulary," *Science 67*, 300 (1928).
16. "The physical pendulum in quantum mechanics," *Phys. Rev. 31*, 891–4 (1928).
17. "Quantum phenomena in the biological action of X-rays" (with H.M. Terrill), *J. Cancer Res. 11*, 324–33 (1927).

18. "Recent developments in quantum mechanics," *Science 68*, 193–5 (1928).
19. "Wave mechanics and radioactive disintegration" (with R.W. Gurney), *Nature 122*, 439 (1928).
20. "Nuclear motions associated with electron transitions in diatomic molecules," *Phys. Rev. 32*, 858–72 (1928).
21. "Quantum mechanics of momentum space," *J. Franklin Inst. 207*, 467–74 (1929).
22. "The critical potentials of molecular hydrogen" (with H.D. Smyth), *Proc. Nat. Acad. Sci. 14*, 871–5 (1928).
23. "Quantum mechanics and radioactive disintegration" (with R.W. Gurney), *Phys. Rev. 33*, 127–40 (1929).
24. "Remarks on uncertainty principles," *Science 69*, 573–4 (1929).
25. *Quantum Mechanics* (with P.M. Morse) (McGraw-Hill, New York, 1929); reprinted 1964.
26. "Complete dissociation of H_2," *Phys. Rev. 35*, 658 (1930) (abstract). See complete paper by W. Bleakney, *Phys. Rev. 35*, 1180–6 (1930).
27. "An interpretation of Pauli's exclusion principle" (with J.E. Mack), *Phys. Rev. 35*, 579–82 (1930).
28. "A cosmological conjecture" (with J.E. Mack), *Nature 125*, 455 (1930).
29. "Predissociation of diatomic molecules from high rotational states" (with D.S. Villars), *Phys. Rev. 35*, 1028–32 (1930).
30. "Singlet-triplet interval ratios in sp, sd, sf, p^5s and d^9s configurations" (with G.H. Shortley), *Phys. Rev. 35*, 1342–6 (1930).
31. "The theory of complex spectra," *Phys. Rev. 36*, 1121–33 (1930).
32. "The theory of complex spectra II" (with G.H. Shortley), *Phys. Rev. 37*, 1025–43 (1931).
33. "Quantum mechanics of collision processes," *Rev. Mod. Phys. 3*, 43–88 (1931).
34. "Quantum phenomena in the biological effects of radiant energy," *J. Franklin Inst. 214*, 105–6 (1932).
35. "Lorentz double refraction in the regular system" (with F. Seitz), *J. Opt. Soc. Am. 22*, 393–401 (1932).
36. "The spin of the neutron" (with R.F. Bacher), *Phys. Rev. 41*, 683–5 (1932).
37. "Production of infra-red spectra with electric fields," *Phys. Rev. 41*, 759–62 (1932).
38. "Note on the velocity of sound," *Am. Phys. Teacher 1*, 18 (1933).
39. "Notes on the Stark effect," *Phys. Rev. 43*, 648–54 (1933).
40. "Food and the theory of probability," *Proc. U.S. Naval Inst. 60*, 75–8 (1934).
41. "Relative multiplet transition probabilities from spectroscopic stability" (with C.W. Ufford), *Phys. Rev. 44*, 740–3 (1933).
42. "The absolute intensity of the nebular lines," *Astrophys. J. 79*, 217–34 (1934).
43. "Where do we live? Reflections on physical units and universal constants," *Am. Phys. Teacher 2*, 63–9 (1934).

44. *The Theory of Atomic Spectra* (with G.H. Shortley) (Cambridge University Press, Cambridge, 1935).
45. "The energy distribution of neutrons slowed by elastic impacts (with G. Breit)," *Phys. Rev. 49*, 229–31 (1936).
46. "Note on electron-neutron interaction," *Phys. Rev. 49*, 459–61 (1936).
47. "Three catch questions," *Am. Phys. Teacher 3*, 85–6 (1935).
48. "The photoelectric effect of the deuteron" (with G. Breit, *Phys. Rev. 49*, 904–11 (1936); correction: (with G. Breit and J.R. Stehn) *51*, 56 (1937).
49. "Interaction between protons as indicated by scattering experiments" (abstract) (with G. Breit), *Phys. Rev. 49*, 866 (1936).
50. "On nuclear forces" (with B. Cassen), *Phys. Rev. 50*, 846–9 (1936).
51. "Theory of scattering of protons by protons" (with G. Breit and R.D. Present), *Phys. Rev. 50*, 825–45 (1936).
52. "Vibration spectra and molecular structure. I. General remarks and a study of the spectrum of the OH group" (with R.B. Barnes and L.G. Bonner), *J. Chem. Phys. 4*, 772–8 (1936).
53. "Ionization and of issociation of molecules by electron impact" (with W. Bleakney and L.G. Smith), *J. Phys. Chem. 41*, 197–208 (1937).
54. "Immersion of the Fourier transform in a continuous group of functional transformations," *Proc. Nat. Acad. Sci. 23*, 158–64 (1937).
55. "A new approach to the Hermite polynomials" (with R. Greenwood), *Phil. Mag. 24* (7), 281–7 (1937).
56. "Interaction between light nuclei" (abstract) (with M. Phillips and L. Eisenbud), *Phys. Rev. 51*, 382 (1937).
57. "Theories of optical rotatory power," *Rev. Mod. Phys. 9*, 432–57 (1937).
58. "One-electron rotatory power" (with W. Altar and H. Eyring), *J. Chem. Phys. 5*, 753–75 (1937)
59. "Mathematical models in modern physics," *J. Franklin Inst. 225*, 255–61 (1938).
60. "The theory of nuclear structure," *J. Franklin Inst. 227*, 801–16 (1939).
61. "A simple derivation of the Maxwell-Boltzmann law," *Phys. Rev. 54*, 937–40 (1938).
62. "Note on the external photoelectric effect of semi-conductors," *Phys. Rev. 54*, 1089–91 (1938).
63. "Electronic generation of electromagnetic oscillations," *J. Appl. Phys. 11*, 502–6 (1940).
64. "Forced oscillations in cavity resonators," *J. Appl. Phys. 12*, 129–32 (1941).
65. "Principles of micro-wave radio," *Rev. Mod. Phys. 14*, 341–89 (1942).
66. "Physics in industry," *Science 96*, 172–4 (1942).
67. "A physicist's peace," *Am. J. Phys. 10*, 96–7 (1942).
68. "The development of high voltage for the production of neutrons and artificial radioactivity," *Ohio J. Sci. 41*, 131–4 (1941).
69. "Space-charge relations in the magnetron with plane electrodes (abstract)," *Proc. Inst. Radio Engrs. 29*, 664 (1941).

70. "Detection of metastable ions with the mass spectrometer" (with J.A. Hipple), *Phys. Rev. 68*, 54–5 (1945).
71. "Nuclear engineering," *Westinghouse Engineer* 1–7, November 1945.
72. "Study of metastable ions with the mass spectrometer" (abstract) (with J.A. Hipple), *Phys. Rev. 69*, 257 (1946).
73. "Metastable ions formed by electron impact in hydrocarbon gases" (with J.A. Hipple and R.E. Fox), *Phys. Rev. 69*, 347–56 (1946).
74. "Foundations of nuclear physics," *Nucleonics 1*, 3–11 (1947).
75. "The Franck-Condon principle and related topics," *Am. J. Phys. 15*, 365–74 (1947). (Retiring address as President of the American Physical Society, 1946.)
76. "Science and the national welfare," *Science, 107*, 2–7 (1948).
77. "Electronics and the future," *Electrical Engineering*, 1–7, April 1947.
78. "Investigation of the attractive forces between the persistent currents in a superconductor and the lattice" (with E. Maxwell), *Phys. Rev. 76*, 578 (1949).
79. "Superconductivity and the Bohr magneton," *Proc. Nat. Acad. Sci. 35*, 488–90 (1949).
80. "Is there a science of Instrumentation?" *Science 110*, 339–42 (1949).
81. "Effect of oscillations of the case on the rate of a watch" (with P.E. Condon), *Am. J. Phys. 16*, 14–6 (1948).
82. "The development of American physics," *Am. J. Phys. 17*, 404–8 (1949).
83. "Present program of the National Bureau of Standards," *Sci. Monthly 73*, 176–82 (1951).
84. "Evolution of the quantum theory," *Sci. Monthly 72*, 217–22 (1951).
85. "Some thoughts on science in the federal government," *Phys. Today 5*, 6–13 (1952).
86. "Physics of the glassy state: I. Constitution and structure," *Am. J. Phys. 22*, 43–53 (1954).
87. "Physics of the glassy state: II. The transformation range," *Am. J. Phys. 22*, 132–42 (1954).
88. "Physics of the glassy state: III. Strength of Glass," *Am. J. Phys. 22*, 224–32 (1954).
89. "Physics of the glassy state: IV. Radiation-sensitive glasses," *Am. J. Phys. 22*, 310–17 (1954).
90. "A half-century of quantum physics," *Science 121*, 221–6 (1955).
91. "Flow of liquid helium through porous Vycor glass" (with K.R. Atkins and H. Seki), *Phys. Rev. 102*, 582 (1956).
92. *Handbook of Physics* (with H. Odishaw) (McGraw-Hill, New York, 1958); 2nd edn., 1967.
93. "What is this new world?" *J. Nat. Assoc. Woman Deans and Counselors 21*, 153–61 (1958).
94. "Graphical representation for unit systems," *Am. J. Phys. 29*, 487–91 (1961).
95. "Intermediate courses in physics," *Am. J. Phys. 30*, 166–71 (1962).

96. "Sixty years of quantum physics," *Phys. Today 15*, 37–48 (1962); *Bull. Phil. Soc. Washington 16*, 82–102 (1962). (Retiring presidential address to Philosophical Society of Washington, Dec. 2, 1960.)

97. "Intensity-dependent absorption of light," *Proc. Nat. Acad. Sci. 52*, 635–7 (1964).

98. "Prof. A.H. Compton" (biographical note), *Nature 194*, 628–9 (1962).

99. "Lyman James Briggs" (biographical memoirs), *Yearbook American Philosophical Society* (New York, 1963), pp. 117–21.

100. "Spin-Orbit Interaction in self-consistent fields" (with H. Odabasi), in *Quantum Theory of Atoms, Molecules, Solid State* (Academic, New York, 1966), pp. 185–201.

101. "William Shipley's barometer (1748)," *Am. J. Phys. 34*, 358 (1966).

102. "Measures for progress," *Appl. Opt. 6*, 9–12 (1967). (Book review.).

103. "Radiative transport in hot glass," *J. Quant. Spectros. Rad. Transfer 8*, 369–85 (1968).

104. "On pair correlation in the theory of atomic structure," *Rev. Mod. Phys. 40*, 872–5 (1968).

105. "The past and the future of the *Reviews of Modern Physics*," *Rev. Mod. Phys. 40*, 876–8 (1968).

106. *Scientific Study of Unidentified Flying Objects* (paperback: Bantam Books, New York, 1969; hardcover: Dutton, New York, 1969).

107. "Self-consistent field calculations for energy levels of 4, 5, 6, 7, 8, 14, 15, and 16 electron isoelectronic sequences" (with H. Odabasi), *J. Opt. Soc. Am. 59*, 658 (1969). (Letter to the Editor calling attention to detailed calculations contained in JILA reports Nos. 95 and 97.)

108. "OFO's I have loved and lost," *Proc. Am. Philos. Soc. 113*, 425–7 (1969).

109. "Self-consistent field calculations of the $1s^22s^22p3p$ configuration in the carbon isoelectronic sequence" (with H. Odabasi), in *In Honor of Philip M. Morse*, H. Fesbach and K.U. Ingard, eds. (MIT Press, Cambridge, Mass., 1969), pp. 5–20.

110. "Education for world understanding," in *Topics in Modern Physics* (Colorado Associated University Press, Boulder, 1971), pp. 349–53.

111. "Reminiscences of a life in and out of quantum mechanics," *Int. J. Quant. Chem. 7*, 7–22 (1973).

112. "Tunnelling—how it all started," *Am. J. Phys. 46*, 319–23 (1978).

113. *Atomic Structure* (with H. Odabasi) (Cambridge University Press, New York, 1980).

114. "Silver anniversary of atomic clocks" (unpublished speech).

Contents

A Theory of Intensity Distribution in Band Systems[*]

EDWARD U. CONDON

Department of Physics
University of California

ABSTRACT

A theory of the relative intensity of the various bands in a system of electronic bands is developed by an extension of an idea used by Franck in discussing the dissociation of molecules by light absorption. The theory predicts the existence of two especially favored values of the change in the vibrational quantum numbers, in accord with the empirical facts as discussed by Birge.

A means of calculating the intensity distribution from the known constants of the molecule is presented and shown to be in semi-quantitative agreement with the facts in the case of the rollowing band systems: SiN, AlO, CO (fourth positive group of carbon), I_2 (absorption), O_2 (Schumann-Runge system), CN (violet system), CO (first negative group of carbon), N_2 (second positive group of nitrogen), and N_2 (first negative group of nitrogen).

In the case of I_2 there is a discrepancy, if Loomis' assignment of n'' values is used, which does not appear if Mecke's original assignment is used. It is suggested that at least some of the lower levels postulated by Mecke are real but that absorption from them always results in dissociation of the molecule and so they are not represented in the quantized absorption spectrum.

1. Introduction

The problem of the explanation of relative or absolute intensities of spectral lines stands out today as one of the most important in the whole field of spectroscopy. As yet the problem has hardly received the necessary attention from experimental workers to make attempts at quantitative theoretical treatment profitable. The experimental difficulties of constructing controllable sources of excited materials seem to be yielding but slowly to the tremendous efforts toward their removal which are now being made. However, as is known, considerable progress has already been made in the discussion of

[*] A preliminary account of this paper was presented at the meeting of the Pacific Coast Section of the American Physical Society at Stanford University in March, 1926.

relative intensities in the fine structure of lines (notably hydrogen Balmer lines in the Stark effect) and more recently in the relative intensities in multiplets in line spectra and of the various lines in the fine structure of individual bands.[1]

When the fine structure of lines or bands is under discussion the experimental problem is considerably easier, for in this case the difficulties of the unknown variation of photographic sensitivity with wavelength do not make themselves felt appreciably. But when relative intensities are desired over long stretches of the spectrum, as between the members of a series of lines in atomic spectra or as in a system of bands in molecular spectra, the difficulties of photometric work present themselves in their full force. The result is that today there are few quantitative data in this field. In the case of the individual bands of a system,—the special subject of this article,—one has only the rough visual estimates of the observers; sometimes nothing better than a knowledge of the presence or absence of the band in question under very roughly described conditions of excitation and observation.

It seems worth while, however, to undertake an attempt at a theoretical correlation of the outstanding characteristics of the observed data with the quantum theory model for the emitter of the band spectra of diatomic molecules. The theory presented here is one of the relative likelihood of the various possible changes in the vibrational quantum number and makes no attempt to explain the distribution of molecules in the initial state. The distribution in the initial state, in cases approaching thermal equilibrium, should be governed by the Maxwell-Boltzmann distribution law. This is apparently the case *within the accuracy of the known data* for absorption spectra, where deviations from thermal equilibrium are least important.[2] The Maxwell-Boltzmann law seems also to hold remarkably well in some cases of violent electrical excitation in discharge tubes where one hardly expects to find thermal equilibrium. Perhaps this is a reflection of the fact that the Maxwell-Boltzmann law is one of a sort of maximum chaos, and it matters not whether the chaos results from disordered heat motions or irregular electrical conditions. To be more concrete, if n' is the vibration quantum number of the excited electronic state of the molecule and n'' that of the normal state of the molecule, then the theory presented here undertakes the discussion of what Birge has called an n' progression for the emission process or an n'' progression for the absorption process.

One expects the considerations of which use is made in the model discussed to be of an approximate nature only. They are based on certain kinematic relationships existing between the vibratory motions in the two electronic states concerned. As the simple mode of discussing the model is one which, without quantum theory, would yield fractional changes in the vibrational

---

[1] Sommerfeld, Atombau und Spektrallinien. 4th ed., Chap. 5.

[2] Chapter IV, Sec. 4, of National Research Council Report on Molecular Spectra (in press). The writer is indebted to Professor R. T. Birge for the opportunity to use the manuscript copy of this report.

quantum numbers, the question arises as to the proper modification of the results of the continuous theory in order to fit the discontinuous quantum theory. There is as yet no known way of treating this problem so the present results are necessarily approximate.

In view of the fact that the latest development in the disentanglement of the quantum enigma is in the direction of the substitution of another kinematics[3] for the description of mechanical systems whose sizes are of the order of Angstrom units it is thought best not to dwell too long on the attempt to formulate this treatment exactly. Rather it is recorded here as giving results in striliing and partially satisfactory agreement with the facts although with the expectation that a better treatment must come through the application of the newer kinematics of Heisenberg, Born, and Jordan.

The modern quantum theory interpretation of electronic bands as presented in Sommerfeld's "Atombau"[4] and in the National Research Council Report on Molecular Spectra (in press) is adopted in this paper. The quantization of the motion of a diatomic molecule leads to energy levels associated with each electronic configuration which are specified by a rotational and a vibrational quantum number, m and n respectively, as given by the following formulas:

$$\frac{E}{hc} = A_n = B_n m^2 = D_n m^4 = F_n m^6 = \cdots$$

$$A_n = \omega^0 n (1 - \alpha n + \cdots)$$

$$B_n = B_0 - \alpha n + \cdots$$

$$D_n = D_0 + \cdots$$

All of the expressions are, in fact, infinite series but the terms indicated suffice for the analysis of most band systems. The derivatives of the potential energy curve at its equilibrium point are given in terms of the empirically determined, ω^0, x, B_0, $\alpha \ldots$ by the formulas:[5]

$$V'(r_0) = 0$$

$$V''(r_0) = c^2 (2\pi\omega^0)^2 \mu$$

$$V'''(r_0) = -\frac{6V''(r_0)\pi\sqrt{2B_0\mu c}}{\sqrt{h}}\left(\frac{\alpha\omega^0}{6B_0^2} + 1\right)$$

$$V^{IV}(r_0) = \frac{2}{V''(r_0)}\left[\frac{5}{6}[V'''(r_0)]^2 - 8\frac{x}{ch\omega_0}[V''(r_0)]^3\right]$$

[3] Heisenberg, Zeits. f. Physik **33**, 879 (1925); Born and Jordan, Zeits. f. Physik **34**, 858 (1925); L. de Broglie, Ann. d. Physique **3**, 22 (1925); Schrödinger, Ann. der Physik **79**, 459 (1926).
[4] Sommerfeld, Atombau, 4th ed., Chap. 9.
[5] The formulas are given by Born and Hückel, Phys. Zeits. **24**, 1 (1925). The nomenclature used here is that of the Report on Molecular Spectra.

in which c is the velocity of light and μ is the harmonic mean of the masses of the two atoms in the molecule.

2. General Theory for Abrupt Structural Changes

The theory of transition probability which will now be developed is an outgrowth of a picture proposed by Franck for a mechanism for the dissociation of molecules by absorption of light.[6] Briefly Franck's picture is as follows:

In a "cold" gas the molecules are not vibrating and are in their lowest electronic level so that the nuclear motions are governed by $V_1(r)$. If now a light quantum is absorbed which is of sumcient energy to bring the molecule into an electronic excited state, it is natural to suppose that in thus changing the electronic energy of the molecule no other specific action is exerted by the light on the molecule. The absorption of light merely substitutes a new law of nuclear interaction, say $V_2(r)$ for the old one $V_1(r)$. But since this new one has a different equilibrium position, the atoms, at the instant after the absorption, will be away from equilibrium and so start to vibrate. Should it happen that

$$V_2(r_{01}) > V_{2\,\text{max}}$$

where $V_{2\,\text{max}}$ is the maximum value of $V_2(r)$ in the range $r_{01} < r < \infty$, then the molecule will tend to execute oscillations of infinite amplitude, i.e., the molecule will dissociate. Thus Franck contemplates the photo-chemical dissociation of a molccule by a simple absorption of light.

It is natural to extend this point of view to the effect of an electron transition in a molecule either in absorption or emission and whether the molecule be vibrating in the initial state or not. The electron transition is supposed to happen in a negligibly short time as compared to the period of the nuclear vibrations. If the transition occurs at the instant when the separation of the atoms is r and the relative momentum is p, then one supposes that the transition does not alter the instantaneous values of r and p, but merely substitutes a new potential energy function, say $V_2(r)$, for the old one, say $V_1(r)$. The values of r and p, at the instant of transition determine exactly a vibrational motion in the final state.

In general, of course, the vibrational motion so determined is not one of the states allowed by the quantum conditions. It becomes necessary, therefore, to suppose that the vibrational motion of the hnal state is not strictly governed by this principle but that it merely tends to take up the quantized vibration nearest to the one indicated by this principle. Here a certain ambiguity appears which is akin to that which occurs in all other cases where attempts are made to reason accurately from properties of the orbits, other than their energies.

It is easy to give a general analytical formulation of the action governing the vibrational transition. Disregarding rotation and writing p for p_r, and q

---

[6] Franck, Trans. Faraday Society (1925).

for r one has for the Hamiltonian functions of the nitial and final states:

$$H_1 = \frac{p^2}{2\mu} + V_1(q)$$

$$H_2 = \frac{p^2}{2\mu} + V_2(q)$$

If the motion in either state is solved by the Hamilton-Jacobi method leading to the introduction of the action and angle variables, $J_1 w_1$ for the initial state and $J_2 w_2$ for the final state one finds relations of the form

$$p = p_1(J_1, w_1)$$

for the initial state,

$$q = q_1(J_1, w_1)$$

and similar relations for the final state. On the assumption underlying the work, if the electron transition akes place at an instant when the initial values of J and w are J_1, w_1 then the values of J and w after the transition, i.e., J_2 and w_2 will be given by the equations:

$$p_2(J_2, w_2) = p_1(J_1, w_1)$$

$$q_2(J_2, w_2) = q_1(J_1, w_1)$$

Or solving these equations for J_2 and w_2 one finds that:

$$J_2 = J_2(J_1, w_1)$$

$$w_2 = w_2(J_1, w_1)$$

so that the final motion is wholly determined by the initial motion. By the quantum conditions J_1 equals $n_1 h$ and J_2 equals $n_2 h$ where n_1 and n_2 are integers. It is thus seen that the solution for J_2 will not, in general, give an integral value. The phase of the final motion as given by w_2 plays no rôle in the spectroscopic theory. Moreover w_1 enters through a function which is periodic, with period 1 with respect to this variable. Thus the full range of possible values of J_2 corresponding to any J_1 is obtained by letting w_1 range from 0 to 1. From the general dynamical theory w_1 increases linearly with the time. Failing any other indications on this point, one may say that the electron transition is just as likely to occur at one instant as at another, and thus there is a definite probability assigned to each range of J_2 to $J_2 + dJ_2$.

It is clear that if all values of w_1 are equally probable then the values of J_2 corresponding to the small values of dJ_2/dw_1 will be stronglyweighted. But $dJ_2/dw_1 = 0$ give the maxima and minima of the possible transitions so that these extremes are to be regarded as more probable than the other allowed jumps.

But the uncertainty of how to treat the fractional n_2 and the inaccuracy of the experimental data make it fruitless to attempt to consider such detailed questions as the accurate relative probabilities of the transitions. For a given J_1 there will be a maximum and minimum value of J_2 given by the theory.

Moreover the most probable transitions, it will appear later, are to the extremes of J_2 if the times of electron transition be taken as equally probable. Therefore in discussing the relationship of the theory to ernpirical facts, only the extremes of J_2 for each J_1 have been computed.

Looking over the theory as just developed, it is to be emphasized that at least the extrema of J_2 for each J_1 are governed entirely by the energy functions $V_1(r)$ and $V_2(r)$. These in turn, as has been pointed out, are fixed bv the positions of the bands in the spectrum. Thus there are no adjustable constants appearing in the theory so that *the vibrational transition probabilities are sharply correlated with the structure of the band system as regards position in the spectrum.*

The actual solution of the equations which present themselves in the case of the non-harmonic oscillator is prohibitively lengthy. In the following section the explicit formulas are developed for the case in which all derivatives of $V_2(r)$ and $V_1(r)$, higher than thesecond, vanish (harmonic oscillator). A graphical mode of treating the non-harmonic case is also presented.

3. Explicit Formulas for the Harmonic Oscillator

In the case in which $V_1(r)$ and $V_2(r)$ are parabolic the modes of vibration are simple harmonic in both initial and final states and the explicit development of the formulas is quite simple. Thus let us assume:

$$V_1(q) = \tfrac{1}{2}K_1(q - q_{01})^2$$
$$V_2(q) = \tfrac{1}{2}K_2(q - q_{02})^2$$

The solution for p and q in terms of J and w for the harmonic oscillator is well-known.[7] One has for the final state:

$$p_2(J_2, w_2) = \sqrt{2\mu v_2 J_2}\cos 2\pi w_2$$
$$q_2(J_2, w_2) = q_{02} + \frac{1}{2\pi}\sqrt{\frac{2J_2}{\mu v_2}}\sin 2\pi w_2$$

in which the frequency v_2 is related to the mass μ and the spring constant K_2 by

$$v_2 = \frac{1}{2\pi}\sqrt{\frac{K_2}{\mu}}$$

Similarly for the initial state:

$$p_1(J_1 w_1) = \sqrt{2\mu v_1 J_1}\cos 2\pi w_1$$
$$q_1(J_1 w_1) = q_{01} + \frac{1}{2\pi}\sqrt{\frac{2J_1}{\mu v_1}}\sin 2\pi w_1$$

[7] Born, Vorlesungen über Atommechanik, pp. 39, 57.

Equating the values of p and q at the instant of the electron transition and writing d for $q_{01} - q_{02}$:

$$\sqrt{2\mu v_2 J_2} \cos 2\pi w_2 = \sqrt{2\mu v_1 J_1} \cos 2\pi w_1$$

$$\frac{1}{2\pi}\sqrt{\frac{2J_2}{\mu v_2}} \sin 2\pi w_2 = d + \frac{1}{2\pi}\sqrt{\frac{2J_1}{\mu v_1}} \sin 2\pi w_1$$

Solving for J_2 in terms of J_1 and w_1 by elimination of w_2:

$$J_2 = 2\pi^2 \mu v_2 d^2 + \frac{1}{2}\left(\frac{v_1}{v_2} + \frac{v_2}{v_1}\right)J_1 + 2d\sqrt{2\pi^2 \mu v_2 \frac{v_2}{v_1}} \sin 2\pi w_1 \cdot \sqrt{J_1}$$

$$+ \frac{1}{2}\left(\frac{v_1}{v_2} - \frac{v_2}{v_1}\right)\cos 4\pi w_1 \cdot J_1$$

The solution for w_2 in terms of J_1 and w_1 is readily obtained but it is not needed. Introducing the quantum integers n_1 and n_2 in place of J_1 and J_2:

$$n_2 = \frac{2\pi^2 \mu v^2}{h} d^2 + \frac{1}{2}\left(\frac{v_1}{v_2} + \frac{v_2}{v_1}\right)n_1 + 2d\sqrt{\frac{2\pi^2 \mu v_2}{h}\cdot\frac{v_2}{v_1}} \sin 2\pi w_1 \cdot \sqrt{n_1}$$

$$+ \frac{1}{2}\left(\frac{v_1}{v_2} - \frac{v_2}{v_1}\right)\cos 4\pi w_1 \cdot n_1$$

For brevity this will be written:

$$n_2 = A + Bn_1 + C\sqrt{n_1}\sin\theta + Dn_1\cos 2\theta,$$

defining the coefficients A, B, C, D by comparison with the preceding equation. The extreme values of n_1 corresponding to any n_2 are found by the roots of $dn_1/d\theta = 0$ in the usual way. If $|C/4D\sqrt{n_1}| > 1$

$$n_2 = A + Bn_1 \pm C\sqrt{n_1} \mp Dn_1$$

and if $|C/4D\sqrt{n_1}| < 1$ in addition to these also secondary maxima, given by $n_2 = (A - C^2/8D) \pm C^2/4D + (B + D)n_1$, the $\pm$ sign being taken the same as the sign of $C/4D$.

As has already been remarked, the extreme allowed transitions are the most probable ones and therefore these equations for the extrema give the bands which one expects to be strongest in a band system.

The formulas for the non-harmonic oscillator can be obtained by successive approximations in a purely analytical way but it is easier to resort to an approximate graphical method. $V_1(r)$ and $V_2(r)$ are supposed to have been determined, in the neighborhood of r_{01} and r_{02} respectively from the analysis of the energy levels of the band system. Suppose these plotted on the same piece of paper. Also suppose the vibrational energy levels marked off as in Fig. 1. Then when the molecule vibrates in either electronic state with a known number of quanta it is easy to see what the amplitude of the motion is, for from the energy integral, $p = 0$ when

$$V(r) = W$$

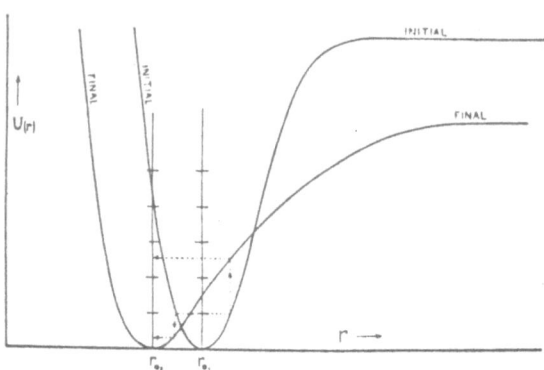

FIGURE 1. Typical relation of the potential energy curves, illustrating graphical method of finding favored transitions.

if W is the energy of the vibratory motion. It is at the two extreme positions that the vibrator spends most of the time, i.e., the electronic transitinn is most likely to occur when the vibrator is in one of these extreme positions. At these extreme values of r for a vibratory motion given by $V_2(r)$, if $V_1(r)$ is suddenly made the law governing the motion then one sees directly from the figure what will be the amplitude and hence the energy of the new motion. Thus two most probable transitions are determined for each value of the quantum number in the initial state.

This graphical method, as will appear later, indicates how very sensitive the theory is to slight uncertainties in the functions, $V(r)$. In the next section are given the details of the application of the theory to all systems for which the necessary data are available.

4. Application to Known Systems

It will now be seen how far the model employed accounts for the main observed features of intensity distribution in band systems. These have been discussed by Birge[8] who points out that it is a general rule that for each value of n' (the quantum number of the initial state for emission) there are two preferred values of $n' - n''$. On a double entry table as normally used, the n' being plotted downuard as ordinates and the n'' to the left as abscissas the locus of the strong bands is a parabolic looking curve; i.e., the strong bands form a locus somewhat like the shaded part in Fig. 2. The size or this general locus uith regard to the coordinate scale varies greatly from system to system, the branches being coincident for SiN, slightly separated for AlO, widely

---

[8] Birge, Phys. Rev. **25**, 240 (1925), Abstract No. 23, also Chap. IV, Sec. 4, Report on molecular spectra.

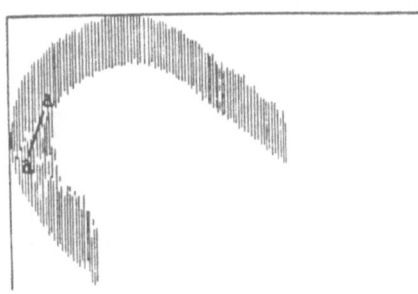

FIGURE 2. Typical distribution of intensity in band system (n' ordinates, downward; n'', abscissas).

separated for CO while the scale for iodine is so great that only the part near the origin (a in the figure) is known. In this section the connection between these observed distributions and the theoretical predictions will be examined.

The band systems to be treated are in order, SiN, silicon nitride; AlO, aluminum oxide; CO, 4th positive group of carbon; I_2, iodine; also O_2, Schumann-Runge system, CN, violet cyanogen; CO^+, first negative group of carbon, N_2, second positive group of nitrogen and N_2^+, first negative group of nitrogen. Of these nine systems, the first rour are typical of the increasing scale of the intensity distribution and are discussed in greater detail. In CO and I_2 the deviations from the harmonic law of force are considerable. For these cases corrections for the deviations are discussed as fully as the data allow. For the remaining five systems the observed intensity data together with the theoretical curve calculated from an assumption or harmonic force law are presented as additional examples or the semi-quantitative correctness of the theory.

In Table I are collected together all of the molecular constants which are needed for a discussion or the systems according to the harmonic force law. In every case the figures are taken from a large table of constants in Chapter IV of the National Research Council Report on Molecular Spectra. In the last four columns of the table are given the four constants A, B, C, D which occur in the formula for n'' in terms of n' as computed for the emission process. C.G.S. units are employed. A quantity with ' refers to the initial state and " to the final state of the emission process. $\omega^{0'}$ is the v_1 of the previous section expressed in cm^{-1} instead of sec^{-1}, similarly for $\omega^{0''}$ and v_2.

Turning now to the first system, one has available the estimated intensities by Mulliken.[9] These are plotted in Fig. 3. Here there is but a single ridge of strong bands. The figure shows the curve of most probable values of n'' as a function of n', calculated from the molecular constants according to the theory. As the theoretically double-branched curve is here scarcely an integer

[9] Mulliken, Phys. Rev. **26**, 319 (1925); Jevons, Proc. Roy. Soc. **89A**, 187 (1913–14).

TABLE I. Summary of molecular constants

Carrier	$\mu \times 10^{24}$	$\nu_e^* $ cm^{-1}	$\omega^{0'}$ cm^{-1}	$\omega^{0''}$ cm^{-1}	$I_0' \times 10^{40}$	$I_0'' \times 10^{40}$	$r_0' \times 10^8$	$r_0'' \times 10^8$	$d \times 10^8$	A	B	C	D	$(C/4D)^2$
SiN	15.417	24234.2	1016.3	1145.0	37.8	37.2	1.567	1.554	+0.013	0.0269	1.007	+ 0.349	−0.119	0.537
AlO	16.60	20635.3	864.8	970.	46.01	43.38	1.665	1.617	+0.048	0.335	1.007	+ 1.23	−0.115	7.1
CO	11.317	64721.	1499.28	2147.74	17.3	14.9	1.24	1.17	+0.07	1.075	1.065	+ 2.485	−0.366	2.88
I$_2$	104.7	15598.3	126.52	213.76	1210. / 1115.	787. / 720.5	3.40 / 3.622	2.74 / 2.622	+0.66 / +0.640	88.0 / 82.8	1.140	+24.4 / +23.62	−0.548	124.
O$_2$	13.17	49359.3	708.	1565.	34.22	19.2	1.61	1.21	+0.40	29.8	1.331	+16.24	−0.879	21.4
CN	10.67	25799.8	2143.9	2055.64	14.140	14.647	1.151	1.172	−0.021	0.875	1.001	− 1.833	+0.043	113.
CO$^+$	11.317	45637.7	1704.42	2197.03	15.4	14.05	1.17	1.11	+0.06	0.809	1.033	+ 2.045	−0.256	3.98
N$_2$	11.558	29653.1	2018.66	1718.40	15.24	16.98	1.149	1.212	−0.063	0.712	1.013	− 1.56	+0.161	5.87
N$_2^+$	11.558	25565.9	2396.0	2187.4	13.35	14.41	1.075	1.117	−0.042	0.403	1.004	− 1.213	+0.092	10.9

* ν_e gives the frequency of the origin of the 0-0 band.

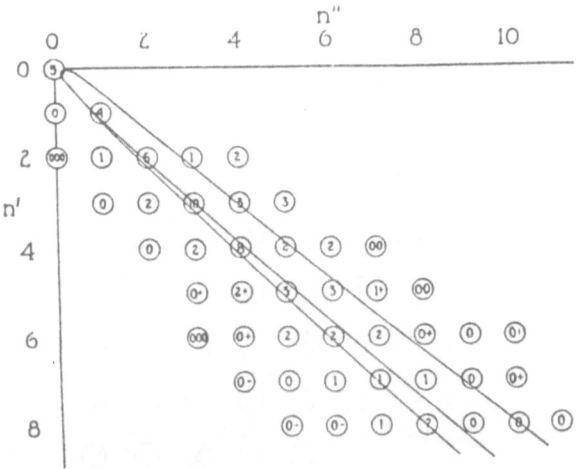

FIGURE 3. The band system of silicon nitride.

in width, it is natural that the double branch should have merged into one for the rough correspondence in question. This molecule is one which shows a doubling of the lower branch due to the cos 2θ term in the formula. The strong bands coincide exactly with this doubled line.

The next typical case is that of aluminum oxide. Here the experimental data both of Birge and also of Eriksson and Hulthén show clearly the double ridge of strong bands.[10] Here the curve computed from the molecular constants for the locus of the strong bands is open enough so that the branches should appear distinct, as in fact they are. The lower branch of the theoretical curve agrees precisely with the data, while the upper branch is slightly high. This distribution of intensity is by far the most common in the systems of bands which have thus far been analyzed on the quantum theory, as can be seen in Table I from inspection of the table of values of A, B, C, and D.

Extending the sequence of increasing scale for the distribution curve, the next example is the great system of the fourth positive group of carbon which is now known to be due to carbon monoxide.[11] In this one system, which extends from about 2650A to 1300A there are known more than 150 different bands. The great extent of this system over the frequency scale (40,000 cm⁻¹) naturally furnishes a severe test of any intensity theory. Unfortunately there are here no even semi-quantitative intensity data. In fact no one observer has ever photographed the entire system. Because of its great extent, however, the main features of the distribution are indicated by a diagram which shows simply which bands are observed. This accordingly has been done in Fig. 6.

---

[10] Birge has made estimates from unpublished spectrograms taken by him at the University of Wisconsin. Also Eriksson and Hulthén, Zeits. f. Physik **34**, 775 (1925).
[11] Birge, Nature **117**, 229 (1926); Lowry, J.O.S.A. **8**, 647 (1924).

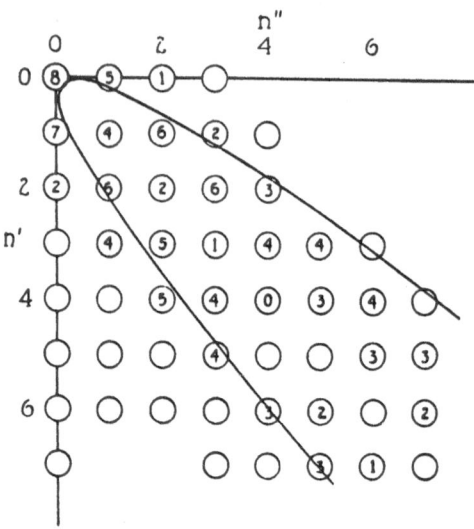

FIGURE 4. The band system of AlO.

The full line gives the curve of the intensity distribution as computed by the simple theory based on the harmonic law of force.

The analysis of the fine structure of the bands is so incomptete that it is only possible in a rough way to find the corrections due to deviations from the harmonic force law. In fact no fine structure work has been carried out on this system. Data on the initial state may be obtained since this, according to the analysis of Birge, is also the final state of the Angstrom CO bands. The moment of inertia in the final state is obtained from the infra-red CO rotation vibration bands. Besides the data given in Table I, the following figures on the 4th positive group are used in this work:

$$x'\omega^{0'} = 17.24 \qquad \alpha' = 0.023$$

$$x''\omega^{0''} = 12.703 \qquad \alpha'' = ?$$

By means of the formulas for the coefficients in the law of force already given, it is possible from these data to compute force law curves for both states, but with some ambiguity and considerable uncertainty in the final state, for which no α value is known. Writing the expansion for the potential energy in the form

$$V = u_2(r - r_0)^2 + u_3(r - r_0)^3 + u_4(r - r_0)^4 + \cdots$$

in which the unit of energy is spectroscopic wave numbers (cm^{-1}) while $(r - r_0)$ is expressed in Angstrom units one readily finds from the formulas previously given the following expressions for u_2 u_3 and u_4 in terms of the band structure constants:

$$u_2 = \frac{10^{-16} V''(r_0)}{2! \, hc} = [0.95641 - 3](\omega^{0^2} \mu)$$

$$u_3 = \frac{10^{-24} V'''(r_0)}{3! \, hc} = -[0.27924 - 1]u_2 \sqrt{B_0 \mu} \left(\frac{\alpha \omega^0}{6 B_0^2} + 1 \right)$$

$$u_4 = \frac{10^{-32} V^{IV}(r_0)}{4! \, hc} = \frac{1}{u_2} \left(\frac{5}{4} u_3^2 - \frac{8}{3} u_2^3 \frac{x}{\omega^0} \right)$$

In these expressions the unit of μ is 10^{-24} gm as in Table I, and the quantities in brackets are the logarithms of the coefficients of the corresponding quantities.

When the formulas are applied to the known data for the initial state in CO one obtains the following values:

$$u_2' = 2.30 \times 10^5$$

$$u_3' = -6.00 \times 10^5$$

$$u_4' = +8.84 \times 10^5$$

For the final state a value of α is lacking. In computing the potential energy curve what seems like a reasonable value of α has been used, namely 0.02. This compares well with the known values for systems which have been analyzed. The coefficients found on this assumption are:

$$u_2'' = 4.718 \times 10^5$$

$$u_3'' = -12.59 \times 10^5$$

$$u_4'' = +25.66 \times 10^5$$

In Fig. 5 are drawn to scale the potential energy curves of the initial and final states in a form suitable for the application of the graphical method of predicting the intensity distribution. In Fig. 6 is given in the curve marked "corrected law" the theoretical position of the two ridges of intense bands as obtained by this method. It is clear that the corrected law improves the agreement for the lower branch while leaving almost unaltered the good fit of the other branch.

Passing now to the band system of iodine,[12] the largest known band system, it is essential to use the higher terms in the law of force since the change in the moment of inertia between the initial and the final states is very large.

---

[12] The analysis is due to Mecke, Ann. d. Physik **71**, 104 (1923), with Loomis' revision of the quantum assignment. The values of B_0[8] and α are due to Loomis (Chapter VI, Report on Molecular Spectra), that of α'', however, having been previously given by Kratzer and Sudholt, Zeits. f. Physik **33**, 144 (1925). As Loomis points out, the values of α are quite uncertain. For new observations on resonance spectra, see Dymond, Zeits. f. Physik, **34**, 553 (1925).

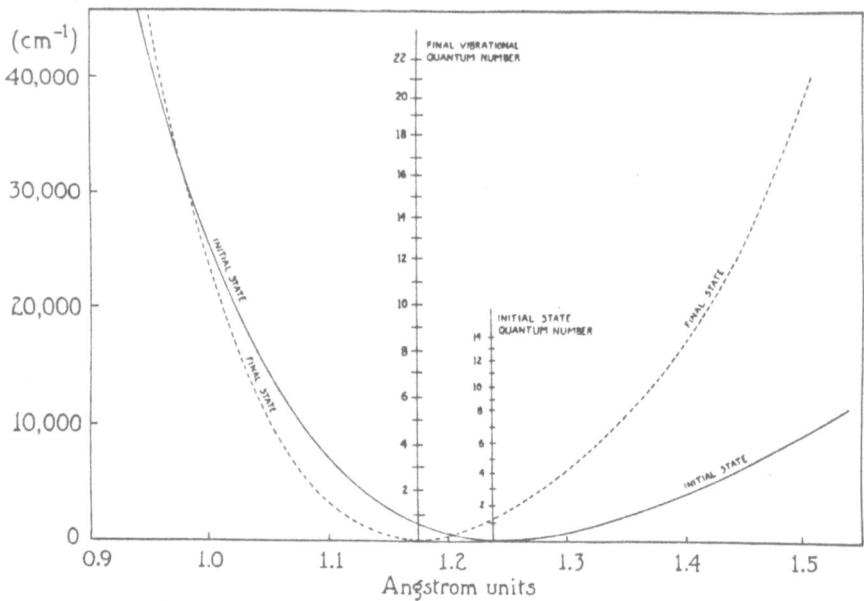

FIGURE 5. The potential energy curves for the two electronic states involved in the fourth positive group of carbon (CO).

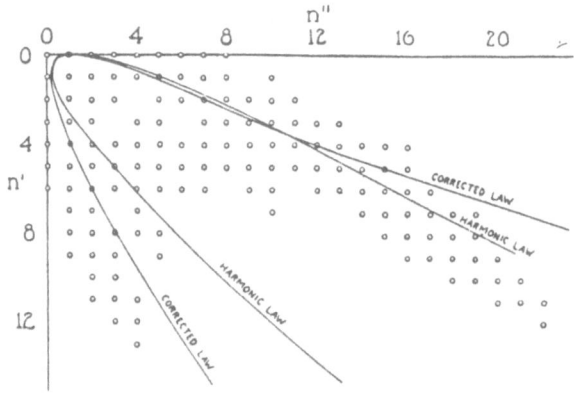

FIGURE 6. The observed bands of the fourth positive group of carbon.

The calculations make use of the following data in addition to that given in the table:

$$\alpha' = 0.00015 \qquad x'\omega^{01'} = 0.85$$

$$\alpha'' = 0.00011 \qquad x''\omega'^{011'} = 0.592$$

The resulting values of the coefficients in the potential energy function are:

	Initial	*Final*
u_2:	$+1.537 \times 10^4$	$+4.313 \times 10^4$
u_3:	-2.906×10^4	-6.002×10^4
u_4:	$+3.580 \times 10^4$	$+4.014 \times 10^4$

As in the case of CO, there are no intensity measurements available for this system of bands. The data are due to Mecke and give merely the presence or absence of the various bands. The system is thus plotted in Fig. 7. Here the data are for the *absorption* spectrum rather than for emission as in the examples previously given. To be consistent with the other examples, however, the terms "initial" and "final" in the foregoing set of constants have been chosen so as to refer to the *emission* process. In the absorption process, n'' becomes the initial quantum number and n' the final. Mecke has already commented on the peculiar intensity distribution in I_2, his Fig. 4 contrasting it with the more typical distribution of intensity of the violet CN-system. The distribution of intensity is, of course, protoundly modified by the distribution of the molecules among the initial states. This accounts for the great observed decrease in intensity of successive n' progressions, for increasing values of n''.

The curve in Fig. 7 is that which is obtained by the graphical process, using the potential energy curves given by the foregoing set of constants. Here evidently is a wide discrepancy between theory and experiment. The

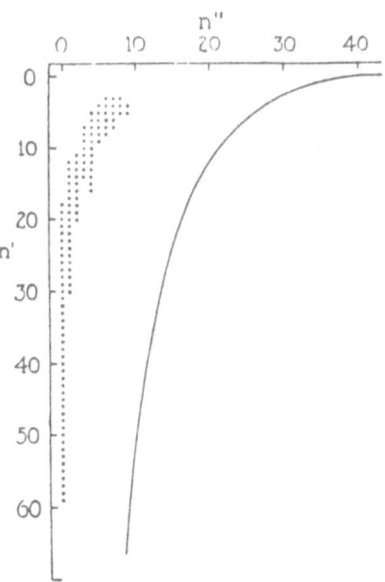

FIGURE 7. The absorption bands or iodine (I_2).

discrepancy may be removed by an alteration in the assignment of vibrational quantum numbers. Mecke's assignment of values of n'' is greater by four than the one, due to Loomis, on which the calculations have been based. Mecke's zero of n'' was chosen to include a place for four anti-Stokes terms in Wood's resonance spectrum. But if the theoretical intensity curve derived from Loomis' assignment is correct it is clear that a molecule in the zero-state initially would favor a transition to an energy state in excess of that required for the dissociation process. This suggests that when the molecule absorbs light from the unexcited non-vibrating state it is always dissociated and no bands are obtained. It is therefore possible that lower vibrational states exist which contribute, by absorption from them, only to the continuous spectrum of iodine. The existence of such lower states is called for, of course, if there are really the so-called anti-Stokes terms in the resonance spectrum of Iodine. Loomis has, however, given an analysis of the resonance doublets which makes it appear that what were believed to be anti-Stokes terms in the $n' = 26$ progression are really due to the excitation of some lines out of other n' progressions. Dymond has recorded observations of anti-Stokes terms in resonance spectra excited by other lines than the green mercury line. It remains to be seen whether Loomis' analysis can be applied to them. It is clear that, since the theory here presented gives the same favored Δn values for absorption as for emission, transitions involving the lower levels in the unexcited state would not appear in the fluorescent spectrum for the same reason that they are not involved in the fully quantized absorption spectrum. Thus it is to be expected that anti-Stokes terms will not appear even though the necessary levels may exist in the molecule.

It is easy to see that a rather small change in the assignment of quantum numbers will make the theory agree with experiment. Full calculations were carried out for the case in which the n'' values were increased by 5 and the n' by 30. It was found that the discrepancy was about the same but in the other direction. Therefore a good fit could be obtained by some reassignment which makes smaller changes in n' and n''. It seems, however, that agreement on the intensities calls for reassignment of *both* n' and n'', since calculations show that good fits are not obtained if n' alone is altered. Iodine, then, may be tentatively regarded as in agreement with the theory.*

This completes the consideration of the four main sizes of the typical intensity distribution. Of the remaining five, the Schumann-Runge system of oxygen is much like that of iodine in that an unusually large change in the moment of inertia occurs during the electron transition. The other four are medium sized systems like that of AlO.

The Schumann-Runge bands of oxygen are quite incompletely known.[13]

---

* See note on p. 1201.

[13] Schumann, Smithsonian Contrib. **29**, no. 1413 (1903); Runge, Physica **1**, 254 (1921); Mulliken, Private communication to Professor Birge; Leifson, Astrophys. J. **63**, 73 (1926); Füchtbauer and Holm, Phys. Zeits. **26**, 345 (1925); Report on Molecular Spectra, Chap. IV, Section 7.

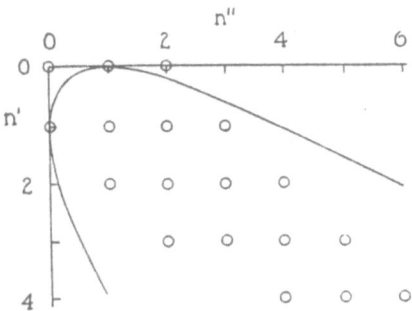

FIGURE 8. The violet cyanogen system (CN).

This name is here given to the system whose $n'' = 0$ progression is known in absorption by the researches or Schumann and others and whose $n' = 0$ progression was photograplled in emission by Runge. The recognition of the fact that the Runge and Schumann bands are part of one system is due to Mulliken. The data on the absorption bands has recently been extended by the work of Leifson as interpreted by Birge. All that can be said here is that in absorption Leifson's strongest bands run from $n' = 9$ to a limit at $n' = 21$ with continuous spectrum beyond. Similarly the emission bands measured by Runge correspond to large values of Δn, the actual bands being $n' = 0, n'' = 11$ to 17. Here is evidently a preference for large values or Δn which is quite in accord with the theory of this paper. The fine structure analysis or the bands thus far permits only the use of parabolic $V(r)$ curves, but these, when computed from the data of Table I show that in absorption the most probable transition is that beyond dissociation. The calculation shows that from the $n'' = 0$ state the molecule in the absorption process tends to take up over 1.2 volts of vibrational energy or 0.4 volts in excess of that needed for dissociation in the excited state for these bands. Similarly the curves indicate for the $n' = 0$ progression or emission bands a large Δn, much larger in fact than that indicated directly by the Runge bands. Here the higller terms in the force law will undoubtedly operate to reduce the theoretical Δn, so that in this system there is as Sood agreement as can be hoped for in the present state of analysis of the bands.

The remaining four band systems to be considered present no new features of special interest. They are shown in Figs. 8, 9, 10, 11 in which the intensities are, when given, simply rough visual estimates from the plates. The theoretical curve in each case is based on the formula which assumes simple harmonic oscillators. The data for CN (Fig. 8), are from Heurlinger.[14]

The intensities given for CO^+ (Fig. 9) as well as the molecular constants used are based on the work or Blackburn and Johnson.[15] It is important to

[14] Kayser's Handbuch, v. 5, p. 190.
[15] Johnson, Proc. Roy. Soc. **108A**, 343 (1925); Blackburn, Proc. Nat. Acad. **11**, 28 (1925).

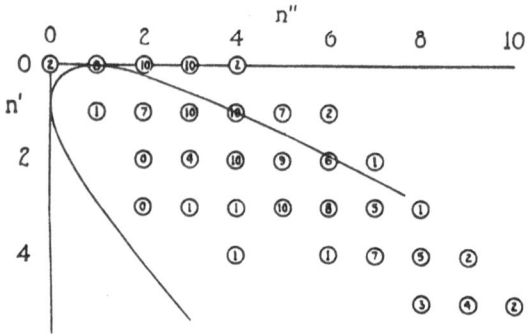

FIGURE 9. The first negative group of carbon (CO$^+$).

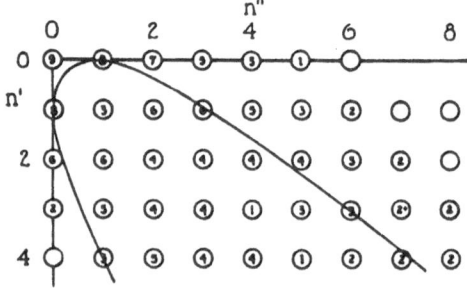

FIGURE 10. The second positive group of nitrogen (N$_2$).

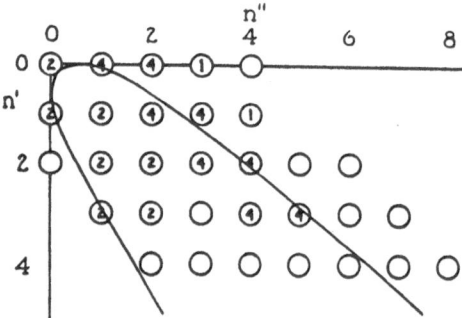

FIGURE 11. The first negative group of nitrogen (N$_2$$^+$).

note that the theory predicts the existence or some strong emission bands in the rar ultraviolet, along the lower branch of the theoretical curve. The single-branched intensity distribution the writer believes is entirely a consequence or the lack of complete experimental data. The intensity estimates for the second positive group of nitrogen (Fig. 10) are due to Birge, being made from inspection of spectrograms taken by E. P. Lewis.[8]

The data on the first negative group of nitrogen (Fig. 11) are due to Merton and Pilley.[16]

In conclusion it seems permissible to claim that the theory outlined here gives a satisfactory semi-quantitative correlation between the intensity distribution in band systems and the positional structure of the systems by means of a definite mechanical picture of the processes governing transition probabilities. The only previous attempt at a theoretical treatment of this topic is that of Lenz[17] who gave, by application of the correspondence principle to an over-simplified molecular model, some general formulas which, however, were incapable of numerical application to specific cases. On the other hand, Lenz' theory does lead to an understanding of the many and seemingly irregular alternations of intensity in the $n' = 26$ progression of the iodine emission spectrum (Wood's resonance spectrum). This is a thing which is wholly unclear in terms of the theory given here.

It is a pleasure to acknowledge my indebtedness to Professor R. T. Birge who has generously given me the benefit of his knowledge of band spectra and their quantum interpretation. It is also appropriate again to call attention to the fact that my work is merely an extension of a leading thought on this subject by Professor J. Franck.

Note added to proof. Since this was written I have learned of the experiments of H. Kuhn (Zeits. f. Physik **39**, 77 (1926)) which seem to show conclusively that the long series of absorption bands which have a convergence limit really do come from the lowest vibration level so that the values of n'' cannot be altered as suggested on page 1198.

Göttingen,
November 8, 1926.

---

[16] Merton and Pilley, Phil. Mag. **50**, 195 (1925).
[17] Lenz, Zeits. f. Phys. **25**, 299 (1924).

Mean Free Paths in a Gas Whose Molecules are Attracting Rigid Elastic Spheres

EDWARD U. CONDON AND E.V. VAN AMRINGE†

Department of Physics
University of California

The beautiful results which the founders of the kinetic theory of gases were able to attain by regarding gas molecules simply as rigid elastic spheres have been considerably refined in several directions by theorists who have imagined a gas molecule which exerts a weak attractive force on its fellows. This concept of attracting molecules lies at the basis of van der Waals' pioneer work on the equation of state, and enabled Sutherland to give a theoretical basis to his remarkably accurate equation for the variation of the coefficient of viscosity with the temperature.

In spite of the success of the attracting elastic sphere molecule in explaining gas behaviour, an evaluation of the mean free path of such molecules which proceeds along classical lines does not seem to have been given. In view of the importance of the free path concept in kinetic theory, it was thought worth while to supply this deficiency. The results are presented in this paper.

On the historical side it may be noted that Sutherland* carried his approximate analysis merely to the point of indicating how the free path should vary with the absolute-temperature. Chapman†, in two papers, has approached the problem from the point of view of finding the solution for the law of distribution of velocities in a gas presenting small deviations from the steady state. James‡ has made certain corrections and extensions in Chapman's work.

1. *Effective Collision Area.* Let us suppose a gas made up of molecules of diameter σ, the density of which is such that the mean distance apart of the molecules is large compared with their size. Suppose that the molecules interact in collision as do rigid elastic spheres. Let it be further supposed that

---

† Communicated by Prof. L. B. Loeb, Ph.D.
* Sutherland, Phil. Mag. [5] xxxvi. p. 507 (1893).
† Chapman, Phil. Trans. Roy. Soc. A, ccxvi. p. 326 (1915).
‡ James, Cambr. Phil. Soc. Proc. xx. p. 447 (1920).

they attract each other through symmetrical fields of force such that the work required to separate them to infinity when their centres are initially at a distance r is $E(r)$. If m is the mass of each molecule, then the equations of motion of the second with respect to the first are

$$m\ddot{x} = +\frac{2x}{r}\frac{\partial E}{\partial r}, \qquad m\ddot{y} = +\frac{2y}{r}\frac{\partial E}{\partial r}. \tag{1}$$

Since the forces are central, these possess an integral of angular momentum. If V is the relative velocity when the molecules are widely separated, and if B is the perpendicular from the origin (centre of first molecule) on the initial line of motion of the second, the angular momentum is equal to VB, i.e.,

$$r^2\frac{d\theta}{dt} = VB. \tag{2}$$

Similarly the energy integral is

$$\tfrac{1}{2}mV^2 - \tfrac{1}{2}mv^2 = -2E(r). \tag{3}$$

Eliminating the time between these integrals we have

$$\frac{1}{2}m\left[\left(\frac{dr}{dt}\right)^2 + \frac{V^2B^2}{r^2}\right] = \frac{1}{2}mV^2 + 2E(r). \tag{4}$$

We are interested in the distance r_0 of minimum approach between the two molecules. This is the value of r in the preceding equation, which corresponds to the condition $dr/dt = 0$. A collision will occur if this value, r_0, is equal to or less than σ. That is, a collision will occur if B is less than the critical value of B given in this equation:

$$B^2 = \sigma^2\left[1 + \frac{2E(\sigma)}{\tfrac{1}{2}mV^2}\right]. \tag{5}$$

This is the result obtained by Sutherland, who obtains his viscosity formula by writing this value of B for σ in the classical free path formulae, and observing that V^2 is proportional to the absolute temperature T.

2. *Maxwell Free Path.* In this section the expression obtained in the preceding section for the effective collision area of the molecules will be employed for the computation of the mean free path according to Clerk Maxwell. In Maxwell's method the mean free path is defined to be the ratio of the mean speed of the molecules, c, to the mean number of collisions per second, Z. In this section the correction to Maxwell's mean path which arises from the attractions of the molecules is evaluated. The method follows closely that used by Jeans* in his treatise on kinetic theory. If $f(u, v, w)$ be written for the distribution function of velocities, v for the number of molecules in unit volume, θ for the angle between the relative velocity and the line of centres

---

* Jeans, 'The Dynamical Theory of Gases,' 3rd ed. (1921).

when at distance B apart, then the mean number of collisions in unit time occurring between molecules of speed components u, v, w and u', v', w' under the conditions described by Jeans as a collision of class α is given by his expression (48):

$$v^2 f(u, v, w) f(u', v', w') V B^2 \cos \theta \, du \, dv \, dw \, du' \, du' \, dw' \, dw, \qquad (6)$$

where B^2 has been written for σ^2 to take into account the increased number of collisions arising from the molecular attractions. The evaluation of the total number of collisions per second is effected by integrating over all the variables with appropriate limits.

If in (6) the value of B given by (5) be substituted, it is evident that the unity term in brackets will contribute exactly the same number of collisions as in the classical elastic sphere theory, and that the correction to this number is given by study of the second term. Thus the analysis may be carried parallel to that in Jeans's book as far as his expression (52), which gives the total number of collisions per second in whic the relative velocity lies between V and $V + dV$. Writing in B^2 for σ^2 in this, one has

$$\sqrt{\frac{\pi h^3 m^3}{2}} v^2 \sigma^2 \left[1 + \frac{4E}{mV^2} \right] e^{-\frac{1}{2}hmV^2} V^3 \, dV. \qquad (7)$$

The total number of collisions per second is obtained on integrating this expression from θ to ∞. The definite integrals involved are well known, and the result is

$$v^2 \sigma^2 \sqrt{\frac{2\pi}{hm}} (1 + 2hE) \qquad (8)$$

as the corrected mean number of collisions per second experienced by each molecule. If R is the gas-constant for a single molecule, we have the well-known relation $RT = \dfrac{1}{2h}$, so that finally the modified form of the mean free path according to Maxwell becomes

$$l = \frac{1}{\sqrt{2} \pi v \sigma^2 \left(1 + \dfrac{E}{RT} \right)},$$

in which E stands for $E(\sigma)$ and represents the work required to separate from contact two molecules to an infinite distance.

Letting $z = E/RT$, this can be written

$$l = \frac{1}{\pi v \sigma^2} \cdot H(z), \quad \text{where} \quad H(z) = \frac{1}{\sqrt{2}(1 + z)}. \qquad (9)$$

3. *Collision Frequency for Molecules of Specified Speed.* As was first pointed out by Tait, the mean collision frequency of the molecules in an assemblage whose speeds are distributed according to Maxwell's law depends on the

speed of the particular group of molecules under consideration. Tait* showed that, as a result, there is a mean free path proper to molecules of each speed, and then computed a mean free path which was the mean of those peculiar to a given speed, weighted according to the Maxwell distribution law for the speeds.

In this section the computation of the correction to this branch of the theory arising from the attractions of the molecules is undertaken. Here again, for brevity, the presentation may be conveniently made to parallel that of the classical work given in Jeans's book. His treatment is contained in Sect. 341 of the third edition. His notation is somewhat encumbered by the fact that it is for the case of a mixture of various kinds of molecules. The reader will find no difficulty in extending a simple treatment for a single kind of molecule to this gas, and so the more concise notation will be used here.

Jeans's expression (706) is applicable to our needs if we write in B^2 in place of S_{12}^2. The transformation yielding his (707) is also directly applicable, so that we have

$$2vB^2\sqrt{\pi h^3 m^3}e^{-hmc'^2}\frac{c'}{c}dc'\,V^2\,dV \qquad (10)$$

for the number of collisions experienced by molecules of speed c with molecules of speeds between c' and $c' + dc'$, when the relative velocity is contained between the limits V and $V + dV$.

Now, as before, substitution of the expression for B given in (5) will contribute a term to which the classical treatment is applicable, and a correction term. The total number of collisions per second experienced by a molecule of speed c is obtained by integrating V from the difference of the speeds to the sum of the speeds, and then integrating c' from 0 to ∞. Let Z be the mean number of collisions on the uncorrected theory and ΔZ the correction arising from the attractions of the molecules. For Z we can quote directly the result given by Jeans:

$$Z = \frac{\sqrt{\pi v \sigma^2}}{hmc}\psi(c\sqrt{hm}), \qquad (11)$$

in which $\psi(x)$ is the following function of its argument:

$$\psi(x) = xe^{-x^2} + (2x^2 + 1)\int_0^x e^{-y^2}\,dy. \qquad (12)$$

Proceeding now to the evaluation of ΔZ, we observe that it is given by

$$\Delta Z = \int_0^\infty \int_{c-c'}^{c+c} 2v\sigma^2\sqrt{\pi hm^3}e^{-hmc'^2}\frac{c'}{c}dc'\cdot\frac{4E}{m}dV. \qquad (13)$$

The integration with respect to V yields a factor

$$2c' \quad \text{when} \quad c' < c,$$

$$2c \quad \text{when} \quad c' > c.$$

---

* Tait, Edin. Trans. xxx. p. 74 (1886).

The correction term ΔZ is therefore given by

$$\Delta Z = \frac{16v\sigma^2 E}{m}\sqrt{\pi h^3 m^3}\left[\frac{1}{c}\int_0^c e^{-hmc'^2}c'^2\,dc' + \int_0^\infty e^{-hmc'^2}c'\,dc'\right]. \quad (14)$$

The first integral may be made to depend on the probability integral by integration by parts, the second integral may be calculated directly. If these reductions be carried out, the correction term assumes the form

$$\Delta Z = \frac{8v\sigma^2 E\sqrt{\pi}}{mc}\int_0^{c\sqrt{hm}} e^{-y^2}\,dy. \quad (15)$$

The correction arising from the attraction is more clearly exhibited when it is expressed as a fractional amount of the total number of collisions. If this be done we arrive at the following expression for the mean number of collisions per second experienced by those molecules which are travelling with a speed c:

$$Z + \Delta Z = \sqrt{\frac{\pi}{hm}}\,v\sigma^2\frac{\psi(x)}{x}\left[1 + z\frac{2\sqrt{\pi}\,\mathrm{Erf}\,x}{\psi(x)}\right], \quad (16)$$

in which $\mathrm{Erf}(x)$ is the error-function,

$$\mathrm{Erf}\,x = \frac{2}{\sqrt{\pi}}\int_0^x e^{-y^2}\,dy,$$

and the argument x is written for $c\sqrt{hm}$.

The manner in which the mean number of collisions per second varies with the speed of the particular molecules under consideration, and the manner in which this is affected by the attractions, are so complicated that a pair of graphs have been made to clarify the situation. In fig. 1 the uncorrected Z is

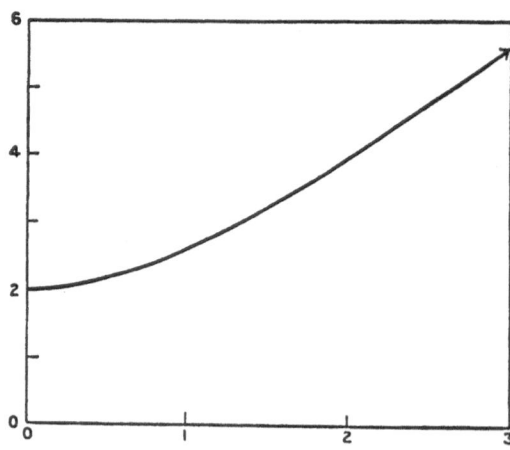

FIGURE 1. $\dfrac{\psi(x)}{x}$ plotted as a function of x.

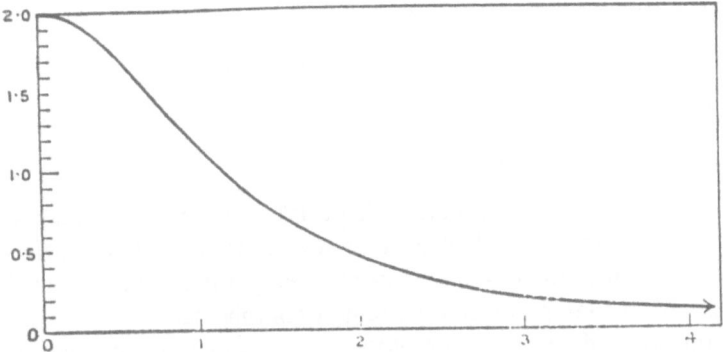

FIGURE 2. $\dfrac{\Delta Z}{Z}$ plotted as a function of x.

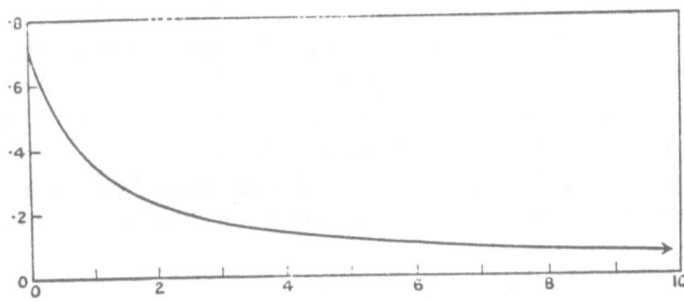

FIGURE 3. $H(z)$ plotted against z.

plotted against x on the supposition that $\sqrt{\dfrac{\pi}{hm}}\, v\sigma^2$ is equal to unity. In fig. 2 the coefficient of z in the correction term has been plotted against x. Values of the error-function are available in many places, while those of $\psi(x)$ are from Tait.

4. *The Tait Free Path.* With the results of the preceding section at hand we are prepared to calculate the modification in the value of the Tait free path which results from molecular attraction. The mean free path of molecules moving with speed c is simply the ratio of the distance gone in a second, c, to the mean number of collisions in a second, that is,

$$\lambda_c = \frac{x}{\sqrt{hm(Z + \Delta Z)}}. \tag{17}$$

Tait's free path is obtained by computing the average of λ_c when the various λ_c are weighted according to the number of molecules whose speeds lie between c and $c + dc$, as given by the Maxwell distribution formula. The corrected Tait free path, therefore, is given by

$$\lambda = \frac{1}{\pi v \sigma^2} \int_0^\infty \frac{4x^4 e^{-x^2}\, dx}{\psi(x)\left(1 + z\dfrac{2\sqrt{\pi}\,\mathrm{Erf}\,x}{\psi(x)}\right)}. \tag{18}$$

This expression is comparable with Jeans's equation (78), where attractive forces are not considered. It is evident at once that, while the Maxwell free path was a simple function of the absolute temperature, the manner in which λ varies with the temperature is extremely complicated.

For the convenient discussion of the variation of λ one may write

$$\lambda = \frac{1}{\pi v \sigma^2} F(z), \quad \text{where} \quad F(z) = \int_0^\infty \frac{4x^4 e^{-x^2}\, dx}{\psi(x)\left[1 + z\dfrac{2\sqrt{\pi}\,\mathrm{Erf}\,x}{\psi(x)}\right]}. \tag{19}$$

The exact evaluation of the Tait free path therefore requires a study of the complicated function $F(z)$. The evaluation of this function can only be effected by quadratures. This has been done in an approximate way which yields values sufficiently accurate for the needs of kinetic theory.

We have at once that $F(0)$ is the integral occurring in Tait's work, the value of which is 0.677, and that $F(\infty) = 0$. Other values of $F(z)$ are contained in Table I.

TABLE I. Values of $F(z)$, $G(z)$, and $H(z)$

z	$F(z)$	$G(z)$	$H(z)$
0.0	.677	.744	.707
.2	.572	.638	.589
.5	.464	.532	.471
1.0	.354	.418	.353
2.0	.245	.295	.236
3.0	.187	.231	.177
10.0	.0696	.0897	.0643

5. *Transfer Theory of Viscosity.* Neither the Maxwell nor the Tait free path is the proper one to use in the development of the kinetic theory of the viscosity coefficient by the method of transfer of momentum, as is well known. Again, it is convenient to parallel the treatment of the subject as given by Jeans in order to avoid lengthy repetitions. The discussion of his Sect. 364–369 is applicable here.

What we require for the transfer theory in the case of attracting elastic sphere molecules* is an exact calculation of the integral (*cf.* Jeans's expression (766))

$$\int_0^\infty f(c)\lambda_c c \, dc, \tag{20}$$

In the case of attracting spheres, the evaluation of this integral requires the use of our equation (17) for the value of the mean free path of a molecule of speed $x/\sqrt{hm}$. The viscosity coefficient η, by the transfer theory, is then given by

$$\eta = \frac{1}{3}\rho\bar{c}l, \quad \text{where} \quad l = \frac{2\sqrt{2\pi}}{\sqrt{2\pi v\sigma^2}} \int_0^\infty \frac{x^5 e^{-x^2}\, dx}{\psi(x)\left[1 + \dfrac{E}{RT}\dfrac{2\sqrt{\pi}\,\text{Erf}\,x}{\psi(x)}\right]}. \tag{21}$$

This, again, is a new kind of mean free path, differing from both the Maxwellian free path and the Tait free path. It is also a very complicated function of the temperature. Inasmuch as it is this latter free path which is appropriate to the theory of viscosity, we see that the Sutherland formula is not a rigorous consequence of the physical assumptions from which it was deduced.

To see how this free path depends on the temperature, one may define the function $G(z)$ to be that part of the free path which is independent of $1/\pi v\sigma^2$:

$$\lambda_v = \frac{1}{\pi v\sigma^2}G(z), \quad \text{where} \quad G(z) = 2\sqrt{\pi}\int_0^\infty \frac{x^5 e^{-x^2}\, dx}{\psi(x)\left[1 + z\dfrac{2\sqrt{\pi}\,\text{Erf}\,x}{\psi(x)}\right]}. \tag{22}$$

Several values of this function have been computed by quadratures. These are included in Table I along with $H(z)$ and $F(z)$.

6. *Discussion of Results.* In fig. 4 are plotted the ratios $F(z)/H(z)$ and $G(z)/H(z)$. If these ratios remained constant, or fairly constant, this would indicate that the functions F and G could be fairly well represented by a Sutherland formula of the form

$$F(z) = F(0)\cdot\frac{1}{1+z}; \quad G(z) = G(0)\cdot\frac{1}{1+z}.$$

However, in the important case of $G(z)$, the ratio shows a 35 per cent variation. One may try to fit a Sutherland form to $G(z)$, introducing another constant c:

$$G(z) = G(0)\cdot\frac{1}{1+cz}. \tag{23}$$

---

* Here, as in Sutherland's theory, we neglect the momentum transfer due to *deflexions* of paths, without collisions.

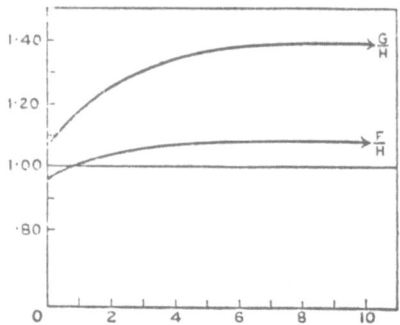

FIGURE 4. $\dfrac{F}{H}$ and $\dfrac{G}{H}$ plotted as functions of z.

If c be computed from the values of G, it is found that c is not constant, but takes these values:

z	c
0.2	0.810
.5	.802
1.0	.774
2.0	.756
3.0	.740
10.0	.728

For rough numerical work this analysis indicates that one may take $c = 0.75$ as representing the correct function over the range of values important in the theory of viscosity. Remembering the definition of z, this means that *the Sutherland constant n the formula for temperature coefficient of viscosity is approximately three-fourths of the ratio of the energy of two molecules at contact to the gas-constant per molecule.*

To summarize, one may say:

1. The Maxwell mean free path is accurately represented by a formula of the Sutherland type, in which Sutherland's constant C is to be interpreted as the ratio of the energy of two molecules in contact to the gas-constant per molecule.

2. The mean free path appropriate to the transfer theory of viscosity is *approximately* represented by a formula of the Sutherland type, in which C is interpreted as about three-fourths of the above ratio.

The results here given lead to a direct method of evaluating the energy of two molecules in contact by means of the published experimental data on temperature coefficient of viscosity. The results so obtained should be of use in the study of other properties of gases. In closing it is desirable to mention a wide discrepancy between this analysis and that given by James*, where he concludes that the Sutherland constant C depends on the form of the law of force, and varies from 0.31 to 0.15 of the ratio of the energy of contact to the gas-constant. James's approach to the problem is an entirely different one, but it is surprising that such a discrepancy should appear. This will require explanation before the results can be used to estimate the energy of the molecules at contact.

---

* James, Cambr. Phil. Soc. Proc. xx. p. 447 (1920).

Coupling of Electronic and Nuclear Motions in Diatomic Molecules

EDWARD U. CONDON

Göttingen, Germany

While for many important problems the quantum mechanical coupling between the motion of the nuclei and of the electrons in a diatomic molecule may be ignored, there are others in which this coupling is the main factor. Among the various phenomena which are conditioned by this coupling and which have recently attracted attention the following may be mentioned:

1. The mechanism of the dissociation of molecules by light absorption. This question has been discussed by Franck who, with his co-workers, is also doing important experimental work in this field.[1]

2. The mechanism of dissociation of molecules as a primary consequence of excitation by electron collision. The ideas of Franck on (1) have here been applied by Birge and Sponer[2] to a discussion of the experimental results of Hogness and Lunn.

3. The relative transition probabilities for the various vibrational transitions associated with a single electron transition in the emission or absorption of light by a molecule; i.e., the problem of intensity distribution in electronic band systems. This question has been treated recently by the writer.[3]

4. The mechanism of the process whereby molecules may be excited to vibration levels by electron couision without simultaneous electronic excitation. That this process actually occurs is proved by the recent experiments of Harries, carried out in Franck's laboratory.[4]

All of these phenomena permit of being understood in terms of the new quantum mechanics, at least in a qualitative way, as it is the purpose of this note to show.

The complete quantum-mechanical problem of a diatomic molecule, from

---

[*] NATIONAL RESEARCH FELLOW.

[1] Franck, *Trans. Faraday Soc.*, **1925**; Dymond, E. G., *Zeits. Phys.*, **34**, 553 (1925); Kuhn, H., *Zeits. Phys.*, **39**, 77 (1926).
[2] Birge and Sponer, *Phys. Rev.*, **28**, 259 (1926).
[3] Condon, E., *Ibid.*, **28**, 1182 (1926).
[4] Harries, W., Preliminary report at meeting of the Deutschen Physikalischen Gesellschaft, Hamburg, Feb. 6, 1927.

the Schrödinger point of view, calls for the solution of a partial differential equation in $3N + 6$ independent variables, if N is the number of electrons in the molecule. The problem is distinguished from the problem of a complicated atom, by the presence of two particles of large mass, instead of one. The usual theory of the diatomic molecule, however, replaces the actual model of two nuclei and N electrons affecting each other with Coulomb forces by one which consists simply of two masses affecting each other by a more or less arbitrary potential energy law. This effective potential energy is supposed to result from the average of the reactions of the rapidly moving electrons on the nuclei together with the mutual Coulomb repulsion of the nuclei.

The known success of this simpler model in explaining the energy levels, in general, and the infra-red transition probabilities, makes it natural to suppose that the heavy masses of the nuclei bring about a partial or approximate separation of variables in the Schrödinger equation. The three coördinates which correspond to the motion of the center of gravity of the system in space are readily separated out. Similarly, if rotation is neglected, the system is described by the nuclear separation, R, and the electronic coördinates, denoted in the aggregate by X. The assumption of partial separation of variables here made requires the two following results:

(*a*) The energy-levels (eigen-wert parameter) are the sum of a function depending only on the electronic quantum numbers, denoted in the aggregate by e, and a function of both the electronic quantum numbers and a vibrational quantum number, n.

(*b*) The amplitude function, ψ, is the product of two factors, one of which depends on all of the quantum numbers and only the nuclear separation, R, while the other depends only on the electronic quantum numbers and the electronic and nuclear coördinates. That is, one has

$$E = E_1(e) + E_2(e, n)$$

$$\psi(e, n, X, R) = \psi_1(e, X, R) \cdot \psi_2(e, n, R).$$

This assumption, of course, needs justification in terms of the theory of the Schrödinger equation. An attempt to do this is being made. But, making the assumption, the four coupling phenomena already listed come within a single picture.

Numbers 1 and 3 are simpler than 2 and 4 for radiative transitions are simpler than those involving electron collisions. For the purpose of computing such transition probabilities one has to consider the matrix component of the electric moment of the molecule with regard to the initial and final state. This matrix component, referring to the initial state e', n' and the final state e'', n'', is given by

$$M(e'e''n'n'') = \iint M(X, R)\psi(e'n'XR)\psi(e''n''XR)\, dX\, dR$$

in which $M(X, R)$ is equal to the electric moment of the molecule in the configuration X, R. The integration is over all values of the coördinates. For the purpose of comparing various vibration transitions associated with the same electronic transition, one may perform the integration over the electronic coördinates. One obtains thereby an effective electric moment function, $m(R)$, for computing the different vibration transitions. The complete matrix component is then the integral of this moment function over the product of the two vibrational factors of the amplitude function.

The vibrational factor in this formula may safely be identified with the amplitude function obtained from the simple anharmonic oscillator treatment of the molecule problem. These amplitude functions have the essential characteristic that they approach a zero value asymptotically, but rapidly, outside of the region of the corresponding classical motion. Inside this region they oscillate, having as many zeros as the order of the quantum state. *Herein lies the quantum mechanical justification of the picture used by Franck* in discussing the dissociation of molecules by light absorption, and which has found application in the theory of transition probabilities in electronic band systems.

The Franck picture is that the favored transitions are those for which the classical vibrational motions cover over-lapping regions of the nuclear separation coördinate. Clearly this comes out of the quantum mechanical treatment by the property of the Schrödinger amplitude function of having its largest values at the coördinate values covered by the classical motion. Hence the larger values of the matrix components are, in general, those for which both amplitude function factors in the integrand have large values for the same values of the independent variable; i.e., those of over-lapping classical motion. The quantum-mechanical formula differs in two respects from the earlier treatment. One is that transitions of small probability corresponding to non-over-lapping classical vibration motions are made possible by the fact that the Schrödinger amplitude functions have values outside the range of the classical motion. Also, since the amplitude function oscillates within the range of the classical motion there is the possibility of a kind of interference in the transition probabilities, reducing the value associated with a transition which corresponds to over-lapping classical motions. This may be the explanation of the irregular alternations of intensity in the Wood's resonance spectrum of iodine, akeady discussed by Lenz in terms of the old correspondence principle for intensities.

To illustrate the semi-quantitative numerical behavior of the new quantum mechanical formula, two cases have been worked out roughly. For the purpose, $m(R)$ was regarded simply as a constant and the vibrational amplitude functions were taken to be the Hermitian polynomial harmonic oscillator solutions.[5] The harmonic oscillator solutions depend on the initial and final electronic state through the vibration frequencies and centers of oscillation associated with the two states. The relative transition probabilities are mea-

---

[5] Schrödinger, *Ann. Phys.*, **79**, 514 (1926).

sured by the product of the square of the matrix component and the fourth power of the quantum frequency involved, in accordance with the classical dipole radiation formula.

Thus, applied to the $n' = 0$ progression of the Schumann-Runge bands of oxygen, one obtains for the relative intensity of the bands corresponding to various final state vibration quantum numbers, n'':

$n'' =$	0	1	2	3	4	5
	8.4×10^{-6}	9×10^{-5}	5×10^{-4}	1.6×10^{-3}	6.6×10^{-3}	0.008

6	7	...	13	15	17
0.019	0.03	...	0.88	1.00	1.13

Experimentally, Runge measured on his plates only the bands $n'' = 11$ to $n'' = 17$, as being the strongest. This is in good semi-quantitative agreement with the foregoing calculations.

On the other hand, SiN offers a band system characterized by a very slight change in moment of inertia and frequency of vibration between the two electronic states.[6] Correspondingly, zero change in the vibration quantum number is the most probable. This comes out of the matrix formula since, had there been no change in moment of inertia or natural frequency, the two amplitude functions in the formula would have been members of the same normal-orthogonal set of functions, so that only the zero vibration change would be allowed. When this is "almost" the case, the functions are "almost orthogonal" thus favoring the zero change in vibrational quantum number.

Turning now to the coupling processes 2 and 4, one may interpret these in essentially the same way as a consequence of Born's quantum mechanical analysis of the problem of the collision of a charged particle with an atomic or molecular system.[7] A free electron colliding with a molecule interacts with each of the electrons and the nuclei in the molecule according to the Coulomb law. If this energy of interaction be developed in negative powers of the distance of the electron from the center of gravity of the molecule, the development begins with an inverse cube term representing the interaction between the free electron and the dipole moment of the atom. The higher powers of the development correspond to the interaction with the quadrupole and higher moments of the electric charge of the molecule. When these are neglected, Born's analysis shows the probabilities of excitation of a molecule by electron impact are proportional to the square of the matrix component of the electric moment of the molecule in regard to the initial and final states in question. This, however, is the same quantity as that which measures the probabilities of transition associated with light emission and absorption. Therefore, to the order of approximation which replaces the electron interac-

---

[6] Compare treatment and discussion in *Phys. Rev.*, **28**, 1182 (1926).
[7] Born, *Zeits. Phys.*, **38**, 803 (1926).

tion with the molecule by that between the electron and equivalent dipole, one has that vibration transitions associated with electronic excitation of the molecule will be the same whether the process is radiative or a result of a collision. This is the justification of the argument of Birge and Sponer in discussing the experiments of Hogness and Lunn.

Using this same analysis of the collision problem, it is clear that the action of a colliding electron on a molecule in exciting vibration transitions without electron excitation, the fourth type of coupling in the list, is a consequence of the non-vanishing of the same matrix components as those which measure the probability of vibration transitions in infra-red, vibration-rotation bands. The correlation is, however, not a sharp one, for in the collision process the electric moments of higher order of the molecule may be active.

Wave Mechanics and the Normal State of the Hydrogen Molecule

EDWARD UHLER CONDON*

Munich, Germany

The problem of the motion of a particle attracted by two fixed centers of force according to the Coulomb force law can be treated by classical mechanics and has been used in quantum theory by Pauli and Niessen for a theory of the hydrogen molecule ion.[1] In the quantum mechanics, where the energy levels are determined as the "eigenwerte" of Schrodinger's equation, the variables are separable and the boundary value problem is easily set up. But thus far a satisfactory treatment of the differential equations involved is lacking. Burrau[2] has recently carried out a numerical integration of the problem for the lowest energy level of an electron moving under the influence of two fixed centers of Coulomb attraction as a function of the distance apart of these centers. In this paper, Burrau's data are used to give a semi-quantitative discussion of the neutral hydrogen molecule. His values are:

Nuclear separation	1.0	1.3	1.6	1.8	2.0	2.2	2.4	2.95
Electronic energy	2.896	2.648	2.436	2.309	2.204	2.109	2.025	1.836

The unit of separation is the Bohr 1_1 orbit radius of hydrogen atom, that of energy is the ionization potential of atomic hydrogen.

In all this work the tacit assumption is made that, because of the large masses of the nuclei, the problem can be solved regarding the nuclei as fixed at a distance which is one of the parameters of the problem. When the energy of the electronic motion as a function of the distance is known, the energy of the Coulomb repulsion of the fixed nuclei is added and so the variation of the total energy of the non-rotating non-vibrating molecule

---

* NATIONAL RESEARCH FELLOW.

[1] For a review of the status of the $H_2{}^+$ and H_2 problems prior to the new quantum mechanics see Van Vleck, *Quantum Principles and Line Spectra*, p. 88. Also discussion by Kemble in last chapter of National Research Council Report on "Molecular Spectra in Gases."

[2] Burrau, *Danske Vidensk. Selskab. Math.-fys. Meddel.*, 7, 14, Copenhagen (1927).

with nuclear distance is found. The minimum of this curve is taken as the "equilibrium" separation of the nuclei and the value of the minimum is taken as the energy of the molecule in that electronic state. (More correctly, the small amount $(1/2)h\nu$ is to be added to the minimum value.) If the nuclei are no longer regarded as fixed this curve is regarded as giving the "law of force" governing the rotational and vibrational motions of the molecule. That this is the correct procedure in the classical mechanics was shown by Born and Heisenberg:[3] that it remains correct in the quantum mechanics has not yet been definitely proved. There is no reason to believe, however, that it is not correct, and it will be used here without further justification.

When the nuclei of a hydrogen molecule ion are far apart one is dealing virtually with a free hydrogen atom and a proton. The electronic energy is then mainly that of the Coulomb interaction between the proton and the electronic charge of the atom. If the atom were not Stark-affected by the proton, this would be just equal to the nuclear repulsion and the total energy would be simply R for all values of the nuclear separation (all values are negative), where R is the Rydberg constant. But the proton induces a polarization of the H-atom and, therefore, the energy of proton-electron interaction is greater than that of proton-proton. On the other hand, when the nuclear separation is zero and the electron moves under the influence of a double central charge, the energy is that of the lowest state of ionized helium. Burrau's numerical integrations supply values of the electronic energy for intermediate electronic separations. When the nuclear repulsive energy (curve b, Fig. 1) is added to Burrau's values there results curve a, figure 2, which is Burrau's curve for H_2^+. The equilibrium separation is 2 units (i.e., 2 times the radius of the Bohr 1_1 hydrogen orbit) and the minimum energy is $1.204R = 16.28$ volts. The heat of dissociation is $0.204R = 2.76$ volts. Burrau checks the value with

---

[3] Born and Heisenberg, *Ann. Physik*, **74**, 1 (1924).

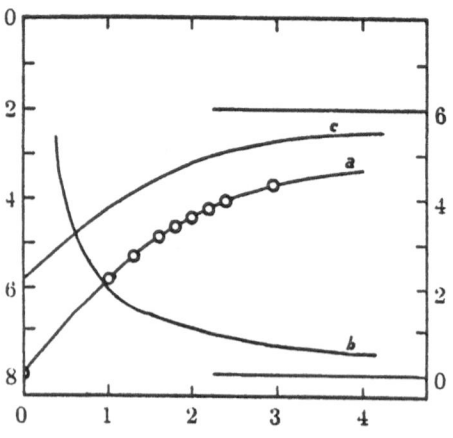

FIGURE 1. Electronic and nuclear energy in H_2. a, Values for non-interacting electrons. b, Coulomb energy of nuclear repulsion. c, Approximate electronic energy curve for interacting electrons. Units: ordinates, 1 = Rydberg constant, abscissas, 1 = radius of first Bohr orbit in hydrogen atom.

FIGURE 2. Resultant energy curves in H_2^+ and H_2.
a, Burrau's curve for H_2^+. b, Curve for H_2 for
non-interacting electrons. c, Approximate curve for
H_2 with interacting electrons. The small circle in
the crook of curve b, represents the equilibrium
position and energy on Hutchisson's classical
crossed-orbit model of H_2. Units: same as figure 1
(note different scales of ordinates for H_2 and H_2^+).

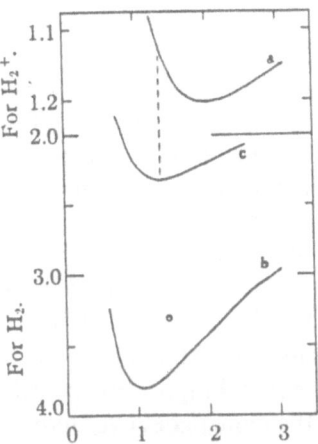

experiment by an indirect comparison with the ionization potential of H_2 as
discussed later in this paper.

Turning now to the neutral molecule one expects, on the Pauli principle
of assigning quantum numbers, that the two electrons will be in equivalent
orbits. The starting point, therefore, for the approximate treatment of the
problem is a model in which the two electrons have no mutual influence
and each moves as it would if alone in the ground state of H_2^+ as given by
Burrau. The electronic energy of this model at each distance is evidently just
twice that for H_2^+. This curve of doubled H_2^+ values is given here as a of
figure 1. Combined with the Coulomb proton-proton energy this yields curve
b, figure 2, for the energy curve of the neutral H_2 molecule with uncoupled
electrons. This gives an equilibrium separation of 1.075 units, i.e., a moment
of inertia of 2.7×10^{-41} gr. cm.^{2} The heat of dissociation is $1.800R = 24.36$
volts.

Naturally, such a model gives only a very rough approximation to the
truth. But it is to be observed that the above model does set a definite lower
limit on the moment of inertia of the molecule. For the electronic interaction, whatever its amount, will be positive and will decrease monotonously
with increasing nuclear separation, since it is the repulsive potential of inter-
acting-like charge. It acts to increase the ordinates of curve b, figure 1, by
decreasing amounts and, therefore, shifts the minimum of the resultant curve
to larger abscissas. This seems to be an important conclusion inasmuch as
*the lower limit here definitely given by quantum mechanics is greater than nine
of the thirteen values obtained on various theories from specific heat data* as
presented in the recent thorough review of the the subject by Van Vleck and
Hutchisson.[4]

---

[4] Van Vleck and Hutchisson, *Physic. Rev.*, **28**, 1022 (1926).

Turning now to the electronic interaction, the analysis of Hund[5] provides the important result that the electronic term of the lowest state of a molecule changes continuously from its value for a neutral atom of equal number of electrons to its value for the dissociated atoms, according to the new quantum mechanics. Herein lies an important difference between the old and the new quantum theory which is essential to the argument of this paper. That unexcited molecules dissociate into two unexcited atoms as a result of vibrations of infinite amplitude has been shown empirically by Birge and Sponer.[6]

The first approximation to the electron interaction in unexcited helium has been computed by Unsöld,[7] by means of the wave mechanics, who finds $5.5R$ for the whole atom, i.e., $1.5R = 20.3$ volts for the ionization potential. Empirically the value is $5.818R$ for the total energy. At large distances the model goes over into two neutral hydrogen atoms. The electronic energy will, therefore, be asymptotically equal to the Coulomb interaction of an electron and a proton, for it is made up of the repulsion of the two electrons and the attraction of each proton for the electron of the other atom. Moreover, the situation is now that of the interaction of two neutral units so that polarization deformation of each atom by the other will be much smaller than in the case of $H_2{}^+$. Inasmuch as Burrau's work shows that at a distance of 2.95 units the electronic part of the $H_2{}^+$ energy differs from the pure Coulomb by about $0.1R$, it is safe to assume for the H_2 molecule a closer approach to Coulomb value for abscissas greater than 3 units. For intermediate points, a natural assumption is to reduce the doubled Burrau value in the ratio $5.818 : 8.00$ in order to secure agreement with helium. If this is done it is found that the resulting curve joins on to the Coulomb curve smoothly. Curve c of figure 1 has been drawn from the theoretical values so reduced up to the value for 2.4 and joined on to a Coulomb curve for abscissa values of 3 and greater.

The resut of combining c and b of figure 1 is to give c of figure 2 as the energy curve for the hydrogen molecule. The minimum of this curve corresponds to a moment of inertia of 4.26×10^{-41} gr. cm.^{2}, and to a heat of dissociation of 4.4 volts. The latter vatue agrees to within 0.1 volt with the band spectrum value of Witmer and of Dieke and Hopfield.[8] These figures should be compared with a moment of inertia of 4.91×10^{-41} gr. cm.^{2} and $1.422R$ heat of dissociation found by Hutchisson[9] from a cross orbit model of H_2 on classical quantum theory.

Another interesting consequence follows from the retation of the H_2 energy curve to that of $H_2{}^+$. According to a principle put forward by Franck, changes involving the electrons in a molecule will affect the nuclei mainly indirectly

[5] Hund, *Zs. Physik*, **40**, 742 (1927).
[6] Birge and Sponer, *Physic. Rev.*, **28**, 259 (1926).
[7] Unsöld, *Ann. Physik*, **82**, 355 (1927).
[8] Witmer, *Physic. Rev.*, **28**, 1223 (1926); Dieke and Hopfield, *Zs. Physik*, **40**, 299 (1926).
[9] Hutchisson, *Physic. Rev.*, **29**, 270 (1927).

through the change in molecular binding.[10] According to this view, the most probable event in an electron collision experiment by means of which an electron is removed from H_2, is the removal of an electron while the nuclei are at a distance of 1.350 units. This requires an amount of energy given by the difference between $2.325R$ and the ordinate of the H_2^+ curve at 1.350, namely $1.125R$. This is the theoretical apparent ionization potential and amounts to $1.2R$ or 16.2 volts in good agreement with the mean experimental value 16.1 given by Franck and Jordan.[11] On the other hand, the true energy of the process $H_2 \rightarrow H_2^+ + e^-$, where H_2^+ is in the lowest state, is the difference between the ordinates of the minima of the H_2 and H_2^+ curves, i.e., $1.12R = 15.2$ volts, comparing favorably with value of Witmer and of Dieke and Hopfield from the ultra-violet band spectrum of H_2.

The theoretical value of the frequency of vibration, depending on the curvature of the curve at its minimum, is naturally more uncertain. Calculation shows that the curve gives a frequency of vibration of 5300 cm.$^{-1}$, about 20% higher than the value 4360 cm.$^{-1}$ from experiment.[8] As for the moment of inertia, while it is larger than most of the values from specific heat theories, it is in accord with the larger values which have been found by Richardson and Tanaka[12] from analysis of the hydrogen bands.

In conclusion it seems proper to emphasize that Burrau's calculation of H_2^+ and the extension here to H_2 constitute the first quantum-theoretic quantitative discussion of the binding of atoms into molecules by electrons—the valence forces of chemistry. The quantitative success of the new quantum mechanics in the face of the classical theory's failure must serve to lend strong support to the new methods.

---

[10] Franck, *Trans. Far. Soc.*, **21**, part 3 (1925). See also discussion by Birge, p. 248 of National Research Council Report on "Molecular Spectra in Gases," and quantum mechanical treatment by Condon, these PROCEEDINGS, **13**, 462 (1927).
[11] Franck and Jordan, *Anregung der Quantensprünge durch Stosse.*
[12] Richardson and Tanaka, *Proc. Roy. Soc.*, **106A**, 663 (1924).

The Zeeman Effect of the Symmetrical Top According to Wave Mechanics

EDWARD U. CONDON†

Zürich, Switzerland

ABSTRACT

The alteration of the quantum-theoretical energy levels of a symmetrical top due to the action of a magnetic field on a charge which is fixed to the top, is investigated by means of the perturbation theory of the wave mechanics. The final formula for the change in the energy levels is given below in Eq. (10).

The model of the rigid rotator with two different moments of inertia (symmetrical top) has proved useful in the theory of molecular spectra. Investigations of the unperturbed energy levels and radiation amplitudes (matrix components of the electric moment) according to wave mechanics have been given by Reiche, Reiche and Rademacher, Kronig and Rabi, and by Manneback.[1] Dennison[2] has also treated the problem by the matrix methods. Reiche has investigated the first-order Stark effect while Manneback has investigated the second-order Stark effect as well.

The Zeeman effect, i.e., the perturbation of the energy levels of the top by a magnetic field, has not been hitherto investigated. The results of the calculation of this perturbation are here communicated.

The notation here used is that of Kronig and Rabi. The top has two of its moments of inertia equal to A and the third, about the axis of symmetry, equal to C. The electrical properties of the top are summarized in the assumption that a charge, e, is fixed to the top at a point whose distance off the axis of symmetry is a, while the distance along the axis of symmetry from the center of gravity of the top to the foot of the perpendicular from this charge on to the axis is c.

---

† National Research Fellow.
[1] Reiche, Zeits. f. Physik **39**, 444, (1926); Reiche and Rademacher, Zeits. f. Physik **42**, 453 (1927); Kronig and Rabi, Phys. Rev. **29**, 262 (1927); Manneback, Phys. Zeits. **28**, 72 (1927).
[2] Dennison, Phys. Rev. **28**, 318 (1926).

The wave equation can be conveniently derived with the aid of a variation principle due to Fock.[3] The Lagrangian function for the particle in a magnetic field is

$$L = T - (e/c)A \cdot v \tag{1}$$

where T is the kinetic energy, A the vector potential, and v the velocity of the charged particle. Using the Eulerian angles θ, ϕ, ψ, one can write the magnetic term as

$$(e/c)A \cdot v = H(\alpha\dot{\theta} + \beta\dot{\phi} + \gamma\dot{\psi}) \tag{2}$$

in which H is the field strength and

$$
\begin{aligned}
\alpha &= -2\pi\mu v \cdot a \cos\phi (c \cos\theta + a \sin\theta \sin\phi) \\
\beta &= 2\pi\mu v \cdot a(a \cos\theta - c \sin\phi \sin\theta) \\
\gamma &= 2\pi\mu v [a^2 \cos^2\phi + (c \sin\theta - a \cos\theta \sin\phi)^2]
\end{aligned}
\tag{3}
$$

where μ is the electronic mass and v the electronic Larmor precession frequency for unit field strength.

The Hamiltonian function is obtained in the usual way by expressing $p_\theta\dot{\theta} + p_\phi\dot{\phi} + p_\psi\dot{\psi} - L$ as a function of the momenta, p_θ, p_ϕ, and p_ψ. The result is

$$H = \frac{1}{2A}(p_\theta + H_\alpha)^2 + \frac{1}{2}\left(\frac{\cos^2\theta}{A \sin^2\theta} + \frac{1}{C}\right)(p_\phi + H_\beta)^2 + \frac{1}{2A \sin^2\theta}(p_\psi + H_\gamma)^2$$

$$- \frac{\cos\theta}{A \sin^2\theta}(p_\phi + H_\beta)(p_\psi + H_\gamma) \tag{4}$$

Applying Fock's variation process to this Hamiltonian function, one finds the following form for the wave-equation on neglecting terms in H^2:

$$DU + \frac{8\pi^2 E}{h^2}U = \frac{2\pi i H}{h}\Delta U \tag{5}$$

Here D is the differential operator occurring in the wave-equation for the unperturbed top while Δ is a linear differential operator of the first order representing the action of the magnetic field:

$$DU \equiv \frac{1}{A \sin\theta}\frac{\partial}{\partial\theta}\left(\sin\theta\frac{\partial U}{\partial\theta}\right) + \left(\frac{\cos^2\theta}{A \sin^2\theta} + \frac{1}{C}\right)\frac{\partial^2 U}{\partial\phi^2} + \frac{1}{A \sin^2\theta}\frac{\partial^2 U}{\partial\psi^2}$$

$$- \frac{2\cos\theta}{A \sin^2\theta}\frac{\partial^2 U}{\partial\phi \partial\psi} \tag{6}$$

---

[3] Fock, Zeits. f. Physik **38**, 242 (1926).

$$\Delta U \equiv \frac{1}{A \sin\theta} \frac{\partial}{\partial\theta}(\alpha U \sin\theta) + \frac{\alpha}{A}\frac{\partial U}{\partial\theta} + \left(\frac{\cos^2\theta}{A\sin^2\theta} + \frac{1}{C}\right)\left(\frac{\partial}{\partial\phi}(\beta U) + \beta\frac{\partial U}{\partial\phi}\right)$$

$$+ \frac{1}{A\sin^2\theta}\left(\frac{\partial}{\partial\psi}(\gamma U) + \gamma\frac{\partial U}{\partial\psi}\right) - \frac{\cos\theta}{A\sin^2\theta}\left(\frac{\partial}{\partial\phi}(\gamma U) + \frac{\partial}{\partial\psi}(\beta U)\right)$$

$$+ \beta\frac{\partial U}{\partial\psi} + \gamma\frac{\partial U}{\partial\phi}\right). \tag{7}$$

The unperturbed equation ($H = 0$) is degenerate in that the energy levels depend only on the quantum numbers, j and n, while the unperturbed characteristic functions are:

$$U_{jmn} = \Theta_{jmn}(\theta)\cdot e^{in\phi}\cdot e^{im\psi} \tag{8}$$

Since ψ does not occur explicitly in the perturbation term, ΔU, this coordinate remains separable and it is thus unnecessary to apply the special apparatus of the perturbation theory for degenerate systems. According to the simple perturbation theory,[4] the amount of the perturbation becomes:

$$\Delta E_{jmn} = \frac{ihH}{4\pi}\int \bar{U}_{jmn}\Delta U_{jmn}\sin\theta\ d\theta\ d\varphi\ d\psi \tag{9}$$

in which it is supposed that the characteristic functions have been normalized and where $\bar{U}_{jmn}$ means the conjugate complex function to U_{jmn}.

The computation of the perturbation energy is readily effected since besides integrals over exponential functions the only integral needed is

$$\int_0^\pi \Theta_{jmn}{}^2(\theta)\cdot\cos\theta\cdot\sin\theta\ d\theta = -nm/j(j+1)$$

The integral is a simple consequence of well-known properties of the hypergeometric function, to which $\Theta_{jmn}(\theta)$ is related.[5] The result, therefore, for ΔE_{jmn} is

$$\Delta E_{jmn} = -mh\nu_L\left[\frac{\mu(c^2 + \frac{1}{2}a^2)}{A} + \frac{n^2}{j(j+1)}\left(\frac{\mu(c^2 + \frac{1}{2}a^2)}{A} - \frac{\mu a^2}{C}\right)\right] \tag{10}$$

in which ν_L is the Larmor precession frequency for a particle of mass μ and charge e in a field of strength H, i.e.,

$$\nu_L = eH/2\mu c$$

The ratio $n^2/j(j+1)$ is to be given the value 1 when $j = n = 0$.

One sees that the alteration of the energy levels for a molecule represented by this model will therefore be small compared with the Zeeman effect for

---

[4] Schrödinger, Ann. d. Physik **80**, 437 (1926).
[5] See e.g. Reiche, Zeits. f. Physik **39**, 453, Eq. (44) (1926).

atoms in the ratio of electron mass to nuclear mass.[6] (The distances a and c are of the same order of magnitude as the distances of the nuclei from the center of gravity of the top-molecule.) The formula deviates from that of the old quantum theory in the appearance of $j(j + 1)$ in place of j^2, the same change as occurs in the formula for the unperturbed energy levels.

The model of a symmetrical top is often used for discussions of band spectra for diatomic molecules having states with a resultant electronic angular momentum about the line joining the two nuclei. In this case the moment of inertia, C, arises solely from the electronic structure of the molecule, hence the $\mu a^2/C$ is of the order of magnitude 1. The other terms may be neglected in comparison with it.

The minus sign which affects this term can then give rise to an inverted Zeeman effect, as already noted by Lenz as a consequence of the old theory. If $n = 0$ in both initial and final states, there is no Zeeman effect of atomic order of magnitude. If $\Delta n = 0$ there is a splitting into the normal Lorentz triplet (when one applies the selection rules for m as given quantum mechanically by Dennison, Kronig and Rabi, and Reiche) except that the spacing decreases with increasing j, i.e., as one moves away from the origin of a band. The spacing varies as $1/j(j + 1)$ or almost inversely as the square of j. But if $n = 0$ in one state and $n = 1$ in the other then an inversion of one set of levels occurs. This has the consequence that a pattern of $2j + 1$ equi-distant lines appears, due to the non-splitting for $n = 0$ and the splitting for $n = 1$. The extreme width of the pattern is $2/(j + 1)$. In other words for such bands the broadening due to the Zeeman effect will vary with the inverse first power of j instead of the inverse second.

This calculation was carried out while the writer was a visitor at the summer semester of the Institute for Theoretical Physics at Munich. I take this opportunity to express to Professor Sommerfeld and his co-workers in Munich my appreciation of friendly advice and discussions during the semester and of the hearty *Münchner Gemütlichkeit*, which pervades the life of the institute.

---

[6] For a review of the present status of theory and experiment concerning Zeeman effect of band spectra, see section on Zeeman effect by Kemble in the Report of the National Research Council on Molecular Spectra in Gases.

The Physical Pendulum in Quantum Mechanics

Edward U. Condon

Department of Physics,
Columbia University

ABSTRACT

It is pointed out that the Mathieu functions of even order are the characteristic functions of the physical pendulum in the sense of Schrödinger's wave mechanics. The relation of various properties of the functions, as known from purely analytical investigations of them, to the pendulum problem is discussed.

The problem of the physical pendulum, that is, of the motion of a mass-point constrained to move in a circle and acted on by a uniform force held, has played such a great role in the study of analytic mechanics that a discussion of the same problem from the standpoint of Schrödinger's wave mechanics cannot be without interest. It turns out that the characteristic functions are certain of the Mathieu, or elliptic cylinder, functions and that the arguments from mechanics serve to illustrate in an interesting way many of the properties of these functions.

Let the mass of the particle be μ and let its position in a circle of radius a be designated by the angle θ. It will be supposed that the particle carries an electric charge e, and that there is a uniform electric held acting in such a way that the potential energy function is $-eEa\cos\theta$. That is, the force is in the direction of $\theta = 0$. Under these circumstances the wave-mechanical equation becomes

$$\frac{d^2\psi}{d\theta^2} + \frac{8\pi^2\mu a^2}{h^2}[W + eEa\cos\theta]\psi = 0,$$

in which W is the energy level parameter. The energy levels are the values of W for which this equation possesses solutions which have the period 2π in θ. Introducing the variable, $x = \frac{1}{2}\theta$, and the abbreviations,

$$\alpha = 8\pi^2\mu a^2 W/h^2, \quad q = 2\pi^2\mu a^2 eE/h^2,$$

the equation appears in the usual form for Mathieu's equation,

$$d^2\psi/dx^2 + (4\alpha + 16q\cos 2x)\psi = 0,$$

where, now, one seeks solutions with the period π in x.

The solutions of this equation which have the period 2π in x are known as Mathieu functions.[1] For $q = 0$, the case of zero field strength, the problem becomes simply that of a free rotator with fixed axis. The characteristic functions are 1, $\sin 2x$, $\cos 2x$, $\sin 4x$, $\cos 4x$... with the associated values of α equal to 0, 1, 1, 4, 4.... The problem is degenerate since there are two distinct characteristic functions associated with each energy level, except the first one. The characteristic function for any degenerate state is an indeterminate linear combination of the two functions associated with that state. In particular, the runctions which correspond to progressive rotatory motion in the two opposing senses are e^{2imx} and e^{-2imx}, rather than $\cos 2mx$ and $\sin 2mx$. These latter functions correspond in some way to equal numbers of rotators turning in opposite senses and these connect in a continuous manner with the non-degenerate characteristic functions for $q \neq 0$, as $q \to 0$. The standard notation for the Mathieu functions is $ce_n(x, q)$ and $se_n(x, q)$ where as $q \to 0$ these become equal to $\cos nx$ and $\sin nx$ respectively. All of these have period 2π, while those in which n is even have the period π in x and so they are the characteristic functions of the pendulum problem.

It will be observed that the zero from which energy is reckoned in the wave equation is from the position at which $\theta = \pi/2$ or $x = \pi/4$. If instead it is reckoned from $\theta = 0$, the minimum of the potential energy curve, one has to add $4q$ to each value of α.

It is of interest now to consider a number of properties of the functions, which have been obtained by purely analytical means, in the light of the pendulum problem. Firstly, it is clear that the functions will bear an invariant relation to the minimum of the potential energy curve, hence yield such relations[2] as

$$ce_{2n}(\tfrac{1}{2}\pi - x, -q) = (-1)^n \cdot ce_{2n}(x, q).$$

Jeffreys[3] has shown that there are no allowed values of α such that $\alpha < -4q$. Physically this means that there are no states for which the total energy is less than the minimum potential energy permissible for the system, and therefore appears natural enough. The relation, $\alpha = 4q$, is a critical one for classical mechanics in that for $\alpha > 4q$ the motion is rotatory while for $\alpha < 4q$ it is oscillatory. This value has also shown itself as a critical one in the analytical theory of the functions.

---

[1] An account of them is given in Whittaker and Watson, A *Course of Modern Analysis*, Chap. 19 (1920). This is, however, quite incomplete now because of the many more recent investigations by British and Scottish mathematicians.

[2] See e.g. Goldstein, Trans. Cambr. Phil. Soc. **23**, 303, (1927) Par. 1.5. This memoir contains a good many of the newer results not given in Whittaker and Watson.

[3] Jeffreys, Proc. London Math. Soc. **23**, 437, (1924–25). This paper and the one preceding it are especially interesting in that the methods of approximate integration which he uses are closely related to those by means of which the connection between classical mechanics and quantum mechanics is established.

For $\alpha > 4q$, Jeffreys finds (approximately) that α must be such that

$$\int_0^{2\pi} (4\alpha + 16q \cos 2x)^{1/2} \cdot dx = 2\pi m,$$

in which m is an integer. Recalling the meaning of α and q, this requirement becomes

$$\int_0^{2\pi} p_\theta \, d\theta = \tfrac{1}{2} mh,$$

so that the condition reduces to the classical quantum condition for even values of m, the ones which correspond to allowed quantum motions.

The most complete tables of the values of α as a function of q are those of Goldstein (loc. cit.). He has given the values of α for $ce_0(x, q)$, $se_2(x, q)$ and $ce_2(x, q)$ besides several others which are not related to the pendulum problem. He gives an asymptotic expansion good for small m and large q as follows:

$$\alpha \sim -4q + (2m + 1)(2q)^{1/2} - \cdots .$$

In this one recognizes that $(\alpha + 4q)$ which is the energy counted up from the minimum of potential energy at $\theta = 0$ goes linearly with the number m in just the way that the energy levels of the harmonic oscillator go in wave mechanics. Moreover, the interval between levels, $2(2q)^{1/2}$, when expressed as energy is exactly equal to $h\nu$ where ν is the frequency of the small oscillations of a pendulum in the field of strength E, reckoned on classical mechanics, i.e.,

$$\nu = (eE/4\pi^2 \mu a)^{1/2}.$$

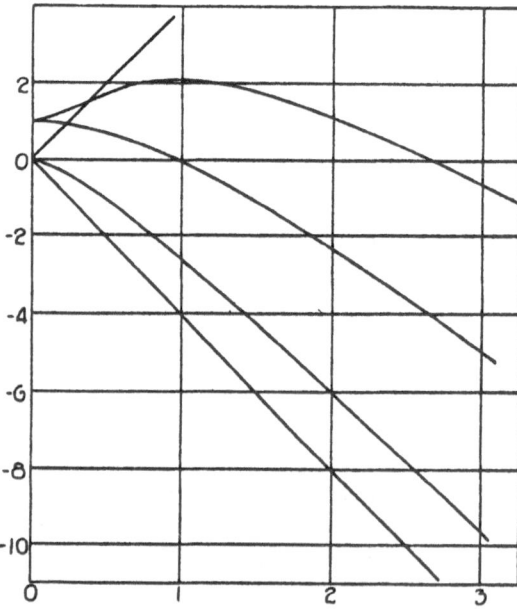

FIGURE 1. The three lowest energy levels of the physical pendulum as a function of field strength (abscissas, q. ordinates, α). The straight line of positive slope divides the region of classical rotatory motions from that of classical oscillatory motions. That of negative slope gives the value of the potential energy minimum which forms a more natural origin from which to count the energy of the small oscillation states.

The rapidity with which the three lowest energy levels approach the values appropriate to the small oscillations theory is indicated in Fig. 1, where α is plotted as a function of q, from Goldstein's tables.

An interesting reature is the way the second and third energy levels cross the line, $\alpha = 4q$, without any discontinuity, this being the critical place at which the associated classical motion changes character from rotatory type to oscillatory by passing through a motion of infinite period. The discontinuous behavior of a pendulum moving with constant energy in a field of slowly decreasing strength caused some trouble in the old mechanics, as noted by Ehrenfest and by Bohr, That this discontinuity is not a feature of the wave mechanical treatment was first recognized by Hund.[4] In a sense this discontinuity still appears in wave mechanics but at $q = 0$ instead of at $q = \alpha/4$, for the characteristic functions, as $q \to 0$, do not join on continuously with those solutions of the equation for the force-free rotator which represent rotatory motion, as already remarked.*

Similarly one expects for small values of the quantum number and large values of q, that the characteristic functions will go over into the characteristic functions of the harmonic oscillator problem. The Mathieu functions do have just such an asymptotic connection with the Hermitian polynomials or parabolic cylinder functions.[5] This behavior is just what one would expect from the relation to the pendulum problem. Thus it makes clear the theorem of Ince (loc. cit.) concerning the clustering of the zeros of the Mathieu functions within a region of the order of magnitude $-Kq^{-1/4} < x < +Kq^{-1/4}$ as $q \to \infty$, where K is a constant. This is because the wave functions in wave mechanics only show a distinctly oscillatory character inside the region of the associated classical motion, where the de Broglie wavelength, h/p, of the system is real. A simple calculation shows that the amplitude of the classical motion for a given quantum state tends to zero as $q^{-1/4}$ for $q \to \infty$.

It thus appears that the principal properties of the Mathieu functions of even order are simply related to the mechanical problem of which they are the characteristic functions in Schrödinger's wave mechanics.

---

[4] Hund, Zeits. f. Phys. **40**, 742, (1927) especially footnote, p. 750.
[5] See e.g., Ince, Jl. London Math. Soc. **2**, 46, (1927).
* *Added in proof:* The ψ function that joins on with e^{i2nx} is, of course,

$$ce_{2n}(x,q)e^{2\pi iE_1t/h} + i\, se_{2n}(x,q)e^{2\pi iE_2t/h}$$

where E_1 and E_2 are the energy levels associated with ce_{2n} and se_{2n} respectively. If one compute the quantum mechanical expression for the current associated with this ϕ, it will be seen that it depends on the time through a factor $\cos 2\pi(E_1 - E_2)t/h$. For small q, $E_1 - E_2 \to 0$ so the current reverses itself with a small frequency which becomes zero for $q = 0$. This behavior reminds one of the "gallows problem" of Ehrenfest and Tolman (Phys. Rev. **24**, 287 (1924)), although here there are no forces due to twist of thread to cause the long period reversal!

Wave Mechanics and Radioactive Disintegration

Ronald W. Gurney, Edward U. Condon

Palmer Physical Laboratory
Princeton University

After the exponential law in radioactive decay had been discovered in 1902, it soon became clear that the time of disintegration of an atom was independent of the previous history of the atom and depended solely on chance. Since a nuclear particle must be held in the nucleus by an attractive field, we must, in order to explain its ejection, arrange for a spontaneous change from an attractive to a repulsive field. It has hitherto been necessary to postulate some special arbitrary 'instability' of the nucleus; but in the following note it is pointed out that disintegration is a natural consequence of the laws of quantum mechanics without any special hypothesis.

It is well known that the failure of classical mechanics in molecular events is due to the fact that the wave-length associated with the particles is not small compared with molecular dimensions. The wavelength associated with α-particles is some 10^5 smaller, but since the nuclear dimensions are smaller than atomic in about the same ratio, the applicability of the wave mechanics would seem to be ensured.

In the classical mechanics, the orbit of a moving particle is entirely confined to those parts of space for which its potential energy is less than its total energy. If a ball be moving in a valley of potential energy and have not enough energy to get over a mountain on one side of the valley, it must certainly stay in the valley for all time, unless it acquire the deficiency n energy somehow. But this is not so on the quantum mechanics. It will always have a small but finite chance of slipping through the mountain and escaping from the valley.

In the diagram (Fig. 1), let O represent the centre of a nucleus, and let $ABCDEFG$ represent a simplified one-dimensional plot of the potential energy. The parts ABC and GHK represent the Coulomb field of repulsion outside the nucleus, and the internal part $CDEFG$ represents the attractive field which holds α-particles in their orbits. Let DF be an allowed orbit the energy of which, say 4 million volts, is given by the height of DF above OX. Approximately, we can say that with this orbit will be associated a wave-function which will die away exponentially from D to B. Again, corresponding to motion outside the nucleus along BM, there will be a wave-function which will die away exponentially from B to D. The fact that these two functions

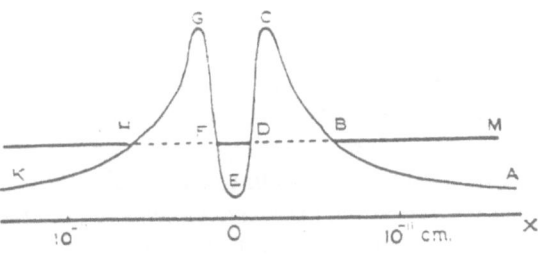

FIGURE 1

overlap in the region *BD* means that there is a small but finite probability that the particle in the orbit *DF* will escape from the nucleus along *BM*, acquiring kinetic energy equal to the height of *DFBM* above *OX*, say 4 million volts. This occurrence will be spontaneous and governed solely by chance.

The rate of disintegration, that is, the probability of escape, depends on the amount of overlapping of the wave-functions in the regions *DB* and *FH*, and this is extremely sensitive to the height to which the potential curve at *C* rises above *BDF*. By varying this height through a small range we can obtain all periods of radioactive decay from a fraction of a second, through the 10^9 years of uranium, to practical stability. (In considering the transmutation of a molecule into its isomer, Hund found a similar vast range of transformation periods, *Zeit. f. P.*, **43**, 810; 1927.) If the potential curves for the interaction of an α-particle with the various radioactive nuclei are similar, we can obtain a qualitative understanding of the Geiger-Nuttall relation between the rate of disintegration and the range of the emitted α-particles. For the α-particles of high energy the wave function for outside motion will overlap that for the inside motion more, and the rate of disintegration will be greater.

Besides obtaining a general idea of the mysterious instability of the nucleus, we can visualise in this way one of the most puzzling results of recent experimental work. An α-particle having the same range (2.7 cm.) as those emitted by uranium should, if fired directly at the uranium nucleus, penetrate its structure; while faster α-particles should do so, even when not fired directly at the nucleus. It was therefore disconcerting when, on examining the scattering of fast α-particles fired at uranium, Rutherford and Chadwick (*Phil. Mag.*, **50**, 904; 1925) could find no indication of any departure from the inverse square laws. But from the model outlined above, this is what would be expected. For if the height of *BM* above *OX* represents the energy of the uranium α-particles, then a faster particle fired at the nucleus will simply run part way up the hill *ABC* and return without having encountered any change in the repulsive field or any nuclear particles (which are describing orbits within the region *GEC*).

The peculiar property of the wave mechanical equations which finds application here has also been applied to the theory of the emission of electrons from cold metals under the action of intense fields (Oppenheimer, *Proc. Nat.*

Acad. Sci., **14**, 363; 1928; and Fowler and Nordheim, *Proc. Roy. Soc.*, A, **119**, 173; 1928). Ordinarily, an atom does not lose its electrons because the attractive field of the atom remains attractive to all distances. But when an intense field is applied, then the attractive field is reversed in sign a short distance from the atom. This makes the resultant potential energy curve similar to that in the diagram, and so the atoms begin to shed their electrons.

Much has been written of the explosive violence with which the α-particle is hurled from its place in the nucleus. But from the process pictured above, one would rather say that α-particle slips away almost unnoticed.

Nuclear Motions Associated with Electron Transitions in Diatomic Molecules

EDWARD U. CONDON

Palmer Physical Laboratory
Princeton University

ABSTRACT

The question of nuclear motions associated with electron transitions is discussed from the standpoint of quantum mechanics. It appears that Heisenberg's indetermination principle gives the clue to the inexactness of the earlier method based on Franck's postulate since its strict application calls for a violation of the principle. The existence of an entirely new type of band spectrum due to the wave nature of matter is predicted and the interpretation of Rayleigh's mercury band at 2476–2482 A.U. as of this type is suggested. Finally it is shown that while Franck's postulate is also true for electron jumps in atoms, it is of but trivial interest because its inexactness is much greater for the electrons than for heavy nuclei.

Two years have elapsed since Franck proposed a mechanism for the direct dissociation of molecules by light absorption[1] and since the extension of that mechanism to a theory of intensity distribution in band systems was made.[2] In the meantime the theory has been applied with gratifying success to the discussion of a number of band systems which have been recently analyzed. It has also been possible to derive the postulate underlying Franck's idea from the new quantum mechanics and thus to bring it into closer relation with the basic principles of quantum physics.

The first publication of this connection with quantum mechanics was very brief.[3] Since the connection provides one of the more easilv visualised applications of quantum mechanics and because of the rather wide applicability of the method in molecular problems it is therefore proposed here to give a fuller account of the connection. Besides providing a justification of Franck's assumption, the quantum mechanical method provides a distinct advance since it gives, in principle at least, an exact method for calculating intensities.

---

[1] Franck, Trans. Faraday Society **21**, part 3 (1925).
[2] Condon, Phys. Rev., **28**, 1182 (1926).
[3] Condon, Proc. Nat. Acad. Sci. **13**, 462 (1927).

This is illustrated particularly neatly in cases where absorption of light results in direct dissociation of the absorbing molecule (Cl_2, Br_2, ICl). It appears that the reason Franck's postulate gives only the most probable transitions and not the range of allowed transitions near the most probable is intimately connected with the so-called uncertainty principle of Heisenberg and Bohr. The quantum mechanical treatment also points to the possible existence of a new type of molecular band structure which is a direct manifestation of the de Broglie wave-length in the spectrum of a molecule.

§1. Franck's Postulate

According to the theory of band spectra the molecule exists in different electronic energy levels. Associated with each of these is an effective law of force which governs the motion of the two nuclei relative to each other and which is most conveniently described by drawing the curve which gives the energy of the molecule in the non-rotating non-vibrating state as a function of the distance of the two nuclei. In Fig. 1 we have such a pair of curves drawn for the two electronic states involved in the emission of the Swan bands of carbon.[4] The curves are inferred from the energy levels of the band spectrum. Thus the moment of inertia in a given state gives the abscissa of the minimum and the frequency of vibration gives the first coefficient in the Taylor's series for the curve around the minimum, etc.

Franck postulated that in an electron transition from a state in which the molecule is not vibrating the heavy nuclei will not be affected "momen-

---

[4] Based on the analysis of Shea, Phys. Rev. **30**, 825 (1927).

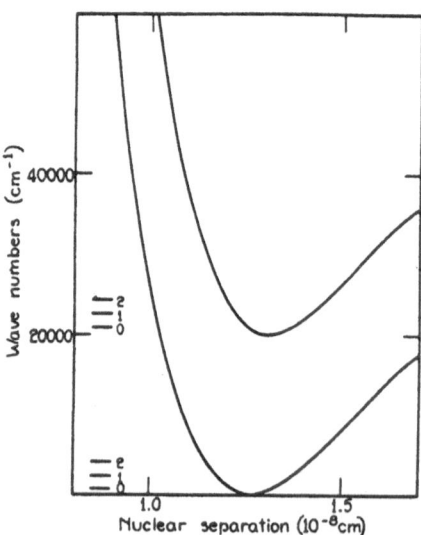

FIGURE 1. Potential energy curves for the electronic states of C_2 involved in the Swan band system.

tarily." Immediately after the transition the nuclei will have the same separation as before. But this will not be, in general, the equilibrium position of the law of force for the new electronic state and therefore the molecule will start to vibrate. In other words the most probable transition from the minimum of the curve for the initial state is to that point on the curve for the final state which has the same nuclear separation. This rule was then extended to make it applicable to transitions in which the molecule is vibrating initially as well as in the final state. One postulates that the electron transition affects neither the position nor the *momentum* directly. Then if the transition occurs while the nuclei are moving relatively, the most probable final state is that in which the position and momentum just after the transition is the same as that just before. This may be supplemented by supposing that the electron transition may occur with equal probability independently of the phase of the vibratory motion. This rule gives a definite probability for each of the possible final energies which can be communicated to the molecule.

It has now to be noted that it may happen that the curves lie as in Fig. 2 (which is approximately to scale for chlorine[5]). In this case absorption of light by the non-vibrating molecule would have as the transition determined by this rule that to energy level A which corresponds to dissociation plus a perfectly definite amount of translational energy in excess of the dissociation work, D. Franck first applied the method to a case of this sort—that of iodine, I_2—to explain the experiments of Dymond.[6] Experiment shows, however, that while the most favored transition is to energy level A, the actual absorption consists of a fairly broad continuous band. This indicates a lack

---

[5] The frequencies of vibration and heats of dissociation are from Kuhn, Zeits. f. Physik **39**, 77 (1926). The relative position of the two curves is not known since the rotation structure of the bands has not been analyzed; their relative position as drawn here was inferred roughly from the known intensity distribution by the theory of this paper.
[6] Dymond, Zeits. f. Physik **34**, 553 (1925).

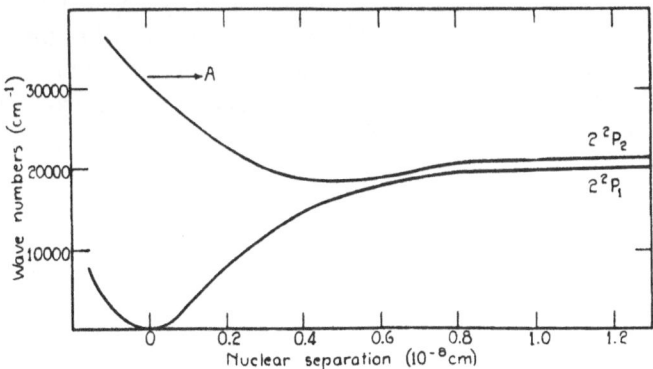

FIGURE 2. Approximate potential energy curves involved in the visible absorption spectrum of Cl_2. Ordinates and abscissas as in Fig. 1.

of sharpness about the principle which requires explanation. Similarly in a case like Fig. 1 where the rule leads to non-dissociating transitions, its strict application leads us to predict transitions to energies in the final state which are not quantum levels. As we know that these cannot occur we are again forced to recognize a certain lack of sharpness in the rule and to suppose that the probability of transition to non-allowed levels must perhaps be assigned to the credit of the nearest allowed level.

Making such adjustments, however, the principle has proven quite useful in giving an easily visualized rule for predicting the nuclear motions set up by electron transitions. The adjustments toward looseness which have to be made in the picture are the result of a too strict application of classical ideas to the problem in hand. For a rational treatment of the problem we turn therefore to the new quantum mechanics.

2. The Principle in Wave Mechanics

Following Schrödinger's method the behavior of a molecule is described in terms of the solution of his wave equation which involves $(3n + 6)$ independent variables if n is the number of electrons in the molecule. Not much headway can be made with the equation without the application of perturbation theory. Born and Oppenheimer have shown how the problem can be handled by successive approximations amounting to an expansion in powers of $(m/M)^{1/4}$ where m is the electronic mass and M is of the order of a nuclear mass.[7] It is because of the great mass of the nuclei that one can regard their motion as determined approximately by the wave-mechanical analogue of the time-mean of the forces of all the swiftly moving electrons on the nuclei. Their work justifies the postulate that the coordinates of the nucleus remain approximately separable from those of the electrons in Schrödinger's theory.

We can ignore the rotation motion of the molecule. Let r be the nuclear separation and x stand for the ensemble of all the electronic coordinates. A definite state of the molecule will be characterized by e, which is the ensemble of all the electronic quantum numbers, and by n, the vibrational quantum number. The fact of approximate separability means that to a good approximation the energy of the state $E(e, n)$ is the sum of an electronic part and a vibrational part,

$$E(e, n) = E_e + E_n$$

and that the wave function for such a state $\psi_{e,n}(x, r)$ is the product of a function of the electronic coordinates and one of the nuclear separation,

$$\psi_{en}(x, r) = X_e(x) \cdot R_{en}(r).$$

[7] Born and Oppenheimer, Ann. d. Physik **84**, 457 (1927).

Moreover the work of Born and Oppenheimer shows that the energies E_n and the function $R_n(r)$ are given by solving the one-dimensional wave equation in the variable r got by using an effective potential energy curve $U(r)$ representing the mean action of all the electrons in the state characterized by $X_e(x)$. Recently Heitler and London have obtained important results for the quantum theory of molecules concerning the curves $U(r)$ by approximations concerning $X_e(x)$. We may take empirically the $U(r)$ curve associated with each state as that one which is consistent with the quantum theory of the empirical energy levels of the band system in question.

Such empirical $U(r)$ curves in turn lead directly to solutions for E_n and $R_n(r)$ from a simple one-dimensional wave equation. The intensities of any given transition $(e', n') \rightarrow (e'', n'')$ have now to be considered. These are given unambiguously by the quantum mechanics in terms of the function $\psi_{en}(x, r)$. The electric moment of the molecule is given classically as the vector sum af the charges multiplied by their distances from a fixed point. Let $M(x, r)$ be this function. If the x are rectangular coordinates M is linear in all of the x and r. The electric moment which functions as a measure of the probability of the foregoing transition is

$$M(e', e'', n', n'') = \iint M(x,r)\psi_{e'n'}(x,r) \cdot \psi_{e''n''}(x,r)\, dx\, dr.$$

One can carry out (in principle) the integration over the electronic coordinates and obtain an intermediate sort of electric moment function which is characteristic of the electron transition $e' \rightarrow e''$ and which depends only on r,

$$M(e', e'', n', n'') = \int M(e'e'', r) \cdot R_{e'n'}(r) R_{e''n''}(r)\, dr$$

Although the range of integration is $-\infty$ to $+\infty$ the only parts which make appreciable contributions are in the range of coordinates in which both $R(e', n', r)$ and $R(e'', n'', r)$ have values not too near zero. In such a range it is probably always sufficient to take $M(e', e'', r)$ as a linear form $A + Br$ where it is supposed that the origin of r has been shifted to the center of the important part of the range of integration. In what follows a little brevity is gained by neglecting all but the constant part A since the reader will have no difficulty in seeing that the argument is not much affected by omitting the linear Br term. Although $M(e', e'', r)$ is unambiguously defined, almost nothing is known about it in the present state of our ignorance of the ψ functions for molecules.

We turn now to the consideration of the relative intensity of the different $n' \rightarrow n''$ transitions, associated with a definite $e' \rightarrow e''$ transition. It is a consequence of Schrödinger's theory that $R_n(r)$ will approach zero very rapidly outside of the region of the classical vibratory motion of the same energy. Inside this region it will be an oscillatory function.

From these properties of the functions $R(e, n, r)$ we read the quantum mechanical justification of the Franck postulate; together with a definite estimate of the uncertainty involved in its application. For example the $n' = 0 \to n'' = 0$ band by the Franck rule is most probable for zero change in moment of inertia. For large change it is an improbable transition. This follows because now the wave functions $R(e', 0, r)$ and $R(e'', 0, r)$ are both Gauss error curves, each one located symmetrically about the minimum of it own potential energy curve. The over-preciseness of the rule arises from the neglect of the fact that in the zero vibration state the particle is not precisely at the minimum of the curve but has a probability of being a short distance on either side of the minimum.

One sees then that when non-oscillatory wave functions come in question the integral has the largest value when the functions overlap most. This corresponds to Franck's rule, but extended to take into account the distribution of positions characteristic of quantum mechanics. We consider next a transition from $n' = 0$ to some large value of n'' for a case in which the change in moment of inertia accompanying the electron jump is small. The wave function for the large value of n'' will now be a rapidly oscillating function of r, and for this reason the integral of the product of two wave functions will be small. This corresponds to the part of the rule of §1 which says that large changes in nuclear momentum at the instant of electron transition are improbable. This comes about because the oscillatory character of the wave function is the wave-mechanical analogue of the fact that the particle has considerable momentum at that part of its orbit. (The spacing of the zeros is given approximately by de Broglie's rule, $\lambda p = h$).

The wave functions of the higher states sink rapidly to zero outside the range of the classical motion. In the neighborhood of one of the turning points they have a broad maximum—broad because here the momentum is small so the quasi-wave-length is long, and maximum because here the particle spends a larger fraction of the time than places where it moves fast. As one goes from either of the classical turning points toward the minimum of the potential energy curve, the function oscillates more rapidly and has decreasing amplitude—both things corresponding to the greater speed of the particles at this part of the motion. We consider next a transition from the $n' = 0$ state accompanied by a large change in moment of inertia. Then the vibration level which is given as most probable by the rule of §1 will be one whose wave function's broad maximum lies in a favorable position with respect to the $n' = 0$ wave function and is therefore one which is favored by the wave mechanics formula. The wave functions for smaller values of n'' will not, overlap the $n' = 0$ wave function, while those for larger values of n'', though they overlap, oscillate rapidly and for this reason the integral measuring the intensity is small.

Thus we see how the quantum mechanical formula agrees with the rule of §1 when that is regarded as approximate. The method of this section goes beyond that of §1 in providing exact intensity predictions and gives appreciable values to transitions in the neighborhood of those given by Franck's rule. This is an evidence of the workings of the basic uncertainty relation of Heisenberg.[8] According to this we cannot reason closely about the simultaneous values of position and momentum. One must admit uncertainties in each quantity such that their product is of the order of h. We were violating this in §1, for example where we spoke of the nuclei as having zero relative momentum when at the extremity of their classical vibratory motions. The conclusion from such a statement must, on the quantum theory, be erroneously over-precise as we have already seen is the case.

§3. The Continuous Spectrum Accompanying Dissociation

One can easily see that the quantum-mechanical formula gives the right order of magnitude for the breadth of the continuous band accompanying dissociation of a molecule. There are several good examples at present of molecules which, on absorbing light from their lowest state, dissociate into two atoms which rush apart from each other with translational energy. We will consider the examples of the halogens for which the data are available.

In Fig. 3 the curves are drawn to scale for I_2.[9] By Franck's rule the most probable transition on absorption from the non-vibrating state is the absorption of light of frequency 19000 cm^{-1}. What is found experimentally is a broad band of absorption which runs from $\lambda 4300$ to $\lambda 5800$. The part from $\lambda 4300$ to $\lambda 4995$ is continuous while from $\lambda 4995$ to $\lambda 5800$ it is banded. This observed range of absorption is indicated near the frequency scale in Fig. 3.

What does the quantum-mechanical method predict for the breadth of this band? The wave function of the non-vibrating initial state is a Gauss error curve. The order of magnitude of the width of the band, as in §2, is given by the breadth of the frequency interval over which the integral for the electric moment has an appreciable value. An exact calculation would require knowl-

---

[8] Heisenberg, Zeits. f. Physik **43**, 172 (1927) and Bohr, Nature **121**, 580 (1928).
[9] Some readers may recall that the outstanding blemish in the original paper[2] was the lack of agreement between theory and experiment for the absorption bands of I_2. This was due to a bad blunder on my part in using the moment of inertia for the 26th vibrational level of the excited state thinking it was the moment of inertia for the non-vibrating molecule. For calling attention to the mistake in private communications I am deeply indebted to Professors R. T. Birge and F. W. Loomis. The curves as drawn are based on the following data:

$r'_0 = 3.010$	$r''_0 = 2.663$
$\omega^{0'} = 127.5$	$\omega^{0''} = 213.67$
$\beta'_0 = 0.02911$	$\beta_0'' = 0.03730$
$\alpha' = 0.00017$	$\alpha'' = 0.00011$
$x'\omega' = 0.85$	$x''\omega'' = 0.592$

58   E.U. Condon

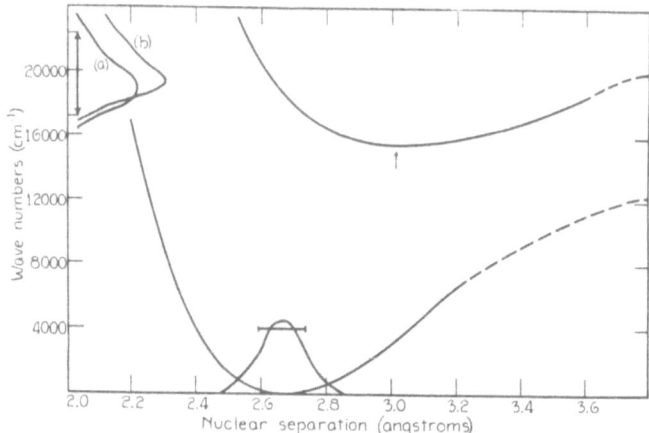

FIGURE 3. Potential energy curves involved in the visible band system of I_2, together with the ψ function for zero vibration in the final state. (a) is the reflection of this ψ function in the upper state energy curve. (b) is (a) multiplied by v^4. The two-headed arrow indicates the approximate extent of the observed absorption from the zero vibration level.

edge of the form of the wave functions for the excited state. The order of magnitude involved one readily sees is about equal to the interval on the frequency scale given by "reflecting" the initial state wave function in the potential energy curve of the final state the result being curve (a) Fig. 3. Such a rough estimate of the quantum mechanical intensity integral has still to be multiplied by v^4 to give intensities. Curve (b) shows the result of doing this. It is seen that the theory predicts the *extent* as well as the position in the spectrum of the absorption quite nicely.

For Br_2, Cl_2, and ICl we do not at present know the moments of inertia so that we are not able to predict theoretically the exact *location* of the continuous bands accompanying dissociation. But if we put the potential energy curve for the excited state at the position which makes theory and experiment agree as to the location of the maximum then the slope comes out so in each case as to predict the observed *width* of the continuous band, at least in order of magnitude.

In Fig. 4 we have the data on the continuous absorption of Cl_2, Br_2, and ICl as a function of frequency and a line showing the theoretical width of the continuous band, as estimated in this rough way.

For Cl_2 the measurements of Halban and Siedentopf were used, for Br_2 those of Kuhn[21] for ICl those of Gibson and Ramsberger[20] while for I_2 we have only Mecke's statement[10] that the continuous begins around $\lambda 4300$. The

---

[10] Mecke, Ann. d. Physik **71**, 104 (1923).

FIGURE 4. The relative absorbing powers from the zero vibration state into the continuous spectrum beyond dissociation for Cl_2, Br_2, and ICl and its approximate extent for I_2 given by the lower line. The shading indicates the location of the $n'' = 0$ progression of bands and their high frequency convergence limits. The horizontal lines under each curve show the quantum mechanical estimate of the width of the continuous absorption. The ordinates are relative values, for each substance, so comparisons between substances are meaningless.

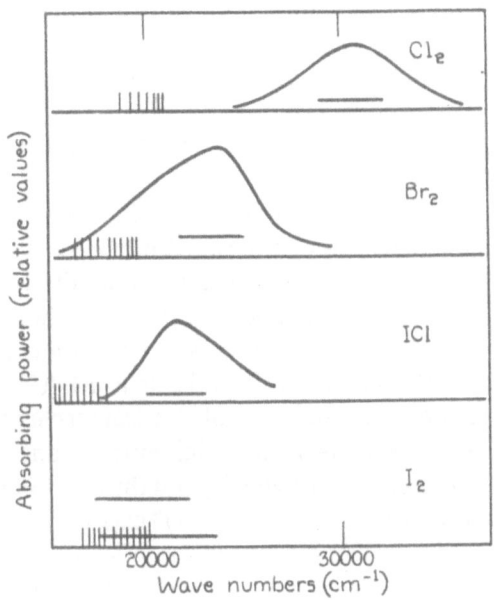

approximate theoretical values were got by regarding the potential energy curve of the excited state as parabolic, the constant being determined by the known frequency of vibration. It was then assumed that this curve is truly so located relatively to that of the ground state as to give the empirical value of the maximum of the continuous correctly by the rule of §1. This determines the slope of the energy curve for the excited state at the equilibrium separation of the unexcited state. One can say that the true slope is certainly greater than the value estimated this way. This slope is then multiplied by the parameter, a, in the wave function of the harmonic oscillator for the normal state to give an estimate of the half-value width of the continuous spectrum on either side of the maximum, as indicated in Fig. 4 by the lines drawn just inside the absorption curves. Naturally such an approximation is quite crude, but it is fair to claim that it agrees well enough with the observations to serve as support of the theory.

While this paper was in preparation Winans and Stueckelberg[11] recognized the application of these ideas to the explanation of the molecular part of the continuous spectrum of hydrogen, making use of Heitler and London's[12] quantum-mechanical potential energy curve for the 1^3S state of the molecule. Their paper serves at the same time to give not only a satisfactory account of this spectrum but also an important application of the methods of this section,

[11] Winans and Stueckelberg, Proc. Nat. Acad. Sci. **14**, (in press) (1928).

[12] Heitler and London, Zeits. f. Physik **44**, 455 (1927).

and the first empirical evidence of the physical reality of the theoretical 1^3S state of the molecule.

§4. Diffraction Bands in the Continuous Spectrum

The argument of the preceding section from the properties of the wave functions can be translated into a justification for Franck's postulate, as we have seen. Moreover it gave a quantitative understanding of the over preciseness of the conclusions drawn from the postulate by showing how the strict use of it violates Heisenberg's uncertainty principle.

The analogy with optics is a helpful one. As is well-known, geometrical optics is the analogue of classical mechanics. In the transition to wave mechanics one can regard Heisenberg's uncertainty principle as being a semiquantitative rule which gives the order of magnitude of diffraction deviations from geometrical optics and classical mechanics respectively. The last stage of refinement is reached in optics where the wave theory is used to predict details of diffraction patterns, and in wave mechanics where Schrödinger's equation gives the exact details of the diffraction phenomena of the de Broglie waves. In this section we shall see that cases may arise in which the quantum mechanical formulas give a rippling fluctuation in intensity in the continuous spectrum accompanying dissociation.

Suppose the potential energy curves lie as in Fig. 5 where the essential point is that one curve has a very gentle slope at the abscissa values near the minimum of the other. Then by the arguments of preceding sections the most probable jump from zero vibration in electron state I to electron state II is in the continuous spectrum. By the uncertainty principle one can estimate roughly the extent of this continuous spectrum. Looking at the matter more in detail, we see that because of the gentleness of slope of curve II the Ψ functions associated with nuclear motion in this electronic state will have a rather long wavelength, so the distance between zeros might be about equal to the breadth of the Gauss error curve which is the ψ function of zero vibration in state I. We consider now the relation of the ψ_{II} functions for state II but different nuclear energies, W, to this error curve ψ_I function of state I.

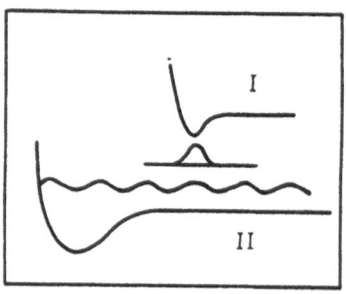

FIGURE 5. Illustrating a possible relation of the two potential energy curves which would give rise to the diffraction bands.

For values of W such that the oscillatory part of ψ lies at the same abscissa values as the maximum of the ψ_I function, the value of the integral of the product, $\psi_I \psi_{II}$ will depend very much on whether a node or a loop of the ψ_{II} function is coincident with the maximum of the ψ_I function. If a node is there, the integral will have a small value, if a loop it will be large.

We have next to consider how ψ_{II} depends on the nuclear energy W for this state. As W increases through its continuous spectrum, the ψ_{II} function will change in such a way that its wave-length gradually shortens about as $W^{1/2}$ while the positions of the zeros shift continuously past any fixed abscissa value. This has for a consequence that the intensity of transition into the continuous spectrum will undergo a rippling variation being large for those values of W which put a loop of ψ_{II} over the center of ψ_I and small for those which put a node there. All of this effect is, of course, contained within the extent of the allowed continuous spectrum as given roughly by the Heisenberg principle; just as in optics the diffraction maxima inside the geometrical shadow of a straight edge are inside the region of diffracted light given by the uncertainty principle. Finally for large values of W the wave-length of ψ_{II} will be so small compared to the breadth of ψ_I that these maxima become less and less clearly resolved while the whole intensity sinks to zero.

With the greatest reserve I will now suggest that the peculiar band recently described by Rayleigh[13] in mercury vapor may be such an intramolecular manifestation of the wave nature of matter. At least this will serve as an illustration on which to make the preceding ideas more precise. The band in question extends from $\lambda 2476$ to $\lambda 2482$ and is found only in emission. It has a sort of head at $\lambda 2476$ and is accompanied by a weaker band with a head at $\lambda 2469$ but this one doesnot show the banded structure. To discuss these as nuclear diffraction bands we will suppose that the head at 2476 corresponds to the transition from the zero vibration state of curve I (Fig. 5) to zero nuclear energy on curve II which will for simplicity be supposed to be absolutely horizontal under the minimum of curve I. On this view curve II is probably that of the ground state of the Hg_2 molecule and curve I one for which the molecule separates into one Hg atom in its normal state and one in the $^3P_2$ state, which is upper state for the forbidden atomic line $\lambda 2270$—but these points are not essential to the present discussion.

The band whose head is at $\lambda 2469$ may be supposed to represent transitions into the continuous range above II from the first vibrational level of curve I. This fixes the vibration frequency in curve I at 106.2 cm^{-1} and therefore the constant, a, giving the breadth of the wave function,

$$a = 8.2/(\omega\mu)^{1/2} = 0.08$$

where a is in Angstrom units, $\omega = 106.2$ and μ is half the atomic weight of Hg. This formula is a convenient way of writing the general formulas

[13] Rayleigh Proc. Roy. Soc. **A119**, 349 (1928), especially p. 353.

$$a^2 = h/2\pi(\kappa\mu)^{1/2} \quad \text{and} \quad \nu = (\kappa/\mu)^{1/2}/2\pi, \quad (C.G.S. \text{ units}),$$

from the wave mechanics of the harmonic oscillator.

On the horizontal part of II the function ψ_{II} will have the form

$$\psi_{II} = \lambda \sin 2\pi(x/\lambda + \delta)$$

in which λ is the de Broglie wave-length, h/p, given by the convenient formula,

$$\lambda(A.U.) = 25.7/(\mu\nu)^{1/2}$$

where ν is the energy of the state in cm^{-1} reckoned up from the horizontal part of II. The factor λ, in front, comes from the normalization of the wave functions for the continuous. As explained before δ, as well as λ, depends on ν. Assuming complete non-polarity for simplicity, the intensity of transition from state I to a level ν cm^{-1} up from the horizontal part of II is therefore proportional to the square of,

$$I_0(\nu) = \int e^{-x/2a} \cdot \lambda \sin 2\pi(x/\lambda + \delta) \, dx$$

$$= (2\pi)^{1/2} a\lambda \sin 2\pi\delta \cdot e^{-2\pi^2 a^2/\lambda^2}.$$

The factor with the exponential function represents the effect of the uncertainty principle, the sine factor gives the banded diffraction structure. Inserting the numerical value of a and the relation of λ to ν the exponential factor becomes $e^{-\nu/52.3}$, so that the intensity for $\nu = 100$ cm^{-1} will be about $e^{-4} = 2$ percent of the value for $\nu = 0$. The Rayleigh band is about 90 cm^{-1} in extent so that the order of magnitude agreement is a good one. But it does not appear to show such an exponential intensity decay.

Next we have to consider roughly the spacing of the diffraction maxima in the spectrum. A rough approximation is to suppose that curve II is simply L-shaped, that is, it extends horizontally a distance d to the left of the minimum of I and then turns up abruptly to infinity. For a range of ν of only 0–100 cm^{-1} this is not as bad an approximation as it might at first seem. At any rate for such a curve ψ_{II} has a node at the place of infinite slope and so δ is simply given by

$$\delta = d/\lambda = d\nu^{1/2}/2.57$$

and the maxima come at $\delta(\nu) = (n + \frac{1}{2})$ with n an integer. This simple argument would then put the nth maximum at

$$\nu^{1/2} = 2.57(n + \frac{1}{2})/d$$

This would make the maxima close together at the short-wave side and more widely spaced toward the long-wave side. Not only is this true but $\Delta\nu^{1/2}$ is fairly constant for Rayleigh's band as Table I based on his data shown.

From the experimental value of $\Delta\nu^{1/2}$ one infers that d is about 6.4A which is a rather large value. But this may well be a consequence of the over simplification of the curve II.

TABLE I. Values of $\Delta v^{1/2}$ for the Hg_2 bands observed by Rayleigh

Wave-number	v	$v^{1/2}$	$\Delta v^{1/2}$
40372.3	0.0	0	
354.0	18.3	4.28	
350.6	21.7	4.66	0.38
346.8	25.5	5.06	0.40
342.6	29.7	5.45	0.39
337.9	34.4	5.87	0.42
332.4	39.9	6.32	0.45
326.1	46.2	6.80	0.48
320.3	52.0	7.22	0.42
313.5	58.8	7.67	0.45
306.7	65.6	8.10	0.43
298.7	73.6	8.58	0.48
291.7	80.6	8.98	0.40
284.4	87.9	9.38	0.40

Another assumption for II is that it is given by a straight-line of small slope, its equation in the neighborhood of the minimum of I being

$$U(x) = -kx$$

For this one finds for the characteristic functions

$$\psi_{II}(W, x) = (x + W/k)^{1/2} J_{1/3}\left[\left(\frac{x + W/k}{b}\right)^{3/2}\right]$$

in which $J_{1/3}(z)$ is the ordinary Bessel function of order 1/3 and

$$b = 3.345/(u_1 \mu)^{1/3}$$

Here u_1 is written for k when the unit of energy is cm^{-1} and of length 10^{-8} cm, and the unit of μ is 1/16 the mass of an oxygen atom. The values of z for which $|J_{1/3}(z)|$ is a maximum are roughly (by inspection from the tables in Watson's Bessel Functions):

$m =$	1	2	3	4	5
$z_m =$	0.92	4.36	7.54	10.70	13.84
$\Delta z_m =$		3.44	3.18	3.16	3.14

For larger values of m, the ordinal number of the maximum, the interval approaches π. For larger m it is therefore sufficient to look for maxima at

$$(v/u_1 b)^{3/2} = (m\pi - 1.89)$$

which gives the law $\Delta v^{3/2} = u_1 \pi b$. This assumption therefore does not fit Rayleigh's data.

The structureless band at $\lambda 2469.7$ has already been supposed to be due to the transition from the first vibration level of state I on to the continuum of II. It may be apparently structureless since the ψ_I function for one vibration unit has a wider range than has ψ_I for zero vibration which tends to blur the diffraction effect. But it is hard to convince oneself that one unit of vibration energy can make so much difference.

The fact that these bands were found only in emission is consistent with the view here presented although this of course, is easily explained on more usual views by saying that their final state is an excited state of the molecule.

In conclusion it should be said that the existence of such diffraction bands either in emission or absorption (if I is a lower state than II) but not in both is a rigorous conclusion from quantum mechanics. It is not unreasonable to suppose that Rayleigh's band is a case of this but the interpretation must as yet be provisional and subject to confirmation, perhaps by the empirical determination of curves I and II through sharp bands coming from their discrete vibration levels.

§5. Additional Applications to Band Systems

Since extension of Franck's postulate to band systems was first made the necessary data for its application to several more band systems has accumulated. This section is devoted to remarks on its application to some new band systems.

As already mentioned[9] the outstanding discrepancy between theory and experiment in the case of the visible bands of iodine has been removed by correction of a blunder in the use of the experimental data.

The iodine spectrum presents an interesting feature which was not amenable to treatment in the original Franck method, namely the somewhat irregular fluctuation of intensity in Wood's long series of resonance doublets. Such fluctuations probably are to be interpreted as arising from diffraction effects akin to those which are the subject of §4, except that we are now dealing with transitions of which both initial and final states are in the region of discrete energy levels. The main resonance doublet series, it is now known[14], arises from transitions from the $n' = 26$ state of the excited molecule to various vibration states of the molecule in the normal electronic state from $n'' = 0$ to $n'' = 27$. Along this series of 28 doublets there are great intensity variations, some of the doublets being quite missing. Lenz[15] showed how this behavior is possible from correspondence principle considerations. From the criterion of the overlapping of the wave functions it is easily seen from Fig. 3 that from $n' = 26$ to $n'' = 0$ to 27 are allowed transitions. The exact intensity of any particular one, however, will depend on the integral of the product of the wave functions of the initial and final states as in §2. And as in §4 this will depend very much on the exact phase relation of the nodes and loops of the two wave

[14] Chapter VI of National Research Council Report on Molecular Spectra in Gases.
[15] Lenz, Zeits. f. Physik **25**, 299 (1924).

functions in question. These finer diffraction effects probably give rise to the intensity fluctuations in the resonance doublet series.

While the principle of their explanation is therefore precisely fomulated, the exact prediction unfortunately calls for a very accurate knowledge of the wave functions. This is lacking since for these large vibration quantum numbers it is quite inacceptable to treat the molecule as a harmonic oscillator or any other model for which ψ functions are at present exactly known.

The band systems which have been recently analyzed and on which the theory has been tested include especially the β bands of nitric oxide,[16] and the blue-green bands of diatomic sodium.[17] The vibrational analysis only for several more band systems has been recently published. In all of these cases the types of intensity distribution are those which this theory explains but lacking analysis of rotational analysis one is unable to use them for detailed verification of the theory. In this class at present may be mentioned the many band systems of the copper halides studied by Ritschl,[18] the absorption and fluorescence spectra of S_2, Se_2, Tl_2, studied by Rosen,[19] the absorption spectrum of iodine monochloride observed by Gibson and Ramsberger,[20] that of chlorine and bromine studied by Kuhn,[21] the ultra-violet system of iodine analyzed by Pringsheim and Rosen,[22] and others.

In a recent article Herzberg[23] has obtained nice measurements on the intensity distribution in the second positive group of nitrogen under different conditions of excitation, one of the systems discussed in the original paper.[2] His criticisms of the paper[2] are already covered by the explicit statement made on p. 1183 that the theory treats only of relative transition probabilities and not of distribution of the molecules in the different excited states. The latter naturally depends on the conditions of excitation and cannot come out of any theory which makes no mention of these.

§6. Electron Jumps and the Franck Postulate

It is natural to ask why the Franck postulate does not apply to electron transitions in atoms, that is, to questions of relative intensity of lines in atomic spectra. The classical mechanical basis for the postulate consisted in saying that the absorbed quanta and the electronic motions involved much smaller amounts of momentum than were necessary to excite the nuclear

---

[16] Barton, Jenkins and Mulliken, Phys. Rev. **30**, 175 (1927).
[17] Loomis and Wood, Phys. Rev. **32**, 223 (1928).
[18] Ritschl, Zeits. f. Physik **42**, 172 (1927).
[19] Rosen, Zeits. f. Physik **43**, 69 (1927).
[20] Gibson and Ramsberger, Phys. Rev. **30**, 598 (1927).
[21] Kuhn, Zeits. f. Physik **39**, 77 (1926); Halban and Siedentopf, Zeits. f. Physik. Chem. **103**, 71 (1922).
[22] Pringsheim and Rosen, Zeits. f. Physik **50**, 1 (1928).
[23] Herzberg, Zeits. f. Physik **49**, 761 (1928).

vibrations. Therefore the principle, which may be said to be one favoring as little discontinuity in the orbits as possible, should also apply to the electron orbits. This is true since one recalls that the momentum of a quantum of visible light is very small compared with the orbital momentum of electrons in the initial and final orbits concerned in the emission or absorption of such light. (Nowadays one remembers this best by recalling that the relation $p = h/\lambda$ holds equally for light and for electrons, together with the fact that λ for visible light is 10^4 times the size of an atom.)

The answer is that the postulate *is* valid to a certain extent for relative probability of electron transitions. For example, the reason why the transition $50_2 \rightarrow 1_1$ (Bohr notation) is so much less probable than a transition $2_2 \rightarrow 1_1$ in atomic hydrogen is simply because the wave functions do not overlap as much in the former case asdn the latter. For atomic spectra the principle does not have so much interest as in molecular spectra, on the other hand, because here the wave-length of the electron is enough greater that the range of the Heisenberg uncertainty principle is so broad that in turn the predictions of Franck postulates become so blurred as to be of almost trivial interest. In principle, however, it is just as true for electrons as for nuclear motions.

Finally it may be said that it is in this direction that the connection of the Franck postulate with the correspondence principle is to be found, although this point today has only historical interest. The transition probabilities are related to the Franck postulate in the way here outlined and also are know to have the appropriate asymptotic connection with the Fourier coefficient of the classical motion which is demanded for the Bohr theory.[24]

In conclusion I should like to say that this paper is an outgrowth of work[3] done while abroad on a National Research fellowship and that it is a pleasure here to express my profound indebtedness to the fellowship board for its generous support of my studies.

---

[24] This fact is built into the very structure of wave mechanics and has been clearly worked out in the special case of the hydrogen atom by Eckart, Zeits. f. Physik **48**, 285 (1928).

Quantum Mechanics of Momentum Space

Edward U. Condon

Assistant Professor of Physics
Princeton University

In Schrödinger's work on quantum mechanics he starts from the Hamiltonian function of the mechanical system under investigation and makes from it an operator by the process of replacing each momentum, p_σ, by the differentiation operator $\dfrac{h}{2\pi i}\dfrac{\partial}{\partial q_\sigma}$. A wave function $\psi(q_1 \ldots q_n)$ is then introduced which is required to satisfy the equation

$$H\left(\frac{h}{2\pi i}\frac{\partial}{\partial q}, q\right)\psi = W\psi, \tag{1}$$

where W is a constant. That is, the energy operator, operating on ψ must give simply a constant into the same function. Only such ψ's as are finite and continuous throughout the range of the q's are allowed. It is then found that such ψ's exist only when W has certain charactcristic values depending on the form of the operator H. These values of W are then taken as the allowed values of the energy. The interpretation given by Schrödinger for his function ψ however is not tenable and must be given up in favor of one due to Born that the quantity $\psi(W, q)$ is related to the probability of finding the system in a little volume element of the coördinate space, dq, when it is known that the system is in the energy state W. This probability is in fact taken equal to

$$\psi(W, q)\bar\psi(W, q)\, dq, \tag{2}$$

where $\bar\psi$ is the conjugate complex function to ψ. This method of finding energy levels and this interpretation of ψ are at the heart of many of the successes of the new quantum mechanics.

As Professor Slater has pointed out, this statistical feature runs throughout the whole of quantum mechanics. The questions which it attempts to answer are to be phrased in the following form: (*A*) What are the numerical values which a physical quantity may assume, and (*B*), In a given set of circumstances what is the relative probability that it does assume any one of these allowed values? In terms of this general formulation one sees that the equation of Schrödinger (1) and the expression (2) form answers to two very special questions of this type. Equation (1) is the means employed for finding what values the particular physical quantity, total energy of the system, may have. It answers a question of the first type. Expression (2) is the answer to a question of the second type. We are supposed to know that the configuration may assume any value represented by the allowed ranges of the q's. The given set of circumstances is the certain information that the system is in the state of energy W and then the expression gives relative probabilities that the system be found in various elements of the configuration space.

It has been the service of Dirac and Jordan to point the way to the generalization of equations (1) and (2) so that the means is at hand to answer any questions of the type (A) and (B) rather than simply the special cases of them covered by the Schrödinger equation. The purpose of these remarks is to illustrate the ideas by the presentation of a few simple cases in which one works with the momenta as independent variables, i.e. in which an equation analogous to (1) is utilized to find a $\psi(W, p)$ which can be used in an expression analogous to (2) to get the relative probability that the momenta will lie in the element of momentum space defined by p_1 between the values p_1 and $p_1 + dp_1$, p_2 between the values p_2 and $p_2 + dp_2$ and so on. It is believed that the results, while containing nothing new of importance to atomic physics will nevertheless be useful in facilitating the study of the general theory.

To each physical quantity can be associated an operator and the laws of association of operators with quantities constitute an important branch of the subject. There are different representations of a physical quantity and the letter denoting the quantity has to be used in two senses. Where written in bold type it will be meant as the quantity in its capacity as an element of the algebra of quantum mechanics. That is, as a matrix in many cases, or more generally as what Dirac calls the q-number ($q \sim$ quantum). In the language of Weyl's recent book,[1] each physical quantity is associated with an operator in the unitary function space. Where ordinary type is used the ordinary real variable which is the arithmetic value of the quantity is meant. The guiding principle in associating operators with the physical quantities is that this must be done consistently with the matrix mechanics of Heisenberg. Thus between coördinate and conjugate momentum one has the rule

$$\mathbf{pq} - \mathbf{qp} = \frac{h}{2\pi i}. \tag{3}$$

In getting Schrödinger's equation (1) we associate each $\mathbf{q}$ with the operation: Multiplication by the arithmetic value q and each $\mathbf{p}$ with the operation: Multiplication by $h/2\pi i$ and partial differentiation with respect to the conjugate q. One sees that this is in fact consistent with (3) since

$$\frac{h}{2\pi i}\frac{\partial}{\partial q}(q\psi) - q\frac{h}{2\pi i}\frac{\partial \psi}{\partial q} \equiv \frac{h}{2\pi i}\psi. \tag{3a}$$

---

[1] Weyl, "Gruppentheorie und Quantenmechanik." Leipzig, Hirzel. 1928.

But another way[2] of associating operators with **p** and **q** which is just as consistent with (3) as the method leading to Schrödinger's equation is the following:

p: Multiplication by p.

q: Operation $-\dfrac{h}{2\pi i}\dfrac{\partial}{\partial p}$.

for this gives

$$p\left(-\frac{h}{2\pi i}\frac{\partial \psi}{\partial p}\right) + \frac{h}{2\pi i}\frac{\partial}{\partial p}(\psi p) \equiv \frac{h}{2\pi i}\psi. \tag{3b}$$

so that this association is consistent with (3).

Before passing to the use of this method in determining energy values it is essential to consider the question: What values may each **p** assume? Because of the kinematical connection between **p** and **q** the allowed values of **p** are determined by the allowed values of **q**, as Jordan[3] has shown. As to the allowed values of **q** these are supposed to be known almost intuitively from the nature of the system. Thus, the allowed values of a Cartesian coordinate are all real values from $-\infty$ to $+\infty$; of a radius vector in polar coordinates are from 0 to $+\infty$; of the azimuthal angle in spherical or cylindrical coördinates, 0 to 2π; of the co-latitude in spherical coördinates, 0 to π. From these allowed ranges in **q** certain results for the allowed values of the associated **p** are found. Thus for a Cartesian coördinate or a radius vector whose range is infinite the **p** may take on all values from $-\infty$ to $+\infty$. But for an azimuthal angle like φ the allowed values of the associated **p** are the discrete range given by $n(h/2\pi)$ where n takes on all *integral* values from $-\infty$ to $+\infty$.

Corresponding to this second mode of associating operators with p and q one finds as the equation analogous to (1),

$$H\left(p, -\frac{h}{2\pi i}\frac{\partial}{\partial p}\right)\psi = W\psi, \tag{4}$$

which will now be considered in some simple instances:

The Free Particle. For the motion of a particle of mass μ in one rectilinear coördinate, x, under no forces the Hamiltonian function is **H** $= (1/2\mu)\mathbf{p}^2$. As it is independent of **q**, (4) takes the simple algebraic form

$$\left(\frac{1}{2\mu}p^2 - W\right)\psi = 0. \tag{4a}$$

---

[2] This method or introducing the operator $-\dfrac{h}{2\pi i}\dfrac{\partial}{\partial p}$ for q when p is multiplication by p was suggested to me after the symposium by my colleague, Dr. H. P. Robertson. It is so neat, that I have taken the liberty of using it rather than the method used in my remarks.

[3] Jordan, Zeitschrift für Physik, **44**, 1 (1927).

It is at once apparent that if p has a value differing from $\pm\sqrt{2\mu W}$ then ψ must be zero. As the range of p is confined to real values one has $W > 0$. But here one strikes a characteristic feature of the theory in that if one adheres to the notion that ψ is to be normalized with respect to the complete allowed range of p, i.e.,

$$\int_{-\infty}^{+\infty} \psi(W,p)\bar{\psi}(W,p)\,dp = 1$$

then ψ must be strongly infinite at $p = \sqrt{2\mu W}$ or $-\sqrt{2\mu W}$ or both in order that the integral shall not vanish. Thus ψ is a function of the type called a δ-function by Dirac:

$$\delta(x) = 0 \quad \text{for } x \neq 0 \quad \text{and} \quad \int_{-\infty}^{+\infty} \delta(x)\,dx = 1.$$

Plainly one does not deal here with a simple requirement of finiteness and continuity for $\psi(W,p)$ as seems usually to be the case with the $\psi(W,q)$ of Schrödinger.

The analogue of (2) is that this $\psi(W,p) \cdot \bar{\psi}(W,p)\,dp$ is equal to the probability that p have a value between p and $p + dp$ when W is known to have the value W. The equation (4a) shows that this vanishes except for

$$\frac{1}{2\mu}p^2 = W,$$

so that we recognize in this an equation of classical mechanics which retains its validity in quantum mechanics.

All values of W are on an equal footing in regard to the nature of the $\psi(W,p)$ associated with them so (4a) would show that all values of W greater than zero are allowed.

The Freely Falling Particle. This is the same as the preceding except that now the Hamiltonian is $(1/2\mu)\mathbf{p}^2 - \kappa\mathbf{x}$ so (4) becomes

$$\left(\frac{1}{2\mu}p^2 + \frac{\kappa h}{2\pi i}\frac{\partial}{\partial p}\right)\psi = W\psi \tag{4b}$$

of which the solution is

$$\psi = \exp\frac{2\pi i}{\kappa h}\left(Wp - \frac{1}{6\mu}p^3\right).$$

This solution is finite throughout the range of values of p no matter what the value of W, therefore all values of W are allowed, both positive and negative, agreeing with the classical mechanics of uniformly accelerated motion. The value of $\psi\bar{\psi}$ is here independent of p which means that all values of p are equally probable. This is in accord with the classical motion if we understand that all values of the time are to be taken as equally probable in reckoning the relative probabilities for the momentum. That this is the correct analogue with classical mechanics is consistent with quantum mechanics for exact specification of the energy W, makes the conjugate quantity, the time, to be wholly indeterminate, as is well known.[4]

The Harmonic Oscillator. Here the Hamiltonian is

$$\frac{1}{2\mu}\mathbf{p}^2 + \frac{1}{2}\kappa\mathbf{x}^2$$

and because it is quadratic in both p and x one gets an equation of the same form here (from (4)) as by Schrödinger's method, (1). It is:

$$-\frac{h^2\kappa}{8\pi^2}\frac{\partial^2\psi}{\partial p^2} + \frac{1}{2\mu}p^2\psi = W\psi. \tag{4c}$$

Requiring that ψ be finite at $p = \pm\infty$ this gives the energy levels

$$W_n = \left(n + \frac{1}{2}\right)hv, \qquad n = 0, 1, 2, \ldots, \qquad v = \frac{1}{2\pi}\sqrt{\frac{\kappa}{\mu}};$$

and

$$\psi_n(p) = \exp(-p^2/p_0{}^2)H_n(p/p_0), \qquad p_0{}^2 = \frac{h}{2\pi}\sqrt{\kappa\mu}.$$

Here $H_n(x)$ is the nth Hermitian polynomial as with Schrödinger. One sees that p_0 is the maximum value of the momentum in classical oscillations of energy $\frac{1}{2}hv$ just as the characteristic length, a, occurring in Schrödinger's functions was the classical amplitude of oscillation in the classical motion of this energy. For various values of n therefore these ψ functions bear the same relation to the momentum values associated with the classical motions of the same energy as they do with Schrödinger when they give the coördinate probabilities instead of momentum probabilities.

Free Rotator with Fixed Axis. The Hamiltonian is $(1/2\mu)\mathbf{p}^2$ as with the free particle moving in a straight line. Here μ is the moment of inertia. But as the coördinate now is one restricted to values from 0 to 2π, it follows that p is restricted to values $n(h/2\pi)$ with n integral. Therefore the allowed values of W are

$$W_n = \frac{n^2h^2}{8\pi^2\mu}$$

with ψ functions of the same type as with the free particle.

The Physical Pendulum.[5] This offers an interesting variation of the preceding, the Hamiltonian being $(1/2\mu)\mathbf{p}^2 - \kappa\cos\theta$. To form equation (4) calls for

---

[4] Heisenberg, Zeitschrift für Physik, **43**, 172 (1927) and Bohr, Nature, April 14, 1928.
[5] For the treatment of this problem by Schrödinger's method see Condon, Physical Review, **31**, 891 (1928).

interpretation of the operator $\cos \theta$. This is clearly

$$\cos \frac{h}{2\pi i} \frac{\partial}{\partial p} = \frac{1}{2}\left[\exp\left(\frac{h}{2\pi i} \frac{\partial}{\partial p}\right) + \exp\left(-\frac{h}{2\pi i} \frac{\partial}{\partial p}\right)\right].$$

Recalling that the terms on the right are a symbolic form of Taylor's theorem it is plain that the operation on $\psi(p)$ gives

$$\cos \theta \psi(p) = \frac{1}{2}\left[\psi\left(p + \frac{h}{2\pi}\right) + \psi\left(p - \frac{h}{2\pi}\right)\right]$$

which remains consistent with the general result that the allowed values of p differ by $h/(2\pi)$ for the momentum conjugate to the coordinate θ of period 2π. Introducing $p = n(h/2\pi)$ the equation (4) for $\psi(W, n)$ becomes

$$\frac{\kappa}{2}[\psi(n + 1) + \psi(n - 1)] = \left(W - \frac{n^2 h^2}{8\pi^2 \mu}\right)\psi(n) \qquad (4d)$$

The allowed energy levels for W are those for which this difference equation for ψ has solutions which remain finite as n takes on integral values between $-\infty$ and $+\infty$. But there seems not to be a very useful mathematical literature on boundary value problems for difference equations.

In these examples we see cases where the equation (4), which is analogous to (1), in momentum space may be algebraic or may be a finite difference equation. In coördinate space, since the Hamiltonian always includes a quadratic form in the momenta a second order differential operator is always generated. This makes (1) always resemble the partial differential equations of wave propagation from elasticity theory. In fact this analogy is fundamental in Schrödinger's approach to quantum mechanics. From the more general standpoint this resemblance to a wave equation is somewhat accidental.

The Critical Potentials of Molecular Hydrogen

E.U. CONDON AND H.D. SMYTH

Palmer Physical Laboratory
Princeton University

Out of a number of experiments in recent years on the impact of electrons with hydrogen molecules, there come certain results which at first sight appear discordant. On the one hand the experiments on ionization show that an electron, when it ionizes a hydrogen molecule, does so without causing dissociation. That is, the production of atomic ions as the primary result of impact of moderate speed electrons with hydrogen molecules is a process which occurs rarely if at all. On the other hand, dissociation into uncharged atoms by electrons with velocities both below and above the ionizing potential, does occur frequently. There is therefore an apparent inconsistency, but the writers believe it is explained by theoretical considerations similar to those advanced by Winans and Stueckelberg[1] in the preceding paper. In fact almost all the critical potential data on hydrogen now seem to fit in the theoretical scheme. However, before presenting the theory, a more precise summary of the experimental results will be given.

Suppose we present the experimental observations as a function of the speed of the impacting electrons. If diatomic hydrogen be bombarded by electrons of any speed below eleven volts the electrons lose a small indeterminate amount of energy; that is, the collisions are not perfectly elastic as in the rare gases, but are very nearly so. No effect of radiation or dissociation is observed. As the speed of the electrons is increased up to twelve volts several effects[2] are observed, first inelastic impact,[3] second radiation photoelectrically effective,[4] third the production of atomic hydrogen (or at least of some form of hydrogen that "cleans up" and is chemically active[4,5]). The exact critical potentials where these processes begin to occur are not very precisely determined but are all in the neighborhood of 11.5 volts.

Now, if the speed of the electrons is further increased the experimental

---

[1] Winans and Stueckelberg, these PROCEEDINGS, **14**, 867, 1928.
[2] We are assuming the weak ionization sometimes reported at about 11 volts, is due to an impurity.
[3] Compton and Mohler, *Critical Potentials*, p. 115.
[4] Hughes and Skellet, *Physic. Rev.*, **30**, pp. 11–35, 1927.

evidence is less definite. Apparently radiation becomes more intense, possibly with discontinuities corresponding to higher critical potentials. In particular a number of investigators report an increase in radiation at about 12.8 volts. As to the production of atoms, that also appears to increase. Thus, Glockler, Baxter and Daltons[5] state that the effectiveness of the impacting electrons increases rapidly between 11.4 volts and the molecular ionizing potential above which the change is slow. Experiments on inelastic impact give no definite results in this region.

When the electrons reach a speed of 13.5 volts, the ionizing potential of atomic hydrogen, weak ionization is always observed, indicating that there is always some atomic hydrogen present. In the past this has been explained by dissociation due to the hot filament.

Somewhat above this voltage but before the molecular ionization potential, are a group of radiating potentials observed by Olsen and Glockler but in no other experiments.

Then at about 15.9 volts strong ionization sets in, which has been shown to be ionization without dissociation. Furthermore, no evidence has been found at this or any higher voltage of the formation of atomic ions from molecules as the immediate result of an electron impact. However, such a process may occur if of very small probability, since it might be masked by the large number of atomic ions formed by secondary dissociation of the molecular ions. This latter process occurs only if the molecular ions have acquired a certain minimum kinetic energy before they make collisions. (See Dorsch and Kallmann, *Zs. Phys.*, **44**, p. 543, 1927.)

As the speed of the impacting electrons is still further increased no new processes are observed by the positive ray or "clean-up" methods. Some experimenters have observed additional ionization setting in at about thirty volts. Of more importance, however, are the results of Franck and Blackett, who showed that the emission of Balmer lines was a primary result of impact of thirty volt or higller speed electrons. In other words, they showed dissociation may occur with excitation of one (or perhaps both) atoms.

We may summarize the principal results in tabular form as given below.

	PROCESS	MINIMUM ENERGY REQUIRED	REMARKS
(a)	$\begin{cases} H_2 \to H + H \\ \text{Radiation produced} \end{cases}$	11.5	
(b)	More radiation	12.8	Not very definite
(c)	$H_2 \to H_2^+ + e$	15.9	The usual ionization
(d)	$H_2 \to H^+ + H + e$		Does not occur
(e)	$H_2 \to H' + H \text{ (or } H')$	30	

It should be pointed out that we have little evidence as to the relative proba-

---

[5] Glockler, Baxter and Dalton, *J. Am. Chem. Soc.*, **49**, pp. 58–65, 1927.

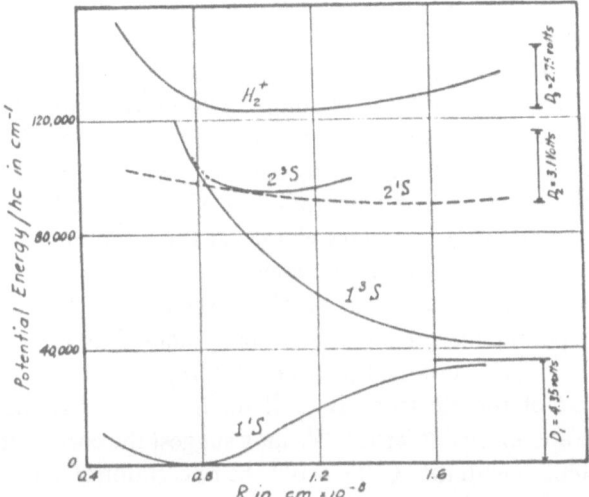

The curves in the above figure represent the potential energy as a function of nuclear separation for five different electronic states of the hydrogen molecule. The values given at the right for the heats of dissociation are obtained from band spectra for 1^1S and 2^1S states and from Burrau's paper for the ion. They indicate the heights which the curves would approach if extended to the right.

bilities of these processes at different voltages, though from the small number of atomic ions observed at low pressures we may infer that (a) is improbable compared to (b) at potentials above 15.9.

In order to interpret these phenomena in terms of energy levels and potential energy curves we will need not only those given in the previous paper but also the potential energy curve for the normal state of the hydrogen molecular ion. This has been calculated by Burrau.[6] In the figure a curve drawn from his data is given. To it have been added a curve for the 2^1S from band spectra data and the curves given by Winans and Stueckelberg for the 1^3S and 2^3S states.

Consider the possible effect of an electron impact on an H_2 molecule. If the electron has less than 11 volts the only possible transition it may cause is to the 1^3S state. But applying the Franck-Condon principle we see that the probability of such a transition will be small except for those nuclear separations where the 1^3S curve nearly coincides with that for the 2^3S term. In other words we might expect a little but only a little atomic hydrogen to be produced below eleven volts since the 1^3S state is unstable.

Slightly above eleven volts the excitation to the 2^1S state becomes possible. Although the nuclear binding in this state is considerably weaker than in the

[6] Burrau, *Kgl. Danske Vid. Selskalb. Math-fys. Med.*, 7, 14, 1927.

normal state and the curve suggests the possibility of dissociation on excitation to this state, the intensity distribution in the absorption bands indicates that this is not usually the case.[7] Return from this state to the normal would, of course, produce radiation that would be photoelectrically effective.

Electrons of somewhat greater speed might do any of three things, (1) excite directly to the 1^3S state producing hydrogen atoms with big kinetic energy, (2) excite to the 2^1S state producing radiation, (3) excite to the 2^3S state which would then revert to the 1^3S state giving off the continuous spectrum and producing atomic hydrogen simultaneously.

The writers believe that some or all of these three processes occur in the various experiments which find a critical potential at about 11.5 both for the production of radiation and of atomic hydrogen.

If the speed of the electrons be still further increased the possibilities become more and more numerous. We may suggest the correlation of the 12.8 radiating potential with the "C" level but the uncertainties of both theory and experiments make further correlations futile. Two points remain to be mentioned for this region of velocities. First, we may expect more and more atomic hydrogen to be produced by excitation to higher triplet levels and reversion to the 1^3S. Second, the presence of this atomic hydrogen probably explains the atomic critical potentials which are so often observed.

Finally, suppose the electrons have sufllcient velocity to ionize the molecule, that is to change it from the lowest to the topmost curve in our figure. If we compare the shapes of the two curves and positions of their minima, we see at once that the most probable transitions are going to give nuclear separations which will not result in dissociation. This is in perfect accord with experiment.

Until we know more of the excited states of the H_2^+ ion we cannot proceed with explanations of the effects that still faster electrons may produce.

---

[7] Hopfield, *Physic. Rev.*, **31**, p. 918. 1928.

Quantum Mechanics and Radioactive Disintegration[1]

R.W. Gurney and E.U. Condon

Palmer Physical Laboratory
Princeton University

ABSTRACT

Application of quantum mechanics to a simple model of the nucleus gives the phenomenon of radioactive disintegration. The statistical nature of the quantum mechanics gives directly disintegration as a chance phenomenon without any special hypothesis. §1 contains a presentation of those features of quantum mechanics which are here used and gives a simple calculation of the disintegration constant. §2 discusses the qualitative application of the model to the nucleus. §3 presents quantitative calculations amounting to a theoretical interpretation of the Geiger-Nuttall relation between the rate of disintegration and the energy of the emitted α-particle. In getting this relation one arrives at the rather remarkable conclusion that the law of force between emitted α-particle and the rest of the nucleus is substantially the same in all the atoms even where the decay rates stand in the ratio 10^{22}. §4 calls attention to the natural way in which the paradoxical results of Rutherford and Chadwick on the scattering of fast α-particles by uranium receive explanation with the model here used. §5 discusses certain limitations inherent in the methods employed.

The study of radioacitivity itself together with the application of it as a working source of high speed helium nuclei and electrons has played a fundamental role in the development of quantum physics. The scattering experiments of Rutherford and his associates gave the picture of the nuclear atom on which all of the success of modern atomic theory depends. Bohr's formulation of quantum postulates to be applied to such a model was a great step in the extension of knowledge of atomic structure and finally culminated in 1925 in the discovery by Heisenberg and by Schrödinger of a reformulation

---

[1] An account of this work was first published in Nature for September 22, 1928. In a number of the Zeitschrift für Physik (**51**, 204, 1928) received here two weeks ago there appears a paper by Gamow who has arrived quite independently at the same basic idea as was presented in our letter and which is here treated in detail. Reports of this paper were also given at the Schenectady meeting of the National Academy of Sciences on November 20, 1928 and at the Minneapolis meeting of the American Physical Society on December 1, 1928.

of mechanical laws which has subsequently proved extremely powerful in handling atomic structure problems. In this development of the last fifteen years little advance has been made on the problem of the structure of the nucleus.

It seems, however, that the new quantum mechanics has had sufficient success to justify the hope that it is competent to carry out an effective attack on the problem. The quantum mechanics has in it just those statistical elements which would seem appropriate to an explanation of the phenomenon of radioactive decay. This is the feature of the general problem with which we are concerned in this paper. We believe that the results provide at last an interpretation of nuclear disintegration which in its fundamental points is very close to the truth although it is necessarily quite incomplete.

The outstanding difficulties in the way of a good theoretical treatment of nuclear structure at present are mainly bound up with our lack of under-standing of the quantum mechanics of the magnetism of the fundamental particles. This question has been much advanced this year by Dirac's exten-sion of Pauli's theory of the spinning electrons[2] but this remains essentially a theory of the behavior of one electron in an electromagnetic field. Not only is it apparently still unsatisfactory as such but this limitation must necessarily be disposed of in principle before the many body nuclear problem can be approached. And with that done there will remain the inevitable analytical difficulties.

Enough is known, however, to teach us that probably the magnetic interac-tion is not to be handled simply by an alteration of a potential energy function depending solely on the coordinates of the several interacting parti-cles. This tends to detract from the value of arguments based simply on the use of quantum mechanics with the positional coordinates of the nuclear constituents. Nevertheless we shall restrict ourselves to the use of such meth-ods in the discussion of the instability or capacity for spontaneous disinte-gration of a very much simplified nuclear model. The simplification to be made will consist in supposing that we can discuss the behavior of any one con-stituent by applying the quantum mechanics to it as a single body moving in a force field due to the rest of the nucleus.

The difference between quantum mechanics and classical mechanics which is here made responsible for the disintegration process is easily stated. In classical mechanics the orbit of a particle is entirely confined to those points in space at which its potential energy is less than its total energy. This is not true in quantum mechanics. Classically if a particle be moving in a basin of low potential energy and have not as much total energy as the maximum of potential energy surrounding the basin, it must *certainly* remain there for all time, unless it acquires the deficiency in energy somehow. But in quantum mechanics most statements of certainty are replaced by statements of prob-ability. And the above statement must now be altered to read "... it may

---

[2] Dirac, Proc. Roy. Soc. **A117**, 610; **A118**, 351 (1928).

remain there for a long time but as time goes on the probability that it has escaped, even without change in its total energy, increases toward unity."

In §1 of this paper the detailed development of the argument leading to the conclusion of the preceding paragraph is given. In §2 we discuss its qualitative application to the nuclear disintegration problem. §3 is devoted to semi-quantitative estimates of the rates of decay.

1. Coupling of Motions of Equal Energy

Consider a particle of mass μ. It is sufficient to consider one degree of freedom; let the coordinate of the particle be x and let the forces be measured by the potential energy function $V(x)$.

In classical mechanics the equations of motion possess the energy integral

$$p^2/2\mu + V(x) = W \tag{1}$$

(p = momentum) which, for values of x such that $W - V(x) < 0$, can only be satisfied by p pure imaginary. Therefore, classically, one had the result that a particle could only be where $W - V(x) \geqq 0$. An important consequence of this was that if there were several ranges of x for which $W - V(x) \geqq 0$ separated by ranges where $W - V(x) < 0$, then there were several different motions possible with the energy level W, each of which was wholly confined to one of these separate ranges. Thus in Fig. 1 for the energy level indicated there would be two distinct types of motion of the same energy W; one is a libration in the range I, and the other a libration in the range II.

These results are modified considerably by the new quantum mechanics. In the first place, Eq. (1) loses its validity and is replaced by an integral theorem, as Born[3] has shown, in which there is no longer a definite correlation between simultaneous value of position and momentum as (1) implies. The quantum mechanical form of (1) is, if $\psi(x)$ is Schroedinger's wave function

$$W = \int_{-\infty}^{+\infty} \left(\frac{h^2}{8\pi^2\mu} \frac{d\psi}{dx} \frac{d\bar{\psi}}{dx} + V(x)\psi\bar{\psi} \right) dx \tag{1a}$$

---

[3] Born, Zeits. f. Physik **38**, 806 (1926).

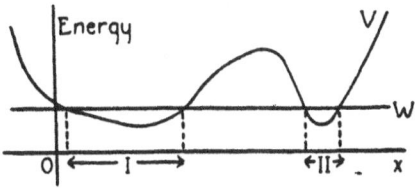

FIGURE 1

The lack of a precise correlation has been much emphasized by Heisenberg and by Bohr,[4] and is a general characteristic of quantum mechanics. From the new standpoint, one has to consider the behavior of Schrödinger's equation for the problem

$$\frac{d^2\psi}{dx^2} + \frac{8\pi^2\mu}{h^2}(W - V(x))\psi = 0. \tag{2}$$

As is well known in some problems there are solutions $\psi(W, x)$ for values of W which are finite and continuous everywhere. These are the "allowed" values of quantum theory. For the $\psi(W, x)$ which comes out of (2) as a by-product, Born has shown that its square may be satisfactorily interpreted as giving the probability that the particle lies between x and $x + dx$ when it is in the state of energy W. This is really the ground for requiring that ψ remain finite. For an energy level, such that $\psi(W, x)$, does not remain finite as $x \to \pm\infty$, the probability that it is not "at infinity" is vanishingly small, and therefore these states do not exist physically. Adopting the probability interpretation of $\psi(W, x)$ one has at once the result that there is a finite probability of being outside the range of the classical motion of that energy.

A simple case is the lowest state of the harmonic oscillator, which has the energy $h\nu/2$. The $\psi(W, x)$ for this state is $e^{-1/2(x/a)^2}$ so $\psi^2 = e^{-(x/a)^2}$ where a is the classical amplitude of motion associated with this energy. The probability of being outside the classical range is therefore

$$\frac{2\displaystyle\int_{a}^{\infty} e^{-(x/a)^2}\,dx}{\displaystyle\int_{-\infty}^{+\infty} e^{-(x/a)^2}\,dx} = 0.157$$

or more than 15 percent.

When one studies the behavior of $\psi(W, x)$ from (2) for a $V(x)$ somewhat like the one in Fig. 1, he finds that, if the W is one for which ψ is finite everywhere, then ψ approaches zero very rapidly (exponential decrease) as $x \to \pm\infty$. In the neighborhood in which $W - V(x)$ is small, the function takes on appreciable values and has oscillatory character where $W - V(x) < 0$, and non-oscillatory character elsewhere. Cases like that of Fig. 1 have been discussed by Hund[5] in connection with his studies of molecular spectra.

An important case is that in which the potential energy curve consists of a single "obstacle" or barrier as in Fig. 2, and the motion is one of insufficient energy, W, to clear the obstacle. In such cases there are *two* finite solutions $\psi_1(W, x)$, and $\psi_2(W, x)$ associated with each energy level, W, and so an arbitrary linear combination of them is also a solution of (2). Born has shown that there

[4] Heisenberg, Zeits. f. Physik **43**, 172 (1927); Bohr, Nature, April 14, 1928.
[5] Hund, Zeits. f. Physik **40**, 742 (1927); Wentzel, Zeits. f. Physik **38**, 518 (1926).

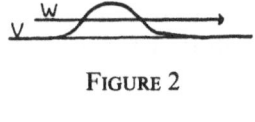

FIGURE 2

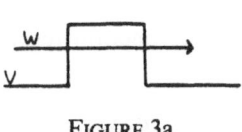

FIGURE 3a

is always a combination of them which depends on x as $e^{+i\sigma x}$ and represents a pure left-to-right progressive wave motion as $x \to +\infty$. Such a solution for x large and negative can then be said to represent an incident left-to-right wave coming from the left side and a reflected wave which is not as strong as the incident wave. The interpretation is that the incident beam of particles is partly reflected and partly transmitted. In the range where $(W - V) < 0$ the de Broglie wave-length h/p becomes imaginary, and so gives rise to an exponential behavior of ψ whose nearest analogue is, perhaps, in optics in the slight penetration of a refracted ray into a rarer medium even beyond the angle of total reflection where the refracted angle is imaginary. In this way, one can find the probability that a particle coming up from the left will get through the wall and escape to the right. The case illustrated in Fig. 3a for which

$$V(x) = 0 \qquad x < -a$$
$$V(x) = V \qquad -a < x < 0,$$
$$V(x) = 0 \qquad x > 0,$$

and for $0 < W < V$, is a simple one with which to illustrate the nature of the calculation. For a given energy level, W, there are two ψ functions satisfying the requirements of finiteness everywhere and of continuity for the ordinates and slopes at the discontinuities in $V(x)$.

These are readily found to be

$$\psi_1(W, x) = \begin{cases} \cosh \sigma_2 a \cdot \cos \sigma_1 (x + a) - (\sigma_2/\sigma_1) \sinh \sigma_2 a \sin \sigma_1 (x + a) & (x < -a) \\ \cosh \sigma_2 x & (-a < x < 0) \\ \cos \sigma_1 x & (0 < x) \end{cases}$$

$$\psi_2(W, x) = \begin{cases} -(\sigma_1/\sigma_2) \sinh \sigma_2 a \cdot \cos \sigma_1 (x + a) + \cosh \sigma_2 a \sin \sigma_1 (x + a) & (x < -a) \\ (\sigma_1/\sigma_2) \sinh \sigma_2 x & (-a < x < 0) \\ \sin \sigma_1 x & (0 < x) \end{cases}$$

(3)

where $\sigma_1 = (2\pi/h)(2\mu W)^{1/2}$ and $\sigma_2 = (2\pi/h)[2\mu(V - W)]^{1/2}$. To find the ψ function corresponding to a beam of particles incident from the left which is partly transmitted and partly reflected, one has to add these together in such a way that to the right of the obstacle there is only the pure left-to-right flow, i.e., one must take

$$\psi_1(W, x) + i\psi_2(W, x).$$

To the left of the obstacle, the ψ function represents the superposition of a left-to-right, or incident beam

$$\psi_{inc} = \left[\cosh \sigma_2 a - \frac{i}{2}\left(\frac{\sigma_1}{\sigma_2} - \frac{\sigma_2}{\sigma_1}\right) \cdot \sinh \sigma_2 a \right] e^{i\sigma_1(x+a)}$$

and a reflected beam

$$\psi_{ref} = \frac{i}{2}\left(\frac{\sigma_1}{\sigma_2} + \frac{\sigma_2}{\sigma_1}\right) \sinh \sigma_2 a e^{-i\sigma_1(x+a)}$$

$$\left. \right\} \quad x < -a \qquad (4)$$

The transmitted beam is simply

$$\psi_{tr} = e^{i\sigma_1 x} \qquad x > 0 \qquad (5)$$

These expressions have, of course, the conservation property

$$(\psi\bar{\psi})_{inc} = (\psi\bar{\psi})_{ref} + (\psi\bar{\psi})_{tr}.$$

The probability that a particle coming up to the wall shall get through to the other side is simply $(\psi\bar{\psi})_{tr} \div (\psi\bar{\psi})_{inc}$ which for $e^{\sigma_2 a} \gg 1$ is clearly equal to

$$P_a(W) = 16(W/V)(1 - W/V)e^{-2\sigma_2 a}. \qquad (6)$$

The controlling factor is the exponential term except when W/V is very near to 0 or 1.

For application to a theory of the pulling of electrons out of metals by electric fields Fowler and Nordheim[6] have derived the probability expression by similar methods for the curve of Fig. 3b, i.e.

$$V(x) = 0 \qquad\qquad x < 0$$
$$V(x) = C - Fx \qquad x > 0$$

The probability that a particle of energy W get through the wall they find to be

---

[6] Fowler and Nordheim, Proc. Roy. Soc. **A119**, 1 (1928).

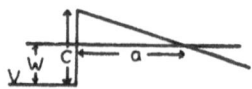

FIGURE 3b

$$P_b(W) = 4[(W/C)(1 - W/C)]^{1/2} \exp(-4k(C - W)^{3/2}/3F)$$

from their Eq. (18) p. 178. The term in the exponent can be written

$$4k(C - W)^{3/2}/3F = 4\sigma_2/3a \quad (k^2 = 8\pi^2\mu/h^2) \tag{7}$$

to exhibit the similarity with the case of the square wall. Here a is the positive value of x for which $V(x) = W$, and σ_2 is defined as

$$\sigma_2 = (2\pi/h)[2\mu(C - W)]^{1/2}.$$

The exponents in each of these cases can be written in the form

$$(4\pi/h) \int [2\mu(V - W)]^{1/2}\, dx$$

the integration extending across the barrier, the limits being the two places where $V(x) - W = 0$.

Application of the method of approximate integration of Schrödinger's wave equation which was first used in quantum mechanics by Wentzel[5] indicates that such a result is quite general. The probability of getting through the wall at a single approach is governed essentially by the factor

$$\exp\left\{-(4\pi/h) \int [2\mu(V - W)]^{1/2}\, dx\right\} \tag{8}$$

being equal to it except for a factor of the order of magnitude of unity.

We have next to consider the case of a potential energy curve of the type shown in Fig. 4. According to classical mechanics there are two modes of motion associated with energy levels below the maximum such as W in the figure. One is a periodic motion in the range I while the other is an aperiodic motion in the range II. By the Bohr-Sommerfeld rule the periodic motions would give a discrete spectrum of allowed energy levels which would overlie the continuous spectrum associated with the aperiodic motions. On the quantum mechanics every energy level is allowed with the essential difference that *there are no energy levels with which two types of motion are associated.* With each energy level there is associated just one wave function $\psi(W, x)$ whose square gives the relative probability of being at different parts of the possible range of x. The $\psi(W, x)$ functions do show traces of the discreteness of the energy levels which the Bohr-Sommerfeld rule associates with the perodic motions in I, in an interesting way. The $\psi(W, x)$ for every W show sinusoidal

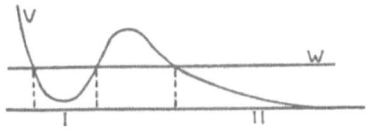

FIGURE 4

oscillations as $x \to \infty$ and also oscillate in the range I. For most energies the amplitude of the oscillations in the range II is overwhelmingly large compared to that in range I, the ratio being of the order of $\exp\{(2\pi/h) \int [2\mu(V - W)]^{1/2} dx\}$ the integration extending across the barrier. This situation is just reversed however for little ranges of W values near those given by the old quantization rules. For these the amplitude in I is large compared to that in II in the same ratio. These then are the "allowed" energy levels. It is not a stationary state for the particle to be in range I and remain there. But for certain energy levels there is an extraordinarily large probability of being in unit length of range I relative to unit length of range II.

We have to find the mean time which a particle remains in the range I before "leaking through" to the outer range II. This can be obtained from the following simple consideration. When the particle is at a place of x large and positive, $V(x) = 0$ (Fig. 4) so the energy is all kinetic and the speed is therefore $(2W/\mu)^{1/2}$. The amount of time which the particle spends in unit length for x large is therefore $(\mu/2W)^{1/2}$. The time spent in a range of length a is therefore $a(\mu/2W)^{1/2}$. Now according to the wave-functions the probability of being in unit length of range I for one of the quasi-discrete energy values relative to the probability of being in unit length of range II is of the order $\exp\{(4\pi/h) \int [2\mu(V - W)]^{1/2} dx\}$. Therefore since the motion is aperiodic and the particle escaping from range I will in the mean only go through unit length of II once, the time T which must be spent in range I before getting through to range II is of the order of

$$T \sim a(\mu/2W)^{1/2} \exp\left\{(4\pi/h) \int [2\mu(V - W)]^{1/2} dx\right\}$$

where a is of the order of the breadth of range I.

Like all of the results of quantum mechanics this is to be interpreted as a probability result. So that if we start with a number of particles in the same allowed energy level in identical regions similar to range I, the number which leak out in time dt is governed by

$$dN = -N\lambda\, dt$$

which gives the usual exponential law of decay $N(t) = N_0 e^{-\lambda t}$ where

$$\lambda = 1/T. \tag{9}$$

The expression for T may be arrived at in a somewhat different way. One can think of the particle as executing its classical motion in range I, but as having at each approach to the barrier the probability of escaping to range II given by expression (8) above. The frequency of the periodic motion in I, which represents the number of approaches to the barrier in unit time, is of the order $a(\mu/2W)^{1/2}$ so the mean time of remaining in range I before escape comes out as the quotient of these two quantities as before. The reader will find it of interest to examine Oppenheimer's formula[7] for the pulling of

---
[7] Oppenheimer, Proc. Nat. Acad. Sci. **14**, 363 (1928).

electrons out of hydrogen atoms by an electric field. His formula for the mean time required for dissociation of the atom by a steady electric field splits naturally into a factor which is the classical frequency of motion in the Bohr orbit multiplied by an exponential probability factor of the type of expression (8) used in this paper.

2. Application to Radioactive Disintegration

After the exponential law in radioactive decay had been discovered in 1902, it soon became clear that the time of disintegration of an atom was as independent of the previous history of the atom as it was of its physical condition. One could not for example suppose that an atom at its birth begins to lose energy by radiation and that its instability is the result of the drain of energy from the nucleus. On such a view it would be expected that the rate of decay would increase with the age of the atoms. When later it was observed that the number of atoms breaking up per second showed the fluctuations demanded by the laws of probability it became clear that the disintegrating depended solely on chance. This has been very puzzling so long as we have accepted a dynamics by which the behaviour of particles is definitely fixed by the conditions. We have had to consider the disintegration as due to the extraordinary conjunction of scores of independent events in the orbitalmotions of nuclear particles. Now, however, we throw the whole responsibility on to the laws of quantum mechanics, recognizing that the behaviour of particles everywhere is equally governed by probability.

From what was said in the preceding section it is clear that the property of the nucleus which we need to know in order to apply the theory is its potential energy curve; and this happens to be a property which we know fairly definitely. Outside a nucleus whose net charge is given by the atomic number we should expect to find a Coulomb inverse-square field of the appropriate strength. And it is well known that in experiments on the scattering of alpha particles from heavy nuclei the proper inverse-square field is found to extend through the whole accessible region. In Fig. 5 the curve AB is a plot of the potential energy of an alpha particle in this field against the distance from the centre of a nucleus of atomic number $Z = 90$. To provide the

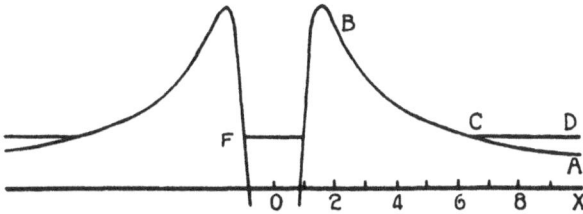

FIGURE 5. The unit of abscissas is 10^{-12} cm. The horizontal line gives the energy of the α-particle emitted by uranium, 6.5×10^{-6} ergs.

attractive field which holds alpha particles in the nucleus it has long been recognized that the potential energy curve must turn over in the way shown in Fig. 5. And it has been shown, for example by Enskog,[8] that curves of this type may be obtained by giving the particle a magnetic moment.

In order to explain the ejection of a particle one has hitherto supposed that the praticle in the internal region received energy sufficient to raise it over the potential barrier. The suggestion that this energy was obtained by absorption of some ultra-penetrating radiation from outside never received wide acceptance. But it was necessary on classical mechanics to suppose that the emitted particle had received energy, if not from outside then from the other nuclear particles. Now the potential barrier which confines particles in the nucleus, i.e. the area under the curve in Fig. 5, is a region where the total energy would be less than the potential energy. And since the quantum mechanics endows particles with the new property of being able to penetrate such regions, this gives us at last a nucleus which can disintegrate without the absorption of energy.

We see that a mere qualitative application of the principles of quantum mechanics seems to account for the principal properties of radioactive atoms, most of which have been familiar for nearly thirty years. We have now to consider the question: How can nearly similar nuclei have periods of decay of anything from a small fraction of a second to over 10^9 years? It has been shown above that in coupling the possible motions of a particle on either side of a potential barrier, the probability of transmission through the barrier is extremely sensitive to the area of the barrier; in fact the relation to it is exponential. In this way we shall show that we can obtain all rates of decay up to practical stability, and that from atoms whose potential curves are almost identical.

3. Quantitative Application

If the height of CD above OX in Fig. 5 gives the energy of the alphaparticle emitted, we have to consider the coupling of the motion along CD with the motion along EF inside the nucleus. We clearly do not have much choice in the area of the potential barrier we may take, since both the point C is fixed and the curve passing through C. For the purposes of numerical calculation we will compare radium A which has a period of 4.4 minutes (half-value period 3.05 minutes) with the extreme cases of uranium and radium C′, which have decay periods of about six thousand million years and a millionth of a second respectively.

An alpha-particle emitted by an element of atomic number Z escapes through the Coulomb field corresponding to $(Z-2)$. Hence the potential energy

---

[8] Enskog, Zeits. f. Physik **45**, 852 (1927).
[9] Blackett, Proc. Roy. Soc. **A107**, 369 (1925).

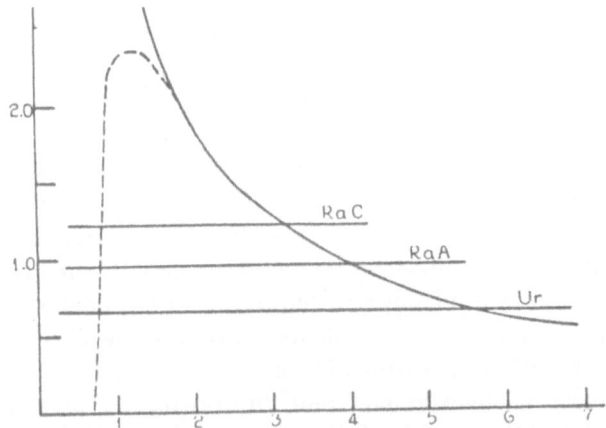

FIGURE 6. The unit of ordinates is 10^{-5} ergs, and the unit of abscissas 10^{-12} cm.

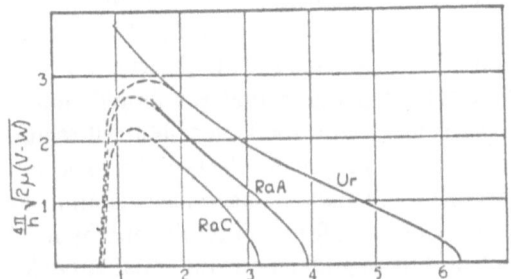

FIGURE 7. Ordinates give the value of $(4\pi/h)[2\mu(V - W)]^{1/2}$ the unit being 10^{13} cm^{-1}. The unit of abscissas is 10^{-12} cm.

for $Z = 82$, which has been plotted in Fig. 6 is appropriate to radium A. The three horizontal lines in Fig. 6 give the energies of the alpha-particles from Ra A, Ra C', and Uranium, which are 1.22×10^{-5}, 9.55×10^{-6}, and 6.5×10^{-6} ergs respectively. It is clear that the factor $(V - W)$, which occurs in the expression (9), given above for the rate of decay, is simply the vertical distance between the horizontal line for W and the potential energy curve for the proper value of Z-2. In Fig. 7 is plotted a curve derived from Fig. 6 giving the value of $(4\pi/h)[2\mu(V - W)]^{1/2}$ for Ra A as a function of the radius. The upper curve is for uranium and the lower for Ra C', derived from curves for the proper atomic numbers.

From these curves we can at once find how large a barrier we have to take in order to obtain any observed rate of decay. For the integral occurring in the exponent of expression (9) is merely the area that we will take under the curve in Fig. 7. Since the unit of abscissas taken is 10^{-12} cm and the unit of ordinates 10^{13} cm^{-1}, each of the squares in Fig. 7 has the dimensionless value 10; so that for an element whose potential barrier has an area of one square on this diagram we should employ the factor e^{-10}. The broken line in Fig. 6 has been drawn so as to give for Ra A in Fig. 7 an area of approximately the value 53.7. For substituting this value in expression (9) together with $W = 9.55 \times 10^{-6}$ and $a = 10^{-12}$ we obtain the decay constant $8.45 \times 10^{20} \times e^{-53.7} = 3.8 \times 10^{-3}$ sec.$^{-1}$, or the decay period $1/\lambda$ is 4.4 minutes in agreement with observation. In the expression for λ the precise value of the first factor is obviously unimportant, for if it were taken five times larger or smaller this would only alter the area of the required barrier by about 1 percent. The general size of the potential barrier in Fig. 6 that we have had to take seems to be a very reasonable one.

Now we reach an unexpected result. In drawing the areas for uranium and Ra C' in Fig. 7 the continuous lines were predetermined, and the broken lines have been derived from the curve in Fig. 6, already used for Ra A. The values of the two areas are found to be 34.4 and 90, though the exact values depend on how the broken line is made to join the Coulomb potential curves. On substituting the values 34.4 and 90 in the expression for T we obtain for Ra C' and uranium decay periods of the order of 10^{-6} sec. and 10^{10} years respectively, in agreement with observation.

It was already clear from Fig. 6 that we should obtain for all elements some qualitative agreement with the Geiger-Nuttall relation: the higher the energy of the alpha-particle the greater the rapidity of decay. But now we have found the unexpected result that the agreement is almost quantitative; that we do not have to choose a different potential energy inside the nucleus for each alpha-particle but having taken one potential curve for the whole series, it is the energy of the emitted alpha-particle which determines its own rate of decay. The mere fact that the velocity of the alpha-particle from Ra A, 1.69×10^9 cm per sec. is a little greater than the 1.4×10^9 cm per sec. of uranium, and a little less than the 1.92×10^9 cm per sec. of Ra C', gives Ra A a decay period 10^{15}; times as short as that of uranium and 10^8 times as long as that of Ra C', in agreement with observation. Questions raised by this agreement with the factor of Geiger and Nuttall will be discussed in the last section of this paper. The radius 2×10^{-12} cm, at which we have taken the deviation from the inverse-square law, seems to be of the magnitude which our knowledge of the nucleus would lead us to expect.

Further we see at once why it is that no slow alpha-particles have been discovered. Although particles of ranges between 2.5 and 7 cm are plentifully distributed, no alpha-particles of energy less than 6.5×10^{-6} ergs have been found. But we now see from Figs. 6 and 7 that for particles of lower energy the area of the potential barrier increases very rapidly; so that for particles of range 2 cm or less the exponential factor would reduce the rate of decay of the element to a value at which its manifestation of radioactivity would be beyond the limits of detection.

Beta-ray disintegration.—It has been customary to assign the central core

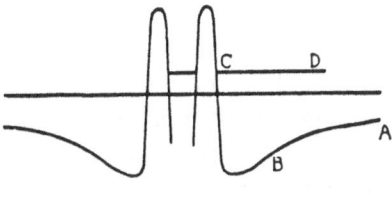

FIGURE 8

of the nucleus as the habitat of the nuclear electrons, with a potential energy curve of the type shown in Fig. 8. The outer slope AB again represents the Coulomb inverse-square field, as in Fig. 5. But since the charge of the electron is $-e$ instead of $+2e$ the potential energy is reversed in sign, and of half the magnitude of that in Fig. 5. There is nothing new in this assumed curve, although it looks somewhat artificial; this type of curve for the nuclear electron was obtained for example by Enskog in the paper referred to above.[8] What is new is the suggestion that an electron in the internal region again has a certain chance of penetrating the barrier, and of escaping at any time along CD with kinetic energy given by the height of CD above the axis.

If we have alpha and beta-particles both with this chance of escaping from the nucleus, it might be thought that every radioactive element should be found to disintegrate part with expulsion of alpha-particles and part with beta-particles. But we would repeat that the chance of escape is extremely sensitive to the height to which the potential energy curve rises above the energy-level in question; and that if the size of this potential barrier be increased by a small factor the probability of escape may be decreased more than a million-fold. There seems then no reason why there should not be the three types of disintegration: that in which the probability of escape is much greater for an alpha-particle than for an electron; that in which it is much greater for an electron than for an alpha-particle; and that in which the probabilities of escape are comparable. The last gives the branching type of disintegration as shown by Ra C, of which 99.97 percent emits beta-particles, and 0.03 percent alpha-particles. By taking this view of the disintegration process, we have raised the question: Does any radioactive element have a unique mode of disintegration, or does it merely appear unique in most cases because the secondary mode is a million times less frequent and escapes detection? The present discussion certainly favours the latter alternative. It need not surprise us then that so few cases of branching disintegration have so far been discovered, since it is unlikely (so far as we know) that the areas of the potential barriers will in many nuclei happen to have just that relative size which will give for alpha and beta-particles comparable probabilities of escape.

Artificial disintegration.—Blackett's cloud-chamber photographs[9] of artificial disintegration in nitrogen showed that the impinging alpha-particle was caught and retained by the nucleus. One is tempted to apply the present theory, using again the fact that the impinging alpha-particle may penetrate the barrier of potential, this time from the outside, instead of passing over the top as required by classical theory. But when we do this we are at once confronted by the fact that instead of approaching the barrier 10^{20} times per second, like a nuclear particle, our alpha-particles will only make one impact apiece. So it would seem that the capture of the alpha-particle could not be due to penetration. There is, however, another consideration; and that is that if the impinging alpha-particle have an energy very near that of an allowed but unoccupied nuclear energy-level, the chance of its penetrating the barrier at a single impact approaches unity. This property has already been referred to in section 1.

4. Experimental Evidence for the Penetration of Potential Barriers

The essential basis of the present theory is the assumed power of particles to pass through regions where their total energy would by classical mechanics be less than their total energy. For this property there is no direct experimental evidence in physics, although it follows from the laws of quantum mechanics. But in applying this to the nucleus we have found that we can actually *obtain* direct experimental evidence. Though on classical mechanics the passage of a particle through such a forbidden region was a manifest absurdity, it was found in 1925 by Rutherford and Chadwick[10] that that is exactly what the alpha-particles from uranium appear to do.

Consider the alpha-particle which the uranium nucleus emits during its disintegration. The alpha-particle will gain energy in escaping through the repulsive Coulomb field outside the nucleus. This energy is given on classical theory as $2Ze^2/r$. Even if the alpha-particle leaves its place in the nucleus with no initial velocity, its energy cannot be less than this amount. The energy with which the alpha-particles leave the disintegrating Uranium atom is observed experimentally to be 6.5×10^{-6} ergs. On referring to Fig. 5, which was drawn for $Z = 90$, we see that this energy corresponds to $r = 6.3 \times 10^{-12}$ cm and if any of the energy was initial energy and not acquired through falling through the repulsive field, the value of r would have to be greater than this value.

It was concluded that the inverse-square law of repulsive field could not possibly hold within this value of r. Consequently if we fire at the uranium nucleus an alpha-particle having slightly more energy than the $6.5 \ 10^{-6}$ ergs, it should penetrate its structure to where the Coulomb law no longer holds; while still faster particles should penetrate, even when not fired directly at the nucleus. It was therefore disconcerting when, on examining the scattering of

---

[10] Rutherford and Chadwick, Phil. Mag **50**, 889 (1975).

fast alpha-particles fired at uranium, Rutherford and Chadwick could find no indication of any departure from the inverse-square laws. The Coulomb field was found to hold inside the radius from which the uranium alpha-particle appeared to come. That is to say, the uranium alpha-particle appeared to emerge from a region where its kinetic energy was negative. To escape this conclusion, Rutherford[11] supposed that the uranium alpha-particles before ejection are electrically neutral, having been neutralised by two electrons which they leave behind when they are ejected. This hypothesis succeeded in circumventing the paradox. But if we abandon classical mechanics, the paradox disappears, yielding us direct experimental evidence in favor of the phenomenon of quantum mechanics in which we are interested.

5. Discussion of Limitations

It must be clearly understood that although the Coulomb part of the potential curve outside the nucleus, represented by AB in Fig. 5 is necessarily common to all particles, the internal part is merely intended to represent the potential energy of a particular alpha-particle. And it must not be taken to represent a general central field common to many particles, such as we are so accustomed to in atomic structure. There is no reason why the internal field should be necessarily symmetrical al)out the center of the nucleus as drawn in Figs. 5 ancl 8. In fact, Rutherford[12] has suggested that the nucleus may have something analogous to a crystalline structure. If this caution is lost sight of, difficulties are encountered.

For the atom of each radioactive element contains within its nucleus not only the alpha-particle which it will itself emit, but also the alpha-particles destined to be emitted by its successors in the radioactive series. Now if the velocity of escape of the alpha-particles from each element were always less than that of those emitted by its predecessors, there would be no serious difficulty; for from an atom loaded with alpha-particles in various allowed energy levels, the particle in the highest level would have the greatest probability of escape. This however is the opposite of what is observed; and we have to account for the subsequent emission of particles of higher energy than that emitted by the parent substance. We may do this by supposing either (a) that the alpha-particles of higher energy have in the parent element been confined bv correspondingly high barriers; or (b) by supposing that the alpha-particles in the nucleus are not permanently in the high energy levels from which they emerge, but are temporarily raised up from lower levels. The latter seems to be a retrograde step, for the principle advantage of the present theory is that it has offered an escape from such processes.

---

[11] Rutherford, Phil. Mag. **4**, 580 (1927); Proc. Phys. Soc. **39**, 370 (1927).
[12] Rutherford, Jour. Franklin Inst. **198**, 743 (1924).

If, however, we accept the former supposition (a), we see that the emission of one alpha-particle must profoundly modify the potential barrier which confines the alpha-particle destined to be emitted next. As we have shown, the Geiger-Nuttall relation seems to require that the barrier through which this alpha-particle emerges be approximately the same in all elements of the series. But until we know how this comes about, it seems inadvisable to discuss the Geiger-Nuttall relation in greater detail. In speaking of the energy of one particle in the nucleus, it must not be forgotten that we are making use of the simplification mentioned in the introduction: that of discussing one nuclear constituent alone.

Remarks on Uncertainty Principles

EDWARD U. CONDON

Palmer Physical Laboratory
Princeton, University

Since the publication of Heisenberg's paper[1] on the "anschaulichen Inhalt" of quantum mechanics, discussions of the fundamental limitations on the accuracy of physical measurements have been much in the foreground. According to Heisenberg, the quantum mechanics implies that it is impossible to measure simultaneous values of a coordinate and its conjugate momentum with unlimited precision. Instead, if Δp be the estimated error or uncertainty in a momentum and Δq that in the associated coordinate one must have the inequality,

$$\Delta p \Delta q > \frac{h}{2\pi} \tag{1}$$

This inequality has come to be known quite generally as Heisenberg's uncertainty relation.

In discussions on this subject it is essential to distinguish two standpoints. One is the annlysis of proposed experiments, whether realizable or ideal, by which it is proposed to make measurements. The other is that of the relation of the uncertainty principles to the laws of quantum mechanics as now formulated. It is only the second standpoint which is considered here.

The origin of the uncertainty relation (1) for p and q lies in the fact that the operators which represent p and q do not commute. Therefore one is tempted to suppose that such an uncertainty relation may be true for any two quantities whose operators do not commute. Such, however, is not the case. These remarks establish by means of specific examples the truth of the following statements:

(*a*) The fact that the operators corresponding to two physical quantities, A and B, do not commute does not imply the existence of an uncertainty relation of the form of (1), namely, that the product of the two uncertainties must be greater than or equal to some lower limit.

(*b*) Even if A and B do not commute, there may be exceptional values of A and B which may be both known simultaneously with no uncertainty.

---

[1] W. Heisenberg, *Zeits. für Physik*, 43: 172. 1927.

(c) There may exist a limited class of states of the system, with regard to which A and B do commute, but in which nevertheless the two quantities A and B can not be known with unlimited precision.

The relation of the uncertainty principle to the quantum mechanics may be formulated as follows. The configuration of a dynamical system of n degrees of freedom is specified by n spatial coordinates, as in classical mechanics. There may appear new coordinates which do not have classical analogs like the electron spin or the permutation variables but these will be left out of account. The particular state of the dynamical system at any instant is then specified by giving a function $\varphi(x_1 \ldots x_n, t)$ which has the property that $\varphi\bar{\varphi}\,d\tau$, where $d\tau$ is the volume element of the configuration space, is the probability that the system be found at the instant, t, with its configuration lying in the volume element $d\tau$ of the configuration space which surrounds the point, $x_1 \ldots x_n$.

Corresponding to each physical quantity there is a linear operator, which when applied to φ gives another function of the coordinates, $x_1 \ldots x_n$. Let A be a physical quantity and at the same time A may stand for the operator which represents A. Then

$$\int \bar{\varphi} A \varphi \, d\tau$$

represents the mean or expected value of A^2 in this state of the system characterized by φ. The integration is over the entire configuration space. Similarly,

$$\int \bar{\varphi} A^2 \varphi \, d\tau$$

represents the mean or expected value of A^2 in this state of the system.

We shall define, the uncertainty in the value of A associated with this state by the equation,

$$(\Delta A)^2 = \int \bar{\varphi} A^2 \varphi \, d\tau - \left(\int \bar{\varphi} A \varphi \, d\tau \right)^2 \qquad (2)$$

where ΔA is written for the uncertainty in A. This evidently corresponds with the classical definition of the uncertainty as the square root of the mean of the square of the deviation from the mean. One observes that if φ is such that $A\varphi = a\varphi$, where a is an ordinary arithmetical number, then ΔA vanishes. In such a case the value of A is precisely a. For example, if A is really the Hamiltonian function and φ is one of the solutions of Schrödinger's equation for the system, one has the proof that the Schrödinger wave functions correspond to states of the system in which the total energy has precisely one of the allowed values of the energy.[2]

---

[2] This formulation corresponds to that of Weyl, "Gruppentheorie und Quantenmechanik," Leipzig, 1928, p. 67.

The examples illustrating propositions (a), (b) and (c) are afforded by considering the different components of angular momentum of a particle about the origin. These will be denoted by M_x, M_y and M_z. As is well known, the operators corresponding to these three quantities do not commute with each other, the operators being,

$$M_x = \frac{h}{2\pi i}\left(y\frac{\partial}{\partial z} - z\frac{\partial}{\partial y}\right) \tag{3}$$

the other two being given by cyclic permutation of x, y and z. These operators satisfy the following commutation rules,

$$M_x M_y - M_y M_x = \frac{h}{2\pi i} M_z, \tag{4}$$

and two others given by cyclic permutation of x, y and z.

As to (a), we observe that the operator for M^2, where

$$M^2 = M^2_x + M^2_y + M^2_x,$$

commutes with the operator for any component, M_x, M_y or M_z. If we consider in particular states where φ is of the form

$$\varphi = R(r)e^{im\phi}P^m_l(\cos\theta), \tag{5}$$

where $R(r)$ is any function of r, and the z axis is the pole of the spherical polar coordinate system, it may be readily verified that such states correspond to precise values for M^2 and M_z given by

$$M^2 = l(l+1)\left(\frac{h}{2\pi}\right)^2 \quad \text{and} \quad M_z = m\frac{h}{2\pi}$$

If one compute the value of ΔM_x or ΔM_y for such states it turns out to be equal to

$$(\Delta M_x)^2 = (\Delta M_y)^2 = \tfrac{1}{2}[l(l+1) - m^2]. \tag{6}$$

This is finite for finite values of l and m. Therefore we have here a class of states of the particle in which, since ΔM_z is zero, the product of the uncertainties in ΔM_x and ΔM_z is zero, in spite of the fact that the operators M_x and M_z do not commute. Hence the truth of (a).

As to (b), we notice that if

$$\varphi = R(r) \quad \text{(independent of } \theta \text{ and } \varphi)$$

this corresponds to the precise values,

$$M_x = 0, \qquad M_y = 0, \qquad M_z = 0,$$

with no uncertainty in any of them. Hence the truth of (b).

As to (c), we observe that if we deal with the class of states in which M_z is known to have the value zero precisely, then from the commutation rule (4), applied to any φ of this class, the operator $M_x M_y$ gives the same result as

$M_y M_x$. But nevertheless the uncertainty in neither M_x nor M_y is zero in such a state. The φ functions for this class of states are evidently of the type of (5) with m set equal to zero, so the preceding calculation of ΔM_x and ΔM_y given in (6) applies here. Hence the truth of (c).

It would appear, therefore, that a general uncertainty principle is not simply to be formulated in terms of the commutativity or lack of commutativity of the operators associated with A and B, as is usually implied in discussions on this subject. What the exact criteria may be, we may not prepared to state.

Although largely of a negative character, these remarks have considerably clarified the situation for me. In working them out, I have derived much benefit from stimulating conversations with my colleagues, especially Drs. J. E. Mack, E. C. G. Stueckelberg, G. P. Harnwell and H. P. Robertson.

An Interpretation of Pauli's Exclusion Principle

E.U. Condon and J.E. Mack*

*University of Minnesota
Minneapolis, Minnesota*

ABSTRACT

I. Pauli's exclusion principle can be understood as an instance of the subjectivity of our knowledge. We are built out of only a particular world constructed according to one of the non-combining patterns possible under the laws of quantum mechanics. Therefore we are capable of having sense perceptions of only that world.

II. Dirac's theory of the proton shows why Pauli's principle governs the world.

I

Probably the most important trend in current physics is that toward the recognition of the fact that our science gives no information about an objective reality wholly independent of ourselves. We are becoming increasingly impressed by the subjectivity of our knowledge. We owe much of this to the work of Bridgman with his emphasis on the unprofitableness of dallying over meaningless questions and his operational formulation of the reality of physical concepts. The rôle played by our subjective limitations on observation of the world about us has been discussed in detail by Heisenberg and by Bohr. We refer to the train of ideas known as Heisenberg's uncertainty principle.

Let us consider what the uncertainty principle did for physics. It came after the fairly complete formulation of the laws of quantum mechanics as we know them to-day. Before the principle the laws were quite definitely stated by Heisenberg, Born, Dirac, Jordan, Schrödinger, Pauli and others. They were stated so definitely that the method of treating special problems was recognised and applied with a good deal of success by a large number of physicists. Therefore, they were valid laws of science, since they met the test of being workable in the hands of physicists in general, and did not rest purely

---

* Nation Research Fellow.

on the genius of the individuals who discovered them. Even without the principle they probably could have attained a universal applicability but we cannot say how that would have been, since, owing to the slowness of diffusion of scientific knowledge, the formal laws were still in the process of becoming the common property of professional physicists when Heisenberg enunciated the principle.

But before he did, even among those who had found the new methods workable and who had contributed to the many successes of the theory, there was a feeling of dissatisfaction, a feeling that the new theory was not understood, perhaps even not understandable. These were disturbing times for we did not know why we felt dissatisfied (or perhaps unsatisfied) and so were quite helpless in attempting to remedy the situation. The remedy was provided by Heisenberg in his uncertainty principle. In a sense his famous paper in *Zeitschrift für Physik*, volume 43, has nothing new in it. For it does not alter or add to the formal statement of the laws of quantum mechanics. Nor does it apply them to any specific problem and produce calculations about some aspect of nature which are triumphantly in agreement with somebody's quantitative measurements. What it did do was to relieve that feeling of dissatisfaction. Many a physicist who had been working with the formalism of quantum mechanics felt that he understood it for the first time after he read that paper. Subsequently this satisfaction of understanding was considerably enhanced by the appearance of Bohr's Como lecture simultaneously in *Nature and Die Naturwissenschaften*. The title of Heisenberg's paper is significant: he called it the "anschaulich Inhalt" of the new quantum mechanics. It was just that: it was the intuitive, clear, evident content of the theory.

The important feature of this clear and intuitive part of the theory is the deflation of the lords-of-creation attitude of physicists. Wholly rational, detached fellows we thought ourselves (in our naive way), studying objectively the true laws of the external world. We had no concern with unfruitful stuff like metaphysics. Now we know otherwise and are healthily aware of our own organic relation to the traditionally inorganic subject matter of our science.

At present there is another law of nature (empirical, like all laws of nature) intimately connected with quantum mechanics, which we feel that we do not understand. This is Pauli's exclusion principle. We refer to the same feeling of dissatisfaction which was felt about quantum mechanics in general before Heisenberg's paper.

Pauli's principle grew out of empirical spectroscopy and was stated in the form that no two electrons in an atom can have the same set of quantum numbers. A formal place was found for this in the equations of quantum mechanics by Heisenberg, Dirac, Wigner and others. The equations of quantum mechanics showed a remarkable property when applied to the description of a dynamical system consisting of several dynamically similar particles. Such properties are naturally of fundamental importance because of the belief that all electrons are dynamically similar and that they are the fundamental stuff of which all substance is built. This remarkable property may be described by saying that such dynamical systems may exist in $N!$ different classes of states such that if the system exists now in one of these classes then it must exist for all time to come in this same class of states. (N is the number of similar particles in the system.) In other words, quantum mechanics provides, when applied to the dynamical system which is the entire universe, not just one solution but $N!$ possible solutions, which we may call worlds, where N is the total number of electrons in the universe. Since it is an exact conclusion from the postulated exact similarity of all the particles that the universe cannot change from one of these classes of states to another, our universe must represent a selection from among these possible worlds. Which one? There is one which has the property that it has no possible states characterized by two of the particles' having the same quantum numbers. That one agrees with the world that we know.

That is fine but it leaves us with a lack of understanding of the matter which is, perhaps, akin to the unsatisfied feeling toward quantum mechanics which preceded the discovery of its "anschaulich Inhalt" by Heisenberg. What of the other $N!-1$ possible worlds?

Now we come to our main thesis: The restriction to a world built in accordance with Pauli's principle can be understood as another and extreme instance of the subjectivity of our knowledge. We, physicists, humans, are creatures who are built out of, and therefore are capable of reacting with and having sense perceptions of, only a particular world constructed according to a particular one of the non-combining patterns possible under the laws of quantum mechanics. Because of our own nature we are aware of, or conscious of, only that part of possible reality that is built on our pattern.

Do worlds really exist that are built (of the same electrons as ours) on one of the other patterns but nevertheless according to the laws of quantum mechanics? The question is meaningless in a sense because of our subjective incapacity to be aware of them. It is meaningless for the same reason as is the question, (sometimes asked after an individual has heard, for the first time, an exposition of Heisenberg's uncertainty principle) "Does a particle really have precise position and momentum simultaneously even though we cannot be experimentally aware of precise values for them?"

The situation turns on the meaning to be ascribed to the word "really" in the two questions. We prefer to associate with it the connotation, within the power of human sense perceptions. In that case the answer to both of them is no, as the "even though" clause makes ridiculously self-evident in the case of the second one. But we are reluctant to dismiss the other possible worlds quite so readily as we do simultaneous position and momentum. This is because the other worlds correspond to solutions of the equations of quantum mechanics, whereas there is no place in the structure of the mathematical formalism of quantum mechanics for simultaneous exact position and momentum.

Objective reality these other worlds cannot have, because of our subjective inability to react to them. (Remember that all objective reality is colored by important subjective elements!) This we feel is the simple, evident substratum of the Pauli exclusion principle which provides us with the right to say we "understand" it.

II

The foregoing shows that we can be conscious of only one of the possible worlds. But it throws no light on a possible reason why the real one is the one which obeys Pauli's exclusion principle.

One feels that the type of symmetry represented in our world is especially simple and that there must be some reason for it. It is another instance of an ancient paradox: out of $N!$ possibilities the chance of the Pauli principle world being the real one seems vanishingly small. But in the matter or building the real world the dice are cast but once and in that one throw the antisymmetric pattern has an equal chance with the others.

Dirac's recent theory of the proton published in the Proceedings of the Royal Society for January, 1930, throws new light on the matter, however. When we spoke in I of there being $N!$ different possible worlds according to quantum mechanics, we were speaking of the non-relativistic theory. But the relativistic theory of the electron must represent a surer foundation on which to build cosmologic hypotheses. According to it, electrons can exist in states of negative energy (less than $-mc^2$), as well as in usual states of positive energy (of the order $+mc^2$). We have no experimental evidence of electrons of this sort and so the prediction of such negative energy electrons was regarded as a blemish in the theory. But now the solution to the difficulty is seen in the fact that Pauli's principle governs our world.

We can do no better than quote the crucial paragraph of Dirac's paper: "The most stable states for an electron (i.e. the states of lowest energy) are those with negative energy and very high velocity. All the electrons in the world will tend to fall into these states with emission of radiation. The Pauli exclusion principle, however, will come into play and prevent more than one electron going into any one state. Let us assume that there are so many electrons in the world that all the most stable states are occupied, or, more accurately, that *all the states of negative energy are occupied except perhaps a few of small velocity.* Any electrons with positive energy will now have very little chance of jumping into negative-energy states and will therefore behave like electrons are observed to behave in the laboratory. We shall have an infinite number of electrons in negative-energy states, and indeed an infinite number per unit volume all over the world, but if their distribution is exactly uniform we should expect them to be completely unobservable. Only the small departures from exact uniformity, brought about by some of the negative-energy states being unoccupied, can we hope to observe."

And now we see why it is that the Pauli principle world is the only one which can be built out of Dirac's relativistic electrons. For it is the only one in which the available states can be filled up. For the others there is no limit to the number of electrons that can occupy a state. The antisymmetric one is the only pattern that has a "bottom." For the others all the electrons in the universe would be drained off into states of indefinitely great negative energy.

Predissociation of Diatomic Molecules from High Rotational States

D.S. Villars and E.U. Condon

University of Minnesota
Minneapolis, Minnesota

ABSTRACT

Instability of molecules in high rotational states as observed especially in HgH and AlH, is a phenomenon whose quantum-mechanical explanation is closely analogous to that of the radioactive disintegration of atomic nuclei. The heat of dissociation of AlH from its normal state is estimated to be 3.07 volts.

The observations of Henri[1] and his school, that certain band systems apparently pass progressively from a discontinuous to a continuous character in such a manner as to indicate that the emitting molecules lose their rotational quantization before they lose their vibrational quantization, has been ascribed to a "predissociation" by various writers. The generally accepted mechanism for this process is that proposed by Bonhoeffer and Farkas,[2] who assume, in accordance with quantum mechanics, that there may occur a radiationless transition of the molecule in a definitely quantized rotational and vibrational state, to another electronic state for which the same total energy places the molecule in the continuous energy range corresponding to dissociation. When the probability of transition from the quantized level to the dissociated state is high, the average life of the molecule may become less than a rotational period of the molecule so the corresponding band is diffuse. But there is an additional mechanism for predissociation which also is a consequence of quantum mechanics which has not been mentioned before, so far as we are aware.

It is known that the set of rotational quantum levels associated with the same vibrational quantum number may extend up to values of the total energy in excess of the amount necessary to dissociate the molecule in that electronic state. Such sets of rotational leveis however do not extend upward indefinitely but are often observed to break off rather suddenly. In at least one

[1] Henri, Photochemie (1919); Structure des Molécules (1925).
[2] Bonhoeffer and Farkas, Zeits. Phys. Chem. **134**, 337 (1927).

instance, that of aluminum hydride, the last two or three levels broaden out so that lines involving them become faint because their energy is spread out more on the photographic plate. This phenomenon has been interpreted by Mulliken,[3] Hulthén,[4] and Ludloff[5] as due to an instability of the molecule arising from the high rotation. What has been lacking is an account of the nature of the instability and the reason that it sets in when it does.

Franck and Sponer[6] have advanced the view that the phenomena both of the sudden termination of such sets of rotational levels and the broadening of the last few levels, are to be interpreted as Bonhoeffer and Farkas have suggested. Oldenberg,[7] from classical ideas alone, has criticized this paper of Franck and Sponer. He points out that the potential energy governing the radial motion of a molecule that is rotating with quantum number, m, is equal to

$$U_m(r) = U_0(r) + \frac{h^2}{8\pi^2\mu} \frac{m(m+1)}{r^2}$$

where $U_0(r)$ is the potential energy for the non-rotating molecule and is determined by the electronic state. Since for large r, $U_0(r)$ becomes constant more rapidly than r^{-2} approaches zero, the function $U_m(r)$ is positive for large values of r and will show, for m not too great, a minimum followed by a maximum at greater nuclear separation. This behavior is illustrated by Fig. 1 which represents the curves (Morse[8]) for HgH.

The similarity of the curves with those that lie at the basis of quantum mechanics of radioactive disintegration[9] is evident. Hence the general properties of the wave equation which were so successful in that field are applicable here. The formula for the mean time of remaining essentially in the range of the periodic classical motion near the minimum, before going over into the range or the aperiodic classical motion extending from beyond the maximum to infinity, there developed for the atomic nucleus, is here applicable to the dissociation of molecules by rotation. We do not expect accurate results from it, both because it is based on an approximate integration of the wave equation, and because it requires an accurate knowledge (which we do not have) of $U_0(r)$ for the larger values of r. In principle, however, we have the means of inferring from the $U_0(r)$ curve, which is inferred from the energy levels, the exact maximum value of the rotational energy which can give levels effectively sharp in radiative transitions. These are the ones for which the mean life, τ, of the molecule before it "leaks" through the hump in the potential

[3] Mulliken, Phys. Rev. **25**, 509 (1925).
[4] Hulthén, Zeits. f. Physik **32**, 32 (1925), **50**, 319 (1928).
[5] Ludloff, Zeits. f. Physik **39**, 526 (1926).
[6] Franck and Sponer, Göttinger Nachr., p. 241 (1928).
[7] Oldenberg, Zeits. f. Physik **56**, 563 (1929).
[8] Morse, Phys. Rev. **34**, 57 (1929).
[9] Gurney and Condon, Phys. Rev. **33**, 127 (1929).

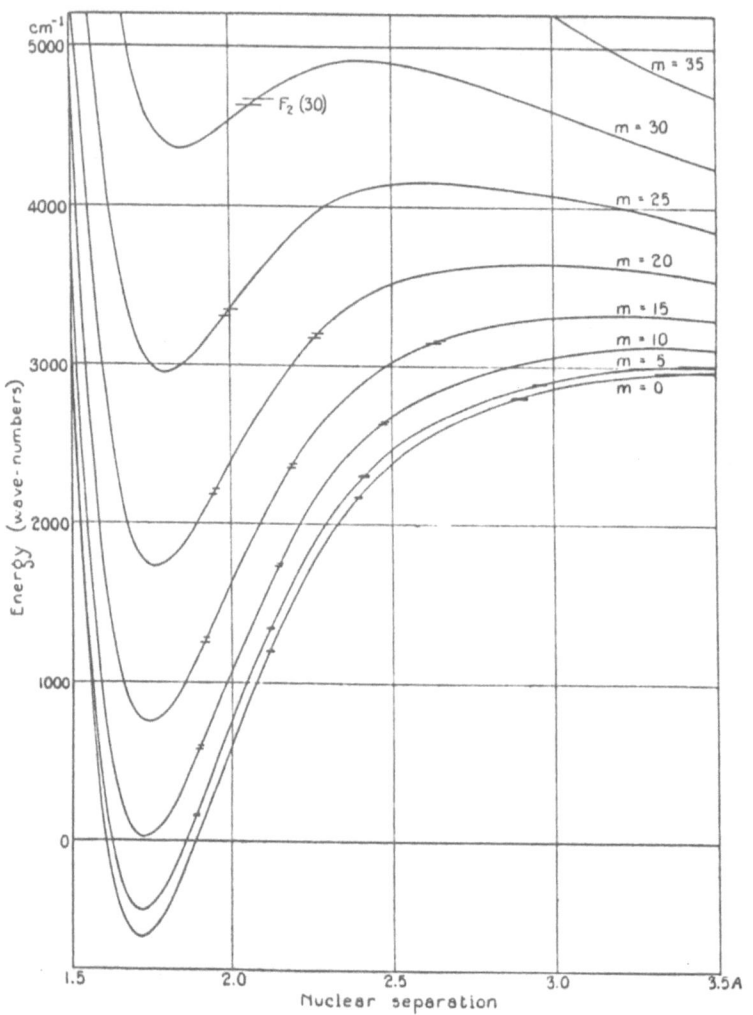

FIGURE 1. Rotational vibrational (Morse) curves for HgH (normal $^2\Sigma$ state.) The designation $F_2(30)$ refers to the upper of the two doublet levels on the $m = 30$ curve.

energy curve is large compared with 10^{-8} sec, which is the mean life needed to provide a sharp line in spectroscopic instruments of the resolving power actually available to experimenters.

Oldenberg has shown that there will be no trace of any discrete level structure for any value of m greater than the one for which $U_m(r)$ has a horizontal inflection point. (An *integral* value of m may not give a horizontal inflection point, m is regarded here as a continuous variable). He also mentions that owing to the existence of the so-called zero-point vibration energy the actual values of m will not extend this far. But there is another factor to consider: the level in question must be sufficiently below the maximum of the

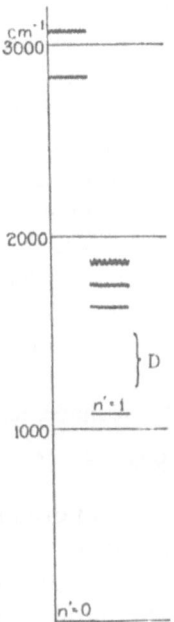

FIGURE 2. Predissociation levels in the excited $^2\Pi$ state of AlH.

$U_m(r)$ that the mean life before leakage is large compared to the mean life before radiation. It is clear why the rotational levels associated with the second vibrational quantum state cannot extend as high as those associated with the first; they are always higher above the minimum of the $U_m(r)$ curve because of the extra vibrational energy and therefore the requirement that they be sufficiently below the maximum of $U_m(r)$ can only be satisfied for smaller values of m. This behavior is nicely illustrated by the levels of HgH plotted by Franck and Sponer in their Fig 2.[6]

This behavior is also shown by the excited electronic state of AlH concerned in the band system studied by Eriksson and Hulthén.[10] Here only two vibrational levels are known so that the heat of dissociation cannot be found by the usual method of extrapolation of vibration levels, due to Birge and Sponer.[11] Fig. 2 shows the range of rotational energy associated with the zero and one vibrational levels of this electronic level. Comparing this figure with the analogous one for HgH, we do not hesitate to say that the dissociation energy for excited AlH must lie below the lowest fuzzy level and above the one quantum vibration level with zero rotation. This puts it probably within the range of the braces around "D" in the figure, that is, it is close to 0.17 ± 0.02 volts. This is the quantity called "y" in Table I of Mulliken's[12] study of band

[10] Eriksson and Hulthén, Zeits. f. Physik **34**, 786 (1925).
[11] Birge and Sponer, Phys. Rev. **28**, 259 (1926).
[12] Mulliken, Phys. Rev. **33**, 130 (1929).

spectra of diatomic hydrides. Hence the heat of dissociation from the normal state is 3.07 volts. This is essentially the conclusion of Franck and Sponer, though from different theoretical considerations.

Oldenberg[13] is of the opinion that the radiationless transfers in the nuclear separation coordinate violate Franck's principle because they call for a fairly large "sudden" alteration in the nuclear separation. We feel that this is an attempt to push the Franck idea beyond the limits of its validity. The principle is simply an easily visualized rule describing approximately what quantum mechanics will predict. The connection with quantum mechanics has already been developed.[14] The principle has to do with the nuclear motions accompanying electron transitions and not with the nuclear motions when there is no change of electronic state.

A rough calculation for HgH was made, using a Morse function for $U_0(r)$,

$$U_0(r) = 2990 + 3592e^{-6.25(r-r_0)} - 7184e^{-3.125(r-r_0)},$$

where energy is measured in cm^{-1} and distance in 10^{-8} cm. The zero-point vibrational energy was arbitrarily taken as one-half the difference between the two lowest vibrational levels. The quantity ω_e was taken as 1528.8 cm^{-1}. These assumptions are somewhat arbitrary but are the best that can be made. The half-breadth of a line is given by

$$\Delta v/c = 1/4\pi\tau c,$$

(compare Pauli, Handbuch der Physik, vol. 23, p. 70). With the formula for mean life from Gurney and Condon[15], the half-breadth of the line with rotation quantum number thirty and zero vibration comes out to be 39 cm^{-1}. Actually this level gives rise to a sharp line, but a slight change in the form of the $U_0(r)$ curve, entering the exponent of the formula for τ will make it be sharp, so we conclude that insufficient knowledge of the $U_0(r)$ function does not justify further calculations at present.

---

[13] Reference 7, p. 573.
[14] Condon, Phys. Rev. **32**, 858 (1928).
[15] Reference 9, p. 133.

Singlet-Triplet Interval Ratios for sp, sd, sf, p^5s and d^9s Configurations[*]

E.U. Condon and G.H. Shortley

University of Minnesota
Minneapolis, Minnesota

ABSTRACT

A systematic comparison of the known data on the singlets and triplets arising from sp, sd, sf, and d^9s configurations with the theory of Houston shows that the theory gives a good account of the deviations from the Landé interval rule which accompany departure from Russell-Saunders coupling. There are numerous significant discrepancies, however. Writing ^1L_l and $^3L_{l+1}$, ^3L_l, $^3L_{l-1}$ with $L = P, D, F$, when $l = 1, 2, 3$ for the term values, we plot as abscissa $(^3L_{l-1} - {}^3L_{l+1})/|{}^3L_l - {}^1L_l|$ and $(^3L_{l-1} - {}^3L_l)/(^3L_l - {}^3L_{l+1})$ as ordinate if $(^3L_l - {}^1L_l)$ is positive, otherwise the reciprocal of this quantity. Houston's equations (12) give functional relations between these interval ratios which are compared with the experimental values.

Houston[1] has worked out an approximate quantum mechanical theory of the relation of the triplet interval ratio to the singlet-triplet interval for two electron configurations in which one of the electrons is in an s state, but the comparison he makes with experimental data gives one very little idea as to just how accurate the theory is. The purpose of the present paper is therefore to make a systematic comparison of the available data with Houston's theory.

If we write the terms of a singlet-triplet system as ^1L_l, $^3L_{l+1}$, ^3L_l, $^3L_{l-1}$, with $L = P, D, F$, when $l = 1, 2, 3$, Houston's equations (12) give the following relative term values in terms of the parameter X, which is the ratio of the exchange perturbation integral to the perturbation theory integral which measures the spin energy. (The first classification applies to $X > 0$, the second to $X < 0$.):

$$^1L_l, {}^3L_l = -\tfrac{1}{2}(X - 1) - \tfrac{1}{2}\{(X - 1)^2 + 4l(l + 1)\}^{1/2}$$

$$^3L_{l+1} = -l$$

$$^3L_l, {}^1L_l = -\tfrac{1}{2}(X - 1) + \tfrac{1}{2}\{(X + 1)^2 + 4l(l + 1)\}^{1/2}$$

$$^3L_{l-1} = l + 1.$$

---

[*] This paper was presented at the Washington meeting of the American Physical Society, April 24, 1930.
[1] W.V. Houston, Phys. Rev. **33**, 297 (1929).

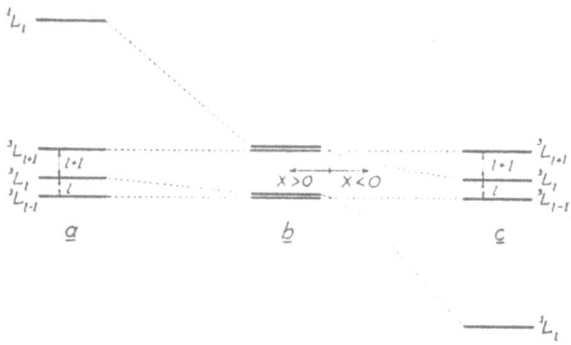

FIGURE 1. Diagram illustrating the behavior of the singlet-triplet intervals. (*a*) X large and positive. (*b*) X very small and positive. (*c*) X large and negative.

Fig. 1 illustrates the behavior of these intervals. With X large and positive the triplet has the Landé interval and the singlet is high above the triplet as at *a*. As X approaches zero the levels approach coincidence in pairs as at *b*; and with X large and negative we again have the Landé interval with the singlet far below the triplet as at *c*.

The theoretical curves of Figs. 2 and 3 are plotted as follows: For abscissa is used $(^3L_{l-1} - \,^3L_{l+1})/|\,^3L_l - \,^1L_l|$. For the lower curve, $X > 0$, the ordinate is $(^3L_{l-1} - \,^3L_l)/(^3L_l - \,^3L_{l+1})$, which goes from $l/(l + 1)$ with abscissa zero, to zero with abscissa one. For the upper curve, $X < 0$, the ordinate is the reciprocal of this, $(^3L_l - \,^3L_{l+1})/(^3L_{l-1} - \,^3L_l)$, which goes from zero with abscissa one to $(l + 1)/l$ with abscissa zero. The fact that the abscissa starts increasing (to $(l + \frac{1}{2})/(l(l + 1))^{1/2}$ at $X = -1$) shows that $^1L_l - \,^3L_l$ becomes less than the whole spread of the triplet until the ordinate equals $1/2l$ at $X = -2$.

A fairly complete search of the literature was made and the points plotted on Figs. 2 and 3. The coordinates are tabulated in the accompanying tables together with a brief reference to the source of the data. No points were plotted for which any of the intervals were less than 2.5 cm^{-1}, since the accuracy of these points did not seem to be sufficient for a fair comparison with theory. Such points are listed in the table with a dagger (†). Those points whose coordinates fell off the graph are listed with an asterisk (*). No *sf* points were plotted because there were just a few clustered near Russell-Saunders coupling.

It is found that the theory works well for p^5s and d^9s configurations, in which we are one *p* or *d* electron short of a closed shell, except that the whole system of lines is inverted. These points all fall on the upper curves.

Pb III[2] and Tl II[3] have *sf* configurations in which $^3F_3$ and $^1F_3$ are on

[2] S. Smith, Phys. Rev. **34**, 397 (1929).
[3] S. Smith, Phys. Rev. **35**, 236 (1930).

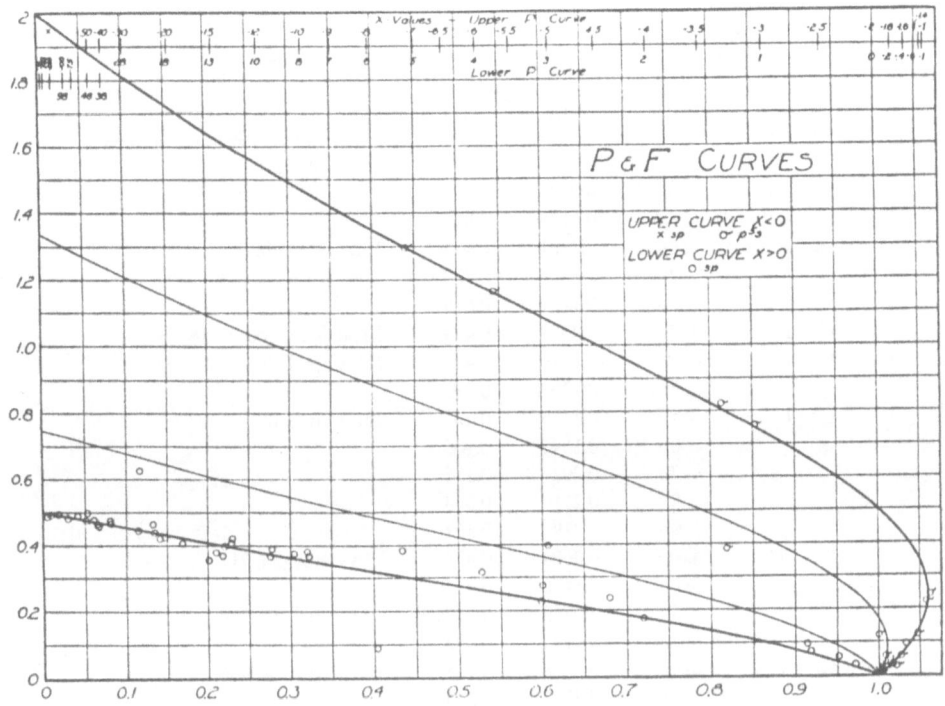

FIGURE 2. *P* and *F* curves showing *sp* and p^5s points. (*F* curve is drawn lightly and no points plotted.)

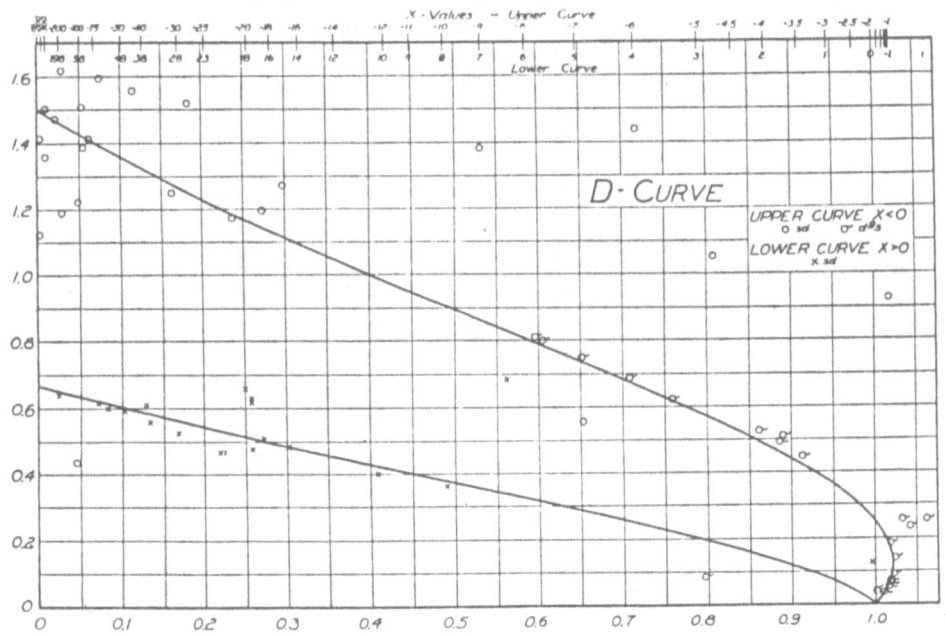

FIGURE 3. *D* curve showing *sd* and d^9s points.

TABLE I. Coordinates.

Elem.	Config.	Albs.	$sp(X > 0)$ Ord.	Sources
Al II	3s3p	0.0084	0.493	Int. Crit. Tab.
	4p	0.0293	0.481	
	5p	0.1145	0.445	
Ba I	6s6p	0.2301	0.422	Int. Crit. Tab.
	7p	0.1410	0.420	
	8p	0.1181	0.619	
C I	3s2p	0.0527	0.499	Proc. Roy. Soc. **A118**, 43
Ca I	4s4p	0.179	0.494	Int. Crit. Tab.
	5p†	0.1253	0.110	
Cd I	5s5p	0.1315	0.462	Int. Crit. Tab.
	6p	0.1695	0.406	
	7p	0.202	0.355	
	8p	0.211	0.378	
	9p	0.219	0.367	
Ga II	4s4p	0.0604	0.478	Phys. Rev. **34**, 714
	5s4p	0.1476	0.424	
Ge I	5s4p	0.0719	0.177	Phys. Rev. **31**, 786
	6s4p	0.914	0.095	
Ge III	4s4p	0.0818	0.465	Phys. Rev. **34**, 698
	5p	0.276	0.366	
Hg I	6s6p	0.436	0.382	Int. Crit. Tab.
	7p	1.035	0.094	
	8p	0.952	0.056	
Mg I	3s3p	0.0046	0.487	Int. Crit. Tab.
N II	3s2p	0.680	0.235	Proc. Roy. Soc. **A114**, 662
	4s2p	0.135	0.437	
	5s2p	0.305	0.373	
O III	3s2p	0.0658	0.458	Proc. Roy. Soc. **A118**, 43
Pb I	7s6p	0.972	0.032	Phys. Rev. **33**, 301
Pb III	6s6p	0.600	0.274	Phys. Rev. **34**, 397
Sb IV	5s5p	0.278	0.388	Phys. Rev. **34**, 402
Si I	4s3p	0.221	0.395	Zeits. f. Phys. **40**, 530
Sn I	6s5p	0.918	0.074	Phys. Rev. **30**, 574
Sn III	5s5p	0.230	0.407	Int. Crit. Tab.
	6p	0.598	0.226	
Sr I	5s5p	0.0808	0.474	Int. Crit. Tab.
	6p	0.607	0.395	
Te V	5s5p	0.323	0.366	Phys. Rev. **34**, 402
Tl II	6s6p	0.528	0.316	Phys. Rev. **35**, 236
	7p	0.403	0.088	
Zn I	4s4p	0.406	0.489	Int. Crit. Tab.
	5p	0.0507	0.475	
	6p	0.0589	0.464	
	7p	0.0666	0.456	
	8p	0.0651	0.464	
	9p†	0.0739	0.514	
Yt II	5s5p	0.321	0.380	Bur. Stand. Jl. of Res. **2**, 738

TABLE I (*continued*)

Elem.	Config.	Albs.	$sp(X < 0)$ Ord.	Sources
Al II	$3s6p*$	0.1045	2.193	Int. Crit. Tab.
Ca I	$4s6p*$	0.0140	1.950	Int. Crit. Tab.
Hg I	$6s9p*$	0.3935	5.356	Int. Crit. Tab.
	$10p*$	0.1273	7.135	
	$11p*$	0.1131	8.73	
Sr I	$5s7p*$	0.0902	2.100	Int. Crit. Tab.

Elem.	Config.	Albs.	$p^5s(X < 0)$ Ord.	Sources
A I	$3p^54s$	0.854	0.756	Zeits. f. Phys. **40**, 839
	$6s$	1.008	0.0578	
	$7s$	1.021	0.0282	
	$9s$	1.002	0.0078	
Ca III	$3p^54s$	0.814	0.823	Phys. Rev. **31**, 501
	$5s*$	1.109	0.269	
K II	$3p^54s$	0.819	0.383	Phys. Rev. **31**, 501
	$5s$	1.063	0.241	
Kr I	$4p^55s$	1.058	0.221	B. S. Jl. Res. **3**, 154
Na II	$2p^53s$	0.442	1.293	Phys. Rev. **31**, 967
Ne I	$2p^53s$	0.544	1.161	Int. Crit. Tab.
	$5s$	1.047	0.122	
	$6s$	1.027	0.0578	
	$7s$	1.017	0.0322	
	$8s$	1.010	0.0211	
	$9s$	1.006	0.0125	
	$10s$	1.003	0.0084	
	$11s$	1.001	0.0061	
Xe I	$5p^56s$	0.999	0.120	R. S. Jl. Res. **3**, 756

Elem.	Config.	Albs.	$sd(X > 0)$ Ord.	Sources
Ba I	$6s5d$	0.258	0.476	Int. Crit. Tab.
Ca I	$4s3d$	0.0237	0.640	Int. Crit. Tab.
	$5d\dagger$	0.0258	0.607	
	$7d\dagger$	0.712	0.750	
Cr I	$4s3d$	0.1678	0.525	Phys. Rev. **33**, 542
Lu II	$6s5d$	0.490	0.363	Bul. Am. Ph. Soc. Apr. 10, 1930, 11
[5]Pb III	$6s5d$	0.220	0.466	Phys. Rev. **34**, 397
Sb IV	$5s5d$	0.258	0.628	Phys. Rev. **34**, 402
	$6d$	0.562	0.683	
Sc II	$4s3d$	0.0718	0.616	Sci. Papers, Bur. Stand. **22**, 329
	$5s3d$	0.301	0.483	
Sn III	$5s5d$	0.250	0.659	Int. Crit. Tab.
	$6d$	0.129	0.610	Phys. Rev. **34**, 402
Sr I	$5s4d$	0.083	0.600	Int. Crit. Tab.
	$6d$	0.409	0.398	
Te V	$5s5d$	0.257	0.616	Phys. Rev. **34**, 402

TABLE I (continued)

Elem.	Config.	Albs.	Ord.	Sources
			$sd(X > 0)$	
Ti III	$4s3d$	0.1031	0.592	Astro. Jl. **66**, 13
V IV	$4s3d$	0.1335	0.558	Phys. Rev. **33**, 542
Yt I	$5s4d$	0.272	0.506	B. S. Jl. Res. **2**, 738
	$6s4d$	0.996	0.124	

Elem.	Config.	Albs.	Ord.	Sources
			$sd(X < 0)$	
Ba I	$6s6d$	0.0485	1.228	Int. Crit. Tab.
	$7d$	0.0465	0.438	
	$8d$	0.0284	1.190	
	$9d$	0.0746	1.595	
	$10d$†	0.0324	4.61	
Ca I	$4s4d$	0.0207	1.474	Int. Crit. Tab.
	$6d$†	0.0544	1.200	
Cd I	$5s5d$	0.1074	1.555	Int. Crit. Tab.
	$6d$	0.0621	1.414	
	$7d$	0.0553	1.388	
	$8d$†	0.0454	6.000	
Ga II	$4s4d$	0.0096	1.360	Phys. Rev. **34**, 714
	$5s4p$	0.0029	1.416	
	$6s4p$	0.0022	1.125	
Ge III	$4s4d$	0.0099	1.505	Phys. Rev. **34**, 697
Hg I	$6s6d*$	1.510	0.585	Int. Crit. Tab.
	$7d$	1.018	0.923	
	$8d$	0.808	1.051	
	$9d$	0.717	1.438	
	$10d$	0.653	0.556	
[5]Pb III	$6s5d*$	0.250	2.145	Phys. Rev. **34**, 397
Sr I	$5s5d$	0.183	1.520	Int. Crit. Tab.
	$7d$	0.273	1.195	
	$8d$	0.163	1.250	
	$9d$	0.298	1.271	
Tl II	$6s6d$	0.532	1.384	Phys. Rev. **35**, 236
	$7d$	0.237	1.173	
	$8d*$	0.118	1.861	
Zn I	$4s4d$	0.0238	1.618	Int. Crit. Tab.
	$5d$†	0.0128	1.818	
Yt II	$5s5d$	0.533	1.509	B. S. Jl. Res. **2**, 738

Elem.	Config.	Albs.	Ord.	Sources
			$d^9s(X < 0)$	
Ag II	$4d^95s$	0.862	0.526	Phys. Rev. **31**, 317
	$6s$	1.022	0.0896	
Au II	$5d^96s$	1.062	0.2569	Phys. Rev. **34**, 19
Cd III	$4d^95s$	0.886	0.4916	Phys. Rev. **31**, 778
Cu II	$3d^94s$	0.605	0.798	Phys. Rev. **29**, 386
	$5s$	1.018	0.1835	
	$6s$	1.018	0.0692	Phys. Rev. **34**, 1128
	$7s$	1.010	0.0348	

TABLE I (*continued*)

Elem.	Config.	Albs.	$d^9s(X < 0)$ Ord.	Sources
Ga IV	$3d^94s$	0.708	0.686	Phys. Rev. **31**, 750
Ge V	$3d^94s$	0.759	0.622	Phys. Rev. **31**, 750
Hg III	$5d^96s$	1.033	0.2568	Phys. Rev. **34**, 19
In IV	$4d^95s$	0.913	0.447	Phys. Rev. **31**, 778
Ni I	$3d^94s$	0.596	0.810	Phys. Rev. **29**, 386
	$5s$	1.024	0.1393	
	$6s$	1.016	0.0519	Phys. Rev. **34**, 828
Pd I	$4d^95s$	0.890	0.510	Phys. Rev. **39**, 386
	$6s$	1.018	0.065	
Pt I	$5d^96s$	0.796	0.0829	Phys. Rev. **34**, 19
	$7s$	1.001	0.0374	Phys. Rev. **34**, 190
Tl IV	$5d^96s$	1.042	0.235	Phys. Rev. **34**, 19
Zn III	$3d^94s$	0.652	0.748	Phys. Rev. **30**, 381

Elem.	Config.	Albs.	$sf(X > 0)$ (not plotted) Ord.	Sources
Al II	$3s4f$†	0.0978	0.750	Int. Crit. Tab.
	$5f$	0.0506	0.783	
	$6f$	0.0573	0.777	
	$7f$	0.0389	0.761	
	$8f$	0.0118	0.657	
	$9f$†	0.0049	0.750	
	$10f$†	0.0030	0.786	
	$11f$†	0.0016	0.667	
	$12f$†	0.0019	0.800	
[6]Ge III	$4s4f$	1.384	0.275	Phys. Rev. **34**, 697
Sn III	$5s4f$	0.0556	0.378	Int. Crit. Tab.
	$5f$	0.0312	0.357	Phys. Rev. **40**, 574
Sr I	$5s4f$†	0.0056	0.630	Int. Crit. Tab.
	$5f$†	0.0113	0.889	
	$6f$†	0.1765	0.049	
	$7f$†	0.314	0.165	

Elem.	Config.	Albs.	$sf(X < 0)$ (not plotted) Ord.	Sources
Ba I	$6s4f$	0.0046	1.014	Int. Crit. Tab.
	$5f$	0.0847	4.34	
	$6f$	0.0092	1.594	

† indicates a point which is not plotted on account of large probable error (see text).
* indicates a point which does not fall within the limit of Figs. 2 and 3.
[5] S. Smith, Phys. Rev. **34**, 397 (1929) has a question as to which of two singlets belong to this configuration. These are both listed, one for $X > 0$, plotted with a question mark and one for $X < 0$, which fell off the graph.
[6] The singlet is within the triplet.

opposite sides and outside of the $^3F_2 - {}^3F_4$ interval. A similar thing happens in the case of the $3p^5\, 8s$ configuration of A I[4]. These partially inverted triplets might have been plotted with negative ordinates on our graphs.

As to the accuracy with which the points fit the theory, it can be said that in general the d^9s and p^5s fit best, sp and sd for $X > 0$ next, and sp and sd for $X < 0$ poorest. Also in general, elements of high atomic number show especially pronounced disagreements. The disagreements are to be regarded provisionally as cases in which the second-order perturbations are not negligible rather than as essential defects in the basic theory.

---

[4] A. Meissner, Zeits. f. Physik **40**, 839 (1927).

The Theory of Complex Spectra

EDWARD U. CONDON

Palmer Physical Laboratory
Princeton University

ABSTRACT

Extending the paper of Slater (Phys. Rev. **34**, 1293, 1929) on complex spectra, it is pointed out that assignment of definite electron configurations to spectral terms is an approximate procedure and only has meaning when the multiplet systems of the several configurations are widely separated. The effect of including spin terms is sketched. Non-diagonal matrix elements for the N-electron problem are reduced to corresponding elements for the 2-electron problem, as Slater did for the diagonal elements. Two-electron jumps occur because of the fact that spectral terms may not be precisely labelled by means of electron configurations.

In a paper of this same name, Slater[1] has given a direct treatment of the application of the first order perturbation theory to the central field approximation to the atom model. He neglects spin forces in the Hamiltonian and so his results correspond to pure Russell-Saunders coupling of the angular momentum vectors, all intervals inside of the multiplets being zero. The electrostatic interaction gives the classification into multiplet levels and the calculations provide definite predictions concerning the intervals between the several multiplets which belong to the same electron configuration.

These intervals are found to be expressible in terms of certain double integrals over the radial factors of the eigen-functions of an electron moving in the central field which is made the starting-point of the perturbation calculation. For a configuration which gives n multiplets, there are thus $(n - 1)$ intervals. Slater's first-order calculation expresses these $(n - 1)$ intervals in terms of a fewer number (say, m) of integrals, usually, so even if one regards all of the integrals as independently adjustable, the theory predicts certain relations between the intervals. One may choose values for the m integrals as if they were independent so as to get the best fit possible in order to obtain a kind of test of Slater's results. Such a test of the results can be made comparatively simply without going into the more difficult question

---

[1] Slater, Phys. Rev. **34**, 1293 (1929).

of determination of the best central field and the radial eigen-functions associated with it. If a good representation of the data is obtained on treating the m integrals as independent, the question still remains whether the m values assigned are compatible with the central field eigen-functions in view of the fact that they are not really independent. However, if a good representation cannot be obtained even by treating the integrals as independent, it certainly will not be improved when allowance is made for the fact that they are not independent

With such considerations in mind an attempt was made to apply Slater's results to a larger number or cases than he has treated in the examples at the end of his paper. It quickly became apparent that the intervals between the multiplets usually disagree badly with the first-order calculations. It is therefore necessary for an adequate theory of complex spectra to extend the calculations to a higher degree of approximation. Some results in that direction are the subject of this paper.

1. Definitions and Notation. The starting point, as with Slater, is a model of the atom in which N electrons each move, without influencing each other, in the same central field which has the potential energy, $-U(r)$. Only eigen-functions that are anti-symmetric in all pairs of electrons are used and so an eigen-function is specified by giving a *complete set of* $4N$ quantum numbers. A complete set consists of N *individual sets* which are called (n, l, m, k) these being the (n, l, m_l, m_s) of Slater. Each electron has four coordinates (x, y, z, s). The first three give its position and the fourth the z-component of spin angular momentum. For short the Greek letters, $\alpha, \beta, \gamma, \delta \cdots$ are written for separate individual sets. Also the capitals $A, B, C, D \cdots$ are written as abbreviations for different complete sets.

By Pauli's exclusion principle all individual sets in a complete set must be different, and two complete sets are not considered as different if they differ merely in regard to the order of listing of the same N individual sets. Nevertheless for definiteness a definite order of writing the individual sets in a complete set is adopted and adhered to during the calculations.

Since anti-symmetric eigen-functions are used there is no one-to-one correspondence between individual electrons and individual sets of quantum numbers. This means that an expression commonly used in spectroscopy such as "the excited electron is in a $4f$ state" refers to the presence of a $4f$ individual set in the complete set of quantum numbers. The electron configuration of a given complete set means the list of n, l values of the individual sets. Thus there are generally a number of different complete sets belonging to each configuration. Since the energy of a particle in a central field depends only on n and l, all of the complete sets belonging to a certain configuration have the same energy in the zero[th] approximation from which the start is made.

For the one-electron eigen-function having the individual set $\propto$, written as a function of the first electron's coordinates, the notation $u_\alpha(1)$ is used, replacing Slater's $u(n_1/x_1)$. ψ is defined as

$$\psi = u_\alpha(1)\cdot u_\beta(2)\cdot u_\gamma(3)\cdots u_\xi(N)$$

so that the normalized eigen-function for the complete set A is

$$\Psi_A = (N!)^{-1/2}\sum_P (-1)^p P\psi$$

in which P stands for a permutation of the indices $1, 2, 3\cdots N$ in ψ relative to the α, β, $\gamma\cdots\xi$. The summation extends over all $N!$ such permutations and p has the parity of P.

For the matrix component of any quantity as H which connects the states having the complete sets A and B the Dirac notation $(A|H|B)$ is used, so that

$$(A|H|B) = \int \bar{\Psi}_A H \Psi_B$$

where $\int$ means integration over the $3N$ position coordinates and summation over the N spin coordinates.

2. Formulation of the energy level problem. The starting point is an exact solution of the quantum mechanical problem for a fictitious atom whose Hamiltonian is E where

$$E = \sum_1 \left[\frac{1}{2\mu}(p_{x1}{}^2 + p_{y1}{}^2 + p_{z1}{}^2) - U(r_1)\right] \qquad (2.1)$$

where Σ_1 means that the same functional form is to be written down successively as depending on the coordinates of all N electrons and the results added together.

A form of the Hamiltonian for real atoms that is much nearer to the truth is

$$H = \sum_1 \left[\frac{1}{2\mu}(p_{x1}{}^2 + p_{y1}{}^2 + p_{z1}{}^2) - \frac{ze^2}{r_1} + V(r_1)M_1\cdot s_1\right] + \frac{e^2}{2}\sum_{1,2}\frac{1}{r_{12}} \qquad (2.2)$$

In this the terms $V(r_1)M_1\cdot s_1$ represent the energy of interaction of each electron's spin with its own orbital angular momentum. $V(r_1)$ is to be chosen in some way along lines of the semi-empirical discussions of the "screening for the spin doublets" of other workers. The question will not be discussed further in this paper. $M\cdot s$ is the scalar product of orbital and spin angular momentum for an electron. $\Sigma_{1,2}$ means a summation over all pairs of electrons, the two indices varying independently so that each pair is counted twice. Since the operators for E and H do not commute with each other, the matrix for H will not be diagonal in terms of the representation that is based on the eigen-functions of E. The problem of finding the energy levels for H is that of finding a transformation to the diagonal form for the matrix for H.

How the function $U(r)$ is to be chosen will not be discussed here. An approximate theoretical treatment, such as that of Thomas and Fermi, or a semi-empirical method of the sort studied especially by Hartree may be used. Slater studied the Hamiltonian (2.2) with omission of the spin term. Houston,[2] Bartlett,[3] and Gaunt,[4] have considered special cases of (2.1) counting the spin term and Goudsmit[5] has recently extended their results by a clever device.

When the spin term is omitted the Hamiltonian H commutes with both the sum of the z-components of orbital angular momentum, and the sum of the z-components of spin angular momentum. Therefore H has no matrix components connecting states of E for which either of these sums is different. In terms of the quantum numbers introduced in §1 the quantities are Σm and Σk, the sums being over all the individual sets belonging to a complete set. Slater makes very good use of these results to calculate the energies of all the multiplets arising from a configuration (except for those configurations in which more than one multiplet of the same kind appears) without having to calculate any non-diagonal matrix components.

If the spin term is not omitted one still has the result that H commutes with the total sum of the z-components of both orbital and spin angular momentum although it no longer commutes with each sum separately. Therefore, even with spin counted there will be no matrix components of H connecting states for which the values of $\Sigma(m + k)$ differ.

There is another important property of the Hamiltonian which arises from its isotropy with regard to different orientations of the coordinate axes. The isotropy means that there is still a degeneracy, that of space quantization, associated with the Hamiltonian, so that the degeneracy of E is not completely removed by the inclusion of the spin and electrostatic repulsion terms which are the essence of the transformation from E a diagonal matrix to H a diagonal matrix. With each eigen-value of H can be associated a maximum value of $\Sigma(m + k)$ which is represented among the eigen-functions belonging to that eigen-value. This number is the quantum number J of that energy level. The isotropy then brings with it the result that there are other eigen-functions for which $\Sigma(m + k)$ has the values $J - 1, J - 2 \cdots, -J$ all of which have the same eigen-value. The proof of this is best obtained by appeal to group theory.

If one uses the perturbation theory to find approximately the transformation from E diagonal to H diagonal, by treating $(H - E)$ as a perturbation, the success of the calculation in the first-order requires that the eigenvalues of H which "grow out of" a particular eigen-value of E remain close together compared with the distance of the particular eigen-values of E from its nearest neighbor in the spectrum of E. This is a well known property of the perturbation theory and shows itself in many particular instances. Perhaps the Paschen-Back effect in the anomalous Zeeman effect is the best known of

[2] Houston, Phys. Rev. **33**, 297 (1929).
[3] Bartlett, Phys. Rev. **35**, 229 (1930).
[4] Gaunt, Proc. Roy. Soc. **A122**, 513 (1929).
[5] Goudsmit, Phys. Rev. **35**, 1325 (1930).

these. In it the effect of a uniform magnetic field on an atomic energy level is required. The first-order calculation is correct only if the spread of the energy levels growing out of the unperturbed level is small compared to the distance from the unperturbed level to its nearest neighbor in the unperturbed scheme. If this condition is not fulfilled then the second order perturbation becomes important. An important second-order correction implies that an important alteration of the eigen-function has taken place so that, when it is expanded in terms of the eigen-functions of E, it begins to have an important component of the eigen-function of the neighboring level of E in addition to the eigen-function of the level from which it grew.

Therefore as the second-order correction becomes more important the quantum numbers which were appropriate to labelling the different eigen-values of E become less and less appropriate for the labelling of the eigen-values of H. They cease to be "good quantum numbers" to use a curiously apt expression introduced by Mulliken[6] in a discussion of the correlation of atomic energy levels with those of diatomic molecules.

3. Validity of configuration assignments. All discussions of complex spectra have hitherto been based on the idea that an electron configuration could be assigned uniquely to each energy level. That this procedure is only of approximate validity is seen at once from the foregoing discussion. The criterion can be so formulated: *The assignment of a definite electron configuration to a group of multiplets is only exact insofar as the spread of the levels belonging to one configuration is small compared to the distance (on the energy level diagram) of the spread of levels belonging to the neighboring configurations.*

It is evident that this criterion for the case of the electron configuration quantum numbers is of the same form as the well known criterion for deciding between "weak" and "strong" magnetic fields in the Paschen-Back effect.

The importance of raising this point lies in the fact that complex spectra have already been analyzed in which the criterion for definite assignments of electron configurations is not fulfilled. A noteworthy instance[7] is that of Ti II as analyzed by Russell.[8]

In Fig. 1 are plotted the levels corresponding to the two lowest configurations in Ti II as assigned by Russell. The arrows in each column give the centers of gravity of all the terms in the column, weights being assigned according to their values of $(2J + 1)$.

It is evident that here, if anywhere, one may expect an appreciable effect of what may be suitably called interaction of neighboring electron configurations. Except for the overlapping of d^3 ^2II and ^2D and of d^3 ^2P and ^4P the distance between multiplets is large compared to intervals inside the multiplets, so that the criterion for Russell-Saunders coupling is fulfilled. The ^2II and ^2D can show no interaction, though, since they have no common value of

---

[6] Mulliken, Reviews of Mod. Phys. **2**, 60 (1930).

[7] I am indebted to my friend, Prof. J.E. Mack, for directing my attention to this case.

[8] Russell, Astrophys. J. **66**, 283 (1927).

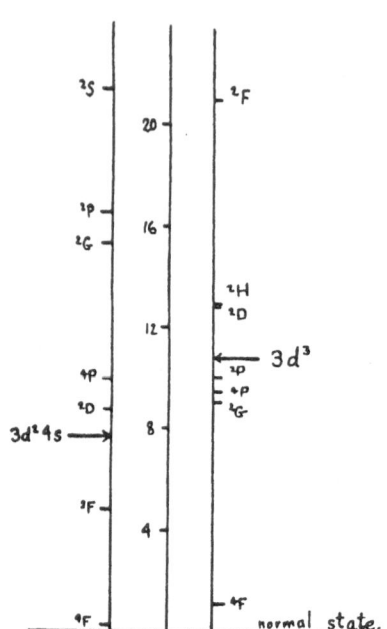

FIGURE 1. *The two lowest configurations in Ti II*. Theory predicts another ^2D for $(3d)^3$ which has not been found; the arrow shows the center of gravity of the known terms, which is therefore not the exact center of gravity of the whole configuration. The center scale is in thousands of cm^{-1}.

J, whereas it is expected that the nearness of ^2P to ^4P will disturb the intervals. This is in fact the case, the ratio of the observed intervals being 3.82 against a theoretical (Landé) value of 1.67.

Perhaps the best way to present the situation is by reference to Slater's diagrams (loc. cit. p. 1301) of the energy matrix. Referring to his Fig. 2, one may suppose that the upper double-shaded square is the matrix for the different complete sets belonging to $(3d)^2 4s$ and the next double-shaded square is the similar matrix for complete sets belonging to $(3d)^3$. Slater says that the terms in the singly-shaded rectangles (these are the matrix components connecting the complete sets in $(3d)^2 4s$ with those in $(3d)^3$) are negligible. This is often the case but the point that is made here is that they are not negligible in a case where the distance between the centers of gravity of the terms coming from the same configuration is small or comparable with the spread of the multiplets arising from the two configurations separately. In case the configurations overlap, as they do in the special instance of Ti II one needs to consider the larger square which includes the two doubly-shaded squares and the two singly-shaded rectangles that border them.

If one next makes use of the exact theorem that there are no matrix components connecting complete sets for which $\Sigma(m + k)$ differs the large square is considerably simplified. It can be rearranged so that complete sets belonging to either configuration but having the same $\Sigma(m + k)$ are in adjacent rows and columns and then the large square corresponding to the two configurations taken together will break up into a series of smaller squares. If one wishes further to neglect the spin terms, as Slater does, then

TABLE I. Sets of quantum numbers for (dds) configuration.

$\Sigma k =$	3/2 (+++)	1/2 (+−+)	(++−)
$\Sigma m = 4$		(220)	
3	(210)	(210) (120)	(210)
2	(200)	(200) (020) (110)	(200)
1	(2−10) (100)	(2−10) (−120) (010) (100)	(2−10) (100)
0	(2−20) (1−10)	(2−20) (−220) (1−10)	(2−20) (1−10)
		(−110) (000)	

TABLE II. Sets of quantum numbers for (ddd) configuration.

$\Sigma k =$	3/2 (+++)	1/2 (++−)
$\Sigma m = 6$		
5		(212)
4		(211) (202)
3	(210)	(210) (201) (2−12) (102)
2	(21−1)	(21−1) (200) (2−11) (2−22) (101) (1−12)
1	(21−2) (20−1)	(21−2) (20−1) (2−10) (2−21) (100) (1−11) (1−22) (0−12)
0	(20−2) (10−1)	(20−2) (2−1−1) (2−20) (10−1) (1−10) (1−21) (0−11) (0−22)

these smaller squares may be broken up into still smaller ones since then both "Σm and Σk separately have to be equal in the quantum numbers labelling rows and columns.

The actual procedure in applying the diagonal sum method will be illustrated in terms of d^2s and d^3 although the numerical application of the formulas to $(3d)^2 4s$ and $(3d)^3$ in Ti II will be reserved to a later paper. One has first to set up the scheme of complete sets of quantum numbers for each configuration, just as Slater does. For example, Table I gives the quantum numbers for the (dds) configuration.

In this table are listed the m values that may be associated with the sets of spin(k) values that head the columns. The corresponding table for d^3 is Table II.

In these tables it is not necessary to list the quantum numbers for negative values of Σm and Σk as these do not give additional information.

The multiplet schemes corresponding to the two configurations are given superposed in the Table III, the first line in each cell being the contribution from d^2s and the second that from d^3.

Now if the interaction between configurations can be neglected the diagonal sum method gives a value for the sum of the energies of all the multiplets lying on each line separately in each cell. But if the interaction is important then the diagonal sums must be taken simply cell by cell, this gives only half as many equations and so the power of the diagonal sum method is greatly diminished (just as it is greatly diminished when spin is not neglected).

TABLE III. Multiplets for d^2s and d^3.

$\Sigma m =$	3/2		1/2								
6											
5			^2H								
4				^2G							
			^2H	^2G							
3	^4F			^2G	^2F					^4F	
	^4F		^2H	^2G	^2F					^4F	
2	^4F			^2G	^2F	^2D				^4F	
	^4F		^2H	^2G	^2F	^2D	^2D			^4F	
1	^4F	^4P		^2G	^2F	^2D		^2P		^4F	^4P
	^4F	^4P	^2H	^2G	^2F	^2D	^2D	^2P		^4F	^4P
0	^4F	^4P		^2G	^2F	^2D		^2P	^2S	^4F	^4P
	^4F	^4P	^2H	^2G	^2F	^2D	^2D	^2P		^4F	^4P

Neglecting the interaction the diagonal sum method in this case is capable of giving the energy of each multiplet except that it can only give the arithmetic mean of the two ^2D's which arise from d^3. Allowing for the interaction it gives much less: now the ^2H and the ^2S are the only ones given directly, also given are the arithmetic means of the two ^4P's, the two ^4F's, the three ^2D's, the two ^2P's, the two ^2F's and the two ^2G's.

In order to get the actual separation of the terms of similar multiplet character whose sums only are given one needs to have the non-diagonal elements of the energy matrix.

With these at hand one can set up the secular determinants for each cell and get the roots of the secular equations. This looks formidable at first sight because the determinants are of the order equal to the number of multiplets in each cell. But the diagonal sum method can be used to depress the order of the secular equations so that one needs only to solve a quadratic where there are two multiplets of the same kind, a cubic if there are three, and so on.

Perhaps it is helpful here to point out how the diagonal sum method works when spin is not neglected. In that case one can not write an equation that the sum of the terms in each cell of the Σm, Σk diagram is equal to the sum of the corresponding diagonal elements of the energy matrix. Also one needs a complete table covering the negative values of Σk as well as the positive. For simplicity the argument will be presented in terms of the sp configuration, which has already been fully treated by Houston.[9] One has for Σm, Σk:
sp

---

[9] Houston, Phys. Rev. 33, 297 (1929).

$\Sigma k =$	$1(++)$	$0(+-)(-+)$	$-1(--)$
$\Sigma m = 1$	(01)	(01)(01)	(01)
0	(00)	(00)(00)	(00)
-1	$(0-1)$	$(0-1)(0-1)$	$(0-1)$

Arranging by values of $\Sigma(m+k)$ one sees that the values of $\Sigma(m+k)$ and the number of complete sets by which each value is realized is

$\Sigma(m+k)$	Realizations	Terms represented			
2	1	$^3P_2$			
1	3	$^3P_2$	$^3P_1$		$^1P_1$
0	4	$^3P_2$	$^3P_1$	$^3P_0$	$^1P_1$
-1	3	$^3P_2$	$^3P_1$		$^1P_1$
-2	1	$^3P_2$			

The third column gives the energy levels whose sum is given by applying the diagonal sum method to all complete sets which have the same value of $\Sigma(m+k)$. Just as the diagonal sums, without spin, fails to give the separation between two multiplets of the same kind, so here with spin it fails to give the separation between two levels having the same J value. To get the separation one would need to solve a quadratic equation, which is what Houston did. And generally, the diagonal sums alone will give simply the arithmetic mean of all the terms in the configuration that have the same J value. To get them separately one has to use the non-diagonal elements and solve an algebraic equation whose degree is equal to the number of times the particular J value occurs. This statement is the starting point of Goudsmit's recent work on the transition from Russell-Saunders to jj-coupling.

Of course it is evident that if perturbations by an external magnetic field are included the method of diagonal sums gives the derivation of the g-sum rules that play such an important role in the theory of the anomalous Zeeman effect. But similarly if one wants to find the individual g's for n terms of the same J value then an algebraic equation of the nth degree has to be solved, and this is true whether the n terms come from the same configuration or not.

4. **Non-diagonal matrix elements.** The diagonal sum method works only with the diagonal matrix elements of the energy and is quite powerful. It gives the energies of each multiplet belonging to a configuration if the interaction between configurations is neglected and if all multiplets are of a different kind. To go beyond this and find the separations between two multiplets of the same kind, or to allow for interaction of configurations and for other questions it is necessary to know the non-diagonal elements.

Two types of symmetrical function of the electron coordinates are of especial importance. One is of the form $F = \Sigma_1 f(1)$, that is, the sum of the same function of each one of the electron's coordinates occurring one at a time. The other is $G = \Sigma_{12} g(1, 2)$, that is, the sum over all possible pairs of electrons, of a symmetrical function of the coordinates of both of them. Slater has carried out the calculation for the diagonal elements, the extension to the non-diagonal elements is very easy and almost exactly like Slater's work so only the results will be stated.

For a quantity of the type F, the matrix components, $(A|F|B)$ connecting states with the complete sets A and B are as follows:

(a) They vanish if B differs from A in regard to more than one individual set.

(b) If B differs from A solely in regard to one individual set then

$$(A|F|B) = \int_1 \bar{u}_\alpha(1) f(1) u_{\alpha'}(1)$$

where α and α' are the only individual sets of A and B respectively which are not equal.

(c) The diagonal element $(A|F|A)$ is worked out by Slater. Its value is

$$(A|F|A) = \sum_\alpha \int_1 \bar{u}_\alpha(1) f(1) u_\alpha(1)$$

where the sum extends over all the individual sets in the complete set A.

An immediate corollary of the results (a) and (b) is that if $f(1)$ is a quantity which is a diagonal matrix in the one-electron problem, then F is a diagonal matrix in the N-electron problem. This is perhaps the simplest way of proving that the total z-component of spin and the total z-component of orbital angular momentum are diagonal matrices in the representation used for this approach to the theory of complex spectra.

Similarly the matrix components, $(A|G|B)$ for a quantity of the type G can be reduced as follows:

(d) They vanish if B differs from A in regard to more than two individual sets.

(e) If B differs from A only in that two of its individual sets which one may call α', β' differ from the individual sets of A, called α, β, then

$$(A|G|B) = \left[2 \int_{1,2} \bar{u}_\alpha(1) \bar{u}_\beta(2) g(1, 2) u_{\alpha'}(1) u_{\beta'}(2) \right.$$

$$\left. - 2 \int_{1,2} \bar{u}_\alpha(1) \bar{u}_\beta(2) g(1, 2) u_{\beta'}(1) u_{\alpha'}(2) \right].$$

(f) If B differs from A only in that its individual set, α', differs from the individual set, α, in A then

$$(A|G|B) = 2 \sum_{\beta} \left[\int_{1,2} \bar{u}_\alpha(1)\bar{u}_\beta(2)g(1,2)u_{\alpha'}(1)u_\beta(2) \right.$$

$$\left. - \int_{1,2} \bar{u}_\alpha(1)\bar{u}_\beta(2)g(1,2)u_\beta(1)u_{\alpha'}(2) \right]$$

where Σ_β means a summation in which β runs over the $N-1$ individual sets that are common to A and B.

(g) The diagonal element $(A|G|A)$ is given by Slater. It is

$$(A|G|A) = \sum_{\alpha,\beta} \left[\int_{1,2} \bar{u}_\alpha(1)\bar{u}_\beta(2)g(1,2)u_\alpha(1)u_\beta(2) \right.$$

$$\left. - \int_{1,2} \bar{u}_\alpha(1)\bar{u}_\beta(2)g(1,2)u_\beta(1)u_\alpha(2) \right]$$

the summation running over all pairs of individual sets.

These results show further simplifications for quantities that are independent of spin. For $u_\alpha(1)$ one has

$$u_\alpha(1) = v_\alpha(1)\delta(s_1,k_\alpha)$$

where $v_\alpha(1)$ is a function of the position coordinates only. In the preceding formulas $\int_1$ means integration over the position coordinates and summation over the spin coordinate. If $f(1)$ or $g(1,2)$ do not depend on spin the summation over s can be carried out at once. The results are:

(b) $(A|F|B) = \delta(k_\alpha,k_{\alpha'}) \displaystyle\int_1 \bar{v}_\alpha(1)f(1)v_{\alpha'}(1)$

(c) $(A|F|A) = \displaystyle\sum_{\alpha} \int_1 \bar{v}_\alpha(1)f(1)v_\alpha(1)$

(e) $(A|G|B) = 2 \left[\delta(k_\alpha,k_{\alpha'})\delta(k_\beta,k_{\beta'}) \displaystyle\int_{1,2} \bar{v}_\alpha(1)\bar{v}_\beta(2)g(1,2)v_{\alpha'}(1)v_{\beta'}(2) \right.$

$$\left. - \delta(k_\alpha,k_{\beta'})\delta(k_\beta,k_{\alpha'}) \int_{1,2} \bar{v}_\alpha(1)\bar{v}_\beta(2)g(1,2)v_{\beta'}(1)v_{\alpha'}(2) \right]$$

(f) $(A|G|B) = 2\delta(k_\alpha,k_{\alpha'}) \displaystyle\sum_{\beta} \left[\int_{1,2} \bar{v}_\alpha(1)\bar{v}_\beta(2)g(1,2)v_{\alpha'}(1)v_\beta(2) \right.$

$$\left. - \int_{1,2} \bar{v}_\alpha(1)\bar{v}_\beta(2)g(1,2)v_\beta(1)v_{\alpha'}(2) \right].$$

(g) $(A|G|A) = \displaystyle\sum_{\alpha,\beta} \left[\int_{1,2} \bar{v}_\alpha(1)\bar{v}_\beta(2)g(1,2)v_\alpha(1)v_\beta(2) \right.$

$$\left. - \delta(k_\alpha,k_\beta) \int_{1,2} \bar{v}_\alpha(1)\bar{v}_\beta(2)g(1,2)v_\beta(1)v_\alpha(2) \right].$$

A consequence of the limitation (d) is that there is no first-order inteaction between two neighboring configurations that differ in regard to more than two electrons. This result is hardly likely to be of much importance for two configurations that differ in regard to as many as three electrons will, in general, lie in widely separated parts of the energy level diagram so the interaction would be negligible anyway.

This completes the reduction of the non-diagonal elements to integrations over the coordinates of only one or two electrons. If in particular $g(1, 2)$ is of the form $1/r$ additional developments may be made in which $1/r_{12}$ is expanded in a series of spherical harmonics and so all the integrals can be expressed, as Slater does for the ones occurring in the diagonal elements, as a sum of certain integrals over the radial eigen-functions multiplied by coefficients that are certain integrals of spherical harmonics. The calculations involve some more general coefficients than Slater's a's and b's. Detailed developments of this part of the reduction will be reserved for a later paper.

5. Two-electron jumps. An immediate application of the results of the preceding section is to the question of "two-electron jumps." These are transitions between energy levels whose configurations differ in regard to two of the sets of nl values. Such transitions are usually weak compared to the more usual one-electron jumps. One may also have "zero-electron jumps," that is, transitions between terms arising from the same configuration. This is the usual way of stating the case, in which one speaks as though each term is uniquely associated with a single configuration of the central field approximation.

The interaction between a light wave whose vector potential is given by $A(x, y, z)$ and an atom consisting of N electrons is measured by a term in the Hamiltonian of the form

$$\sum_1 A(x_1, y_1, z_1) \cdot p_1$$

where p_1 is the momentum of the first electron. This form embraces quadrupole and all multipole radiations. The dipole radiation which is the first approximation is obtained by replacing $A(x, y, z)$ by its value at the center of the atom and writing

$$A(0, 0, 0) \cdot \sum_1 p_1$$

which is valid if the wave-length is great compared to the size of the atom so that A does not vary much over the size of the atom. Now even the exact interaction is of the form of the sum of the same operator function of the coordinates of each electron summed over each of the electrons. Therefore one sees that its matrix components connecting complete sets which differ in regard to more than one individual set vanish.

In other words, if assignment of electron configurations to energy levels were an exact procedure there would be no two-electron jumps. Thus the existence of two electron jumps is an indication of a break-down of exact configuration assignments.

This point is similar to the one that the existence of inter-system combinations, i.e. transitions between states of different multiplicities, is an indication of the break down of exact assignment of L and S values of the terms, i.e. break-down in the Russell-Saunders coupling scheme.

There is an interesting question of language involved here. If the expression "electron jump" is to be translated into quantum mechanics as meaning a change in an (nl) individual set in going from an initial state to a final state, then, strictly, only one-electron jumps occur. But the exact eigen-function of each energy level has in its make-up components belonging to several different configurations of the central field model, whereas it has been the custom to assign to the energy level one configuration which is presumed to be the one that has the largest component in the expansion of the exact eigen-function. One-electron jumps are the only ones really occurring but since the exact eigen-function has other configurations in it than the principal one from which it derives its configuration name, there will appear to be two-electron jumps when the transitions are described solely by reference to the single configuration name that is assigned to each level by the customary procedure. It is the same with inter-system combinations: the selection rule $\Delta S = 0$ is exact, but the actual energy levels have components in their eigen-functions corresponding to more than one value of S; then when one persists in labelling the terms simply with the value of S most strongly represented in the eigen-function he is confronted with apparent violations of this selection rule.

The break-down of exact configuration assignments also has important implications for the theory of relative intensity of spectral lines. Ornstein and Burger[10] have already shown that when inter-system combinations have appreciable intensity that they must be appropriately reckoned in the application of the intensity sum rules. This point is brought out clearly also in recent work by Harrison.[11] That is, the sum rules have to be applied to all the lines arising from transitions between all the terms of the initial and final configurations. Evidently all this has to be extended one step further for cases in which the interaction between two configurations is important. For simplicity suppose the initial state levels can be all given a fairly excact configuration assignment but that for the final state there are two configurations in interaction. Then the sum rules for intensities will have to be extended to summations over all lines terminating on-any of the levels belonging to both of the interacting configurations. Detailed consideration of the intensity relations will be postponed to a later paper.

This paper was written at Stanford University during the summer quarter. The writer takes pleasure in this opportunity to thank the Stanford physicists for their cordiality and especially Professor G.R. Harrison for stimulating discussions on spectroscopy. He is also indebted to Professor J.E. Mack for discussions last winter at the University of Minnesota.

[10] Ornstein and Burger, Zeits. f. Physik **40**, 403 (1926).
[11] Harrison, J. Opt. Soc. Amer. **19**, 109 (1929).

The Theory of Complex Spectra II[*]

E.U. Condon and G.H. Shortley

Palmer Physical Laboratory
Princeton University

ABSTRACT

Formulas for the relations between the energies of multiplets arising from the same electron configuration for all two-electron cofigurations up to ff and several cases of three-electron configurations are worked out following Slater's method; Slater's table of a's and b's being extended to cover f electrons. A systematic comparison of the known data with this first order perturbation theory shows poor agreement in many cases and good agreement in many. The theory predicts the observed alternation in the relative positions of singlet and triplet through S, P, D, F, etc. in the pp, pd, and pf triads, and the dd and df pentads. In general the p electron configurations fit very poorly; a uniform trend with atomic number is observed for p^3 and good fits are obtained for $4p3d$ in Ti III, V IV, and Cr V. For d electrons the theory fits very well in the first long period of the periodic table, and fairly well in the second. The ^1S of d^2 and the ^2D of d^2s are predicted much higher than the levels assigned to those multiplets when such an assignment is made. d^3 fits well except for ^2P. An energy level table of La II is given as recently analysed by Meggers and Russell. Here we have complete $5d4f$ and $4f^2$ configurations which fit the theory very well, these calculations having assisted in the assignment of some of the singlets and resulted in a rearrangement of singlet lines.

§1. Introduction

This paper is a sequel to one[1] published last fall in which the first steps were taken toward working out the second approximation for atomic spectra with Russell-Saunders coupling. Before going on with that work it was thought desirable to make a careful study of the application of the first approximation formulas, given by Slater's method,[2] to all of the known data. That is the subject of this paper.

---

[*] This paper was presented at the New York Meeting of the American Physical Society, February 27,1931.

[1] Condon, Phys. Rev. **36**, 1121 (1930).
[2] Slater, Phys. Rev. **34**, 1293 (1929).

It will be recalled that the first-order calculation gives formulas for the energy of each of the multiplets arising from a given electron configuration in terms of certain integrals taken over the radial factor of the wave function for an electron in the central force field that lies at the basis of the calculations. These integrals represent the perturbation energy due to the electrostatic repulsion of the electrons. It is inconvenient to work out these integrals for they involve the unknown wave functions of the screened average force field in which the electrons move. Instead these integrals are treated as adjustable (except for restrictions such as that certain of them are essentially positive, etc.) in order to see how well the data can be represented. If a good fit is obtained that is, therefore, only a partial confirmation of the theory, for the question still remains open whether the relative magnitudes assumed for the several integrals are really compatible with their definition as integrals.

Since Meggers and Russell[3] have recently obtained for the first time, in La II, complete sets of multiplets involving the two-electron configurations pf, df, f^2 we have thought it worthwhile to extend Slater's tables of a's and b's to provide the necessary coefficients for applying the method to configurations involving f electrons. These results are presented in §2. In §3 and §4 the explicit formulas are given for the first-order energies in a number of important configurations and in §5, §6, and §7 comparison of the formulas with the data is made.

§2. Slater's Coefficients for f Electron Configurations

In Slater's paper[2] there are a few details connected with normalization that need to be straightened out. Slater, in his manuscript, had normalized his wave functions in an unusual way: namely, so that the normalizing integral over the spherical harmonic factor was set equal to 4π instead of 1. This requires the normalizing integral over the radial coordinate to be set equal to $1/4\pi$. When his paper went through the Physical Review office one of us (E.U.C.) thought that a mistake had been made in the normalizing factor and inserted a $(2\pi)^{-1/2}$ to normalize the $\Phi(m_l/\phi)$ on page 1308 in the usual way. As Slater was in Europe at the time he did not have an opportunity to set the matter straight again. Therefore this factor should be removed from the $\Phi(m_l/\phi)$ on page 1308 and then it should be borne in mind, what Slater does not mention, that the radial wave-function is to be normalized to $1/4\pi$ instead of 1. The usual normalization of each factor to 1, is the one we prefer. To have this one needs to leave the Φ factor as printed on page 1308, to insert $2^{-1/2}$ on the right side of the equation defining $\Theta(lm_l/\theta)$ on that page, to remove the factor 4π in the equation for $I(nl)$ on page 1310 and the factor $(4\pi)^2$ in the equations for $F^k(nl; n'l')$ and $G^k(nl; n'l')$ on page 1311.

We have also found it convenient to treat the a's and b's of page 1311 as

---

[3] See §7 of this paper.

integers by associating the denominator of the a's and b's as they occur in the tables of page 1312 with the corresponding F. Therefore we write

$$F_k(nl; n'l') = \frac{1}{D_k} F^k(nl; n'l')$$

where F^k is Slater's F and D_k is the denominator of the fractional value for $a^k(l, m_l; l'm_l')$ as given on page 1312. The corresponding definition of G_k is also made.

Having detected an error in Slater's table of b's by reaching an inconsistency in deriving the energy levels for the pd configuration it was thought worthwhile

TABLE I. Extension of table of $a^k(lm_l; l'm_l')$

Electrons	l	m_l	l'	m_l'	$k = 0$	2	4	6
sf	0	0	0	±3	1			
				±2	1			
				±1	1			
				0	1			
pf	1	±1	3	±3	1	5/75		
	1	±1	3	±2	1	0		
	1	±1	3	±1	1	− 3		
	1	±1	3	0	1	− 4		
	1	0	3	±3	1	−10		
	1	0	3	±2	1	0		
	1	0	3	±1	1	6		
	1	0	3	0	1	8		
df	2	±2	3	±3	1	10/105	3/693	
	2	±2	3	±2	1	0	− 7	
	2	±2	3	±1	1	− 6	1	
	2	±2	3	0	1	− 8	6	
	2	±1	3	±3	1	− 5	−12	
	2	±1	3	±2	1	0	28	
	2	±1	3	±1	1	3	− 4	
	2	±1	3	0	1	4	−24	
	2	0	3	±3	1	−10	18	
	2	0	3	±2	1	0	−42	
	2	0	3	±1	1	6	6	
	2	0	3	0	1	8	36	
ff	3	±3	3	±3	1	25/225	9/1089	1/7361.64
	3	±3	3	±2	1	0	−21	− 6
	3	±3	3	±1	1	−15	3	15
	3	±3	3	0	1	−20	18	− 20
	3	±2	3	±2	1	0	49	36
	3	±2	3	±1	1	0	− 7	− 90
	3	±2	3	0	1	0	−42	120
	3	±1	3	±1	1	9	1	225
	3	±1	3	0	1	12	6	−300
	3	0	3	0	1	16	36	400

Note: In cases with two ± signs, the two can be combined in any of the four possible ways.

TABLE II. Extension of table of $b^k(lm; l'm_l')$

Electrons	l	m_l	l'	m_l'	$k=0$	1	2	3	4	5	6
sf	0	0	3	±3				1/7			
	0	0	3	±2				1			
	0	0	3	±1				1			
	0	0	3	0				1			
pf	1	±1	3	±3			45/175		1/189		
	1	±1	3	±2			30		3		
	1	±1	3	±1			18		6		
	1	±1	3	0			9		10		
	1	0	3	±3			0		7		
	1	0	3	±2			15		12		
	1	0	3	±1			24		15		
	1	0	3	0			27		16		
	1	±1	3	∓3			0		28		
	1	±1	3	∓2			0		21		
	1	±1	3	∓1			3		15		
df	2	±2	3	±3		15/35		10/315		1/1524.6	
	2	±2	3	±2		5		20		5	
	2	±2	3	±1		1		24		15	
	2	±2	3	0		0		20		35	
	2	±1	3	±3		0		25		7	
	2	±1	3	±2		10		15		24	
	2	±1	3	±1		8		2		50	
	2	±1	3	0		3		2		80	
	2	0	3	±3		0		25		28	
	2	0	3	±2		0		0		63	
	2	0	3	±1		6		9		90	
	2	0	3	0		9		16		100	
	2	±2	3	∓3		0		0		210	
	2	±2	3	∓2		0		0		126	
	2	±2	3	∓1		0		10		70	
	2	±1	3	∓3		0		0		84	
	2	±1	3	∓2		0		25		112	
	2	±1	3	∓1		0		15		105	
ff	3	±3	3	±3	1		25/225		9/1089		1/7361.64
	3	±3	3	±2	0		25		30		7
	3	±3	3	±1	0		10		54		28
	3	±3	3	0	0		0		63		84
	3	±2	3	±2	1		0		49		36
	3	±2	3	±1	0		15		32		105
	3	±2	3	0	0		20		3		224
	3	±1	3	±1	1		9		1		225
	3	±1	3	0	0		2		15		350
	3	0	3	0	1		16		36		400
	3	±3	3	∓3	0		0		0		924
	3	±3	3	∓2	0		0		0		462
	3	±3	3	∓1	0		0		42		210
	3	±2	3	∓2	0		0		70		504
	3	±2	3	∓1	0		0		14		378
	3	±1	3	∓1	0		24		40		420

Note: In cases where there are two ± signs, the two upper or the two lower signs must be taken together.

to check these tables by a complete recalculation using Gaunt's formulas[4] for the integrals. These formulas were also used to extend the tables to the pairs, *sf*, *pf*, *df*, *ff*. This straightforward but laborious computation makes us now feel confident that there are no errors in Slater's table of page 1312 except the one originally detected. The value

$$b^3(1, \pm 1; 2, \mp 2) = 45/245$$

is correct, instead of 9/245 as printed.

The extension to *f* electron values is covered in Tables I and II.

§3. The Energy Levels in Two-Electron Configurations

Slater has treated in detail the method whereby the energy levels of the several multiplets are to be found in terms of the *F* and *G* perturbation integrals and has given some examples. He has shown that the electrons in closed shells are without direct effect on the perturbation theory, although of course, they have the indirect effect of determining the nature of the best central field on which to base the approximation. Therefore, we do not need to give the details of the calculations but merely summarize the results.

Slater's F^k integrals are necessarily positive and decreasing with increasing *k*, from their definition. Therefore, since the denominator *D* in the definition of our F_k increases rapidly, F_k necessarily decreases very rapidly with increasing *k*. Since the *G*'s are not essentially positive by definition, no definite statement can be made concerning their relative magnitudes; however, in every instance we have found, the G_k's have been positive and rapidly decreasing with *k*. Although not consistent with its definition as an integral, it is convenient to measure F_0, which occurs in the formula for each multiplet, from an arbitrary low level of the spectrum. (Slater's theory provides an integral *I*, dependent only on the configuration, to locate the height of the whole multiplet.)

If one electron is in an *s* state the result is simply a singlet and triplet whose *L* is the *l* of the other electron outside closed shells, as Slater shows on page 1315, duplicating by this method a result of Heisenberg.

For *pp*, non-equivalent *p* electrons, Slater gives the triplet intervals. The complete formulas are, if ^3P is written for "relative energy of the center of gravity of the ^3P terms,"

$$^1S = F_0 + 10F_2 + G_0 + 10G_2$$
$$^3S = F_0 + 10F_2 - G_0 - 10G_2$$
$$^1P = F_0 - 5F_2 - G_0 + 5G_2$$
$$^3P = F_0 - 5F_2 + G_0 - 5G_2 \qquad (pp)$$
$$^1D = F_0 + F_2 + G_0 + G_2$$
$$^3D = F_0 + F_2 - G_0 - G_2$$

---

[4] Gaunt, Phil. Trans. Roy. Soc. **A228**, 151 (1929).

We note that the arithmetic mean of the corresponding singlet and triplet energies is independent of the G's while corresponding singlet-triplet intervals are independent of F's. Also since $G_2 \ll G_0$ usually we have $^1S > {}^3S$ and $^1P < {}^3P$ and $^1D > {}^3D$; an alternation which is quite a general prediction of the theory.

For p^2, equivalent p electrons, the ^3S, ^1P and ^3D are ruled out by the exclusion principle and the formulas for what is left are the same as those for the arithmetic means of singlet and triplet in pp, namely

$$
\begin{aligned}
^1S &= F_0 + 10F_2 \\
^3P &= F_0 - 5F_2 \\
^1D &= F_0 + F_2.
\end{aligned}
\qquad (p^2)
$$

The formulas for dd and d^2, also ff and f^2, show similar relationships. For dd we have

$$
\begin{aligned}
^1S, {}^3S &= F_0 + 14F_2 + 126F_4 \pm (G_0 + 14G_2 + 126G_4) \\
^1P, {}^3P &= F_0 + 7F_2 - 84F_4 \mp (G_0 + 7G_2 - 84G_4) \\
^1D, {}^3D &= F_0 - 3F_2 + 36F_4 \pm (G_0 - 3G_2 + 36G_4) \\
^1F, {}^3F &= F_0 - 8F_2 - 9F_4 \mp (G_0 - 8G_2 - 9G_4) \\
^1G, {}^3G &= F_0 + 4F_2 + F_4 \pm (G_0 + 4G_2 + G_4)
\end{aligned}
\qquad (dd)
$$

where the upper sign is for the singlet and the lower for the triplet. For d^2 the multiplets are ^1S, ^3P, ^1D, ^3F, ^1G and their energies are given by the same formulas upon omitting the terms involving G integrals.

For pd the formulas are

$$
\begin{aligned}
^1P, {}^3P &= F_0 + 7F_2 \pm (G_1 + 63G_3) \\
^1D, {}^3D &= F_0 - 7F_2 \mp (3G_1 - 21G_3) \\
^1F, {}^3F &= F_0 + 2F_2 \pm (6G_1 + 3G_3).
\end{aligned}
\qquad (pd)
$$

This shows the same alternation in sign of the leading G integral but differs from the preceding ones in that the F and G parts are not similar.

The formulas for pf are

$$
\begin{aligned}
^1D, {}^3D &= F_0 + 12F_2 \pm (3G_2 + 36G_4) \\
^1F, {}^3F &= F_0 - 15F_2 \mp (15G_1 - 9G_4) \\
^1G, {}^3G &= F_0 + 5F_2 \pm (45G_1 + G_4).
\end{aligned}
\qquad (pf)
$$

The formulas for df are

$$
\begin{aligned}
^1P, {}^3P &= F_0 + 24F_2 + 66F_4 \pm (G_1 + 24G_3 + 330G_5) \\
^1D, {}^3D &= F_0 + 6F_2 - 99F_4 \mp (3G_1 + 42G_3 - 165G_5) \\
^1F, {}^3F &= F_0 - 11G_2 + 66F_4 \pm (6G_1 + 19G_3 + 55G_5)
\end{aligned}
\qquad (df)
$$

$$^1G, {}^3G = F_0 - 15F_2 - 22F_4 \mp (10G_1 - 35G_3 - 11G_5)$$

$$^1H, {}^3H = F_0 + 10F_2 + 3F_4 \pm (15G_1 + 10G_3 + G_5). \tag{df}$$

Finally the formulas for two non-equivalent f electrons are

$$^1S, {}^3S = F_0 + 60F_2 + 198F_4 + 1716F_6 \pm (G_0 + 60G_2 + 198G_4 + 1716G_6)$$

$$^1P, {}^3P = F_0 + 45F_2 + 33F_4 - 1287F_6 \mp (G_0 + 45G_2 + 33G_4 - 1287G_6)$$

$$^1D, {}^3D = F_0 + 19F_2 - 99F_4 + 715F_6 \pm (G_0 + 19G_2 - 99G_4 + 715G_6)$$

$$^1F, {}^3F = F_0 - 10F_2 - 33F_4 - 286F_6 \mp (G_0 - 10G_2 - 33G_4 - 286G_6)$$

$$^1G, {}^3G = F_0 - 30F_2 + 97F_4 + 78F_6 \pm (G_0 - 30G_2 + 97G_4 + 78G_6)$$

$$^1H, {}^3H = F_0 - 25F_2 - 51F_4 - 13F_6 \mp (G_0 - 25G_2 - 51G_4 - 13G_6)$$

$$^1I, {}^3I = F_0 + 25F_2 + 9F_4 + F_6 \pm (G_0 + 25G_2 + 9G_4 + G_6), \tag{ff}$$

from which the values for f^2 can be obtained by ignoring the part involving G integrals and remembering that the allowed terms are ^1S, ^3P, ^1D, ^3F, ^1G, ^3H and ^1I.

The most striking thing about these results perhaps is the uniform way in which an alternation of the relative height of singlet and triplet is predicted, since in most cases the G of lowest index will be enough larger than the others to dominate the whole expression in the G's. Russell and Meggers[5] called attention to this alternation in 1927 and "commended it to the attention of theoretical investigators." Its explanation by the quantum mechanics must be counted as an important success for the theory.

§4. Three-Electron Configurations

It would be a waste of time to work out all possible three-electron configurations at present, therefore we confine ourselves to cases for which we have been able to find experimental data with which to check the results. The addition of an s electron to p^2, d^2 or pd gives three important cases.

According to the vector coupling viewpoint the addition of an s electron to p^2 gives the results $^1S \to {}^2S$, $^3P \to {}^2P$ and ^4P, and $^1D \to {}^2D$. In the formulas for p^2s the F integrals are of the type $F(np^2)$ while there now appears a G integral to represent the action of the s electron, which is $G_1(np, n's)$:

$$^2S = F_0 + 10F_2 - G_1$$

$$^2P = F_0 - 5F_2 + G_1$$

$$^4P = F_0 - 5F_2 - 2G_1 \tag{p^2s}$$

$$^2D = F_0 + F_2 - G_1$$

[5] Russell and Meggers, Sci. Papers Bur. Stand. **22**, 364 (1927).

where $F_0 = 2F_0(np, n's) + F_0(np^2)$. Thus $3G_1$ is the $^2P - {}^4P$ interval and the quantities ^2S, $(^2P + 2{}^4P)/3$ (which is the center of gravity of this combination), and ^2D have the same intervals as ^1S, ^3P and ^1D in p^2.

Similar results hold for d^2s. Here too the singlets of d^2 become doublets and the triplets split into doublets and quartets:

$$^2S = F_0 + 14F_2 + 126F_4 - \quad G_2$$

$$^2P = F_0 + \quad 7F_2 - \quad 84F_4 + \quad G_2$$

$$^4P = F_0 + \quad 7F_2 - \quad 84F_4 - 2G_2$$

$$^2D = F_0 - \quad 3F_2 + \quad 36F_4 - \quad G_2 \qquad (d^2s)$$

$$^2F = F_0 - \quad 8F_2 - \quad 9F_4 + \quad G_2$$

$$^4F = F_0 - \quad 8F_2 - \quad 9F_4 - 2G_2$$

$$^2G = F_0 + \quad 4F_2 + \quad\quad F_4 - \quad G_2$$

in which

$$F_0 = F_0(nd^2) + 2F_0(nd, n's)$$

$$G_2 = G_2(nd, n's).$$

The relation to d^2 is evident on comparison. Further we see that the doublet-quartet interval is the same for the P and the F multiplets, being equal to $3G_2$.

In the case of pds we encounter the first instance in which it is impossible to get complete formulas by Slater's diagonal sum method since this configuration gives two different ^2P, ^2D and ^2F. This comes about because pd gives $^{1,3}P, D, F$ and the added s electron makes the singlets into $^2P, D, F$, and splits each triplet into $^{2,4}P, D, F$. In such a case the method gives simply the arithmetic mean of the two doublets of similar L value:

$$(^2P) = F_0 + 7F_2$$

$$^4P = F_0 + 7F_2 - (G_1{}^{pd} + 63G_3{}^{pd}) - G_1{}^{sp} - G_2{}^{sd}$$

$$(^2D) = F_0 - 7F_2$$

$$^4D = F_0 - 7F_2 + (3G_1{}^{pd} - 21G_3{}^{pd}) - G_1{}^{sp} - G_2{}^{sd} \qquad (spd)$$

$$(^2F) = F_0 + 2F_2$$

$$^4F = F_0 + 2F_2 - (6G_1{}^{pd} + 3G_3{}^{pd}) - G_1{}^{sp} - G_2{}^{sd}$$

in which (^2P) indicates the mean of the two ^2P's and

$$F_0 = F_0(ns, n'p) + F_0(ns, n''d) + F_0(n'p, n''d)$$

$$F_2 = F_2(n'p, n''d)$$

$$G_1{}^{sp} = G_1(ns, n'p)$$

$$G_2{}^{sd} = G_2(ns, n''d)$$

$$G_1{}^{pd} = G_1(n'p, n''d)$$

$$G_3{}^{pd} = G_3(n'p, n''d).$$

Slater has worked out p^3, the result being that the energies increase in the order ^4S, ^2D, ^2P and the interval $(^2P - {}^2D)$ is to $(^2D - {}^4S)$ as $2:3$.

For the configuration d^3 we find the energy levels to be,

$$^2P = 3F_0 - 6F_2 - 12F_4$$

$$^4P = 3F_0 \qquad\quad - 147F_4$$

$$(^2D) = 3F_0 + 5F_2 + 3F_4$$

$$^2F = 3F_0 + 9F_2 - 87F_4 \qquad\qquad (d^3)$$

$$^4F = 3F_0 - 15F_2 - 72F_4$$

$$^2G = 3F_0 - 11F_2 + 13F_4$$

$$^2II = 3F_0 - 6F_2 - 12F_4$$

where (^2D) indicates the mean of the two ^2D's.

§5. Comparison with Experimental Data, Configurations with p Electrons

The simplest case for comparison with the data is p^2, where the theory predicts that the multiplets come in the order ^3P, ^1D, ^1S, as Slater noted. From §3 we see that the ratio $(^1S - {}^1D)/(^1D - {}^3P) = 3/2$ according to theory. Slater gives the normal configuration of Si I as an example, and we find several more, as given in Table III.

From this it would appear that some influence depresses the ^1S in general, and that the doubtful ^1S given by Fowler and Selwyn for N II is probably wrong. Pb I and Bi II are so far from Russell-Saunders coupling that a good fit would hardly be expected.

The configurations p^3 and p^4 may be discussed here because they are similar to p^2. For p^3 the theory says that the ratio $(^2P - {}^2D)/(^2D - {}^4S)$ equals $2/3$. Table IV gives the known instances.

The continued increase in this ratio as the total quantum number increases is to be noted particularly. The closeness of the ratio for N I and O II would indicate that the ratio for the unreliable F III should perhaps be closer to 0.51.

For p^4 the theory gives the same result as for p^2. Table V gives the examples.

The new data of Frerichs show that O I is not in as good agreement with the theory as indicated in one of Slater's examples, his ratio being 1.55. Slater took his data from a remarkable energy-level diagram by McLennan, McLeod and Ruedy, Phil. Mag. **6**, 565 (1928), in which the wave number difference for $^1S - {}^1D$ is given as 39,500, although the main point of the paper is the identification of this transition with the auroral green line!

TABLE III. p^2 configurations

Element	Config.	$(^1S - {}^1D)/(^1D - {}^3P)$	Reference
Theory		1.500	
C I	$2p^2$	1.13	1
N II	"	5. (? on ^1S)	2
O III	"	1.14	2
Si I	$3p^2$	1.48	3
Ca I	$4p^2$	−0.01	4
Ge I	"	1.50	5
Sn I	$5p^2$	1.39	5
Pb I	$6p^2$	0.62	6
Bi II	"	0.51	7

[1] Paschen and Kruger, Ann. d. Physik **7**, 1 (1930).
[2] Fowler and Selwyn, Proc. Roy. Soc. **A118**, 42 (1928).
[3] Slater, Phys. Rev. **34**, 1317 (1929).
[4] Russell, Astrophys. J. **66**, 190 (1927).
[5] Rao, Proc. Roy. Soc. **A124**, 475 (1929).
[6] Gieseler and Grotrian, Zeits. f. Physik **39**, 377 (1926); Sur, Phil. Mag. **3**, 736 (1927).
[7] McLennan, McLay, and Crawford, Proc. Roy. Soc. **A129**, 584 (1930).

TABLE IV. p^3 configurations

Element	Config.	$(^2P - {}^2D)/(^2D - {}^4S)$	Reference
Theory		0.67	
N I	$2p^3$	0.50	1
O II	"	0.51	2
F III*	"	0.46	3
S II	$3p^3$	0.65	4
As I	$4p^3$	0.72	5
Sb I	$5p^3$	0.91	6
Bi I	$6p^3$	1.12	6

[1] Compton and Boyce, Phys. Rev. **33**, 147 (1929); Ekefors, Zeits. f. Physik **63**, 442 (1930).
[2] Russell, Phys. Rev. **31**, 27 (1928).
[3] Dingle, Proc. Roy. Soc. **A122**, 144 (1929).
[4] Ingram, Phys. Rev. **32**, 172 (1928); L. and E. Bloch, C. R. **188**, 160 (1929).
[5] Rao, Proc. Roy. Soc. **A125**, 240 (1929).
[6] Charola, Phys. Zeits. **31**, 458 (1930).
* No intercombinations are found between the quartet and doublet systems, and the relative term values are probably quite inaccurate.

E.U. Condon and G.H. Shortley

TABLE V. p^4 configurations

Element	Config.	$({}^1S - {}^1D)/({}^1D - {}^3P)$	Reference
Theory		1.50	
O I	$2p^4$	1.14	1
Se I	$4p^4$	1.71	2
Te I	$5p^4$	1.71	2

[1] Frerichs, Phys. Rev. **36**, 398 (1930); Hopfield, Phys. Rev. **37**, 160 (1931).
[2] McLennan and Crawford, Nature **124**, 874 (1929).

We have found two complete p^2s configurations, which should be similar to p^2 as indicated in §4. The ratio $({}^2S - {}^2D)/({}^2D - P)$, where $P = (2{}^4P + {}^2P)/3$, which should be 1.50, as 0.58 in As I[6] and 3.58 in Sb III.[7]

pp, pd, and pf give similar triads, in which the means of singlet and triplet should lie in the corresponding orders: $P\ D\ S$, $D\ F\ P$, $F\ G\ D$, lowest energy first, with the ratios $(S - D)/(D - P) = 9/6$; $(P - F)/(F - D) = 5/9$; $(D - G)/(G - F) = 7/20$, respectively.

We have found the pp complete only for C I[8], N II[9], and O III[10], but since these all occur with the means in the wrong order we do not give the details.

This failure of C I, N II, and O III to agree with the theory appears also in the known pd configurations, the mean of the F's being low in every case.[8,9,10] Yt II,[11] La II,[12] and Ge I[13] $4p5d$, also have the pd means in the wrong order. The means of Ge I $4p4d$, and Zr III $5p4d$[14] come in the right order with the ratio $(P - F)/(F - D)$, which is theoretically 0.555, having the value 0.28 and 3.58, respectively. More interesting is the behavior of $4p3d$ in the isoelectronic sequence Ca I[15], Sc II[16], Ti III[17], V IV[18] and Cr V[18]. The first two come in the wrong order, giving $(P - F)/(F - D) = -0.15$ and -0.06, respectively, but the last three agree satisfactorily, the ratios being $+0.45$, 0.49, and 0.548. Since $P - F = 10F_2(4p3d)$ and $F - D = 18F_2(4p3d)$, we can get two values of F_2 for each of these last three ions. Taking the average of these two, we have $F_2(4p3d)$ for Ti III, 427; V IV, 569; Cr V, 707 cm^{-1}. These are perfectly linear,

[6] Rao, Proc. Roy. Soc. **A125**, 240 (1929).
[7] Lang, Phys. Rev. **35**, 445 (1930).
[8] Paschen and Kruger, Ann. d. Physik **7**, 1 (1930).
[9] Ingram, Phys. Rev. **34**, 427 (1929).
[10] Fowler and Selwyn, Proc. Roy. Soc. **A118**, 42 (1928).
[11] Meggers and Russell, B. S. Jl. of Res. **2**, 733 (1929).
[12] Meggers and Russell, see §7.
[13] Rao, Proc. Roy. Soc. **A124**, 467 (1929).
[14] Kiess and Lang, B. S. Jl. of Res. **5**, 311 (1930).
[15] Russell, Astrophys. J. **66**, 190 (1927).
[16] Russell and Meggers, Sci. Papers Bur. Stand. **22**, 329 (1927).
[17] Russell and Lang, Astrophys. J. **66**, 19 (1927).
[18] White, Phys. Rev. **33**, 542 (1929).

TABLE VI. $3p\,4d$ singlet-triplet separations

	Ti III		V IV		Cr V	
	obs.	calc.	obs.	calc.	obs.	calc.
$^1P - {}^3P$	2810	2785	4761	4873	6894	7208
$^1D - {}^3D$	-2056	-1970	-1588	-1935	-826	-1704
$^1F - {}^3F$	5270	5310	6784	6590	8161	7648
G_1		435		534		614
G_3		15.2		30.2		47.4
G^1		6520		8000		9200
G^3		3720		7400		11600

as shown by the plot in Fig. 2, together with corresponding F's for d^2. We have a further check on the three singlet-triplet separations in terms of the two parameters G_1 and G_3, which it is fruitless to apply to the cases in which the means fitted poorly, but interesting in the other three cases. Ti III, V IV and Cr V all alternate properly, as pointed out in §3. Table VI shows the results of fitting the data to the formulas for the separations by least squares.

It is seen that, in the opposite order from the means, Ti III fits best and V IV next. The G's are again approximately linear functions of the ionic charge. G^3 becomes greater than G^1, which is allowed.

For pf in La II[12] the ratio $(D - G)/(G - F)$ is 1.05 instead of the theoretical 0.35.

There is one more instance of a configuration with p electrons in which one gets a determined ratio, namely pds. In this case (see §4) nothing can be predicted concerning the quartets, ^4P, ^4D, ^4F; but the theory predicts a constant ratio between the means of the two ^2P's, the two ^2D's and the two ^2F's which occur. If we designate these means by $P\,D\,F$, we have the order $D\,F\,P$ as in pd, with the same ratio $(P - F)/(F - D) = 5/9$. In Sc I[16] these means have the wrong order, $D\,P\,F$; and in Yt I[11] and Zr II[19] the above ratio has the values 1.86 and 2.08, respectively, much too large.

In general the predictions of the theory have been seen to fit very poorly for configurations with p electrons, a uniform trend having been observed in p^3, and good fits having been obtained for pd in the higher members of the CaI isoelectronic sequence.

§6. Comparison with Experimental Data, d Electrons

For d^2 the theory predicts (see §3) the energies of the five multiplets ^1S, ^3P, ^1D, ^3F, ^1G in terms of three integrals, the ^1S being predicted extremely high. In no instance, where a ^1S is reported, is it anywhere nearly high enough, and so

---

[19] *Kiess and Kiess*, B. S. Jl. of Res. **5**, 1210 (1930).

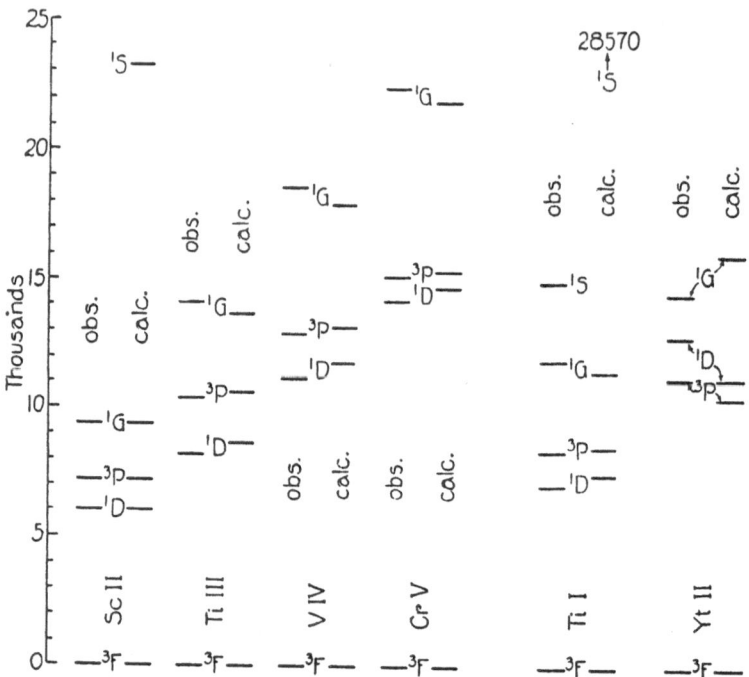

FIGURE 1. The configuration d^2.

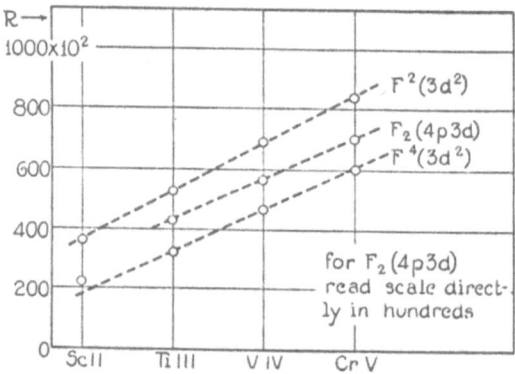

FIGURE 2. Values of some integrals for the Ca I isoelectronic sequence.

we have investigated the other four levels, making a least-squares fit of the separations of ^3P, ^1D, ^1G from the low ^3F, which three separations the theory gives in terms of the two parameters F_2 and F_4.

In the isoelectronic sequence Sc II[16], Ti III[17], V IV[18], and Cr V[18], it was found possible to make the excellent fits plotted in Fig. 1 for $3d^2$. The predicted height of ^1S in Sc II is shown, the other ^1S's being correspondingly high. Russell and Meggers reported a ^1S between the ^1D and ^3P of Sc II, but later, in a note to their Yt paper[11] they ascribe this not to $3d^2$ but to $4s^2$. The $3d^2$ ^1S as predicted would be in the midst of the configuration $4p3d$ with which they get their strong combinations, and so difficult to find. In Ti III, Russell and Lang report a ^1S just under the ^1G, but with a question mark. This undoubtedly does not belong to $3d^2$. In V IV and Cr V, White reports a ^1S just above the ^1G, but since these spectra were analyzed practically by extrapolation from Sc II and Ti III these levels also are probably not part of $3d^2$. The values of F^2 and F^4 as determined in this way show a striking linearity as plotted in Fig. 2. Slater's $F^2(3d^2)$ and $F^4(3d^2)$ are plotted in place of our F_2 and F_4 in order to show their relative magnitudes. Table VII gives these values ($D_2 = 49$, $D_4 = 441$):

TABLE VII

	Sc II	Ti III	V IV	Cr V
$F^2(3d^2)$	36180	52840	69300	84330
$F^4(3d^2)$	22740	32520	47020	60880

These values are in good accord with Slater's rough estimate that F^4 is approximately half of F^2.

Russell[20] has completed the $3d^2$ also in Ti I, the levels being shown in Fig. 1, together with the theoretical levels with $F_2(3d^2) = 899.5$ and $F_4(3d^2) = 66.15$. This is again an excellent fit except for ^1S, as was pointed out by Slater. Meggers and Russell[11] have found all but the ^1S in the $4d^2$ of Yt II, but here, see Fig. 1, not a very good fit is obtained of the rest of the levels, although the perturbations are not great. This plot is made with $F_2(4d^2) = 625.2$ and $F_4(4d^2) = 55.1$.

In contrast to these reasonable fits we have the Zr III $4d^2$ of Kiess and Lang,[14] in which not only is the ^1S much too low, but the *two* intervals between *no three* of the other levels may be fitted with possible values of the *two* parameters F_2 and F_4; either it is necessary to assume that one of the integrals is negative or that F^4 is many times greater than F^2. In the $5d^2$ of La II (see §7 for energy levels) we have a somewhat different situation. No ^1S is reported, and of the four remaining levels there are three, and just three, which may be fitted with possible values of the parameters. These are ^3P, ^3F, ^1G with $F_2 = 495$ and $F_4 = 35.4$. (The two parameters fit the two intervals exactly, of course.) This leaves the ^1D almost 5000 cm^{-1} too high.

In d^2s we have exactly the same situation as in d^2 when we take the weighted means of ^2P and ^4P, and ^2F and ^4F, as analogous to ^3P and ^3F, respectively. Here again we can get good least squares fits of the F, ^2D, P, ^2G levels of $3d^24s$ of Ti II[21] and $4d^25s$ of Yt I[11]. These are plotted in Fig. 3. In Ti II a ^2S is

[20] Russell, Astrophys. J. **66**, 347 (1927).
[21] Russell, Astrophys. J. **66**, 283 (1927).

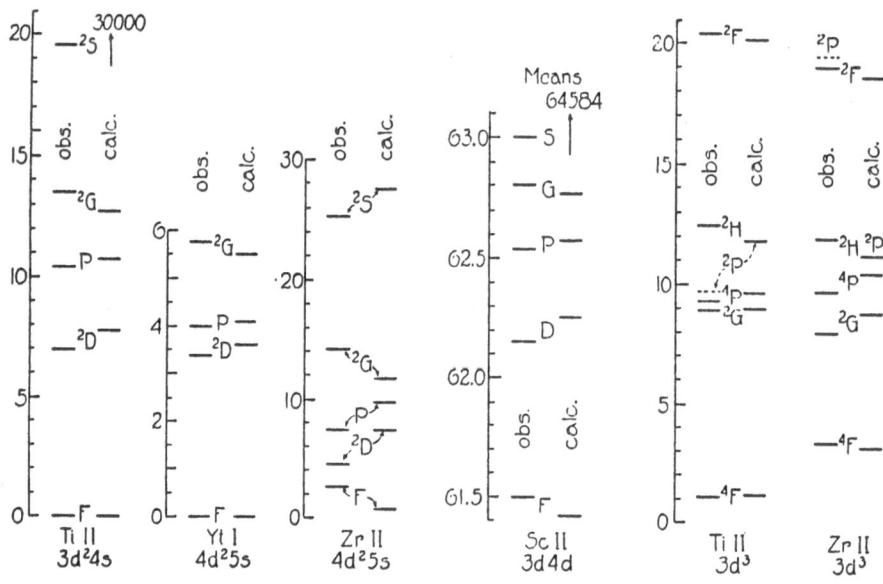

FIGURE 3. The configurations d^2s, dd, and d^3. (Scale in thousands of wave numbers.)

observed at about the same relative position as the ^1S of Ti I, while calculated it should be much higher. Again in Yt I no ^2S is observed. The values of the parameters are, for Ti II, $F_2 = 1014.7$, $F_4 = 59.2$ and for Yt I, $F_2 = 433.6$, $F_4 = 32.1$. We have $F^2 = 1.91F^4$ and $1.51F^4$ respectively in these two cases. In Zr II[19], $4d^2 5s$, we have the first instance in which a ^2S is found reasonably high. In this configuration, although we cannot get a very good fit, in contrast to the $4d^2$ of Zr III almost any three of the five levels will give reasonable values of the parameters, and we can get an approximate fit of all five by taking $F_2 = 905$, $F_4 = 55$ as indicated in Fig. 3 ($F_0 - G_2 = 8000 \text{ cm}^{-1}$ measured from Kiess and Kiess' low $^4F_{1\frac{1}{2}}$, see §3). The fit would not be much improved by omitting any one of the mulltplets from consideration so here we have probably a generaly large second order perturbation. We have a further check in this configuration on the doublet-quartet separations, which should be the same for P and F terms, being equal to $3G_2$. We have for Ti II, $^2P - {}^4P = 6620$, $^2F - {}^4F = 4558$; and for Yt I, $^2P - {}^4P = 3964$, $^2F - {}^4E = 4357$. These show reasonable agreements, but Zr II is badly off, for here we have $^2P - {}^4P = -1877$, and $^2F - {}^4F = +5404$.

The dd configuration gives us a pentad, $^{1,3}SPDFG$, with three F's to fit the means and three G's to fit the separations. The first and best instance of this configuration is the Sc II $3d4d$ of Russell and Meggers.[16] Here we can fit the $PDFG$ means as shown in Fig. 3 using $F_2(3d4d) = 107$, $F_4(3d4d) = 6$ ($F_0 = 62330$). The S mean is observed some 1600 cm^{-1} too low, which could be caused by the ^1S being about 3000 cm^{-1} too low. When we consider the separations (*singlet-triplet*) we find that we can obtain a fairly good fit of the

TABLE VIII. Sc II $3d\,4d$ separations

	Obs.	Calc.
$^1S - {}^3S$	3872	6368
$^1P - {}^3P$	-4275	-4385
$^1D - {}^3D$	4384	4492
$^1F - {}^3F$	-3935	-3810
$^1G - {}^3G$	4864	4760

$P\,D\,F\,G$ separations, with the observed S separation much too small, as shown by Table VIII.

These are calculated using $G_0 = 2230$, $G_2 = 36.7$ and $G_4 = 3.5$, making $G^0(3d4d) = 2230$, $G^2(3d4d) = 1800$, $G^4(3d4d) = 1540$. (The G^9, unlike F^0, carries here its full meaning as an integral.) The S separatin is about 2500 cm^{-1} too small, which would indicate a ^1S about that much too low. This is in good agreement with the predictions of the calculation of the means and indicates that this ^1S is perhaps incorrect, the correct one being some 2500–3000 cm^{-1} higher. An error in the assignment of a singlet term, especially a ^1S, is quite likely to occur, since the identification of such terms is very difficult. An interesting example of this sort will be discussed in §7.

The other instances of the *odd* configuration are not as good. Of the $4d\,5d$ of Yt II[11], no *three* of the means will fit with reasonable values of F_2 and F_4, the D and the S in particular being very low. Since the S separation was also large and negative where the theory, by comparison with the other separations, says it should be positive, Professor Russell assigned the 61367 ^1S to $4d5d$, discovering that the 59615 level he had assigned was not real; but this still is perhaps not the right level, since we have now a separation of only $+167$ whereas the S separation should probably be a great deal large. The other levels do not fit well enough to say anything definite. In the $4d5d$ of Zr III[14], Kiess and Lang do not find a ^1G, but of the means of the $S\,P\,D\,F$ singlets and triplets, no three will fit with positive values of F_2. Of the La II[12] $5d6d$ means that $P\,D\,F\,G$ will fit very well, with $F_2 = 115$, $F_4 = 4$ $(F_0 = 54430)$, with the S much too low, as in the case of Sc II, but this fit is not borne out by the separations, which are entirely skew, the P separation even being -2176 when it should probably be positive.

The other d configuration which we have found almost complete is d^3. Here we get ^2PFGH, ^4PF, and two ^2D's. Russell[21] in Ti II $3d^3$, has found all but one ^2D, and Kiess and Kiess have found all the multiplets of the $4d^3$ of Zr II.[19] The theory predicts, surprisingly, that ^2H and ^2P should have the same energy. Of these, in our two instances, the ^2H fits well, but the ^2P is considerably separated from it. If the two ^2D's only the mean is given by the theory, but in Zr II this comes far from fitting well. However, in both cases we can fit the other five levels, ^2FGH and ^4PF surprisingly well with our three parameters, as indicated in Fig. 3. For Ti II we have $F_2(3d^2) = 845$, $F_4(3d^2) = 54$

$(3F_0 = 17750)$; corresponding to $F^2 = 1.74F^4$. The mean of the ^2D's is predicted at 22140, which would throw the ^2D not found at about 31000, very high indeed. For Zr II we have $F_4 = 36$ $(3F_0 = 16000)$; corresponding to $F^2 = 2.11F^4$. The mean of the ^2D's is predicted at 19523, observed at 14214 (multiplets at 13869 and 14559).

We have seen that the theory works much better for d electrons than for p. It has been good in every instance in the first long period of the periodic table, and better in the second than the third.

§7. La II and the f Electron Configurations

W.F. Meggers and H.N. Russell have recently completed the analysis of La II, obtaining the first complete pf, df and f^2 configurations. Through their kindness in allowing us to use these data we have been able to obtain two beautiful fits of the df and f^2 configurations, and to prove the actual service of this theory to spectroscopists by helping to straighten out the analysis at two or three points.

Since these data are unpublished as yet, Meggers and Russell have kindly allowed us to publish a preliminary energy table, the levels being measured up from the low $5d^2$ $^3F_2$. We give only the centers of gravity of the multiplets in Table IX.

Of these the p^2, pd, pf, d^2, and dd have already been discussed, and were not found to agree particularly well with the theory. The new $5d4f$ and $4f^2$ remain to be considered, which will be done in some detail.

As we received the data from Professor Russell, the df ^1H was placed at 18169, rather close to ^3H, instead of having the extremely great separation of about 10,000 units and being the highest level in the configuration. An attempt to fit the means by the method illustrated in Fig. 4 was made. This is incidentally the method used on many of the previous configurations. The formulas to which we are fitting these means are as follows:

Calc.

$$P = 25214 = F_0 + 24F_2 + 66F_4 \qquad 25216$$
$$D = 20534 = F_0 + 6F_2 - 99F_4 \qquad 20506$$
$$F = 21467 = F_0 - 11F_2 + 66F_4 \qquad 21191$$
$$G = 19038 = F_0 - 15F_2 - 22F_4 \qquad 19323$$
$$H = 18502 = F_0 + 10F_2 + 3F_4 \qquad 22598$$

Since F_0 occurs uniformly, we plot in Fig. 4 the value of these means against the coefficients of F_2. Then the line determining F_2 must be so drawn that its separations from these points are as closely as possible proportional to the coefficients of F_4 and in the direction shown by the arrows. It is seen at once

TABLE IX. La II

$6s^2$		^1D	40458	^1P	27424
^1S	7473	^3D	38835	^3P	23003
		^1F	37210	^1D	18895
$6s6p$		^3F	37034	^3D	22174
^1P	45692	^1G	39221	^1F	24523
^3P	32699	^3G	37479	^3F	18411
				^1G	16599
$6s5d$		$5d^2$		^3G	21478
^1D	1394	^1S	—	^1H	28525
^3D	2760	^3P	5949	^3H	18835
		^1D	10095		
$6s4f$		^3F	1183	$4f^2$	
^1F	15773	^1G	7473	^1S	69505
^3F	14888			^3P	63960
		$5d6d$		^1D	59900
$6p^2$		^1S	54794	^3F	57939
^1S	66592	^3S	55230	^1G	59528
^3P	61779	^1P	56037	^3H	56080
^1D	62026	^3P	53861	^1I	62408
		^1D	55184[1]		
$6p5d$		^3D	53067		
^1P	30353	^1F	52138?		
^3P	28833	^3F	54819[1]		
^1D	24462	^1G	56036		
^3D	27538	^3G	53659		
^1F	32201				
^3F	27477	$5d4f$			
$6p4f$					

[1] The $^1D_2$ and $^3F_2$ of the $5d6d$ may possibly be interchanged. ($^3F_4 = 55321$, $^3F_3 = 54840$, $^3F_2 = 53885$.)

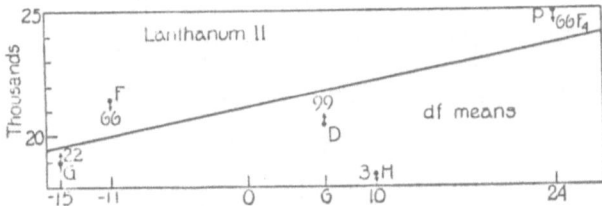

FIGURE 4. Illustrating the method of locating poorly fitting multiplets and a method of estimating the values of the integrals.

that we can thus fit G, F, D, and P quite well, but that H is far too low. From the slope of the line we get $F_2 = 115$, from the average separations $F_4 = 16$ (and from the height at zero abscissa, $F_0 = 21400$). From the level diagram in Fig. 5 we see that changing none of these values will tend to improve the general fit, so that these are approximately as good as can be found. A more

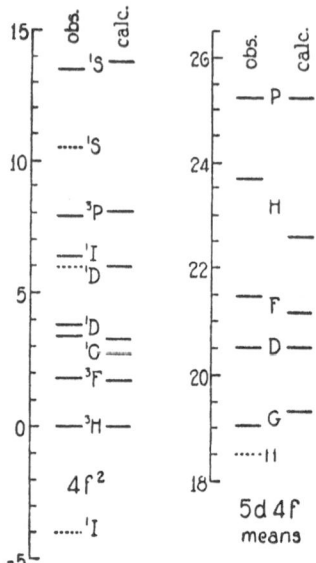

FIGURE 5. The configurations df and f^2 of La II. (Scale in thousands of wave numbers.)

exact determination of these values, such as a least squares fit, gives an accuracy which is meaningless. The calculated means are shown in the above table. From the calculated H mean it would seen that the ^1H should be about 8200 units higher.

When we consider the separations we have the following formulas to determine the G's

				Calc.
$^1P - {}^3P =$	$-4421 =$	$-2($	$G_1 + 24G_3 + 330G_5)$	$- 4638$
$^1D - {}^3D =$	$3279 =$	$2($	$3G_1 + 42G_3 - 165G_5)$	3397
$^1F - {}^3F =$	$-6112 =$	$-2($	$6G_1 + 19G_3 + 55G_5)$	$- 5837$
$^1G - {}^3G =$	$4879 =$	$2($	$10G_1 - 35G_3 - 11G_5)$	4987
$^1H - {}^3H =$	$+ 666 =$	$-2($	$15G_1 + 10G_3 + G_5)$	-11330

It is inconvenient here to use such a method of fitting as describd above, but should be very large and negative. A least squares fit of the $P\,D\,F\,G$ separations gives $G_1 = 357.6$, $G_3 = 29.7$, $G_5 = 3.78$, and the calculated separations show in the table, a good fit. These values correspond to $G^1(5d4f) = 12,500$, $G^3 = 9350$, $G^5 = 5750$. This calculation shows that the ^1H should be about 12,000 units higher.

In the data as received from Professor Russell, the $4f^2\,{}^1I$ was placed at 52,052 instead of the value noted in the table, and two possibilities given for ^1S and ^1D, as noted below. The f^2 consists of seven levels to be fitted with four parameters as follows:

$$^1S = \begin{Bmatrix} 69505 \\ 66592 \end{Bmatrix} = F_0 + 60F_2 + 198F_4 + 1716F_6$$

$$^3P = \quad 63963 \quad = F_0 + 45F_2 + \quad 33F_4 - 1287F_6$$

$$^1D = \begin{Bmatrix} 59900 \\ 62026 \end{Bmatrix} = F_0 + 19F_2 - \quad 99F_4 + \quad 715F_6$$

$$^3F = \quad 57939 \quad = F_0 - 10F_2 - \quad 33F_4 - \quad 286F_6$$

$$^1G = \quad 59528 \quad = F_0 - 30F_2 + \quad 97F_4 + \quad 78F_6$$

$$^3H = \quad 56080 \quad = F_0 - 25F_2 - \quad 51F_4 - \quad 13F_6$$

$$^1I = \quad 52052 \quad = F_0 + 25F_2 + \quad 9F_4 + \quad F_6.$$

Now from a diagram such as Fig. 4, it may be seen that F_6 will be extremely small, for a good fit of the high ^1S, the low ^1D, the ^3P, ^3F, ^1G, and ^3H may be obtained using just the parameters $F_2 = 94.0$, $F_4 = 22.1$. The ^1I is definitely observed 10000 units too low. The correct ^1S and ^1D are at once determined, since the other possibilities will not fit under any circumstances.

Thus we have seen that the theory predicted the ^1H of df and the ^1I of f^2 both about 10,000 units higher, and when this was called to the attention of Professor Russell, he discovered that he could make this shift by rearranging his lines as shown in Fig. 6. Upon doing this he immediately was able to

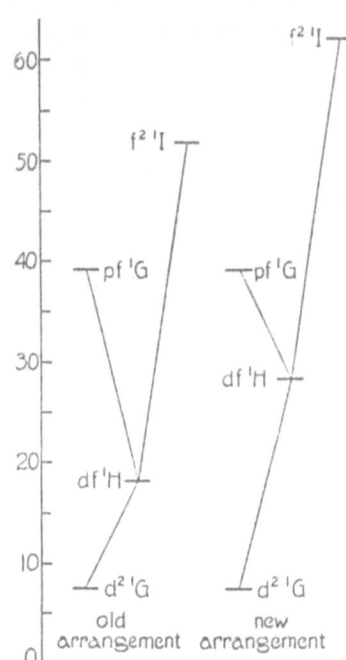

FIGURE 6. Russell's rearrangement of the La II singlets according to the predictions of the theory. (Scale in thousands of wave numbers.)

check each of these levels by faint cross-combination lines, thus definitely proving the correctness of this arrangement. Both of these singlets were raised 10,356 units to the places shown in Table IX. In Fig. 5, the old levels are shown broken and the corrected levels by full lines. The df diagram is merely the old diagram determined from the means of P D F G, with the H mean put in its proper place. The $^1H - {}^3H$ separation becomes now 9690 against the calculated 11330 from the other four separations. The ^1H was raised just about the average of the predictions from the means and from the separations. It is supposed that the $3d4d\,^1S$ of Sc II would behave in about this fashion. For the f^2 the ^1I is placed within 300 units of its prediction using the two coefficients above. Fig. 5, however, is a recalculation using least squares to fit the three coefficients F_2, F_4, F_6 to the separations of all the levels from the low ^3II. This gives a surprisingly good fit of the 6 intervals in terms of the 3 integrals, with the value of $F^2 = 21{,}000$, $F^4 = 23{,}500$, $F^6 = 1930$. Unfortunately F^4 comes out a little larger than F^2 whereas it must be smaller.

Thus we have obtained quite pleasing results with these two f-electron instances, the theory having stood the test of prediction and been of actual service in the analysis of the spectrum, both in the assignment of ^1S and ^1D, f^2, the former of hich Professor Russell says he had little possibility of assigning definitely, and in the rearrangement of the levels in both df and f^2.

Kiess and Lang[14] have completed the $4d4f$ configuration of Zr III except for the ^3H, but when the other four means are plotted as in Fig. 4, it is seen that no three of them will fit in any fashion. The P, D, and F separations alternate but the G does not.

In conclusion we may say that this first approximation seems to be most accurate for those configurations with lowest total quantum numbers in comparison with the angular momentum quantum numbers, the $3d$ and $4f$ in particular giving good results.

We wish to thank Professor H.N. Russell for his helpful interest in this work, and him and Dr. Meggers for permission to publish their lanthanum data.

Quantum Mechanics of Collision Processes

I. Scattering of Particles in a Definite Force Field

By E.U. Condon

Appendix, Electron Diffraction

By P.M. Morse

Palmer Physical Laboratory
Princeton University

TABLE OF CONTENTS

By a collision will be understood the interaction of two or more of the entities with which atomic physics is concerned under such conditions that before and after the collision they are widely separated in space but during the collision they are close together. The entities which may take part in collisions include electrons, protons, photons, atoms and molecules. If photons were to be cnsidered the report would also need to treat of the entire problem of the interaction of radiation and matter. This is too much and so it is arbitrarily excluded. Certain other topics such as the Ramsauer effect, the collisions of two atoms and the polarization of the light emitted by atoms excited by an incident unidirectional electron beam are also left out of consideration. The report is thus intended simply to serve as an introduction to the simpler problems and to the method by which the more complicated problems are attacked.

§1. The Laws of Quantum Mechanics

The general principles of the theory are now available in a number of accounts. Of the elementary grade may be mentioned the articles of Kemble and of Kemble and Hill[1] in this journal, and the books of Condon and Morse,[2] Sommerfeld,[3] Frenkel,[4] deBroglie,[5] all of which cover approximately the same ground. More advanced are the books of Born and Jordan,[6] Weyl[7] and Dirac.[8] This last presents a general formulation of the theory in a form most suitable for our purposes so that reference to it will be the main source for fundamental ideas.

Although duplicating Dirac, it is perhaps worthwhile to preface the treatment of collision problems with a condensed statement of the fundamental ideas. Page references to Dirac follow the essential statements.

In the next few pages the principles of the theory will be set down in terms of the properties of certain abstract symbols, this corresponding to the purely symbolic treatment of vector analysis which is independent of any coordinate system. There follows the translation of the same ideas into a notation which makes reference explicitly to coordinate systems but where, of course, all the results must remain invariant under all the allowed changes of coordinate system. The latter method, of course, is more analogous to the spirit of tensor analysis as used in relativity theory while the former is the analogue of the treatment of vectors by Gibbs.

The state (p. 7) of a system is described by a quantity called ψ (not to be confused with Schrödinger's ψ-function) which is analogous to a vector in a space of many (perhaps an infinite number) of dimensions (p. 18).

In any given coordinate system the components of ψ may be complex numbers. Associated with each ψ will be a ϕ whose components in any coordinate system are the conjugate complex numbers to the components of ψ in that same coordinate system.

One can add two different ψ's to get another ψ. Similarly one can add two ϕ's to get a ϕ. But there is no place in the algebra for addition of a ϕ and a ψ.

A given state is thus described by a ψ and the associated ϕ. The symbol $\phi\psi$ is defined as the analogue of the scalar product of two vectors in vector analysis and is therefore an ordinary number (p. 21). One never multiplies two ψ's nor two ϕ's nor a ϕ and a ψ in the order $\psi\phi$, but always in the order $\phi\psi$.

---

[1] Kemble and Hill, Rev. Mod. Phys. **1**, 157 (1929) and **2**, 1 (1930).

[2] Condon and Morse, Quantum Mechanics, New York, McGraw-Hill, 1929.

[3] Sommerfeld, Wave Mechanics, New York, E.P. Dutton, 1930.

[4] Frenkel, Einführung in die Wellenmechanik, Berlin, J. Springer, 1929.

[5] de Broglie, Introduction à l'étude de la mécanique ondulatoire, Paris, Hermann et Cie, 1930.

[6] Born and Jordan, Elementare Quantenmechanik, Berlin, J. Springer, 1930.

[7] Weyl, Gruppentheorie und Quantenmechanik, Leipzig, S. Hirzel, 1928.

[8] Dirac, Quantum Mechanics. Oxford University Press, 1930.

If ϕ_r, ψ_r, refer to one state and ϕ_s, ψ_s, refer to another, then the ordinary number $\phi_r \psi_s$ is the conjugate complex number to $\phi_s \psi_r$, and moreover $\phi_r \psi_r$ is real and positive.

In observing a system experimentally we build an apparatus on a macroscopic scale which acts on and is acted upon by the system by a certain set of operations, and a scale or pointer reading results. The essential feature of classical physics has been that we have expected to be able to formulate the laws of physics in terms of unctial relations between the pointer-readings given by one set of observing apparatus and those given by the pointer-readings of another set, or of other sets. All of physics and exact natural science has proceeded along such lines hitherto. Quantum mechanics does not do this. This departure from the ordinary way in which mathematics is employed in natural science is so fundamental that it is at first hard to grasp. But once grasped the formalism of the theory is easily understood.

Any set of experimental apparatus and operations is therefore not going to appear in theory theory simply as the source of certain pointer-readings which bear a direct functional relationship with other sets of pointer-readings. Instead it appears as a quantity of a more complicated sort about to be described. Thus we are dealing not merely with a new set of laws of nature but with an entirely new mathematical canvas on which to represent these laws. In this respect the quantum mechanics is a much more far-reaching advance than was the theory of relativity.

There are two parts to the study of the theory. One is the weaving of the new canvas of purely mathematical relations on which the picture of nature is to be painted. The other is the painting of the picture.

Any given set of experimental operations leading to numerical results of observation of the system, $i.e.$ pointer-readings, will be called an observable (p. 25). In the mathematical framework an observable is considered as a quantity which acts on a ψ and converts it into another ψ. It is analogous to a tensor of the second rank or to the linear vector function of Gibbs (p. 26). Because of experimental difficulties we are not always able to observe the observables directly as pointer-readings and much of importance remains to be said on the side of the relation of observing apparatus to the observed system. But we pass over that. Thus most actual observations in atomic physics are quite indirect. Nevertheless we speak of ordinary classical concepts as observables, e.g. the position of an electron.

An observable in the mathematical theory is a rule for acting on any ψ and converting it into another ψ. If α denotes the observable, one has $\alpha\psi$ as another ψ and the α's are assumed to be linear so that $\alpha(\psi_1 + \psi_2) = \alpha\psi_1 + \alpha\psi_2$. Similarly an observable acting on a ϕ is written $\phi\alpha$ and the result is another ϕ. The sum of two observables, α_1 and α_2, is defined by $(\alpha_1 + \alpha_2)\psi = \alpha_1\psi + \alpha_2\psi$, where ψ is arbitrary. The product of two observables is defined by $(\alpha_2\alpha_1)\psi = \alpha_2(\alpha_1\psi)$ and so in general $\alpha_2\alpha_1 \neq \alpha_1\alpha_2$. One requires the assumption $\phi(\alpha\psi) = (\phi\alpha)\psi$ and hence such an expression can be written $\phi\alpha\psi$.

We now make the important physical postulate (p. 30) that a state ψ for which $\alpha\psi = a\psi$, where a is an ordinary number, is characterized by α having precisely the numerical value a. For a given state ϕ, ψ the quantity $\phi\alpha\psi$ is defined to be the average value of α in that state if the auxiliary condition, $\phi\psi = 1$, is satisfied. This is an important notion: that there exist states in which an observable does not have a precise value but instead may have various values weighted with different probabilities so that many observations of α made on the same state will yield different values, leading to an average α characteristic of the state. $\phi\alpha\psi$ cannot be regarded as the value of α associated with that state because then if $\phi\alpha_1\psi$ is the value of α_1 and $\phi\alpha_2\psi$ that of α_2, we ought to have $(\phi\alpha_1\psi)(\phi\alpha_2\psi)$ equal to $\phi(\alpha_1\alpha_2)\psi$. But this is not the case, and this agrees with the ordinary behavior of averages where the average of the product of two quantities is not in general equal to the product of the averages (p. 32).

The conjugate complex of α, written $\bar{\alpha}$, is defined by its satisfying the equation

$$\phi_s\bar{\alpha}\psi_r = \overline{\phi_r\alpha\psi_s}, \tag{1.1}$$

where the two states r and s are arbitrary (p. 28).

An observable in general is not capable of assuming all values *precisely*. The values which it can take on are the ones for which the equation in ψ,

$$\alpha\psi = \alpha'\psi \tag{1.2}$$

where α' is an ordinary number, has solutions (p. 35). These are the allowed values of α and may form a discrete set of numbers or a continuous set. Of course $\phi\alpha\psi$ can have any value between the least α' and the greatest. Weyl calls a state ψ satisfying this equation a pure state (reine Fall) for this observable. Dirac calls the allowed values eigenvalues and the ψ's which satisfy this equation eigen-ψ's of α and speaks of a given eigen-ψ belonging to an α' if it satisfies this equation with that particular α'. The eigen-ψ belonging to α' will be denoted by $\psi(\alpha')$. Similarly one has the eigen-ϕ associated with α' satisfying the equation

$$\phi\alpha = \alpha'\phi$$

and denoted by $\phi(\alpha')$ when desired.

Theorem:

$$\phi(\alpha')\psi(\alpha'') = 0, \qquad \text{if } \alpha' \neq \alpha'', \tag{1.3}$$

i.e. the eigenstates are orthogonal.

Proof:

$$\phi(\alpha')\alpha\psi(\alpha'') = \alpha''\phi(\alpha')\psi(\alpha'') \qquad \text{(Since } \psi \text{ is an eigen-}\psi\text{)}$$

$$= \alpha'\phi(\alpha')\psi(\alpha'') \qquad \text{(Since } \phi \text{ is an eigen-}\phi\text{)}$$

Therefore: $(\alpha' - \alpha'')\phi(\alpha')\psi(\alpha'') = 0$, hence the theorem (p. 36).

Assume that any ψ can be expanded in terms of the eigen-ψ's of any observable. This amounts to assuming the whole of a kind of generalized

Strum-Liouville theory at one step; hence mathematicians will recognize that much needs to be filled in here by studying exactly what class of ψ's can be so expanded (p. 38). One may write the expansion coefficient of $\psi(\alpha')$ as $(\alpha'|)$, i.e.

$$\psi = \sum_{\alpha'} \psi(\alpha')(\alpha'|) \tag{1.4}$$

where the space to the right of $|$ can be used to put indices to distinguish the expansion coefficients of different ψ's with respect to the eigen-ψ's of α (p. 73). $(\alpha'|)$ is a function of α' which need be defined only for all the allowed values of α. From the orthogonality property of the eigen-ψ's just as with Fourier series, multiplying (1.4) by $\phi(\alpha')$,

$$(\alpha'|) = \phi(\alpha')\psi \tag{1.5}$$

provided $\phi(\alpha')\psi(\alpha') = 1$. Similarly an arbitrary ϕ can be expanded,

$$\phi = \sum_{\alpha'} (|\alpha')\phi(\alpha'), \qquad \text{with } (|\alpha') = \phi\psi(\alpha').$$

The set of eigenstates can be thought of as a set of unit orthogonal vectors in which case $(\alpha'|)$ becomes the components of ψ on the coordinate system formed by them, found by taking the scalar product of ψ with the appropriate eigenstate.[9]

Consider $\phi\alpha\psi$ when ϕ, ψ are so expanded, in terms of eigenstates of α. One has

$$\phi\alpha\psi = \left\{ \sum_{\alpha'} (|\alpha')\phi(\alpha')\alpha' \right\} \left\{ \sum_{\alpha''} \psi(\alpha'')(\alpha''|) \right\} = \sum_{\alpha'} \alpha'(|\alpha')(\alpha'|), \tag{1.6}$$

by the orthogonality property for the eigenstates. Here one has the average of α expressed as a sum of the allowed values of α, multtiplied by weighting factors. Hence the interpretation of $(|\alpha')(\alpha'|)$ as the probability that α have the value α' in the state ϕ, ψ (p. 82). Since one has $(|\alpha') = \overline{(\alpha'|)}$ these probabilities are necessarily real and positive as they should be.

In case $\alpha\beta = \beta\alpha$, states may exist, $\psi(\alpha', \beta')$, which are simultaneous eigenstates of both α and β, so that

$$\alpha\psi(\alpha', \beta') = \alpha'\psi(\alpha', \beta')$$

$$\beta\psi(\alpha', \beta') = \beta'\psi(\alpha', \beta')$$

and similarly for the associated ϕ's (p. 44). To describe a system one may

---

[9] In the preceding, and in what follows, where the symbol $\sum_{\alpha'}$ is used, it is tacitly supposed that all the allowed values of α' form a discrete set. In many important cases the allowed values form a set that is wholly continuous or consists of discrete and continuous portions. In these cases generalizations of the above treatment are required which are analogous to Kemble I, p. 188. It is convenient to write the formulas as if only discrete values were involved with the understanding that the appropriate change has to be made where continuous sets present themselves. (Compare Dirac §25, p. 70).

choose any set of commuting observables such that there is no other observable that is not a function[10] of them commutes with all of them. Such a set is
said to be complete (p. 47). It is convenient to denote all the members of a
complete set by α for short, meaning thereby the ensemble $\alpha_1, \alpha_2, \ldots$.

Consider now an observable β which does not commute with all the
members of the set α. Then generally an eigenstate of β will not be one of the
set α. Suppose $\psi(\beta')$ belongs to β'. One can expand $\psi(\beta')$ in terms of the $\psi(\alpha')$
and get

$$\left.\begin{aligned}
\psi(\beta') &= \sum_{\alpha'} \psi(\alpha')(\alpha'|\beta'). \\
\phi(\beta') &= \sum_{\alpha'} (\beta'|\alpha')\phi(\alpha').
\end{aligned}\right\} \tag{1.7}$$

Similarly,

Since the state belonging to β' is one in which β has precisely the value β' we
have the result that

$$(\beta'|\alpha')(\alpha'|\beta') = (\alpha'|\beta')\overline{(\alpha'|\beta')} \tag{1.8}$$

is the probability that the observables $\alpha_1 \ldots \alpha_n$ have the values $\alpha_1' \ldots \alpha_n'$ when
β is known to have precisely the value β' (p. 83).

Perhaps it will help the reader who is already familiar with Schrödinger's
work to be told that a solution of Schrödinger's equation, as $\psi_W(x, y, z)$, in the
Schrödinger notation, is a special case of this result where the positions x, y,
z are the observables symbolized here by α and the total energy W is to be
identified with β.

Consider the action of β on an eigenstate of the α's. $\beta\psi(\alpha')$ will be another
ψ and so can be expanded in terms of the eigenstates of the α's. For this we
adopt the notation

$$\beta\psi(\alpha') = \sum_{\alpha''} \psi(\alpha'')(\alpha''|\beta|\alpha'). \tag{1.9}$$

The quantity β is thus described by the double array of coefficients $(\alpha''|\beta|\alpha')$
obtained when α'' and α' range independently over all the allowed values of
the α's. This array of coefficients is called the matrix of β in the α scheme (p.
58 and p. 74).

If ψ is arbitrary it is given by $(\alpha'|)$ in the α scheme. Then $\beta\psi$ is

$$\beta\psi = \sum_{\alpha'} \beta\psi(\alpha')(\alpha'|) = \sum_{\alpha'\alpha''} \psi(\alpha'')(\alpha''|\beta|\alpha')(\alpha'|)$$

and therefore the component of $\beta\psi$ relative to the α'' eigenstate in the α scheme
is

---

[10] A function of an observable α need be defined only for the allowed values of α since
no other values of α have any significance in the theory (p. 39). γ is said to be a function
of α if its value is specified for each allowed value of α. As to functions of several
observables, these are only defined when the several observables appearing as arguments commute with each other, and are then defined similarly (p. 46).

$$\phi(\alpha'')\beta\psi = \sum_{\alpha'} (\alpha''|\beta|\alpha')(\alpha'|), \tag{1.10}$$

by the orthogonality property.

In the α scheme the matrix of any one of the α's is especially simple: $\alpha_1\psi(\alpha') = \alpha_1'\psi(\alpha')$ and therefore $(\alpha''|\alpha_1|\alpha') = \alpha_1''\delta_{\alpha''\alpha'}$, where $\delta_{\alpha''\alpha'} = 0$ if $\alpha'' \neq \alpha'$ and equals 1 if $\alpha'' = \alpha'$, and the α's have discrete values. This needs some special consideration for the case in which the allowed α values form a continuous set, but that will not be given here.

Suppose now β stands for another set of commuting observables. An arbitrary ψ will be given by

$$\psi = \sum_{\alpha'} \psi(\alpha')(\alpha'|).$$

One has further that

$$\psi(\alpha') = \sum_{\beta'} \psi(\beta')(\beta'|\alpha').$$

Hence

$$\psi = \sum_{\beta'\alpha'} \psi(\beta')(\beta'|\alpha')(\alpha'|) = \sum_{\beta'} \psi(\beta')(\beta'|).$$

So

$$(\beta'|) = \sum_{\alpha'} (\beta'|\alpha')(\alpha'|), \tag{1.11}$$

which gives the relation between the components of a ψ in two different coordinate systems.

Similarly one has, if γ is another observable,

$$\gamma\psi(\alpha'') = \sum_{\beta'\beta''} \psi(\beta')(\beta'|\gamma|\beta'')(\beta''|\alpha''),$$

$$\psi(\beta') = \sum_{\alpha'} \psi(\alpha')(\alpha'|\beta'),$$

so

$$\gamma\psi(\alpha'') = \sum_{\alpha'\beta'\beta''} \psi(\alpha')(\alpha'|\beta')(\beta'|\gamma|\beta'')(\beta''|\alpha'').$$

But

$$\gamma\psi(\alpha'') = \sum_{\alpha'} \psi(\alpha')(\alpha'|\gamma|\alpha''),$$

so by the orthogonality property

$$(\alpha'|\gamma|\alpha'') = \sum_{\beta'\beta''} (\alpha'|\beta')(\beta'|\gamma|\beta'')(\beta''|\alpha''), \tag{1.12}$$

which gives the rule for changing a matrix from one coordinate system, or scheme, to another.

The transformation functions $(\alpha'|\beta')$ satisfy certain identities since

$$(\alpha'|) = \sum_{\beta'} (\alpha'|\beta')(\beta'|) = \sum_{\beta'\alpha''} (\alpha'|\beta')(\beta'|\alpha'')(\alpha''|).$$

Hence, comparing coefficients, one must have

$$\sum_{\beta'} (\alpha'|\beta')(\beta'|\alpha'') = \delta_{\alpha'\alpha''},$$

and similarly

$$\sum_{\alpha'} (\beta'|\alpha')(\alpha'|\beta'') = \delta_{\beta'\beta''}.$$

$$\text{(1.13)}$$

This completes the statement of the formal rules.

This is the mathematical pattern in terms of which we seek to formulate the laws of atmic physics. The remainder of the theory consists in the recognition of the properties of the operators which are to represent various observables. To a particular mode of observation with certain apparatus is to be associated a certain operator. The laws of nature are not, as before, the functional relations between the numerical values given by certain experiments, but relations between the operaors that stand for various modes of observation. The recognition of what operator is to be associated with each set of experimental operations has been carried out thus far partly by appeal to the correspondence principle (as with coordinate position and conjugate momentum) and partly by appeal to experiment (as with electron spin). Of course the correspondence principle itself is a broad generalization from experiment so the known relations between operators for physical quantities all spring from experiment. Essentially the laws are as follows (Dirac, Chap. VI):

The quantities $q_1 \ldots q_n$ which are analogous to Cartesian coordinates of particles are capable of taking on all values from $-\infty$ to $+\infty$. The quantities $p_1 \ldots p_n$ which are analogous to the Cartesian components of momentum similarly take all values from $-\infty$ to $+\infty$.

The quantities q and p satisfy the following quantum-theoretic laws of nature,

$$q_i q_j - q_j q_i = 0$$
$$p_i p_j - p_j p_i = 0$$
$$p_i q_j - q_j p_i = (h/2\pi i)\delta_{ij}.$$

$$\text{(1.14)}$$

Analogous to the total energy of the system is a Hamiltonian function H which is represented by the same functional form of the p's and q's as on the classical theory for the analogous classical dynamical system. The importance of H on the classical theory lay in the fact that through Hamilton's equations of motion it determined the time variation of the state. That continues to be its imortance here, the dependence of the state ψ on the time being given by

$$-\frac{h}{2\pi i}\frac{\partial\psi}{\partial t} = H\psi.$$

$$\text{(1.15)}$$

The eigenstates for the Hamiltonian are stationary states in the sense that the probability that any quantity α have a value α' is independent of the time. One has for the eigenstate belonging to H',

$$H\psi = H'\psi,$$

and therefore the time dependence of such a ψ is,

$$\psi_t(H') = \psi_0(H')e^{-2\pi iH't/h}.$$

Similarly

$$\phi_t(H') = \phi_0(H')e^{+2\pi iH't/h}.$$

The average value of α is independent of the time in such a state.[11] The proof is as follows:

$$\phi_t(H')\alpha\psi_t(H') = \phi_0(H')e^{+2\pi iH't/h}\alpha\psi_0(H')e^{-2\pi iH't/h}$$

so that

$$\phi_t(H')\alpha\psi_t(H') = \phi_0(H')\alpha\psi_0(H'),$$

the time dependence having just cancelled out.

The place of Schrödinger's equation in this scheme (pp. 103, 104) is that it is a special case of the equation $\alpha\psi = a\psi$. Suppose we are dealing with a system which is specified by Cartesian coordinates and momenta $q_1p_1 \ldots q_np_n$. Then the coordinates $q_1 \ldots q_n$ form a complete set of commuting observables whose allowed values are all values from $-\infty$ to $+\infty$. Let $\psi(q')$ be the eigen-ψ belonging to the set of values $q_1' \ldots q_n'$. Then an eigen-ψ for the total energy can be written, as usual,

$$\psi(H') = \sum_{q'} \psi(q')(q'|H').$$

The equation $H\psi = H'\psi$ can then be written

$$H\psi(H') = \sum_{q'} H\psi(q')(q'|H')$$

$$= \sum_{q'q''} \psi(q'')(q''|H|q')(q'|H')$$

$$= \sum_{q''} H'\psi(q'')(q''|H').$$

Hence, by the orthogonality property of the $\psi(q')$,

$$\sum_{q'} (q''|H|q')(q'|H') = H'(q''|H').$$

Schrödinger's great discovery consisted in he observation that the operation on $(q'|H')$ which is represented by the left-hand side of the equation could be replaced by a differential operator, so that this equation became a differential equation for $(q|H')$.

The differential operator for H is to be found[12] by replacing each p by the corresponding $(h/2\pi i)(\partial/\partial q)$ in the classical expression for $H(p,q)$. The equation for $(q|H')$ is then the partial differential equation,

---

[11] If α does not involve the time explicity.

[12] Compare Kemble I, Sec 4, "The Energy Operators."

$$H\left\{\frac{h}{2\pi i}\frac{\partial}{\partial q}\right\}(q|H') = H'(q|H'). \qquad (1.16)$$

Hence the $\psi_W(q)$ of Schrödinger is to be identified with the $(q|H')$ of the present notation, where $W = H'$ and is an allowed energy level.

In fact, more generally, any quantity, such as angular momentum, which classically is expressed as a function of the q's and p's may be made into an operator for the q-scheme by writing $(h/2\pi i)(\partial/\partial q)$ for the corresponding p.

§2. One Dimensional Collisions

The best way to become familiar with the workings of the theory is by consideration of simple one-dimensional problems in which a particle moves under the influence of a potential energy function $V(x)$ so that the classical Hamiltonian is

$$H = \frac{1}{2\mu}p^2 + V(x). \qquad (\mu = \text{mass}).$$

It is of importance to know the behavior of the function $(x|H')$ which in Schrödinger's notation is the $\psi(x)$ belonging to the energy level H'.

In order to speak of a particle colliding with a force-field it is necessary to suppose that the forces become negligible at large distances from the origin. Hence, since the zero of V is arbitrary we suppose that $V \to 0$ as $x \to -\infty$ and $V \to V_0$ as $x \to +\infty$ and for definiteness suppose $V_0 > 0$.

The Schrödinger equation for $(x|H')$ is then

$$\left\{-\frac{h^2}{8\pi^2\mu}\frac{\partial^2}{\partial x^2} + V(x) - H'\right\}(x|H') = 0.$$

This has to be solved with the understanding that $(x|H')$ remains finite. (More accurately, so that $\int_a^b (H'|x)(x|H')\,dx$ is finite if $(b - a)$ is finite and such that $\int_a^b (H'|x)(x|H')\,dx/(b - a)$ is finite as $b \to \infty$ and $a \to -\infty$). By easy studies of this ordinary differential equation one can establish the following:

(a) H' not less than V_{min} where V_{min} is the absolute minimum of $V(x)$.

(b) If $V_{min} < 0$ then the allowed H' in the range $V_{min} < H' < 0$ form a discrete set, and for these $\int_{-\infty}^{+\infty} (H'|x)(x|H')\,dx$ is finite and can therefore be set equal to unity by a proper normalization of $(x|H')$. Also $(x|H')$ is real so that

$$(H'|x) = (x|H').$$

(c) The allowed H' in the range $0 < H' < V_0$ form a continuous set, i.e. all values are allowed. The $(x|H')$ is real and

$$\frac{1}{b - a}\int_a^b (H'|x)(x|H')\,dx$$

tends to a finite limit (not zero) for b fixed and $a \to -\infty$ and tends to zero for a fixed and $b \to +\infty$.

(d) For $H' > V_0$ all values are allowed and associated with each H' are two different $(x|H')$ which will be denoted by $(x|H', 1)$ and $(x|H', 2)$. Since an arbitrary linear combination $c_1(x|H', 1) + c_2(x|H', 2)$ is a solution of the Schrödinger equation one sees that $(x|H')$ is here to a certain extent indeterminate. For either solution one has

$$\frac{1}{b-a} \int_a^b (H'|x)(x|H') \, dx$$

equal to a finite limit for either $b \to +\infty$ or $a \to -\infty$ or both.

The states of class (b) correspond to the classical periodic vibratory motion inside the range of values of x in which $H' - V(x) > 0$. In this range $(x|H')$ is an oscillatory function and in the range where $H' - V(x) < 0$, it sinks asymptotically and without oscillations to zero roughly at the same rate as an exponential function. Since the particle has a negligible probability of being at infinity for these states they are not of interest for collision problems.

The states of class (c) correspond to the classical aperiodic motion in which a particle comes from $x = -\infty$ with kinetic energy H', goes as far as the least value of x for which $H' - V(x) = 0$ and returns to $x = -\infty$ with kinetic energy H'. For them $(x|H')$ is an oscillatory function in the range from $-\infty$ to the least value of x for which $H' - V(x) = 0$. For x greater than the greatest value of x for which $H' - V(x) = 0$, the function sinks asymptotically to zero, roughly like an exponential function. A detailed consideration of its behavior between the least and greatest values of x for which $H' - V(x) = 0$ reveals some striking and important phenomena in which quantum mechanics gives very different results from classical mechanics. The particle has some chance of penetrating to positions where, classically speaking, its potential energy alone is greater than its total energy.

The states of class (d) correspond to classical aperiodic motions, which classically are of quite different nature according as $H' > V_{max}$ or $< V_{max}$ where V_{max} is the absolute maximum of $V(x)$. In the former case one type of classical motion is that in which the particle comes from $x = -\infty$ with kinetic energy H' and goes, without ever changing its direction to $x = +\infty$ with kinetic energy $H' - V_0$. The other type would be a similar motion carried out in the reverse direction. Similarly for $H' < V_{max}$ one classical motion is approach from $x = -\infty$ to the least value of x for which $H' - V(x) = 0$ and return to $x = -\infty$. The other is approach from $x = +\infty$ to the greatest value of x for which $H' - V(x) = 0$ and return to $x = +\infty$. We shall see that in quantum mechanics the sign of $(H' - V_{max})$ does not play the role of a sharp dividing line between two distinct classes of motions.

To remove the arbitrariness about the two $(x|H')$ in class (d) we will suppose that $(x|H', 1)$ has been so chosen that it has the asymptotic form

$$(x|H', 1) \to e^{ik_0 x} \qquad \text{as } x \to +\infty$$

$$k_0 = \frac{2\pi}{h} [2\mu(H' - V_0)]^{1/2},$$

and similarly $(x|H', 2)$ will be taken so that

$$(x|H', 2) \to e^{-ikx} \qquad \text{as } x \to -\infty.$$

$$k = \frac{2\pi}{h}(2\mu H')^{1/2}.$$

We shall see that this choice of $(x|H', 1)$ and $(x|H', 2)$ is a canonical one, the former corresponding to a state of affairs in which particles in the neighborhood of $x = +\infty$ are certainly moving from left to right, the second corresponding to a state in which the particle when in the neighborhood of $x = -\infty$ is certainly moving from right to left.

In accordance with that consequence of quantum mechanics known as Heisenberg's uncertainty principle[13] we cannot speak of a precise value of the momentum of a particle at a precisely given position (because p and x do not commute). But we can ask about the value of the momentum when the particle is at $x = \pm\infty$ because this virtually amounts to no restriction on the position of the particle. This point is akin to an approximation that is always made in optics which needs for its correction the theory of the resolving power of optical instruments. A light wave is strictly monochromatic only if it is of infinite extent both in the wave-front and perpendicular to it—and yet we admit only a small portion of the wave into the narrow slit of a spectroscope and continue to talk of it as essentially monochromatic! Since $(x|H', 1)$ becomes for large x asymptotically the same as the wave function of a free particle of precisely known momentum, $(hk_0/2\pi)$, it is safe to interpret this as a state of the system in which the particle's momentum certainly has the value $+(hk_0/2\pi)$ when at $x = +\infty$. Similarly $(x|H', 2)$ represents a state in which the particle's momentum certainly has the value $-(hk/2\pi)$ or $-(2\mu H')^{1/1}$ when at $x = -\infty$.

Another property of $(x|H')$ of which we now make use is that the asymptotic expansion of $(x|H', 1)$ as $x \to -\infty$ is of the form

$$(x|H', 1) \to A(H')e^{ikx} + B(H')e^{-ikx}.$$

which means that $(x|H', 1)$ represents a state in which particles are coming from $x = -\infty$ with momentum, $+(hk/2\pi)$, in numbers proportional to $A(H')\bar{A}(H')$ and are going to $x = -\infty$ in numbers proportional to $B(H')\bar{B}(H')$ with momentum $-(hk/2\pi)$. Therefore we shall interpret $(x|H', 1)$ as a state of motion in which $A(H')\bar{A}(H')(hk/2\pi)$ particles come from $x = -\infty$ in unit time, and of these $hk_0/2\pi$ in unit time get through to $x = +\infty$ and $B(H')\bar{B}(H')(hk/2\pi)$ are reflected or scattered back to $x = -\infty$. What will happen to any particlar particle cannot be stated. Evidently for this view to be tenable one must have the sum of the numbers transmitted and scattered back equal to the number

---

[13] Heisenberg's book, *The Physical Principles of the Quantum Theory*, published in July, 1930 by the University of Chicago Press gives a full account of this.

coming from $x = -\infty$. That this is true quite generally may be readily proven as follows.

We write down the Schrödinger equation for $(x|H')$ and for $(H'|x)$, its complex conjugate; multiply the former by $(H'|x)$ and the latter by $(x|H')$, substract and integrate with regard to x. The result is

$$(x|H')\frac{d}{dx}(H'|x) - (H'|x)\frac{d}{dx}(x|H') = \text{constant}.$$

Evaluating this expression for $x \to +\infty$ and for $x \to -\infty$ and equating the results we have the theorem of the conservation of the number of particles. (Compare Weyl, p. 63.)

We can thus define the probablity of transmission as

$$\text{tr.} = k_0/kA(H')\bar{A}(H'),$$

and of reflection as

$$\text{ref.} = B(H')\bar{B}(H')/A(H')\bar{A}(H'),$$

where

$$\text{tr.} + \text{ref.} = 1,$$

as it should. The interpretation of $(x|H', 2)$ is exactly analogous, this representing a state of motion in which a particle coming from $x = +\infty$ has a chance of being reflected back and a chance of being transmitted to $x = -\infty$.

The interpretation of states of class (c) is now quite evident. Asymptotically for $x \to +\infty$ one has $(x|H') \to 0$ so that no particles move off to $x = +\infty$. For $x \to -\infty$, $(x|H')$ is asymptotically real and oscillatory so that the intensity of the incident and reflected beams are equal, as they should be.

The essential thing in the study of problems of the type of class (d) is to be able to compute $A(H')$, for from it the transmission and reflection probabilities can be obtained at once. This calls for a solution of Schrödinger's equation for the problem in question. As stated before, classically we have tr.$(H') = 0$ for $H' < V_{max}$ and tr.$(H') = 1$, for $H' > V_{max}$. This is replaced in quantum mechanics by a gradually changing function with the properties: tr.$(H') \to 0$ for $H' \ll V_{max}$ and tr.$(H') \to 1$ for $H' \gg V_{max}$. Instead of proving this in general, the main features of the theory will now be illustrated by working out a few cases where $V(x)$ has a mathematically tractable form.

Such cases are afforded by supposing $V(x)$ to be constant except for a finite number of finite discontinuities. For our Case I suppose $V(x) = 0$, for $x < 0$ and $V(x) = V_0$ for $x > 0$. In this type of problem it is assumed that $(x|H')$ and $(d/dx)(x|H')$ are continuous at the discontinuities in $V(x)$. These assumptions were first made by Faxen and Holtsmark.[14] Of course, it is to be understood that no problems of real physical interest will show mathematical discon-

---

[14] Faxen and Holtsmark, Zeits. f. Physik **45**, 311 (1927).

tinuities in $V(x)$ and so boundary conditions are not needed for real problems. But if $V(x)$ changes by a considerable amount in a space small compared to a de Borglie wave-length, $2\pi/k$, we expect that the solution for such a case can be found by treating the change as an actual discontinuity with approximately chosen boundary conditions. Faxen and Holtsmark's paper purports to show that these are the appropriate conditions in this sense. On this point see also H. A. Wilson, Phys. Rev. **35**, 948 (1930), Eckart, Phys. Rev. **35**, 1298 (1930) and Wilson, Phys. Rev. **35**, 1586 (1930).

For $0 < H' < V_0$ one has,

$$x > 0 \qquad (x|H') = e^{-k_0 x} \qquad\qquad k_0 = \frac{2\pi}{h}[2\mu(V_0 - H')]^{1/2},$$

$$x < 0 \qquad (x|H') = Ae^{ikx} + \bar{A}e^{-ikx} \quad k = \frac{2\pi}{h}(2\mu H')^{1/2}$$

where, in order to have the required continuities:

$$Re(A) = \tfrac{1}{2}, \qquad Im(A) = k_0/2k,$$

where $Re(A)$ and $Im(A)$ mean real and imaginary parts of A.

Hence there is some probability of being at places where $x > 0$, which is impossible on classical mechanics. The total probability of x being between 0 and $+\infty$ is proportional to

$$\int_0^\infty e^{-2k_0 x}\, dx = 1/2k_0.$$

The probability of being in unit length at $x < 0$ is proportional to

$$\lim_{a \to \infty} \frac{1}{a} \int_{-a}^0 (H'|x)(x|H')\, dx = \frac{1}{2}[1 + (k_0/k)^2].$$

The ratio of these two quantities gives a kind of mean depth of penetration of the particle into the non-classical region. It is

$$1/k_0[1 + (k_0/k)^2].$$

It varies from 0 at $H' = 0$, to ∞ as $H' \to V_0$.

For $H' = V_0$ the same treatment holds, if we write $k_0 = 0$. For $x > 0$, $(x|H') = 1$ and for $x < 0$, $(x|H') = \cos kx$. This is a peculiar transition case since the flow at $x > 0$ vanishes with k_0 but nevertheless the penetration is infinite.

For $H' > V_0$, we have

$$x > 0 \qquad (x|H', 1) = e^{ik_0 x} \quad k_0 = \frac{2\pi}{h}[2\mu(H' - V_0)]^{1/2}$$

$$x < 0 \qquad (x|H', 1) = Ae^{+ikx} + Be^{-ikx}, \quad k \text{ as before,}$$

where, to satisfy the continuity requirements,

$$A = \tfrac{1}{2}(1 + k_0/k), \qquad B = \tfrac{1}{2}(1 - k_0/k).$$

FIGURE 1. Probability of reflection at a
sudden drop in the potential energy.

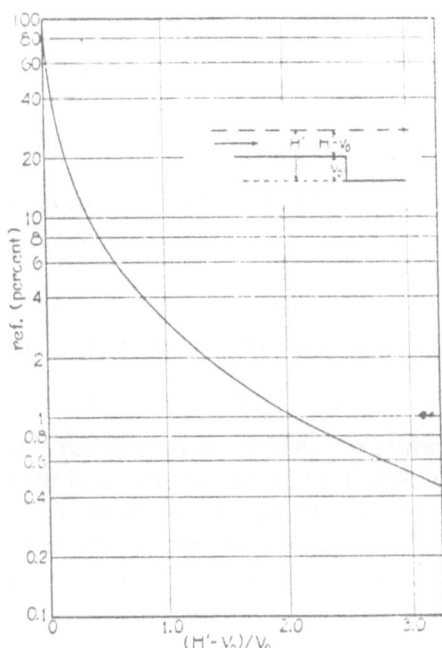

Hence the transmission and reflection probabilities are,

$$\text{tr.} = \frac{4k_0/k}{(1 + k_0/k)^2}, \qquad \text{ref.} = \frac{(1 - k_0/k)^2}{(1 + k_0/k)^2}.$$

Since tr. + ref. = 1, it is sufficient to consider the dependence of tr. on H'. At
$H' = V_0$, tr. = 0, since $k_0 = 0$. At $H' = 4V_0/3$, $k_0/k = \frac{1}{2}$ and tr. = 8/9 so we see
that the rise of tr. to the value unity is rapid when $(H' - V_0)$ becomes an
appreciable fraction of V_0. By considering $(x|H', 2)$ in the analogous way we
may see that the probability of transmission in the right-to-left motion is zero
for $H' = V_0$ but rapidly rises to unity for $H' > V_0$. In Figure 1 is plotted the
logarithm of ref. against $(H' - V_0)/V_0$ for this case. One can see that the particle
approaches rapidly to its classical behavior as H' becomes appreciably than
V_0.

Case II will be defined by $V(x) = 0$, for $x < 0$, $V(x) = V_1$ for $0 < x < a$ and
$V(x) = V_0$ for $a < x$ where $V_1 > V_0$. For $H' < V_0$ we get certain reflection as
in Case I but with a somewhat more complicated calculation of the mean
depth of penetration. The result, it can easily be foreseen, will be a somewhat
greater mean depth for a given V_1 than if V_0 were equal to V_1. We pass to
$(x|H', 1)$ for $V_0 < H' < V_1$. We have

$a < x$ $\qquad (x|H', 1) = e^{ik_0 x}$ $\qquad\qquad k_0$ as before

$0 < x < a$ $\qquad (x|H', 1) = Ae^{k_1 x} + Be^{-k_1 x}$ $\quad k_1 = \dfrac{2\pi}{h}[2\mu(V_1 - H')]^{1/2}$

$x < 0$ $\qquad (x|H', 1) = Ce^{ikx} + De^{-ikx},$ $\quad k$ as before, where the A, B, C and D are determined by the continuity requirements at $x = 0$ and $x = a$. The results are,

$$A = \tfrac{1}{2}e^{-k_1 a}(1 + ik_0/k_1)e^{ik_0 a}$$

$$B = \tfrac{1}{2}e^{k_1 a}(1 - ik_0/k_1)e^{ik_0 a}.$$

$$C = \tfrac{1}{4}e^{ik_0 a}[(1 - ik_1/k)(1 + ik_0/k_1)e^{-k_1 a} + (1 + ik_1/k)(1 - ik_0/k_1)e^{+k_1 a}]$$

$$D = \tfrac{1}{4}e^{ik_0 a}[(1 + ik_1/k)(1 + ik_0/k_1)e^{-k_1 a} + (1 - ik_1/k)(1 - ik_0/k_1)e^{+k_1 a}],$$

the details being left to the reader. Hence the transmission coefficient is

$$\text{tr.} = k_0/kC\bar{C} = (8k_0/k)\{1 + 4k_0/k + k_0^2/k^2 - k_0^2/k_1^2 - k_1^2/k^2$$

$$+ (1 + k_0^2/k_1^2 + k_1^2/k^2 + k_0^2/k^2)\cosh 2k_1 a\}^{-1},$$

which is a somewhat complicated function of H'. If $2k_1 a \gg 1$ the second term dominates the denominator and the transmission coefficient is small, of the order $e^{-2k_1 a}$. Hence for a potential wall of finite height and finite thickness there is always some chance of penetration and escape to $x = +\infty$, contrary to classical mechanics. For $H' > V_1$, $(x|H', 1)$ is of the same form except in the middle portion, where it is

$$0 < x < a \qquad (x|H', 1) = Ae^{ik_1 x} + Be^{-ik_1 x} \qquad k_1 = \frac{2\pi}{h}[2\mu(H' - V_1)]^{1/2}.$$

This amounts merely to a substitution of ik_1 for k_1 in the preceding expressions. Hence the transmission coefficient now is

$$\text{tr.} = 8(k_0/k)\left\{1 + 4\frac{k_0}{k} + \frac{k_0^2}{k^2} + \frac{k_0^2}{k_1^2} + \frac{k_1^2}{k^2}\right.$$

$$\left. + \left(1 - \frac{k_0^2}{k_1^2} - \frac{k_1^2}{k^2} + \frac{k_0^2}{k^2}\right)\cos 2k_1 a\right\}^{-1}.$$

One observes that as $H' \to \infty$ the transmission probability tends to certainty

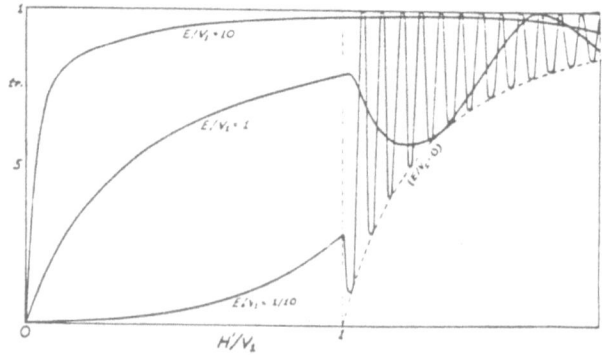

FIGURE 2. Probability of transmission at a potential wall of height V_1. (E is the energy of a particle of de Broglie wave-length equal to the thickness of the wall.)

but not monotonically because of the oscillations of the cosine term. This term is a maximum for $2k_1 a = 2n\pi$ where n is an integer, that is, when an integral number of half de Broglie wave-lengths is contained in a (the wave-length corresponding to the classical momentum in the range 0 to a is meant). Because the other terms vary with H', the transmission minima or maxima will not come precisely at these values but only approximately so.

The behavior of tr. as a function of H'/V_1 is shown in Figure 2 where curves for three proportions of the wall are shown, for the case $V_0 = 0$.[15] The curves are labelled by the value of the ratio, $h^2/2\mu a^2 V_1$, which expresses completely the characteristics of the wall for this problem, being the ratio of the energy of a particle of de Broglie wave-length a to V_1.

Case III will be defined by

$$V(x) = 0 \qquad x < 0$$

$$= V_1 \qquad 0 < x < a$$

$$= V_2 \qquad a < x < b$$

$$= V_0 \qquad b < x$$

where it is supposed that $V_1 > 0$, $V_2 < V_1$, and $V_0 \gg V_1$. If $V_0 \gg V_1$ the penetration of the particles into $b < x$ may be neglected for $H' \ll V_0$ and $(x|H')$ set equal to zero at $x = b$ for this case. We shall consider in detail only the case for which $V_2 < H' < V_1$ as the other range of H' will present phenomena not differing essentially from the preceding cases. Since $(x|H') = 0.$ at $x = b$ one can write

$$a < x < b \qquad (x|H') = \sin k_2(x - b) \qquad k_2 = \frac{2\pi}{h}[2\mu(H' - V_2)]^{1/2}$$

$$0 < x < a \qquad \qquad = Ae^{k_1 x} + Be^{-k_1 x} \quad k_1 \text{ as before}$$

$$x < 0 \qquad \qquad = Ce^{ikx} + De^{-ikx} \quad k \text{ as before.}$$

A, B, C, and D are determined by the continuity requirements as before, and $D = \bar{C}$. It is here somewhat more convenient to write

$$x < 0 \qquad (x|H') = E \sin k(x - \delta).$$

One finds for E,

$$E^2 = (1 + k_1^2/k^2)(A^2 + B^2) + 2(1 - k_1^2/k^2)AB,$$

where

$$A = \frac{1}{2}\left[\frac{k_2}{k_1}\cos k_2(b - a) - \sin k_2(b - a)\right]e^{-k_1 a},$$

$$B = \frac{1}{2}\left[-\frac{k_2}{k_1}\cos k_2(b - a) - \sin k_2(b - a)\right]e^{+k_1 a}.$$

---

[15] Kindly prepared for this report by Professor J.E. Mack.

The probability of being in the range $a < x < b$ is $\int_a^b \sin^2 k_2(x-b)\,dx$ which is of the order of $\frac{1}{2}(b-a)$. The probability of being in unit length of the range $x < 0$ is $\frac{1}{2}E^2$. Ordinarily, if $k_1 a \gg 1$, then E^2 is a very large number because of the quantity B^2 in the formula for E^2, which makes E^2 of the order of $e^{2k_1 a}$ if the bracket in the formula for B is of the order of unity. For such energy levels one has the result that particles coming from $x = -\infty$ are reflected without appreciable penetration into the region $a < x < b$ which might have been expected by analogy with Case I. The important new result is that for values of H' such that

$$(k_2/k_1)\cos k_2(b-a) + \sin k_2(b-a)$$

is of the order of $e^{-2k_1 a}$, then E^2 is very small, of the order $e^{-2k_1 a}$, so that the probability of being in unit length in $x < 0$ is almost vanishingly small compared with that in the region $a < x < b$. This means that, on the average, particles coming from $x = -\infty$ with such energies will not only penetrate into the region $a < x < b$ but will remain there a long time before going out to $x = -\infty$. It does not mean, of course, that *every* particle coming from $x = -\infty$ will penetrate into $a < x < b$. Some may be reflected in the region $0 < x < a$; if there are such then the mean stay in $a < x < b$ of those that do penetrate must be correspondingly longer. There is no way of telling by means of the $(x|H')$ function *for a single energy level*[16] how many penetrate into $a < x < b$ and hence what the mean duration in this region is for those which pentrate. But we can set a lower limit on the mean duration. In a length $(b-a)$, where $x < 0$ any one particle spends the time $2(b-a)/(2H'/\mu)^{1/2}$ since it traverses the distance twice, once going in and once coming out, with almost precisely the classical velocity. The time spent inside must bear the ratio to the time spent outside given by the relative values of $(H'|x)(x|H')$, which is E^2, hence the mean time spent inside by those that penetrate must be

$$T = \frac{2(b-a)}{(2H'/\mu)^{1/2}}(1/E^2),$$

if each particle penetrates, and longer if some do not.

These three cases will suffice to illustrate some of the striking characteristics of the theory. Naturally result, semi-quantitatively the same, hold for $V(x)$ a continuous function of x of the same general shape as those considered here.[17] In particular, Case III illustrates the quantum theory of radioactive disintegration by alpha-particle emission in the form due to Gurney and Condon.[18] The theory was developed independently by Gamow[19] along somewhat different lines.

---

[16] For elucidation of this point, see §6 of this report.
[17] An interesting example in treated by Eckart, Phys. Rev. **35**, 1303 (1930).
[18] Gurney and Condon, Phys. Rev. **33**, 127 (1929).
[19] Gamow, Zeits. f. Physik **51**, 204 (1928).

Approximation Methods

It is now clear how we have to interpret a rigorous solution for $(x|H')$ if one has been found for the particular potential energy curve $V(x)$ describing the problem. We next consider what approximate methods are available for finding $(x|H')$ when the rigorous solution cannot be obtained. Born has developed a method of successive approximations for this problem. For convenience let $u(x)$ stand for $(x|H', 1)$ and suppose $V(x) \to 0$ as $x \to \pm\infty$. We write

$$u(x) = e^{ikx} + u_1(kx) + u_2(kx) + \cdots$$

where $u_1, u_2, \ldots$ vanish for $x \to +\infty$ and are determined by the equations:

$$u_1'' + u_1 = Ve^{i\xi}/H'$$

$$\cdots\cdots\cdots\cdots\cdots\cdots\cdots \quad (\xi = kx).$$

$$u_n'' + u_n = Vu_{n-1}/H'.$$

These simple equations can be solved with the condition, $u_n(\infty) = 0$, the result being,

$$u_n(\xi) = \frac{1}{H'} \int_\xi^\infty u_{n-1}(\eta) V\left(\frac{\eta}{k}\right) \sin(\eta - \xi)\, d\eta.$$

For $x \to -\infty$ this gives,

$$u_n(\xi) = (i/2H')\left[e^{i\xi} \int_\xi^\infty u_{n-1}(\eta) V(\eta/k) e^{-i\eta}\, d\eta - e^{-i\xi} \int_\xi^\infty u_{n-1}(\eta) V(\eta/k) e^{i\eta}\, d\eta \right],$$

which becomes

$$u_n(\xi) = (i/2H') e^{i\xi} \int_{-\infty}^{+\infty} u_{n-1}(\eta) V(\eta/k) e^{-i\eta}\, d\eta$$

$$- (i/2H') e^{-i\xi} \int_{-\infty}^{+\infty} u_{n-1}(\eta) V(\eta/k) e^{i\eta}\, d\eta,$$

so that asymptotically each $u_n(\xi)$ behaves like a sum of constant coefficients into $e^{i\xi}$ and $e^{-i\xi}$. Hence the sum of them behaves this way, hence the possibility of making an interpretation of incident and reflected streams of particles as in the preceding paragraphs.

Born's[20] published convergence proof for the method is erroneous. Weyl has given a proof of convergence in his book[21] which imposes the restriction

$$k \int_{-\infty}^{+\infty} \left| \frac{V(x)}{H'} \right| dx < 1,$$

[20] Born, Zeits. f. Physik **38**, 803 (1926).
[21] Weyl, Gruppentheorie und Quantenmechanik, p. 61.

which for a given $V(x)$ can always be satisfied if H' is taken large enough. One can easily see that this restriction greatly narrows the range of important physical equations to which the method is applicable. It is therefore desirable to find a better form of the successive approximations method. Thus in Case II, if $V_0 = 0$, we have

$$k \int_{-\infty}^{+\infty} \left| \frac{V(x)}{H'} \right| dx = kaV_1/H' = (k_1 aV_1/H')(k/k_1)$$

so that Born's method would not converge in the case where $H' < V_1$ and $K_1 a \gg 1$, since $k/k_1 \sim 1$, if Weyl's restriction is necessary.

In any approximation method one starts from a problem whose solution is rigorously known, and it is desirable to have this starting problem correspond as closely as possible to the one whose solution is sought. In Born's method the starting point problem is that of the absolutely free particle. Instead let us suppose $V(x) = V_0(x) + V_1(x)$ where $V_0(x)$ is the potential energy function of a problem whose rigorous solution is known and which is to chosen that $V_1(x)$ is small compared to $V_0(x)$ everywhere. The equation for $(x|H')$ can be written

$$\frac{d^2u}{d\xi^2} + [1 - U_0(\xi)]u = U_1(\xi)u$$

where

$$k^2 = \frac{8\pi^2\mu H'}{h^2}, \quad \xi = kx, \quad U_0(\xi) = V_0(x)/H', \quad U_1(\xi) = V_1(x)/H',$$

and $u(\xi)$ is $(x|H')$. Let $u(\xi)$ be in particular $(x|H', 1)$ and suppose that $u_a(\xi)$ is $(x|H', 1)$ for the potential energy function $V_0(x)$ while $u_b(\xi)$ is any linearly independent solution of the Schrödinger equation for $u_a(\xi)$.

For $u(\xi)$ we write

$$u(\xi) = u_a(\xi) + u_1(\xi) + u_2(\xi) + \cdots$$

where the sum of all terms after the first must vanish for $\xi \to +\infty$. These terms will be determined as solutions of the equations,

$$\frac{d^2u_1}{d\xi^2} + (1 - U_0(\xi))u_1 = U_1(\xi)u_a(\xi),$$

$$\cdots\cdots\cdots\cdots\cdots\cdots\cdots\cdots\cdots\cdots\cdots\cdots\cdots$$

$$\frac{d^2u_n}{d\xi^2} + (1 - U_0(\xi))u_n = U_1(\xi)u_{n-1}.$$

Letting

$$D(\xi) = u_a(\xi)u_b'(\xi) - u_b(\xi)u_a'(\xi)$$

the solutions are found to be,

$$u_1(\xi) = \int_\xi^\infty K(\xi,\eta)U_1(\eta)u_a(\eta)\,d\eta$$

$$\dots\dots\dots\dots\dots\dots\dots\dots\dots\dots\dots\dots$$

$$u_n(\xi) = \int_\xi^\infty K(\xi,\eta)U_1(\eta)u_{n-1}(\eta)\,d\eta$$

where,

$$K(\xi,\eta) = [u_b(\xi)u_a(\eta) - u_a(\xi)u_b(\eta)]/D(\eta).$$

From this it is evident that $u(\xi)$ satisfies the integral equation

$$u(\xi) = \int_\xi^\infty K(\xi,\eta)U_1(\eta)u(\eta)\,d\eta.$$

This is a Volterra equation and the method of solving it here given is a standard one.

Dirac[22] has treated one-dimensional collisions by an approximate method, which is equivalent to the first order of Born's successive approximations, using the momentum representative of the states instead of the coordinate representative which characterizes the preceding treatment. The calculations are considerably complicated by the fact that the momentum representatives, $(p|)$, are discontinuous and involve the δ-function.

§3. Special One-Dimensional Calculations

Detailed evaluation of reflection and transmission coefficients for various one-dimensional potential energy walls have their chief application in the theory of thermionic emission of metals worked out by Nordheim and Fowler. According to the present ideas in the electron theory of metals, the free electrons behave like a gas at such a high density that the Fermi equation of state must be applied in discussing its properties.[23] At the boundary between metal and free space the potential energy of an electron increases by some 20 volts, as experiments on the diffraction of de Broglie waves show. (Appendix). Of those electrons which attain a velocity component normal to the surface whose energy equivalent is in excess of this amount only a certain fraction will actually escape because of the existence of a reflection coefficient at the wall, as was discussed in Sec. 2.

Nordheim and Fowler have shown that if I is the saturated electronic thermionic current then A in the formula

---

[22] Dirac, Zeits. f. Physik **44**, 585 (1927).
[23] Compare Darrow, Phys. Rev. Suppl. **1**, 115 and 123 (1929).

$$I = AT^2 e^{-\chi/kT}$$

(T = absolute temperature, k = Boltzmann constant, χ = work function) is given by

$$A = \frac{2\pi m e k^2}{h^3} g\bar{D}.$$

Here $g = 2$ and is the weight factor brought in by the two spin orientations for each ordinary phase-space cell and $\bar{D}$ is a mean of the transmission coefficient for the surface potential energy function weighted according to the distribution of normal components of electronic translational energy. Therefore thermionic emission depends on the mean value of the transmission coefficient and hence on the form of the potential energy law in the neighborhood of the metal surface. This has given rise to the calculation of $\bar{D}$ for a number of special assumptions concerning the potential energy $V(x)$, where x is a coordinate running from metal to vacuum, normal to the metal-vacuum interface.

Nordheim[24] first worked out the case which is called Case II of Sec. 2 but made an algebraic error which was corrected by Fowler.[25] But Fowler's Fig. 2 is quite wrong in that it fails to show the interference effects in the transmission coefficient for the energies greater than the potential energy maximum. The correct form is that given by Fig. 2 of this report.

Nordheim[26] has also made the calculation for the case in which the square wall of our Case I, Sec. 2 is rounded off by the Schottky image force, to give a better approximation to actual conditions. He uses

$$V(x) = C - e^2/4x, \qquad x > x_0$$

$$= 0 \qquad\qquad x < x_0$$

where x_0 is given by $e^2/4x_0 = C$. Naturally this makes the transmission coefficient a rather complicated function of the energy, W (called H' in Sec. 2). The important effect of rounding off the sharp discontinuity in V is that now the transmission coefficient does not tend to zero as $W \to C$, but takes on a limiting value 0.927 in case the particles are electrons and $C = 12$ electron volts. Therefore the non-classical reflection when $W > C$ is almost inappreciable. Therefore the mean value $\bar{D}$ is close to unity so that with the weighting factor, $g = 2$, the thermionic constant A comes out to be 120 amp./cm.^{2} Empirically A has a value nearer half of this.

The theory of A has had its ups and downs! The theoretical expression, omitting $g\bar{D}$, is due to Dushman and agrees with experimental values. The electron spin makes $g = 2$ instead of assigning unit weight to each phase cell as implied in Dushman's derivation, but at first Fowler and Nordheim

[24] Nordheim, Zeits. f. Physik **46**, 833 (1928).
[25] Fowler, Proc. Roy. Soc. **A122**, 39 (1929).
[26] Nordheim, Proc. Roy. Soc. **A121**, 626 (1928).

FIGURE 3. Form of the potential energy function used by
Georgeson.

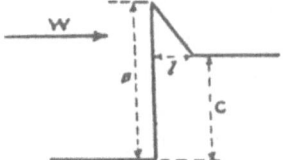

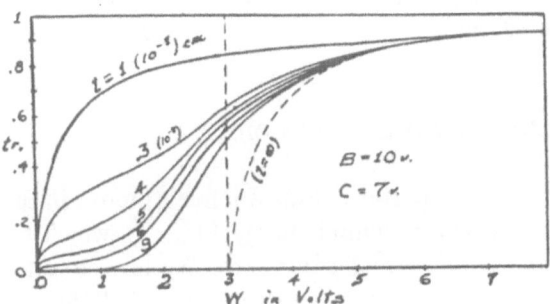

FIGURE 4. Transmission probabilities for the potential barrier of Fig. 3.

thought $\bar{D} \sim \frac{1}{2}$ so that the agreement of theory and experiment remained
unimpaired. But now Nordheim's analysis shows that $\bar{D} \sim 1$ so theory gives
a value about twice too large. Fowler[27] has recently discussed several possible
ways of removing this difficulty.

Georgeson[28] has worked out the transmission coefficient for the case in
which the potential energy function is of the form given in Fig. 3. He finds for
tr. for $W > B$

$$\text{tr.} = \frac{4W^{1/2}(W - B)^{1/2}}{\{(W - B)^{1/2} + W^{1/2}\}^2 - (BE/\sigma)\sin 2\lambda}$$

in which $F = (B - C)/l$, $\sigma = 4k(W - C)^{3/2}$, $\lambda = (2k/3F)\{(W - C)^{3/2} -$
$(W - B)^{3/2}\}$ and $k^2 = 8\pi^2\mu/h^2$, where μ is the mass of the particle.

For $W < B$ but not very near to C, he finds

$$\text{tr.} = \frac{4(B - W)^{1/2}W^{1/2}e^{-2Q}}{B + \dfrac{F}{k(B - W)^{1/2}} + \dfrac{F^2}{3k^2(B - W)^2} + \dfrac{1}{3^{1/2}}\dfrac{FW^{1/2}}{k(B - W)}},$$

in which

$$Q = \frac{2k}{3F}(B - W)^{3/2}.$$

[27] Fowler, Proc. Roy. Soc. **A122**, 36 (1929).
[28] Georgeson, Proc. Camb. Phil. Soc. **25**, 175 (1929).

Georgeson also works out special forms valid when $W \sim C$ and $W \sim B$. He gives three figures showing the transmission coefficient as a function of W for various values of B, C and l. These are probably only meant to show the general trend since they do not show the interference effects for $W > B$ which are implied by the sine term in the denominator. One of Georgeson's figures is reproduced in our Fig. 4.

Eckart[29] has published an interesting special case in which the penetration of electrons through a potential barrier that is without discontinuities in value or slope is worked out.

§4. Three-Dimensional Collisions

We shall suppose that a particle of mass μ free to move in three dimensions is acted on by a field of potential energy $V(x, y, z)$, which becomes rapidly equal to zero as $r = (x^2 + y^2 + z^2)^{1/2} \to \infty$. We have to study the possible stationary states of the motion for positive values of H' through the representative $(x, y, z|H')$ which for simplicity will be written $u(x, y, z)$. The essential difference between the three-dimensional and the one-dimensional case is that we now have an ∞^2 set of stationary states associated with each H' instead of two as in the one-dimensional problem. This is because there is now an ∞^2 continuum of directions of motion possible when far from the origin, instead of just two as in the one-dimensional case. It is imperative therefore that we have some means of classifying this wealth of solutions and that we have some knowledge of their properties.

This is best done in analogy with the one-dimensional problems by making a preliminary study of the $(x, y, z|H')$ for a free particle in three dimensions. Then $V = 0$ everywhere and the equation for $(r|H')$ is

$$\Delta(r|H') + k^2(r|H') = 0 \qquad k^2 = \frac{8\pi^2 \mu H'}{h^2}$$

of which a canonical form of solution is

$$(r|H') = e^{ik \cdot r}$$

where k is any vector whose magnitude is k; hence this form implies an ∞^2 family of solutions. Since $H' = (h^2/8\pi^2\mu)k^2$ it is best to use the vector k to label the canonical form, i.e.,

$$(r|k) = e^{ik \cdot r}$$

This function is that of an infinitely extended plane wave and corresponds, according to Sec. 1, to a precisely given value of the linear momentum vector. The operator for the linear momentum, p, is $(h/2\pi i)grad$ and this operator

---

[29] Eckart, Phys. Rev. **35**, 1303 (1930).

applied to $(r|k)$ gives $(kh/2\pi)(r|k)$; hence $(r|k)$ is the representative of a state in which p has precisely the value $kh/2\pi$.

Since we are going to be concerned with the scattering of particles by a localized field of force it will also be of interest to learn the representatives of states in which the angular momentum about some fixed origin has a precise value. A set of solutions appropriate for answering this question is obtained by solving the Schrödinger equation in polar coordinates. The equation is known to have solutions of the type,

$$(r|k, l, m) = \frac{R_l(kr)}{kr} \Theta_{lm}(\theta)\Phi_m(\phi).$$

Here the factors Θ_{lm} and Φ_m are the normalized factors of a surface harmonic

$$\Theta_{lm}(\theta) = \left(\frac{2l + 1}{2} \frac{(l - |m|)!}{(l + |m|)!}\right)^{1/2} \sin^{|m|}\theta \frac{d^{|m|}}{d(\cos\theta)^{|m|}} P_l(\cos\theta),$$

$$\Phi_m(\phi) = e^{im\phi}/(2\pi)^{1/2},$$

and $R_l(\xi)$ is a solution of the equation

$$\left(\frac{d^2}{d\xi^2} + 1 - \frac{l(l + 1)}{\xi^2}\right) R_l(\xi) = 0, \qquad (\xi = kr)$$

which vanishes at the origin. Functions which satisfy this equation arise in many problems of physics and many notations have been used for them (Watson, Bessel Functions, p. 55). We shall be interested in the general solution of this equation. It is of the form

$$y = c_1 e^{i\xi} f_l(-i\xi) + c_2 e^{-i\xi} f_l(+i\xi)$$

where $f_l(i\xi)$ is defined by the terminating series

$$f_l(i\xi) = \sum_{r=0}^{l} \frac{(l + r)!}{r!(l - r)!(2i\xi)^r}.$$

If the solution is to be finite at the origin one must have $c_2 = (-1)^{l+1}c_1$. We shall define $R_l(\xi)$ by the equation

$$R_l(\xi) = e^{i\xi} f_l(-i\xi) - (-1)^l e^{-i\xi} f_l(+i\xi).$$

With this definition one has the relation

$$R_l(\xi) = (+i)^{l+1}(2\pi\xi)^{1/2} J_{l+1/2}(\xi)$$

where $J_{l+1/2}(\xi)$ is the usual Bessel function of half an odd integer.

Since the operator for the component of angular momentum along the pole of a spherical polar coordinate system is $(h/2\pi i)(\partial/\partial\phi)$, one sees that in $(r|k, l, m)$ we have a pure state for this component and that m labels the precise value of it in Bohr units $(h/2\pi)$. Similarly the operator for the squared resultant angular momentum is

$$-\left(\frac{h}{2\pi}\right)^2\left[\frac{1}{\sin\theta}\frac{\partial}{\partial\theta}\left(\sin\theta\frac{\partial}{\partial\theta}\right)+\frac{1}{\sin^2\theta}\frac{\partial^2}{\partial\phi^2}\right]$$

so that it can be seen that $(r|k, l, m)$ refers to a pure state for squared resultant angular momentum, the precise value being $l(l + 1)(h/2\pi)^2$. The interpretation of k, as proportional to resultant linear momentum, we already know.

To find the distribution of angular momentum in the infinite unidirectional beam we must expand $(r|k)$ in terms of the $(r|k, l, m)$. Suppose the pole of the polar coordinates has the direction of k, then $k \cdot r = kr \cos\theta$ and the expansion desired as a well-known one,

$$e^{ikr\cos\theta} = (2\pi/kr)^{1/2} \sum_{l=0}^{\infty} (l + \tfrac{1}{2})i^l J_{l+1/2}(kr)P_l(\cos\theta)$$

or in our notation,

$$(r|k) = -i \sum_{l=0}^{\infty} [2\pi(l + \tfrac{1}{2})]^{1/2}(r|k, l, 0).$$

The absence of terms for $m \neq 0$ shows that the component of angular momentum in the direction of the beam is certainly zero, which corresponds with classical mechanics.

Let us now consider the scattering of particles by a spherically symmetric field of force. First we study the solutions of the equation for the motion of the particles,

$$\Delta u + \frac{8\pi^2\mu}{h^2}(H' - V(r))u = 0.$$

Since V is spherically symmetric this will have solutions of the form

$$u = \frac{R_l(kr)}{kr}\Theta_{lm}(\theta)\Phi_m(\phi)$$

where the angle factors are the same as for the free particle but $R_l(\xi)$ is a new function whose properties we need to study. It satisfies the equation

$$\left[\frac{d^2}{d\xi^2} + 1 - \frac{l(l + 1)}{\xi^2} - U(\xi)\right]R_l(\xi) = 0.$$

in which $U(\xi) = V(\xi/k)/H'$. The boundary conditions on $R_l(\xi)$ are:

$$R_l(0) = 0, \quad \text{and} \quad R_l(\infty), \quad \text{finite}.$$

It will now be supposed that for $\xi \to \infty$, we have $U(\xi) \ll l(l + 1)/\xi^2$.

We observe that the equation for $R_l(\xi)$ is of the same form as that of a one-dimensional problem in which $U(\xi)$ is infinite for $\xi < 0$, if ξ is regarded as a coordinate which can range from $-\infty$ to $+\infty$. Hence by the arguments used in Sec. 2, there will be one solution associated with each positive value of H', which will be essentially real and which asymptotically will have the form,

$$R_l(\xi) \to A_l[e^{i(\xi-\gamma_l)} - (-1)^l e^{-i(\xi-\gamma_l)}], \qquad \xi \to \infty.$$

Here A is in general a complex number to be determined by normalization or other requirements and γ_l is a phase which is fully determined by the fact that $R_l(\xi)$ is the particlar solution that vanishes for $\xi = 0$.

For a given H' there is thus the same wealth of solutions for the particle in the force field as for the free particle, namely $(2l + 1)$ corresponding to the different values of m associated with each l, with l taking on all integral values from 0 to ∞. To find the scattering we proceed to build up a superposition of these fundamental solutions which will represent an incident plane wave plus outgoing waves only, at points far from the origin. To do this we have to choose the coefficients A_l in such a way that

$$A_l[e^{i(\xi-\gamma_l)} - (-1)^l e^{-i(\xi-\gamma_l)}] = [2\pi(l+\tfrac{1}{2})]^{1/2}(-i)[e^{i\xi} - (-1)^l e^{-i\xi}] + B_l e^{i\xi}.$$

The first term on the right is that needed to make up a plane wave, as shown by the expansion of $(r|k)$ in terms of $(r|k, l, m)$, while the second corresponds to outgoing waves. Hence, equating coefficients of $e^{i\xi}$ and $e^{-i\xi}$, we find

$$A_l = -i[2\pi(l+\tfrac{1}{2})]^{1/2} e^{-i\gamma_l}$$
$$B_l = -2[2\pi(l+\tfrac{1}{2})]^{1/2} e^{-i\gamma_l} \sin\gamma_l.$$

The solution with these values of A_l and $m = 0$ is then

$$u = \sum_{l=0}^{\infty} (2\pi)^{-1/2} A_l(R_l(\xi)/\xi)\Theta_l(\theta)$$

$$\to e^{ikz} + \sum_{l=0}^{\infty} (2\pi)^{-1/2} B_l(e^{i\xi}/\xi)\Theta_l(\theta) \qquad \text{for } \xi \to \infty.$$

Hence at very large distances from the origin one has a unit intensity per unit volume of particles moving in the positive z-direction, and particles moving outward away from the origin. The number of outward moving particles in the volume element $r^2 \sin\theta \, dr \, d\theta \, d\phi$ is proportional to

$$\frac{e^{i\xi}e^{-i\xi}}{2\pi\xi^2} r^2 \sin\theta \, dr \, d\theta \, d\phi \left(\sum_l B_l\Theta_l\right)\left(\sum_l \bar{B}_l\Theta_l\right)$$

$$= \frac{1}{2\pi k^2} dr \, d\omega (\sum B_l\Theta_l)(\sum \bar{B}_l\Theta_l).$$

Hence the number crossing the surface element bounded by the differential of solid angle $d\omega$ in unit time is

$$v \cdot \frac{1}{2\pi k^2} (\sum B_l\Theta_l)(\sum \bar{B}_l\Theta_l) \, d\omega$$

where $v = (2H'/\mu)^{1/2}$.

The number of incident particles per unit normal cross-section area of beam per unit time is simply v. The ratio of the number going out in unit time in solid angle $d\omega$ to this is therefore a quantity of the dimensions of an area and is the cross-section of the incident beam needed to contribute the number of particles scattered off in the solid angle $d\omega$. One speaks therefore of the scattering power of the field of force in terms of this effective cross-section of the force field for producing the scattering in question. The coefficient $1/2\pi k^2$ can be written $\lambda^2/8\pi^3$ where λ is the de Broglie wave-length for the particle being scattered. The cross-section for scattering into the element of solid angle $d\omega$ is thus,

$$\alpha \, d\omega = \frac{\lambda^2}{8\pi^3} \left(\sum_l B_l \Theta_l \right) \left(\sum_l \bar{B}_l \Theta_l \right) d\omega.$$

The total cross-section for scattering in all directions is

$$\alpha = \int_0^{2\pi} \int_0^{\pi} \alpha_\theta \, d\theta \, d\phi = \frac{\lambda^2}{4\pi^2} \sum_l B_l \bar{B}_l$$

because of the orthogonality of the Θ_l for different l. Recalling the definition of B_l this can be written

$$\alpha = \frac{\lambda^2}{\pi} \sum_l (2l + 1) \sin^2 \gamma_l.$$

In the corresponding classical motion particles of momentum p whose line of motion at infinite distance passes at distance d from the origin would have angular momentum pd. The area of the beam in which the particles have angular momentum between l and $(l + 1)$ Bohr units is therefore $\alpha_l = (2l + 1) \lambda^2/4\pi$ where $\lambda = h/p$. The scattering formula can thus be written

$$\alpha = 4 \sum_l \alpha_l \sin^2 \gamma_l,$$

to bring out the fact that each term in the summation is of the order of magnitude of a corresponding classical term.

For a particle of energy W and mass M the value of α_l in terms of the area, πa^2 of the first Bohr orbit in hydrogen, is given by

$$\alpha_l/\pi a^2 = (2l + 1)(R/W)(\mu/M)$$

in which R is the ionization energy of atomic hydrogen and μ is the mass of the electron.

The preceding calculation shows that the scattering arises essentially from the fact that the phases of the asymptotic solution in the force field are not the same as in the case when no forces act. The scattering power is thus referred back completely to the shift in phase of the wave function of the particle in the force field relative to that of the free particle. The convergence of the series for α is insured if $\sin \gamma_l \to 0$ sufficiently rapidly as $l \to \infty$. For large l the term $U(\xi)$ becomes negligible relative to $l(l + 1)/\xi^2$ in the equation for $R_l(\xi)$ which tends to bring this about, but the exact criteria for the convergence are not known.

The foregoing rigorous theory calls for an exact solution for $R_l(\xi)$ or at least an exact calculation of the phase shifts, γ_l, relative to the free particle solutions. There are not very many functions $V(r)$ for which such an exact solution is possible. The Coulomb law, $V(r) \sim r^{-1}$, requires special treatment (Sec. 5) since it falls off less rapidly than the $l(l + 1)/\xi^2$ term due to centrifugal force, whereas the preceding developments imply the opposite behavior. As in the one-dimensional case, various features of the theory can be illustrated by supposing $V(r)$ to consist of a finite number of constant portions connected by finite discontinuities. The calculations follow the pattern of the one-dimensional calculations with Bessel functions appearing in the former role of the exponential and trigonometric functions. This makes them considerably less susceptible to numerical treatment because the necessary tables of Bessel functions are not available.

Successive Approximations Method

The method of successive approximations was extended to the three-dimensional case by Born in his original paper. One seeks a particular solution of

$$\Delta u + \frac{8\pi^2 \mu}{h^2}(H' - V)u = 0$$

which at large distances from the origin consists of a plane wave and outgoing scattered waves. Writing $k^2 = 8\pi^2 \mu H'/h^2$ one can write $\xi = kx, \eta = ky, \zeta = kz$ and $U(\xi, \eta, \zeta) = V/H'$ so the equation is

$$\Delta u + (1 - U(\xi, \eta, \zeta))u = 0$$

where now ξ, η, ζ are the independent variables in the Laplacian. We may write,

$$u = u_0 + u_1 + u_2 + \cdots$$

and find $u_0, u_1 \ldots$ etc. from the equations,

$$\Delta u_0 + u_0 = 0$$

$$\Delta u_1 + u_1 = u_0 U$$

$$\cdots\cdots\cdots\cdots\cdots\cdots$$

$$\Delta u_n + u_n = u_{n-1} U.$$

$$\cdots\cdots\cdots\cdots\cdots\cdots$$

We may take u_0 to be the incident plane wave, $e^{i\xi}$, and seek the particular solutions for $u_1, u_2, \ldots$ which represent outgoing waves only. Such a solution of the equation for u_n may be found, by an application of Green's theorem, to be

$$u_n(\xi, \eta, \zeta) = -\frac{1}{4\pi} \int u_{n-1}(\xi', \eta', \zeta') U(\xi', \eta', \zeta') \frac{\exp(i|\varrho - \varrho'|)}{|\varrho - \varrho'|} \, d\xi' \, d\eta' \, d\zeta',$$

where $\varrho = \xi i + \eta j + \zeta k$ and similarly for ϱ' and the integration extends over all space. Thus each u_n may be found, in particular u_1 being,

$$u_1(\xi, \eta, \zeta) = -\frac{1}{4\pi} \int e^{i\zeta'} U(\xi', \eta', \zeta') \frac{\exp(i|\varrho - \varrho'|)}{|\varrho - \varrho'|} d\xi' \, d\eta' \, d\zeta'.$$

Asymptotically for large ρ one has $|\varrho - \varrho'| = \rho - \rho' \cos \theta'$ where θ' is the angle between ϱ and ϱ', so that

$$u_n(\xi, \eta, \zeta) = -\frac{1}{4\pi} \frac{e^{i\rho}}{\rho} \int u_{n-1}(\xi', \eta', \zeta') U(\xi', \eta', \zeta') e^{-i\rho' \cos \theta'} d\xi' \, d\eta' \, d\zeta',$$

which is an outgoing wave, whose amplitude is a function of the direction of ϱ, which may be specified as usual by polar angles, θ, ϕ, the z-axis being the pole for which $\theta = 0$.

Hence, if the series converges, we have

$$u = e^{i\zeta} + A(\theta, \phi)e^{i\rho}/\rho \qquad \text{for } \rho \to \infty,$$

where

$$A(\theta, \phi) = -\frac{1}{4\pi} \sum_{n=1}^{\infty} \int u_{n-1}(\xi', \eta', \zeta') U(\xi', \eta', \zeta') e^{-i\rho' \cos \theta'} d\xi' \, d\eta' \, d\zeta'.$$

$A(\theta, \phi)$ is thus theoretically known although the alculation of A accurately is not feasible. The interpretation is as before: the number of outward moving particles in a volume element at distance r in solid angle $d\omega$ is proportional to

$$\frac{A\bar{A}}{k^2} d\omega = \left(\frac{\lambda}{2\pi}\right)^2 A\bar{A} \, d\omega$$

where λ is the de Broglie wave-length. This is the effective area for scattering particles in the particular solid angle $d\omega$. The total area for scattering in all directions can then be obtained by integrating over all directions.

It may be remarked that the particular solution here found satisfies the integral equation

$$u(\xi, \eta, \zeta) = e^{i\zeta} - \frac{1}{4\pi} \int u(\xi', \eta', \zeta') U(\xi', \eta', \zeta') \frac{\exp(i|\varrho - \varrho'|)}{|\varrho - \varrho'|} d\xi' \, d\eta' \, d\zeta',$$

so the question of convergence of the successive approximations process may be referred back to the theory of such an integral equation. It is remarkable that simply the first term in the series for $A(\theta, \phi)$ can give a fair approximation to the solution in many instances.

As in the one-dimensional case there are many interesting physical cases for which the method probably does not converge. It would be possible to modify it by basing the approximations on known rigorous solutions for a $V_0(r)$ approximating to the actual $V(r)$ as was done in detail for one dimension but the method has not been hitherto used in the literature. When applicable, the successive approximations method does not require $V(x, y, z)$ to be spherically symmetrical.

First approximation: Central force.

The formulas for the Born successive approximation method appear to be so different from those of the preceding rigorous theory that it is instructive to show the connection of Born's first approximation with the formula of page 69, where the scattering is related to the phase-shifts in the radial factor of $(r|H')$ relative to the corresponding representative for the free particle. We assume $U(\xi', \eta', \zeta')$ is a function of ρ' alone. For large ρ one has asymptotically $|\varrho - \varrho'| = \rho - \rho' \cos \omega$ where ω is the angle between ϱ and ϱ'. If θ, ϕ and θ', ϕ' give the directions of ϱ and ϱ' then

$$\cos \omega = \cos \theta \cos \theta' + \sin \theta \sin \theta' \cos(\phi - \phi')$$

For $u_1(\xi, \eta, \zeta)$ we have therefore,

$$-\frac{e^{i\rho}}{4\pi\rho} \int_0^\infty U(\rho')\rho'^2 \, d\rho' \int\int e^{i(\zeta' - \rho' \cos \omega)} \sin \theta' \, d\theta' \, d\phi'.$$

The integral over the unit $\theta'\phi'$ sphere may be evaluated by making use of the developments,

$$e^{i\zeta'} = (\pi/2)^{1/2} \sum_{l'=0}^\infty (+i)^{l'}(2l' + 1)\rho'^{-1/2} J_{l'+1/2}(\rho') P_{l'}(\cos \theta')$$

$$e^{-i\rho' \cos \omega} = (\pi/2)^{1/2} \sum_{l=0}^\infty (-i)^l (2l + 1)\rho'^{-1/2} J_{l+1/2}(\rho') P_l(\cos \omega).$$

Multiplying these two together term by term and integrating, all terms vanish in which two Legendre polynomials of differing l appear. For those of equal l, a well-known result (e.g. MacRobert, Spherical Harmonics, p. 137) gives,

$$\int\int P_l(\cos \omega) P_l(\cos \theta') \sin \theta' \, d\theta' \, d\phi' = \frac{4\pi}{2l + 1} P_l(\cos \theta)$$

so the expression for $u_1(\xi, \eta, \zeta)$ for large ρ is

$$u_1(\xi, \eta, \zeta) = -\frac{\pi e^{i\rho}}{2\rho} \sum_{l=0}^\infty (2l + 1) P_l(\cos \theta) \int_0^\infty U(\rho') J^2_{l+1/2}(\rho')\rho' \, d\rho'.$$

so that the amplitude $A(\theta, \phi)$ which determines the distribution in angle of the scattering is the coefficient of $e^{i\rho}/\rho$ in this equation. The first order scattering area, for all directions, is obtained by integrating $A\bar{A}$ over all directions giving a result which may be written, in order to bring out the correspondence with the rigorous theory,

$$\sum_{l=0}^\infty \alpha_l \left[\int_0^\infty U(\rho)((\pi/\rho)^{1/2} J_{l+1/2}(\rho))^2 \rho^2 \, d\rho \right]^2$$

where α_l is the cross-section for classical angular momentum between l and $l + 1$ Bohr units as before. One sees therefore that the Born first approximation amounts to a particular approximation to the phase shifts γ_l which occur in the rigorous theory.

The theory for central force fields as developed here is due to Faxen and Holtsmark (Zeits. f. Physik, **45**, 307 (1927)). This comparison of the first approximation of Born's method with the exact theory has not been published before although Mott (Proc. Camb. Phil. Soc. **25**, 304 (1928)) has approached the problem in a similar manner.

§5. Scattering in a Coulomb Field

The scattering of a beam of particles moving in the field of a Coulombian force center is one of the problems which can be treated rigorously by quantum mechanics. When an attempt is made to apply Born's successive approximations method (Sec. 4) to this problem a divergent integral presents itself. Wentzel[30] first avoided this difficulty by supposing the potential energy to vary as e^{-kr}/r instead of $1/r$. This made the integrals converge even when k was set equal to zero *after* the calculation! The result was a formula for the total intensity and angular distribution of scattered particles agreeing with that of Rutherford which was based on classical dynamics. But Wentzel's calculation is an approximate one. Later Oppenheimer[31] also treated the problem by Born's method.

Mott[32] and Gordon[33] gave the first rigorous proofs that the exact quantum mechanical solution is exactly in accord with the classical Rutherford formula. Later Temple[34] provided a much simpler proof of the same fact.

The mathematical methods employed are rather advanced in any case so the details will not be considered here. The aim, as in Sec. 4, is to find a particular solution of Schrödinger's equation which corresponds to incident particles in a plane wave and outgoing particles only. The feature of this problem of special interest is that one cannot find a solution which behaves asymptotically like a plane wave. This is connected with the fact that the Coulomb field falls off so slowly as the distance increases.

Classically we consider the scattering of a stream of particles all of which are moving with velocity v parallel to the axis of x when at large distances from the force center. The trajectory of each particle is a hyperbola and those of all the particles form a family of hyperbolas. By the general laws of correspondence between classical and quantum mechanics the wave fronts of the Schrödinger wave function should approximate to the surfaces orthogonal to this family of trajectories. Although each hyperbola has an asymptote

[30] Wentzel, Zeits. f. Physik **40**, 590 (1927).
[31] Oppenheimer, Zeits. f. Physik **43**, 413 (1927).
[32] Mott, Proc. Roy. Soc. **A118**, 542 (1928).
[33] Gordon, Zeits. f. Physik **48**, 180 (1928).
[34] Temple, Proc. Roy. Soc. **A121**, 673 (1928).

parallel to the x-axis, the orthogonal surfaces nevertheless are not plane, but instead become the surfaces given by

$$x - \frac{ZZ'e^2}{mv^2} \log(r - x) = \text{const.}$$

where $Z'e$ is the charge on the particles being scattered and m their mass while Ze is the charge on the force center. (Mott's Eq. (16), p. 546, reference 32, should have ε^2, not ε.) Therefore we must find a particular solution of the wave equation which consists asymptotically of waves having a wave front of this form plus outgoing waves only. Mott finds such a solution, its asymptotic expansion being

$$e^{i[x - \rho \log(r-x)]} + R \cdot \left(\frac{1}{r}\right) e^{i[r + \rho \log(r-x) + \alpha]}$$

in which the unit of length is so chosen that $2\pi\rho/h = 1$ and where, in the ordinary units of length,

$$R = \frac{ZZ'e^2}{2mv^2} \csc^2 \frac{\theta}{2}.$$

α is a phase shift which is without effect on the intensity of the scattered particles. The intensity of scattered particles per unit incident intensity is then $R^2 \, d\omega$ where $d\omega$ is the differential of solid angle between θ and $\theta + d\theta$ so this gives the Rutherford formula exactly for all velocities of the incident particles and all angles of scattering.

Temple's contribution consists in the observation that this result may be obtained more simply if paraboloidal coordinates are used, as in the Stark effect of atomic hydrogen, instead of spherical polar coordinates.

§6. Non-stationary States. Wave-packet Methods

Thus far we have only worked with stationary states, i.e. states in which the total energy of the system has a precise value. Such states are *stationary* in the sense that the probabilities for them do not vary with time. The equations of motion for ψ and ϕ are

$$-\frac{h}{2\pi i} \frac{\partial \psi}{\partial t} = H\psi \qquad \text{and} \qquad +\frac{h}{2\pi i} \frac{\partial \phi}{\partial t} = \phi H.$$

If ψ is an eigen-ψ of H so that $H\psi = H'\psi$, then the variation of ψ with time is given by

$$\psi_t = \psi_0 e^{-2\pi i H' t/h}.$$

Similarly,

$$\phi_t = \phi_0 e^{+2\pi i H' t/h},$$

which means that each component of ϕ and ψ varies according to these equations in any representation. Hence the product of any component of ϕ with the corresponding component of ψ, such as $(H'|\alpha')(\alpha'|H')$, which gives the probability that α have the value α' when H is known to have the value H', is independent of the time. All probabilities are therefore independent of time in the eigenstates of total energy.

Because of this fact we cannot follow the course of events in any situation by confining ourselves to states in which energy has a precise value. In this section, we shall consider therefore the way in which the course of events may be followed in a collision by employing non-stationary states. Any non-stationary state and its time variation are conveniently studied in terms of its H representative, $(H'|)_t$, at a particular time, t. Because of the fundamental equation of motion the different components, $(H'|)$, change at different rates and thus the resultant ψ changes in time. We have

$$\left(\frac{h}{2\pi i}\frac{\partial \psi}{\partial t} + H\psi\right) = \sum_{H'} \psi_0(H')\left[\frac{h}{2\pi i}\frac{\partial(H'|)}{\partial t} + H'(H'|)\right] = 0.$$

Hence by the orthogonality property of the $\psi_0(H')$ each bracketed coefficient must vanish. Therefore,

$$(H'|)_t = (H'|)_0 e^{-2\pi i H't/h}$$

$$(|H')_t = (|H')_0 e^{+2\pi i H't/h}.$$

Hence ψ at any time is given, in terms of the coordinate system formed by the eigen-ψ's of energy at the initial time, by the relation,

$$\psi_t = \sum_{H'} \psi_0(H')(H'|)_0 e^{-2\pi i H't/h}.$$

Thus although the magnitudes of the various components of ψ do not change in this representation, their phases relative to each other do change, which is sufficient to make the probabilities vary with the time for dynamical quantities that do not commute with H.

In any other representation the time variation is more complicated. If at $t = 0$ the state is described by $(\alpha'|)_0$, then the initial value of $(H'|)$ is, from equation (1.11),

$$(H'|)_0 = \sum_{\alpha''} (H'|\alpha'')(\alpha''|)_0$$

and at time t we have

$$(\alpha'|)_t = \sum_{H'} (\alpha'|H')(H'|)_t.$$

Hence

$$(\alpha'|)_t = \sum_{\alpha''}\left(\sum_{H'} (\alpha'|H')e^{-2\pi i H't/h}(H'|\alpha'')\right)(\alpha''|)_0.$$

We observe that the quantity in the bracket is simply the matrix component, in the α-scheme, of the operator, $e^{-2\pi iHt/h}$, which contains the time parametrically. The variation of the state with the time is thus given by a linear operation performed on the initial representatives.

Some simple illustrations of the variation of the state in time have been worked out by Kennard and by Darwin.[35] They have considered the variation in time of $(r|)$ for a particle in various simple fields of force, such as no forces, constant field, charge in uniform magnetic field, etc. They show that if initially $(r|)_0$ represents a particle in the neighborhood of the position whose vector is a, then the place of maximum probability will move quite closely according to classical laws. Generally speaking, the uncertainty in position of the particle increases with the time in accordance with the fact that a finite initial uncertainty in position implies an uncertainty in initial momentum and so the analogue is to a family of classical motions rather than to a particular one. Usually in classical mechanics the positions of particles tracing out such a family of motions become farther and farther apart. In exceptional cases, as in that of the harmonic oscillator, this does not happen.

Such calculations with non-stationary states are often described as wave-packet methods. This is connected with the fact that the representatives, $(r|H')$, occurring in the formula for $(r|)_t$ in terms of $(r|)_0$ are the solutions of Schrödinger's "wave" equation, and the summation over H' involved is analogous to a superposition of various wave patterns to give a moving group of waves. This face is a great help in actually working with the theory because it gives a quantum mechanical meaning to much of the classical diffraction and interference theory.[36] The first wave packet problem to be worked out was that of the harmonic oscillator, due to Schrödinger,[37] who showed that by superposing the $\psi(H')$ for various energy levels of the harmonic osicilator, the amplitudes and phases being properly chosen, a wave packet could be obtained in which the place of maximum probability density oscillated back and forth with a simple harmonic motion having the classical frequency. Debye[38] has attempted a general treatment of the motion of a wave packet representing a particle whose position is fairly accurately known in a general one-dimensional force field. Other investigations in this direction are those of Ehrenfest and of Ruark.[39] These studies are all based on the use of approximate solutions of Schrödinger's equation for $(r|H')$ and content themselves with showing the approximate validity of the Newtonian laws of motion for the place of maximum probability density. It would be of interest to pursue the matter further, for example, to construct a wave packet enabling one to watch the probabilities while a particle is in the act of slipping through a potential wall where it wouldn't be allowed by classical mechanics.

---

[35] Kennard, Zeits. f. Physik **44**, 326 (1927); Darwin, Proc. Roy. Soc. **117A**, 258 (1927).

[36] Compare e.g., Slater, Phys. Rev. **31**, 895 (1928).

[37] Schrödinger, Naturwiss. **14**, 664 (1926); Markoff, A. Zeits. f. Physik **42**, 637 (1927).

[38] Debye, Phys. Zeits. **28**, 170 (1927).

[39] Ehrenfest, Zeits. f. Physik **45**, 455 (1927): Ruark. Phys. Rev. **32**, 1133 (1928).

184 P.M. Morse

We shall not present the details of any of the wave packet calculations as they have not played a great role in collision theory thus far. The main object in mentioning them is simply to indicate the nature of the calculations to be done if one cares to follow the course of a collision in time.

The standard form (recommended by its analytical simplicity and *not* having any connection with the appearance of the same function as an approximate representation of Bernoulli's theorem in classical probability theory) for representing a situation in which the particle is known to be in the neighborhood of the place whose position vector is a, with an uncertainty measured by σ, and with a mean momentum, p_0, is to take for $(r|)$, the expression,

$$(r|) = \pi^{-3/4}\sigma^{-3/2}\exp\{-(r-a)^2/2\sigma^2 + 2\pi i p_0 \cdot r/h\}$$

This is so normalized that the integral of $(|r)(r|)$ over all space is unity. This is an expression which will be used in later sections. To start with such a state at $t=0$ and to trace its development as time goes on is the nearest thing that one can do in quantum mechanics to the corresponding process of finding the motion belonging to given initial conditions. One can, of course, use such an initial state no matter what the Hamiltonian governing the particle. But the subsequent change of $(r|)_t$ will depend on the Hamiltonian because of the occurrence of the functions, $(r|H')$, in the equation for the change of $(r|)_t$ with time.

The $(p|)$ representative of this same state is readily found to be

$$(p|) = \pi^{-3/4}\tau^{-3/2}\exp\{-(p-p_0)^2/2\tau^2 + 2\pi i a \cdot (p-p_0)/h\}$$

where $\tau = h/2\pi\sigma$. This brings out clearly the fact that the mean momentum for the state is p_0 and that the product of the uncertainties in position and momentum is equal to $h/2\pi$ for this state.

Appendix

Scattering by a Crystal Lattice

By P.M. Morse

When we consider the behavior of electrons inside a crystal, we encounter a much more complex problem than any of the previous examples. Inside the crystal, atoms are arranged in a regular three dimensional geometric pattern. An electron is never free from the influence of one or another atom, but is continually battering its way through the lattice, disrupting some of the bound atomic electrons, changing atomic energies, in turn being temporarily held by an atom, and, again, absorbing some atomic energy. In order to deal with this complicated process we are forced to make several drastic simplifications.

In the first place the atoms are much heavier bodies than the wandering electron, and are only slightly disturbed from their equilibrium position by electronic impact; and so the nuclei will be considered as fixed at their equilibrium points. But this simplification is not enough, for each atom is too complicated a system in its interactions with the electrons for us to be able to deal with, and a whole lattice of atoms would be still more impossible to handle.

In other words we first consider the behavior of a single electron in the potential field caused by the fixed atoms; then consider the vibration of the crystal atoms when undisturbed by electronic motions; and lastly we must calculate the effect on these motions of the interaction between electron and vibrating atoms and between electron and electron by approximate perturbation methods. These interactions must be taken into account when we wish to discuss the crystal's electrical conductivity, or its magnetic or mechanical properties; but the interaction calculations are very involved, and in the scattering of electrons from crystals they presumably introduce only slight corrections, so we shall neglect them here.

The problem discussed in this section is therefore that of the behavior of an electron traversing a fixed, three-dimensionally periodic potential field. This has been attacked in a number of different ways, and a number of different approximate forms of the function $(x'|H')$ have been used.

The simplest approximation is to consider the periodic potential variation as being negligible. In this case the crystal is merely a uniform depression in the potential, an amount $h^2 V_0/8\pi^2\mu$ less than the potential outside. If the crystal is infinite in extent, the function $(r'|W')$ for an electron of kinetic energy $h^2 W/8\pi^2\mu$ is $e^{i(r \cdot p)}$, where the magnitude of p is $(W)^{1/2}$ and its direction is in the direction of the electronic motion.

This rough approximation was used by Sommerfeld, Houston and others[40] to explain many phenomena of metallic conduction, etc. However this approximation cannot deal with such experiments as those of Davisson and Germer,[41] where electrons are shot from the outside at a crystal and are reflected from this surface. Since the electron stream can be represented as a plane wave of wave-length inversely proportional to the electronic momentum, we should expect strong reflection for certain wave-lengths, similar to the Bragg beams for x-rays. Davisson and Germer obtain such strong reflections. Since the solution mentioned above neglects the atom-grating entirely we cannot expect it to deal with such an experiment. To this order of approximation the crystal acts as a homogeneous medium of index of refraction $(1 + V_0/E)^{1/2}$, where $h^2 E/8\pi^2\mu$ is the electron's kinetic energy outside the crystal,[42] and the reflection from such media has been discussed in an earlier section.

---

[40] Pauli, Zeits. f. Physik **41**, 81 (1927); Sommerfeld, Houston, Eckart, Zeits. f. Physik **47**, 1 (1928); Houston, Zeits. f. Physik **48**, 449 (1928).
[41] Davisson and Germer, Proc. Nat. Acad. **14**, 619 (1928).
[42] Bethe, Naturwiss. **15**, 787 (1927).

Perhaps a better approximation can be obtained by approaching the problem from a different viewpoint. Instead of considering the electron as approximating a completely free electron, we can consider it as approximating an atomic electron.[43] Since the electrons are under the influence of the nuclear fields their behavior will be somewhat like that of an electron in an isolated atom, and the lower the electronic energy, the better is this approximation.

In order words, the crystal can be considered as a large molecule composed of similar atoms, and the electron behavior can be determined by methods used in discussing electrons in molecules.

For instance in the case of an electron in a diatomic molecule of similar atoms,[44] since the potential barrier between the two nuclei is finite, there is a possibility that an electron originally about one nucleus can get through to the other nucleus. Therefore in equilibrium conditions the function describing the electronic behavior will best be approximated by a linear combination of the wave function about one nucleus, Φ_1, and that about the other, Φ_2, i.e.,

$$(x'|W') = a\Phi_1 + b\Phi_2 = \Psi.$$

The values of a and b are determined by the average energy change due to the proximity of the two nuclei, by the usual methods of dealing with degenerate systems. If the perturbing energy V is the change in potential about one nucleus due to the proximity of the other, then the average energies used are

$$E_1 = \int \bar{\Phi}_1 V \Phi_1 \, dv = E_2; \qquad E_{12} = \int \bar{\Phi}_1 V \Phi_2 \, dv = E_{21}$$

and the values of a and b are determined by the equations

$$(E_1 - E)a + E_{12}b = 0$$

$$E_{21}a + (E_2 - E)b = 0$$

where E is the difference between the atomic energy level and the corresponding molecular level. The secular determinant determining E from these equations is

$$(E_1 - E)^2 = E_{12}^2$$

and so $E = E_1 \pm E_{12}$. In other words the proximity of the two similar nuclei splits the single atomic level into two levels. The two wave functions corresponding to the two levels are

$$\Psi_1 = (\Phi_1 + \Phi_2)/2^{1/2}; \qquad \Psi_2 = (\Phi_1 - \Phi_2)/2^{1/2}.$$

If now we build up a one-dimensional crystal by stringing N similar nuclei in a line an equal distance d_x apart, we find that the original atomic energy

[43] Heisenberg, Zeits. f. Physik **49**, 619 (1928); Bloch, Zeits. f. Physik **52**, 555 (1928); Slater, Phys. Rev. **35**, 509 (1930).
[44] Morse and Stueckelberg, Phys. Rev. **33**, 932 (1929).

level is split up into N different levels, each level corresponding to one of the linear combinations

$$\Psi_n = \sum_{r=1}^{N} a_{nr}\Phi_r \qquad (n = 1, 2, \ldots, N)$$

where Φ_r is the atomic wave function about the rth nucleus.

If the perturbation between adjacent nuclei only be considered, then the equations determining the a's will be the set

$$a_{n,r-1}E_{12} + (E_1 - E_n)a_{n,r} + E_{12}a_{n,r+1} = 0 \qquad (r = 1, 2, \ldots, N)$$

similarly to the simple case above. A solution of the secular determinant arising from these equations shows that

$$E_n = E_1 - 2E_{12}\cos \pi n/(N + 1).$$

If the ratio between the successive coefficients is taken equal

$$a_{n2}/a_{n1} = a_{n3}/a_{n2} = \cdots = a_{nN}/a_{nN-1} = x_n.$$

Then each set of equations above becomes

$$x_n^2 + 2\cos[\pi n/(N + 1)]\cdot x_n + 1 = 0$$

or

$$x_n = \exp \pm \lceil \pi i n/(N + 1)\rceil$$

and therefore

$$\Psi_n = \frac{1}{N^{1/2}} \sum_{r=1}^{N} \Phi_r \exp[\pi i n r/(N + 1)] \qquad \text{(for the plus sign).}$$

The value of Ψ_n near the r'th nucleus, i.e., when $x = rd_x$, if the origin be placed at the first nucleus, will be nearly entirely due to the term with Φ_r, since the Φ's become very small at distances from their nucleus greater than $d_x/2$. This means that the function Ψ_n can be quite closely approximated by the function

$$\exp[\pi i n x/(N + 1)d_x]\cdot U(x)$$

where

$$U(x) = \sum_{r=1}^{N} \Phi_r/N^{1/2}$$

and is a function periodic in x with period d_x.

If the crystal is very large, N is large, and the quantity $n/2(N + 1)$ can be considered as having a continuous range of possible values between zero and $1/2$, and if U be expanded into a Fourier series,

$$\Psi = \exp[ik_x\alpha/x] \sum_{l=-\infty}^{\infty} b_{xl}\exp[il\alpha x]$$

where $\alpha = 2\pi/d_x$, and where the value of k_x, the variable corresponding to $n/2(N + 1)$, determines the particular wave function chosen. It is seen also that the original single atomic energy level is now spread into a continuous band of allowed levels, which may or may not be separated from the band corresponding to the next atomic level by a band of forbidden energies.

For small values of k_x the energy varies linearly with the square of k_x, and so $k_x\alpha$ is analogous to the p in the free electron function.

The discussion so far has been for but one dimension, but the generalization to three dimensions is obvious, and for a simple cubic, orthorhombic or tetragonal lattice the wave function is

$$\Psi = \exp i(k_x\alpha x + k_y\beta y + k_z\gamma z)\cdot \sum_{l,m,n=-\infty}^{\infty} B_{lmn}\exp i(l\alpha x + m\beta y + n\gamma z).$$

This can be considered as a free electron, multiplied by a Fourier series representing the distortion due to the presence of the nuclei.

This approximation is somewhat better than the first mentioned one, and Bloch[43] has obtained fairly good results for conductivity with it. However it is a good approximation only for the electrons with the lowest energies. For higher energies the atomic wave functions used in the function U are not particularly good approximations to the actual wave function, for the distortion due to the presence of the neighboring nuclei is relatively large.

Perhaps the best way to treat the problem is to attack it directly, by solving for the motion of an electron in the three-dimensionally periodic field of the nuclei.[45] This should give results which will approximate that of the free electron for high energies, and that of the Bloch combination of atomic electrons for low energies. And it should predict the results of Davisson and Germer.

Any three dimensionally periodic potential function can be represented by the Fourier series

$$V = \frac{h^2}{8\pi^2\mu}\sum_{l,m,n} A_{l,m,n}\exp[i(l\alpha + m\beta + n\gamma)\cdot r]$$

where the summation extends from minus to plus infinity for all three indices, and where the A's are chosen so that V is everywhere real. r is the vector distance from some origin, and α, b and γ are vectors parallel to each of the crystal axes (four would be needed for the hexagonal system, but the generalization is apparent), of lengths equal to 2π divided by the respective lattice spacings.

This potential function is inserted in the usual Schrödinger equation and the function $(r|H')$ is to be found which is finite, single valued and continuous over all space. The resulting equation is a generalized form of Hill's equation,[46]

[45] Morse, Phys. Rev. **35**, 1310 (1930).
[46] Whittaker and Watson, A Course of Modern Analysis, Cambridge University Press (1915), Chapter 19. An approximate solution has been found by Peierls, Ann. d. Physik **4**, 121 (1930). Bethe, Ann. d. Physik **87**, 55 (1928) discussed the equation and obtained a number of the properties of the solution.

and solutions satisfying the boundary conditions can be found if the series $\sum A_{l,m,n}$ is absolutely convergent.

For a working example a cubic, tetragonal, or orthorhombic crystal will be assumed, and the potential function will be simplified into the form

$$V = -\frac{h^2}{8\pi^2\mu}\left[\sum_l \alpha^2 A_{x,l}e^{il\alpha x} + \sum_m \beta^2 A_{y,m}e^{im\beta y} + \sum_n \gamma^2 A_{z,n}e^{in\gamma z}\right]$$

where α, β and γ are the scalar magnitudes of $\boldsymbol{\alpha}$, $\boldsymbol{\beta}$ and $\boldsymbol{\gamma}$; and, to make the average energy equal to zero and V real, $A_{x,0} = A_{y,0} = A_{z,0} = 0$, and $A_{x,l} = A_{x,-l}$ etc. This form of potential function is not general enough to express every sort of lattice, but the electronic behavior in such a field will be sufficiently illustrative.

The resulting Schrödinger equation,

$$\Delta u - [W + \alpha^2 \sum A_{x,l}e^{il\alpha x} + \beta^2 \sum A_{y,m}e^{im\beta y} + \gamma^2 \sum A_{z,n}e^{in\gamma z}]u = 0$$

where u is the function $(x, y, z|H')$ and where W is $8\pi^2\mu/h^2$ times the electronic energy, can be broken up into three simple equations if $u = X(x) \cdot Y(y) \cdot Z(z)$. These equations are

$$\frac{d^2X}{dx^2} - (Wa^2 + \sum \alpha^2 A_{x,l}e^{il\alpha x})X = 0$$

$$\frac{d^2Y}{dy^2} - (Wb^2 + \sum \beta^2 A_{y,m}e^{im\beta y})Y = 0$$

$$\frac{d^2Z}{dz^2} - (Wc^2 + \sum \gamma^2 A_{z,n}e^{in\gamma z})Z = 0$$

where $(a^2 + b^2 + c^2) = 1$. The resulting solution, u, represents a stream of electrons travelling with a velocity equal to $(h^2W/4\pi^2\mu^2)^{1/2}$ in the direction given by the direction cosines a, b and c.

Since these equations are similar, an investigation of one of them will indicate the solutions to all three. The solution[47] of the first equation is

$$X = \sum_{r=-\infty}^{\infty} B_{x,r}e^{i(k_x+r)\alpha x}$$

where the B's are determined by the equations

$$\left[\frac{Wa^2}{\alpha^2} - (k_x + r)^2\right]B_{x,r} + A_{x,l}B_{x,r-1} = 0 \qquad (r = \ldots, -1, 0, 1, 2, \ldots)$$

and the constant k_x is determined by the equation

---

[47] Van der Pol and M.J.O. Strutt, Phil. Mag. **5**, 18 (1928).

$$\sin^2(\pi k_x) = \Delta(0) \cdot \sin^2(\pi a(W)^{1/2}/\alpha)$$

where $\Delta(0)$ is the determinant

$$
\begin{vmatrix}
\cdots & 1 & \dfrac{-A_{x,1}}{16-W'} & \dfrac{-A_{x,2}}{16-W'} & \dfrac{-A_{x,3}}{16-W'} & \dfrac{-A_{x,4}}{16-W'} & \cdots \\[2ex]
\cdots & \dfrac{-A_{x,1}}{4-W'} & 1 & \dfrac{-A_{x,1}}{4-W'} & \dfrac{-A_{x,2}}{4-W'} & \dfrac{-A_{x,3}}{4-W'} & \cdots \\[2ex]
\cdots & \dfrac{-A_{x,2}}{1-W'} & \dfrac{-A_{x,1}}{1-W'} & 1 & \dfrac{-A_{x,1}}{1-W'} & \dfrac{-A_{x,2}}{1-W'} & \cdots \\[2ex]
\cdots & \dfrac{-A_{x,3}}{-W'} & \dfrac{-A_{x,2}}{-W'} & \dfrac{-A_{x,1}}{-W'} & 1 & \dfrac{-A_{x,1}}{-W'} & \cdots \\[2ex]
\cdots & \dfrac{-A_{x,4}}{1-W'} & \dfrac{-A_{x,3}}{1-W'} & \dfrac{-A_{x,2}}{1-W'} & \dfrac{-A_{x,1}}{1-W'} & 1 & \cdots
\end{vmatrix}
$$

where $W' = Wa^2/\alpha^2$.

This solution is finite everywhere in a crystal of infinite extent for all real values of k_x. A study[47] of the behavior of k_x for values of the A's of the same order of magnitude or less than W shows that k_x is only complex for values of $a^2 W$ close to the set of values $\alpha^2/4$, α^2, $9\alpha^2/4$, $4\alpha^2$, ... This indicates that for any given electronic energy, there are certain values of a, b and c, i.e., certain directions of electron motion inside the crystal, which are barred.

Notice that this form of wave function is of the Bloch form, and will equal his solution for low energy values. For high energy values, unless the k's are complex, only B_{x0} is large, and the solution is that of the free electron.

If the crystal is not infinite in extent, but is bounded by surfaces, then these directions will not be barred, for u can then be finite everywhere even for complex values of the k's. However the real exponential term due to the imaginary part of the k's will insure that the amplitude of u for these directions is negligible except near the surface of the crystal. This indicates that for a beam of electrons of energy and direction such that $a^2 W = l^2\alpha^2/4$, $b^2 W = m^2\beta^2/4$, and $c^2 W = n^2\gamma^2/4$ (l, m, n integers) the electron wave will be damped out as it penetrates the crystal, and therefore such beams will be strongly reflected from the crystal surface. The above relations between the energy and direction of strongly reflected beams and the grating constants are those defining the analogues of Laue beams in x-rays.

The relation between $a^2 W/\alpha^2$ and k_x can be represented by the equation

$$\frac{a^2 W}{\alpha^2} + k_x^2 + f(k_x)$$

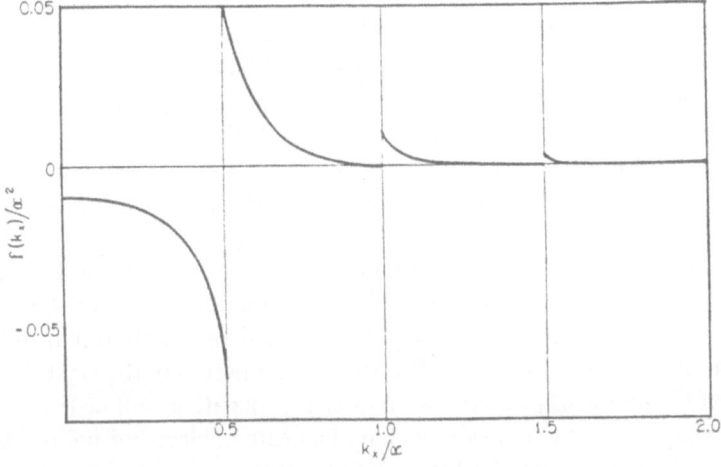

FIGURE 5. Values of $f(k_x)/\alpha^2$ for the potential function $V = (h^2/8\pi^2\mu)(\cos \alpha x/2)$

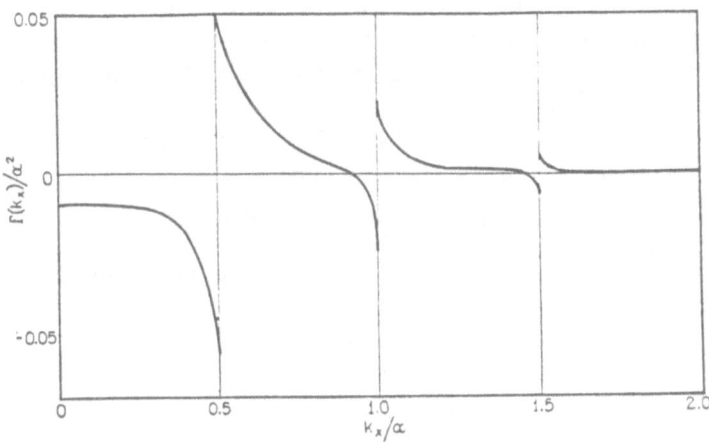

FIGURE 6. Values of $f(k_x)/\alpha^2$ for the potential function $V = (h^2/8\pi^2\mu) [3 \cos(\alpha x)/4 - 3 \cos(2\alpha x)/10 + \cos(3\alpha x)/20]$

where f only has values appreciably different from zero when k_x is near $n\alpha/2$. The values of $f(k_x)$ are shown in Figs. 5 and 6 for two different forms of potential function, for real values of k_x. Note that the form of f when k_x is near $n/2$ is sensitive to changes in the form of V, and so if f is known, the general form of V can be estimated.

Thus we see immediately that some of the results of Davisson and Germer and others are predicted. Their experiments have shown that an electron beam scatters from a crystal in a manner strikingly analogous to a beam of x-rays of wave-length equal to the de Broglie "wave-length" of the electron beam, if the crystal were considered to have an "index of refraction" equal to $[(E + V_0)/E]^{1/2}$, where E is the energy of the electron beam and V_0 is approximatily equal to the work function of the metal. We have already shown that electrons will be scattered from crystal surfaces in directions analogous to Laue beams, and we shall proceed to show that an application of the theory outlined above explains the other experimental results.

For simplicity, let the crystal surface be the plane, $x = 0$, and let the space, x negative, be field-free. Since the average potential inside the crystal has been set equal to zero, and the average potential inside is less than that outside by the amount ϕ, where ϕ can be called the work function of the crystal, then the value of the constant potential in the space, x negative, will be ϕ.

The simplest case to consider is that of a beam of electrons falling normally on the surface: we wish to find the relative intensity of the beam reflected back on the primary beam as a function of the energy of the primary beam.

Since for the present we have ruled out electronic loss of energy by omitting the interaction between the electron and the nuclear vibration, we cannot deal with the electrons heterogeneously scattered with loss of energy: and since we are not for the present interested in the homogeneous scattering to the side, we can neglect the Y and Z factors, and impose the boundary conditions on the X factor alone.

Since the primary and reflected beams are in field-free space, use can be made of the results of section 2, and the function X will be

$$X_0 = Ce^{i(E)\frac{1}{2}x} + De^{-i(E)\frac{1}{2}x}$$

outside the crystal. E is $8\pi^2\mu/h^2$ times the electronic energy outside the crystal. Therefore the electronic energy inside the crystal will be $W = E + V_0$, where $V_0 = 8\pi^2\mu\phi/h^2$, and so the electron beam inside the crystal will be represented by the function

$$X_i = \sum_r B_{x,r} e^{i(k_x + r)\alpha x}.$$

At $x = 0$ this must equal X_0 in slope and magnitude. The B's can be calculated for any assumed type of potential function, by the equations given previously, and so the two boundary conditions

$$C + D = \sum B_{x,r}$$

$$C - D = \sum \frac{\alpha(k_x + r)B_{x,r}}{(E)^{1/2}}$$

serve to determine C and D. The ratio $J = D\bar{D}/C\bar{C}$ will then be the relative intensity of the reflected beam. Analysis of the properties of the B's indicates that J will be unity for values of E which make k_x complex. Figure (7) shows a simple form of potential function, and Figure (8) shows the values of J for different values of E. Actually, since a number of electrons are scattered

FIGURE 7. Assumed form of potential energy function.

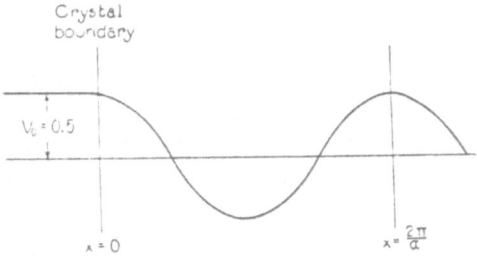

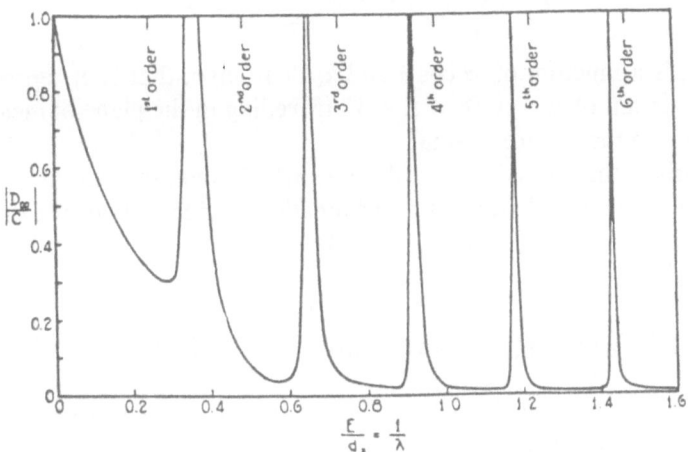

FIGURE 8. Relative intensity of the regularly reflected beam as a function of the electronic wave number. The incident beam is normal to the 111 surface of a nickel crystal. Lines marked 1st order, etc., indicate positions of strong x-ray reflection for the same crystal.

heterogeneously in their passage through the crystal, the maximum values of J will not be unity, and since more electrons are so scattered at high energies than at low, these peaks will diminish in height for increasing E. Approximately, the energy E_n for each peak is given by the equation

$$E_n = \frac{n^3\alpha^2}{4} - V_0$$

which would be the values found for simple Bragg wave reinforcement if the crystal has an index of refraction equal to $[(E + V_0)/E]^{1/2}$.

The exact value of E_m is given by the equation

$$E_m + V_0 = \frac{n^2\alpha^2}{4} + \alpha^2 G_n$$

where G_n is a quantity, usually small compared to $\frac{1}{4}$, whose value depends on the surface conditions.

Thus the only information given by this exact treatment, additional to that given by the simpler Bragg theory, is that the difference between the energy values E_n and the values $n^2\alpha^2/4$ is not exactly V_0, but that this difference varies with change of surface conditions and with different n values. A glance at the experimental curves[48] shows that they bear out these additional refinements.

The case where the primary beam is in the x, y plane, and strikes the surface at an incident angle θ to the surface normal, corresponds to the usual Bragg method of x-ray analysis. In this case only the Z factor of the u function can be neglected. The function inside the crystal will be then

$$U_i = \sum_m B_{y,m} e^{i(k_y+m)\beta y} \cdot \sum_n B_{x,n} e^{i(k_x+n)\alpha x}$$

where k_x is a function of $W \cos \phi$ and k_y of $W \sin \phi$; that is, it represents an electron stream of energy $W = E + V_0$ travelling in the plane of incidence at an angle ϕ to the surface normal.

To satisfy the boundary conditions, which require that at $x = 0$ the amplitude and normal gradient of the functions inside and outside the crystal be the same, the function outside must be

$$U_0 = C \exp\{i((E)^{1/2} \sin \theta y + (E)^{1/2} \cos \theta x)\}$$
$$+ \sum_r D_r \exp\{i[((E)^{1/2} \sin \theta + r\beta)y - (E - ((E)^{1/2} \sin \theta + r\beta)^2)^{1/2}x]\}$$

where $(E)^{1/2} \sin \theta = k_y \beta$. From the previous discussion we have seen that when k_y is not near $m/2$, then

$$\beta k_y = (W)^{1/2} \sin \phi \text{ approximately}$$

and since $\beta k_y = (E + V_0)^{1/2} \sin \theta$, this shows that the index of refraction, $\sin \theta / \sin \phi$, is approximately equal to $(E + V_0)^{1/2}/(E)^{1/2}$, except for values of E and θ such that k_y is near one of the values $m/2$.

A more correct relationship between E, θ, V_0 and ϕ is obtained by means of the equation for $g(k)$, and is

$$(E + V_0) \sin^2 \phi = E \sin^2 \theta + \beta^2 \frac{((E)^{1/2} \sin \theta)}{\beta^{1/2}}$$

where g is only appreciably different from zero when $(E)^{1/2} \sin \theta$ is near one of the values $m\beta/2$.

The intensity of the regularly reflected beam can be found by equating the terms of U_0 and U_i, and $\partial U_0/\partial x_i$ and $\partial U_i/\partial x$ which have the y part of their exponential equal, at $x = 0$. These equations reduce to

$$C + D_0 = B_{y,0} \sum_n B_{x,n}$$

$$C - D_0 = B_{y,0} \sum_n \frac{\alpha(k_x + n)B_{x,n}}{(E)^{1/2} \cos \theta}.$$

[48] Davisson and Germer, Proc. Nat. Acad. **14**, 622 (1928).

These equations are similar to those for the one dimensional case of normal incidence, and therefore the values of E and θ giving a strong regularly reflected beam are determined by the equation

$$(E_n + V_0)\cos^2 \theta_n = \frac{n^2\alpha^2}{4} + \alpha^2 G_n$$

where G_n is small and depends on the surface conditions. To relate this to θ, use is made of the previous equation, and we finally find that the values E_n and θ_n for a maximum regular reflection are given by

$$E_n \cos^2 \theta_n = \frac{n^2\alpha^2}{4} - V_0 + G_n + \beta^2 \frac{((E)^{1/2} \sin \theta)}{\rho}.$$

The Bragg analysis, for an external wave-length equal to $2\pi/(E)^{1/2}$ would give $E_n \cos^2 \theta = n^2\alpha^2/4$ for index of refraction unity, and $E_n \cos^2 \theta_n = (n^2\alpha^2/4) - V_0$ for index of refraction $[(E + V_0)/E]^{1/2}$.

If $1/(E)_n^{1/2}$, which is proportional to the de Broglie wave-length of the electron beam, is plotted against $\cos \theta_n$ for maximum regular reflection, the Bragg analysis would require straight lines all going through the origin, $1/(E)^{1/2} = 0$, $\cos \theta = 0$. The curves required by the exact analysis above differ from the Bragg lines appreciably only when $(E)^{1/2} \sin \theta$ is near $m\beta/2$, i.e., when f is large.

A curve of values of $1/(E)_n^{1/2}$ and $\cos \theta_n$ for maximum regularly reflected beams is given in Figure 9. The x and y parts of the potential function are

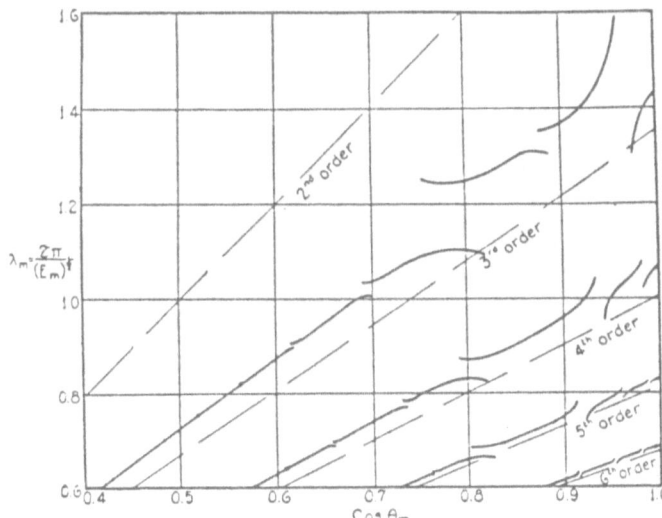

FIGURE 9. Values of electronic wave number and angle of incidence of electron beam for strong regularly reflected beam. The crystal is nickel, the surface the 111 plane. Broken lines indicate positions of analogous x-ray reflection.

taken to be simple sinusoidal variations, for simplicity. The dotted lines, marked 1st order, 2nd order, etc., are those given for index of refraction unity, and the curved lines give the correct relation. Note the "breaks" in the curves whenever $(E_n)^{1/2} \sin \theta_n$ equals $m\beta/2$.

Of course the shape of the curves near the breaks will be different for different forms of the potential function: but the breaks will always occur where $E_n \sin \theta_n = m\beta/2$.

A glance at the experimental curve[49] shows that such breaks are present, the most marked one near $(E_n)^{1/2} \sin \theta_n = 2.96$, and another near $(E_n)^{1/2} \sin \theta_n = 2.83$. The value of $\beta/2$ for this face of nickel and for the azimuth used is 1.46, so that the breaks occur at the proper places for $m = 2$.

Thus the simple examples we have worked out show that the analysis is not only capable of accounting for the general effects of electron scattering from crystal surfaces, but also of explaining the small peculiarities in the results. It appears probable that when the shape of the $1/(E_n)^{1/2}$, $\cos \theta_n$ curves are better known near the breaks it will be possible to make an empirical estimate of the form of the potential function inside a crystal.

---

[49] Davisson and Germer, Proc. Nat. Acad. **14**, 624 (1928).

The Spin of the Neutron

R.F. Bacher[1] AND E.U. Condon

Massachusetts Institute of Technology

From some of the recent information reported about neutrons it seems possible to find out whether the neutron has a spin and if so what its value is. In order to do this it is necessary to make several assumptions about the structures of light nuclei: (1) They are built as far as possible out of alpha-particles. (2) The nuclear spin is determined from the spin of its component spins alone. (3) From the evidence that neutrons may be obtained from Li, it is assumed that at least one of the isotopes contains a neutron in its nucleus. (4) The possibilities considered for the spins of the various particles in the nucleus are proton $0, \frac{1}{2}$; neutron $0, \frac{1}{2}, 1$; electron $0, \frac{1}{2}$; alpha-particle 0. The first assumption is one commonly made and is supported by studies on the artificial disintegration of the light elements. The second assumption seems reasonable for the light elements in view of what is known about their nuclear moments. (If orbital moments quantized to integer values are assumed for the nucleus, deductions about whether the nuclear moment should be integer or half integer are not disturbed but those utilizing the actual value are.)

The spin of the free proton is determined from hydrogen to be $\frac{1}{2}$ and that of the extra-nuclear electron is well known to be also $\frac{1}{2}$ but the possibility of both of these losing their spin in a nucleus should not be overlooked. The various possibilities given above will be divided into two groups for which the proton spin is considered to be 0 and $\frac{1}{2}$ respectively.

As nuclei on which to try these various possibilities the two isotopes of Li, and N^{14} are considered. Curie and Joliot[2] have recently discovered that the highly penetrating radiation emitted from lithium when it is bombarded with alpha-particles from polonium consists of neutrons. They are similar to the particles emitted by Be and B recently discussed by Chadwick,[3] though having less energy. It thus seems probable that one of the two isotopes of lithium contains a neutron. Under the above assumptions we have the following possibilities for the structure of Li^6, Li^7 and N^{14}:

---

[1] National Research Fellow.
[2] Curie and Joliot, Nature **130**, 57 (1932).
[3] J. Chadwick, Proc. Roy. Soc. **A136**, 692 (1932).

$$\begin{aligned}
\text{Li}^6 \quad (A) &\quad 1\alpha + 2p + 1e \\
(B) &\quad 1\alpha + 1p + 1n \\
\text{Li}^7 \quad (C) &\quad 1\alpha + 3p + 2e \\
(D) &\quad 1\alpha + 2p + 1n + 1e \\
(E) &\quad 1\alpha + 1p + 2n \\
\text{N}^{14} \quad (F) &\quad 3\alpha + 2p + 1e \\
(G) &\quad 3\alpha + 1p + 1n.
\end{aligned}$$

Let us first consider that the proton has zero spin when it is a part of a complex nucleus. Let us denote the spin of the proton, electron, and neutron by s_p, s_e, and s_n, respectively. The possible resultants may be presented in Table I.

The Li^7 nucleus is known to have a resultant moment of $1\frac{1}{2}$. This possibility appears only once in the table and to obtain it s_e must be $\frac{1}{2}$ and $s_n = 1$. These

TABLE I

Nucleus	Case	Structure			$s_e = 0$			$s_e = \frac{1}{2}$		
		p	n	e	$s_n = 0$	$s_n = \frac{1}{2}$	$s_n = 1$	$s_n = 0$	$s_n = \frac{1}{2}$	$s_n = 1$
Li^6	A	2	0	1	0	0	0	$\frac{1}{2}$	$\frac{1}{2}$	$\frac{1}{2}$
	B	1	1	0	0	$\frac{1}{2}$	1	0	$\frac{1}{2}$	1
Li^7	C	3	0	2	0	0	0	0, 1	0, 1	0, 1
	D	2	1	1	0	$\frac{1}{2}$	1	$\frac{1}{2}$	0, 1	$\frac{1}{2}, 1\frac{1}{2}$
	E	1	2	0	0	0, 1	0, 2	0	0, 1	0, 2

TABLE II

Nucleus	Case	Structure			$s_e = 0$			$s_e = \frac{1}{2}$		
		p	n	e	$s_n = 0$	$s_n = \frac{1}{2}$	$s_n = 1$	$s_n = 0$	$s_n = \frac{1}{2}$	$s_n = 1$
Li^6	A	2	0	1	0, 1	0, 1	0, 1	$\frac{1}{2}, 1\frac{1}{2}$	$\frac{1}{2}, 1\frac{1}{2}$	$\frac{1}{2}, 1\frac{1}{2}$
	B	1	1	0	$\frac{1}{2}$	0, 1	$\frac{1}{2}, 1\frac{1}{2}$	$\frac{1}{2}$	0, 1	$\frac{1}{2}, 1\frac{1}{2}$
Li^7	C	3	0	2	$\frac{1}{2}, 1\frac{1}{2}$	$\frac{1}{2}, 1\frac{1}{2}$	$\frac{1}{2}, 1\frac{1}{2}$	$\frac{1}{2}, 1\frac{1}{2}, 2\frac{1}{2}$	$\frac{1}{2}, 1\frac{1}{2}, 2\frac{1}{2}$	$\frac{1}{2}, 1\frac{1}{2}, 2\frac{1}{2}$
	D	2	1	1	0, 1	$\frac{1}{2}, 1\frac{1}{2}$	0, 1, 2	$\frac{1}{2}, 1\frac{1}{2}$	0, 1, 2	$\frac{1}{2}, 1\frac{1}{2}, 2\frac{1}{2}$
	E	1	2	0	$\frac{1}{2}$	$\frac{1}{2}, 1\frac{1}{2}$	$\frac{1}{2}, 1\frac{1}{2}, 2\frac{1}{2}$	$\frac{1}{2}$	$\frac{1}{2}, 1\frac{1}{2}$	$\frac{1}{2}, 1\frac{1}{2}, 2\frac{1}{2}$
N^{14}	F	2	0	1	0, 1	0, 1	0, 1	$\frac{1}{2}, 1\frac{1}{2}$	$\frac{1}{2}, 1\frac{1}{2}$	$\frac{1}{2}, 1\frac{1}{2}$
	G	1	1	0	$\frac{1}{2}$	0, 1	$\frac{1}{2}, 1\frac{1}{2}$	$\frac{1}{2}$	0, 1	$\frac{1}{2}, 1\frac{1}{2}$

same conditions however for Li^6, give a resultant of either $\frac{1}{2}$ or 1 and it is known that the resultant moment is 0. This contradiction means that the spin of the nuclear proton must not be 0.

Let us now consider that the proton has spin $\frac{1}{2}$ in the nucleus. The possible resultant spins for Li^6, Li^7, and N^{14}, may be presented in Table II.

Using the facts that the resultant nuclear spins of Li^6, Li^7, and N^{14} are 0, $1\frac{1}{2}$ and 1, respectively, one sees that if $s_n = 0$ then s_e must be 0 and hence neither Li^6, Li^7, or N^{14} may contain a neutron. But since we consider the fact that neutrons are observed from lithium to indicate that they are present in the nucleus of either or both Li^6 and Li^7, this is not possible and the neutron must have a spin. If $s_n = 1$ then s_e must be 0 and cases A, C, E, and G satisfy the requirements. If $s_n = \frac{1}{2}$, then s_e may be either 0 (all cases acceptable) or $\frac{1}{2}$ (B, C, E, and G acceptable).

Three possibilities remain:

$$\text{I} \quad s_p = \tfrac{1}{2}, \qquad s_n = \tfrac{1}{2}, \qquad s_e = 0$$

$$\text{II} \quad s_p = \tfrac{1}{2}, \qquad s_n = \tfrac{1}{2}, \qquad s_e = \tfrac{1}{2}$$

$$\text{III} \quad s_p = \tfrac{1}{2}, \qquad s_n = 1, \qquad s_e = 0.$$

If $s_n = \frac{1}{2}$ and $s_e = \frac{1}{2}$ then Li^6 must contain a neutron. If this neutron is the one which is dislodged by bombardment with alpha-particles then either with or without capture of the alpha-particle the products give isotopes which are not known and which in addition being to a class of which an isotope has never been bound to exist, namely with more than twice as many protons as electrons:

$$Li^6 + He^4 = n_1 + B^9$$

$$Li^6 + He^4 = n_1 + Li^5 + He^4.$$

Of course there could be complete disintegration but it is doubtful whether this would furnish sufficient energy to eject the neutron. Also if $s_n = \frac{1}{2}$ and $s_e = \frac{1}{2}$, N^{14} must contain a neutron but there is no evidence for this though there is ample evidence that it disintegrates with the emission of a proton when bombarded with alpha-particles. Finally, if $s_e = \frac{1}{2}$ then it must be assumed that its magnetic moment is diminished to about 0.001th the value which it has for an extra-nuclear electron to account for the small size of hyperfine structure separations. This all seems improbable.

Consider the possibility $s_n = 1$ and $s_e = 0$. There are over fifteen isotopes of various elements which have odd mass numbers and odd nuclear moments which are quite certain. If $s_n = 1$, then neutrons can exist in the nuclei of any of these only in pairs, which seems to be a peculiar limitation.

The possibility $s_n = \frac{1}{2}$, $s_e = 0$, meets with none of these objections as far as is known. In addition it is more satisfying than either of the other two possibilities if one thinks of the neutron as composed of a proton and an electron in some way, since II, which gives both electron and proton a spin of $\frac{1}{2}$ in the nucleus, would require one or the other to lose its spin in the neutron and III, which gives only the proton a spin, would require both to have spin in the neutron.

One may conclude that, under the assumptions mentioned at first, the neutron must have a spin and that the proton must have a spin of $\frac{1}{2}$ in the nucleus. Of the three possibilities which present themselves the case with $s_p = \frac{1}{2}$, $s_n = \frac{1}{2}$, and $s_e = 0$ seems most probable.

Notes on the Stark Effect

E.U. Condon

Princeton University

The theory of the Stark effect in atomic spectra is discussed in general terms and it is shown that large effects arise when two terms which are related to each other in such a way as to satisfy the optical combining rules come close together in the spectrum. Fairly good values of the Stark displacements in complicated atoms may be obtained simply by using the hydrogenic values of the matrix components involved in the theoretical formulas. The ideas are illustrated by discussion of the existing data on the spectra of nickel, lithium, carbon spark, and argon.

There exists in the literature of spectroscopy a large amount of experimental data referring to the Stark effect in atomic spectra which has not been hitherto brought into any kind of relation to the modern theory of atomic structure. Of course the effect, and its theoretical interpretation, for atomic hydrogen is very well known and has played an important rôle in development of the theory. Likewise the work of Foster[1] and of Dewey[2] on the effect in helium has brought the theory into satisfactory relation with experiment there. Some work has also been done on the small quadratic effect on the resonance lines of sodium and potassium, but in addition to these contributions a glance for example at the Stark effect chapter in the *Handbuch der Experimental Physik* (Volume 21) will show a considerable amount of uncorrelated experimental data. In this paper the aim is to show how the theory may be used to throw light on some of this material and to point the way for further work in this field.

§1. Theory of the Stark Effect

If H_0 is the Hamiltonian of the unperturbed atom and if $\mathbf{P}$ is the electric moment of the atom and $\mathbf{E}$ the applied electric field, then the Hamiltonian for the atom in the field is

---

[1] Foster, Proc. Roy. Soc. **A114**, 47 (1927) and **A117**, 137 (1927).
[2] Dewey, Phys. Rev. **28**, 1108 (1926) and **30**, 770 (1927).

$$H = H_0 - \mathbf{E} \cdot \mathbf{P}. \tag{1}$$

The alteration of energy levels and proper states is therefore governed, in accordance with usual theory, by the matrix components of $\mathbf{P}$. Moreover the only component of $\mathbf{P}$ that is effective is that along $\mathbf{E}$ so we may choose axes so this is the z-component. Now P_z is the same quantity which governs that part of the dipole radiation of atoms which is polarized with its electric vector parallel to the z-axis. Therefore the selection rules and other calculations of its matrix components which have been made for the theory of line intensities are applicable here. Writing $(A|P|B)$ for the matrix component of P_z which connects two unperturbed energy states A and B we know that $(A|P|B)$ vanishes unless:

(a) A and B are of opposite parity (Laporte rule), (b) $J_B = J_A$ or $J_A \pm 1$ where J_B and J_A are the resultant angular momentum quantum numbers of B and A respectively,

(c) $M_{JA} = M_{JB}$ where M_{JA} and M_{JB} are the z-components of resultant angular momentum of the atom in the two states measured with $h/2\pi$ as unit.

These three properties are rigorous. In addition, insofar as it is accurate to assign electronic configuration labels to the terms we may say:

(d) Configuration A may differ from configuration B at most in regard to one electronic n, l value.

Likewise in case of Russell-Saunders coupling in which resultant spin angular momentum S and resultant orbital angular momentum L are quite accurately diagonal when H_0 is diagonal we may say that $(A|P|B)$ vanishes unless

(e) $S_A = S_B$.

(f) $L_B = L_A, L_A \pm 1$.

In addition for Russell-Saunders coupling the actual dependence on L, S, J and M_J is the same as that which in the intensity theory leads to the well-known Kronig-Russell-Hönl formulas.

So far as orders of magnitudes are concerned we may expect that the nonvanishing components will be of the same order as if calculated with hydrogen-like eigenfunctions for the individual electrons of the atom. For this reason we note that Schrodinger's calculations for hydrogen give

$$(n, l-1, m|EP|n, l, m) = \frac{3}{2} e E n a_0 \left[\frac{(n^2 - l^2)(l^2 - m^2)}{4l^2 - 1} \right]^{1/2} \tag{2}$$

for the matrix component connecting two states of equal n value and of l value differing by unity. Here a_0 is the first Bohr radius. Expressing energy in cm^{-1} and taking 100 kv/cm as the unit of E the coefficient of the irrational expression in (2) becomes

$$6.43n \text{ cm}^{-1} \text{ per 100 kv/cm.}$$

The maximum value assumed by the radical for fixed n and varying l and m is about $n/3^{\frac{1}{2}}$ so far n not greater than 10 the matrix component is under 200 cm^{-1} in value in all cases for fields up to 100 kv/cm. This field is about the largest which has been used in ordinary Stark effect investigations.

Suppose now we put such a perturbing term into the Hamiltonian of an atom and consider what may be said in general about the result. The perturbed states will become linear combinations of such of the unperturbed states as are joined by nonvanishing matrix components of the perturbation. Therefore quantities which were quantum numbers in the unperturbed problem lose their significance after the perturbation. Thus it is no longer possible to speak of a state as being definitely odd or even, also the J value loses its precision, but M_J retains its status as an accurate quantum number.

The extent to which this tendency is realized the perturbation theory shows to be related to the nearness of the unperturbed terms. If W_A and W_B are the unperturbed energies of states A and B and $(A|EP|B)$ is the perturbation component connecting them then the perturbed states will be appreciable linear combinations of A and B if $(A|EP|B)$ is comparable with or large compared to $(W_A - W_B)$. If $(A|EP|B) \ll (W_A - W_B)$ then the alteration in the perturbed states is relatively small as is also the change in energy. In this case so far as the change in energy is concerned due simply to the matrix component connecting A and B it is such as to displace the upper state upward and the lower state downward each by the same amount, namely

$$|(A|EP|B)|^2/|W_A - W_B|. \tag{3}$$

The states A and B thus seem to "repel" each other.

As to the amount of this "repulsion" of two terms it is generally rather small. For weaker fields a common value of $(A|EP|B)$ would be 15 cm^{-1} whereas the terms (A and B) may be separated by 1000 cm^{-1} so the perturbation of each would amount to 0.22 cm^{-1}. The small quadratic Stark effect of any state A is the sum of the actions of this type of all other states B of the atom which have matrix components connecting it with A. Most of the theoretical work on the Stark effect of alkalis, like that of Unsöld[3] and of Kirkwood,[4] is concerned with attempts to sum up, for particular states A, like the normal state, or the first resonance state, the whole effect of all other states which perturb it.

But such effects are never comparable with the relatively large effect in atomic hydrogen. Nevertheless such large Stark effects are observed in other atomic spectra. Large Stark shifts may occur in other atoms if two terms connected by a matrix component $(A|EP|B)$ are closer together in the unperturbed energy scheme than the energy value $(A|EP|B)$. In particular if A and B have zero interval in the unperturbed scheme then they will give rise to two levels in the perturbed scheme, one of which is moved up by $(A|EP|B)$ and the other down by the same amount. This makes it clear why large Stark effect displacements are relatively rare in spectra other than hydrogen. The matrix components $(A|EP|B)$ are rather severely restricted by the selection rules. Even when nonvanishing they are quite small so that it is something of an

[3] Unsöld, Ann. d. Physik **82**, 355 (1927).
[4] Kirkwood, Phys. Zeits. **33**, 521 (1932).

exceptional case when two states A and B which are connected by the perturbation matrix are close together relative to the magnitude of $(A|EP|B)$.

The condition for large Stark effect just set out is more general than the one given by Bohr[5] before the new quantum mechanics which is usually used by experimental workers in qualitative discussions of their data. Bohr pointed out that we may expect large effects for a spectral term that is nearly hydrogen-like, i.e., one which if equated to Rh/n^{*2} leads to a nearly integral value of n^*. This was regarded as indicating that the electronic orbit was effectively in a Coulomb field and hence the hydrogen-like behavior was to be expected. But it did not give a means of estimating the magnitude of the effect. The account of the theory we have given shows clearly the justification of Bohr's rule. If a particular term is almost hydrogen-like then we may be fairly sure that terms arising from an electron configuration in which one of the electronic l values is increased by unity will also be hydrogen-like. If both are nearly hydrogen-like terms they will be near each other and so produce a large effect. The essential thing is nearness of W_A and W_B for nonvanishing $(A|EP|B)$ and Bohr's rule gives us a convenient way of noticing what is in point of fact a large class of the cases in which large Stark effects are observed. The rule is sufficient though not necessary so we may expect to find cases in which the effect is large even though the perturbed states are not hydrogen-like in their energy values.

The rule was enunciated before the development of Pauli's principle with its sound basis for getting true n values. In those days if n^* was nearly an integer it was thought that the nearest integer was the true value of n. Now we know that $n^* - n$ may become as great as several units so that a term may be apparently hydrogen-like through having $(n^* - n)$ be almost exactly equal to an integer instead of zero. This is in fact the case in silver, where there is a good-sized Stark effect that comes about from the nearness of the $6d^2D$ to the $4f^2F$, both being very close to the hydrogenic value for $n = 4$.

In what follows attention will be paid to the application of the perturbation theory to the discussion of the existing data for the spectra of several elements. The general standpoint will be that the effect has to be calculated by finding the allowed values of $H_0 - \mathbf{E} \cdot \mathbf{P}$ which can be done in view of the smallness of the matrix components of $\mathbf{E} \cdot \mathbf{P}$ by solving the finite secular equation which represents the interaction of just those interacting terms which are within a few hundred wave numbers of each other on the energy scale.

§2. Nickel

To illustrate how the principles of the preceding section may be used to throw some light on the experimental data, even in a case that is too complicated for detailed discussion at present, let us consider the Stark effect in nickel. The

---

[5] Bohr, Proc. Phys. Soc. London **35**, 275 (1923) and Ann. d. Physik **71**, 228 (1923).

only observations apparently are those of Takamine,[6] which were made a decade before the analysis of the spectrum by Russell.[7] Of the fifty lines for which Takamine records Stark displacements, twenty-nine may be found in Russell's list of classified lines. These are all transitions from levels of the high even configurations d^8s5s, d^94d, d^96s, d^95d and d^8s4d to the intermediate odd configurations d^94p and d^8s4p. The final states have term values between 25,000 and 33,000 cm^{-1} above the normal state. As there are no even configurations in this region we conclude that the Stark displacement of these terms is very small.

The initial states involved in the identified lines fall into two groups, one between 49,000 and 51,000 cm^{-1}, and the other between 54,000 and 57,000 cm^{-1}. Taking the ionization limit to be 61,579 cm^{-1} we find that the hydrogenic terms are 49,400 cm^{-1} for $n = 3$ and 54,730 cm^{-1} for $n = 4$. For the lower group the configurations are $(3d)^94d$ and $(3d)^44s5s$, so these are not $n = 3$ states for valence electrons. But these lower even configurations occupy the same region in the energy diagram as does the odd d^95p which is known from Russell's analysis. So it is more in accord with our general principles to regard the Stark effect as an interaction of d^94d and d^8s5s with d^95p which is induced by the applied field and to regard the nearness to a hydrogenic value as accidental. Similarly the upper group of even terms is in the correct position to interact with d^96p but as this configuration has not been identified in the spectrum as yet we must confine our attention to the lower group of initial states.

This further restriction reduces the number of identified lines for which Takamine gives Stark displacements to twelve, listed in Table 1. The first column gives Takamine's wave-length, the second that of the identified line in Russell's paper which I take to be the same line (there is a systematic difference of about 0.2A). The third column gives the identification in terms of Russell's multiplet analysis, the fourth and fifth are the shifts in wave numbers of the parallel and perpendicular polarized components as observed by Takamine for the values of the field strength given in the sixth column of the table. The line 5142.77 is given two alternative identifications by Russell both of which are listed in the table.

It will be noticed that e^3F_4 occurs as initial state for the first, second, fifth and eleventh lines in the table. The nearness of the equality of their Stark displacements, when these are reduced to the same field strength assuming a variation with E^2, is an indication of the accuracy to which the whole effect may be attributed to the upper state. In the twelve lines there are eight different initial states represented. All the terms show a displacement upward in energy except e^3P_1 and e^3F_2. According to the ideas of §1 we therefore expect that the nearest odd combining term to each of these terms will be below the perturbed term, except for these two for which the nearest combining term is

[6] Takamine, Astrophys. J. **50**, 1 (1919).
[7] Russell, Phys. Rev. **34**, 821 (1929).

TABLE I. Stark effect in nickel I

(Takamine)	(Russell)	Identification	$\Delta\nu$ (cm^{-1}) $\parallel$	$\perp$	Field (kv/cm)
4410.66	4410.50	$z^5D_3{}^0 - e^3F_4$	+6.1	+4.1	39
4937.45	4937.33	$z^5F_4{}^0 - e^3F_4$	+5.0	+5.5	38.5
5018.48	5018.30	$z^3D_1{}^0 - e^3F_2$	−2.8	0.0	38.5
5082.55	5082.38	$z^3P_1{}^0 - e^3P_1$	−1.7	−0.97	38.5
5084.20	5084.07	$z^3D_3{}^0 - e^3F_4$	+5.0	+5.4	38.5
5142.91	5142.77	$\begin{cases} z^3D_2{}^0 - f^3D_2 \\ z^5F_3{}^0 - f^3D_3 \end{cases}$	+0.34	+0.26	21.8
5146.61	5146.48	$z^3D_2{}^0 - e^5F_4$	+0.44	+0.26	21.8
5155.90	5155.76	$z^1D_2{}^0 - e^1F_3$	+3.0	+3.0	21.8
5176.72	5176.56	$z^1D_2{}^0 - f^1D_2$	+0.56	+0.34	21.8
5184.78	5184.59	$z^3D_2{}^0 - e^3P_1$	−1.1	−1.1	21.8
5462.69	5462.48	$z^1F_3{}^0 - e^3F_4$	+2.9	+1.9	21.8
5588.09	5587.85	$a^3P_2 - y^3D_3{}^0$	+0.48	+0.35	21.8

TABLE II

Perturbed term	Configuration	Nearest perturbing term	Interval (cm^{-1})
e^3F_4	d^94d	$3D_3{}^0d^95p$	−5.2
e^3F_2	d^94d	$3D_1{}^0d^95p$	+16.7
e^1F_3	d^94d	$^3F_4{}^0d^85p$	−42.5
e^3P_1	d^94d	$^3D_2{}^0d^95p$	+12.0
e^5F_4	d^8s5s	$6^0J = 3$	−53.3
f^3D_2	d^94d	$^3D_3{}^0d^95p$	−0.5
f^3D_3	d^94d	$\begin{cases} ^3D_2{}^0d^95p \\ ^3D_3{}^0d^95p \end{cases}$	−86.5 +56.1
f^1D_2	d^94d	$^1D_2{}^0d^95p$	−65.0
$y^3D_3{}^0$	d^8s4p	$^3D_3d^95s$	−15.11

above. This is in fact the case as Table II shows. In Table II are listed in the successive columns: the name of the term, its configuration, the name and configuration of the nearest term which could perturb it, and the interval between the perturbed term and this nearest term in cm^{-1} counted negative if the perturbing term is below the perturbed term and positive if it is above. Of course the designations perturbed term and perturbing term are simply relative to the particular lines under discussion; actually the perturbation is mutual and (if it existed) Stark effect data on lines involving the terms called "perturbing' here should show that they are perturbed by amounts equal and opposite to their perturbing action on the terms here called "perturbed."

The table shows that we should expect f^3D_2 to show a strong upward perturbation whereas f^3D_3 would probably show a very small perturbation since the nearest terms are relatively far from it and there are two, one above and one below of roughly the same distance whose effects tend to cancel. Since the Stark effect for 5142.77 is actually quite small this tells us unambiguously that this line is to be identified with the second of the two alternatives offered by Russell's analysis. It is believed that this is the first time that a doubtful point in an analysis of a complex spectrum has been settled by reference to the expected Stark effect of the lines.

In the absence of more complete experimental data it is not thought worthwhile to attempt more precise discussion of the relations involved in the theory. It is felt, however, that even these rough agreements suffice to show that the Takamine data are in accord with the general outlines of the theory of the Stark effect.

§3. Lithium

The experimental data for lithium are summarized in the *Handbuch der Experimental Physik* (Volume 21) article by Stark, page 479. All of it refers to combinations in which 2s or 2p are the final states. Either of these should have extremely small effects owing to their great separation from the nearest terms of opposite parity. Therefore we expect the line shifts to be essentially those of the corresponding upper states. Evidence for this is found in comparing the recorded shifts of 2s–5p and 2p–5p, assuming these to be quadratic effects. The former is -5.5 cm^{-1} at 0.26 unit (1 unit $= 10^5$ volt/cm) while the latter is -45.5 $^{-1}$ at 0.8 unit. Writing $\Delta v = kE^2$ corresponding values of k are 81. and 72. respectively.

In this spectrum the doublet separations are negligibly small. Also the excited states are fairly hydrogen-like. Let us first consider the effect of the field on the states with $n = 4$. The unperturbed energy values relative to the 4d term are:

4s	-1611.7 cm^{-1}	4d	0.0
4p	-154.7	4f	$+7.4$

The matrix components or electric interaction we may expect to be close to the values given by (2) for hydrogen. In any case the relative magnitudes will be given quite accurately by (2) and this is all that matters for a description of the pattern. We may expect a fairly strong interaction between 4d and 4f, a smaller effect on 4p and a quite small effect on 4s. The secular equation for $m = 3$ is linear and tells us that this substate of the 4f term is not affected by the field. Owing to the selection rule on m this does not show up in combinations with 2p. For $m = 2$ it is a quadratic equation connecting these substates or 4d and 4f. Using the matrix components for (2) and the empirical 4d–4f separation we find the roots to be

$$\lambda = 3.7 \pm [3.7^2 + 25.8^2 E^2]^{1/2}$$

when measured from the 4d level.

Likewise for $m = 1$ the secular equation is a cubic connecting corresponding substates of the $4p$, $4d$ and $4f$ terms:

$$\begin{vmatrix} -154.7 - \lambda & -39.8E & 0 \\ -39.8E & -\lambda & -32.5E \\ 0 & -32.5E & 7.4 - \lambda \end{vmatrix} = 0.$$

For $m = 0$ the secular equation is a quartic but here the $4s$ term is so far away compared to the size of the perturbation that its effect may be neglected. The matrix component connecting $4s$ and $4p$ has the value 56.9 for unit field strength so the perturbation due to interaction of these terms is $(56.9)^2/1457.0 = 2.2$ cm^{-1} which is negligible compared to the main effect. Even the $4p$–$4d$ interaction is quite small so that all of the secular equations may with good accuracy be taken simply as quadratics connecting the $4d$ and $4f$ levels. If we do this we obtain as the levels

$$\lambda_2 = 3.7 \pm [(3.7)^2 + 25.8^2 E^2]^{1/2}$$

$$\lambda_1 = 3.7 \pm [(3.7)^2 + 32.5^2 E^2]^{1/2}$$

$$\lambda_0 = 3.7 \pm [(3.7)^2 + 34.5^2 E^2]^{1/2}.$$

In Fig. 1 these levels are plotted against E. The experimental points are from the date of Snyder[8] on $\lambda 4602$ which is the combination with $2p$. The agreement is quite good. This calculation makes no reference to the doublet character of the spectrum and so is not in accord with Snyder's suggestion concerning the line's behavior.[9]

Similar remarks may be made concerning the group of levels for $n = 5$. In combination with $2p$ only the substates with $m = 0$, 1 and 2 are effective. The energies relative to $5d$ are:

---

[8] Snyder, Phys. Rev. **33**, 354 (1929).
[9] Reference 8, bottom of page 359.

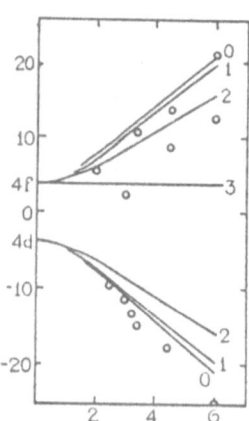

FIGURE 1. Stark displacements due to interaction of $4d$ and $4f$ terms in lithium. Ordinates, displacement in cm^{-1}; abscissas, field strength in 10^4 volt/cm. The points are observed displacements of the lines $4d - 2p$ and $4f - 2p$.

$5s$	-798.2	$5f$	$+7.8$
$5p$	-84.0	$5g$	$7.8 < ? < 10$
$5d$	0.0		

where probable limits for the unknown $5g$ are merely suggested. In the field the perturbed eigenstates become linear combinations of the almost coincident $5d$, $5f$, $5g$ terms. The $5p$ is not so far away but that its displacement is appreciable. With the second-order perturbation formula this term must be pushed up by $5s$ and down by $5d$ giving as its displacement in unit field -57 cm^{-1} which is in fair agreement with the values -81 and -72 obtained experimentally. The displacement of the line $2p$–$5s$ corresponds to a downward displacement in unit field or the $5s$ term of 28 cm^{-1}. The theoretical value is 12. cm^{-1}.

The group $2p - (5d + 5f + 5g)$ may be discussed easily by omitting the $5p$ interaction and neglecting the small and empirically uncertain intervals between these initial state terms. Doing this the energy levels are given by

$$\lambda_2 = \pm 64.2E, \qquad \lambda_1 = \pm 77.5E, \qquad \lambda_0 = \pm 81.4E.$$

In Fig. 2 the data of Snyder are plotted together with the straight lines given by the theory.

Similarly the line $2p$–$6s$ shows a shift corresponding to a displacement of $6s$ of -32 cm^{-1} in unit field whereas the theoretical push from $6p$ is -45 cm^{-1}. The displacement of the $6p$ term in unit field as inferred from the $2p$–$6p$ line is -163 cm^{-1}. The theoretical value is the difference between the upward push from $6s$ and the downward push of $6d$ which comes out to be -200 cm^{-1}. Since the experimental data are rough and the theoretical values depend on the square of matrix components which are taken from hydrogen it is felt that the agreement here presented is good enough to indicate that the ideas of §1 are adequate for a discussion of the main effect.

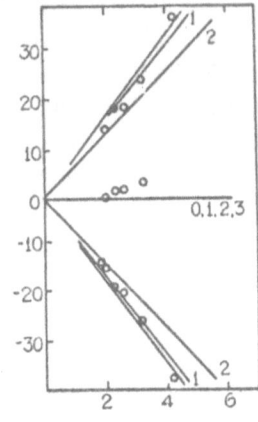

FIGURE 2. Stark displacements due to interaction of $5d$, $5f$ and $5g$ terms in lithium. Ordinates, displacement in cm^{-1}; abscissas, field strength in 10^4 volt/cm. The points are observed displacements of the combinations with the $2p$ term.

§4. Carbon Spark

The data are some observations made by Ishida and Fukushima[10] and are believed to be the only data on a nonhydrogenic ion spectrum. The displacements, as is to be expected, correspond to smaller values of the matrix components than in arc spectra. The matrix components are in fact roughly half the hydrogenic values which is the factor of reduction to be expected for orbits in a Coulomb field with $Z = 2$.

Their Fig. 2a gives the results of observation on $\lambda 2747.31$ and $\lambda 2746.50$ which is the transition, $s^2 3p\,^2P - s^2 4d\,^2D$, in CII. The shift is to longer wavelengths and corresponds to $k = -0.004$ in the equation $\Delta v = kE^2$ where Δv is in cm^{-1} and E is in 10^4 volt/cm. Likewise their Fig. 2b gives the data on $\lambda 4267.27$ and $\lambda 4267.02$ which is $s^2 3d - s^2 4f$. Here the k is about $+0.002$ in the same units. Attributing these shifts to the upper states we see that they are due to the interaction of the $s^2 4d$ term and the $s^2 4f$ term. The $4d$ term is 855 cm^{-1} below the $4f$ term and so the $4d$ term is pushed down in energy and the $4f$ term is pushed up, which is in accord with the facts.

An interesting feature of their Fig. 2b is the fact that there is an unshifted component in the $\perp$ polarization which is absent in the $\parallel$ polarization. This is due to the fact that the $m = 3$ substates in ^2F are unperturbed since there is no $m = 3$ substate in ^2D to perturb them. In combining with ^2D to form a radiative transition the $m = 3$ substate of ^2F must jump into an $m = 2$ substate of ^2D which accounts for the presence of the unshifted component in the $\perp$ but not in the $\parallel$ polarization.

The hydrogenic matrix component for $m = 0$ connecting the $4d$ and $4f$ states for a field of 10^4 volt/cm is 3.44 cm^{-1} from Eq. (2). The value of the matrix component corresponding to empirical k of 0.002 and Δv of 855 is 1.37 cm^{-1} or roughly half of the hydrogenic value. This is largely due to the fact that we are dealing with an ion rather than a neutral atom, also to the fact that the pictures are blends of components for $m = 1$ and $m = 2$ for which the matric components are smaller.

Their data on $\lambda 2992.63$ furnish an excellent example of the interaction of terms as sketched in §1. This is the transition $s^2 3d - s^2 5f$. The experimental value of k is -0.018. At first sight this appears to be an exception to our general rules as the field makes it go down in energy and yet the $5d$ term, below it, should push it up.

This is due to the presence of a $5g$ term. This term is not known experimentally but we will not be far wrong if we place it at the exact hydrogenic value for $n = 5$. The $5f$ term is 144 cm^{-1} lower than this and the $5d$ is 462 cm^{-1} lower than $5f$. Assuming hydrogen-like matrix components (divided by 2 because we have an ion) and these empirical term separations we calculate for k the value

---

[10] Ishida and Fukushima, Sci. Papers Inst. Phys. Chem. Research **14**, 123 (1930).

$$k = -(2.43)^2/144 + (3.45)^2/462 = -0.015.$$

Here the first term is the downward push due to $5g$ and the second the upward push due to $5d$; the final value is in good agreement with Lhe experimental, -0.018.

These are all the data on this spectrum which are available. Theory agrees with experiment as accurately as could be expected and, in particular, explains the difference in the behavior of the $4f$ and $5f$ terms.

§5. Argon

The effect in argon has been investigated recently by Ryde.[11] He measured the displacements of a large number of lines and has given in his Table VII the displacement of the energy levels produced by a field of 100 kv/cm. The configuration $p^5 5d$ is given with some completeness and his Fig. 3 shows how the displacement of a term is greater the smaller its difference from the hydrogenic value for $n = 5$. In argon the states may be separated into two sets, one built on the state $p^5\,^2P_{3/2}$ of the ion, the other on $p^5\,^2P_{1/2}$ the ion. Combinations between the two sets are weak.

Looking at the data from the viewpoint of this paper, we see that the $p^5 5d(^2P_{3/2})$ terms are perturbed downward because of a repulsion from the $p^5 5f(^2P_{3/2})$ terms which lie in a close group at just the hydrogenic value reckoned down from $^2P_{3/2}$ of A II. Hence distance from hydrogenic value is here synonymous with distance from nearest perturbing term. At present we

---

[11] Ryde, Zeits. f. Physik **77**, 516 (1932).

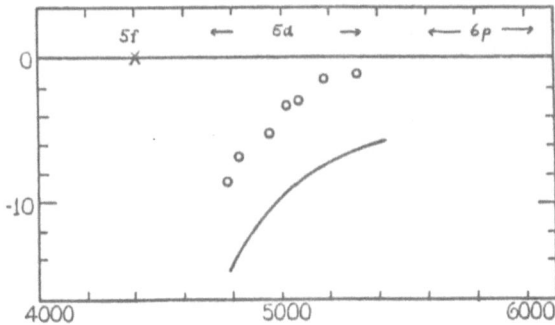

FIGURE 3. Stark displacements of the several $p^5 5d(^2P_{3/2})$ terms in argon, showing relation of the displacement to nearness to the perturbing $p^5 5f$ and $p^5 6p$ groups of terms. Ordinates, displacement in energy in field of 10^5 volt/cm; abscissas, term values of the terms in cm^{-1}. The hyperbola is drawn, using hydrogenic values of the matrix component connecting $5d$ and $5f$.

do not have very good knowledge of the coupling relations in argon so it is hard to give a detailed discussion of the interaction of the $5d$ with the $5f$ terms. Neglecting the separation between the $5f$ terms, the perturbation of any particular $5d$ term due to the $5f$ terms is

$$\Delta v = \sum_{\beta} \frac{|(5d_{\alpha}|E \cdot P|5f_{\beta})|^2}{(W_{5d}{}^{\alpha} - W_{5f}{}^{\beta})},$$

where the sum extends over all the $5f$ terms in the group. If the numerator is about the same for all the d terms the Δv values should lie on a hyperbola whose axis is the location of the $5f$ group, i.e., about 4400 cm^{-1}. Using the hydrogenic value with $m = 0$ of the $5d$–$5f$ interaction from Eq. (2) for the numerator we have 5770. In Fig. 3 is plotted the curve $(5770)/(W_{5d} - W_{5f})$ together with points showing Ryde's experimental values for the corresponding term displacements in a field or 100 kv/cm. It is seen that the observed values agree with this curve in order of magnitude but drop to zero more rapidly than does the hyperbola representing interaction with $5f$. This might be due to a variation of the numerator with the $5d$, but another cause which is certainly acting is the upward push on the $5d$ terms from the $6p$ group which is not so very far below the $5d$ terms.

Relative Multiplet Transition Probabilities from Spectroscopic Stability

E.U. CONDON AND C.W. UFFORD

Palmer Physical Laboratory
Princeton University

A method analogous to the diagonal sum rule for term values is given for calculating the relative transition probabilities of the different multiplets in Russell-Saunders coupling. The method is based on the principle of spectroscopic stability. The relative multiplet transition probabilities are given for the transitions $p^2d - p^3$, $dp - d^3$, $p^3d - p^4$ and $p^4d - p^5$.

This paper will give a method, based on the principle of spectroscopic stability, of calculating the relative transition probabilities of different multiplets in Russell-Saunders coupling. This principle allows one to use a method of finding the transition probabilities in Russell-Saunders coupling in terms of the transition probabilities between the zero order states analogous to the diagonal sum rule used by Slater[1] for the energy values. This method gives the same results as that used by Ufford[2] but avoids using the eigenfunctions in Russell-Saunders coupling.

We have the theorem[3] analogous to the principle of spectroscopic stability, which states that the sum of the squares of the matrix elements connecting a set of states is invariant when these states are subjected to a unitary transformation. Since the transformation from zero order states to Russell-Saunders coupling states is unitary, this theorem will give us equations from which the matrix elements of the electric moment may be found in Russell-Saunders coupling. In cases where more than one term of a kind occurs in a configuration, this method will give only the sums of the squares of the matrix elements and not the elements themselves. The transition probabilities are proportional to the squares of these matrix elements of the electric moment. The transition probability for the entire multiplet may be determined from the transition probability for a transition between a single initial and final state by using the summation rules.

---

[1] J.C. Slater, Phys. Rev. **34**, 1293 (1929).
[2] C.W. Ufford, Phys. Rev. **40**, 974 (1932).
[3] M. Born, W. Heisenberg and P. Jordan, Zeits. f. Physik **35**, 557 (1926); J.H. Van Vleck, Phys. Rev. **29**, 740 (1927).

1. Calculation of the Matrix Elements of the Electric Moment in Russell-Saunders Coupling

We are to calculate the matrix elements of the electric moment between a set of states belonging to a first electron configuration and a set belonging to a second electron configuration differing from the first in regard to just one nl value with a change in l equal to unity. Call the first configuration αnl and the second $\alpha n'l'$. First, one draws up the zero-order scheme for each configuration, classified by M_L and M_s values; and the LSM_LM_s scheme for each configuration also classified by M_L and M_s values. As in the case of energies, only positive values of M_L and M_s need be considered.

The matrix components of er will vanish between states of different M_s, so one may take each value of M_s separately, starting with the largest, and draw up the matrix for the different M_L values in each scheme. The matrix components of er in the zero-order scheme may be obtained from the formulas given by Condon.[4] The only nonvanishing matrix components are those in which but one individual set of quantum numbers changes, so that this must be the individual set referring to $(nl) - (n'l')$, where $l' = l \pm 1$. We want

$$|(nlm_l|e\mathbf{r}|n'l'm_l')|^2 = |(nl|er|n'l')(lm_l|\mathbf{r}|l'm_l')|^2, \qquad (1)$$

where $(lm_l|\mathbf{r}|l'm_l')$ is the numerator of the factor given by Condon and Morse[4] in Eqs. (35) which contains the dependence of $(nlm_l|e\mathbf{r}|n'l'm_l')$ on m_l and m_l'. Therefore $(nl|er|n'l')$ must be defined to include the denominator of the factor given by Condon and Morse, thus

$$(nl|e\mathbf{r}|n'l') = [(2l + 1)(2l' + 1)]^{-1/2} \int_0^\infty erR(nl)R(n'l')\,dr. \qquad (2)$$

Again in the LSM_LM_s scheme, we must separate from each matrix element $(LSM_LM_s|er| \times L'SM_L'M_s)$ the part, $(LM_L|\mathbf{r}|L'M_L')$, depending on M_L, so that the remaining part $(^{2S+1}L|er|^{2S+1}L')$ will be independent of M_L, and hence the same in each rectangle characterized by a different value of M_L. It is the squares of these factors independent of M_L which we are then able to calculate by the invariance of the sum of their squares. The dependence of $(LSM_LM_S|er|L'SM_L'M_S)$ on M_L may be obtained from formulas given by Güttinger and Pauli.[5]

Now in each rectangle characterized by a given value of M_SM_L and $M_{L'}$ we write an equation

[4] E.U. Condon, Phys. Rev. **36**, 1121 (1930). For the matrix elements between single electron states see for example, E.U. Condon and P.M. Morse, *Quantum Mechanics*, p. 100, McGraw-Hill (1929).

[5] P. Güttinger and W. Pauli, Zeits. f. Physik **67**, 743 (1931); p. 761, Eq. (25). The $a_{l'}^l$ of this equation is $(^{2S+1}L|r|^{2S+1}L')$.

$$|(nl|er|n'l')|^2 \sum_{\substack{\text{zero order} \\ \text{states}}} |(lm_l|\mathbf{r}|l'm_l')|^2$$

$$= \sum_{L\cdot S\,\text{state}} |(LSM_L M_S|er|L'SM_L'M_S)|^2$$

$$= \sum_{L\cdot S\,\text{states}} |(^{2S+1}L|er|^{2S+1}L')|^2 |(LM_L|\mathbf{r}|LM_L')|^2. \tag{3}$$

The factor $|(nl|cr|n'l')|^2$ remains the same throughout the entire matrix so that, as we are interested only in relative values, it may be omitted. Also, the squares of the matrix elements contain the factor $\frac{1}{4}$ for the x and y components since $x = \frac{1}{2}(x + iy + x - iy)$ and the matrix element of either $x + iy$ or $x - iy$ is zero for a given transition. Since this factor, $\frac{1}{4}$, will occur in both the zero order and $LSM_L M_S$ schemes, on each side of Eq. (3), it also may be omitted. Then, beginning in the rectangle with the largest M_L, we solve the equation of each rectangle for the $|(^{2S+1}L|er|^{2S+1}L')|^2$'s using the values obtained from the rectangles of higher M_L. Thus we obtain an equation for each rectangle with only one unknown as long as only one multiplet of the same kind occurs in a configuration. An equation for the sum of the squares of the matrix components of the multiplets of the same kind is obtained. In some rectangles it often happens that all the elements are known, so that a check is obtained on the work up to this point.

The relative transition probability for the entire multiplet is then obtained from

$$|(^{2S+1}L|er|^{2S+1}L')|^2$$

by multiplying by the factors given by Güttinger and Pauli[5] in Eq. (24) for the sum over the final states and by $(2S + 1)(2L + 1)$ for the sum over the initial states.

2. Detailed Calculation of Quartet Transitions in $d^2p - d^3$

The method of calculating relative multiplet transition probabilities is shown in Table I. To obtain, for example, the equation in the rectangle labeled $d^2pM_L = 3$, $d^3M_L = 3$, one has for the zero order states:[4]

$$|(|2^+0^+1_p{}^+||z|2^+1^+0^+|)|^2 = |(1_p{}^+|z|1^+)|^2 = (1 + 1 + 1)(1 - 1 + 1) = 3,$$
$$|(|2^+1^+1_p{}^+||z|2^+1^+0^+|)|^2 = |(0_p{}^+|z|0^+)|^2 = (1 + 0 + 1)(1 + 0 + 1) = 4. \tag{4}$$

For the $LSM_L M_S$ states:[5]

$$|(d^2p4, 3/2, 3, 3/2|ez|d^33, 3/2, 3, 3/2)|^2 = (4^2 - 3^2)|(d^2p\,^4G|er|d^3\,^4F)|^2$$
$$= 7|(d^2p\,^4G|er|d^3\,^4F)|^2,$$
$$|(d^2p3, 3/2, 3, 3/2|ez|d^33, 3/2, 3, 3/2)|^2 = 3^2|(d^2p\,^4F|er|d^3\,^4F)|^2 \tag{5}$$
$$= 9|(d^2p\,^4F|er|d^3\,^4F)|^2.$$

TABLE I. Calculation of the matrix elements of the electric moment in Russell-Saunders coupling for the quartets of $d^2p - d^3$ with $M_S = 3/2$

				d^3
	M_L			3
		Zero order states		$|2^+1^+0^+|$
			LSM_LM_S states	^4F
	4	$|1_p{}^+2^+1^+|$	^4G	$2 = 56|(d^2p\,{}^4G|er|d^3\,{}^4F)|^2$ $|(d^2p\,{}^4G|er|d^3\,{}^4F)|^2 = \frac{1}{28}$
d^2p	3	$|1_p{}^+2^+0^+|$ $|0_p{}^+2^+1^+|$	^4G ^4F	$3 + 4$ $= 7|(d^2p\,{}^4G|er|d^3\,{}^4F)|^2$ $+ 9|(d^2p\,{}^4F|er|d^3\,{}^4F)|^2$ $7 = \frac{7}{28}$ $+ 9|(d^2p\,{}^4F|er|d^3\,{}^4F)|^2$ $|(d^2p\,{}^4F|er|d^3\,{}^4F)|^2 = \frac{3}{4}$
	2	$|1_p{}^+2^+ - 1^+|$ $|1_p{}^+1^+0^+|$ $|0_p{}^+2^+0^+|$ $|-1_p{}^+2^+1^+|$	^4G ^4F ^4D ^4D	$2 + 12 + 6$ $= 2|(d^2p\,{}^4G|er|d^3\,{}^4F)|^2$ $+ 6|(d^2p\,{}^4F|er|d^3\,{}^4F)|^2$ $+ 30|(d^2p2\,{}^4D|er|d^3\,{}^4F)|^2$ $20 = \frac{2}{28} + 6(\frac{3}{4})$ $+ 30|(d^2p2\,{}^4D|er|d^3\,{}^4F)|^2$ $|(d^2p2\,{}^4D|er|d^3\,{}^4F)|^2 = \frac{18}{35}$

Equating the sum of the elements of Eq. (4) to the sum of the elements of Eq. (5) gives the required equation. The value of $|(d^2p\,{}^4G|er|d^3\,{}^4F)|^2$ from the previous rectangle $d^{2p}M_L = 4$, $d^3M_L = 3$ is then substituted in this equation and the equation solved for $|(d^2p\,{}^4F|er|d^3\,{}^4F)|^2$. Now the relative multiplet transition probabilities are obtained from these elements by multiplying by the factors to sum over the initial and final states. Thus[5]

$$(d^2p\,{}^4G, d^3\,{}^4F) = 1/28(4 \cdot 9)(4 \cdot 7) = 36,$$

$$(d^2p\,{}^4F, d^3\,{}^4F) = 3/4(4 \cdot 7)(3 \cdot 4) = 252, \qquad (6)$$

$$(d^2p2\,{}^2D, d^3\,{}^4F) = 18/35(4 \cdot 5)(3 \cdot 7) = 216.$$

$(d^2p2^2D, d^3\,{}^4F)$ is the sum of the relative transition probabilities of the two $^4D - {}^4F$ multiplets.

3. Relative Multiplet Transition Probabilities

The relative multiplet transition probabilities are shown in the tables as follows: $p^2d - p^3$ in Table II, $d^2p - d^3$ in Table III, $p^3d - p^4$ in Table IV, and $p^4d - p^5$ in Table V.

TABLE II. Relative multiplet transition probabilities in Russell-Saunders coupling for the transitions $p^2d - p^3$

		p^3		
		^4S	^2D	^2P
p^2d	^4F	y	x	x, y
	^4D	y	x	x
	^4P	24	x	x
	^2G	x, y	y	y
	$2\,^2F$	x, y	42	y
	$3\,^2D$	x, y	15	25
	$2\,^2P$	x	3	9
	^2S	x	y	2

TABLE III. Relative multiplet transition probabilities in Russell-Saunders coupling for the transitions $d^2p - d^3$

		d^3						
		^4F	^4P	^2H	^2G	^2F	$2\,^2D$	^2P
d^2p	^4G	180	y	x	x	x	x, y	x, y
	^4F	1260	y	x, y	x	x	x	x, y
	$2\,^4D$	1080	520	x, y	x, y	x	x	x
	^4P	y	240	x, y	x, y	x, y	x	x
	^4S	y	320	x, y	x, y	x, y	x, y	x
	^2H	x, y	x, y	396	44	y	y	y
	$2\,^2G$	x	x, y	1584	576	180	y	y
	$3\,^2F$	x	x, y	y	1000	420	470	y
	$3\,^2D$	x	x	y	y	660	625	215
	$3\,^2P$	x, y	x	y	y	y	705	255
	^2S	x, y	x	y	y	y	y	70

It must be remembered that a transition such as $d^2p3\,^2D - d^3 2\,^2D$ in Table III represents the sum of the relative transition probabilities of six $d^2p\,^2D - d^3\,^2D$ multiplets. As before,[2] the intersystem transitions forbidden by the selection rule for the spin quantum number, $\Delta S = 0$, are marked x; and those forbidden by the selection rule for the total orbital angular momentum, $\Delta L = 0, \pm 1$, are marked y. Thus the method of spectroscopic stability is seen to aid effectively in the theoretical calculation of relative multiplet transition probabilities in Russell-Saunders coupling.

TABLE IV. Relative multiplet transition probabilities in Russell-Saunders coupling for the transitions $p^3d - p^4$

		p^4		
		^3P	^1D	^1S
	^5D	x	x	x, y
	^3G	y	x, y	x, y
	$2\,^3F$	y	x	x, y
	$3\,^3D$	120	x	x, y
	$2\,^3P$	45	x	x
p^3d	^3S	15	x, y	x
	^1G	x, y	y	y
	$2\,^1F$	x, y	63	y
	$2\,^1D$	x	30	y
	$2\,^1P$	x	7	20
	^1S	x	y	0

TABLE V. Relative multiplet transition probabilities in Russell-Saunders coupling for the transitions $p^4d - p^5$

		p^4d							
		^4F	^4D	^4P	^2G	$2\,^2F$	$3\,^2D$	$2\,^2P$	^2S
p^5	^2P	x, y	x	x	y	y	19	9	2

The Energy Distribution of Neutrons Slowed by Elastic Impacts

E.U. Condon and G. Breit

Palmer Physical Laboratory
Princeton University

The problem of finding the distribution in energy of particles of mass m, initially of the same energy, which have made n impacts with particles of mass M all initially at rest, is solved. It is supposed the impacts are elastic and the distribution in angle igotropic in a coordinate system in which the center of mass is at rest. If x is the ratio of the energy after n impacts to the initial energy then the chance that x lie in dx at x is $(\log 1/x)^{n-1}/(n-1)!$ for $m = M$. For unequal masses the expression is more complicated but easy to calculate. The results have some interest in connection with the slowing of neutrons by elastic impacts with other nuclei, especially with hydrogen nuclei.

In this note we work out the energy distribution of neutrons which, starting with the same initial energy, have made n impacts with other nuclei all initially at rest. We suppose the impacts are elastic and the scattering isotropic in a coordinate system in which the center of mass is at rest. The result is of some interest in connection with current researches on "slow" neutrons. The work grew out of a desire to understand a statement due to Fermi[1] that "It is easily shown that an impact of a neutron against a proton reduces, on the average, the neutron energy by a ractor $1/e$."

Let the nuclei of the medium be all at rest and of mass M while the incident neutron is of mass m and energy, E_0. Then by a simple application of the conservation laws it is found that the neutron energy after an impact is given by $E_1 = E_0(1 - \alpha x)$ where

$$\alpha = 4mM/(m + M)^2 \qquad \text{and} \qquad \cos \varphi = 1 - 2x,$$

φ being the angle of scattering of the neutron in a coordinate system in which the center of mass of m and M is at rest. For elastic spheres on classical mechanics or for short range forces of any type in quantum mechanics the chance that φ lie between φ and $\varphi + d\varphi$ is proportional to $d(\cos \varphi)$ so we see that x may take each value between 0 and 1 with equal probability.

---

[1] Amaldi, D'Agostino, Fermi, Pontecorvo, Rasetti and Segrè, Proc. Roy. Soc. **A149**, 522 (1935). See p. 524.

After n collisions the energy will be E_n where

$$E_n = E_0(1 - \alpha x_1)(1 - \alpha x_2) \ldots (1 - \alpha x_n).$$

Let $x = E_n/E_0$. We have to find the probability that x lie in dx at x being given that each x_i has equal chance of having any value between zero and unity. Hence x may vary between $(1 - \alpha)^n$ and unity. Write

$$(1 - \alpha x_i) = e^{-ui} \qquad \text{and} \qquad u = \sum u_i \qquad \text{so} \qquad x = e^{-u}.$$

Each u_i varies between 0 and $a = \log(1 - \alpha)^{-1}$. The chance of a given set of u_i values is

$$dx_1 \, dx_2 \ldots dx_n = e^{-u} \, du_1 \, du_2 \ldots du_n.$$

The chance that the sum u, lie in du at u is therefore the factor e^{-u} multiplied by the chance that u lie in du at u assuming each u_i to be equally likely to have any value between 0 and a. The evaluation of this turns out to be a very old problem, apparently first considered by Laplace,[2] and the result is

$$f_n(u) \, du = \frac{1}{a^n(n - 1)!} \left\{ u^{n-1} - \binom{n}{1}(u - a)^{n-1} \right.$$

$$\left. + \binom{n}{2}(u - 2a)^{n-1} - \cdots + (-1)^{n-1}\binom{n}{n - 1}[u - (n - 1)a]^{n-1} \right\}$$

with the understanding that the term involving $(x - ka)^{n-1}$ is to be assigned the value zero for $x < ka$. Thus the distribution function has a discontinuous $(n - 1)$st derivative with discontinuities at the places $x = ka$ with $k = 0, 1, 2, \ldots, n$. A modern derivation of the result has been given by Rietz.[3]

Before we looked it up in the library we had worked out a solution of the problem which is enough different from those we have seen to make it worth communicating briefly. The first step is to consider the recursion relation connecting the distribution function for $n + 1$ with that for n. Let $u = \sum_{i=1}^n u_i$ and $v = u + u_{n+1}$. Let $I_k{}^n(u)$ be the expression for $f_n(u)$ in the range $(k - 1)a < u < ka$. The chance of u being in du at u and u_{n+1} in du_{n+1} at u_{n+1} is then $I_k{}^n(u) \, du \, du_{n+1}$. Changing the variables to v and u and summing over the values of u which lead to a fixed value of v we find the recursion relation

$$I_k{}^{n+1}(v) = \int_{v-a}^{(k-1)a} I^n{}_{k-1}(u) \, du + \int_{(k-1)a}^{v} I_k{}^n(u) \, du$$

with the understanding that $I_0{}^n(u) \equiv 0$.

Now let us consider the geometrical situation. We have to find the $(n - 1)$ dimensional measure of the intersection of the hypercube $0 \le u_i \le a$ with the two hyperplanes $\sum u_i = u$ and $\sum u_i = u + du$. Plainly it will be proportional

[2] Laplace, *Théorie Analytique des Probabilités* (1820), pp, 257–263.
[3] Rietz, Proc. Int. Math. Congress, Toronto 2, 795 (1924). The problem is also discussed by H.P. Lawther, Jr., Annals of Math. Statistics 4, 241 (1933) whose Fig. 1 graphs the distribution function for $n = 1, 2, 4, 8, 16, 32$.

to $u^{n-1} du$ for $0 < u < a$ by a dimensional argument. As u becomes a little greater than a the hyperplane passes n corners and so the expression $u^{n-1} du$ has to be corrected by subtracting off $n(u - a)^{n-1} du$ to allow for the part of the hyperplane that is outside the hypercube. When u becomes just greater than $2a$ the hyperplane passes $\binom{n}{2}$ corners so correction by $\binom{n}{2}(u - 2a)^{n-1} du$ is necessary. That the correction has to be added this time is readily seen by inspection in the three-dimensional case where the figure is easily visualized. This argument suggests the general form for $I_k{}^n(u)$ and it is easily verified that it satisfies the recursion relation and is therefore correct when properly normalized so that its integral from 0 to na will give the correct volume, a^n.

For $u > na$, $I_n{}^n(u)$ is a sum of terms in $u^{n-1}, (u - a)^{n-1} \ldots$ which is easily seen to vanish identically. In fact this polynomial may be written as $[1 - e^{-D}]^n u^{n-1}/(n - 1)!$ which must vanish because $[1 - e^{-D}]^n$ expanded in D contains only D^n and higher powers. Here D is the differentiation operator. The ratio of the coefficients of $(u - ka)^{n-1}$ is fixed because the n independent ratios must satisfy $n + 1$ conditions. It is thus not even necessary to verify the recurrence relation for $I_k{}^n(u)$.

The case of most interest experimentally is that corresponding to the slowing of neutrons by protons. Here $m = M$ and $\alpha = 1$ so $a = \infty$ and the complications due to passing corners of the hypercube do not arise. All finite values of u are now less than a so only $I_1{}^n(u)$ plays a role. The distribution function for u is therefore $e^{-u} u^{n-1} du/(n - 1)!$ or for x,

$$F_n(x)\, dx = (\log 1/x)^{n-1}\, dx/(n - 1)!$$

Thus the probability that x have a value between 0 and ξ is given by

$$P_n(\xi) = \int_0^\xi F_n(x)\, dx.$$

The integral involved here is an incomplete gamma function for which very complete tables exist.[4] We have prepared Fig. 1 in which $P_n(\xi)$ is plotted against a logarithmic scale of ξ for several values of n. There is not much point

---

[4] Karl Pearson, ed. *Tables of the Incomplete Gamma Function* (London, H. M. Stationery Office, 1922).

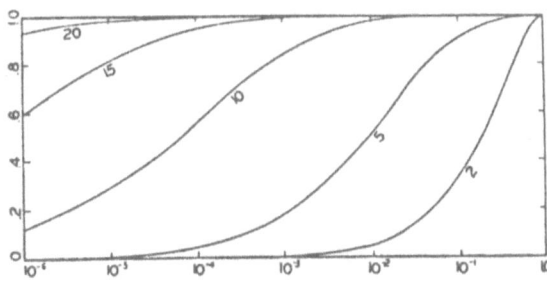

FIGURE 1. Showing the probability as ordinate that a neutron have a fraction of its initial energy less than or equal to the abscissa after a number of impacts with prot ns that is marked on each curve.

in extending the curves below $\xi = 10^{-6}$ since, for ordinary initial energies of neutrons, a value of ξ much less than this brings the neutron energies down to thermal values where the assumption that the protons are initially at rest is no longer valid. The curves give a good idea of the way in which the energy is rapidly reduced by a moderate number of collisions.

Although one can calculate the average of any power of x from the distribution function, it is simpler to do it from the expression for x as a product of the n factors $(1 - \alpha x_i)$. Then we have

$$\overline{x^s} = \left[\int_0^1 (1 + \alpha x_i)^s \, dx_i \right]^n = \left[\frac{1 - (1 - \alpha)^{s+1}}{\alpha(s + 1)} \right]^n.$$

In particular for the case of equal masses, $\alpha = 1$, and the ordinary arithmetic mean, $s = 1$ we have $\bar{x} = 2^{-n}$. The statement of Fermi referred to above is verified for the logarithmic mean, $\exp(\overline{\log x})$.

The function $f_n(u)$ can be represented around its maximum approximately by $e^{-6(u-na/2)^2/na^2}$ as may be found by considering $\overline{x^2} - \bar{x}^2$ according to the above formulas or by empirical fitting of numerical computations. The latter indicate that 5.7 gives a somewhat better approximation than 6.

Calculation with the distribution function for unequal masses is quite simple. In Fig. 2 we give, as an illustrative example, the integrated energy distribution of neutrons which have made 10 and 20 collisions with carbon nuclei ($M = 12$). Comparison of Figs. 1 and 2 affords a striking indication of how little the carbon nuclei in paraffin contribute to the slowing down of the neutrons.

The above discussion is not intended to give the distribution of neutrons slowed down by passing through a given thickness of paraffin—some of the emerging neutrons obviously perform more collisions than others. The distribution function considered here is nevertheless useful for approximate estimates when most neutrons can be considered to have performed the same number of collisions.

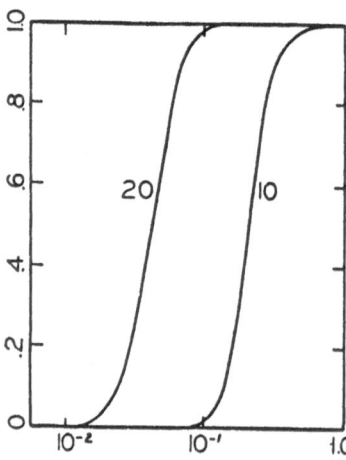

FIGURE 2. Same as Fig. 1 except that it is for collisions with carbon nuclei instead of with protons.

Note on Electron-Neutron Interaction

EDWARD U.CONDON

Palmer Physical Laboratory
Princeton University

Assuming a possible short-range interaction between electrons and neutrons of the form $V = K\delta(\mathbf{r}_N - \mathbf{r}_e)$ where δ is the Dirac delta function and $\mathbf{r}_N$ and $\mathbf{r}_e$ are the vector positions of electron and neutron, respectively, it is shown that probably $|K| < 30mc^2(e^2/mc^2)^3$ from consideration of the effect of the interaction on slow neutron scattering cross sections. If K is positive and a little less than this upper limit this interaction could be responsible for the observed isotope displacement or spectral energy levels.

In most theoretical speculations of nuclear physics it is usual to omit consideration of terms in the Hamiltonian corresponding to short range forces between electrons and neutrons. In this note the situation is examined more closely.

Dee[1] showed that fast neutrons do not produce more than one ion pair in three meters of air path at normal conditions by interaction with the electrons. As there are 15 electrons in an air molecule and 2.7×10^{19} molecules/cm^3 this result says only that the cross section between electrons and neutrons (of about 5 MEV energy) is less than 8.3×10^{-24}. But this is not a very small upper limit being in fact larger than the cross section for scattering of fast neutrons by protons.

It is generally felt that the actual cross section must be very much smaller than this. If we suppose the interaction between the neutron and electron to be represented by a potential well (*or* wall) of depth (*or* height) V and radius of action a then the wave function of an electron's motion relative to the neutron is given by

$$\Delta\psi + (2m/\hbar^2)[W - V(r)]\psi = 0$$

and the approximate solution due to Born for the scattering is

$$\psi = e^{ikz} - \frac{1}{4\pi}\frac{2m}{\hbar^2}\int \exp(i\Delta\mathbf{k}\cdot\mathbf{r})V(r)\,dv\cdot e^{ikr}/r$$

---

[1] Dee, Proc. Roy. Soc. **A136**, 727 (1932).

where $\hbar\Delta\mathbf{k}$ is the vectorial change in momentum of the electron in the scattering. The second term becomes $(\pm 2ma^3 V/3\hbar^2)e^{ikr}/r$ so in the usual way this gives for the total cross section

$$\sigma = 4\pi(2ma^3 V/3\hbar^2)^2$$

on the assumption that the electron wave-length associatecl with the change of momentum is large compared to the range of the interaction forces. A "reasonable" value for V is mc^2 and for a is (e^2/mc^2) so on this basis the expected "reasonable" cross section for scattering of electrons by neutrons is $\sigma = (16/9)\alpha^4\pi a^2$, where $\alpha = e^2/\hbar c$ or numerically, $\sigma = 1.24 \times 10^{-32}$ cm^2. To get a cross section as large as Dee's upper limit one would need to have V equal to $2.6 \times 10^4 mc^2$ if one keeps to the same range. This is "unreasonable" nor would it be "reasonable" to get the larger cross section by retaining mc^2 for V and increasing the range of action of the forces. Of course this calculation is really for a free electron rather than an electron bound in an atom but since for a 5 MEV neutron the energy of relative motion of electron and neutron (2500 volts) is large compared to the binding energy it is probable that the order of the result is not much diflerent from the result of a calculation which does not neglect the binding.

Such considerations have been more or less generally known for some time. They make it desirable to have another approach which gives a smaller upper limit to the possible interaction. This is provided by consideration of the possible effect of electron-neutron interaction on the scattering of slow neutrons.

When the neutron energy is small, of the order of thermal energy, there is no possibility of excitation of the atom by the neutron impact so we are dealing with an elastic collision. The usual molecular considerations apply here to indicate that the slow moving neutron will move in an effective potential field given by considering the normal electronic level of the "molecule" of atom + neutron. If we assume a force law of negligible range (alongside of atomic dimensions) we may write $V = K\delta(\mathbf{r}_N - \mathbf{r}_e)$ where K is the $4\pi V a^3/3$ of the square potential well, and the other is the Dirac δ function in three dimensions normalized so its volume integral is unity.

Then if $\rho(r_e)$ is the electron density in the atom and the neutron interaction is supposed weak enough not to affect it the potential energy of the system atom + neutron differs by $V(r) = K\rho(r)$ from the energy with the neutron absent. The scattering of slow neutrons by the atom will be governed by the wave equation

$$\Delta\psi + (2M/\hbar^2)[W - K\rho(r)]\psi = 0,$$

in which M is the reduced mass of neutron and atom and W is the energy of the internal or relative motion.

The elastic scattering is given by an expression of the same form as before, with these differences, that now we are dealing with a long range force determined by the distribution in space of the bound electrons and the mass is here M instead of the electron mass m. The wave-length is now comparable with the range of the forces so the integral is a little more complicated. However it is the same function of electron density as the structure factor used in x-ray scattering work. Calling the coefficient of e^{ikr}/r in the scattered wave $f(\theta)$ we have

$$f(\theta) = -(4\pi)^{-1} \int \exp ik(\mathbf{n_0} - \mathbf{n}) \cdot \mathbf{r} K\rho(r)(2M/\hbar^2)\, d\tau$$

$$= (2M/4\hbar^2) \int_0^\infty \frac{\sin \kappa r}{\kappa r} K\rho(r)r^2\, dr,$$

where $\kappa = (4\pi \sin \theta/2)/\lambda$ and λ is the de Broglie wave-length of the relative motion, $h/(2MW)^{1/2}$. By introducing the structure factor $F(\kappa)$ as defined on p. 140 of Compton and Allison, X-Rays and Electrons, this is

$$f(\theta) = -(2M/\hbar^2)K(4\pi)^{-1}ZF(\kappa)$$

so, as usual the differential cross section for scattering into $d\omega$ is

$$|f(\theta)|^2\, d\omega = (K^2M^2Z^2/4\pi^2\hbar^4)F^2(\kappa)\, d\omega.$$

The total cross section is then

$$\sigma = \sigma_0 \frac{2}{\kappa^2} \int_0^K F^2(\kappa)\kappa\, d\kappa$$

with $\kappa = 4\pi/\lambda$ and $\sigma_0 = 4\pi(2MZK/\hbar^2 4\pi)^2$. Since $F^2(\kappa) \to 1$ for $\kappa \to 0$ it follows that σ_0 is the limiting value of the cross section.

The value of σ_0 in this case is larger than for the scattering of fast, free electrons by the factor $Z^2(M/m)^2$. Thus for scattering of slow neutrons by atomic hydrogen one need have K only 1/920 as great to get the same cross section as for the previous work.

We have next to consider the value of the integral over the form factor occurring in the cross section expression. For a first orientation the electron distribution is satisfactorily given by the Fermi-Thomas statistical method and for this distribution the form factor may be found in Compton and Allison, p. 148. From this the integral was calculated as a function of J where J is a complicated expression which reduces to $J = 6.6V^{1/2}Z^{-1/2}$ where V is the value of the relative energy in electron volts. We find for $G(J)$ where $\sigma = \sigma_0 G(J)$

J	$G(J)$	J	$G(J)$	J	$G(J)$	J	$G(J)$
0.0	1.00	0.6	0.18	0.3	0.38	0.9	0.11
.1	.74	.7	.15	.4	.30	1.0	.09
.2	.53	.8	.13	.5	.23	1.5	.05

The average relative energy Ve is equal to $\frac{1}{2}Mv^2$ where v is the average relative velocity of the neutron and the atom, if the atoms are at rest and the neutrons have room temperature thermal energy, $V \sim 1/40$. For hydrogen then $J \sim 1$ so the correction owing to the structure factor amounts to a factor of about 0.1. In addition there is another factor which undoubtedly plays a role here— the zero-point energy of oscillation of the atoms in a molecule.[2] For hydrogen this may amount to $V \sim \frac{1}{4}$ volt or ten times the thermal energy. This tends to cut down the cross section still more. In comparing hydrogen and deuterium the zero-point energy will be less in the latter so this would tend to make the deuterium cross section greater. Also the difference in the reduced masses relative to the neutron introduces a factor of 16/9 in the ratio of the cross sections. Both work however to make the hydrogen cross section less than that for deuterium whereas Dunning[3] finds 35×10^{-24} cm^2 for hydrogen and 4×10^{-24} cm^2 for deuterium.

For heavier atoms the zero-point energy is negligible and owing to the appearance of $Z^{-1/2}$ in J the form factor average $G(J)$ does not change by more than a factor of two. So the electron-neutron interaction would vary smoothly as Z^2. As there is no trace of such a "Moseley law" for the slow neutron cross sections we must conclude that this possible manifestation of the interaction contributes less than 10^{-24} cm^2 to the slow neutron scattering cross section of hydrogen.

If this be so one can set the limit

$$|K| < 30mc^2(e^2/mc^2)^3,$$

which is much lower than the limit

$$|K| < 2.6 \times 10^4 mc^2(e^2/mc^2)^3,$$

which one sets from Dee's study or the upper limit of the ionization produced by fast neutrons.

Another way in which electron-neutron interaction would manifest itself is in the isotopc displacement of spectral terms. If the interaction energy is of very short range type as we have been supposing then the energy difference between a heavier and a lighter isotope due to this interaction will be K times the total electron density at the origin for each neutron added to the nucleus. The particle density at the origin is zero exccept for s electrons, and the only contribution that will be observable in the shift of frequency of a spectrum line will be that from the particular s electron that is involved directly in the optical transition which produces the line.

Accordingly the energy is raised by $K\psi^2(0)$ for each neutron if the electron-neutron interaction is repulsive, where $\psi^2(0)$ is the particle density at the origin

[2] This idea that the zero-point oscillations of the hydrogen in a molecule may affect the slowing of neutrons by matter was suggested by Professor Rabi in a conversation last fall. See also Halban and Preiswerk, Nature **136**, 951 (1935) for experimental indications of such an effect.
[3] Dunning, Pegram, Fink and Mitchell, Phys. Rev. **48**, 265 (1935).

standpoint of the effect of departure from the Coulomb law associated with a change in nuclear radius for the different isotopes The formulas occurring in that discussion also require a knowledge of $\psi^2(0)$ so we may use the values that he calculates to estimate how large K must be to explain the entire effect as due to electron-neutron interaction. Here the levels are observed to be raised in going to heavier isotopes indicating electron-neutron *repulsion* if this interaction is mainly responsible for the isotope shift. The sign of K is not determined by the scattering considerations so this is a new bit of information.

In the $6s$ term of Hg II there is a difference of 0.52 cm^{-1} between Hg^{204} and Hg^{202}, that is, 0.26 cm^{-1} per neutron assuming the effect on the lines is all due to a shift of the s term. Breit calculates $\psi^2(0) = 1.45 \times 10^{-26}$ cm^{-3} for the $6s$ electron. Writing $K = kmc^2(e^2/mc^2)^3$, the whole shift is accounted for if $k = 20$, which is consistent with the upper limit obtained from consideration of the slow neutron cross sections. The same value is obtained for Tl I $7s$ using $\psi^2(0) = 0.17 \times 10^{-26}$ cm^{-3} and an observed $\Delta v = 0.03$ cm^{-1} per neutron. The data for Hg I $7s$ indicate $k = 5.5$ so that there is no quantitative consistency except as to order of magnitude.

In conclusion, the foregoing discussion shows that electron-neutron interaction if of the form $K\delta(\mathbf{r}_N - \mathbf{r}_e)$ must have $K < 30mc^2(e^2/mc^2)^3$ to be in accord with slow neutron scattering data; and if K is about of this magnitude, say from $\frac{1}{6}$ to $\frac{2}{3}$ of this upper limit, and is positive, such interaction could be the main source of the isotope displacement of spectral lines.

---

[4] Breit, Phys. Rev. **42**, 348 (1932).

The Photoelectric Effect of the Deuteron

G. Breit and E.U. Condon

*Institute for Advanced Study
and Princeton University*

Theoretical cross sections for the dissociation of the deuteron by absorption of γ-rays, the Chadwick-Goldhaber effect, have been calculated, by using a square well law of potential, both of the ordinary type and of the Majorana type. The curves of cross section as a function of energy for various assumed widths are given. For widths less than 2×10^{-13} cm they are quite similar in shape for either type of interaction. For greater widths the ordinary, potential shows a fairly sharp peak at 4.7 MEV (for a width of 4×10^{-13} cm) whereas the Majorana shows for the same width a much flatter maximum at 6.2 MEV. It is pointed out that suitable measurements of relative cross sections would give a means of telling which type of interaction is obeyed and give an approximate figure for the range of interaction.

The problem of the nature of the interaction between proton-neutron and proton-proton is basic to all theoretical work in nuclear physics and therefore is deserving of detailed consideration from every point of view. There seem to be at present three possible ways of getting interaction laws: (1) Study of angular distribution and magnitude of scattering, such as the scattering of protons and neutrons in hydrogen, (2) interpretation of mass defects of atoms, and (3) the study of the Chadwick-Goldhaber effect[1] or photodissociation of nuclei. Already there is quite an extensive literature devoted to the problems presented by the first two methods. In this paper we examine the situation with regard to the information about proton-neutron interaction that is obtainable from studies of the cross section for the photodissociation of deuterons by γ-rays of more than 2.2 MEV energy.

The deuteron is the simplest of nuclei heavier than H^1 and occupies among them a position analogous to that of the hydrogen atom in the theory of atomic spectra. One may expect that a study of its properties will be essential for the development of views about the structure of nuclei and that the relative simplicity of the mathematical concepts involved in its treatment will make the information derivable from its study more definite than that obtainable

---

[1] Chadwick and Goldhaber, Nature **134**, 237 (1934); Proc. Roy. Soc. **A151**, 479 (1935).

from studies of heavier nuclei. We have thought it of interest, therefore, to calculate the theoretical cross section for photodissociation in some detail so as to make it possible to plan experiments which can throw light on the type of interaction involved as well as on the spatial extension of the forces between neutrons and protons.

Calculations of the cross section of the deuteron for this process have already been made by Bethe and Peierls[2] for interaction forces of extremely short range. Independently, Massey and Mohr[3] have made some calculations for finite ranges of interaction, using both the square potential well and a law of force of exponential type. They found the cross section to be substantially the same for the two types of force law and they also found that an increase in the range of force brings about an increase in the expected cross section. Their calculations are the most complete of those that have been published so far but they are not complete enough inasmuch as only the Wigner types of forces are used and the variation of the cross section with radius is not discussed in much detail. Later, Mamasachlisof[4] published an extension to the Bethe-Peierls calculations which takes into account the nuclear radius to the first order but unfortunately the first term of his formula is too large by a factor 2 and the second by a factor 4. The first-order correction is correctly given by Hall[5] whose results are in agreement with those of Massey and Mohr as well as those presented in this paper.

This first-order correction is entirely due to the effect of the range of interaction on the shape of the S wave function of the bound state. The discussions of this state by the Wigner and Majorana interactions are identical and it is thus impossible to distinguish between these types of interaction by considerations of the first-order effect. Higher order effects involve also changes in the p state with range and for this state the use of Majorana's operator brings about effectively a repulsion as contrasted with the attraction which would be used in Wigner's treatment. It will be seen as the result of the calculations given below that the higher order effects are large for sufficient energies of the γ-rays and that it should be possible to make use of them in distinguishing between the two types of interaction.

Interpretation of observed cross sections in terms of the force-law is complicated by the fact that dissociative transitions caused by magnetic dipole interaction of the γ-rays with the deuteron are of importance comparable with the more familiar electric dipole interaction. The importance of the magnetic dipole contribution for the inverse effect—capture of neutrons by protons with emission of γ-rays—has already been emphasized by Fermi[6] who also gives a formula for the magnetic dipole cross section of the photo-dissociation

[2] Bethe and Peierls, Proc. Roy Soc. **A148**, 146 (1935).
[3] Massey and Mohr, Proc. Roy. Soc. **A148**, 206 (1935); Nature **133**, 211 (1934).
[4] Mamasachlisof, Physik. Zeits. Sowjetunion **8**, 206 (1935).
[5] Hall, Phys. Rev. **49**, 401 (1936).
[6] Fermi, Phys. Rev. **48**, 570 (1935).

of deuterons by γ-rays. The magnetic dipole effect has also been investigated by Bethe, Peierls, Teller and Wigner in a paper which, unfortunately, is not being published. The magnitude of the magnetic dipole cross section may be one-half or one-fourth the electric dipole effect, according to whether the ^1S level of the deuteron (which interpretation of scattering cross sections places within 50 kv of the dissociation limit) is unstable or stable.

Now the experimental value for the cross section obtained by Chadwick and Goldhaber is actually smaller than that given by theory for the electric dipole effect alone, using extremely short range neutron-proton forces. So inclusion of the magnetic dipole effect only makes matters worse. The situation is not that of a definite contradiction between theory and experiment since the experimental measure of the proton energy is apparently uncertain by 80 kv in 240 kv. If the proton energy is really only 160 kv instead of 240 kv, this would correspond to a reduction by a factor $\frac{2}{3}$ of the corresponding theoretical electric dipole cross section.

When sources of much harder γ-rays become available, however, these difficulties become less important. The magnetic dipole effect drops off rapidly with increasing energy so that for 6-MEV γ-rays it contributes only about 3 percent of the cross section. For such γ-rays the energy of the protons would be about 2 MEV and so could be accurately determined by measuring their range if the γ-ray energy were not already known from other work. In the hope of stimulating the interest of experimentalists in this problem we have therefore worked out the theoretical cross sections in their dependence on energy and the assumed force-law in some detail.

The calculation is arranged below so as to have explicit formulas for the most general type of interaction laws. Closed expressions for contributions to the matrix elements due to regions outside the interaction region will be given. These are often the main parts of the matrix elements. The formulas will then be specialized for the case of a "square well." The notation used is as follows.

σ = collision cross section for the incident photon;
M = mass of proton;
E = sum of kinetic energies of proton and neutron after dissociation;
ε = absolute value of binding energy of the deuteron,
V = potential energy;
D = constant depth of potential hole if it is "square";
a = nuclear radius = distance between proton neutron beyond which their mutual potential energy vanishes;
$h\nu$ = energy of photon;
v = relative velocity of proton and neutron after dissociation;
$\gamma = h\nu/\varepsilon$; $\alpha = (M\varepsilon)^{1/2}/\hbar$; $\beta_2 = (ME)^{1/2}/\hbar$.

For "square" hole:

$\beta = M^{1/2}(D - \varepsilon)^{1/2}/\hbar$; $\beta_1 = M^{1/2}(D + E)^{1/2}/\hbar$;
r = distance between proton and neutron; $z = \beta r$, $z_1 = \beta_1 r$; $z_2 = \beta_2 r$;

u = regular solution of radial wave equation for s terms:

$$\frac{d^2u}{dr^2} + \frac{2}{r}\frac{du}{dr} + (M/\hbar^2)(E - V)u = 0;$$

N = normalization factor for s terms defined by

$$4\pi N^2 \int_0^\infty r^2 u^2 \, dr = 1;$$

$F = \sin z_2/z_2 - \cos z_2$; $G = \cos z_2/z_2 + \sin z_2$;

$\bar{F}/r$ = regular solution of radial wave equation for p terms normalized so as to make $\bar{F}$ asymptotic to a sine wave of unit amplitude at infinity;

F_i/r = any (not normalized) regular solution of the radial equation for p terms in $0 < r < a$;

$F' = dF/dz_2$, $F_i' = dF_i/dz_2$, etc.; unless otherwise specified the ' stands always for d/dz_2;

h = Planck's constant; $\hbar$ = Dirac's constant.

For any shape of "hole" one finds by standard methods making use of Einstein's absorption probability expressed in terms of matrix elements:

$$\sigma = (16\pi^3/3)(e^2/\hbar c)(v/v)N^2 \left| \int_0^\infty r^2 u \bar{F} \, dr \right|^2. \qquad (1)$$

This formula should hold for γ-rays having a wave-length large in comparison with the dimensions of the deuteron provided it is correct to represent the effect of the γ-ray by a term $-e\mathbf{E} \cdot \mathbf{r}$ in the Hamiltonian. Here $\mathbf{E}$ is the electric vector of the γ-ray and $\mathbf{r}$ is the displacement of the proton. In fact such a term gives the above formula in Schrödinger's treatment of absorption. Its use is correct and logical if the interaction between neutrons and protons arises from an ordinary potential. For Majorana forces one may formally also use such a term. The complete consequences of doing so do not appear to have been investigated from a theoretical point of view. One may give arguments in favor of using such a procedure. Thus it gives correct results for cases in which the binding between the proton and neutron is weak and the use of the same form of the interaction energy for cases of finite coupling between protons and neutrons appears to be the simplest mathematical procedure. For finite binding the effect of a constant field is very reasonably represented by the same interaction term. It should be noted, however, that the use of this or any other type of interaction with radiation is speculative and that the usual connection of quantum theory with the classical is established with more difficulty for forces of the Majorana type than usually. Thus a wave packet formed out of Majorana wave functions will describe a condition in which the proton and the neutron parts of the wave packet change places at a rate determined by the binding between the proton and neutron. Under such conditions the classical analogy breaks down and one does not have a real justification for the use of standard interaction energies. According to an observation of

Feenberg's kindly communicated by him to us the f sum rule of Thomas and Kuhn does not hold for systems with Majorana's forces because the classical relations $\dot{x} = p_x/m$ are violated in such systems. This fact throws additional doubt on the use of $-e\mathbf{E}\cdot\mathbf{r}$ because the interaction of photons having energies high in comparison with the binding is as a consequence definitely non-classical. There is thus no point at which a close connection with classical theory can be established in the same sense as for forces of the ordinary potential type.

It has been suggested by Massey and Mohr following a related discussion of Taylor and Mott that dipole radiation should disappear altogether for Majorana systems. This is a too stringent point of view since for loosely bound systems one may discuss conditions satisfactorily by usual means.

The use of an interaction energy $-e\mathbf{E}\cdot\mathbf{r}$ is equivalent for long wave-lengths to using $-(e/c)\mathbf{A}\cdot\dot{\mathbf{r}}$ where $\mathbf{A}$ is the vector potential. The latter form of the interaction energy is correct whenever $e\dot{\mathbf{r}}$ can be identified with the electric current. For systems obeying ordinary interaction laws such an identification is very reasonable. For systems governed by exchange forces it is not so immediate because the operators $e\dot{\mathbf{r}}$ represent only the part of the current due to the motion of the separate particles. It is conceivable that the exchange of particles also contributes to the electric current in a manner not directly describable by following the motion of the charged particle. A complete understanding of this question presumably requires a better insight into the nature of heavy particles and their interactions than that which we have at present. Being unable to see the situation more fully we use Eq. (1) even though one cannot be absolutely sure of its validity. For a "hole" of any shape and for either the Wigner or Majorana type of interaction

$$\bar{F} = N_p F_i(r)/F_i(a) \qquad (r < a), \tag{2a}$$

$$\bar{F} = N_p[F_a'G - G_a'F + (F_i'/F_i)_a(FG_a - F_aG)] \qquad (r > a), \tag{2b}$$

where N_p for p states is given by

$$N_p = [1 - z_2^{-2} + z_2^{-4} + 2z_2^{-3}(F_i'/F_i) + (1 + z_2^{-2})(F_i'/F_i)^2]_a^{-1/2}. \tag{2c}$$

The correctness of (2a), (2b) to within the common factor N_p is verified most easily using the relation $F'G - FG' = 1$. The form of the factor N_p has been specialized to p states and to a force free condition in $r > a$. In the above formulas the suffix a indicates that the quantity is evaluated at $r = a$. As a rule the most important part of the integral in (1) is from a to ∞. A general form of this integral is obtained with (2b) and is given by

$$\int_a^\infty r^2 u\bar{F}\, dr = N_p\beta_2^{-2}(1 + \alpha^2\beta_2^{-2})^{-2}u(a)\{2z_2^{-2} + \alpha^2\beta_2^{-2} - 1$$

$$+ 2\alpha\beta_2^{-1}z_2^{-1} + \alpha\beta_2^{-1}(1 + \alpha^2\beta_2^{-2})z_2$$

$$+ (F_i'/F_i)[z_2 + 2z_2^{-1} + \alpha^2\beta_2^{-2}z_2 + 2\alpha\beta_2^{-1}]\}_a. \tag{3}$$

To this one must add the integral from 0 to a which depends on the shape and size of the hole and which can usually be estimated for low energies and radii with sufficient accuracy without precise calculation.

For a square hole the s state is described by

$$ru = \sin \beta r \quad (r < a); \quad ru = \sin \beta a \cdot e^{-\alpha(r-a)} \quad (r > a). \quad (4)$$

The boundary conditions at $r = a$ give

$$z \cot z = -\alpha a. \quad (5)$$

The normalizing factor N on using (4) and (5) can be put into the form

$$2\pi(1 + a\alpha)N^2 = \alpha. \quad (6)$$

For ordinary interactions one has as a result of calculation making use of Eq. (5):

$$\int_0^a r^2 \bar{u} \bar{F}\, dr$$

$$= N_p \sin z \left\{ \frac{2\beta_1{}^2}{(\beta_1{}^2 - \beta^2)^2} - \frac{\alpha a}{\beta_1{}^2 - \beta^2} - \frac{z_1 \sin z_1}{F_i} \cdot \frac{2(1 + a\alpha) + a^2(\beta_1{}^2 - \beta^2)}{a^2(\beta_1{}^2 - \beta^2)^2} \right\} \quad (7)$$

all quantities on the right side being taken for $r = a$. Adding this to the integral given by (2) the expressions simplify on using

$$\beta_1{}^2 - \beta^2 = \alpha^2 + \beta_2{}^2. \quad (8)$$

By substituting the value of N_p and by using in it the expressions for F_i appropriate to the square hole it is found that

$$\sigma = \sigma_{BP} \frac{[(\alpha a)^2 + (\beta a)^2](\beta a)^2}{(1 + \alpha a)[((z_1 \sin z_1)/F_i - z_2{}^2)^2 + z_2{}^2((z_1 \sin z_1)/F_i)^2]_a}, \quad (9a)$$

where $\qquad \sigma_{BP} = (8\pi/3)(e^2/\hbar c)\alpha^{-2}\gamma^{-3}(\gamma - 1)^{3/2} \qquad (9b)$

is the value of σ for $a = 0$ obtained by Bethe and Peierls. In applying (9a) it is convenient to use

$$z_1{}^2 = (\beta a)^2 + \gamma(\alpha a)^2; \quad z_2{}^2 = (\alpha a)^2(\gamma - 1); \quad F_i = \sin z_1/z_1 - \cos z_1. \quad (9c)$$

For $a = 0$ both βa and z_1 approach $\pi/2$ and σ approaches σ_{BP} as it should. Eq. (5) determines $z = \beta a$ for a given αa most easily graphically by plotting z cot z or else by an expansion.[7] Eqs. (9c) give z_1, z_2, and F_i.[8] Numerical calculations can be checked by using the sum rule

$$\int_0^\infty \sigma(v)\, d(hv) = \pi e^2 h/2Mc. \quad (10)$$

---

[7] Wigner, Zeits. f. Physik **83**, 253 (1933).
[8] The function F_i is tabulated in Yost, Wheeler and Breit, J. Terr. Magn. and Atoms. Elec. **40**, 443 (1935). See Table I_1 and Table I_0 for z cot z.

For the Majorana potential there is effectively repulsion in the p state which tends to push the p wave function out of the region $r < a$. The kinetic energy E may be either smaller or greater than the potential energy in this region. Only the first case is considered here in detail because it has the greater practical interest. For this case the quantity

$$\alpha_1 = M^{1/2}(D - E)^{1/2}/\hbar \tag{11}$$

is real. It is convenient to use the abbreviation:

$$Q = \alpha_1 a \sinh \alpha_1 a[(\sinh \alpha_1 a)/\alpha_1 a - \cosh \alpha_1 a]^{-1}. \tag{12}$$

It is then found that:

$$\sigma/\sigma_{BP} = \frac{1}{4}\gamma^4(\gamma - 1)^{-2}\alpha^4(1 + \alpha a)^{-1}\left|\int_0^\infty r^2 u\bar{F}\, dr\right|^2, \tag{13}$$

$$\alpha^2 \int_a^\infty r^2 u\bar{F}\, dr = N_p \sin \beta a \cdot \gamma^{-2}\{-2(\gamma-1)+\alpha a\gamma - Q[\gamma+2(1+\alpha a)(\alpha a)^{-2}]\}, \tag{13'}$$

$$\alpha^2 \int_0^a r^2 u\bar{F}\, dr = N_p \sin \beta a(2D/\varepsilon - \gamma)^{-2}\{(2D/\varepsilon)(\alpha a - 1) + 2(\gamma - 1) - \gamma\alpha a$$

$$+ Q[2(1 + \alpha a)(\alpha a)^{-2} + \gamma - 2D/\varepsilon]\}, \tag{13''}$$

$$N_p^2 = z_2^4[(Q + z_2^2)^2 + z_2^2 Q^2]^{-1},$$

which give on substitution into (1)

$$\sigma/\sigma_{BP} = \frac{1}{4}\sin^2 \beta a(1 + \alpha a)^{-1}(1 - \gamma\varepsilon/2D)^{-4}[(Q + z_2^2)^2 + z_2^2 Q^2]^{-1}$$

$$\cdot \{(\gamma\varepsilon/2D)[(\alpha a)^2(3\gamma - 4 - \gamma\alpha a) + Q(\gamma\alpha^2 a^2 + 4(1 + \alpha a))]$$

$$+ (\alpha a)^2[-2(\gamma - 1) + \alpha a\gamma] - Q[\gamma(\alpha a)^2 + 2(1 + \alpha a)]\}^2. \tag{14}$$

Another form is

$$\sigma/\sigma_{BP} = \frac{1}{4}(\alpha a)^4 \sin^2 \beta a(1 + \alpha a)^{-1}(A + B)^2/C, \tag{15}$$

where

$$A = \gamma^2(2D/\varepsilon - \gamma)^{-1}\{\alpha a - 2\alpha_1^2(\beta^2 + \alpha_1^2)^{-1}$$

$$- Q[1 - 2(1 + \alpha a)(\alpha a)^{-2}(2D/\varepsilon - \gamma)^{-1}]\};$$

$$B = 2(1 + \alpha a)(\alpha a)^{-2} + 2 - \gamma + \gamma\alpha a + P\beta_2 a[2(1 + \alpha a)(\alpha a)^{-2} + \gamma];$$

$$C = 1 - z_2^2 + z_2^4 + 2Pz_2 + (1 + z_2^2)P^2 z_2^2;$$

$$P = F_i'/F_i.$$

Here A is proportional to the contribution to the integral from 0 to a, while B is proportional to the contribution from a to ∞. Numerical calculations can be made either by means of (13) combined with (13') and (13'') or else by

FIGURE 1. Dependence of
cross section on range and
type of interaction law at
the photoelectric threshold,
$E = 0$. Ratio σ/σ_{BP} is
plotted against a in 10^{-13}
cm. P refers to ordinary, M
to Majorana, interaction.

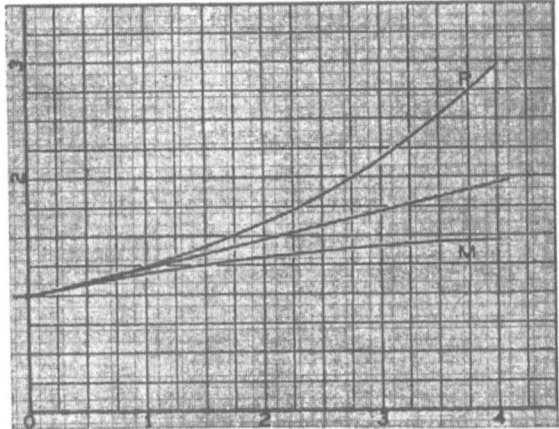

(14) or by (15). The forms (13) and (15) keep track of the contributions to the integral due to $r < a$ and due to $r > a$.

As the formulas are too complicated for one to be able to recognize their properties by inspection we have calculated the cross section for the range of values likely to be of interest experimentally, namely, for γ-ray energies from 2.2 MEV to 11 MEV and for the widths $a = 1, 2, 3$ and 4×10^{-13} cm. The results are shown in Figs. 1 to 5. All calculations for these graphs were made with the same value of α (0.231×10^{13}) and of ε (2.2 MEV). It is easy to change these graphs to other values of ε or universal constants by remembering that σ/σ_{BP} is a function only of γ and αa since αa determines βa and D/ε. Thus if ε is changed to ε' each graph gives the right value of σ/σ_{BP} for a radius $a' = a(\varepsilon/\varepsilon')^{1/2}$ and the γ-ray energy for each point is $h\nu' = (\varepsilon'/\varepsilon)h\nu$.

Fig. 1 shows the effect of the range of the interaction force on the cross section in the limit of energies near the "photoelectric threshold," $E = 0$. The curves are the ratio σ/σ_{BP} plotted against a, at $E = 0$, the one marked P being for ordinary potential, that marked M for the Majorana law. Both are tangent to the line, $1 + \alpha a$, at small values of αa. This straight line corresponds to the values given by Hall.

Figs. 2a and 2b are graphs of σ/σ_{BP} as a function of E in MEV for various values of a, Fig. 2a referring to ordinary and Fig. 2b to Majorana potential. These bring out clearly that the two kinds of interaction behave quite differently if the range of interaction exceeds 2×10^{-13} cm but for narrow ranges of interaction the results are pretty much the same.

Figs. 3a and 3b are graphs of the electric dipole cross section, σ, as a function of E in MEV for various values of a, Fig. 3a referring to ordinary and Fig. 3b to Majorana interaction. For wide ranges of interaction the curves are quite different for the two forms.

Perhaps the experimental possibilities are best brought out by considering what information could be obtained by a measurement of the relative cross

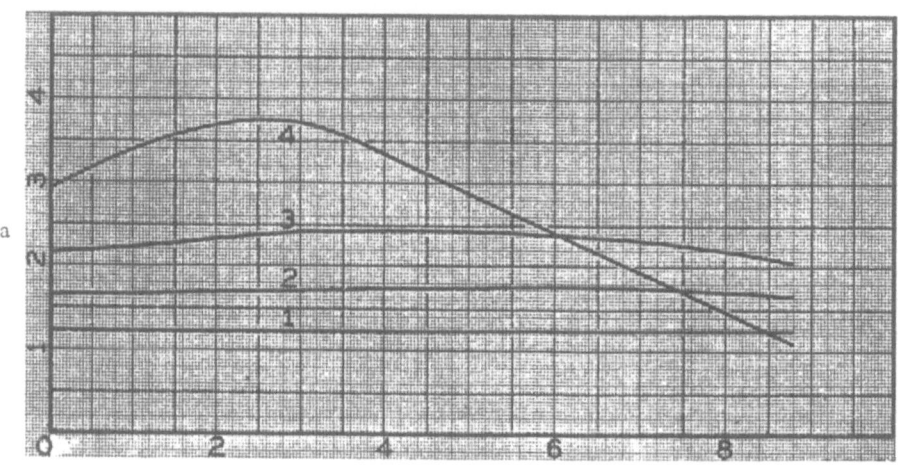

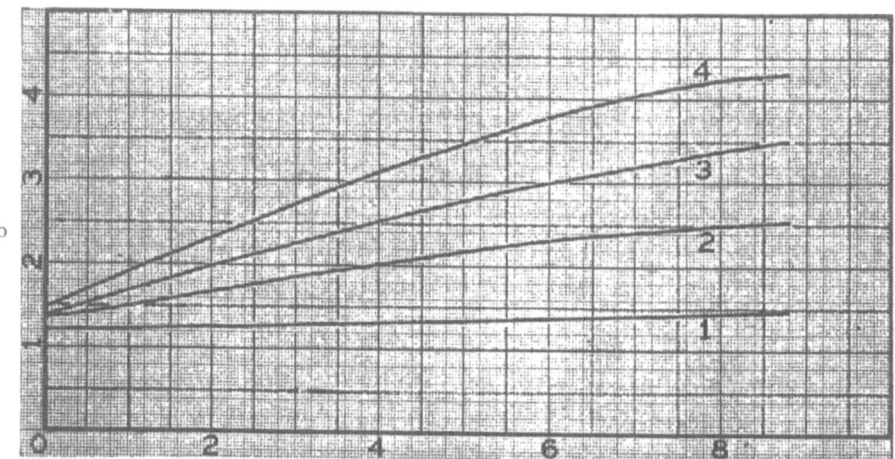

FIGURE 2. Dependence of cross section on range and type of interaction law for energies up to $E = 8.8$ MEV. Ratio σ/σ_{BP} is plotted against E in MEV. Curves are labeled by value of a in 10^{-13} cm. Fig. 2a is for ordinary, 2b for Majorana, interaction.

section for γ-rays of quite different amounts. In Fig. 4 we have plotted the ratio[9] of the cross section for 8.8-MEV γ-rays to that for 4.4-MEV γ-rays, against the range of interaction, a. Clearly if the experimental value of the ratio came out definitely greater than 0.65 one could conclude that the interaction is of Majorana type and could get an estimate of a. If the experimental value were close to 0.65 the conclusion would not be so definite but

---

[9] Of course similar results will be obtained for any two γ-ray energies of this general order; these values were picked simply for convenience, not because of any special properties.

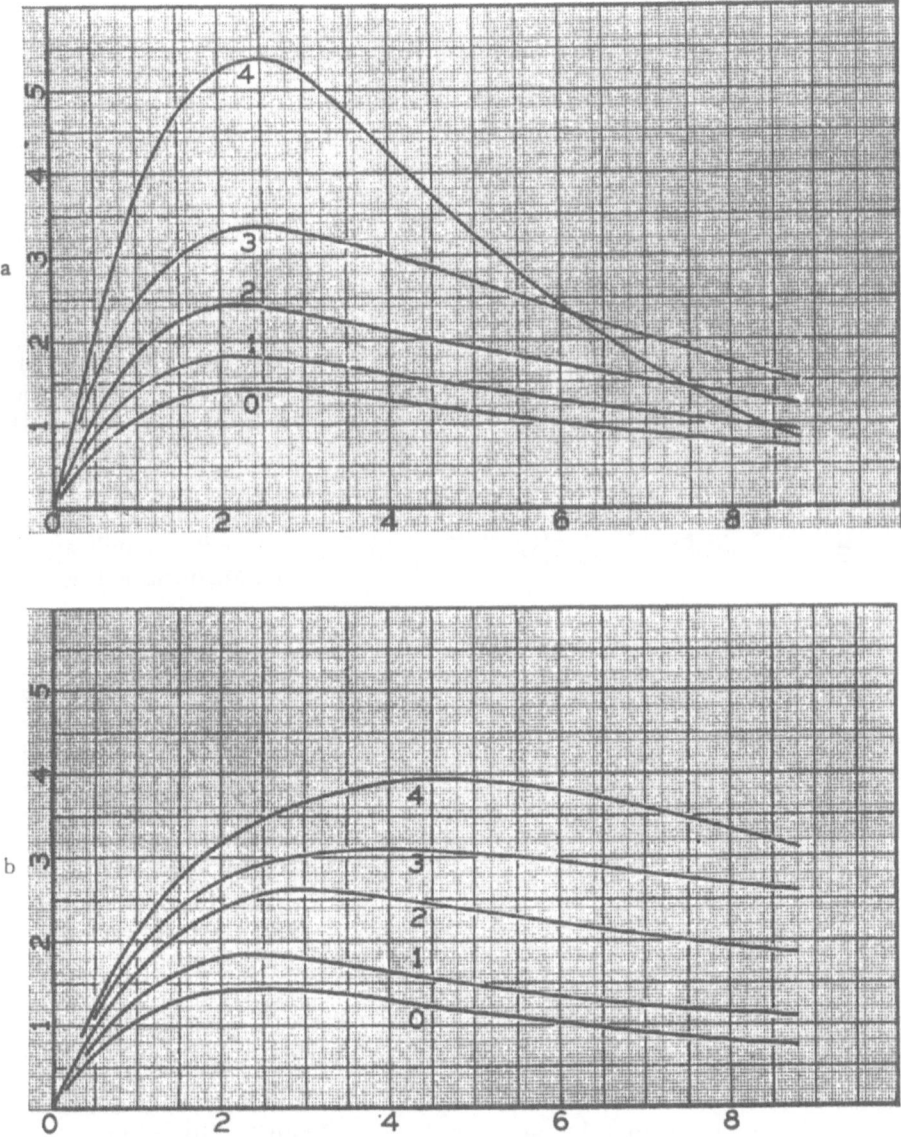

FIGURE 3. Dependence of cross section on range and type of interaction for energies up to $E = 8.8$ MEV. Absolute electric dipole cross section is plotted with 10^{-27} cm^2 as unit, against E in MEV. Curves are labeled by value of a in 10^{-13} cm. Fig. 3a is for ordinary, 3b for Majorana, interaction.

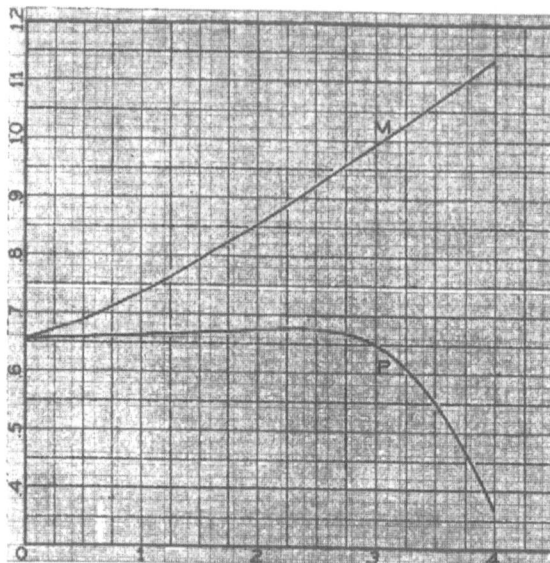

FIGURE 4. Showing dependence on range and type of interaction of ratio of cross section at $h\nu = 8.8$ MEV to that at $h\nu = 4.4$ MEV. Ratio is plotted against a in 10^{-13} cm. Curve P is for ordinary, M for Majorana, interaction.

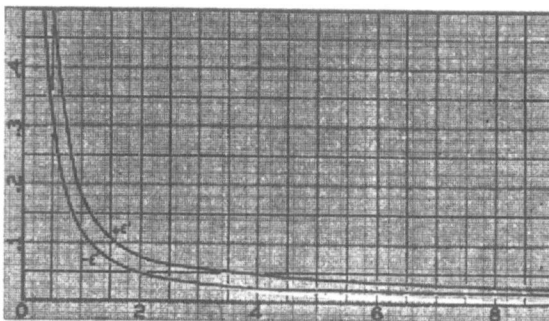

FIGURE 5. Ratio of magnetic dipole cross section to electric dipole cross section at $a = 0$, plotted against E in MEV. Curve $+\varepsilon'$ refers to unstable, $-\varepsilon'$ to stable, level for ^1S level of deuteron at 46 kev from dissociation limit.

again if it were definitely lower one could conclude in favor of the ordinary potential and know that the range was between 3 and 4×10^{-13} cm.

Finally, Fig. 5 shows the ratio of the cross section for magnetic dipole to that for electric dipole in the limit $a = 0$ calculated from a formula given in the unpublished paper of Bethe, Peierls, Teller and Wigner. The formula is

$$\sigma_m/\sigma_{BP} = \frac{(g_p - g_n)^2}{4Mc^2/\varepsilon}\gamma^2(\gamma - 1)^{-1}\frac{[1 \pm (\varepsilon'/\varepsilon)^{1/2}]^2}{\varepsilon'/\varepsilon + \gamma - 1},$$

where ε' is the magnitude of the energy of the deuteron in its lowest ^1S level. The positive sign of $(\varepsilon'/\varepsilon)^{1/2}$ is to be used for the case in which the ^1S level is unstable, the negative sign in the case of stability. Fig. 5 is constructed for the case $\varepsilon' = 46$ kev and $g_p - g_n = 5$. The positive and negative signs of the square root correspond respectively to curves marked $+\varepsilon'$ and $-\varepsilon'$. The relative importance of the magnetic effect diminishes rapidly with γ so there is a considerable advantage in working with 4.4-MEV γ-rays instead of the 2.6 MEV γ-rays. For 10 MEV the γ-ray wave-length is 120×10^{-13} cm, so even a nuclear size of 4×10^{-13} cm, a probable upper limit, is only 1/30 of the γ-ray wave-length and corrections for retardation are probably not important here.[10] Potentials with long ranges can be made to influence p states even for low energies. By means of them it would be possible to decrease the theoretically expected cross sections.

We are indebted to Mr. F.L. Yost and Mr. L. Eisenbud for checking some of the arithmetical calculations.

Correction to the Photoelectric Effect of the Deuteron

Two of us[11] published formulas and graphs for the collision cross section of a photon with a deuteron. Miss Katharine Way of the University of North Carolina has been comparing the calculations with some of her own and she has found arithmetical mistakes in the graphs. These mistakes were made in the calculations which involved the Majorana interaction. The calculations have been repeated and checked. The corrected graphs are reproduced below. They are numbered in the same way as in the paper.[1] Instead of Fig. 3b one should use the following values of $10^{27}\sigma$ (accurate to approximately 1 percent):

---

[10] In discussing retardation effects for such problems it is convenient to use a somewhat more general method than is customary for atomic spectra. The interaction energy between light and matter is proportional to $\mathbf{A} \cdot \dot{\mathbf{r}}$. The vector potential $\mathbf{A}$ contains in it the factor e^{ikz} where it has been supposed that the light wave propagates in the z direction. This factor can be expanded

$$e^{ikz} = \sum_{0}^{\infty} i^n (2n+1) P_n(\cos\theta)(\pi/2kr)^{1/2} J_{n+1/2}(kr),$$

where the P_n are Legendre functions of order n and the $J_{n+1/2}$ are Bessel functions of order $n + \frac{1}{2}$. Each term in the sum when multiplied by $\dot{x}$ or $\dot{y}$ gives rise to a linear combination of two spherical harmonics of order $n \pm 1$. The first term of the sum gives rise only to a spherical harmonic of order 1. It can cause only transitions between states obeying the selection rule $\Delta L = \pm 1$ in accordance with the triangle rule for Gaunt's integrals of products of three spherical harmonics. The values of the matrix elements are modified by the presence of $J_{1/2}(kr)$ in the integrand of the matrix elements rather than the first term of its power series expansion. The ratio of the second term to the first is $-k^2r^2/6 = (2\pi^2/3)(r^2/\lambda^2)$, where λ is the wave-length of the γ-ray. For $r/\lambda = 1/30$ this quantity is $\sim -1/150$ and the correction for retardation to the dipole effect due to this cause is still of little importance. The only other term in the sum for e^{ikz} which need be considered for an $s - p$ transition is that corresponding to $n = 2$ because $n \pm 1$ cannot be equal to 1 for any other term. The ratio of this term to the main one is found to be, on performing the angular integrations, $k^2r^2/15$ to the first order of k^2r^2 and it may also be neglected.

[11] G. Breit and E.U. Condon, Phys. Rev. **49**, 904 (1936).

TABLE I. Values of $10^{27}\sigma$ (Majorana)

E	$a = 1 \times 10^{-13}$ cm	$a = 2 \times 10^{-13}$ cm	$a = 3 \times 10^{-13}$ cm	$a = 4 \times 10^{-13}$ cm
0.275 Mev	0.45	0.57	0.72	0.92
0.55	0.92	1.17	1.47	1.86
1.1	1.51	1.90	2.39	2.97
2.2	1.78	2.26	2.82	3.41
3.3	1.68	2.12	2.60	3.14
4.4	1.50	1.89	2.31	2.78
6.6	1.16	1.45	1.78	2.17
8.8	0.92	1.14	1.38	1.72

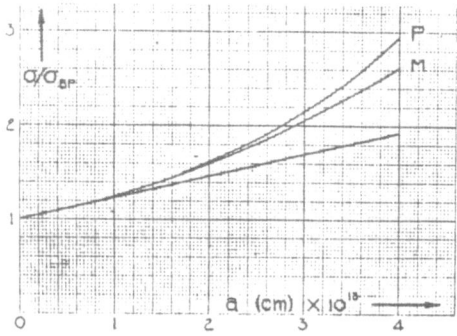

Corrected FIGURE 1. Dependence of cross section on range and type of interaction law at the photoelectric threshold, $E = 0$. Ratio σ/σ_{BP} is plotted against a in 10^{-13} cm. P refers to ordinary, M to Majorana interaction.

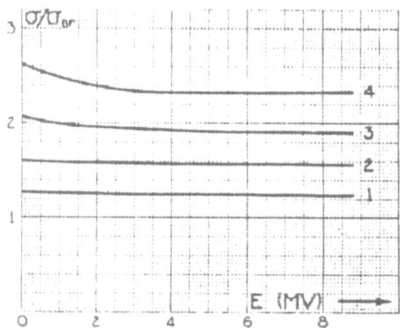

Corrected FIGURE 2b. (Majorana interaction.) Dependence of cross section on range and type of interaction law for energies up to $E = 8.8$ Mev. Ratio σ/σ_{BP} is plotted against E in Mev. Curves are labeled by values of a in 10^{-13} cm.

In Fig. 4 the curve M should be replaced by a straight horizontal line corresponding to $\sigma (h\nu = 8.8 \text{ Mev})/\sigma (h\nu = 4.4 \text{ Mev}) = 0.64$. It is seen that the photoelectric effect of the deuteron in the region $h\nu = 0 \rightarrow 10$ Mev is not very suitable for distinguishing between exchange and nonexchange interactions, the collision cross sections being nearly the same in the two cases. The determination of the radius of the deuteron from experimental material on the photoelectric cross section is simpler, taking into account the above corrections, because nearly the same values for the radius are obtained whether one uses Wigner, Heisenberg or Majorana forces.

G. Breit
J.R. Stehn
University of Wisconsin,
Madison, Wisconsin,

E.U. Condon
Princeton University,
Princeton, New Jersey

On Nuclear Forces

B. Cassen and E.U. Condon

*Harper Hospital, Detroit, Michigan
and Palmer Physical Laboratory,
Princeton, New Jersey*

The various types of exchange forces that are being used in current discussions of nuclear structure may all be simply expressed in terms of a formalism which attributes five coordinates to each "heavy" particle and applies the Pauli exclusion principle to all the particles in the system. The simplest assumption for the interaction law is that which implies equality of proton-proton and neutron-neutron forces and also equality with the proton-neutron forces of corresponding symmetry. This is in accord with the empirical knowledge of these interactions at present.

In this paper we show how the use of a coordinate having two proper values which tells whether a particle is a proton or a neutron, together with the assumption of the Pauli exclusion principle for all the particles, gives a unified description of the various types of exchange forces used in nuclear structure theories. Such a coordinate was first introduced by Heisenberg[1] and also plays a role in the Fermi-Konopinski-Uhlenbeck[2] theory of beta disintegration.

We suppose that each heavy particle (proton or neutron) is described by five coordinates. These are three for its position in space, a spin coordinate σ giving the component of its angular momentum along some direction in space, and a fifth coordinate, τ, which can have the values ± 1. If τ has the value $+1$ the particle is a proton, while the value -1 indicates that it is a neutron.

The spin angular momentum is a vector equal to $\frac{1}{2}\hbar$ times the vector, σ, which is represented by

$$\sigma = \begin{pmatrix} \mathbf{k} & \mathbf{i} - i\mathbf{j} \\ \mathbf{i} + i\mathbf{j} & -\mathbf{k} \end{pmatrix},$$

the rows and columns referring to states which are labeled by precise values of the z component of σ. This nonrelativistic description of the spin was introduced by Pauli and by Darwin.

---

[1] Heisenberg, Zeits, f. Physik **77**, 1 (1932).
[2] Fermi, Zeits. f. Physik **88**, 161 (1934); Konopinski and Uhlenbeck, Phys. Rev. **48**, 7 (1935).

In the same way τ can be considered purely formally like the z component of a vector. The analogy is purely formal in that the three "components" of τ do not refer to directions in space. Formally we may write

$$\tau = \begin{pmatrix} \mathbf{n} & \mathbf{l} - i\mathbf{m} \\ \mathbf{l} + i\mathbf{m} & -\mathbf{n} \end{pmatrix},$$

where $\mathbf{l}$, $\mathbf{m}$ and $\mathbf{n}$ behave algebraically like the three unit vectors $\mathbf{i}$, $\mathbf{j}$ and $\mathbf{k}$. The third component of τ may be called the character coordinate and the whole expression τ the character vector.

We postulate that in an assembly of heavy particles the wave function has to be antisymmetric in all particles with regard to exchange of all five of their coordinates. We want to show that this gives a convenient formalism for working with nuclear problems.

Let us first consider any attribute of a single heavy particle such as its mass, its charge or its magnetic moment. If A is the arithmetic mean of the two values for proton and neutron and B is half the difference, proton value minus neutron value, then that attribute will appear in the equations as a term involving,

$$(A + B\tau).$$

For example, the electrostatic charge will appear as $\frac{1}{2}e(1 + \tau)$ where e is the electronic charge.

Next, let us consider the scalar product $\tau_1 \cdot \tau_2$ of the character vectors associated with two particles. We have

$$(\tau_1 + \tau_2)^2 = \tau_1{}^2 + \tau_2{}^2 + 2\tau_1 \cdot \tau_2,$$

since the operators for two different particles commute. Now as defined τ is formally like twice an angular momentum vector of magnitude $\frac{1}{2}$. Therefore, the possible values of the vector sum are twice 1 and zero. Letting $2T$ stand for the magnitude of the resultant we have

$$4T(T + 1) = 3 + 3 + 2\tau_1 \cdot \tau_2,$$

so the allowed values of $\tau_1 \cdot \tau_2$ are $+1$ and -3. The value $+1$ corresponds to the case of parallel character vectors and so to a wave function that is symmetric in τ_1 and τ_2 while the value -3 corresponds to resultant zero of the two character vectors and hence to an antisymmetric dependence of the wave function on τ_1 and τ_2.

Therefore, the expression

$$\tfrac{1}{2}(1 + \tau_1 \cdot \tau_2)$$

has the allowed values $+1$ and -1, the positive value going with wave functions symmetric in τ_1 and τ_2, while the negative value has for its proper function a wave function antisymmetric in these two character coordinates.

These results are, of course, exactly analogous to the well-known results for the vector sum of two spin angular momenta and their connection with the symmetry properties of the wave function with regard to exchange of σ_1 and σ_2.

The applicability of the Pauli exclusion principle to a dynamical system requires that the Hamiltonian function for the system be a symmetric function of the coordinates of the particles. In looking for possible interaction laws we therefore have to confine ourselves to symmetric functions.

So far, four types of exchange forces have been proposed for description of the interaction between heavy particles. These are:

1. Ordinary (Wigner) potential.[3] This is the familiar kind and is simply a function of the distance between the two particles.

2. Heisenberg potential.[4] This is of the form of a function of the distance multiplied by an operator H. This operator is defined as having the value $+1$ when applied to a wave function that is symmetric with regard to exchange of both position and spin coordinates of the two particles whose interaction is being considered, and the value -1 for the antisymmetric case.

3. Bartlett potential.[5] This is a function of the distance multiplied by an operator B. This operator is defined as having the value $+1$ when applied to a wave function that is symmetric in the spin coordinates alone and -1 for the antisymmetric case.

4. Majorana potential.[6] This is a function of the distance multiplied by an operator M. This operator is defined as having the value $+1$ when applied to a wave function that is symmetric with regard to exchange of the positional coordinates only of the two particles in question, and -1 when applied to an antisymmetric function in the positional coordinates.

Evidently the Majorana type can be expressed in terms of the preceding two:

$$M = HB = BH.$$

Since the operator H exchanges both position and spin, and the Bartlett operator B exchanges spin only, the product will be equivalent to an exchange of position only, for the double exchange of spin provided by the combined action of H and B cancels out and is the same as no exchange of spin.

We now point out that the four operators, 1, H, B and M are readily expressible in terms of the spin and character vectors, σ and τ of the two particles. This follows from the requirement of over-all antisymmetry of the wave functions in the five coordinates of each of the particles. Let the letters a, b, c, d ... stand for the different particles and consider a general wave function ψ that is a function of all five of the coordinates of each particle. More explicitly

$$\psi = \psi(\mathbf{r}_a, \tau_a, \sigma_a; \mathbf{r}_b, \tau_b, \sigma_b; \mathbf{r}_c, \tau_c, \sigma_c; \dots).$$

[3] Wigner, Phys. Rev. **43**, 252 (1933).
[4] Heisenberg, Zeits. f. Physik **77**, 1 (1932)
[5] Bartlett, Phys. Rev. **49**, 102 (1936).
[6] Majorana, Zeits. f. Physik **82**, 137 (1933).

Whatever the functional form of ψ this can be written

$$\psi = \tfrac{1}{2}[\psi(\mathbf{r}_a, \tau_a, \sigma_a; \mathbf{r}_b, \tau_b, \sigma_b; \ldots) + \psi(\mathbf{r}_b, \tau_a, \sigma_b; \mathbf{r}_a, \tau_b, \sigma_a; \ldots)]$$
$$+ \tfrac{1}{2}[\psi(\mathbf{r}_a, \tau_a, \sigma_a; \mathbf{r}_b, \tau_b, \sigma_b; \ldots) - \psi(\mathbf{r}_b, \tau_a, \sigma_b; \mathbf{r}_a, \tau_b, \sigma_a; \ldots)],$$

that is, as the sum of a function symmetric in the position and spin coordinates of particles a and b and one antisymmetric in these same coordinates. As we require ψ to be antisymmetric in all five coordinates of a and b we know that the first term here must be antisymmetric in τ_a and τ_b and the second term must be symmetric in τ_a and τ_b. Therefore the operator H has the value -1 for symmetry in τ_a and τ_b, and $+1$ for antisymmetry in τ_a and τ_b. Using the earlier calculation of $\tau_1 \cdot \tau_2$ we have

$$H_{ab} = -\tfrac{1}{2}(1 + \tau_a \cdot \tau_b)$$

which expresses the Heisenberg exchange operator in terms of the two character vectors.

Similarly it is easy to see that

$$B_{ab} = +\tfrac{1}{2}(1 + \sigma_a \cdot \sigma_b)$$

and therefore, in view of the relation, $M = HB$, we have

$$M_{ab} = -\tfrac{1}{4}(1 + \sigma_a \cdot \sigma_b)(1 + \tau_a \cdot \tau_b),$$

which completes the expression of each of the exchange operators in terms of symmetric functions of the coordinates of the two particles.

With the different types of exchange operators written in this simple way it suggests itself that the general law of interaction for the specifically nuclear forces can be written in the form:

$$U = V + V_h H + V_b B + V_m M.$$

Here the four V's may be quite different functions of the separation distance but the simplest assumption is that the entire dependence of the interaction on σ and τ is contained in the operators 1, H, B and M.

Of course, this simple result is not required by the formalism. It is simply the simplest form for the exchange operators. The mere requirement of a symmetric function would be met if any one or all of the V's were replaced by

$$A + B(\tau_1 + \tau_2) + C\tau_1 \tau_2,$$

where A, B and C are functions of the distance of separation. In fact, this more general form is necessary even for the description of the Coulomb interaction between the particles which for two particles is expressed as

$$\tfrac{1}{4}(e^2/r)(1 + \tau_1)(1 + \tau_2).$$

The expression above, involving A, B and C, has the value $(A + 2B + C)$ for two protons, the value $(A - C)$ for a proton and a neutron, and the value $(A - 2B + C)$ for two neutrons. If proton-proton forces are the same as neutron-neutron forces we may conclude that $B = 0$, and if like-particle forces are the same as proton-neutron forces in states of corresponding symmetry then we can conclude that $C = 0$. With both B and C equal to zero the dependence on the components τ_1 and τ_2 is gone and we are reduced to the simpler original form.

The assumption that B and C are zero seems to be in accord with the facts about nuclear interactions as far as these are known.* The assumption makes the unification that there are only four different force laws, corresponding to the four possible types of symmetry in σ and τ. These four types are describable in terms of more usual notation by giving the symmetry in position and spin, since this determines the symmetry in character. A state that is symmetric in spin is called a triplet, one antisymmetric a singlet. Symmetry for exchange of position will be denoted by S and antisymmetry by P, since these are the standard notations for states of least orbital angular momentum in the two-body problem which have these positional symmetry properties. Here, however, we use S and P in a more general sense.

The distinct laws of interaction are given in Table I. Using the values of the operators 1, H, B and M we can write for the four interaction laws:

$$U(^3S) = V + V_h + V_b + V_m,$$

$$U(^1S) = V - V_h - V_b - V_m,$$

$$U(^3P) = V - V_h + V_b - V_m,$$

$$U(^1P) = V + V_h - V_b - V_m.$$

---

* The consequences of assuming equality of the various specifically nuclear forces for like and unlike particles are considered in detail in a paper by Feenberg and Breit in this issue which we had the pleasure of seeing in manuscript after this paper was sent in.

TABLE I

Symmetry in			Nota-tion	Occurrence
Position	Spin	Char-acter		
s	s	a	^3S	proton-neutron
s	a	s	^1S	proton-neutron proton-proton neutron-neutron
a	s	s	^3P	proton-neutron proton-proton neutron-neutron
a	a	a	^1P	proton-neutron

These are readily solved for explicit expressions for each V in terms of the four empirically occurring combinations.

We shall only make a few brief remarks about the empirical facts as they are known as these have been recently reviewed by Bethe and Bacher.[7] Ideally one would like to learn all eight force laws (there are eight if we do not make the simple formal assumption of the previous section) from studies based wholly on the two-body problem. So far this is not possible.

The situation with regard to the two-body problems is this:

Proton-Neutron

$U(^3S)$: Normal state of deuteron. Observed binding energy gives a relation between depth and width of a potential well.

Scattering of neutrons by protons: This involves all four laws in principle, but in fact owing to short range of the forces only the two S laws enter in an important way for neutron energies less than some tens of millions of volts. Slow neutron scattering cross section indicates a ^1S level of deuteron near to zero binding energy, according to Wigner.

Photodissociation of the deuteron: Electric dipole effect involves transitions from bound ^3S normal state to continuum of ^3P, hence these two laws. In addition to Bethe and Bacher, the problem is discussed by Breit and Condon.[8] Magnetic dipole effect produces transitions from ^3S normal state to ^1S continuum. This is important near the photoelectric threshold.

Radiative capture of neutrons by protons: Here the important effect is for show neutrons principally by action of magnetic dipole radiation from ^1S continuum to ^3S normal state, according to Fermi.[9]

None of these involve the ^1P law in an essential way. Apparently this can only be studied by scattering very high energy neutrons with protons.

Proton-Proton

Here a little evidence comes from the probable nonexistence of He^2. But mainly the knowledge comes from the recent work of Tuve, Heydenburg and Hafstad as analyzed by Breit, Condon and Present.[10] The analysis indicates that up to 1 MeV the departures from coulomb scattering may be described entirely in terms of effects of the ^1S law and gives strong indication that this is the same as the ^1S law in the deuteron.

---

[7] Bethe and Bacher, Rev. Mod. Phys. **8**, 82 (1936).

[8] Breit and Condon, Phys. Rev. **49**, 904 (1936). See also Morse, Fisk and Schiff, Phys. Rev. **50**, 748 (1936).

[9] Fermi, Phys. Rev. **48**, 570 (1935).

[10] Breit, Condon and Present, Phys. Rev. this issue.

Neutron-Neutron

No positive evidence from two-body interactions. Absence of a double neutron is in accord with assumption of the same ^1S law as in protoneutron since the ^1S level is now supposed to be virtual (see reference 10 for details).

All other knowledge of the force laws comes from approximate calculations of binding energies of many-body nuclei as fully reviewed by Bethe and Bacher. These are in accord with assumption of equality of the interaction laws for various kinds of particles so far as specifically nuclear forces are concerned.

This paper grew out of association at the 1936 summer symposium on theoretical physics of the University of Michigan. We wish to express here to Professor H.M. Randall our deep appreciation of the opportunity of working in the stimulating atmosphere of the Michigan laboratory.

Theory of Scattering of Protons by Protons[*]

G. Breit, E.U. Condon, and R.D. Present

University of Wisconsin, Madison, Wisconsin;
Princeton University,
Princeton, New Jersey;
Purdue University,
Lafayette, Indiana

The experiments of Tuve, Heydenburg and Hafstad and those of White are discussed by means of the standard theory of scattering in central fields. The theoretical formulas are presented in a form convenient for numerical computation and are supplemented by tables. These are arranged so as to enable an experimentalist to compute the effect of phase shifts due to angular momenta $L = 0$, $\hbar$, $2\hbar$, and to infer these phase shifts from the experimental material (Tables I, II, III, IV, V, VI, VII, VIII, IX). Tables of necessary Coulomb wave functions are also given for zero angular momentum. By means of these the interaction energy can be computed from the experimental material (Tables X, XI, XII, XIII).

Statistical fluctuations make conclusions drawn from White's data somewhat uncertain. The experiments of Tuve, Heydenburg and Hafstad are comparatively free of statistical effects and their comparison with theory shows that (a) There is an unmistakable difference between the observed scattering and that to be expected according to Mott's formula which uses the inverse square law. (b) This difference can be explained by using practically entirely effects of the phase shift in the partial wave having $L = 0$ (head on collisions; s wave distortions). The distortion of p and d waves ($L = \hbar, 2\hbar$) is secondary and the experimental accuracy does not yet suffice to enable their quantitative determination. (c) The variation of the scattering anomaly with proton energy is in approximate agreement with that to be expected from an interaction potential independent of the energy. (d) For a given range of nuclear forces the interaction potential is accurately determined by the data. The values obtained are in good agreement with those found by Feenberg and Knipp and by Bethe from the mass defects of H^2, H^3, He^4 provided the mass defect calculations are made on the basis of a proton-neutron interaction which depends on the relative orientation of the spins of proton and neutron in accordance with Wigner's explanation of the large scattering of slow neutrons by hydrogen. Mass defect calculations based on a proton-neutron interaction indicated by the binding energy of H^2 without

[*] A paper delivered at the Tercentary Conference of Arts and Sciences at Harvard University, September, 1936.

dependence on the spin orientation give a much lower value for the interaction between like particles than that obtained from the proton-proton scattering experiments. The "like-particle" interaction for a Gauss error potential is $39mc^2e^{-17r^2}$ with 8.97×10^{-13} cm as the unit of length and the interaction energy is 11.1 mev for a potential which is constant (except for its Coulombian part) within a distance $e^2/mc^2 = 2.82 \times 10^{-13}$ cm. (e) The interaction between protons as derived from the scattering experiments is found to be very nearly equal to that between a proton and a neutron in the corresponding condition of relative spin orientation and angular momentum (^1S state). The proton-neutron values which come closest to being equal to the proton-proton values are those obtained by Fermi and Amaldi from the scattering and absorption of slow neutrons.

The close agreement between the empirical values of the proton-proton and proton-neutron interactions in ^1S states suggests that aside from Coulombian and spin effects the interactions between heavy particles are independent of their charge and that the apparent preference for equal numbers of protons and neutrons in the building up of nuclei is conditioned more by the operation of the exclusion principle than by the greater values of proton-neutron forces.

1. Introduction

Observations on the anomalous scattering of protons by protons have been made by Wells,[1] White[2] and Tuve, Heydenburg and Hafstad.[3] In the experiments of Wells a cloud chamber was used. On account of the small number of observed collisions the accuracy of his experiments was insufficient to make theoretical conclusions possible. White's experiments were made by the same method. A greater number of tracks was observed and White concluded that there is a large discrepancy between the actual scattering and that to be expected on the assumption of the inverse square law of force between protons. Such a discrepancy will be referred to below as a *scattering anomaly*. The theoretical interpretation of White's results has been considered in more detail by Present.[4] It was found impossible to account for White's angular distribution by any ordinary theory with a potential describing the interaction between protons. Nevertheless the order of magnitude of the scattering anomaly turned out to be in approximate agreement with the attractive potential which Feenberg and Knipp[5] found from the binding energies of $H_1{}^2$, $He_2{}^3$, $He_2{}^4$ by

[1] W.H. Wells, Phys. Rev. **47**, 591 (1935).

[2] M.G. White, Phys. Rev. **47**, 573 (1935); **49**, 309 (1935). We are very grateful to Dr. White for communicating to us his complete data.

[3] M.A. Tuve, N.P. Heydenburg and L.R Hafstad, Phys. Rev. **49**, 402 (1936), **50**, 806 (1936) (Preceding paper). We are very grateful to Messrs. Tuve, Heydenburg and Hafstad for making their results available to us before publication and for their wholehearted cooperation in answering by experiment the questions which came up in the interpretation of their earlier results.

[4] R.D. Present, Phys. Rev. **48**, 919 (1935).

[5] E. Feenberg and J.K. Knipp, Phys. Rev. **48**, 906 (1935).

using a mixed Heisenberg-Majorana operator for the interaction between protons and neutrons. This operator gives an interaction between protons and neutrons in singlet states that is somewhat weaker than in triplet states, as has been suggested by Wigner from evidence on the scattering of slow neutrons by hydrogen.[6] The difficulties with the angular distribution and energy dependence in White's experiments make, however, conclusions about the interaction potential obtained from his data very uncertain.

The work of Tuve, Heydenburg, and Hafstad was done with electrical counters. Large numbers of scattered particles were observed and their data are comparatively free of statistical fluctuations due to an insufficiency of such particles. The scattering anomaly is smaller in their experiments than in those of White and it starts at a higher energy. It will be seen below that their final results are in reasonably good agreement with a simple form of scattering theory with respect to both the angular distribution and the energy dependence. There is no consistent evidence for the presence in the scattering of anything but an s wave (head-on collisions, $L = 0$) which is in agreement with expectation for interaction forces confined to distances smaller than 10^{-12} cm. Although there were indications in the early experiments of THH of a too rapid variation of scattering anomaly with energy, later and more accurate data are in approximate agreement with theory also in this respect. The observations available at present are not yet precise enough to determine accurately the range of the proton-proton forces. Nevertheless, they appear to be good enough to eliminate strong long range forces. Thus interaction energies of constant magnitude through a distance of $3e^2/mc^2 = 8.5 \times 10^{-13}$ cm are in disagreement with the energy dependence of the scattering anomaly while a constant interaction potential through $e^2/mc^2 = 2.8 \times 10^{-13}$ cm is in fair accord with observation. The magnitude of the interaction potential corresponds to $D = 10.3$ mev through $e2/mc^2$ if one uses square wells and to $A = 39mc^2$ for the potential $Ae^{-\alpha r^2}$ with $\alpha = 17$ and $\hbar(Mm)^{-1/2}c^{-1} = 8.97 \times 10^{-13}$ cm as the unit of length. This value is in good agreement with the result of Feenberg and Knipp who obtain $A = 41mc^2$ for the same α. The sensitivity of the expected scattering anomaly to the magnitude of the interaction potential is great and the above values of A and D are determinable to about $0.1mc^2$ from the scattering data at any given energy aside from uncertainties due to the possible effect of higher phase shifts.

In their present form the scattering experiments of THH give information about the force between two protons ($\pi - \pi$ force) when they collide head on (S state) and when they have antiparallel spins (^1S state). According to Wigner a similar state is of importance for the scattering of slow neutrons by hydrogen

---

[6] Dunning, Pegram, Fink and Mitchell, Phys. Rev. **47**, 970 (1935). Bjerge and Westcott, Proc. Roy. Soc. **A150**, 790 (1935). Fermi and Amaldi, La Ricerca Scientifica **1**, 1 (1936); Fermi, ibid., July, 1936. We are very grateful to Professor Fermi for informing us of his last results before publication. The value of 130 kv used for the position of the virtual level by us is too high. The effect of changing it to 110 kv is scarcely noticeable in Table XIV.

and the magnitude of the proton-neutron attraction in ^1S states ($\pi - \nu$ force) can be determined from scattering experiments combined with measurements of the absorption of slow neutrons in hydrogen. The $\pi - \pi$ and $\pi - \nu$ attractions will be compared for the ^1S states and it will be seen that the more careful experiments indicate a practically exact equality of the $\pi - \pi$ and $\pi - \nu$ forces. Although this comparison has been made only in the ^1S state the agreement is so striking as to suggest that the interactions between heavy particles are *universally equal*, i.e., that the only essential difference in the interactions between like and unlike particles is due to the exclusion principle. The magnetic moment of the proton is different from that of the neutron and, therefore, the force between protons cannot be expected to be exactly equal to that between neutrons or to that between protons and neutrons. The energy due to the magnetic interaction between two protons is, however, of the order $9(e\hbar/2Mc)^2/(e^2/mc^2)^3 \sim 0.012mc^2$ which is very much smaller than the main part of the interaction energy ($\sim 20mc^2$). As a tentative hypothesis we may consider the interactions between heavy particles to be universally equal except for the Coulombian effects between protons and the spin effects between all the heavy particles. Whether this interaction has actually one of the forms used by Feenberg and Knipp and by Bethe still remains to be seen. Since the form

$$[(1 - g)P^M + gP^H]J(r),$$

which has been used successfully by them (P^M = Majorana exchange operator, P^H = Heisenberg exchange operator) gives a value of the "like particle" interaction agreeing with that arrived at from the scattering experiments on protons, one may regard this form as the most likely. Since in H^3, He^3 and He^4 like particles can be considered as nearly in ^1S states the above operator would give for them the same effect as for the ^1S state of unlike particles. We should like to acknowledge our indebtedness to Dr. Feenberg who noticed independently that the above operator could be applied both to like and to unlike particles and kindly communicated his considerations to us. It is possible that the universal operator contains as a part of it a small Wignerian term.

The observations of Bjerge and Westcott and of Dunning[6] on the scattering of slow neutrons by hydrogen indicate that there is either a virtual or a stationary ^1S level of the deuteron at ± 43 kv. If the level is stationary no agreement whatsoever is obtained between the $\pi - \pi$ and $\pi - \nu$ interactions. Assuming it to be virtual, the $\pi - \nu$ interaction as derived from these experiments corresponds to $A = 42mc^2$ and is thus somewhat larger than the proton-proton interaction. The difference of $3mc^2$ is still too large to consider the interactions as the same because this difference will be seen to have significant and quite observable consequences for the scattering amounting to approximately a factor of 3 for neutrons. More recently improved measurements on neutrons were made by Fermi and Amaldi.[6] Their observations on the absorption of slow neutrons lead them to the conclusion that the level is virtual.

Secondly the scattering cross section in hydrogen has been found by them to be 12×10^{-24} cm^2 instead of the previous larger values and the position of the virtual level has been raised to about 130 kv. *With these changes the difference between the interactions of like and unlike particles is insignificant and may be considered as lying within the limits of error of the experiments.*

Recent experiments of Goldhaber[7] raised doubts concerning Wigner's explanation of the scattering of neutrons in hydrogen. According to Goldhaber the mean free path of 200 kv neutrons is about three times greater than that to be expected theoretically. If Goldhaber's result were correct all of the conclusions arrived at in this paper would have to be changed because the $\pi - \nu$ interaction would be modified and because Feenberg and Knipp's value of the like-particle interaction would be altered. In view of these radical implications of Goldhaber's experiment it was repeated by Tuve and his colleagues using carbon bombarded with deuterons as a source. The energy of the deuterons is between 600 kv and 1200 kv. Its precise value does not matter for the interpretation of the experiments. They find a mean free path in paraffin of 2.2 cm which is to be compared with a theoretical mean free path of 2.4 cm using 30 kv for the position of the virtual level and 600 kv for the neutron energy and 2.2 cm using 140 kv for the position of the virtual level and again 600 kv for the neutron energy. Their mean free path is smaller than Goldhaber's even though the energy of the neutrons is higher. Goldhaber's result thus implies a minimum in the scattering at about 200 kv of neutron energy and is improbable. We are very grateful to Dr. Tuve and his colleagues for permission to quote their experiments in this connection.

2. Phase Shift Analysis

The scattering of charged particles by an inverse square field of force has been considered by Gordon and Mott[8] while the effect of symmetry due to the identity of particles has been worked out by Mott.[9] The scattering anomaly

---

[7] M. Goldhaber, Nature **137**, 824 (1936). Cf. reference to M.A. Tuve in text. M.A. Tuve and L.R. Hafstad, Phys. Rev. **50**, 490 (1936).

[8] W. Gordon, Zeits. f. Physik **48**, 180 (1928). N.F. Mott, Proc. Roy. Soc. **A118**, 542 (1928).

[9] N.F. Mott, Proc. Roy. Soc. **A126**, 259 (1930). Cf. J.R. Oppenheimer, Phys. Rev. **32**, 361 (1928).

In applications of the solutions of Gordon and Mott the meaning of the physical condition represented by the solution is usually not clearly stated. This solution is given by Eqs. (1) (1') in the text. It represents the wave inside a very large screening sphere of radius R. For $r > R$ there is supposed to be no field of force while for $r < R$ the field is given by the Coulombian potential. For elastic collisions between a proton and a hydrogen atom or between two hydrogen atoms one is interested in the state of the system before and after the atoms have collided, i.e. in the state in a force free region. Thus rigorously it is the solution "outside" the screening sphere $(r > R)$ that matters. Such a solution is complicated and for ordinary applications it is not con-

due to deviations from an inverse square field has been discussed by Taylor[10] who applied the theory to the scattering of alpha-particles in helium and in hydrogen. The general form of the theory is well established in these papers, and its systematic presentation is given by Mott and Massey.[11] It will suffice here to suppose the methods as known and it will not be necessary to give the derivations of the formulas.

Notation

The following notation will be used:

M = mass of proton.

$\mu = M/2$ = reduced mass in the collision of two protons.

v = relative velocity of the two protons before the collision.

E = kinetic energy of incident protons = $\frac{1}{2}Mv^2$.

E' = energy in frame of center of gravity = $\frac{1}{2}E$

$\Lambda = h/\mu v$ = de Broglie wave-length

$k = 2\pi/\Lambda$.

$a = \hbar^2/\mu e^2$.

$\eta = 1/ka = (e^2/\hbar c)ZZ'c/v$.

r = distance between proton and neutron or proton and proton.

$\rho = kr$, $y = \rho\eta = r/a$.

$L\hbar$ = angular momentum of proton and neutron or proton and proton around common center of gravity.

P_L = Legendre function.

θ = scattering angle in the reference system of center of gravity.

$\Theta = \theta/2$ = scattering angle in the reference system of the laboratory.

---

sidered explicitly for the following reasons. According to the asymptotic expansion given by Eq. (3) in the text the solution inside the screening sphere is the sum of two parts represented by the two exponentials. The second of these is of special interest because it represents the scattered wave. Over a small area of the screening sphere the "spherical wave" represented by this term may be approximated by a plane wave which may be regarded as subject to reflection and refraction at $r = R$. Actually the screening is taking place through a region large compared with the wave-length and therefore the reflection may be neglected. There will also be no refraction as long as the wave may be considered as plane. For very small scattering angles, however, the wave cannot be considered as plane because the second term in the curly brackets in the spherical wave part of Eq. (3) is not negligible in comparison with unity. Thus at small angles one may expect the solutions used here to give incorrect results, in agreement with the Born method calculation of Coulomb scattering due to Wentzel. In order that the second term in curly brackets should become ~ 1 for 1 Mev protons at 0.53×10^{-8} cm it is necessary to have $2\Theta = \theta \sim 1°$. The magnitude of the term decreases with Θ^{-2} at small angles and therefore no serious effect due to this cause is expected in practical applications.

[10] H.M. Taylor, Proc. Roy. Soc. **A134**, 103 (1931); **A136**, 605 (1932).

[11] N.F. Mott and H.S.W. Massey, *The Theory of Atomic Collisions* (Oxford University Press).

$c = \cos \Theta.$ $s = \sin \Theta.$

Γ = the gamma function.

$\sigma_L = \arg \Gamma(L + 1 + i\eta).$

F_L = regular solution of the differential equation for r times the radial wave function in a Coulomb field normalized so as to be asymptotic to a sine wave of unit amplitude at ∞. The asymptotic given by $F_L \sim \sin(\rho - \frac{1}{2}L\pi - \eta \ln 2\rho + \sigma_L)$ and the diff. eq. is $[d^2/d\rho^2 + 1 - 2\eta/\rho - L(L + 1)/\rho^2]F_L = 0.$

G_L = irregular solution of the same differential equation normalized so that its asymptotic form is given by $G_L \sim \cos(\rho - \frac{1}{2}L\pi - \eta \ln 2\rho + \sigma_L).$

K_L = phase shift defined by the asymptotic form of $\mathfrak{F}_L = r$ times the radial wave function in the actual field. This form shall be $\mathfrak{F}_L \sim e^{iK_L}\sin(\rho - \frac{1}{2}L\pi - \eta \ln 2\rho + \sigma_L + K_L).$

The above notation for the Coulomb wave functions is the same as that used by Yost, Wheeler Breit[12] in tabulations of these wave functions.

The plane wave e^{ikz} is changed by the Coulomb field into

$$\psi^c = \sum_0^\infty i^L(2L + 1)P_L(\cos \theta)e^{i\sigma_L}F_L/\rho. \tag{1}$$

An alternative form is

$$\psi^c = e^{-\frac{1}{2}\pi\eta + ikz}\Gamma(1 + i\eta)F - (i\eta; 1; ik(r - z)) \tag{1'}$$

where F is the confluent hypergeometric series. If the field is Coulombian at large distances but not at small ones then the same plane wave is changed into

$$\psi = \sum_0^\infty i^L(2L + 1)P_L e^{i\sigma_L}\mathfrak{F}_L/\rho. \tag{2}$$

At large distances the asymptotic form of ψ^c is

$$\psi^c \sim \left\{1 + \frac{\eta^2}{ik(r - z)} + \cdots\right\}\exp\{i[kz + \eta \ln k(r - z)]\} - \frac{\eta}{k(r - z)}$$

$$\times \left\{1 + \frac{(1 + i\eta)^2}{ik(r - z)} + \cdots\right\}\exp\{i[kr - \eta \ln k(r - z) + 2\sigma_0]\}. \tag{3}$$

The collision cross section per unit solid angle of the laboratory reference system is, on the classical theory, for Coulombian fields:

$$\sigma_{Cl} = 4c(s^{-4} + c^{-4})(e^2/2\mu v^2)^2. \tag{4}$$

The effect of taking into account the symmetry of the wave functions is according to Mott such as to change this into

---

[12] F.L. Yost, John A. Wheeler and G. Breit, Phys. Rev. **49** 174 (1936); Journal of Terrestrial Magnetism and Atmospheric Electricity, December, 1935, p. 443. See also T. Sexl, Zeits. f. Physik **56**, 62 (1929) for discussion of $L = 0$.

$$\sigma_{\text{Mott}} = 4c[s^{-4} + c^{-4} - s^{-2}c^{-2}\cos(\eta \ln s^2 c^{-2})](e^2/2\mu v^2)^2. \qquad (4')$$

If the field deviates from the Coulombian the collision cross section per unit solid angle is

$$\sigma = 4cP, \qquad (4'')$$

where

$$P = \tfrac{1}{4}|f(\theta) + f(\pi - \theta)|^2 + \tfrac{3}{4}|f(\theta) - f(\pi - \theta)|^2$$

in the notation of Mott and Massey.[11] The quantity P and hence σ can be computed using

$$P = (e^2/2\mu v^2)^2 \left\{ s^{-4} + c^{-4} - s^{-2}c^{-2}\cos(\eta \ln s^2 c^{-2}) - \frac{2}{\eta}\sum_L g_L(2L+1) \right.$$

$$\times [s^{-2}\cos\varphi_L{}^s + (-)^L c^{-2}\cos\varphi_L{}^c] + \frac{4}{\eta^2}\sum_L g_L(2L+1)^2 P_L{}^2 \sin^2 K_L$$

$$+ \frac{8}{\eta^2}[\sum_e + 3\sum_o]_{L'>L}(2L+1)(2L'+1)P_L P_{L'}$$

$$\left. \times \sin K_L \sin K_{L'}\cos(\varphi_L - \varphi_{L'}) \right\}, \qquad (5)$$

where g_L has the value 1 for even L and 3 for odd L. The last term in braces contains two sums, the first referring to even L, as indicated by suffix e, and the second referring to odd L as indicated by suffix o. Both of these sums are taken over pairs of unequal values of L. No cross products between even and odd L occur. As before, the argument of the Legendre functions P_L is $\cos\theta$. The other quantities needed in this formula are:

$$\varphi_L = K_L + 2(\sigma_L - \sigma_0); \quad \varphi_L{}^s = \varphi_L + \eta \ln s^2; \quad \varphi_L{}^c = \varphi_L + \eta \ln c^2, \quad (5.1)$$

$$\sigma_1 - \sigma_0 = \tan^{-1}\eta; \quad \sigma_2 - \sigma_1 = \tan^{-1}(\eta/2); \quad \sigma_L - \sigma_{L-1} = \tan^{-1}(\eta/L). \quad (5.2)$$

If all K_L beyond K_2 vanish

$$P = P_M + (\Delta P)_0 + (\Delta P)_1 + (\Delta P)_2, \qquad (6)$$

with

$$(2\mu v^2/e^2)^2 P_M = s^{-4} + c^{-4} - s^{-2}c^{-2}\cos(\eta \ln s^2 c^{-2}), \qquad (6.1)$$

$$(2\mu v^2/e^2)^2(\Delta P)_0 = -\frac{2}{\eta}(s^{-2}\cos\varphi_0{}^s + c^{-2}\cos\varphi_0{}^c)\sin K_0 + \frac{4}{\eta^2}\sin^2 K_0, \qquad (6.2)$$

$$(2\mu v^2/e^2)^2(\Delta P)_1 = -\frac{18}{\eta}(s^{-2}\cos\varphi_1{}^s - c^{-2}\cos\varphi_1{}^c)P_1 \sin K_1$$

$$+ \frac{108}{\eta^2}P_1{}^2\sin^2 K_1, \qquad (6.3)$$

$$(2\mu v^2/e^2)^2(\Delta P)_2 = -\frac{10}{\eta}(s^{-2}\cos\varphi_2{}^s - c^{-2}\cos\varphi_2{}^c)P_2 \sin K_2$$

$$+ \frac{100}{\eta^2}P_2{}^2 \sin^2 K_2 + \frac{40}{\eta^2}\sin K_0 \sin K_2 \cos(\varphi_2 - \varphi_0)P_2.$$

$$(6.4)$$

By means of (4″) and (6.1) one obtains Mott's value as is seen from (4′). The additions to **P** due to K_0, K_1, K_2 are given by (6.2), (6.3), (6.4), respectively. It will be noted that the effect of K_2 depends on K_0. The above formulas are convenient for computation if one is not interested in many values of the phase shifts. For such cases it is more convenient to expand (6.2), (6.3), (6.4) as follows

$$(2\mu v^2/e^2)^2(\Delta P)_0 = -\frac{2}{\eta}\left(\frac{\cos\alpha_0}{s^2} + \frac{\cos\beta_0}{c^2}\right)\sin K_0 \cos K_0$$

$$+ \left(\frac{4}{\eta^2} + \frac{2}{\eta}\frac{\sin\alpha_0}{s^2} + \frac{2}{\eta}\frac{\sin\beta_0}{c^2}\right)\sin^2 K_0,$$

$$\alpha_0 = \eta \ln s^2, \quad \beta_0 = \eta \ln c^2, \qquad (6.5)$$

$$(2\mu v^2/e^2)^2(\Delta P)_1 = -\frac{18}{\eta}P_1\left(\frac{\cos\alpha_1}{s^2} - \frac{\cos\beta_1}{c^2}\right)\sin K_1 \cos K_1$$

$$+ \left[\frac{108}{\eta^2}P_1{}^2 + \frac{18}{\eta}\left(\frac{\sin\alpha_1}{s^2} - \frac{\sin\beta_1}{c^2}\right)P_1\right]\sin^2 K_1,$$

$$\alpha_1 = \alpha_0 + 2(\sigma_1 - \sigma_0), \quad \beta_1 = \beta_0 + 2(\sigma_1 - \sigma_0), \quad (6.6)$$

$$(2\mu v^2/e^2)^2(\Delta P)_2 = -\frac{10}{\eta}P_2\left(\frac{\cos\alpha_2}{s^2} + \frac{\cos\beta_2}{c^2}\right)\sin K_2 \cos K_2$$

$$+ \left[\frac{100}{\eta^2}P_2{}^2 + \frac{10}{\eta}P_2\left(\frac{\sin\alpha_2}{s^2} + \frac{\sin\beta_2}{c^2}\right)\right]\sin^2 K_2,$$

$$+ \frac{40}{\eta^2}\sin K_0 \sin K_2 \cos[K_2 - K_0 + 2\sigma_2 - 2\sigma_0],$$

$$\alpha_2 = \alpha_0 + 2(\sigma_2 - \sigma_0), \quad \beta_2 = \beta_0 + 2(\sigma_2 - \sigma_0). \quad (6.7)$$

In these equations, the coefficients of $\sin K_L \times \cos K_L$, $\sin^2 K_L$ are functions only of the energy and of the scattering angle. In formula (6.7) the cross product term in K_0, K_2 is conveniently computed directly. Values of the coefficients and other quantities for the computation of **P** are given in Tables I, II, ..., IX. In Table I are given values of $(2\mu v^2/e^2)^2\mathbf{P}_M$. In the first column are listed values of η and the second column gives the approximate value of the energy of the incident proton for this η. The succeeding columns give the values of $(2\mu v^2/e^2)^2\mathbf{P}_M$ for $\Theta = 15°$, $20°$, etc. as indicated at the top of each column. The values of $(2\mu v^2/e^2)^2\mathbf{P}_M$ are seen to vary slowly with E and interpolation can be easily made in Table I. In Tables II, III are given

TABLE I. Values of $(2\mu v^2/e^2)^2 \mathbf{P}_M$

η	E in kv	$\Theta = 15°$	$20°$	$25°$	$30°$	$35°$	$40°$	$45°$
0.4775	108.8	219.1	68.84	27.74	13.15	7.19	4.694	4
.3821	170.0	215.4	67.42	27.14	12.90	7.10	4.673	4
.2867	302.4	212.3	66.25	26.65	12.70	7.02	4.657	4
.2069	580	210.3	65.52	26.35	12.58	6.98	4.648	4
.1751	810	209.7	65.27	26.25	12.53	6.97	4.644	4
.1591	981	209.4	65.18	26.21	12.52	6.96	4.644	4
.1017	2400	208.6	64.88	26.10	12.47	6.94	4.638	4

TABLE II. Values of coefficients of $-\sin K_0 \cos K_0$ for $\mathbf{P}/\mathbf{P}_M$

η	E in kv	$\Theta = 15°$	$20°$	$25°$	$30°$	$35°$	$40°$	$45°$
0.4775	108.8	0.0994	0.339	0.758	1.426	2.380	3.441	3.96
.3821	170.0	.212	.541	1.088	1.938	3.128	4.426	5.05
.2866	302.4	.385	.854	1.609	2.755	4.339	6.032	6.84
.1989	628	.665	1.376	2.484	4.15	6.42	8.84	9.95
.1830	741	.741	1.511	2.720	4.53	7.02	9.64	10.84
.1671	890	.830	1.679	3.004	4.99	7.69	10.53	11.88
.1512	1088	.936	1.873	3.350	5.54	8.54	11.69	13.16
.1432	1211	.995	1.990	3.54	5.86	9.06	12.34	13.88
.1332	1400	1.081	2.155	3.83	6.33	9.71	13.25	14.95
.1174	1800	1.244	2.468	4.37	7.21	11.04	15.06	16.98
.1017	2400	1.455	2.872	5.07	8.35	12.78	17.43	19.61

TABLE III. Values of coefficients of $\sin^2 K_0$ for $\mathbf{P}/\mathbf{P}_M$

E in kv	$\Theta = 15°$	$20°$	$25°$	$30°$	$35°$	$40°$	$45°$
108.8	-0.213	-0.194	$-0.004(3)$	0.493	1.380	2.469	3.025
170.0	$-.185$	$-.0828$	.332	1.244	2.769	4.579	5.48
302.4	$-.117$	.215	1.114	2.971	5.819	9.124	10.79
628	.114	1.000	3.11	7.12	13.37	20.5	23.9
741	.198	1.276	3.80	8.58	15.99	24.4	28.4
890	.309	1.643	4.71	10.49	19.42	29.6	34.4
1088	.460	2.130	5.94	13.07	24.02	36.4	42.4
1211	.565	2.438	6.71	14.68	26.96	40.6	47.4
1400	.700	2.907	7.87	17.11	31.3	47.3	55.0
1800	1.009	3.90	10.36	22.31	40.6	61.2	71.2
2400	1.471	5.40	14.07	30.07	54.6	82.1	95.3

TABLE IV. Values of coefficients of $-\sin K_1 \cos K_1$ for P/P_M

η	E in kv	$\Theta = 15°$	20°	25°	30°	35°	40°	45°
0.4775	108.8	1.946	3.23	4.14	4.19	3.06	1.100	0
.2866	302.4	3.50	5.48	6.85	6.84	4.92	1.739	0
.1989	628	5.14	7.92	9.85	9.75	6.99	2.471	0
.1671	890	6.15	9.44	11.69	11.60	8.32	2.91	0
.1432	1211	7.19	10.98	13.60	13.52	9.70	3.39	0
.1174	1800	8.80	13.45	16.61	16.48	11.78	4.13	0
.1017	2400	10.17	15.54	19.18	19.02	13.59	4.77	0

TABLE V. Values of coefficients of $\sin^2 K_1$ for P/P_M

E in kv	$\Theta = 15°$	20°	25°	30°	35°	40°	45°
108.8	0.635	3.21	6.63	9.00	7.51	2.927	0
302.4	3.68	10.88	20.03	25.94	22.15	8.63	0
628	8.78	23.94	42.6	54.4	46.1	17.89	0
890	12.91	34.1	60.5	77.3	65.1	25.22	0
1211	17.93	46.8	82.7	105.2	88.9	34.3	0
1800	27.21	70.1	123.7	157.1	132.2	51.1	0
2400	36.6	93.7	165.1	209.5	176.4	68.0	0

TABLE VI. Values of coefficients of $-\sin K_2 \cos K_2$ for P/P_M

E in kv	$\Theta = 15°$	20°	25°	30°	35°	40°	45°
628	2.36	2.74	1.48	−2.48	−9.68	−18.36	−22.62
890	2.82	3.29	1.79	−3.02	−11.81	−22.48	−27.68
1211	3.32	3.88	2.10	−3.58	−14.10	−26.80	−33.02

TABLE VII. Values of coefficients of $\sin^2 K_2$ for P/P_M

E in kv	$\Theta = 15°$	20°	25°	30°	35°	40°	45°
628	4.90	6.17	1.840	2.19	34.0	104.0	147.0
890	6.89	8.54	2.426	3.50	49.9	150.8	214.3
1211	9.34	11.40	3.14	5.10	69.5	208.6	293.4

TABLE VIII. Values of $(40/\eta^2)P_2/[s^{-4} + c^{-4} - s^{-2}c^{-2}\cos(\eta \ln s^2 c^{-2})]$.

E in kv	$\Theta = 15°$	20°	25°	30°	35°	40°	45°
628	3.00	5.88	4.60	−10.05	−47.0	−99.1	−126.3
890	4.27	8.35	6.54	−14.29	−66.7	−140.4	−179.0
1211	5.82	11.39	8.93	−19.48	−91.0	−191.2	−243.7

TABLE IX. Values of $2\sigma_2 - 2\sigma_0$

E in kv	= 108.8	170.0	302.4	580	1088	1800	2400
$2\sigma_2 - 2\sigma_0 =$	77.9°	63.5°	48.3°	35.4°	25.8°	20.1°	17.4°

values of the coefficients of $-\sin K_0 \cos K_0$ and $\sin^2 K_0$ for the calculation of $(\Delta P)_0/P_M$.

The values of η listed in Table II were used in the computations of the coefficients for Tables II and III. The values of E are not as accurate as those of η because their calculation involves the somewhat uncertain values of the fundamental physical constants. The same applies to Table I and to Tables III, IV, V, VI. The numbers tabulated above, where they are plotted against E, form smooth curves. The graphs for the coefficient of $\sin^2 K_0$ are practically straight lines. By means of such graphs the coefficients can be easily obtained for the needed values of E. The same applies to the coefficients for the effect of K_1 tabulated in Tables IV, V and the coefficients for the effect of K_2 tabulated in Tables VI, VII. In the computation of the cross product term in K_0, K_2 contained in (6.7) it is also convenient to have values of

$$(40/\eta^2)P_2/[s^{-4} + c^{-4} - s^{-2}c^{-2}\cos(\eta \ln s^{-2}c^{-2})]$$

which gives the coefficient of the trigonometric functions for P/P_M. These are given in Table VIII. Values of $2(\sigma_2 - \sigma_0)$ for different E which are needed for the last trigonometric function in (6.7) can be computed by means of Eq. (5.2). Some values are given in Table IX. It will be noted that the coefficients of $\sin^2 K_1$, $\sin^2 K_2$ are practically linear functions of E for most of the energies covered here.

When the values of the energy and of the scattering angle are fixed P/P_M turns out to be a smooth and almost linear function of K_0 and it is thus possible to interpolate for most values of K_0 having made calculations for a few of them.

Since it is probable that these scattering experiments will be repeated and since it is advisable for the experimentalists to have a ready means of testing their data for agreement with a possible analysis in terms of the phase shifts K_0, K_1, K_2 we summarize the procedure for using the above tables. From the experiments one obtains the collision cross section per unit solid angle. This is σ of Eq. (4''); it corresponds to the total number of protons observed and it thus includes the effect of recoil protons. Then P is obtained by means of Eq. (4''). By means of Table I one obtains P_M and hence P/P_M. Fixing the voltage the curve of P/P_M against energy has to be fitted by means of K_0, K_1, K_2. The effect of K_0 is obtained by the following procedure. The coefficients given in Tables II, III are plotted for each scattering angle as a function of E. Using the graphs the coefficients for the needed voltage E are found. Usually no great accuracy is required in these coefficients. The coefficients of Table II are then multiplied by $-\sin K_0 \cos K_0$, those of Table III are multiplied by $\sin^2 K_0$ and the results are added. The result is the contribution to P/P_M due to K_0. When added to unity it gives the expected P/P_M for this K_0 if the other K are zero. These contributions are plotted as functions of K_0 keeping Θ fixed. The graphs give then the values of K_0 which are needed to fit the actual P/P_M for any scattering angle as well as values of P/P_M that correspond to this K_0 for other Θ. The same procedure may be used for the calculation of the effects of K_2 and K_2.

3. Calculation of Phase Shifts for Given Laws of Interaction

The nature of the forces between protons is not yet known. They may be partly describable by means of exchange potentials and partly by means of ordinary potentials. In spite of this apparent complication all kinds of interactions that have been seriously considered so far in nuclear theory give in effect a simple potential which is a function only of the distance r, at any fixed value of the relative angular momentum $L\hbar$. Thus exchange forces of the Majorana type give interaction potentials of the same absolute value but of opposite signs for states with odd and with even L. We will therefore suppose that there is some interaction potential for any L which may be different for different L. On account of the spins of the two protons they may be either in singlet or in triplet states. If the state is a singlet the orbital wave function is symmetric and therefore contains only terms with even L. If the state is a triplet the orbital wave function is antisymmetric and contains only odd L. Conversely the even L occur only in singlets and the odd only in triplets. For each L there is thus no necessity of considering interactions for singlets and triplets separately.

It will be supposed that for any L the potential is practically Coulombian beyond a certain distance r_0. The radial wave equation determines the function $\mathfrak{F}_L$ for $r < r_0$ to within a constant factor (see list of notation). In order to indicate that this solution involves only calculation inside r_0 it will be written as F_i. The suffix L will be omitted for the present so as not to complicate the formulas. The derivatives of F_i, F, G with respect to ρ will be written as F_i', F', G'. The relations between F'/F and K are

$$\tan K = (F'F_i - FF_i')/(GF_i' - F_iG')$$

$$= (F^2\delta)/(1 - FG\delta), \tag{7.1}$$

$$\delta = F'/F - F_i'/F_i, \tag{7.2}$$

$$(F_i'/F_i) = (F' + G'\tan K)/(F + G\tan K). \tag{7.3}$$

Here all quantities are supposed to be taken at $r = r_0$. The calculation of F, G, F', G' can be made by means of the formulas given by Yost, Wheeler, Breit[12] in terms of the series Φ, Φ^*, Ψ, Ψ^*. The necessary relations are

$$F_L = C_L \rho^{L+1} \Phi_L; \quad \rho F_L'/F_L = \Phi_L^*/\Phi_L; \quad G_L = D_L \rho^{-L} \Theta_L;$$

$$\Theta_L = \Psi_L + \rho^{2L+1}(p_L \ln 2\rho + q_L)\Phi_L. \tag{7.4}$$

Using the last form in (7.1) one does not need G'. The tabulations of Yost, Wheeler, Breit do not cover the range of values needed here. The necessary numbers are given in Tables X, XI, XII, XIII. From the phase shifts determined by means of the angular distribution $\rho F_i'/F_i$ can be determined by means of (7.1) or (7.3). This is necessary for example if one wishes to determine the

TABLE X. Coulomb functions for $y = 0.02445$; $r_0 \cong e^2/2mc^2$

$2\pi\eta$	$E \cong$	Φ_0^*/Φ_0	$\Phi_0\Theta_0$	$C_0{}^2\rho\Phi_0{}^2$
0.95	1088	1.0156	0.9276	0.1008
1.00	981	1.0165	.9268	.0931
1.05	890	1.0172	.9258	.0862
1.10	810	1.0178	.9248	.0800
1.15	741	1.0184	.9237	.0743
1.20	681	1.0188	.9227	.0692
1.30	580	1.0197	.9206	.0602
1.50	436	1.0208	.9163	.0461

TABLE XI. Coulomb functions for $y = 0.0489$; $r_0 \cong e^2/mc^2$

$2\pi\eta$	Φ_0^*/Φ_0	$\Phi_0\Theta_0$	$C_0{}^2\rho\Phi_0{}^2$
0.95	1.014	0.8847	0.2062
1.00	1.017	.8861	.1910
1.05	1.020	.8869	.1772
1.10	1.022	.8873	.1646
1.15	1.025	.8872	.1533
1.20	1.027	.8866	.1427
1.30	1.030	.8851	.1258
1.50	1.034	.8804	.0959

TABLE XII. Coulomb functions for $y = 0.0978$; $r_0 \cong 2e^2/mc^2$

$2\pi\eta$	Φ_0^*/Φ_0	$\Phi_0\Theta_0$	$C_0{}^2\rho\Phi_0{}^2$
0.95	0.9525	0.7663	0.4084
1.00	.9674	.7802	.3823
1.05	.9799	.7918	.3580
1.10	.9904	.8008	.3353
1.15	.9998	.8085	.3143
1.20	1.0079	.8139	.2945
1.30	1.0213	.8218	.2593
1.50	1.0400	.8277	.2025

TABLE XIII. Coulomb functions for $y = 0.1467$; $r_0 \cong 3e^2/mc^2$

$2\pi\eta$	Φ_0^*/Φ_0	$\Phi_0\Theta_0$	$C_0^2\rho\Phi_0^2$
0.95	0.798	0.586	0.560
1.00	.837	.620	.535
1.05	.869	.649	.509
1.10	.895	.673	.484
1.15	.919	.694	.459
1.20	.939	.711	.435
1.30	.971	.738	.390
1.50	1.016	.769	.312

magnitude of the interaction when its shape as a function of distance is known. For small radii and $L = 0$ the use of (7.1) is not advisable because the numerical accuracy is then poor. In such a case it is better to use

$$\frac{\rho F_i'}{F_i} = y\frac{X + (2\ln 2y + f)\Phi_0^*}{\Psi_0 + y(2\ln 2y + f)\Phi_0}, \tag{7.5}$$

where $y = \rho\eta = r/a$;

$$f = -2\ln\eta + q_0/\eta + (C_0^2/\eta)\cot K_0, \tag{7.6}$$

$$X = (\Psi_0^* + 2y\Phi_0)/y = 2 - \left(4 + \frac{1}{\eta^2}\right)y - 4y^2$$

$$+ \left(\frac{1}{6\eta^4} + \frac{37}{27\eta^2} - \frac{32}{27}\right)y^3 + \cdots,$$

$$\Phi_0 = 1 + y + \left(\frac{1}{3} - \frac{1}{6\eta^2}\right)y^2 + \left(\frac{1}{18} - \frac{1}{9\eta^2}\right)y^3 + \cdots,$$

$$\Phi_0^* = 1 + 2y + \left(1 - \frac{1}{2\eta^2}\right)y^2 + \left(\frac{2}{9} - \frac{4}{9\eta^2}\right)y^3 + \cdots,$$

$$\Psi_0 = 1 - \left(3 + \frac{1}{2\eta^2}\right)y^2 + \left(\frac{1}{9\eta^2} - \frac{14}{9}\right)y^3$$

$$+ \left(\frac{1}{24\eta^4} + \frac{43}{108\eta^2} - \frac{35}{108}\right)y^4 + \cdots,$$

$$q_0 = 2\eta\left[\gamma - \frac{1}{1 + \eta^2} + (s_3 - 1)\eta^2 - (s_5 - 1)\eta^4 + \cdots\right],$$

$$s_3 = 1.2021; \quad s_5 = 1.0369; \quad s_7 = 1.00835.$$

If r_0 is not very small it is more convenient to use

$$\frac{\rho F_i'}{F_i} = \frac{\Phi_0^*}{\Phi_0} - \frac{1}{\Phi_0\Theta_0 + C_0^2\rho\Phi_0^2\cot K_0}. \tag{7.8}$$

For $r = 0$ it is found from Eqs. (7.1) and (7.5) that

$$\tan K_0 = \frac{C_0^2/\eta}{(\bar{C}_0^2/\bar{\eta})\cot \bar{K}_0 + 2\ln \eta/\bar{\eta} + \bar{q}/\bar{\eta} - q/\eta},\tag{7.9}$$

where the barred quantities refer to that energy at which $K_0 = \bar{K}_0$. Some useful values are given in Tables X, XI. The values of η in Tables X, XI, XII, XIII correspond to those in Tables I, ..., VIII wherever the listed E are the same. As before η is the actual quantity used in the calculations while E was computed from η using values of the fundamental constants. Similarly the values of y given in the headings of the tables are accurate while the values of r_0 are approximate. The values tabulated are those needed for Eq. (7.8) in order to calculate $\rho F_i'/F_i$ from K_0 as well as for the calculation of K_0 from $\rho F_i'/F_i$ by means of the second form of Eq. (7.1).

The calculation of K_0 is now reduced to finding $\rho F_i'/F_i$. If the interaction potential is constant within r_0 the expression for $L = 0$ is

$$\rho F_i'/F_i = z \cot z;$$
$$z = [2\mu\hbar^{-2}(D + E')]^{1/2}r_0 = 0.439(rmc^2/e^2)(D + E')^{1/2}\text{ mv.}\tag{8}$$

Here D is the negative of the potential energy in $0 < r < r_0$. For attractive forces D is positive. Estimates show that K_1, K_2, are probably very small if the range of the nuclear forces is of the order of magnitude arrived at from nuclear mass defects. However, nuclear mass defect calculations are not sensitive to small interactions at large distances. The magnitude of the expected phase shifts can be estimated using

$$K_L \cong -\int (V/E')F_L^2\, d\rho.\tag{8.1}$$

Graphs of F_L for $L = 0, 1, 2$ for energies $E = 0.4, 0.6, 0.8, 1.0$ Mev are given

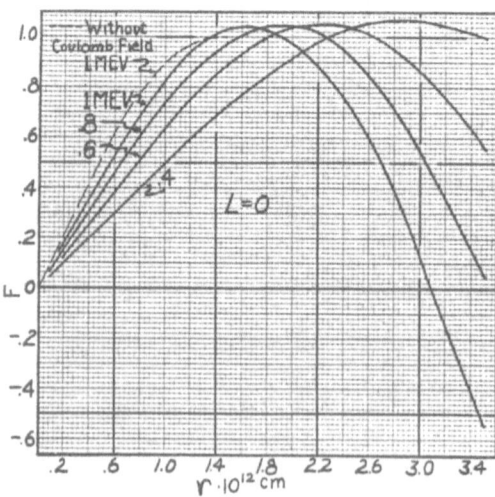

FIGURE 1. Coulomb wave functions for $L = 0$. Normalization is such as to give unit amplitude of sine wave at ∞.

FIGURE 2. Coulomb wave functions for $L = 1, 2$. Normalization is such as to give unit amplitude of sine wave at ∞.

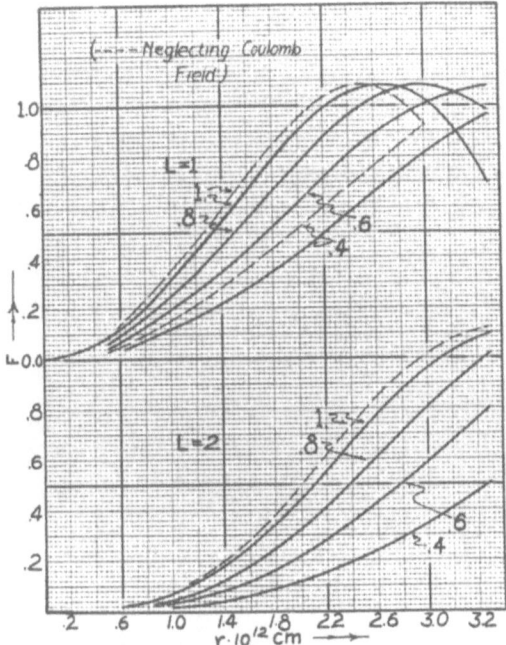

in Figs. 1, 2. For 1 Mev these functions are compared with corresponding functions in the absence of a Coulomb field. The validity of the approximation implied in Eq. (8.1) was tested by an explicit numerical integration for $L = 1$ using the Gauss error potential $Ae^{-\alpha r^2}$ with values of A and α that correspond to those found by Feenberg and Knipp. The agreement is good and the use of Eq. (8.1) appears to be justified for such estimates. If resonance is approached on account of a sufficiently large V the equation becomes unreliable. Direct calculation for K_1 using accurate formulas with a square well having a radius $2e^2/mc^2$ and a depth of 2 Mev gives $K_1 \cong 0.3°$. Since this radius is too great one may expect the principal part of the interaction potential to give rise only to negligible higher phase shifts.

4. Discussion of Experiments

In Fig. 3 are given values of $\sigma \sin \Theta$ from White's experiments. Crosses mark the experimental points observed for energies of the incident protons ranging from 600–750 kv. The dotted curve was calculated for a pure Coulomb field from Mott's formula. The full curve was obtained by adjusting the phase shift K_0 for the distorted s wave so as to give agreement with experiment in the neighborhood of 45°. It is to be noticed that if the two points at 37.5° and 42.5° are rejected, there is surprisingly good agreement with exact Coulomb scattering. However, there is apparently no reason except for statistical fluctuations specially to doubt the reliability of these two points. If they are correct

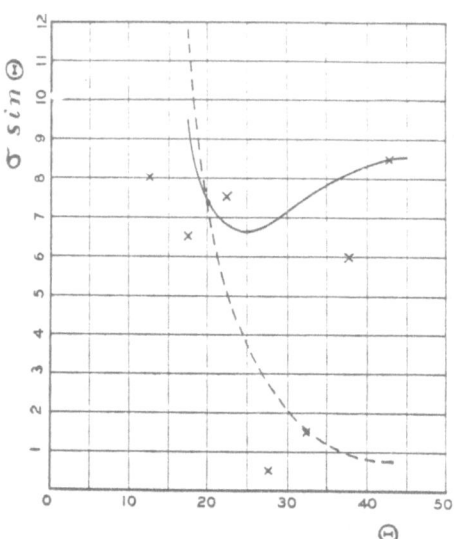

FIGURE 3. White's data. Number of collisions of 600–750 kv protons within a 5° range plotted against angle. The experimental points are as communicated to us by Dr. White from his more complete data.

it becomes necessary to explain the strange deficiency at 30°. The effect of K_1 cannot reasonably explain this condition because it increases relatively to the effect of K_0 at smaller angles. Since this point is in contradiction with the observations of Tuve, Heydenburg, Hafstad it appears simplest to attribute it to statistical fluctuations in White's experiments. Conclusions about the magnitude of the interaction potential from his work appear to be somewhat unsafe in view of this erratic angular dependence. Nevertheless his values at 45° for the 600–750 kv range give an interaction potential which is in approximate agreement with that obtained from the experiments of THH. This happens essentially because the observed scattering anomaly is so large that its explanation calls for approximate *resonance*. In more detail the situation is as follows.

There are in general two values of K_0 that will account for the scattering at 45°. The smallest positive K_0 turns out to be the most probable. It can be explained in terms of an attractive potential agreeing closely with the calculation of Feenberg and Knipp. The other value would either require repulsive forces or a definitely stronger attraction than that obtained by Feenberg and Knipp with the mixed operator for neutron-proton interactions. Additional strong evidence against this K_0 will be seen to be contained in the experiments of THH. Combined evidence from all sources thus indicates only one of the two essentially different values of K_0 to be probable. In order to account for this value it becomes necessary to use approximate resonance of the s wave with the potential hole because the scattering anomaly is great. The expected scattering is then sensitive to the depth of the hole and nearly the same value of the depth is obtained for different values of scattering. By using the interaction energy $Ae^{-(r/a)^2}$ White's values are fitted with $A = 45mc^2$ and

$a = 2.2 \times 10^{-13}$ cm. Between 450 and 600 kv his experiments show agreement with Coulomb scattering. The potential determined from the 600–750 kv range was used to calculate the scattering for energies between 450 and 600 kv. A value roughly six times Mott's was obtained. The disagreement between theory and experiment is here very definite and it is hard to account for it by any simple modification of the theory. Briefly White's angular distribution and voltage dependence of scattering do not allow of a simple theoretical explanation. The effects observed at a scattering angle of 45° for energies between 600 and 750 kv give nevertheless an interaction energy which is in approximate agreement with that found by THH, as will be seen presently.

The observations of THH will now be discussed and it will be seen that most of the scattering anomaly observed by them can be accounted for by the distortion of the s wave. At small scattering angles there are effects calling for p and perhaps d wave distortions. These effects are at present not decided enough to be regarded as definitely real. Nevertheless, they will be considered so as to give an idea of the reliability of the conclusions regarding the magnitude of the interaction potential. It will be seen that, aside from uncertainties having to do with the higher phases (p and d wave distortions), one can obtain very accurate values of the interaction energy from the experiments on the scattering of protons by hydrogen; the uncertainties due to higher phase shifts will be seen to be relatively small. The interaction energy derived from scattering experiments is in good agreement with that obtained from mass defect calculations.

The results of THH for P/P_M at 900 kv are plotted in Fig. 4. The broken curve is drawn smoothly through their points. The full curves give the theoretical dependence of the scattering anomaly on the scattering angle when 30° and 31° are used for the phase shift K_0. Experiment and theory are seen to agree nicely at this energy. In Fig. 5 observations at 800 kv are compared with theory. It will be noted that here the agreement is less satisfactory if one uses only the s wave phase-shift K_0. At small angles there is relatively too much

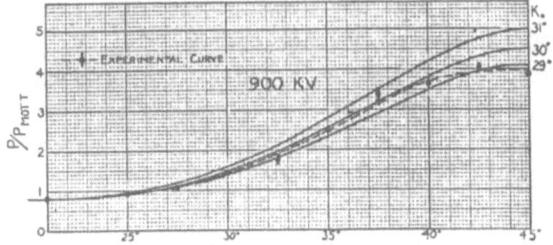

FIGURE 4. Angular distribution of scattering anomaly at 900 kv according to THH. Statistical error as estimated in the observations is indicated where it exceeds size of dot. Points at small angles may be in error on account of difficult angle measurement. Some points in this and in Figs. 5 and 6 do not correspond to latest revision of data from which they differ by amounts insignificant for interpretation.

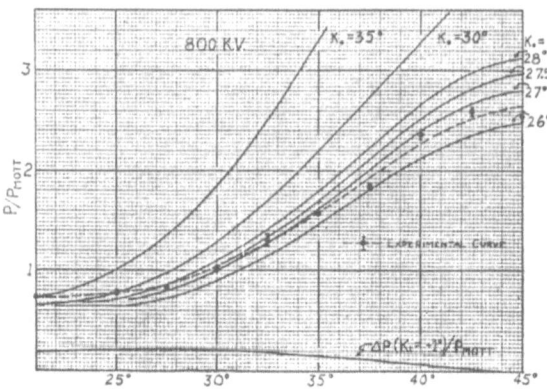

FIGURE 5. Same as Fig. 4 but for 800 kv. Curve at bottom gives effect of K_1.

scattering. Thus if the experimental data are fitted in the neighborhood of 45° then one expects about 15 percent less scattering at 25° than is actually observed. It will be noted that this discrepancy is independent of the precise value which is used for P/P_M at $\Theta = 45°$. Thus at $\Theta = 20°$ and 15° the curves for $K_0 = 26°$ and 28° give nearly the same P/P_M while at $\Theta = 45°$ the values of P/P_M which correspond to these K_0 differ by about 0.6 in a total of 3. The estimated error of the observations is shown on the same graph for some of the points. It is at the most 0.1 in the neighborhood of $\Theta = 45°$. There is also an uncertain error in the measurement of the scattering angle which is important for small Θ. It is not clear, however, why this error should matter at 800 kv and not at 900 kv. Curves for $K_0 = 30°$ and $K_0 = 35°$ are drawn in in order to show what happens when one attempts to fit the data at small Θ. There is then no indication of agreement between theory and experiment from $\Theta = 30°$ on to higher values. At the bottom of the graph is shown the contribution to P/P_M which may be expected on account of a distortion of the p wave by a phase shift $K_1 = -1°$. This has a relatively insignificant effect for values of Θ between 35° and 45° and it raises the theoretically expected values by the necessary amount to agree with experiment from $\Theta = 15°$ to 30°. The data in their present form thus indicate $K_0 = 26°$ and $K_1 = -1°$ at 800 kv. Similarly in Fig. 6 comparisons between theory and experiment are made for 700 kv and 600 kv. At these energies and at high scattering angles the observations are supposedly more dimcult on account of the increased importance of the stopping power of the window in the electrical counter. Thus at 600 kv and $\Theta = 40°$ the experimental point is known to be definitely too low and for this reason the experimental curve is drawn in by THH somewhat higher than the number of observed particles at this angle would indicate. It is dimcult to be sure of the angular distribution curves sufficiently to make a definite phase angle analysis possible. The difference between the experimental and theoretical P/P_M at 800 kv amounts to roughly 0.1 at $\Theta = 20°$ which when attributed

FIGURE 6. Same as Fig. 4
for 600 kv and 700 kv.

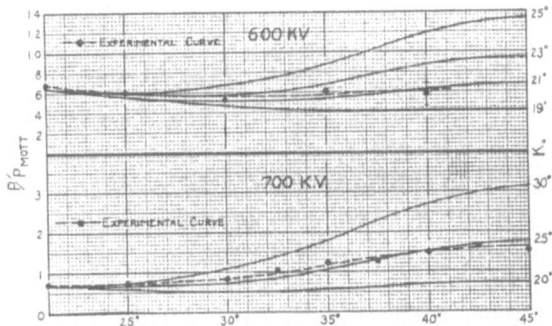

FIGURE 6. Same as Fig. 4
for 600 kv and 700 kv.

to the distortion of the p wave gives roughly $-\frac{1}{2}^\circ$ for K_1. There is thus an
indication that the phase shift K_1 is present from 600 kv to 800 kv and that
it is negligible at 900 kv. Such a variation of K_1 is contrary to all expectation
for forces of such spatial extension as is usually assumed in theories of nuclear
structure. Thus according to Fig. 2 and Eq. (8.1) the distortion of the p wave
would have to be attributed to potentials extending to 3×10^{-12} cm. Otherwise Fig. 2 shows that $F_1{}^2$ will increase with E much too rapidly to make such
a behavior of K_1 possible. In order to account for $K_1 = -1^\circ$ at 1 mev one
would need roughly an interaction energy of 10 kv extending through a
distance of 10^{-12} cm. The Coulomb energy at 3×10^{-12} cm is about 50 kv.
There appears to be at present no other evidence of such long range forces
that is at all definite.

Quantitative conclusions about K_0 and K_1 are sensitive to possible effects
of K_2. If K_1 is due to an interaction extending as far out as the above
estimates would indicate then appreciable values of K_2 would also be expected. This is seen again from Fig. 2 by comparing the wave functions for
$L = 2$ with those for $L = 1$. Fig. 7 shows qualitatively the effect of combining
the effects of $K_0 = 35^\circ$ and $K_2 = 2.2^\circ$. This curve should be compared with
that representing the effect of $K_0 = 29^\circ$ and $K_2 = 0$ shown in the same figure.
The difference in shape is seen to be relatively slight and it will be noted that
it corresponds to the difference in shape between the experimental curves for

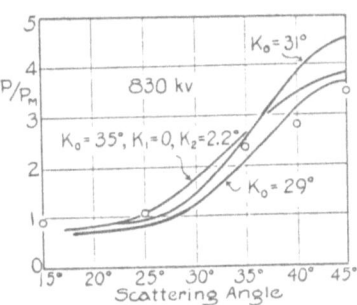

FIGURE 7. Effect of K_2 on angular distribution.
Circles represent experimental points as obtained by THH in their first set of experiments.
These points should be lowered by 13 percent
on account of a geometrical correction.

900 kv and 800 kv and the corresponding theoretical curves in Fig. 4 and Fig. 5 in that slightly higher values of P/P_M are obtained in the region of 30–35°. The same applies to Fig. 6 for 700 kv while at 600 kv the angular distribution is well represented by K_0 alone. Less weight should be given to 600 kv because the number of observations is smaller and none were possible at 45°. The present data are seen to agree better with a combination of K_0, K_1, K_2 than with K_0 alone. To some extent this is doubtless due to the larger number of available parameters. The signs of K_1, K_2 suggested by the above discussion of the experiments are such as to correspond to attractive forces for $L = 2$ and to repulsive forces for $L = 1$. These are the signs which would be expected if the interaction were representable by a pure Majorana exchange operator or by the linear combination of Wigner and Majorana potentials which is expressible as a spin-spin interaction.[5,13] Agreement in sign between the empirical and the expected phase shifts is an argument in favor of their reality. This argument is not very strong because the interaction potential for $L = 0$ may change sign between the long range region of 3×10^{-12} cm and the short range region 3×10^{-13} cm. The value of K_0 derived from scattering at 45° does not depend on K_1 but is quite sensitive to the presence of small amounts of K_2. It is seen from Fig. 7 that inclusion of $K_2 = 2.2°$ in the theoretical analysis makes it necessary to change K_0 from 29° to about 36°. Such a change in K_0 will be seen to have serious consequences on the possible interpretation of the change of K_0 with energy as well as on the comparison of the proton-proton with proton-neutron interactions.

In the above discussion it was supposed that the main effect is due to K_0 which is a reasonable hypothesis on present views regarding the range of nuclear forces. As has been already noted in connection with White's experiments there are for any scattering angle essentially two values of K_0 which account for a given experimental value of P. In the discussion of angular distributions due to K_0 alone the effect of an addition of π to K_0 cannot be noticed. To every K_0 in the first quadrant there corresponds another possible K_0 in the third and to every K_0 in the second quadrant there corresponds another possible K_0 in the fourth. For the present purpose one can consider values of K_0 differing by π as equivalent. They are also equivalent for the purpose of drawing conclusions about the interaction potential for $L = 0$ using K_0, because only $\tan K_0$ enters into the expression for $\rho F_i'/F_i$. Aside from this duplicity it is possible to fit the experimental values by means of K_0 lying either in the first or in the fourth quadrant. The possibilities in the fourth quadrant were not considered above. For such K_0 the values of P/P_M remain consistently above unity while according to the experimental points presented in Figs. 4, 5, 6 the actual P/P_M drop below unity for small scattering angles in all cases. At 600 kv P/P_M remains in fact below 1 for all angles at which observations were made. In addition the dependence of P/P_M on the scattering

---

[13] J.H. Van Vleck, Phys. Rev. **48**, 367 (1935).

angle for all voltages experimentally excamined is represented poorly by means of such K_0 even for those angles for which $P/P_M < 1$. If values of K_0 in the fourth quadrant were to be seriously considered one would need to use large values of the phase shifts for higher L in order to bring about agreement with the observed angular distributions. The difference between the first and fourth quadrant for K_0 can be understood qualitatively as follows. The interaction potential which exists in addition to that representing the inverse square law may be imagined to be either increased or decreased by small amounts starting with zero. If it is increased one gets repulsive forces and values of K_0 in the fourth quadrant; if it is decreased the forces are attractive and K_0 is positive. In the first case the repulsive forces reinforce the Coulombian effect and a larger scattering is to be expected. For attractive forces the Coulombian effect is partly counteracted and a smaller scattering should be found. If, however, the attractive force is made sufficiently great then the Coulombian effect may be practically entirely overcome and the scattering will become nearly zero. As the attractive force is increased further the scattering becomes due primarily to the attraction and may exceed that which would exist if only the Coulombian force were acting. Qualitatively this corresponds to the condition of the theoretical curves shown for 900 kv in Fig. 4 for most of the scattering angles. At small scattering angles the effect of the inverse square field on the wave function becomes great and is sufficient to partly neutralize the effect of attraction so that P/P_M again becomes <1. As the energy of the incident protons decreases the effect of the attraction becomes less pronounced since the proton penetrates into the attractive region with greater difficulty. The region of $P/P_M < 1$ thus moves towards higher scattering angles. These qualitative features of the attractive potentials which correspond to the values of K_0 in the first quadrant are in good agreement with the data which appear to give very *direct evidence against repulsions and for attraction inside the nucleus*. Further evidence in favor of this view will be found in a quantitative discussion of the variation of the scattering anomaly with energy.

In Figs. 8, 9, 10, 11 calculations with "square wells" are compared with experiment. The interaction potential is here constant for $0 < r < r_0$ and its value will be referred to as $-D$. For $r > r_0$ the potential is supposed to be Coulombian. The curves represent the theoretical dependence of K on the energy. In Fig. 8, r_0 was taken to be $e^2/2mc^2 \cong 1.4 \times 10^{-13}$ cm. The three curves correspond to $D = 47.9, 47.0, 46.3$ mev. The ovals mark the values of K_0 derived from the data of Figs. 4, 5, 6 using scattering close to $45°$ and neglecting possible effects of K_2. The dotted curve gives the theoretical dependence of K_0 on the energy when $r_0 = 0$. This curve may be raised or lowered by approaching the limit of $r_0 = 0$ in different ways. Its shape does not vary greatly when this is done. The same curve is reproduced in Figs. 9, 10, 11 so as to give a standard of comparison. No relativistic corrections and no spin forces were taken into account in the calculation of the curves. The experimental points are seen to be in good agreement with $D = 47.0$ Mev. It

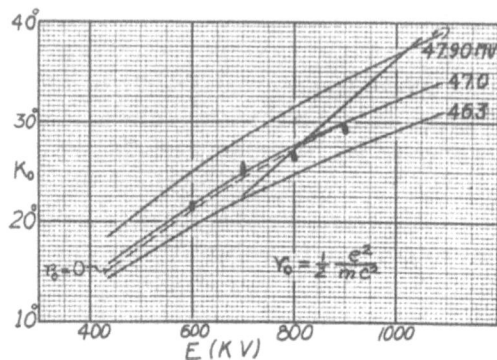

FIGURE 8. Theoretical variation of K_0 with proton energy for an interaction energy constant within $r_0 = e^2/2mc^2 = 1.4 \times 10^{-13}$ cm. Coulombian potential is supposed to be inoperative for $r < r_0$. Ovals represent values of K_0 derived from the newer experiments of THH using data near $\Theta = 45°$. Straight line gives average of the first set of data of THH. Dashed line represents theoretical behavior for $r_0 = 0$. Numbers like 47.0 refer to negative of interaction energy in mev.

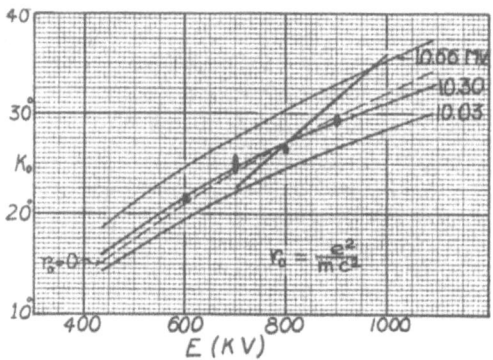

FIGURE 9. Same as Fig. 8 but for $r_0 = e^2/mc^2 = 2.8 \times 10^{-13}$ cm.

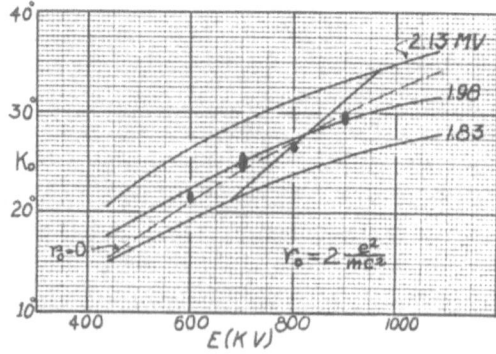

FIGURE 10. Same as Fig. 8 but for $r_0 = 2e^2/mc^2$.

FIGURE 11. Same as Fig. 8 but for $r_0 = 3e^2/mc^2$. Note insufficient slope of theoretical curve.

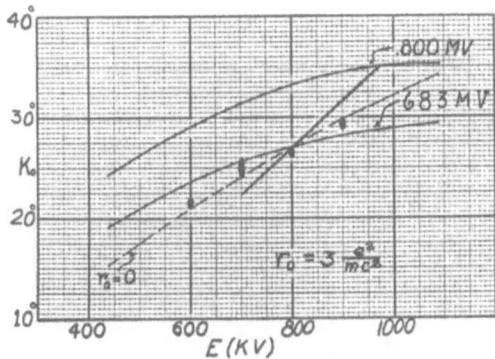

will be noted from the figure that D is determined by this fit with an apparent accuracy of about 0.2 Mev. In order to illustrate this sensitivity the differences in the values of D were used accurately even though the absolute values are perhaps not quite accurate for each r_0. These depend on the correctness of numerical conversion factors such as that occurring in Eq. (8). (Comparisons with neutron-proton forces will be made without demanding great accuracy of the conversion factors and the absolute values of D.) The heavy straight line cutting obliquely across the three curves gives an average of the dependence of K_0 on E in an earlier set of data taken by THH. This is drawn in because it illustrates how hard it would be to fit these earlier data by means of a simple theory. It should be noted that with the plausible potentials used here such steepness cannot be attained and that, therefore, the rate at which K_0 varies with the energy can be used to rule out some kinds of interactions. Fortunately the newer data represented by ovals are free of this troublesome feature. In Figs. 9, 10, 11 similar comparisons of empirical and theoretical calculations for K_0 are made for $r_0 = e^2/mc^2$, $2e^2/mc^2$, $3e^2/mc^2$, respectively. The theoretical dependence for $r_0 = 3e^2/mc^2$ is seen to fit experiment poorlty.

The theoretical curves for $r_0 = 0$, $e^2/2mc^2$, e^2/mc^2, $2e^2/mc^2$ are seen to be in fair agreement with observation. The interval from $E = 700$ kv to 900 kv appears to be too small to make it possible to determine r_0 to a higher accuracy. Inspection of the curves shows that *in a larger energy interval more precise information about r_0 should be obtainable*. It should be noted that the interaction energy $-D$ is determinable with great accuracy in all the cases considered as is obvious from the graphs. It would nevertheless be premature to claim at present an absolutely precise determination of the depth of the potential well because the question of the possible presence of the higher phase shifts has not been settled. Although it appears probable that K_2 is not the largest phase shift it is seen that a small positive K_2 of less than 2° would necessitate using a K_0 larger than what has been used by about 5°. The whole difference between the curves corresponding to $D = -47.9$ and 47.0 Mev amounts to roughly 4° in K_0. The graphs of Figs. 8, 9, 10, 11 thus give an

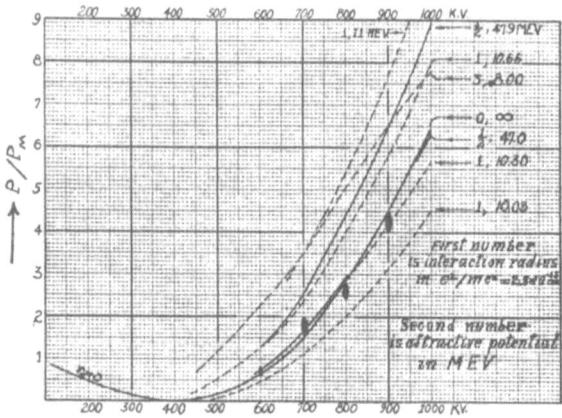

FIGURE 12. Theoretical variation of P/P_M with energy at a scattering angle of 45°. Curves are labeled by two numbers. First number gives the interaction radius of square well (r_0) in units of $e^2/mc^2 = 2.82 \times 10^{-13}$ cm. Second number gives the negative of interaction potential in mv. The ovals are obtained from the second set of data of THH using theoretical angular distribution curves fitted to experiment at higher scattering angles.

exaggerated impression of accuracy if K_2 is of importance, i.e., if the forces extend to large distances, even to a small degree.

In Fig. 12 are given theoretical and experimental values of P/P_M as a function of E for scattering at 45°. The experimental values are indicated by ovals. The full lines refer to $r_0 = 0$ and $r_0 = e^2/2mc^2$. All of the theoretical curves are labeled by means of two numbers such as 1, 10.03. The first of these gives the radius r_0 in units of e^2/mc^2. The second gives the depth in mev. For $r_0 = 0$ the graph is extended to low energies. It will be noted that around 400 kv the scattering at 45° should become very small as a consequence of dealing with an attractive potential. As the energy is decreased towards 100 kv the penetration through the Coulombian barrier becomes small and the attractive potential ceases to be effective. From 100 kv down one may expect the scattering to obey Mott's formula closely, in agreement with the experiments of Gerthsen.[14] However, even the relatively low energy of 200 kv is definitely of interest in drawing conclusions about forces between protons, since the scattering can be expected to be roughly $\frac{1}{2}$ of Mott's value in this region. The sensitivity of P/P_M to the magnitude of the interaction energy is even more striking than the sensitivity of K_0. Thus the difference between $r_0 = e^2/mc^2$, $D = 10.30$ Mev and $r_0 = e^2/mc^2$, $D = 10.03$ Mev is quite unmistakable, and the latter value is seen to be definitely excluded by comparison with experi-

---

[14] C. Gerthsen, Ann. d. Physik **9**, 769 (1931).

ment. Comparison of observation at 800 kv with that at 900 kv favors values of $r_0 < e^2/mc^2$ while comparison of 700 kv with 800 kv is in better agreement with somewhat larger ranges. This comparison is of course theoretically equivalent to that made in Figs. 8, 9, 10, 11, but it puts relatively more emphasis on the 900 kv point on account of the sensitivity of P/P_M to K_0 in this region. As before, the apparent precision in the determination of the depth of interaction may be deceptive on account of the possible presence of higher phase shifts.

Dr. J.A. Wheeler in his work on the scattering of alpha-particles has developed a criterion which makes it possible to eliminate certain kinds of potentials. This criterion will now be used to give an additional argument against the second possibility for K_0 which is due to the fact that it is related to **P** by an equation having two roots. The point of Wheeler's criterion is that the quantity $\rho F_i'/F_i$ must decrease with energy as a consequence of Green's theorem. According to Figs. 8, 9, 10, 11, 12, the K_0 which was used varies approximately as would be expected. For the other possible K_0 the following numbers are obtained at $r_0 = 2e^2/mc^2$ using Eq. (7.8). At 700 kv, $K_0 = -4.1°$, $\rho F_i'/F_i = 1.30$; at 800 kv, $K_0 = -7.3°$, $\rho F_i'/F_i = 1.54$; at 900 kv, $K_0 = -11.3°$, $\rho F_i'/F_i = 1.97$. These numbers show that $\rho F_i'/F_i$ would have to increase with energy if these K_0's were true. In the light of this, combined with the angular dependence as well as the presence in the experimental data of regions in which $P/P_M < 1$ *one must consider the choice of K_0 made here as correct* provided higher phase shifts do not interfere with the analysis. If there were even an infinite repulsive interaction through e^2/mc^2 one would expect K_0 to be approximately $-11.3°$ and it is thus *impossible to use repulsive interactions* in *s* states for the explanation of the data both on account of the wrong energy dependence and on account of the difficulty of obtaining a sufficiently large scattering anomaly. In addition the angular dependence of the scattering anomaly would be wrong. It is satisfactory to have the data point so definitely to one rather than two possibilities of interpretation.

In Fig. 13 graphical comparisons are made between the observed scattering and the Gauss error potential $-Ae^{-\alpha r^2}$. This potential is supposed to be present in addition to the potential e^2/r which represents the Coulombian interaction. With mc^2 as the unit of energy and $\hbar(Mm)^{-1/2}c^{-1} = 8.97 \times 10^{-13}$ cm as the unit of length the equation for $\mathfrak{F}$ is

$$\left[\frac{d^2}{dr^2} + E' - \frac{0.312}{r} + Ae^{-\alpha r^2}\right]\mathfrak{F} = 0.$$

By introducing $x = \alpha^{1/2}r$ and letting α, A have the probable values 17, 40 this becomes

$$\left[\frac{d^2}{dx^2} + \frac{E'}{17} - \frac{0.075(7)}{x} + 2.35(4)e^{-x^2}\right]\mathfrak{F} = 0. \tag{9}$$

The quantity $-0.0757/x + 2.354e^{-x^2}$ can be regarded as the negative of the

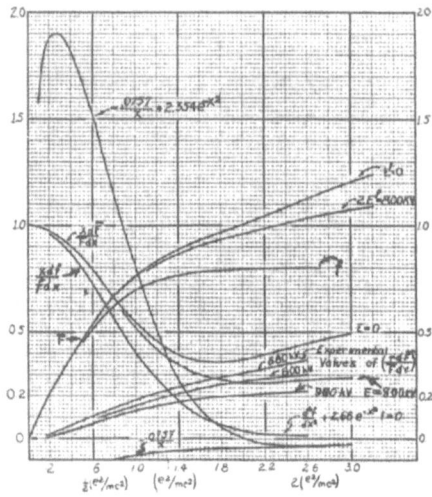

FIGURE 13. Comparison of observation with the potential Ae^{-ar^2}. Curve marked $-0.0757/x + 2.354e^{-x^2}$ represents negative of potential energy in units of $17mc^2$. Coulombian part is shown by $-0.0757/x$. Note approximate agreement of experimental value of $r\,d\mathfrak{F}/\mathfrak{F}\,dr$ for 800 kv with corresponding theoretical curve for $x\,d\mathfrak{F}/\mathfrak{F}\,dx$. In the figure $\mathfrak{F}$ is written as F with a bar over it.

potential energy in units of $17\ mc^2$. It is plotted against x in the figure. The quantity $-0.0757/x$ corresponds to the Coulombian potential and is also plotted for comparison in the same figure. The potential is seen to be very nearly Coulombian at $x = 2.6$ which corresponds to $r_0 \cong 2e^2/mc^2$. There are three curves in the figure going through the origin which represent wave functions. Two of them are marked $E' = 0$, $2E' = 800$ kv. They are regular solutions of Eq. (9) corresponding to these values of E'. The third is marked f and is a regular solution of

$$d^2f/dx^2 + 2.66e^{-x^2}f = 0. \tag{9.1}$$

This equation corresponds approximately to the condition of having $df/dx = 0$ at large distances which makes the stationary level of the two particles in each other's field fall at $E' = 0$. The function f will be used later in order to obtain the value of A for the proton-neutron interaction as well as in order to check on calculations using Eq. (9). The value of the constant multiplying factor for $\mathfrak{F}$ was not determined here by joining to the Coulomb wave functions. This factor cancels out in the applications. Corresponding to these three curves there are three other graphs starting from the point 2.0 on the axis of ordinates which represent $x\,d\mathfrak{F}/\mathfrak{F}\,dx$ and $x\,df/f\,dx$. The uppermost of these carries the label $E = 0$ at its right-hand end. It corresponds to the solution of Eq. (9) for $E = 0$. The lowest gives values of $x\,df/f\,dx$ and is labeled by Eq. (9.1) at its lowest right-hand end. The intermediate curve of the set corresponds to $E = 2E' = 800$ kv. It should be compared with the curve marked 800 kv among the three pointed out as "experimental values of $r\,d\mathfrak{F}/\mathfrak{F}\,dr$" on the figure. These "experimental values of $r\,d\mathfrak{F}/\mathfrak{F}\,dr$" are obtained from empirical values of K_0 by means of Eq. (7.8), neglecting higher phase shifts, for the energies of 680 kv, 800 kv, 980 kv, respectively. For any r these curves give the value of $r\,d\mathfrak{F}/\mathfrak{F}\,dr$ which corresponds to an experimental K_0 provided the field is Coulombian at all distances greater than this r.

Comparison of the theoretical and experimental curves for $r\,d\mathfrak{F}/\mathfrak{F}\,dr = x\,d\mathfrak{F}/\mathfrak{F}\,dx$ at 800 kv shows, as is clear from the graph, that the two curves are very nearly the same for $x = 2.6$ on towards larger r and that therefore the values of A and α used here are nearly right. The values of Feenberg and Knipp are $\alpha = 17$, $A = 41$ and the agreement is seen to be satisfactory. It is to be noted that Feenberg and Knipp considered two possibilities. In one of these the neutron-proton interaction was taken to be represented by a Majorana exchange operator. It gave $A \sim 26$ for the proton-proton and neutron-neutron interactions using the same α. In the other possibility Wigner's suggestion of using different interactions in the singlet and triplet states of the deuteron was used by regarding the proton-neutron interaction as a linear combination of a Majorana and a Heisenberg exchange operator. This view gave rise to the proton-proton and neutron-neutron interactions being represented by the Gauss error potential with $A = 41$ and $\alpha = 17$. There is on the whole a remarkable qualitative consistency in the way in which the proton-neutron scattering, the mass defects of H^2, He^3, H^3, He^4 and the proton-proton scattering fit in with each other. *The data on the scattering of protons in hydrogen are seen to speak definitely in favor of using different neutron-proton interactions in singlet and triplet states and make a pure Majorana force between neutrons and protons improbable.*

5. More Accurate Determinations and Comparisons with the Neutron-Proton Potential

It has been pointed out that, once definite values are assigned to the higher phases and to the range of the forces then accurate information can be obtained about the magnitude of the interaction potential. This is due to the fact that the experimental scattering anomaly is rather large and requires for its explanation interaction energies giving rise to approximate resonance. Thus the depth of the "square well" which was used to fit the data for $r_0 = e^2/mc^2$ is 10.3 Mev while the depth of the "square well" required to give a virtual level at $E = 0$ is 12.8 Mev. Qualitatively this condition is similar to that in the interaction of a proton and neutron in their singlet S state. In order to account for the large scattering of slow neutrons in hydrogen it is necessary to suppose that there is either a virtual or a stationary ^1S level of the proton and neutron in their mutual field. Evidence as to whether the level is stationary or virtual is very scant. It appears of interest to see whether it is possible to consider the proton-proton and proton-neutron interactions to be identical in the ^1S states. We are indebted to Dr. L.A. Young, who made estimates analogous to and qualitatively agreeing with those presented here, for pointing out to us that one should not neglect to correct the depth of the

TABLE XIV. Comparison of proton-proton and proton-neutron singlet S interactions

$r_0 mc^2/e^2 =$	1/2	1	2
$D_{\pi\pi}$ =	47.0	10.3	1.98
$D_{\pi\pi}{}^c$ =	48.7	11.1	2.42
$(D)_{E=0}$ =	51.2	12.8	3.20
$(D_{\pi v})_{43}$ =	49.3	11.9	2.73
$(D_{\pi v})_{130}$ =	48.0	11.2	2.38

"well" for the position of the stationary or virtual level in the neutron-proton potential.

In addition one must consider with greater care the possible effect of a Coulomb potential within the "square well." If it is supposed that the Coulombian force acts everywhere, a closer agreement is obtained between proton-proton and proton-neutron potentials than on the assumption that it acts only outside the square well $(r > r_0)$. Since the range of interaction is not definitely determined by either the proton-proton or the proton-neutron scattering experiments, comparisons are listed below in Table XIV, for three values of the interaction radius r_0. The depth of the square well is given in Mev. The first row in the table gives the radius in units of e^2/mc^2. The second row $D_{\pi\pi}$ gives the depth that is obtained if the Coulomb potential does not act inside the well. The third row gives similarly the depth, if the potential inside the well is taken to be $-D_{\pi\pi}{}^c + e^2/r$. In order to counterbalance the effect of the repulsion the depth has to be increased by roughly mc^2. The fourth row gives the depth required to give a virtual or stationary level at $E = 0$ taking the potential energy to be $-D$ inside the well and zero outside. In the fifth row $D_{\pi v}$ gives the depth for the proton-neutron interaction required to give a virtual level at 43 kv which corresponds to a collision cross section of 30×10^{-24} cm^2 for slow neutrons scattered by protons. The sixth row gives $D_{\pi v}$ using 130 kv for the position of the virtual level. The third and fifth rows agree fairly well indicating that the proton-proton force acting in addition to the Coulombian may be equal to the proton-neutron force in states with antiparallel spins. The third and sixth rows agree still better.

Although the numbers in the above comparison suggest very strongly that they are actually the same, caution in drawing this conclusion should be exercised for the following reasons: (a) It was not known until recently whether the neutron-proton interaction should be taken so as to give a virtual or a stationary level. If the level were stationary rather than virtual the depth for $r_0 = e^2/mc^2$ would be 13.7 Mev and there would be then no agreement between $D_{\pi v}$ and $D_{\pi\pi}{}^c$ for this r_0. However, the recent experiments of Fermi and Amaldi[6] show that the level is virtual and indicate that $D_{\pi v} = D_{\pi\pi}{}^c$. (b) The theoretical value of the neutron-proton scattering cross section is in disagreement with the experiments of Goldhaber.[7] Goldhaber's experiments are, however, contradicted by those of Tuve and may be considered as probably incorrect. (c) The values of $D_{\pi\pi}$ and $D_{\pi\pi}{}^c$ are sensitive to the value of K_2 which is used in the interpretation of proton-proton scattering experiments. (d) The numbers given in the third row of Table XIV for $D_{\pi\pi}{}^c$ would give a smaller scattering cross section of slow neutrons in hydrogen than 30×10^{-24} cm^2 by roughly a factor of four and the values of $D_{\pi v}$ given in the fifth row would give larger cross sections for proton-proton scattering at 45° by roughly a factor of 1.3. The sixth row in Table XIV corresponds to the measurements of Fermi and Amaldi and indicates a practically perfect agreement with $D_{\pi v}$ for reasonably large values of r_0. The agreement is poorer for $rmc^2/e^2 = \frac{1}{2}$ but this is a smaller range of force than is usually considered probable.

In order to be sure of the differences $\delta D = D_{\pi\pi}{}^c - D_{\pi\pi}$ the calculations were carried out by two methods one of which is a direct calculation using Coulomb wave functions inside the "well." The other is a perturbation calculation using $D_{\pi\pi}$ as a starting point. For the direct calculation only the regular wave function inside the "well" need be known. The main part of this function is the same as though there were no Coulomb field and it is convenient to arrange the series so as to take this into account. The functions Φ_0, $\Phi_0{}^*$ of Eq. (7.4) can be expressed as

$$\Phi_0 = \sin z/z + c_1 y + c_2 y^2 + \cdots;$$
$$\Phi_0{}^* = \cos z + 2c_1 y + 3c_2 y^2 + \cdots, \tag{10}$$

where z is given by Eq. (8) and the coefficients are

$$c_1 = 1, \quad c_2 = \frac{1}{3}, \quad c_3 = \frac{1}{18} - \frac{1}{9\eta^2}, \quad c_4 = \frac{1}{180} - \frac{1}{36\eta^2},$$

$$c_5 = \frac{1}{2700} - \frac{1}{270\eta^2} + \frac{23}{5400\eta^4},$$

$$c_6 = \frac{1}{56700} - \frac{1}{3240\eta^2} + \frac{7}{8100\eta^4}.$$

The perturbation calculations were made using the formula

$$\delta D = \frac{e^2}{r_0} \frac{z \int_0^z (\sin^2 z/z)\, dz}{\int_0^z \sin^2 z\, dz}$$

$$= \frac{e^2 z}{r_0} \frac{\ln 2z + 0.5772 - Ci(2z)}{z - \sin z \cos z} \tag{10.1}$$

with

$$Cix = -\int_x^\infty \frac{\cos u}{u}\, du.$$

The two methods of calculation agree to the desired accuracy. This fact is of practical interest if exact calculations are more dimcult to perform than those with "square wells." Use of the validity of the perturbation method will now be made for an improvement on calculations with the Gauss error potential. The general relations needed for extensions of the perturbation method are as follows. A given differential equation

$$[(d^2/dr^2) - E' + \lambda\chi(r) + A\varphi(r)]\mathfrak{F} = 0 \tag{10.2}$$

of the type considered here has, to within an arbitrary constant factor, one and only one solution which is regular at $r = 0$. For such a solution $r\, d\mathfrak{F}/\mathfrak{F}\, dr$ is, therefore, uniquely defined. It may be considered as a function of E', A and λ. Using Green's theorem one obtains

$$\mathfrak{F}_1 \mathfrak{F}_2 \left[\frac{1}{\mathfrak{F}_1} \frac{d\mathfrak{F}_1}{dr} - \frac{1}{\mathfrak{F}_2} \frac{d\mathfrak{F}_2}{dr} \right] + (A_1 - A_2) \int_0^r \varphi \mathfrak{F}_1 \mathfrak{F}_2 \, dr = 0$$

and similar equations with E' and λ. In the limiting cases $A_1 = A_2$, $\lambda_1 = \lambda_2$, $E_1' = E_2'$ they become:

$$\frac{\partial}{\partial \lambda} \left(\frac{\partial \mathfrak{F}}{\mathfrak{F}\partial r} \right) = -\frac{1}{\mathfrak{F}^2} \int_0^r \mathfrak{F}^2 \chi \, dr;$$

$$\frac{\partial}{\partial A} \left(\frac{\partial \mathfrak{F}}{\mathfrak{F}\partial r} \right) = -\frac{1}{\mathfrak{F}^2} \int_0^r \mathfrak{F}^2 \varphi \, dr; \tag{10.3}$$

$$\frac{\partial}{\partial E'} \left(\frac{\partial \mathfrak{F}}{\mathfrak{F}\partial r} \right) = \frac{1}{\mathfrak{F}^2} \int_0^r \mathfrak{F}^2 \, dr.$$

These formulas are useful for the following applications: (a) The determination of the change which must be made in A in order to compensate for a given change in λ so as to leave the phase shift unaltered. This is possible by means of Eqs. (10.3) as long as the rates of change of $\partial\mathfrak{F}/\mathfrak{F}\partial r$ with λ and A are nearly constant. For all that is required is to leave $\partial\mathfrak{F}/\mathfrak{F}\partial r$ unchanged at the boundary of the "well." Eq. (10.1) can be obtained by this procedure by regarding $\chi(r)$ as arising from the Coulombian energy. (b) From the values of the phase shift K_0 one can obtain $\partial\mathfrak{F}/\mathfrak{F}\partial r$ for definite energies and hence from the last Eq. (10.3) the quantity $(1/r\mathfrak{F}^2) \int_0^r \mathfrak{F}^2 \, dr$ can be determined. It may be considered as a rough form factor of the wave function. By means of the form factor information can also be obtained about the character of the interaction potential. Thus the first set of data of THH indicated a variation of K_0 such as is shown by the straight line in Figs. 8, 9, 10, 11. Computing $(1/r\mathfrak{F}^2) \int_0^r \mathfrak{F}^2 \, dr$ from the variation of $\partial\mathfrak{F}/\mathfrak{F}\partial r$ with energy gives then values >1 for the form factor which cannot be explained by the simple potentials used here but would have required other less probable possibilities. This is in agreement with the fact that the older data of THH and the data of White gave a more rapid variation of K_0 with energy than would be expected for $r_0 = 0$. The same effect is shown by the two curves for "experimental values of $r\partial\mathfrak{F}/\mathfrak{F}\partial r$" of Fig. 13 which are marked by 680 kv and 980 kv. These were computed for the old data of THH which were represented by straight lines in Figs. 8, 9, 10, 11.

Comparing them with the theoretical curves for $E = 0$ and $E = 800$ kv the experimental variation of $r\partial\mathfrak{F}/\mathfrak{F}\partial r$ is found to be too great. It will be seen presently, however, that the newer data which were taken at 900, 800, 700 and 600 kv give an energy dependence of K_0 and of $r\partial\mathfrak{F}/\mathfrak{F}\partial r$ which is in approximate agreement with expectation for the Gauss error potential used for Fig. 13.

In order to improve the comparison of the Gauss error potential obtainable from proton-proton scattering with those derivable from mass defects and from neutron-proton scattering, calculations were made in which the field was supposed to become Coulombian for $r > 2e^2/mc^2$ while for $r < 2e^2/mc^2$ the potential was taken as in Eq. (9). By means of Eq. (10.3) the value of A used in Eq. (9) was then corrected so as to give the experimental value of $r\partial\mathfrak{F}/\mathfrak{F}\partial r$ at $r = 2e^2/mc^2$ for $E = 800$ kv. This calculation gave $A = 38.5$ in units of mc^2 which is slightly lower than Feenberg and Knipp's value of $A = 41$. It should be noted that if a small positive K_2 is present this value of A should be raised and that therefore the agreement with Feenberg and Knipp may be better than $A = 38.5$ would indicate. For comparison with neutron-proton interactions it is desirable to eliminate cumulative errors which might be present in the numerical integration. This was done by using the solution for a stationary level at $E = 0$ in the absence of a Coulomb field as a starting point for the determination of both the $\pi\pi$ and the $\pi\nu$ potentials. For $A_{\pi\pi}$ this method of calculation is not very accurate because the shape of the wave function changes appreciably between the initial condition and the final one. The value of A obtained for a stationary level at $E = 0$ is 45.7. The result of applying the perturbation method using the wave function in this state (see curve marked f in Fig. 13) is to give $A_{\pi\pi} = 40.9$. Using the wave function in the final state (marked $2\,E' = 800$ kv in Fig. 13) we get $A_{\pi\pi} = 37.8$. The mean from the two perturbation calculations is 39.3 which is somewhat higher than that obtained by direct calculation. The mean of the direct determination and the perturbation method is $A_{\pi\pi} = 38.9$.

For the $\pi\nu$ interaction the perturbation calculation should be much more accurate. With the solution f as a starting point and by solving for $A_{\pi\nu}$ so as to have a virtual level at 43 kv. $A_{\pi\nu} = 42.0$. The difference $A_{\pi\nu} - A_{\pi\pi}$ is seen to be positive just as $D_{\pi\nu} - D_{\pi\pi}$ and it is of the same order of magnitude. Use of $A_{\pi\nu}$ for calculations on $\pi\pi$ scattering gives too high values and use of $A_{\pi\pi}$ for $\pi\nu$ scattering gives values too small by approximately the same amounts as for square wells with $r_0 = e^2/mc^2$. As expected, this feature of the comparison does not depend critically on the shape of the "well." Use of Fermi and Amaldi's position of the virtual level (130 kv), lowers the value of $A_{\pi\nu}$ to 39.2 which is in practically perfect agreement with $A_{\pi\pi} = 38.9$.

The rate of change of $\partial\mathfrak{F}/\mathfrak{F}\partial r$ with energy was calculated for $\alpha = 17$ by means of Eq. (10.3) and hence also the rate of change with energy of K_0. It was found that K_0 varies approximately in the same way at 800 kv as for a square well with $D = 10.3$ Mev and $r_0 = e^2/mc^2$. It definitely varies more rapidly than for $D = 1.98$ Mev and $r_0 = 2e^2/mc^2$.

Measurements at voltages below and above those used so far will be valuable in determining the range of nuclear forces as is clear from Figs. 8, 9, 10, 11 and they should be helpful in establishing the effects due to higher angular momenta. For an attractive interaction energy of 10 Mev through 2.8×10^{-13} cm the phase shift K_1 should be 0.2° and 1.5° at incident energies of 2 and 9 Mev, respectively; K_1 is roughly twice as great for $D = 2$ Mev, $r_0 = 5.6 \times 10^{-13}$ cm as for $D = 10$ Mev, $r_0 = 2.8 \times 10^{-13}$ cm at 2 Mev of incident proton energy. Attractive and repulsive potentials for $L = 1$ can be distinguished by the sign of K_1.

Summary. The experiments of THH indicate an interaction potential between protons equivalent to -11.1 Mev in a distance of 2.82×10^{-13} cm acting in addition to the Coulombian repulsion. The potential agrees closely with that obtained from mass defect calculations which use a neutron-proton interaction depending on spin orientation. Higher phase shifts than those for $L = 0$ are not called for sufficiently definitely to make their existence certain.

The magnitude of the interaction between like particles in ^1S states is arrived at here with a relatively high precision. It is compared with the proton-neutron interaction in the corresponding state as derived from the experiments of Fermi and Amaldi. The proton-proton and proton-neutron interactions in ^1S states are found to be equal within the experimental error. This suggests that interactions between heavy particles are equal also in other states.

In addition to our indebtedness to Messrs. Tuve, Heydenburg, Hafstad, Wheeler and Young which was mentioned in the text and footnotes we should like to acknowledge our gratitude to Mr. L.R. Eisenbud for conscientious help in checking the numerical calculations.

Vibration Spectra and Molecular Structure

R.B. Barnes,* L.G. Bonner** and E.U. Condon

Palmer Physical Laboratory
Princeton University

A brief survey is given of the ideas underlying the interpretation of molecular vibration spectra in connection with molecular structure problems. This is followed by a discussion of the reasons for the appearance or nonappearance in the spectrum of the characteristic frequency at 3400 cm^{-1} for various molecules containing the OH group. In certain alcohols and glycols this frequency has not been reported in the Raman effect, but it is concluded that this must be due to experimental difficulties. In the carboxylic acids, on the other hand, as well as in certain aromatic compounds containing OH, the characteristic frequency is definitely absent from both the Raman and infrared spectra, and the role of the hydrogen bond in association and chelation is discussed in connection with these cases.

I. General Remarks and a Study of the Spectrum of the OH Group

Introduction

Study of molecular structured by means of infrared and Raman spectra has progressed in two directions. One attack may be said to derive from the highly developed subject of diatomic molecular spectroscopy. Here triatomic molecules and the simpler molecules of four, five and six atoms are studied with high resolution and carefully analyzed in terms of accurate solutions of the complicated vibration problems involved. Progress in this direction is necessarily slow and beset with many difficulties.

Another attack corresponds to the one which has received attention for many years—the semi-quantitative, half-empirical correlation of observed frequencies to individual valence bonds in the molecule. In spite of the vast amount of work that has been put on this side of the subject we believe that there is still a great deal to be learned about molecular structure by a critical

---

* Now with the American Cyanamid Company.
** National Research Fellow.

discussion of the data in terms of these less precise methods. Not only can many questions of molecular structure be answered in this way now, but a careful study of the empirical material from this standpoint will be a necessary preliminary to the extension of precision methods to complicated molecules. Therefore it is our intention in the series of papers of which this is the first, to make a careful survey of the empirical material in terms of the simpler methods of interpretation.

This paper includes a survey of the general ideas used and a discussion of factors affecting the appearance of the band due to the hydroxyl group in various compounds.

Valence Bond Frequencies

A well-known fact of infrared spectroscopy and of its lusty nephew, Raman spectroscopy, is that certain frequencies can be correlated closely with the occurrence in the molecule of certain valence bonds. Curiously enough such correlations are often more constant and uniform in complicated molecules, the notable departures occurring in the lowest members of homologous series. This is connected with the fact that symmetry restrictions usually affect the simpler molecules while in more complicated structures the symmetry elements are lost. Also, in a series of complex molecules the environment of a given bond is more likely to be constant.

If the situation were so simple that there would always be associated a vibration frequency with each type of bond in the molecule, then obviously the molecular vibration method of analysis would provide a very powerful means of investigating the structures of molecules. The situation is in fact almost that simple and a large part of the spectrum in the fundamental vibration region is understandable at a glance. In a recent paper by one of us[1] there is given a brief survey of the type of problem to which these methods are applicable, and of the mode of attack used in some cases. Suffice it to say that important information has already been obtained in the fields of deutero-chemistry, polymerization, isomerism, chelation, and many others and that the possible extensions in these and other fields appear to be almost unlimited. In the present paper we wish to point out that exceptions to the principle of constant bond frequencies frequently occur, and that from these exceptions information is obtained which is just as valuable as that obtained from the regularities. When such exceptions occur it is often possible to give a simple dynamical explanation in terms of the vibration forms of the molecule. If such an explanation is not possible it can frequently be shown that the classical structural formula of the molecule under investigation is not entirely correct.

As regards the dynamical effects mentioned above, it should be pointed out that it is the bond force constant and not the position of the infrared

---

[1] R.B. Barnes, Rev. Sci. Inst. 7, 265 (1936).

absorption or Raman frequency that is characteristic of the specific bond. Therefore a knowledge of these bond force constants is essential if the effects on the positions of the molecular frequencies caused by such phenomena as the interaction between two neighboring similar bonds is to be correctly explained. For this reason frequent reference will be made in what follows to the values of these constants and to the methods used in their calculation.

Following the ideas of Mecke, Dennison, Sutherland and others[2-4] we suppose the principal forces involved are those of a valence force system, namely forces between those atoms which are chemically bonded, obeying Hooke's law for small displacements from equilibrium, and a system of torques tending to preserve the bond angles at the tetrahedral value for carbon and at the appropriate other angles for other polyvalent atoms. In the more accurate analyses other terms are inserted to represent interaction forces between different valence bonds but their effects are usually small.

The next point to make is that accurate analysis of simpler cases has shown the forces preserving valence angles to be much weaker than those available for preserving bond distances. This fact together with the fact that "straight-chain" compounds are actually zigzag in form tends to reduce the amount of dynamical interaction between different valence bond oscillations, as was pointed out by Bartholomé and Teller[5] and recently by Bauermeister and Weizel.[6] Because of the weakness of the directed valence forces those frequencies in which angles vary are generally lower in value than the valence oscillations, being generally less than 1000 cm^{-1}. Also as a result of this weakness the deformation frequencies are affected more strongly by neighboring atoms or groups and do not show such simple empirical regularities as the valence oscillations. They will, therefore, receive almost no consideration in what follows.

To the approximation that Hooke's law forces are obeyed, every theoretical formula relating a vibration frequency to the force constants and masses in a molecule will have a dimensional similarity with that for the simple harmonic oscillator,

$$\tilde{v} = (2\pi c)^{-1}(k/m)^{1/2}. \tag{1}$$

Here $\tilde{v}$ is the frequency in cm^{-1}, m is the mass and k is the force constant in the equation connecting potential energy, V and displacement x from the equilibrium position,

$$V = \tfrac{1}{2}kx^2. \tag{2}$$

It is convenient to standardize the units, once and for all. All known values

[2] R. Mecke, Zeits. f. physik. Chemie **B16**, 409, 421; **B17**, 1 (1932).
[3] D.M. Dennison, Rev. Mod. Phys. **3**, 280 (1931).
[4] Sutherland and Dennison, Proc. Roy. Soc. **A148**, 250 (1935).
[5] Bartholomé and Teller, Zeits. f. physik. Chemie **B19**, 366 (1932).
[6] Bauermeister and Weizel, Physik. Zeits. **37**, 169 (1936).

of the force constant are of the order of 10^5 dyne/cm so we shall adopt this as a convenient unit for k. We have:

$$10^5 \text{ dyne/cm} = 10^5 \text{ erg/cm}^2 = 6.284 \ eV/\text{A}^2$$

$$= 509.5 \text{ cm}^{-1}/(10^{-9} \text{ cm})^2$$

$$= 1.449 \text{ (cal/mole)}/(10^{-9} \text{ cm})^2, \tag{3}$$

which enables conversions of k to other energy and length units to be readily made. For the masses it is convenient to use atomic weights on the oxygen = 16 scale. Using Birge's values for the universal constants the formula that is the dimensional model for all normal modes solutions of vibration problems becomes

$$\tilde{\nu} = 1307 \ (k/m)^{1/2}. \tag{1'}$$

We turn now to a rapid survey of the fundamental frequencies and force constants characteristic of bonds of the type commonly occurring in organic molecules.

Single Bonds with Hydrogen: CH, OH, NH, SH

Nearly every molecule considered contains one or more CH bonds. All such show infrared absorption and Raman lines between 2600 and 3050 cm^{-1}, the strongest frequencies in aliphatic molecules being generally close to 2900. In aromatic molecules the frequency is generally higher being near 3050 as a rule. The simplest view of the matter is to suppose the CH bonds capable of vibration pretty much as independent units in spite of their being part of the larger structure. To that approximation the frequency is given by (1).

With increasing complexity of structure the number of frequencies attributable to CH increases. Here, however, the emphasis is on the comparatively narrow range of these frequencies. The frequency 2918 in methane corresponds to symmetric contraction-expansion of the tetrahedron: here the carbon stands still and the four hydrogen atoms move with alteration of the four bond distances. The frequency may be calculated from (1') with $m = 1$, and k the constant for the CH bond. This yields a value of $k = 5.00$ which agrees well with the value 5.02 obtained by Bonner[7] from a detailed analysis for ethylene. For further comparison we note that the frequency in the normal electronic state of CH in the diatomic band system is 2859, which together with the reduced mass value 0.924 gives 4.44 for k. Provisionally we postulate that k has the value 5.00 for all organic CH bonds.

Characteristic of all compounds containing the OH group is a frequency close to 3400. Since this group has many interesting special properties and since there is a good deal of confusion about it in the literature it is dealt with

---

[7] L.G. Bonner, J. Am. Chem. Soc. **58**, 34 (1936).

in detail in the latter part of this paper. In diatomic OH from electronic bands, the normal state frequency is 3735 corresponding to $k = 7.72$ while in H_2O vapor the two normal modes which are essentially valence oscillations are 3600 and 3756, and the force constant for OH found n Bonner's[8] analysis of the water spectrum is 8.233. The value obtained from 3400 assuming the H to be vibrating against a large mass (reduced mass = 1) is $k = 6.8$.

Frequencies characteristic of the NH bond are not greatly different from those for OH, being in the region around 3300. Here the force constant from the diatomic frequency is 5.22 while that calculated from the frequency 3320 using for the reduced mass $\mu = 1$ is 6.4.

These three commonest types of bonds involving hydrogen have therefore frequencies in the range from 2800 to 3400 and may give rise to some overlapping in infrared studies with low dispersion, as with a single prism rocksalt instrument. Nevertheless it is often possible to distinguish between the bands due to these bonds as will appear later.

Characteristic of SH is a remarkably constant frequency which stays between 2570 and 2580 in the mercaptans. This is a convenient location as it happens that there are usually no other bands in this immediate neighborhood. This frequency corresponds to the much weaker force constant $k = 3.85$.

Single Bonds: C—C, C—O, C—N

After CH, the CC linkage occurs most frequently in organic chemistry. For the single bond CC linkage the general range of frequencies is 800 to 1150. Here there is a tendency to develop a fairly open structure of increasing complexity with growing length of carbon chain. This interaction splitting continues as the length of the chain increases, and leads to the often observed differences between the spectra of the first and higher members of homologous series. This effect has been treated recently by Bauermeister and Weizel.[6]

In ethane, for example, we find a single CC frequency at 993, whereas in propane we find two frequencies characteristic of this bond, one at 867 and the other at 1055. This splitting is an example of the type of dynamical interaction mentioned above, resulting from the presence in the molecule of adjacent similar bonds. Further examples are found in the spectra of the C—O—C and the C=C=C groups.

At this point it is perhaps timely to say a few words concerning a useful approximate method of force constant calculation introduced by Mecke,[2] and used with considerable success by him and by Sutherland and Dennison.[4] Since the hydrogen atoms, which are found on practically all molecules investigated to date, are very light, and their vibration frequencies as have been shown are high compared with those of neighboring groups of heavier atoms, it is frequently possible to neglect the effect of these hydrogens and to

---

[8] L.G. Bonner, Phys. Rev. **46**, 458 (1934).

treat the groups containing hydrogen as rigid bodies. To this approximation, ethane would be treated as a diatomic molecule, each body being of mass 15, while propane would be triatomic, with two mass 15 bodies and one of mass 14. The applications and limitations of this rigid group approximation will be illustrated for a few cases in what follows, and will be taken up in considerable detail in later papers of this series. As an example of such a simplified calculation, ethane with a reduced mass of 7.5 yields a force constant $k = 4.32$, which is an entirely reasonable value.

The single bond CO frequency is not always recognizable since it is so close to the single bond CC frequency. However, in methyl alcohol the frequency 1030 is readily associated with this vibration. Regarding it as a vibration of OH against CH_3 this gives $k = 4.96$ for the CO force constant.

Since what has been said above illustrates well the behavior of bond frequencies and the approximate constancy of the force constants deduced therefrom, we feel it to be unnecessary to discuss further cases in detail. At this point we shall point out simply the ranges of the frequencies and force constants associated with some of the remaining principal bond types. For a detailed compilation of the frequency values the reader is referred to standard works, such as those of Schaeffer and Matossi,[9] Kohlrausch[10] and Hibben.[11]

To repeat, for purposes of comparison, the single bond values given above, we find that, for single bonds between light atoms the frequencies all lie in the range 800–1150 cm^{-1} and the force constants between 4 and 5 (in units of 10^5 dynes/cm). Double bond frequencies are generally between 1500 and 1800 corresponding to a force constant in the neighborhood of 10–12. The triple bond, again, is still higher, ranging from 1900–2300, with a force constant range of 15–20. In the methyl halides frequencies probably to be associated with the C—Cl, C—Br, and C—I bonds are found at 710, 600, and 530, respectively. It must be emphasized that these values apply principally to the aliphatic or open chain compounds and may differ appreciably from the frequencies for the same combinations existing in ring structures.

It is hoped that with this brief introduction we have made clear the empirical background underlying these studies and have outlined the aims, methods and terminology of these papers.

The Spectrum of the OH Group

Since the assumptions implied above are that any molecule containing an OH group will have a frequency in the neighborhood of 3400 cm^{-1} and since many molecules which are known, by chemical evidence, to have such a group do not exhibit this frequency, one must look for an explanation of this apparent

---

[9] Schaeffer and Matossi, *Das Ultrarote Spektrum.*
[10] Kohlrausch, *Der Smekal Raman Effekt.*
[11] J.H. Hibben, Chem. Rev. **18**, 1 (1936).

anomaly. In what follows we propose to point out some of the principal exceptions to this rule of constant bond frequencies and to draw attention to some of the factors that may lead to the disappearance of this frequency.

Raman Effect

Careful study of the original literature and of the reviews mentioned above reveals the remarkable fact that throughout the period of development through which the Raman effect has passed since its discovery in 1928 the overhelming majority of observers failed consistently to record frequencies in the 3400 region. In view of the fact that the strong band which occurs in the infrared spectra of all alcohols at about 2.9μ (3400 cm^{-1}) has, since te days of Julius and Aschkinass been identified with the internal vibration of the OH group, it is extremely surprising that this Raman frequency should continue to be conspicuous by its absence. On page 17 of his review, Hibben states, with regard to the 3400 shift, "Unfortunately ... this can only be detected in less than half the alcohols studied." On page 20 one sees that neither propylene glycol, glycerol nor glucose shows an OH frequency at 3400, if the list of frequencies reported is to be taken as complete. Wood and Collins[12] recently reported Raman measurements upon a beautifully complete series of eleven normal alcohols but failed to record any frequency higher than 3000. Venkateswaran and Bhagavantam[13] presented an extensive table of frequencies observed for alcohols; none, however, higher than 3000 cm^{-1}. In a footnote they remarked that "The Raman spectrum of methyl alcohol shows, besides the lines given, ..., a prominent broad band corresponding to a shift of about 3μ ...." The implication is that the remaining alcohols did not contain even this broad band. So it is with numerous other observers and in the cases of many other compounds containing OH. In view of the fact that all of the compounds herein mentioned do exhibit strong OH absorption in the infrared, as shown in part in curves 1–5 of the figure, and since most of them are reasonably complex molecules so that the possibility of the corresponding Raman frequencies being forbidden by selection rules based on symmetry is very small, the absence of these frequencies is puzzling. Recent Raman measurements in this laboratory have indicated the presence of the weak broad band at about 3400 cm^{-1} characteristic of the OH group, in two of the compounds mentioned above, namely propylene glycol and glycerol. As a result it is clear that caution must be used in basing analyses of the type mentioned in the introduction solely on the results of Raman measurements. The authors have found that a good general principle to be applied in the interpretation of Raman results is that, although the appearance of a line is definite, the nonappearance may be due to any one of a number of factors and does not necessarily indicate that the frequency is absent in the molecule.

[12] Wood and Collins, Phys. Rev. **42**, 386 (1932).
[13] Venkateswaran and Bhagavantam, Ind. J. Phys. **5**, 129 (1930).

Infrared

As already mentioned, all the alcohols heretofore measured have shown a strong OH absorption in the infrared, and, in fact, practically all compounds containing OH exhibit this band. There are, however, a few cases in which this absorption is definitely absent, and it is to these that we shall turn our attention.

Molecules Containing the Carboxyl Group

Sappenfield,[14] Eichmann,[15] Roth,[16] and Errera and Mollet[17] have all reported on infrared absorption of the fatty acid series, and are all in agreement that the overtones of the 3400 frequency are missing in the lower members. In addition measurements in this laboratory on formic, acetic, propionic and valeric acids in the fundamental region show that this frequency is missing for the lower members and appears only very weakly for some of the higher members of the series. This effect is exemplified by curves 6 and 7 of the figure.

The fatty acids are in general highly associated, some fifty percent of the molecules of the lower members existing as dimers even in the vapor state at an elevated temperature, according to Drucker and Ullmann.[18] This gives us the explanation for the disappearance of this band. According to Sidgwick,[19] this association takes place through the OH group of each interacting molecule, as given by the structural formula

$$
\begin{array}{ccccc}
R & & R & & R \\
| & & | & & | \\
H-O-H & -O-H & -O.
\end{array}
$$

Pauling and Brockway,[20] however, have shown by electron diffraction experiments that the dimer of formic acid has a ring structure of the form

$$
\begin{array}{c}
O-H-O \\
H-C \qquad\qquad C-H. \\
O-H-O
\end{array}
$$

In either case, since a hydrogen bond is formed between the oxygen atoms of two separate molecules, it seems clear that the bonding between H and O must be different from that in normal OH, with a corresponding effect on the

---

[14] J.W. Sappenfield, Phys. Rev. **33**, 37 (1929).
[15] O. Eichmann, Zeits. f. Physik **82**, 461 (1933).
[16] A. Roth, Zeits. f. Physik **87**, 192 (1933).
[17] J. Errera and P. Mollet, J. de physique **7**, 281 (1935).
[18] Drucker and Ullman Zeits. f. physik. Chemie **74**, 604 (1910).
[19] Sidgwick, *The Electronic Theory of Valence* (Oxford, 1929).
[20] L. Pauling and L. Brockway, Proc. Nat. Acad. Sci. **20**, 336 (1934).

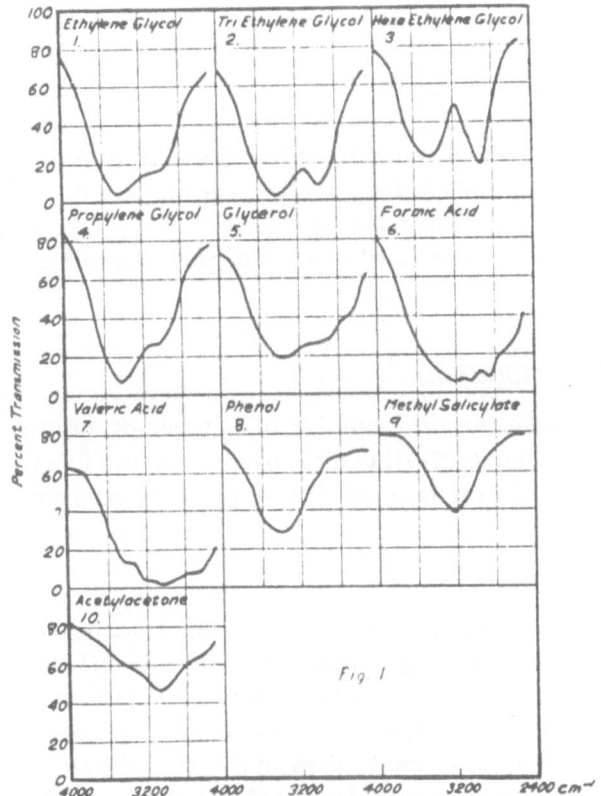

FIGURE 1. These infrared transmission curves cover, for several compounds containing the hydroxyl group, the regions of the fundamental OH and CH absorptions at about 3400 and 3000 cm^{-1}, respectively. The absence of the OH absorption is to be noted in formic acid, methyl salicylate and acetylacetone, curves 6, 9, and 10.

vibration frequency. In fact, Pauling and Broackway showed that the O—O distance in the formic acid dimer is 2.66A, indicating that the O—H distance is perhaps 1.33A, rather than the normal 0.96A.

Association through linkage of the OH appears to a certain extent in most hydroxyl containing molecules, and the relative intensity of the 3400 absorption region provides an indication as to the extent to which this has taken place. Even the alcohols, which were cited above as showing the OH frequency in the infrared, are known to be slightly associated. Kinsey and Ellis[21] found that the characteristic band, for the case of methyl alcohol, was shifted slightly, sharpened, and increased in intensity by passing from the liquid to a solution in CCl$_4$. Dilution with a solvent tends, of course, to diminish association,

---

[21] E.L. Kinsey and J.W. Ellis, Phys. Rev. **49**, 105 (1936).

and Kinsey and Ellis found that the absorption in solution bore a close resemblance to that of the vapor, which is almost certainly unassociated. As might be expected, an increase in temperature also has a tendency to decrease the degree of association, as was shown by experiments of Bloch and Errera.[22]

Molecular association is affected by the weight and volume of the interacting molecules, as illustrated by curves 6 and 7. In formic acid the OH absorption is completely lacking, while in valeric there does appear a slight trace of absorption in this region. This same effect has been found by Errera and Mollet,[17] who investigated a series of acids, including caproic, in the overtone region.

Chelated Compounds

It has been shown that in a number of compounds, principally aromatic compounds containing an OH in the *ortho* position to a C=O or N=O, the possibility exists of forming a ring. This ring, usually six membered, is closed by a hydrogen bond, again involving the hydrogen of the OH group. Both Sidgwick and Errera and Mollet present much evidence of a chemical and physical nature justifying the ring structure for methyl salicylate, salicylaldehyde and many other similar compounds. The final proof seems to be supplied by the fact that none of these compounds exhibit any OH absorption, as was shown by Errera and Mollet and by Hilbert, Wulf, Hendricks and Liddel[23] by measurements in the overtone region. Curves 8 and 9, taken in this laboratory, show how, in the fundamental region, the strong OH band of phenol is completely wiped out when the C=O is placed in the *ortho* position to it in methyl salicylate. Acetylacetone, which is known from other evidence to exist almost entirely in the enol form, is also capable of forming a chelate ring, as is shown by the absence of the 3400 band in curve 10.

In the case of these compounds, as also for the fatty acids, it is interesting to speculate as to the fate of this OH absorption band. Certainly, even though the frequency is greatly affected, there must still be some vibration characteristic of the motion of H against O, though it has not yet been located. It seems reasonably certain that the strength of such a hydrogen bond is only a small fraction of that of the normal OH bond. That is to say, the dissociation energy of the O—H—O link formed in associated or chelated compounds is of the same order of magnitude as that of an ordinary OH. In formic acid Pauling and Brockway found, however, that the distance between two oxygen atoms bound by a hydrogen bond was about 2.67A, which agrees well with previous crystal structure values, and with distances calculated by Errera and Mollet from known angles and distances alone. If, then, the H atom remains at approximately the normal distance of 0.96A from the oxygen to which it was originally bonded it will be 1.71A from the other. It is known that

[22] B. Bloch and J. Errera, J. de physique **6**, 154 (1935).
[23] Hilbert, Wulf, Hendricks and Liddel, Nature **135**, 147 (1935).

interatomic forces decrease very rapidly with distance so the vibration in this case would be essentially H against one O and the energy consideration given above would indicate that the frequency would be almost unchanged. Since this is obviously not so, let us turn to the perhaps more reasonable case of the hydrogen atom midway between the two oxygens, with the force so distributed that each OH bond is the same. Assuming that the force constant varies with distance in the inverse cube fashion suggested by Badger[24] it is possible to calculate the OH vibration frequency for this case. Assuming further that the oxygen atoms are stationary and that the equilibrium O—H distance is 1.33A we find, that the 3400 cm^{-1} frequency shifts to nearer 2400. However, it is not necessary that the H lie on the O—O line and the O—H separation may be larger than that used above. Below are given frequency values corresponding to several O—H separations, taking the O—O distance constant at 2.67A.

$$r \quad 1.33 \quad 1.40 \quad 1.50 \quad 1.60 \quad 1.70$$

$$\bar{\nu} \quad 2390 \quad 2050 \quad 1680 \quad 1390 \quad 1170.$$

In a recent paper Gillette and Sherman[25] seem to favor a structure for the formic acid dimer in which the hydrogen atom, while located on the line of centers of the oxygen atoms, is neither in the center nor at the normal O—H distance from one. This would lead to a vibration frequency between the above 2400 and 3400 cm^{-1}. It would appear from this that almost any value of the OH frequency in compounds of this type would be reasonable, and it is unfortunate that as yet no spectrum has been analyzed carefully enough to permit of its identification.

---

[24] R.M. Badger, J. Chem. Phys. **2**, 128 (1934).
[25] R.H. Gillette and A. Sherman, J. Am. Chem. Soc. **58**, 1135 (1936).

Ionization and Dissociation of Molecules by Electron Impact[1]

W. BLEAKNEY, E.U. CONDON, AND L.G. SMITH

Palmer Physical Laboratory
Princeton University

Among the methods available for experimental study of molecular structure
an important place is occupied by that one which uses a mass spectrograph
to study the kind of products formed when molecules are struck by electrons
of known energy. In the following paper we wish to give some illustrations of
work done in this field including a brief account of some new results recently
obtained which will be published in detail elsewhere.

By a mass spectrograph is meant any apparatus embodying a combination
of electric and magnetic fields which sorts out the ions formed in a source
according to the value of the ratio of mass to charge of the ion, enabling a
measurement of the ratio to be made and also a measurement of the relative
number of ions of each kind as conditions in the source are varied. Research
in this field started about twenty years ago, and was at first directed to study
of atomic ionization processes and to study of isotopes. In a few years the
importance of the mass spectrograph for studies of molecular structure was
recognized independently by H.D. Smyth in Princeton nnd Hogness and
Lunn in Berkeley.

A good review of the field up to about five years ago is found in a paper
by Smyth (14). For a rational classification of the various types of spectrographs with illustrations of their applications reference may be made to
a recent paper by Bleakney (3). More recent reviews of the experimental
material have been prepared by de Groot and Penning (5) and also by
Sponer (15).

I. Results Obtained with Diatomic Molecules

Most of the ideas involved in the application of mass spectroscopy to studies
of molecular structure are illustrated by considering in detail the well-studied
case of molecular hydrogen. In figure 1 there are plotted the potential energy

---

[1] Presented at the Symposium on Molecular Structure, held at Princeton University,
Princeton, New Jersey, December 31, 1936 to January 2, 1927, under the auspices of
the Division of Physical and Inorganic Chemistry of the American Chemical Society.

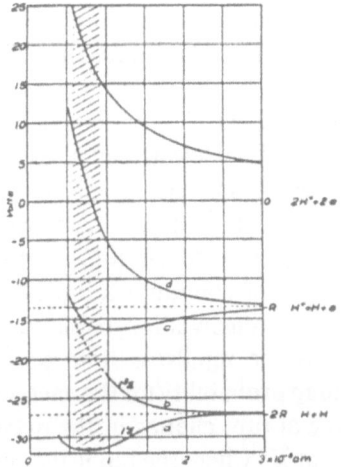

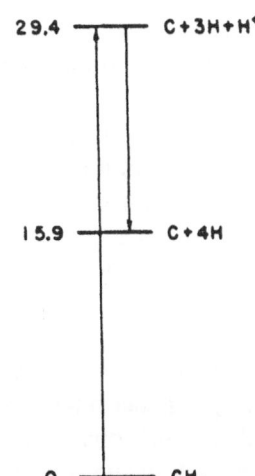

FIGURE 1. Potential energy curves for a number of electronic states of the hydrogen molecule.

FIGURE 2. Energy level deduced from the 29.4 volt appearance potential of H^+.

curves for a number of electronic states of the hydrogen molecule, neutral and singly and doubly ionized. Of these we are most sure of the uppermost, that for doubly ionized H_2, for this is merely two protons and therefore curve e is simply the function e^2/r representing the Coulomb interaction of the two particles. Curve a is for the normal state of neutral H_2, as obtained from band spectra using the moment of inertia, the vibration frequency, and the heat of dissociation to fix the main features of the curve. Curve b is for the repulsive $^3\Sigma$ state of neutral H_2, first calculated by Heitler and London, and recently calculated more exactly by James, Coolidge, and Present (9). This new calculation puts the curve several volts lower than the first-order perturbation calculation of Heitler and London, in the region of nuclear separations near the nuclear separation for the normal state. Curves c and d correspond, respectively, to the stable and unstable electronic states of H_2^+ that are based on combination of a proton and a normal hydrogen atom (12, 18). In addiition, a complete diagram would show a large number of curves corresponding to various electronically excited states of H_2 and H_2^+, but these are not of interest here.

In considering the processes which may be expected to occur when electrons strike normal hydrogen molecules, we are guided by an idea which has come to be known as the Franck-Condon principle (6, 4). This is simply a recognition of the fact that light quanta or electrons transport a negligibly small amount of momentum for a given amount of energy as compared with the momentum which the massive nuclei possess when they have the same amount of energy. Therefore thc distribution of momentum of the heavy particles of a nucleus and their instantaneous positions will be essentially the same after an electron impact (or light quantum emission or absorption) as it is before. Hence among transitions that are energetically possible, the most probable will be those that call for least change in the positions and momenta of the massive nuclei.

This is indicated on the diagram by the shaded vertical region extending up from the minimum of the curve for the normal electronic level. The width of this region is the amplitude of the zero-point vibrational motion. We consider the effect of striking hydrogen molecules with electrons whose energy is gradually increased. To excite the molecule to vibration without making an electronic change would call for direct communication of momentum to the nuclei by the electron, which is extremely improbable. As the voltage is increased above 4.4, transitions to the repulsive state become energetically possible, but at first are of vanishing probability for the energetically possible transitions would call for a large abrupt change in the nuclear separation. Only as the voltage gets up to about 10, where the repulsive curve enters the shaded area, do favored transitions become energetically possible. Molecules making such transitions find themselves in an unstable electronic state with some 6 volts more energy than necessary for dissociation, so they dissociate at once into two normal hydrogen atoms each having about 3 volts of kinetic energy. A critical potential is observed in this region, which may be interpreted in this way, but of course the mass spectrograph cannot check the details since the products are not charged. Chemically such atoms should be extremely reactive, for they have a translational energy of the order of 75 kg-cal. per mole.

Between 15 and 16 volts the shaded region is crossed by the curve for the normal electronic state of H_2^+, so this is where we have to expect appearance of the molecular ion. Since the minimum of the curve does not lie in the shaded region, we expect that as the energy of the electrons is increased above 16 volts the most probable mode of formation of H_2^+ will be by transitions which give the ions some vibrational energy. As the repulsive part of curve c is still within the shaded area at a level of 18 volts, which is enough for simultaneous ionization and dissociation of the molecule into a proton and a normal hydrogen atom, we may expect to observe a slight yield of atomic H^+ ions at this voltage.

An interesting confirmation of these views is afforded by some hitherto unpublished results obtained by Mr. J.A. Hipple in this laboratory. For electron voltages just above 18 volts be compared the yield of monatomic and diatomic ions in hydrogen and in deuterium. The electronic potential energy curves for the two molecules are the same to within the accuracy needed here. But the greater mass of deuterium reduces the zero-point energy of vibrations in the normal electronic state, with corresponding reduction in the zero-point amplitude of vibration. Therefore the shaded region should be drawn but $2^{-1/4}$ as wide for deuterium as for hydrogen. This should considerably reduce the yield of atomic ions as compared to molecular ions from transitions to curve c. This is exactly what was found experimentally. Using 22-volt electrons Hipple found the ratio $H_2^+/H^+ = 104$, while in heavy hydrogen the corresponding ratio is $D_2^+/D^+ = 292$.

As the voltage is increased to about 30 volts, we come to the place where curve d enters the shaded region. At this place transitions to curve d may occur, which result in formation of atomic hydrogen in the normal state and protons, each with about 7 volts of kinetic energy. Curiously enough this energy happens to coincide with the energy needed for complete dissociation of the hydrogen molecule (4.4 plus twice 13.5 volts), and so this critical potential was at first attributed to such a process. That our interpretation is the correct one was shown by Bleakney (1) by experiments which demonstrated that the protons produced in this process have kinetic energy. More exact studies of the processes in which the ions are formed directly with kinetic energy have been made by Lozier (10) and, for corresponding processes in nitrogen, oxygen, and carbon monoxide, by Tate and Lozier (16, 11). Finally at some 45 volts energy for the impacting electron we come to the place where the simple Coulomb potentinl energy curve for the H_2^{++} molecule (!) crosses the shaded area. At these voltages electron impact will be able to strip off both electrons from the hydrogen molecule, leaving the two protons at the same distances as they were in the original molecule in the first instant. They then move out, changing their approximately 18 volts of potential energy into kinetic energy, which is equally divided-between them because of their equal mass.

This account of what happens in hydrogen has been given in full, for it represents most of the ideas and is so satisfyingly complete in its correlation of theory and experiment. Moreover, processes of this kind gave the first definite experimental evidence for the reality of the repulsive potential energy curves that were obtained from quantum-mechanical calculation of electronic states.

Similar studies have been made for other simple diatomic molecules with, however, the difference that in the other cases one cannot say so much about the potential curves from theory, owing to the difficulty of the calculations involved. The method gives us information supplementary to that obtained from molecular spectra about the potential energy curves leading to stable molecular states, and also additional rough information about the location of repuliive potential energy curves. A recent compilation of collision data has been given by Sponer (15).

II. Polyatomic Molecules

As is the case with every method of molecular structure study, the problems presented by studies of polyatomic molecules with the mass spectrograph are considerably more complicated than for diatomic molecules.

The method of electron collisions is not a universal one since it, like all the other methods, tells only a small part of the story of molecular structure, but this small part is nevertheless a significant and useful one. It is primarily concerned with the number of ways in which a molecule may be broken up, the energies necessary for these transformations, and the probabilities of their occurrence. Since a review of this whole field is not feasible at this time, we will confine our discussion to a few carbon compounds. It is a remarkable fact that all or almost any combination of the valence bonds of a molecule may be broken by a single electron impact if the transformation is energetically possible. As an illustration we cite the case of benzene. Nier[2] found that as the result of a single impact all the hydrogens may be stripped off, leaving a C_6^+ ion. This result is somewhat surprising, since, according to the Franck-Condon principle, the energy (~ 35 volts) must first go into an electronic excitation.

The chief difficulty with the interpretation of the results when the mass spectrograph is applied to the study of polyatomic molecules is the lack of information on what happens to the neutral particles. In addition to the discrete energy levels to which they may be excited, there is the possible continuum of translational kinetic energies which serve to obscure the final potential energies of the products. The situation is not hopeless, however, as the analysis given below will indicate. Fortunately the states of the hydrogen atom and molecule are well known, and the first excited state of the hydrogen atom is so far above the ground state that excited states of this atom in dissociated products of molecules containing hydrogen atoms are seldom a cause of misinterpretation. It is to be remembered that the ionization potentials represent the minimum electron energy at which the particular ion in question appears. It is therefore natural to assume, if there is no evidence to the contrary, that the products of ionization are in the lowest electronic states. The question of the vibrational state is however a different matter, since it involves nuclear motions which respond slowly to changes in bond energies.

Although it is undoubtedly true that the Franck-Condon principle restricts the vibrational transitions which accompany an electronic change in a poly-atomic molecule, the situation is here so much more complicated that the principle is not of much help. In a polyatomic molecule containing N atoms there are $3N - 6$ internal degrees of freedom for the nuclear motion, so in place of a potential energy curve as in the diatomic case we shall have a potential energy surface giving the energy as a function of the $3N - 6$ internal coordinates of the nuclear frame of the molecule. There will be one such surface for each electronic state of the molecule. At present we know almost nothing about such surfaces except in the few case, which have been studied by Eyring, Polányi, and others in connection with problems of chemical kinetics. The analysis of Raman and infra-red vibration spectra gives us

---

[2] Private communication from Dr. A.O. Nier.

information about the shape of the surface of the lowest electronic state in the neighborhood of its minimum, and we may hope to learn about excited states from analysis of electronic band systems of polyatomic molecules.

The picture then is this: By appropriate choice of coordinates, in a $(3N - 6)$ dimensional configuration space the motions of the nuclear frame will be the same as the motion of a single mass point in the configuration space moving under the forces described by the potential energy surface. Before being struck by an electron the representative mass point is moving on the potential energy surface of the normal electronic state. This means it is performing small oscillations around the minimum of this surface. The electron impact changes the electronic state, substituting a new potential energy surface to govern the mass point's motions. The mass point then begins to move on the new surface starting from the initial configuration that corresponds to the initial state before impact. The ensuing motion has much greater possibilities for complexity than in the diatomic case, with the general result that a process of dissociation that is energetically possible is much more likely to occur.

It is tempting to try to make progress by assuming some gross oversimplifications of the surfaces in the absence of better information. Thus in methane the H—H distance is greater than in normal H_2 or H_2^+. Does this mean that if H_2^+ is formed by electron impact in methane it will necessarily be formed with vibrational energy? At first sight it might seem that we could reason thus: From the fact that H_2^+ appears at all we know that two C—H bonds are broken in the excited electronic state of the CH_4^+ complex from which it is produced. If the bonds are broken we might neglect the forces between C and H altogether and assume the force between the two hydrogen atoms to be the same as in normal H_2^+. In that case we could use the curves of figure 1 to make a definite estimate of the amount of vibrational energy in the H_2^+ produced.

In a first draft of this paper we had analyzed the data for methane that are discussed below, from this point of view. It appears on going over the matter more carefully that such arguments are too inaccurate to be of value. This is emphatically brought out below in the striking contrast between the processes of forming C_2^+ in C_2H_2 and C_2N_2, respectively.

Experimental Results for Methane

The ions produced by electron impact in methane were studied with a mass spectrograph of the type described by Bleakney (2). The positive ions observed, together with their relative intensities and their appearance potentials, are shown in table 1. In addition, several negative ions were found, which will be described in a later paper.

The positive ions were all observed using first a tungsten filament and later an oxide-coated, equipotential cathode, run at about 1000°C., as source of electrons, the relative intensities of the first five ions being the same in both cases, while the intensities of the last three were not recorded with the tungsten

TABLE 1. Positive ions formed in methane

m/e	ion	Percent of total ionization (electron energy = 50 volts)	Appearance potentials
			volts
16	CH_4^+	50.7	13.1 ± 0.4
15	CH_3^+	39.5	14.4 ± 0.4
14	CH_2^+	4.2	15.7 ± 0.5
13	CH^+	1.7	23.3 ± 0.6
12	C^+	0.6	26.7 ± 0.7
3	H_3^+	0.005	25.3 ± 1.0
2	H_2^+	0.3	27.9 ± 0.5
1	∎	∎	$\begin{cases} 22.7 \pm 0.5 \\ 29.1 \pm 0.6 \end{cases}$

filament. The relative intensities, furthermore, were all found to be unchanged by changes in pressure, electron current, and filament temperature over wide ranges. Hence we may say that each ion is produced by the single impact of an electron with a methane molecule, and that none is the result of primary thermal dissociation of methane at the cathode, as was thought to be the case in the experiments of Hogness and Kvalnes (5).

The errors quoted in the last column of table 1 are estimated probable errors which we believe to be quite conservative. Argon, introduced into the instrument simultaneously with the methane, was used to correct the observed electron energy scale for contact potentials, etc., in measuring the appearance potentials of CH_4^+, CH_3^+, CH_2^+, CH^+, and C^+, while helium and a slight impurity of hydrogen were used as calibrating gases for H_3^+, H_2^+, and H^+.

Interpretation of Data on Methane

Our chief guide in selecting most probable processes to explain the observed appearance potentials is the heat of dissociation of methane into atoms in the gaseous state $D(CH_4)$. This quantity is obtainable from thermochemical data, using the following cycle:[3]

$$C(s) + 2H_2 \rightarrow CH_4 + \quad 18.24 \quad \text{kg-cal.} = 0.79 \text{ volts}$$

$$4H \rightarrow 2H_2 + 207.6 \quad \text{kg-cal.} = 9.00 \text{ volts}$$

$$C(g) \rightarrow C(s) + 140 \pm 30 \text{ kg-cal.} = 6.1 \pm 1.3 \text{ volts} \equiv \lambda_c$$

$$C(g) + 4H \rightarrow CH_4 + 366 \pm 30 \text{ kg-cal.} = 15.9 \pm 1.3 \text{ volt} \equiv D(CH_4)$$

---

[3] Values are taken from Bichowsky and Rossini's *Thermochemistry of Chemical Substances*, Reinhold Publishing Corp., New York (1936). This provides an excellent summary of the values obtained for the heat of sublimation of carbon (λ_c) up to January 1, 1934.

The very large uncertainty in the value of the heat of sublimation of carbon λ_c is estimated by Bichowsky and Rossini from data up to 1934, and most values that have been proposed since that time (11, 7, 13)[4] fall within it.

H^+ at 29.4 volts

On the assumption that 29.4 volts is the minimum energy required for the reaction $CH_4 \rightarrow C + 3H + H^+$, we obtain $D(CH_4) = 15.9 \pm 0.6$ volts or 366 ± 14 kg-cal. by subtracting the ionization potential of H (13.5 volts) from 29.4 volts (figure 2). This gives $\lambda_c = 6.1 \pm 0.6$ volts or 140 ± 14 kg-cal. We could assume that CH_4 molecules do not dissociate completely in yielding H^+ ions at this minimum energy. In this case we should also have to assume that the products of dissociation have some kinetic or excitational energy in order not to arrive at values of $D(CH_4)$ and λ_c which are too high. Since the reaction first proposed does not require the latter assumption it is the preferred explanation. If in this reaction the products possess kinetic energy when formed, the values obtained for $D(CH_4)$ and λ_c represent upper limits to the true values of these quantities.

H^+ at 22.7 volts

If we are correct in taking 29.4 volts as the minimum energy required to dissociate completely the methane molecule with ionization of one of the hydrogen atoms, the only energetically possible reactions which could account for the formation of an H^+ ion at 3 minimum energy of 22.7 volts are:

$$CH_4 \rightarrow CH_3 + H^+; \quad CH_4 \rightarrow CH_2 + H + H^+; \quad CH_4 \rightarrow CH + H_2 + H^+$$

The energy required for the corresponding reactions where H^+ is replaced by H is $22.7 - 13.5 = 9.2$ volts. Since it is known that it takes about 4.0 volts to break a C—H bond this means that the kinetic or excitational energies of the products of dissociation are respectively: $9.2 - 4.0 = 5.2$ volts; $9.2 - 8.0 = 1.2$ volts; $9.2 - (12.0 - 4.5) = 1.7$ volts. The first of these values seems too high, so either the second or third reaction is probably the correct one. The reasoning by which one obtains $D(CH_4) = 17.2$ volts, assuming the second reaction, is shown in figure 3. If the products of dissociation have 1.2 volts of excitational or kinetic energy this value is in accord with the 16.0 volt value.

H_2^+ and H_3^+

The simplest interpretation of the observed appearance potential of H_2^+ ions is that 27.9 ± 0.5 volts is the minimum energy required for the reaction

---

[4] In the two papers referred to in reference 13, it is suggested that $C(s) \rightarrow C(g)$ ^5S at 7.57 volts or 177 kg-cal. Since the ^5S state is 4.3 volts or 100 kg-cal. above the normal ^3P state of the carbon atom, this would mean that $C(s) \rightarrow C(g)$ ^3P at 3.34 volts or 77 kg-cal. Unless the transition $^5S \rightarrow ^3P$ by collisions even at high pressure has an extraordinarily low probability, it is highly improbable that in the determinations by the vapor pressure method the energy measured is that for the process $C(s) \rightarrow C(g)$ ^5S, but rather for $C(s) \rightarrow C(g)$ ^3P.

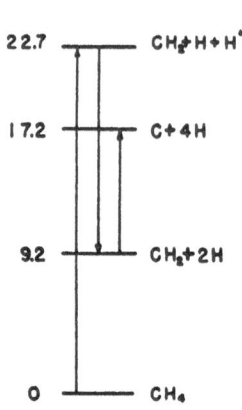

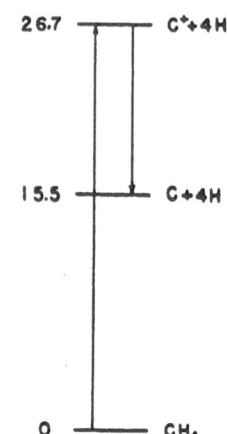

FIGURE 3. Energy levels deduced from the 22.7 volt appearance potential of H^+.

FIGURE 4. Energy level deduced from the appearance potential of C^+.

$CH_4 \rightarrow H_2^+ + C + 2H$. Subtraction of the ionization potential of H_2 (15.4 volts) and addition of the heat of dissociation of H_2 (4.5 volts) yields $D(CH_4) = 17.0 \pm 0.5$ volts or 389 ± 12 kg-cal., whence $\lambda_c = 7.1 \pm 0.5$ volts or 163 ± 12 kg-cal. provided the products are formed in their normal states without kinetic energy. These values being somewhat above the upper limits for $D(CH_4)$ and λ_c set by the 29.4 volt appearance potential of H^+, we are inclined to attribute about a volt of kinetic or excitational energy to the products of dissociation $H_2^+ + C + 2H$.

Since H_3^+ ions are apparently formed along with negative ions they will not be discussed here but will be treated elsewhere.

<h2 style="text-align:center">C^+</h2>

C^+ ions are well explained by the assumption that 26.7 ± 0.7 volts is the minimum energy necessary to dissociate completely a methane molecule and to ionize the carbon atom. If the C^+ ion is unexcited and none of the products acquires kinetic energy as a result of a dissociation, we get $D(CH_4) = 26.7 - 11.2 = 15.5 \pm 0.7$ volts or 357 ± 16 kg-cal., where 11.2 volts is the ionization potential of carbon (figure 4). This gives $\lambda_c = 5.7 \pm 0.7$ volts or 131 ± 16 kg-cal. If the products dissociate with energy these are upper limits to the true values.

<h2 style="text-align:center">CH^+</h2>

Values of $D(CH_4)$ and the ionization potential of CH may be obtained simultaneously by combining the observed appearance potential of CH^+ ions formed from methane with that of these ions formed from acetylene. The latter quantity has been measured by Tate, Smith, and Vaughan (17) as 22.2 ± 0.5 volts. Let us denote the former quantity by $A(CH^+)$ and the latter by $A'(CH^+)$, and let us assume the reaction responsible for the appearance of CH^+ ions at these minimum energies to be $CH_4 \rightarrow CH^+ + 3H$ and $C_2H_2 \rightarrow CH^+ + C + H$. Let us also denote a heat of dissociation by D, an ionization potential by I, and the energy evolved in the formation of a gram-mole of a substance from its elements in their standard states by Q. If the products of dissociation are formed without kinetic or excitational energy in both reactions we have:

$$I(CH) = A(CH^+) + D(CH) - D(CH_4)$$
$$= A'(CH^+) + D(CH) - D(C_2H_2)$$

also

$$D(CH_4) = Q(CH_4) + 2D(H_2) + \lambda_c$$

and

$$D(C_2H_2) = Q(C_2H_2) + D(H_2) + 2\lambda_c$$

Hence

$$A(CH^+) - A'(CH^+) = D(CH_4) - D(C_2H_2)$$
$$= Q(CH_4) - Q(C_2H_2) + D(H_2) - \lambda_c$$

or

$$\lambda_c = Q(CH_4) - Q(C_2H_2) + D(H_2) + A'(CH^+) - A(CH^+)$$
$$= 0.79 + 2.34 + 4.50 + 22.2 - 23.3$$
$$= 6.5 \pm 0.8 \text{ volts or } 150 \pm 18 \text{ kg-cal.}$$

This gives $D(CH_4) = 16.3 \pm 0.6$ volts or 376 ± 18 kg-cal. and also gives $I(CH) = 10.9 \pm 0.8$ volts. Since these values of $D(CH_4)$ and λ_c are reasonable, it appears that the assumed reactions are probably correct, and that if the products of dissociation have excitational or kinetic energies when formed they have about the same amount in the two cases.

CH_2^+, CH_3^+, and CH_4^+

Little can be said by way of correlating the appearance potentials of these ions with known quantities. Simple considerations show that at a minimum energy of 14.4 ± 0.4 volts a CH_3^+ ion must be formed in accordance with the reaction $CH_4 \rightarrow CH_3^+ + H$. This means that the ionization potential of CH_3 is less than or equal to $14.4 - 4.0 = 10.4$ volts, according as the CH_3^+ ion and H atom are formed with or without energy. Similarly, the observed appearance potential of CH_2^+ ions leads to the conclusion that the ionization potential of CH_2 is less than or equal to $15.7 - 8.0 = 7.7$ volts if a CH_2^+ ion is accompanied by two H atoms or is less than or equal to $15.7 - (8.0 - 4.5) = 12.2$ volts if it is accompanied by an H_2 molecule.

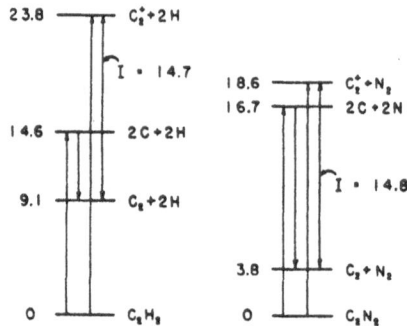

FIGURE 5. Deduction of the ionization potential (I) of C_2 from the appearance potentials of $C_2{}^+$ in C_2H_2 and C_2N_2.

Acetylene and Cyanogen

A few remarks will now be made concerning possibilities of interpretation of the data of Tate, Smith, and Vaughan (17) for acetylene and cyanogen. They find that $C_2{}^+$ appears at 23.8 ± 0.3 in acetylene and at 18.6 ± 0.5 in cyanogen. From Bichowsky and Rossini, using $\lambda_c = 5.8$ volts, we get 14.6 volts and 16.7 volts for the atomic heats of formation of acetylene and cyanogen respectively. The diagram, figure 5, will show the relations involved. The simplest interpretation in each case is to suppose that $C_2{}^+$ is accompanied by two hydrogen or two nitrogen atoms, dissociated, unexcited, and without kinetic energy in each case. But that leads to discordant values of the ionization potential of C_2 from the two sets of data, namely 14.7 from acetylene and 7.3 from cyanogen. The difference between these two values, 7.3 volts, agrees almost exactly with the 7.4 volt heat of dissociation of N_2. If we suppose $C_2{}^+$ is accompanied by N_2 in the case of nitrogen, but by 2H in the case of hydrogen, we get concordant values of 14.7 and 14.8 for the ionization potential of $C_2{}^+$ in the two cases.[5]

Conclusion

The preceding remarks have shown that information of value for molecular structure problems can be obtained by the mass spectrograph. The importance of a great deal of further work is obvious. All of the simpler organic molecules lend themselves to studies of the type here reported for methane, and a program of such work is being planned in this laboratory.

---

[5] This agrees with the rough value 12 ± 2 obtained by Kallmann and Rosen (Z. Physik **61**, 332 (1930)) from a study of electron transfer processes involving the $C_2{}^+$ ion formed in C_2N_2 and absorbed in other gases.

References

(1) BLEAKNEY: Phys. Rev. **35**, 1180 (1930).

(2) BLEAKNEY: Phys. Rev. **40**, 496 (1932).

(3) BLEAKNEY: Am. Physics Teacher **4**, 12 (1936).

(4) CONDON: Phys. Rev. **28**, 1182 (1926); **32**, 858 (1928).

(5) DE GROOT AND PENNING: Handbuch der Physik, Vol. 23/1. J. Springer, Berlin (1933).

(6) FRANCK: Trans. Faraday Soc. **21**, part 3 (1925).

(7) HERZBERG: Nature **137**, 620 (1936).

(8) HOGNESS AND KVALNES: Phys. Rev. **32**, 942 (1928).

(9) JAMES, COOLIDGE, AND PRESENT: J. Chem. Physics **4**, 187 (1936).

(10) LOZIER: Phys. Rev. **36**, 1285 (1930).

(11) LOZIER: Phys. Rev. **46**, 268 (1934).

(12) MORSE AND STUECKELBERG: Phys. Rev. **33**, 932 (1929).

(13) SCHMID, R.: Z. Physik **99**, 274 (1936); **99**, 626 (1936).

(14) SMYTH: Rev. Modern Phys. **3**, 347 (1931).

(15) SPONER: Molekülspektren. J. Springer, Berlin (1936).

(16) TATE AND LOZIER: Phys. Rev. **39**, 254 (1932).

(17) TATE, SMITH, AND VAUGHAN: Phys. Rev. **48**, 524 (1935).

(18) TELLER: Z. Physik **61**, 458 (1930).

Theories of Optical Rotatory Power

EDWARD U. CONDON

Palmer Physical Laboratory
Princeton University

1. Introduction

By optical rotatory power is meant the property of a medium to rotate the plane of polarization of linearly polarized light that is transmitted through it. The first effect of this kind was discovered by Arago in 1811. He found that quartz had this property in the direction of the optical axis, the direction along with ordinary double refraction vanishes.

The sense of rotation bears a fixed relation to the direction of propagation of the light, so if the light traverses the same medium once in each of two opposite directions—as when it is returned through the active medium by reflection at a mirror—the net rotation just vanishes. A substance is said to show positive rotation if the plane of polarization is turned in a clockwise sense as viewed by an observer into whose eye the light is entering. The rotation is proportional to the thickness of the active medium traversed and the rotatory power is defined as the angle through which the plane of polarization is turned per unit path in the medium.

The discovery of optically active liquids is due to Biot.[1] Here, since there is no preferred orientation of the molecules the effect must be due to a structural peculiarity of the individual molecules. The modern theories which relate this property of a fluid to the structure of the individual molecules is the subject of this report.[2]

Conventionally the rotatory power of a medium is given in degrees/decimeter. This quantity will be denoted by φ. In the c.g.s. system it is measured in radian/cm, and evidently one has to multiply the value in degree/cm by $\pi/1800$ to convert to radian/cm.

---

[1] Biot, Bull soc. philomath. **190** (1815).
[2] The book by T.M. Lowry, *Optical Rotatory Power* (Longmans, Green, 1935), contains a very thorough account of the experimental and empirical side of the subject. In this report the general principles are briefly reviewed and the main emphasis is devoted to a review of the applications of dispersion theory to our understanding of the phenomenon.

Another measure of rotatory power called the specific rotatory power is also in common use. It is the rotatory power divided by the density of the active material in gram/cm^3. It will be denoted by $[\varphi]$. It is more common in the experimental literature to denote these quantities by α and $[\alpha]$, respectively, but this departure from the common notation is introduced here because it is desired to reserve the letter α for molecular polarizability corresponding to another well-established usage.

Still another measure of rotatory power is in common use which is called the molecular rotatory power. It is defined as being the specific rotatory power $[\varphi]$ in degree/decimeter per gram/cm^3, multiplied by one one-hundredth of the molecular weight and is denoted by $[M]$. That is,

$$[M] = [\varphi]/100; \qquad [\varphi] = \varphi/\rho, \tag{1}$$

in which ρ is the density and M the molecular weight.

In the theoretical formulas the combination $\varphi M/\rho$ often occurs where φ is the rotatory power in radian/cm. It is convenient to remember the relation,

$$\varphi M/\rho = (\pi/18)[M]. \tag{2}$$

The basic feature of propagation in an optically active medium that is responsible for rotating the plane of polarization is circular double refraction. This was recognized by Fresnel.[3] A substance is said to be double refracting if in a given direction the phase velocity of propagation of light waves is different for two different states of polarization. In the case of optical activity the velocity is different for right and left circularly polarized waves respectively.

Although Fresnel's work greatly antedates the electromagnetic theory of light it is convenient to discuss the situation at once in terms of the modern theory. In the electromagnetic theory there are two vectors associated with the wave that are transverse to the direction of propagation. These are the electric induction, $\mathbf{D}$, and the magnetic induction, $\mathbf{B}$. Suppose the wave is traveling in the direction of the unit vector $\mathbf{k}$ with a velocity c/n where n is the index of refraction. If we introduce unit vectors, $\mathbf{i}$ and $\mathbf{j}$, mutually orthogonal and orthogonal to $\mathbf{k}$ in such a way that $(\mathbf{i}, \mathbf{j} \; \mathbf{k})$ form the basis of a right-handed coordinate system, then $\mathbf{D}$ and $\mathbf{B}$ may be written in the forms,

$$\mathbf{D} = R\{\mathbf{D}_0 e^{i\psi}\}, \quad \mathbf{B} = R\{\mathbf{B}_0 e^{i\psi}\}, \quad \psi = 2\pi v(t - n\mathbf{k}\cdot\mathbf{r}/c), \tag{3}$$

where $\mathbf{D}_0$ and $\mathbf{B}_0$ are constant vectors, expressible in terms of $\mathbf{i}$ and $\mathbf{j}$, and ψ is the phase of the wave at time t and place $\mathbf{r}$, it being supposed that v is the frequency of the wave. The symbol $R\{\ \}$ means that the real part of the complex expression is to be taken.

For a right circularly polarized wave the constant amplitude will be of the form of a constant multiplying into $(\mathbf{i} + i\mathbf{j})$, say $D(\mathbf{i} + i\mathbf{j})$ for, on taking the real part of this expression, we have

---

[3] Fresnel, Ann. Chim. Phys. **28**, 147 (1925); *Oeuvres complètes* **1**, 731, Paris (1866).

$$D(\mathbf{i}\cos\psi - \mathbf{j}\sin\psi).$$

When $\psi = 0$ the vector $\mathbf{D}$ is parallel to $\mathbf{i}$ and as times goes on ψ increases and the vector $\mathbf{D}$ rotates in the clockwise sense as viewed by an observer faced in the $-\mathbf{k}$ direction, that is, faced so the light enters his eyes. Similarly a left circularly polarized wave is represented by a constant multiplying into $(\mathbf{i} - i\mathbf{j})$.

These results contain expression of the fact that circularly polarized light can be regarded as the superposition of two plane polarized waves having the proper phase relation. The factor i in $(\mathbf{i} + i\mathbf{j})$ may be written $e^{i\pi/2}$ from which it is clear that the phase of the linearly polarized constituent along $\mathbf{j}$ is a quarter-cycle ahead of the linearly polarized constituent along $\mathbf{i}$. Similarly in the left circularly polarized wave the component along $\mathbf{j}$ lags a quarter-cycle behind the component along $\mathbf{i}$.

Linearly polarized light may likewise be regarded as arising from the superposition of two circularly polarized waves. Consider the wave made up of a superposition of a right circularly polarized wave with a phase $e^{i\delta}$ and a left circularly polarized wave with a phase $e^{-i\delta}$. The expression for $\mathbf{D}$ assumes the form,

$$\mathbf{D} \approx R\{(\mathbf{i} + i\mathbf{j})e^{i\delta} + (\mathbf{i} - i\mathbf{j})e^{-i\delta}\}$$
$$= 2(\mathbf{i}\cos\delta - \mathbf{j}\sin\delta). \tag{4}$$

For $\delta = 0$ this represents a linearly polarized wave whose plane of polarization is given by the vector $\mathbf{i}$ and for $\delta > 0$ the plane of polarization is turned clockwise through an angle δ with respect to the $\mathbf{i}$ axis.

Now let us suppose that the medium has different refractive indices for right and left circularly polarized waves which may be denoted by n_r and n_l, respectively. Suppose the light enters at the plane $\mathbf{k}\cdot\mathbf{r} = 0$ and leaves at the plane $\mathbf{k}\cdot\mathbf{r} = d$, so d is the length of path traversed. Further suppose that the light enters linearly polarized along $\mathbf{i}$. At the exit plane the phases of the two components will be

$$\psi_r = 2\pi v(t - n_r d/c),$$
$$\psi_l = 2\pi v(t - n_l d/c),$$

so we can write $(\psi_r, \psi_l) = \psi \pm \delta$, where

$$\psi = 2\pi v(t - \tfrac{1}{2}(n_r + n_l)d/c)$$

is the phase corresponding to the mean index of refraction and

$$\delta = \pi(n_l - n_r)d/\lambda \quad (\lambda = c/v) \tag{5}$$

arises from the difference of the two indices of refraction.

But as we have seen, an advance in phase δ for the right and a retardation in phase δ of the left circularly polarized components results in their superposing to produce linearly polarized light whose plane of polarization is turned through an angle δ. Hence the rotation per unit path length is δ/d or the rotatory power φ is expressed directly in terms of the difference of the two indices of refraction:

$$\varphi = (\pi/\lambda)(n_l - n_r). \tag{6}$$

In this equation if φ is expressed in degree/decimeter, then the vacuum wave-length λ must be expressed in decimeters and $\pi = 180$.

Owing to the fact that λ is very small compared to macroscopic values of d, the quantity d/λ in (5) is large compared to unity and so appreciable rotations are produced in spite of $(n_l - n_r)$ being small compared to unity. It is convenient to remember that the sense of the rotation is that of the circularly polarized component that travels most rapidly.

Just as optical activity is produced as an indirect effect of the difference in velocity of propagation of right and left circularly polarized waves, so also there is an indirect effect, known as circular dichroism, arising from differential absorption of the two kinds of waves. It was discovered by Cotton[4] in 1896. It is well known that in the case of ordinary refraction the refractivity is closely associated with the absorption bands, strong absorption bands producing more refractivity. In the case of optically active liquids the difference $(n_l - n_r)$ is of the order of a few parts in a million. The general connection between refraction and absorption suggests that a difference of this order in the absorption coefficients is all that can be expected. In view of the difficulties surrounding intensity measurements it cannot be expected that such a small difference in absorption can be detected by measuring separately absorption coefficients for the two kinds of circularly polarized light.

Instead we have to look for a differential effect whose very existence is due to the difference in the two absorption coefficients. This is provided by studying the propagation of a linearly polarized wave through the absorbing medium. Let ε_l and ε_r be the absorption coefficients for left and right circularly polarized waves and write

$$(\varepsilon_l, \varepsilon_r) = \varepsilon \pm \varepsilon', \tag{7}$$

where ε is the mean absorption coefficient and ε' is half the difference. Then since the intensity of light varies as $\mathbf{D}^2$, the amplitude of $\mathbf{D}$ is subject to exponential decrease by the factors $e^{-\varepsilon_r d/2}$ and $e^{-\varepsilon_l d/2}$ for right and left circularly polarized waves respectively after traversing a thickness d of the medium. Hence the two components which initially unite to give a linearly polarized wave are of unequal amplitude after traversing the medium. As a result they recombine to give elliptically polarized light the ellipticity of which is connected with the difference of the two absorption coefficients.

Each amplitude will be reduced by the common factor $e^{-\varepsilon d/2}$ corresponding to the mean absorption coefficient, but after going through a thickness d the

---

[4] Cotton, Ann. Chim Phys. **8**, 347 (1896).

amplitude will in addition to this factor be given by

$$e^{\varepsilon' d/2}(\mathbf{i} + i\mathbf{j})e^{i\delta} + e^{-\varepsilon' d/2}(\mathbf{i} - i\mathbf{j})e^{-i\delta}.$$

Denoting by $\mathbf{i}(\delta)$ and $\mathbf{j}(\delta)$ the unit vectors obtained by making a clockwise rotation through δ of the unit vectors $\mathbf{i}$ and $\mathbf{j}$, this can be written

$$\mathbf{i}(\delta)\cosh \varepsilon' d/2 + i\mathbf{j}(\delta)\sinh \varepsilon' d/2. \tag{8}$$

This represents, for $\varepsilon > 0$, a right elliptically polarized wave whose major amplitude is rotated by δ from the original direction of the linear polarization. The ellipticity is conventionally measured by an angle Ψ whose tangent is equal to the ratio of the minor to the major amplitudes, that is,

$$\tan \Psi = \tanh \varepsilon' d/2, \tag{9}$$

if we adopt the convention that the ellipticity is positive for a right elliptically polarized wave and negative for left elliptic polarization.

The observable phenomenon known as circular dichroism consists in this appearance of elliptic polarization when linearly polarized light is partially absorbed in passing through an active medium. In view of the fact that ε' will be small compared to ε and that we must choose d such that εd is not very great in order to have an appreciable amount of light transmitted, we shall always have $\varepsilon' d \ll 1$ and therefore

$$\Psi = \tfrac{1}{2}\varepsilon' d = \tfrac{1}{4}(\varepsilon_l - \varepsilon_r)d, \tag{10}$$

so the ellipticity will be proportional to the thickness d. Bruhat[5] has shown that the relative error in measurement of δ and Ψ is a minimum when such a thickness of material is chosen that the amplitude is reduced by the factor e^{-1}, that is the energy reduced by e^{-2}, or to 13.5 percent of its initial value.

2. Electromagnetic Theory

After the general introduction of the preceding section we now turn to the problem of generalizing the ordinary elementary treatments of the electromagnetic theory of light in such a way as to give circular double refraction. The starting point is, of course, Maxwell's equations:

$$\text{div}\,\mathbf{D} = 0, \qquad\qquad \text{div}\,\mathbf{B} = 0,$$

$$\text{curl}\,\mathbf{E} = -(1/c)\dot{\mathbf{B}}, \qquad \text{curl}\,\mathbf{H} = (1/c)\dot{\mathbf{D}}, \tag{11}$$

$$\mathbf{D} = E + 4\pi\mathbf{P}, \qquad\qquad \mathbf{B} = H + 4\pi\mathbf{I}.$$

The properties of the medium find their expression entirely in that they give rise to $\mathbf{P}$, the electric moment per unit volume and to $\mathbf{I}$, the magnetic moment

---

[5] Bruhat, Ann. de physique 3, 232 (1915).

per unit volume. The theory of wave propagation, underlying all kinds of dispersion effects, needs, in addition to Eqs. (11), a connection which relates $\mathbf{P}$ and $\mathbf{I}$ to $\mathbf{E}$ and $\mathbf{H}$. This connection will be provided by a detailed application of electrodynamics to a particular model of the material of the medium.

In the simple theory of isotropic media we have

$$\mathbf{P} = \kappa\mathbf{E}, \quad \mathbf{I} = \kappa'\mathbf{H},$$

where κ and κ' are scalars. Then

$$\mathbf{D} = (1 + 4\pi\kappa)\mathbf{E} = \varepsilon\mathbf{E},$$
$$\mathbf{B} = (1 + 4\pi\kappa')\mathbf{H} = \mu\mathbf{H},$$

(12)

where ε and μ are the usual dielectric constant and magnetic permeability. This simple connection leads, as is well knoun, to the propagation of waves with an index of refraction given by $n^2 = \varepsilon\mu$ As a general rule $\kappa' \approx 10^{-4}\kappa$ except for strongly magnetic substances which absorb light anyway, so it is generally permissible to write, $n^2 = \varepsilon$.

The theory of crystal optics, so far as ordinary double refraction is concerned, is given as a generalization in which κ and κ' are replaced by tensors, although usually κ' is neglected in comparison with κ and the whole refraction regarded as due to the electric polarization.

To get a theory of optical activity we need a different kind of generalization of the material connections. It turns out that the essential point is that there be a part of $\mathbf{P}$ that is proportional to $\dot{\mathbf{B}}$ and a part of $\mathbf{I}$ that is proportional to $\dot{\mathbf{D}}$. We shall assume for the moment that we have a molecular theory which leads to connections of the form

$$\mathbf{D} = \varepsilon\mathbf{E} - g\dot{\mathbf{H}}, \qquad \mathbf{E} = \varepsilon^{-1}\mathbf{D} + \varepsilon^{-1}g\dot{\mathbf{B}},$$
$$\text{or}$$
$$\mathbf{B} = \mathbf{H} + g\dot{\mathbf{E}} \qquad \mathbf{H} = \mathbf{B} - \varepsilon^{-1}g\dot{\mathbf{D}}$$

(13)

and proceed to find solutions of the Maxwell equations using these connections. Every vector may be assumed to be a constant amplitude multiplied into $e^{i\psi}$ where ψ is the phase as in (3). The equations $\text{div } \mathbf{D} = 0$ and $\text{div } \mathbf{B} = 0$ lead to $\mathbf{k} \cdot \mathbf{D} = 0$ and $\mathbf{k} \cdot \mathbf{B} = 0$ so $\mathbf{D}$ and $\mathbf{B}$ are transverse to the direction of propagation.

Next consider the curl equations. These give

$$n\mathbf{k} \times \mathbf{E} = \mathbf{B},$$
$$n\mathbf{k} \times \mathbf{H} = -\mathbf{D}$$

and the material connections (13) give

$$\mathbf{E} = \varepsilon^{-1}\mathbf{D} + i\gamma\mathbf{B},$$
$$\mathbf{H} = \mathbf{B} - i\gamma\mathbf{D},$$

where

$$\gamma = 2\pi\nu g\varepsilon^{-1}.$$

With these we may eliminate **E** and **H** from the curl equations and obtain

$$n(\varepsilon^{-1}\mathbf{k} \times \mathbf{D} + i\gamma\mathbf{k} \times \mathbf{B}) = \mathbf{B},$$

$$n(-i\gamma\mathbf{k} \times \mathbf{D} + \mathbf{k} \times \mathbf{B}) = -\mathbf{D}.$$

Written out in terms of two components for **B** and **D** these are four homogeneous equations for the four unknown components. To make them consistent the determinant of their coefficients must vanish. This condition gives an equation for n whose roots give the possible values of the index of refraction. Writing

$$\mathbf{D} = D_1\mathbf{i} + D_2\mathbf{j},$$

$$\mathbf{B} = B_1\mathbf{i} + B_2\mathbf{j},$$

the equations are

$$-n(\varepsilon^{-1}D_2 + i\gamma B_2) = B_1 \qquad n(i\gamma D_2 - B_2) = -D_1$$

$$n(\varepsilon^{-1}D_1 + i\gamma B_1) = B_2 \qquad n(-i\gamma D_1 + B_1) = -D_2.$$

The permissible values of the index of refraction are readily found to be

$$n^{-2} = (\varepsilon^{-1/2} \pm \gamma)^2. \tag{14}$$

The two negative roots correspond to propagation in the direction $-\mathbf{k}$ and are not of interest. The two positive roots give the indices for propagation in the direction $+\mathbf{k}$. The root with the positive sign for γ is easily seen to correspond to a solution for **D** and **B** of the form of a right circularly polarized wave, the other to left circular polarization. Since γ is a small quantity compared with unity, it follows therefore that

$$n_r = \varepsilon^{1/2} - 2\pi v g,$$

$$n_l = \varepsilon^{1/2} + 2\pi v g. \tag{15}$$

This result combined with (6) gives a connection between the rotatory power of the medium and the parameter g introduced in (13) as follows:

$$\varphi = (2\pi/\lambda)^2 c g. \tag{16}$$

The next problem is to see what kind of response of the individual molecules to the fields of the light wave is needed to give terms in the macroscopic field equations of the type introduced in (13). The effective electric field acting on a molecule is not only the **E** vector of the macroscopic field theory but an average field due to the neighboring molecules. Lorentz[6] showed that for a medium in which the molecules are distributed at random we have to take for the effective field, **E′**,

$$\mathbf{E'} = \mathbf{E} + (4\pi/3)\mathbf{P}. \tag{17}$$

Similar considerations hold in principle for the magnetic field but as the

---

[6] Lorentz, *Theory of Electrons*, p. 305.

intensity of magnetization of the medium is practically negligible in comparison with H this need not be considered.

Now let us suppose that the theory of the response of individual molecules to the external fields leads to formulas,

$$\mathbf{p} = \alpha \mathbf{E}' - (\beta/c)\dot{H},$$
$$\mathbf{m} = +(\beta/c)\dot{E}',$$
(18)

in which $\mathbf{p}$ is the induced electric moment and $\mathbf{m}$ is the induced magnetic moment of an individual molecule. The essential point is the introduction of the term involving the parameter β. Here α is the usual polarizability term giving an induced electric moment proportional to the applied electric field.

The total electric and magnetic moments in unit volume are

$$\mathbf{P} = N_1 \mathbf{p}, \qquad \mathbf{I} = N_1 \mathbf{m},$$
(19)

where N_1 is the number of molecules in unit volume. In case the medium consists of a simple mixture of different kinds of molecules there will be different coefficients α_i and β_i for each species and then $\mathbf{P}$ and $\mathbf{I}$ will be given by a sum of terms one for each species each $\mathbf{p}_i$ and $\mathbf{m}_i$ being multiplied by the corresponding N_i, the number of molecules of the ith species present in unit volume.

Combining (19), (18) and (17) one can easily find that the implied connections of $\mathbf{D}$ and $\mathbf{B}$ with $\mathbf{E}$ and $\mathbf{H}$ are those given by (13) as already assumed. The new result is the connection between the individual molecular parameters α and β and the molar parameters ε and g, which turn out to be

$$4\pi N_1 \alpha/3 = (\varepsilon - 1)/(\varepsilon + 2),$$
(20)

which is familiar from the ordinary theory of dispersion, and the analogous expression for the rotatory parameter, g,

$$4\pi N_1 (\beta/c)/3 = g/(\varepsilon + 2).$$
(21)

Using the familiar expression $n^2 = \varepsilon$ where n is the mean index of refraction and combining (21) with (16) one may arrive at a formula giving a direct connection between the rotatory power and the molecular parameter β:

$$\varphi = \frac{16\pi^3 N_1 \beta}{\lambda^2} \cdot \frac{n^2 + 2}{3}.$$
(22)

This is the main result of the electromagnetic theory in that it refers the activity of the medium back to the parameter β. This parameter has to be explained in terms of detailed molecular theories as will be seen in later sections.

From its definition in (18) we see that molecules with a nonvanishing β have the property that an increasing electric field produces a magnetic moment in them, and an increasing magnetic field produces an electric moment. Let us try to visualize how this can come about.

The main effect when the molecule is put in an electric field is that measured by the polarizability α; positive charges are displaced in the direction of $\mathbf{E}$ and negative charges the other way, the amount being proportional to the field strength, resulting in the production of an induced dipole moment in the molecule. If the electric field is increasing the charges are moving to provide the increasing displacement necessary to go with the increasing dipole moment. Suppose now the molecular structure is such that these flowing charges are not allowed to move directly from their initial to their final positions but are constrained to move in somewhat helical paths so that there is a circulatory component of motion around $\dot{E}$, accompanying the general forward motion in the direction of $\mathbf{E}$.

The currents associated with the circulatory component of the motion give rise to a magnetic moment that is proportional to the amount of $\dot{E}$ and in the same direction as $\dot{E}$. This is a simple pictorial view of the mechanism underlying the term involving β in the equation for the induced magnetic moment.

Conversely, suppose the molecule is in a changing magnetic field. The changing flux through the molecule sets up induced currents in the molecule, that is, induces a flow of charges around the direction of $\dot{H}$ in the sense given by Lenz' law. The same constraints which previously required a circulatory motion to accompany a general displacement, now will require a displacement of positives one way and negatives the other to accompany the induced circulatory currents. Thus there will be produced a separation of positive and negative charges as a result of the action of the induced currents. This is the simple pictorial view of the production of electric moment by a changing magnetic field, the minus sign relative to the other effect being a simple consequence of Lenz' law for induced currents.

At this point it is well to emphasize that the pictorial description may with equal validity be stated in other terms. Thus in Maxwell's equations a changing electric field is invariably associated uith an inhomogeneity of the magnetic field. It is therefore meaningless to say whether the magnetic moment we have formally and pictorially associated with changing electric field is really "due to" the changing electric field rather than "due to" the inhomogeneities of the magnetic field. A similar remark holds, of course, for the electric moment associated here with a changing magnetic field.

In writing this section it has seemed best to put all references to the original literature at the end. It is not really worth uhile to trace back the history completely for it leads into phenomenological extensions of the old elastic solid theory of light which today are only of antiquarian interest. Apparently the first theory of the kind given in this section is due to Gibbs.[7] Others who contributed to the application of the electromagnetic theory to the problem of optical activity are Drude,[8] Lorentz[9] and Livens.[10]

[7] Gibbs, *Collected Works*, Vol. 2, p. 195. Originally published in Am. J. Science **25**, 460 (1882).
[8] Drude, *Göttinger Nachrichten* (1892), p. 366.
[9] Lorentz, *Versuch einer Theorie* ... (Leipzig, 1906).
[10] Livens, Phil. Mag. **25**, 817 (1913); **26**, 362, 535 (1913); **27**, 468, 994 (1914); **28**, 756 (1914); Physik. Zeits. **15**, 385 (1914).

3. The Parameter β and Rotatory Dispersion

In the next section we shall develop the quantum-mechanical theory of dispersion in a form which includes the theory of the parameter β. Before doing this it will be of interest to consider the empirical data in relation to the theory. It will be shown that β is given by

$$\beta_a = \frac{c}{3\pi h} \sum_b \frac{R_{ba}}{v_{ba}{}^2 - v^2}. \tag{23}$$

Here β_a is the value of β appropriate to molecules in the quantum state a, v_{ba} is the frequency of the light absorbed in the jump $a \rightarrow b$ and R_{ba} is a constant characteristic of that particular absorption line which we shall call the *rotational strength* of the line v_{ba}.

If the molecules are distributed over various states in the thermal equilibrium appropriate to the temperature of the medium, the number per unit volume in the state a is $N_1(a)$ where this will be given by a Boltzmann distribution formula. Then the effective value of β to be used in (22) is

$$N_1 \beta = \sum_a N_1(a)\beta_a. \tag{24}$$

Equation (23) for β_a is analogous to the better-known equation for the polarizability α_a of molecules in the quantum state a. This formula, due to Kramers and Heisenberg,[11] is

$$\alpha_a = \frac{2}{3h} \sum_b \frac{v_{ba} S_{ba}}{v_{ba}{}^2 - v^2}, \tag{25}$$

in which S_{ba} is another characteristic of the line v_{ba} which is called the *strength* of the line.[12] The strength of the line should not be confused with the *oscillator strength* which is another measure of the intensity of a line, important in dispersion theory. The oscillator strength is usually denoted by f_{ba}. It is dimensionless and is defined by the equation,

$$f_{ba} = (8\pi^2 \mu / 3e^2 h) v_{ba} S_{ba}, \tag{26}$$

in which e and μ are the charge and mass of the electron. In terms of the oscillator strengths the formula (25) for the polarizability becomes

$$\alpha_a = \frac{e^2}{4\pi^2 \mu} \sum_b \frac{f_{ba}}{v_{ba}{}^2 - v^2}. \tag{27}$$

The oscillator strengths satisfy an important sum rule,

$$\sum_b f_{ba} = n. \tag{28}$$

---

[11] Kramers and Heisenberg, Zeits. f. Physik **31**, 681 (1925); Ladenburg, Zeits. f. Physik **4**, 451 (1921).
[12] Condon and Shortley, *The Theory of Atomic Spectra* (Cambridge, 1935), p. 98.

316

E.U. Condon

which was discovered independently by Thomas and Kuhn.[13] In (28) the letter n stands for the total number of electrons in the molecule. Eq. (28) is true for each state a of the molecule, the sum extending over all other states.

Instead of the rotational strengths of the lines, Kuhn[14] has introduced another measure of the importance of a line in contributing to the rotatory power. This he calls the anisotropy factor for the line. In the notation that is being used here, Kuhn's anisotropy factor, g_{ba}, is

$$g_{ba} = R_{ba}/S_{ba}. \tag{29}$$

The anisotropy factor is easily seen to be a pure number.

Let us now turn to the empirical data in its relation to these equations. When one first approaches the study of optical activity he is apt to be overwhelmed at the large amount of data that exists. But as soon as one tries to use it in connection with the theory, however, he finds that very little of the data is complete enough to be of any real use. In the first place many of the measurements have been made for a single wave-length, usually the sodium D lines, so it is insuffficient to determine the constants R_{ba} and v_{ba} of a dispersion formula for β. This deficiency has been emphasized by Lowry[2] to whose efforts is largely due the recent tendency of experimentalists to secure more complete dispersion data. In the second place, the index of refraction is usually not measured so that one is forced to make an estimate of the $(n^2 + 2)/3$ factor which appears in (22).

Ordinarily we may assume that all of the molecules are in the lowest electronic state so an averaging over initial states as in (24) is not needed, in which case β_a refers to the normal electronic state of the molecule. Combining (23) and (22) we have

$$\frac{\varphi M/\rho}{\frac{1}{3}(n^2 + 2)} = \frac{16\pi^2 N}{3hc} \sum_b \frac{R_{ba}v^2}{v_{ba}^2 - v^2}, \tag{30}$$

in which M is the molecular weight, ρ the density and N is Avogadro's number. In this equation φ is the rotation in radian/cm. Referring to (2) one may write this as an equation for the molecular rotatory power $[M]$ in the conventional units of §1. The result is

$$\frac{[M]}{\frac{1}{3}(n^2 + 2)} = \frac{96\pi N}{hc} \sum_b \frac{R_{ba}v^2}{v_{ba}^2 - v^2}. \tag{31}$$

An ordinary transparent liquid will have its absorption frequencies v_{ba} in the ultraviolet and ordinarily the data on $[M]$ will be confined to the visible part of the spectrum. As a consequence if there are a great number of different absorption bands in the same general region of the ultraviolet, this will not be recognizable in the data which refer simply to the visible spectrum. Instead

---

[13] Thomas, Naturwiss. **13**, 627 (1925); Kuhn, Zeits. f. Physik **33**, 408 (1925).
[14] Kuhn, Trans. Faraday Soc. "Discussion on Optical Rotatory Power," p. 299 (1930).

all of these frequencies will lump together to produce a single effective term having for its effective v_{ba} some kind of average of the individual v_{ba} of the actual lines, and having an effective rotational strength, R_{ba} that is essentially the sum of the separate rotational strengths. This is a situation that is quite familiar in the discussion of the dispersion formula for ordinary refractivity.

A good example of the use of (31) and the related formulae just developed is provided by discussing the data of Hunter[15] on d-sec octyl alcohol which has the rare virtue of including refractive index measurements. His data are given in Table I, which he represents by these empirical formulas,

$$n^2 = 1.6913 + 0.313\lambda^2/(\lambda^2 - 0.0283),$$

$$[\varphi] = 3.14/(\lambda^2 - 0.0283),$$

in which λ is in microns.

According to (31) it is not $[\varphi]$ or $[M]$ which satisfy a formula of the type used empirically but these quantities divided by $(n^2 + 2)/3$. But in this example the total variation in this factor from one end of the table to the other is from 1.34 to 1.36 so we may treat the factor as constant and equal to 1.35 without appreciable error. The molecular weight is $M = 130$ and therefore by the use of Hunter's empirical formula one obtains

$$[M]/\tfrac{1}{3}(n^2 + 2) = 107\sigma^2/(35.6 - \sigma^2),$$

where σ is the wave number in reciprocal microns (10^4 cm^{-1}). Since σ is proportional to v and the formula (31) is homogeneous in the frequency it follows that

$$96\pi N R_{ba}/hc = 107,$$

---

[15] Hunter, J. Chem. Soc. **123**, 1671 (1923).

TABLE I. Index of refraction and specific rotation of d-sec octyl alcohol. (Hunter.)

λ	n	$[\varphi]$
6438	1.4238	8.12
5896	1.4256	9.86
5461	1.4273	11.65
5086	1.4292	13.58
4800	1.4311	15.46
4678	1.4320	16.42
4358	1.4349	19.49
4251		20.6
3969		24.2
3790		27.3
3650		29.9

where R_{ba} is the effective rotational strength of the group of ultraviolet bands whose effective wave number is 58,800 cm^{-1} that is, whose effective wavelength is 1700A. The combination of universal constants occurring here is

$$96\pi N/hc = 0.943 \times 10^{42}$$

and therefore the effective rotational strength from the data is

$$R_{ba} = 1.13 \times 10^{-40}.$$

There are not many cases for which the data on rotatory dispersion are complete enough to permit a calculation of this sort. Even so this merely represents the effect of lumping all the active absorption bands at one effective average frequency in the ultraviolet.

The quantum-mechanical theory of the next section gives a formula for the rotational strength, R_{ba}, of an individual line in terms of matrix components of the unperturbed molecule. There it is shown that

$$R_{ba} = \text{Im}\{(a|\mathbf{p}|b)\cdot(b|\mathbf{m}|a)\}, \tag{32}$$

in which $(a|\mathbf{p}|b)$ and $(b|\mathbf{m}|a)$ are, respectively, the matrix components connecting states b and a of the electric and magnetic dipole moments of the molecule. The symbol Im$\{\ \}$ means "imaginary part of" in the sense

$$\text{Im}\{u + iv\} = v,$$

if u and v are real. The electric moment of a molecule is

$$\mathbf{p} = \sum e\mathbf{r}_i,$$

where $\mathbf{r}_i$ is the position vector of the ith electron. Nonvanishing matrix components of $\mathbf{p}$ may then be expected to be of the order of the electron charge times the radius of the first Bohr orbit, a_H, that is, of the order,

$$ea_H = 2.53 \times 10^{-18} \text{ c.g.s.}$$

Similarly nonvanishing matrix components of the magnetic dipole moment may be expected to be of the order of the Bohr magneton, or roughly,

$$e\hbar/2\mu c = 0.92 \times 10^{-20} \text{ c.g.s.}$$

Hence on this crude estimation the expected order of magnitude of a rotational strength, R_{ba}, is

$$ea_H \cdot e\hbar/2\mu c = 2.32 \times 10^{-38} \text{ c.g.s.}$$

The value just found for octyl alcohol is about one two-hundredth of this crude estimate which is very reasonable, considering that the estimate was too large for a number of reasons: (a) generally the scalar product of the two matrix components will be smaller than the product of their magnitudes as they will usually not be parallel, (b) generally the product of the two matrix components will not be purely imaginary so part of the quantity is lost on taking the imaginary part and (c) the empirical value is the sum of the values referring to different absorption bands in the molecule some of which have negative and others positive values, so there is a tendency for the sum of many such terms to be smaller than the average individual term.

It is of interest to continue the discussion of the illustrative example and to handle the refractive index data in a manner comparable to what has been done for rotatory dispersion. The combination of (25) and (20) yields the formula,

$$\frac{(n^2 - 1)M/\rho}{\frac{1}{3}(n^2 + 2)} = \frac{8\pi N}{3hc} \sum_b \frac{\sigma_{ba} S_{ba}}{\sigma_{ba}^2 - \sigma^2}, \tag{33}$$

in which σ_{ba} and σ are the wave number equivalents of the corresponding frequencies ν_{ba} and ν, and ρ is the density. The refractivity data on octyl alcohol already given then lead to

$$S_{ba} = 83.7 \cdot 10^{-36} \text{ c.g.s.,}$$

as the effective strength of the ultraviolet absorption bands.

By the standard dispersion theory, the formula for S_{ba} which is the analog of (32) is

$$S_{ba} = |(a|\mathbf{p}|b)|^2, \tag{34}$$

so the expected order of magnitude of an S_{ba} term is

$$(ea_H)^2 = 6.40 \cdot 10^{-36} \text{ c.g.s.}$$

The experimental value just found is 13.1 times this natural atomic unit, which is reasonable, for none of the reasons given for smallness of R_{ba} are applicable here.

We conclude the discussion of the example by calculating the effective f_{ba} (oscillator strength) and g_{ba} (anisotropy factor) for the ultraviolet bands lumped at 1700A. From (26) we have, in general,

$$f_{ba} = (4.79 \times 10^{29})\sigma_{ba} S_{ba}$$

and hence using the values of σ_{ba} and S_{ba} we have

$$f_{ba} = 2.36.$$

This shows that transitions of more than one electron are involved in the totality of bands which contribute to the dispersion in the empirical formula given by Hunter. Kuhn's anisotropy factor for these bands is readily calculated from (29) and found to be

$$g_{ba} = 1.35 \times 10^{-6}.$$

This value is typical of the values obtained for g_{ba} in the case of strong absorption bands for which f_{ba} is of the order of unity.

In discussing the example we have referred rather casually to the calculated R_{ba} as being an effective value representing the total contribution of a group of bands. To be more precise, suppose a group of bands to be represented by a single rotatory dispersion term. Then, say

$$\frac{R_1}{v_1{}^2 - v^2} = \sum_b \frac{R_{ba}}{v_{ba}{}^2 - v^2}, \tag{35}$$

where the equality holds for small values of v. To get agreement of both sides in the first two terms of a power series development in v one must have

$$v_1{}^{-2} = (\sum R_{ba}/v_{ba}{}^4) \div (\sum R_{ba}/v_{ba}{}^2),$$
$$R_1 = (\sum R_{ba}/v_{ba}{}^2)^2 \div (\sum R_{ba}/v_{ba}{}^4). \tag{36}$$

Similarly, if one is representing a whole group of ultraviolet bands on the ordinary dispersion formula by a single lumped term, then the analogs of (35) and (36) are easily seen to be

$$\frac{v_1 S_1}{v_1{}^2 - v^2} \approx \sum \frac{v_{ba} S_{ba}}{v_{ba}{}^2 - v^2}, \tag{37}$$

where, to get agreement in the first two terms in a power series in the frequency, we must have

$$v_1{}^{-2} = (\sum S_{ba}/v_{ba}{}^3) \div (\sum S_{ba}/v_{ba}),$$
$$S_1{}^2 = (\sum S_{ba}/v_{ba})^3 \div (\sum S_{ba}/v_{ba}{}^3). \tag{38}$$

As the formulas (36) and (38) are quite different there is no reason to expect that the effective frequency v_1 in the ordinary dispersion formula will be the same as that in the rotatory dispersion formula, even though, of course, the individual frequencies, v_{ba}, are the same. An interesting illustration of this point is provided by data obtained by Volkmann.[16] He found for limonene the critical wave-length in a single term formula for the refractivity to be 998A while that for a corresponding single term formula for rotatory power was 1878A, almost twice as great. This comes about because the R_{ba} are not all of the same sign so the contribution of various bands deep in the ultraviolet tends to cancel out. However, the S_{ba} are all positive so in the refractivity the deep ultraviolet bands are fully effective.

The case of ethyl tartrate is interesting in that it requires two terms in its rotatory dispersion formula. The empirical data are from measurements by Lowry and Cutter[17] and may be represented by the empirical formula

$$[\varphi] = \frac{25.005}{\lambda^2 - 0.03} - \frac{20.678}{\lambda^2 - 0.056}$$

corresponding to absorption bands of opposite rotatory power at 1730A and at 2360A. As a result of the joint action of these two different groups of

---

[16] Zeits. f. physik. Chemie **B10**, 161 (1930).
[17] Lowry and Cutter, J. Chem. Soc. London **121**, 532 (1922).

absorption bands, the rotatory power of ethyl tartrate actually passes through zero with reversal of sign in the neighborhood of 4250A, although there is no characteristic frequency of the molecule at this place in the spectrum. The refractive index data are not given so we merely estimate $(n^2 + 2)/3$ to be 1.30 for want of anything better. Making the calculations exactly as for the previous calculation we arrive at the results:

Ethyl Tartrate

$$\lambda_1 = 1730A, \qquad \lambda_2 = 2360A,$$

$$R_1 = 12.1 \cdot 10^{-40}, \qquad R_2 = 5.42 \cdot 10^{-40}.$$

The critical frequencies appearing in the empirically determined formulas for rotatory dispersion all lie in the ultraviolet or in the visible part of the spectrum if the substance absorbs in the visible. Such frequencies are associated with electronic transitions in the molecule. All of these compounds possess infrared absorption spectra corresponding to changes in the state of nuclear vibration so the question arises: is there any contribution to the rotatory power associated with the infrared absorption frequencies? Several investigations[18] bear on this question, the rotatory power of several substances having been measured out to a wave-length of 2.14 microns. No detectable irregularity in the rotation could be found that could be associated with the infrared absorption.

This is not surprising for we know from ordinary dispersion theory that the contribution to ordinary refractivity of the infrared bands is of the order of 2000 times smaller than electronic bands (of the order of the mass ratio), even for the fundamental vibrations in which the quantum number changes by but one unit. These investigations only went into the harmonic infrared region where the vibrational quantum number changes by two units which would make contributions from such bands extremely weak.

4. Quantum Mechanics of Rotatory Dispersion

Modern work on the problem of deriving the parameter β from a molecular model dates from the independent discovery by Born, Oseen and Gray[19] in 1915 that a calculation of β depends essentially on taking into account the finite ratio of the molecular diameter to the wave-length of light. In other

[18] Meyer, Ann. d. Physik **30**, 607 (1909); Ingersoll, Phil. Mag. **11**, 41 (1906); Phys. Rev. **23**, 489 (1906); Phys. Rev. **9**, 257 (1917); Lowry and Coude-Adams, Phil. Trans. **A226**, 391 (1927), Lowry and Snow, Proc. Roy. Soc. **A127**, 271 (1930).
[19] Born, Physik. Zeits. **16**, 251 (1915); Ann. d. Physik **55**, 177 (1918); Oseen, Ann. d. Physik **48**, 1 (1915); Gray, Phys. Rev. **7**, 472 (1916); Landé, Ann. d. Physik **56**, 225 (1918); Gans, Zeits. f. Physik **17**, 353 (1923); **27**, 164 (1924); Ann. d. Physik **79**, 548 (1926).

words the fact that the phase of the light wave is different for different parts of a molecule is essential.

The molecular model used was that of a spatial distribution of coupled oscillators and corresponds to a natural extension of the form of electron theory of dispersion then in vogue. More or less independently of this work the same general view was also developed by Thomson, de Malleman and Boys.[20] Kuhn[21] has also contributed greatly to the problem by a detailed consideration of the most simple special case of the coupled oscillator model to show activity. His work on this has been very stimulating to the recent development of the subject.

The quantum-mechanical calculation of β was first formulated by Rosenfeld.[22] His calculations lead to the formulas (23) and (32) which have already been discussed in the preceding section. This work unquestionably provides an approach that is much superior to that of the older coupled oscillator models so that one may hope that future work will be built on the quantum-mechanical theory rather than on further study of the coupled oscillator models.

In the rest of this section we give an account of the quantum-mechanical theory for β following essentially Rosenfeld's work but differing somewhat in the details of the calculations. Readers who are not interested in the details of quantum-mechanical calculations will find great pleasure in skipping the rest of this section, for all that will be accomplished will be the derivation of Eqs. (23) and (32).

The fields of a light wave traveling in the direction of the unit vector $\mathbf{k}$ are all derivable from the vector potential $\mathbf{A}$:

$$A = R\{\mathbf{A} \exp(i(t - \mathbf{k} \cdot \mathbf{r}/c)E/\hbar)\}. \tag{39}$$

Here $E = h\nu$ represents the quantum energy associated with the wave. The electric and magnetic vectors of the wave are given by

$$\mathbf{E} = -A/c = -(E/\hbar c)R\{i\mathbf{A}e^+\},$$
$$\mathbf{H} = \operatorname{curl} \mathbf{A} = -(E/\hbar c)R\{i\mathbf{k} \times \mathbf{A}e^+\}, \tag{40}$$

where the exponents of e are the same as in (39).

We may neglect the direct interaction of the atomic nuclei in a molecule with the light wave owing to their comparatively great mass. Hence the interaction gives rise to a perturbation term

$$\mathbf{H} = -(e/mc) \sum_1 [\mathbf{p}_i \cdot \mathbf{A}_i + \mathbf{S}_i \cdot (\operatorname{curl} A)_i]. \tag{41}$$

Here e/mc refers to the charge and mass of the electron and the subscript i on

[20] J.J. Thomson, Phil. Mag. **40**, 713 (1920); de Malleman, Rev. gen des sci. **38**, 453 (1927); Boys, Proc. Roy. Soc. **144**, 655 (1934); Kirkwood, J. Chem. Phys. **5**, 479 (1937).
[21] Kuhn, Zeits. f. physik. Chemie **B4**, 14 (1929).
[22] Rosenfeld, Zeits. f. Physik **52**, 161 (1928). An account will also be found in Born and Jordan, *Elementare Quantenmechanik* (1930), p. 250.

A and curl **A** means that they are to be evaluated at the position of the ith electron.

We have to find how the wave function for the molecule in a particular state labeled by quantum numbers a is affected by the perturbing action of the light wave. For the perturbed wave function we may write

$$\Psi = \psi(a)e^{-iW_a t/h} + \psi_1(a), \qquad (42)$$

where $\psi_1(a)$ has to be determined from the dynamical equation of quantum mechanics

$$i\hbar\partial\Psi/\partial t = (H_0 + H)\Psi, \qquad (43)$$

in which H_0 is the Hamiltonian for the unperturbed molecule. This leads to the following equation for $\psi_1(a)$

$$(H_0 - i\hbar\partial/\partial t)\psi_1(a) = -H\psi(a)e^{-iW_a t/h}. \qquad (44)$$

This will now be solved in the usual way. The right-hand side is expanded in terms of the unperturbed wave functions so

$$-H\psi(a)e^{-iW_a t/h} = \frac{1}{2}\sum_b \psi(b)[(b|H_+|a)e^{i(E-W_a)t/h} + (b|H_-|a)e^{-i(E+W_a)t/h}], \qquad (45)$$

where the coefficients are

$$(b|H_\pm|a) = \frac{e}{mc}\left\{ \bar{\psi}(b)\left(\sum_i \mathbf{p}_i \exp(\mp i\mathbf{k}\cdot\mathbf{r}_i E/\hbar c))\psi(a)\cdot\mathbf{A} \right.\right.$$
$$\left.\left. \mp \frac{iE}{\hbar c}\bar{\psi}(b)\left(\sum_i \mathbf{S}_i\right)\psi(a)\cdot(\mathbf{k}\times\mathbf{A}) \right\}. \qquad (46)$$

Here it is to be understood that $\bar{A}$ stands in place of **A** when the lower sign is used and the approximation has been made of neglecting the retardation factor in the small spin term on the second line. Integration of the wave functions over the configuration space of the molecule is also implied in (46).

If we now expand the retardation factor and save only the first two terms then by some easy reductions

$$(b|H_\pm|a) = \frac{i}{\hbar c}W_{ba}(b|\mathbf{p}|a)\cdot\mathbf{A} \pm \frac{EW_{ba}}{2\hbar^2 c^2}\mathbf{k}\cdot(b|\mathfrak{R}|a)\cdot\mathbf{A} \mp \frac{iE}{\hbar c}(b|\mathbf{m}|a)\cdot(\mathbf{k}\times\mathbf{A}), \qquad (47)$$

where again $\bar{A}$ is to be written for **A** when the lower sign is used. In (47) the following abbreviations have been introduced:

$$W_{ba} = W_b - W_a, \qquad \mathbf{p} = e\sum_i \mathbf{r}_i, \qquad \mathfrak{R} = e\sum_i \mathbf{r}_i\mathbf{r}_i, \qquad (48)$$

$$\mathbf{m} = \frac{e}{2mc}\sum_i (\mathbf{r}_i\times\mathbf{p}_i + 2\mathbf{S}_i),$$

so **p**, $\mathfrak{R}$ and **m** are respectively the electronic contributions to the electric dipole moment, the electric quadrupole moment and the magnetic moment of the molecule.

The terms in the electric quadrupole moment do not introduce any new type of propagation of light through the medium. They give a small correction, of the order of a few parts in a million, to the ordinary connection between mean refractive index and the electric dipole moment and so will be neglected in what follows. Therefore when (47) is used in later calculations the terms in $\mathfrak{R}$ will simply be dropped.

Following the usual procedure one next assumes an expansion for $\psi_1(a)$ in terms of unperturbed wave functions and determines coefficients in the expansion by equating coefficients of both sides of (44). The resulting formula for $\psi_1(a)$ is

$$\psi_1(a) = \frac{1}{2} \sum_b \psi(b) \left[\frac{(b|H_+|a)e^{i(E-W_a)t/\hbar}}{W_{ba} + E} + \frac{(b|H_-|a)e^{-i(E+W_a)t/\hbar}}{W_{ba} - E} \right]. \qquad (49)$$

The first-order correction to the diagonal involving matrix element of any observable F referring to atoms in the state a is then

$$2R\{\bar{\psi}(a)F\psi_1(a)e^{iW_a t/\hbar}\} = R \sum_b \left\{ \frac{(a|F|b)(b|H_+|a)}{W_{ba} + E} e^{iEt/\hbar} \right.$$

$$\left. + \frac{(a|F|b)(b|H_-|a)}{W_{ba} - E} e^{-iEt/\hbar} \right\}. \qquad (50)$$

What is needed are the special cases of (50) in which the values (47) are used for $(b|H_\pm|a)$ and F is identified with **p** and with **m**. If we denote the induced values of **p** and **m** by $\mathbf{p}_1$ and $\mathbf{m}_1$ we have

$$\mathbf{p}_1 = \sum_b R \left\{ \frac{(a|\mathbf{p}|b)(b|H_+|a)}{W_{ba} + E} e^{iEt/\hbar} + \frac{(a|\mathbf{p}|b)(b|H_-|a)}{W_{ba} - E} e^{-iEt/\hbar} \right\} \qquad (51)$$

and a corresponding expression for $\mathbf{m}_1$. Substituting the expressions (47), with neglect of the quadrupole terms as already mentioned, this can be written

$$\mathbf{p}_1 = 2 \sum_b R \left\{ \frac{W_{ba}}{W_{ba}^2 - E^2} (a|\mathbf{p}|b)(b|\mathbf{p}|a) \cdot \mathbf{E} + i \frac{\hbar W_{ba}^2/E^2}{W_{ba}^2 - E^2} (a|\mathbf{p}|b)(b|\mathbf{p}|a) \cdot \dot{\mathbf{E}} \right.$$

$$\left. + \frac{W_{ba}}{W_{ba}^2 - E^2} (a|\mathbf{p}|b)(b|\mathbf{m}|a) \cdot \mathbf{H} + i \frac{\hbar}{W_{ba}^2 - E^2} (a|\mathbf{p}|b)(b|\mathbf{m}|a) \cdot \dot{\mathbf{H}} \right\} \qquad (52)$$

and the corresponding expression for $\mathbf{m}_1$ becomes

$$\mathbf{m}_1 = 2 \sum_b R \left\{ \frac{W_{ba}}{W_{ba}^2 - E^2} (a|\mathbf{m}|b)(b|\mathbf{p}|a) \cdot \mathbf{E} + \frac{iW_{ba}^2/E^2}{W_{ba}^2 - E^2} (a|\mathbf{m}|b)(b|\mathbf{p}|a) \cdot \dot{\mathbf{E}} \right\}. \qquad (53)$$

Each of these expressions contains a term involving

$$\frac{W_{ba}^{2}/E^{2}}{W_{ba}^{2} - E^{2}} = E^{-2} + \frac{1}{W_{ba}^{2} - E^{2}}.$$

The contribution to $\mathbf{p}_1$ resulting from the E^{-2} part of this can be written

$$R\left\{\frac{2i\hbar}{E^{2}} \sum_{b} (a|\mathbf{p}|b)(b|\mathbf{p}|a) \cdot \dot{E}\right\} = \frac{2\hbar}{E^{2}} R\{i(a|\mathbf{pp}|a) \cdot \dot{E}\}$$

by the law of matrix multiplication. Now $(a|\mathbf{pp}|a)$ is a diagonal matrix element of a real observable so it is real; hence the expression whose real part is to be taken is purely imaginary, so the real part vanishes. A similar reduction can also be made in the case of the corresponding term in (53). Since the real part of iX is $-\text{Im}\{X\}$ where $\text{Im}\{X\}$ denotes the imaginary part of X the induced moment expressions can finally be written

$$\mathbf{p}_1 = 2 \sum_{b} \frac{W_{ba}}{W_{ba}^{2} - E^{2}} R\{(a|\mathbf{p}|b)(b|\mathbf{p}|a)\} \cdot \mathbf{E}$$

$$- \frac{\hbar}{W_{ba}^{2} - E^{2}} \text{Im}\{(a|\mathbf{p}|b)(b|\mathbf{p}|a)\} \cdot \mathbf{E}$$

$$+ \frac{W_{ba}}{W_{ba}^{2} - E^{2}} R\{(a|\mathbf{p}|b)(b|\mathbf{m}|a)\} \cdot \dot{\mathbf{H}}$$

$$- \frac{\hbar}{W_{ba}^{2} - E^{2}} \text{Im}\{(a|\mathbf{p}|b)(b|\mathbf{m}|a)\} \cdot \dot{H},$$

$$\mathbf{m}_1 = 2 \sum_{b} \frac{W_{ba}}{W_{ba}^{2} - E^{2}} R\{(a|\mathbf{m}|b)(b|\mathbf{p}|a)\} \cdot \mathbf{E}$$

$$- \frac{\hbar}{W_{ba}^{2} - E^{2}} \text{Im}\{(a|\mathbf{m}|\mathbf{r})(b|\mathbf{p}|a)\} \cdot \dot{E}.$$

(54)

These expressions give the coherent induced electric and magnetic moments which are needed for the calculation of the relations of $\mathbf{D}$ and $\mathbf{B}$ to $\mathbf{E}$ and $\mathbf{H}$ in the medium. They can now be further simplified by averaging over all orientations of the molecules in space. In the ordinary case this means that all orientations are equally likely. If they are not equally likely and the partial lining up of molecules is produced say by an external electric field then special effects may arise, such as that which has apparently been discovered by Kunz and Babcock[23]—the effect of an external electric field on the rotatory power of an active liquid.

The averaging over all orientations is carried out by the following argument. One has to average expressions of the type $\mathbf{pm} \cdot \mathbf{H}$ for all orientations of $\mathbf{p}$ and $\mathbf{m}$ keeping fixed the magnitudes of $\mathbf{p}$ and $\mathbf{m}$ and the angle between them. Now

[23] Kunz and Babcock, Phil. Mag. 22, 616 (1937). See, however, Nature 140, 194 (1937) in which further work is reported showing that the earlier results were due to suspended materials.

pm · H is a vector in the direction of **p**. As **p** takes on all possible directions consistent with a fixed direction of **m** the average value of **pm · H** will be the component of this along **m**, that is $\mathbf{p} \cdot \mathbf{m_0 m} \cdot \mathbf{Hm_0}$ where $\mathbf{m_0}$ is a unit vector in the direction of **m**. Next we average over all directions of **m** by considering first those directions making a fixed angle θ with **H**. The result will be a vector along **H** of magnitude $\mathbf{p} \cdot \mathbf{m} \cos^2 \theta \mathbf{H}$. Finally averaging over all directions of θ weighting them per unit solid angle the end result is $\mathbf{p} \cdot \mathbf{m} (\cos^2 \theta)_{Av} \mathbf{H} = (1/3)\mathbf{p} \cdot \mathbf{mH}$.

Hence after averaging over all orientations of the molecules the expressions for $\mathbf{p_1}$ and $\mathbf{m_1}$ simplify to

$$\mathbf{p_1} = \alpha_a \mathbf{E} + \gamma_a \mathbf{H} - (1/c)\beta_a \dot{\mathbf{H}},$$
$$\mathbf{m_1} = \quad\quad \gamma_a \mathbf{E} + (1/c)\beta_a \dot{\mathbf{E}}, \tag{55}$$

which are almost exactly like Eqs. (18) assumed in the phenomenological discussion of §2. Here, however, we have the additional results connecting α, β and γ with the quantum-mechanical description of the molecular model that is given in the following equations,

$$\alpha_a = \frac{2}{3h} \sum_b \frac{v_{ba} |(a|\mathbf{p}|b)|^2}{v_{ba}^2 - v^2}, \tag{56α}$$

$$2\pi v \beta_a / c = \frac{2}{3h} \sum_b \frac{v \, \mathrm{Im}\{(a|\mathbf{p}|b) \cdot (b|\mathbf{m}|a)\}}{v_{ba}^2 - v^2} \tag{56β}$$

$$\gamma_a = \frac{2}{3h} \sum_b \frac{v_{ba} R\{(a|\mathbf{p}|b) \cdot (b|\mathbf{m}|a)\}}{v_{ba}^2 - v^2}. \tag{56γ}$$

Here v is written for the frequency of the light wave and $v_{ba} = W_{ba}/h$. The subscript a has been written on the coemcients α, β and γ to indicate that these are attributes of molecules in the state a.

The end result of the calculation is contained in (55) and (56). These equations with neglect of the terms in β and γ correspond to the ordinary theory of the refractive index in an isotropic medium. It is easy to see by work analogous to that of §2 that the term in γ has only a second order effect on the mean refractive index for right and left circularly polarized light.[24] There-

---

[24] The details are as follows: corresponding to introduction of g by (13) and (21), the quantity γ leads to introduction of an $f = 4\pi N\gamma(1 - 4\pi N\alpha/3)^{-1}$ where N is the number of molecules per cm^3 so in place of (11) one has

$$\mathbf{D} = \varepsilon \mathbf{E} + f\mathbf{H} - g\dot{\mathbf{H}},$$
$$\mathbf{B} = \mathbf{H} + f\mathbf{E} + g\dot{\mathbf{E}}.$$

Using Maxwell's equations as in Section 4 and assuming

$$\mathbf{D} = d_1(1 - im) + d_2(1 + im),$$
$$\mathbf{B} = b_1(1 - im) + b_2(1 + im),$$

fore its effect may be neglected and (55) reduces precisely to (18). We see that a calculation of the rotatory power of a given substance involves evaluation of the numerators,

$$R_{ba} = \text{Im}\{(a|\mathbf{p}|b)\cdot(b|\mathbf{m}|a)\}, \tag{32}$$

which occur in the formula (56) for β. This part of the problem is discussed in later sections.

We observe that v_{ba} appears in the formula for α at the place where v appears in that for β, in the numerators. The appearance of v_{ba} is what gives rise to the phenomenon of negative dispersion shown by substances having an appreciable number of atoms in electronically excited states.[25] This raises the question: is there an analogous negative rotatory dispersion? The question is rather of pure theoretic interest as it is unlikely that a sufficient concentration of excited optically active molecules could be obtained to study the question experimentally. The answer is in the affirmative, and consists simply in the theorem:

$$R_{ba} = -R_{ab}.$$

Since R_{ba} is the strength associated with resonance to the virtual jump from state a to state b by molecules actually in state a, it follows that R_{ab} is the strength associated with resonance to the virtual jump from state b to state a by molecules actually in state a. The equation just stated is obvious from (32), for interchanging a and b replaces the matrix components by their complex conjugates, which reverses the sign of the imaginary part.

---

one finds that the curl equations give

$$(in^{-1}d_1 - b_1 + \bar{F}d_1)(1 - im) + (-in^{-1}d_2 - b_2 + \bar{F}d_2)(1 + im) = 0,$$
$$(in^{-1}b_1 + \varepsilon^{-1}d_1 - Fb_1)(1 - im) + (-in^{-1}b_2 + \varepsilon^{-1}d_2 - Fb_2)(1 + im) = 0.$$

One can equate to zero separately the coefficients of $(1 - im)$ and $(1 + im)$ in these equations to find the index of refraction for left and right circularly polarized light, respectively. Here

$$F = \varepsilon^{-1}(f - 2\pi ivg)$$

and $\bar{F}$ is the complex conjugate. The resulting equations give

$$\varepsilon^{1/2}n_e^{-1} = (1 - f^2/\varepsilon)^{1/2} - 2\pi vg\varepsilon^{-1},$$
$$\varepsilon^{1/2}n_r^{-1} = (1 - f^2/\varepsilon)^{1/2} + 2\pi vg\varepsilon^{-1},$$

which shows that f enters only as a term in f^2 and does not affect the difference $(n_l - n_r)$. One has therefore

$$n_l = \varepsilon^{1/2} + 2\pi vg \qquad \text{and} \qquad n_r = \varepsilon^{1/2} - 2\pi vg$$

as in (15).

[25] Ladenburg, Rev. Mod. Phys. 5, 243 (1933).

5. General Properties of Rotatory Strengths

By means of (31) the rotatory power of a medium is expressed entirely in terms of the rotatory strengths of the absorption lines, where the rotatory strengths are defined in terms of electric and magnetic dipole matrix components by (32). We wish now to consider certain properties of the R_{ba} which can be derived without specialization of the molecular model.

First of all there is a sum rule, analogous to (28),

$$\sum_b R_{ba} = 0, \tag{57}$$

which is true for all states, a, where the sum is extended over all other states, b. This rule was discovered by Kuhn in connection with the coupled oscillator model. It is easy to give a general quantum-mechanical proof in one line:

$$\sum_b R_{ba} = \mathrm{Im}\left\{\sum_b (a|\mathbf{p}|b)\cdot(b|\mathbf{m}|a)\right\} = \mathrm{Im}\{(a|\mathbf{p}\cdot\mathbf{m}|a)\} = 0,$$

the equality to zero following from the fact that any diagonal matrix element of a real observable is real and therefore its imaginary part vanishes.

Because of the sum rule the optical activity of all substances must vanish in the limit in which $v \gg$ all v_{ba} as is readily seen from (23) or (31). From (31) it is also evident that the rotatory power vanishes as $v \to 0$ because of the v^2 factor in the numerator. Hence rotatory power is a property which tends to zero at both ends of the spectrum.

Next we may consider the symmetry properties of optical activity. This can be done independently of any special theory of the phenomenon and, in fact such considerations which are due to Pasteur, van't Hoff and Le Bel, lie at the foundation of modern stereochemistry. The basic result is: the quantity β is a pseudoscalar which means that it reverses sign on passing from a right-handed coordinate system to a left-handed system. This means that two molecules which are mirror images of each other will have equal and opposite rotatory strengths. This is in accord with (32) for electric dipole moment is a polar vector whereas magnetic dipole moment is an axial vector so their scalar product is a pseudoscalar rather than a true invariant.[26]

The generality of the symmetry argument is also its weakness. It tells us

---

[26] Perhaps these remarks should be made more explicit since so little attention is paid to these points in the usual presentations of electromagnetic theory. Let $\mathbf{i}$, $\mathbf{j}$, and $\mathbf{k}$ be the basic vectors of a right-handed system and $\mathbf{i}'$, $\mathbf{j}'$, $\mathbf{k}'$ be those of the left-handed system that is related to it by inversion,

$$\mathbf{i}' = -\mathbf{i}, \qquad \mathbf{j}' = -\mathbf{j}, \qquad \mathbf{k}' = -\mathbf{k}.$$

The point P whose coordinates are (x, y, z) in the r system has coordinates $(-x, -y, -z)$ in the l system. The coordinates of a point reverse sign on changing systems: any vector whose components do this is called a polar vector.

With regard to the vector products we have, in the r system,

$$\mathbf{i} \times \mathbf{j} = \mathbf{k}, \ldots, \ldots,$$

that two molecules related as mirror images will have equal and opposite rotatory powers, but it does not give us the slightest clue as to what structural feature of the molecule is responsible for the activity—any pseudoscalar associated with the structure might be responsible for the activity and the symmetry argument would be unable to distinguish between them. The attempts to relate optical activity to structure without a detailed theory have been guided to this extent but no more.

Thus Crum Brown[27] proposed a formula for the case of a molecule in which a single asymmetric carbon atom was linked to four different atoms or radicals, A, B, C, D. If k is any scalar attribute of each of the four radicals then the quantity,

$$K = (k_A - k_B)(k_A - k_C)(k_A - k_D)(k_B - k_C)(k_B - k_D)(k_C - k_D),$$

is evidently a pseudoscalar, for we may pass to the mirror image by interchanging any two of the groups and this reverses the sign of K. Of course, it vanishes if any two of the k's are equal. Crum Brown identified the attribute k with the mass of the attached group and achieved some success in correlating empirical data with it but it is now discredited. For example Walden showed that propyl-isopropyl-cyano-acetic acid,

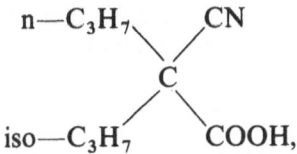

is a polar vector and that the curl of a polar vector field is an axial vector field while
so in the l system, $\mathbf{i}' \times \mathbf{j}' = -\mathbf{k}'$ on using the transformation equations. From this it is evident that the vector product of two polar vectors has components which do not reverse sign on transforming from the r system to the l system. Such a vector is called an axial vector.

It is easy to see that the differential operator

$$\nabla = \mathbf{i}\frac{\partial}{\partial x} + \mathbf{j}\frac{\partial}{\partial y} + \mathbf{k}\frac{\partial}{\partial z} = \mathbf{i}'\frac{\partial}{\partial x^1} + \cdots + \cdots$$

is a polar vector and that the curl of a polar vector field is an axial vector field while the curl of an axial vector field is a polar vector field. Hence if the curl equations of the electromagnetic field are to be invariant on passage from an r system to an l system, it must be that $\mathbf{B}$ and $\mathbf{E}$ are of opposite character and that $\mathbf{D}$ and $\mathbf{H}$ are of opposite character.

It is conventional to assume that $\mathbf{D}$ and $\mathbf{E}$ are polar vectors and $\mathbf{B}$ and $\mathbf{H}$ are axial vectors but all that the theory requires is that the two pairs be of opposite character. The choice is fixed by the assumption that charge density is a true scalar and that current density is a polar vector. If we have a simple connection whereby a vector of one type is equated to a scalar times a vector of the other type, it is evident that that scalar cannot be a true scalar but must reverse sign on passing from an r system to an l system. Such a scalar is called a pseudoscalar. Since $\mathbf{D}$ and $\mathbf{H}$ are of opposite type it follows that g in (13) must be a pseudoscalar and hence that β in (18) is also a pseudoscalar.

[27] Crum Brown, Proc. Roy. Soc. Edinburgh 17, 181 (1890).

is optically active in spite of the equal masses of the propyl and isopropyl groups. Later workers[28] have tried using other attributes of the groups in this type of formula but without much success.

6. The Coupled Oscillator Model

We now come to the problem of calculating the rotatory strengths for particular molecular models. These are of two classes: the coupled oscillator model of Born, Oseen[19] and Kuhn,[21] and the single oscillator model recently proposed by Condon, Altar and Eyring.[29] This section will be devoted to a discussion of the main results for the coupled oscillator model and the single oscillator model will be discussed in the next section.

It is convenient to consider first the extremely simple version of the coupled oscillator model which was devised by Kuhn as this shows all the essential features of this type of model.[30] Suppose we have two particles, the coordinates of which are $(x_1, 0, -d/2)$ and $(0, y_2, +d/2)$ where d is a constant. The arrangement in space is indicated in Fig. 1. Let the charges and masses be e_1, e_2 and m_1, m_2, respectively, and suppose each particle to be bound elastically to its own equilibrium position. Also let there be a quadratic interaction term so the potential energy of the system is given by

$$U = \tfrac{1}{2}k_1 x_1^2 + k_{12} x_1 y_2 + \tfrac{1}{2}k_2 y_2^2, \qquad (58)$$

while the kinetic energy is given by

$$T = \tfrac{1}{2}m_1 \dot{x}_1^2 + \tfrac{1}{2}m_2 \dot{y}_2^2. \qquad (59)$$

The motion is expressed in terms of normal coordinates, ξ_1 and ξ_2, defined as follows:

---

[28] Bose, Zeits. f. physik. Chemie **65**, 695 (1909); Physik. Zeits. **9**, 680 (1908); Bose and Wellers, Zeits. f. physik. Chemie **65**, 702 (1909); Walker J. Phys. Chem. **13**, 574 (1909).
[29] Condon, Altar and Eyring, J. Chem. Phys. **5**, 753 (1937).
[30] The presentation here follows closely that given by Kuhn and Freudenberg, *Hand und Jahrbuch der chemischen Physik*, Vol. 8, part 3 (1932), p. 47.

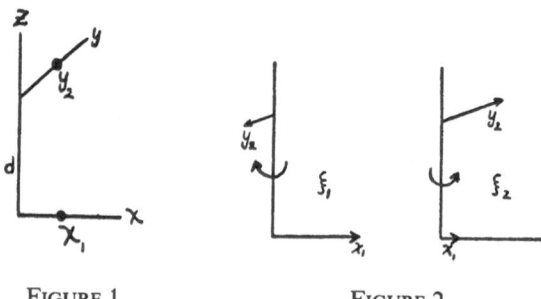

FIGURE 1   FIGURE 2

$$x_1(m_1)^{1/2} = \xi_1 \cos \alpha + \xi_2 \sin \alpha,$$

$$y_2(m_2)^{1/2} = -\xi_1 \sin \alpha + \xi_2 \cos \alpha,$$

(60)

where the parameter α has to be so chosen that the potential energy (58) transforms into a sum of squares in ξ_1 and ξ_2. The value of α is then easily seen to be given by

$$(k_1/m_1 - k_2/m_2) \sin 2\alpha + k_{12}/(m_1 m_2)^{1/2} \cos 2\alpha = 0,$$

(61)

in which case the expressions for kinetic and potential energy become

$$T = \tfrac{1}{2}(\dot{\xi}_1^2 + \dot{\xi}_2^2), \qquad U = 2\pi^2(v_1^2 \xi_1^2 + v_2^2 \xi_2^2),$$

in which

$$(2\pi v_1)^2 = (k_1/m_1) \cos^2 \alpha - 2(k_{12}/(m_1 m_2)^{1/2}) \sin \alpha \cos \alpha + (k_2/m_2) \sin^2 \alpha,$$

$$(2\pi v_2)^2 = (k_1/m_1) \sin^2 \alpha + 2(k_{12}/(m_1 m_2)^{1/2}) \sin \alpha \cos \alpha + (k_2/m_2) \cos^2 \alpha.$$

(62)

Hence the general motion is a superposition of a simple harmonic motion of ξ_1 with frequency v_1 and a simple harmonic motion of ξ_2 with frequency v_2. It is interesting to consider some qualitative features of the motion. Suppose that

$$k_{12}/(m_1 m_2)^{1/2} \ll (k_1/m_1 - k_2/m_2),$$

then (61) shows that α will be a small angle. Hence by (60) the motion in the ξ_1 mode is largely a motion of x_1 with a much smaller amplitude for y_2 (assuming the masses to be of the same order of magnitude). Similarly in the ξ_2 mode the y_2 motion is large compared to the x_1 motion. Qualitatively the two modes have vibration patterns as shown in Fig. 2. On the diagrams are indicated the opposite screw senses characteristic of the two modes. Here the screw sense is the direction of turn around the z axis associated with a displacement toward $+z$ needed to make the direction of the first particle's displacement coincide with that of the second particle.

The next step in the classical discussion of the model consists in determining the forced oscillations set up by the fields of right and left circularly polarized waves, respectively. These forced oscillations give rise to coherent scattering which determines the index of refraction. This is given in full detail by Kuhn and Freudenberg[30] and so need not be repeated here. Instead we shall derive the formula for the rotatory power of this model by applying to it the quantum-mechanical theory of §4. This will provide an interesting variant of their treatment and at the same time give a proof that there is no difference between the classical and quantum-mechanical treatment of this model.

We need an expression for the variable part of the electric dipole moment. This is readily seen to be

$$\mathbf{p} = e_1 x_1 \mathbf{i} + e_2 y_2 \mathbf{j} = (e_1/(m_1)^{1/2} \cos \alpha \mathbf{i} - e_2/(m_2)^{1/2} \sin \alpha \mathbf{j}) \xi_1$$

$$+ (e_1/(m_1)^{1/2} \sin \alpha \mathbf{i} + e_2/(m_2)^{1/2} \cos \alpha \mathbf{j}) \xi_2.$$

(63)

Similarly the magnetic moment of the orbital motion of the two charges is given by

$$\mathbf{m} = (1/2c)[e_1\mathbf{r}_1 \times \mathbf{v}_1 + e_2\mathbf{r}_2 \times \mathbf{v}_2],$$

where $\mathbf{r}_1$ and $\mathbf{r}_2$ are the position vectors and $\mathbf{v}_1$ and $\mathbf{v}_2$ the velocities of the particles. When one uses

$$\mathbf{r}_1 = (-d/2)\mathbf{k} + x_1\mathbf{i}, \qquad \mathbf{r}_2 = (+d/2)\mathbf{k} + y_2\mathbf{j},$$
$$\mathbf{v}_1 = \dot{x}_1\mathbf{i}, \qquad\qquad \mathbf{v}_2 = \dot{y}_2\mathbf{j},$$

this can be written

$$\begin{aligned}
\mathbf{m} &= -(d/2c)(e_1\dot{x}_1\mathbf{j} + e_2\dot{y}_2\mathbf{i}) \\
&= -(d/2c)(e_1/(m_1)^{1/2}\cos\alpha\mathbf{j} - e_2/(m_2)^{1/2}\sin\alpha\mathbf{i})\dot{\xi}_1 \\
&\quad -(d/2c)(e_1/(m_1)^{1/2}\sin\alpha\mathbf{j} + e_2/(m_2)^{1/2}\cos\alpha\mathbf{i})\dot{\xi}_2.
\end{aligned} \tag{64}$$

Next we have to calculate the matrix components of $\mathbf{p}$ and $\mathbf{m}$ in order to get the rotatory strengths given by (32). The energy levels will be labeled by two harmonic oscillator quantum numbers, n_1 and n_2, corresponding to the normal coordinates, ξ_1 and ξ_2. The energies are given by

$$W(n_1, n_2) = (n_1 + \tfrac{1}{2})h\nu_1 + (n_2 + \tfrac{1}{2})h\nu_2 \tag{65}$$

and the corresponding wave functions are

$$\psi(n_1, n_2) = \varphi_{n_1}(\xi_1/a_1)\varphi_{n_2}(\xi_2/a_2), \tag{66}$$

where

$$\varphi_n(z) = \frac{1}{[2^n n!(\pi)^{1/2}]^{1/2}} H_n(z)e^{-z^2/2}, \tag{67}$$

in which $H_n(z)$ stands for the nth Hermitian polynomial and

$$a_i = (1/2\pi)(h/\nu_i)^{1/2} \qquad (i = 1, 2). \tag{68}$$

The necessary matrix components are those for ξ_1 and ξ_2, $\dot{\xi}_1$ and $\dot{\xi}_2$ appearing in (63) and (64). These are well known from the quantum mechanics of the harmonic oscillator:

$$(n_1 n_2|\xi_1|n_1'n_2') = a_1(\bar{n}_1/2)^{1/2}\delta(n_2, n_2'). \tag{69}$$

Here $n_1' = n_1 \pm 1$ and $(\bar{n}_1)$ stands for the larger of n_1 and n_1'. Here the effective mass of the oscillator is unity so $\dot{\xi}_1$ is the same as the momentum, p_{ξ_1}, so the matrix components of $\dot{\xi}_1$ are

$$(n_1'n_2'|\dot{\xi}_1|n_1 n_2) = \pm i(\hbar/a_1)(\bar{n}_1/2)^{1/2}\delta(n_2, n_2'), \tag{70}$$

in which $(\bar{n}_1)$ has the same meaning as before and the $\pm$ sign has the same value as occurs in $n_1' = n_1 \pm 1$. The formulas for matrix components for ξ_2 and $\dot{\xi}_2$ are obtained from (69) and (70) by obvious changes.

If we are calculating the parameter β for molecules in the state (n_1, n_2) there will be four other states for which there are nonvanishing values of the rotatory strength, namely,

$$(n_1 + 1, n_2) \quad \text{and} \quad (n_1 - 1, n_2).$$

Also
$$(n_1, n_2 + 1) \quad \text{and} \quad (n_1, n_2 - 1).$$

The pair in the first line will have the same resonance denominator in the formula (23) and the same is true of the pair on the second line, so although there are four quantum jumps which can contribute there will only be two critical frequencies in the dispersion formula.

If we write R_1 and R_2 for the two rotatory strengths, it is easy to calculate by a combination of (63), (64), (69) and (70) that

$$R_1 = -R_2 = \hbar(d/4c)\frac{e_1 e_2}{(m_1 m_2)^{1/2}} \sin \alpha \cos \alpha. \tag{71}$$

This result is in accord with the sum rule of §5 in giving $R_1 + R_2 = 0$. In calculating this result it will be noticed that the upward virtual jump contributes an amount to R_1 that is proportional to $(n_1 + 1)$, while the downward virtual jump contributes an amount proportional to $(-n_1)$ so the resultant effect of both contributions is an amount which is independent of the quantum number, n_1. This, it uill be recalled, is exactly analogous to the quantum-mechanical theory of the ordinary dispersion by a harmonic oscillator. There the positive dispersion due to the upward virtual jump increases with the quantum number, but so also does the negative dispersion due to the downward virtual jump, and in fact in exactly the same way as here so that the difference is independent of the quantum number.

If we substitute these values in (23) and use the resulting value of β in (22), the final result for the rotatory power of a medium containing N_1 of these models in unit volume, oriented at random, is

$$\varphi = \frac{2\pi N_1}{3} \cdot \lambda^{-2} \cdot \tfrac{1}{3}(n^2 + 2) \cdot d \sin \alpha \cos \alpha \cdot \frac{e_1 e_2}{(m_1 m_2)^{1/2}} \left[\frac{1}{v_1^2 - v^2} - \frac{1}{v_2^2 - v^2} \right]. \tag{72}$$

This corresponds exactly with the result given by Kuhn and Freudenberg[31] except for the factor $(1/3)$ in (72) which arises because (72) is the final formula applying for random orientations of the molecules whereas the equations (46) and (46a) of Kuhn and Freudenberg have still to be averaged. (Compare reference 30, pp. 72–76.)

It is of interest to note how (72) exhibits the two essential features of the model: (l) the oscillators must be coupled, for if k_{12} is zero then by (61), $\sin 2\alpha = 0$ so the rotation vanishes; and (2) the oscillators must be offset by

---

[31] Kuhn and Freudenberg, reference 30, p. 59, Eqs. (46) and (46a).

the distance d as is evident from the proportionality of φ to d that occurs explicitly in (72).

As the foregoing calculations illustrate all the essential features of the coupled oscillator model, it will not be necessary to present the detailed calculations of the general case here. This will be found in the original papers of Born and Oseen[19] and also in the Kuhn and Freudenberg article[30] already referred to, pp. 69–72.

The results of such calculations may be stated as follows: Let the model consist of s particles, the charge and mass of the κth particle being e_κ and m_κ and let $x_\kappa, y_\kappa, z_\kappa$ be the equilibrium coordinates of the κth particle uith respect to a coordinate system fixed in the molecule. The potential energy will be assumed to be a general quadratic form in the displacements of the particles from equilibrium, $u_\kappa, v_\kappa, w_\kappa$ being the displacements of the κth particle. This means that in general there will be $3s$ different proper frequencies and $3s$ normal modes of vibration, described by $3s$ normal coordinates ($a = 1, 2 \ldots 3s$). Analogous to (60) we may express the connection between the displacements and the normal coordinates by

$$\xi_a = \sum_{k=1}^{s} (m_\kappa)^{1/2}(\alpha_{\kappa a}u_\kappa + \beta_{\kappa a}v_\kappa + \gamma_{\kappa a}w_\kappa) \tag{73}$$

for $a = 1, 2, \ldots 3s$. Here the coefficients $\alpha_{\kappa a}, \beta_{\kappa a}, \gamma_{\kappa a}$ define an orthogonal transformation which is determined as usual by the requirement that the potential energy be expressed as a sum of squares in terms of the ξ_a coordinates. The coefficients therefore depend on the coupling terms in the potential energy.

The end result of the calculations is to show that the rotatory power is given by

$$\varphi = \frac{2\pi N_1}{3} \cdot \lambda^{-2} \cdot \tfrac{1}{3}(n^2 + 2) \sum_{a=1}^{3s} \frac{L_a \cdot M_a}{v_a^2 - v^2}. \tag{74}$$

Here v_a is the frequency associated with the normal coordinate ξ_a, and the vectors L_a and M_a are defined by

$$L_a = \sum_{\kappa=1}^{s} \frac{e_\kappa}{(m_\kappa)^{1/2}}(\alpha_{\kappa a}i + \beta_{\kappa a}j + \gamma_{\kappa a}k),$$

$$M_a = \sum_{\kappa=1}^{s} \frac{e_\kappa}{(m_\kappa)^{1/2}}(\alpha_{\kappa a}i + \beta_{\kappa a}j + \gamma_{\kappa a}k)(x_\kappa i + y_\kappa j + z_\kappa k). \tag{75}$$

It is easy to see that the vector L_a is related to the electric dipole moment belonging to the ξ_a mode of vibration and that M_a is related to the corresponding magnetic dipole moment. Hence this classical formula is related to the quantum-mechanical formula in a manner analogous to that revealed by our detailed calculations of the Kuhn model.

7. The Single Oscillator Model

Drude[32] proposed a model for optical activity in which a single electron was constrained to move on a helix while being elastically bound to an equilibrium position on the curve. This model stood in the literature for many years until Kuhn[33] in 1933 pointed out an error in the calculations. Kuhn showed that when correctly treated this model has no rotatory power. This result was extremely interesting in that it provided a case in which the rotatory power vanished in spite of the proper kind of dissymmetry being present. From this result he was inclined to conclude that coupled oscillators of the type discussed in the preceding section were essential to optical activity. The statement that rotatory power *requires* the existence of coupled oscillators for its explanation has been made repeatedly in the literature in recent years.

This view can no longer be held since the work of Condon, Altar and Eyring[29] has shown that it is possible to build a model of a single charged particle moving in a dissymmetric field which has rotatory power. This section will be devoted to a brief summary of that work: for further details the original paper should be consulted.

Most of our thinking about molecular electronic structure nowadays is done in terms of the Hartree approximation. In this each electron is regarded in the first approximation as moving in a static potential field due to the average charge distribution of the nuclei and of the other electrons in the molecule. This has been shown to be an accurate way of accounting for the larger part of the interaction between electrons. The dynamic coupling of the electrons only comes in in higher approximations. Now the coupled oscillator model of Born, Oseen, Kuhn and others takes account of this dynamic coupling of the electrons, but the question arises: If we are proceeding in the sense of a Hartree development, does optical activity make its appearance at the outset, where each electron is moving in the *static* average dissymmetric field of the rest of the molecule, or does it vanish in this approximation and first make its appearance when the *dynamical* coupling of electrons is considered? Kuhn's elimination of the Drude helix model had tended to make it appear that dynamic coupling of more than one electron was a necessary condition, until the single oscillator model was developed.

The single oscillator model assumes that an electron moves in a dissymmetric potential field in which the potential energy as a function of Cartesian coordinates, x_1, x_2, and x_3, is given by

$$V = \tfrac{1}{2}k_1 x_1{}^2 + \tfrac{1}{2}k_2 x_2{}^2 + \tfrac{1}{2}k_3 x_3{}^2 + A x_1 x_2 x_3. \tag{76}$$

The term in A is what produces the necessary dissymmetry.

It is easy to see that an equipotential surface of (76) is qualitatively like what one would get if he took an ellipsoid of three unequal axes and subjected it

[32] Drude, *Göttinger Nachrichten* (1892), p. 400.
[33] Kuhn, Zeits. f. physik. Chemie **B20**, 325 (1933).

to a torsional stress. An equipotential surface has elliptical cross sections when cut by any of the planes parallel to the basic planes of the coordinate system. For definiteness think of the section of the surface $V = $ constant by a plane $x_3 = $ constant. The section by the plane $x_3 = 0$ is an ellipse whose principal axes are the x_1 and x_2 axes. For positive x_3 the section is an ellipse whose principal axes have turned around in the sense of a right-handed screw if $k_1 > k_2$ and $A > 0$. The same screw sense holds for $x_3 < 0$. Hence there is a right-handed screw sense associated with the x_3 axis under the conditions stated.

The screw sense associated with the other coordinate axes is easily determined in the same way. It turns out that in every case there is a screw sense associated with each axis and that two of these are alike while the third one is different. Thus in the special case

$$k_1 > k_2 > k_3 \qquad \text{and} \qquad A > 0,$$

it turns out that one has a right-hand screw sense associated with the first and third axes and a left-hand screw sense associated with the second axis.

In the actual calculations the $Ax_1x_2x_3$ term is handled by perturbation theory based on a starting solution of the anisotropic oscillator problem represented by the quadratic terms in the potential. Thus the stationary states are labeled by harmonic oscillator quantum numbers $(n_1 n_2 n_3)$, corresponding to the energy levels,

$$W(n_1 n_2 n_3) = (n_1 + \tfrac{1}{2})hv_1 + (n_2 + \tfrac{1}{2})hv_2 + (n_3 + \tfrac{1}{2})hv_3, \qquad (77)$$

which are not affected by the perturbation energy in at least the first two approximations. Here v_1, v_2, v_3 are the frequencies determined by the force constants, k_1, k_2, k_3, in the usual way, $v_i = (2\pi)^{-1}(k_i/\mu)^{1/2}$. By the standard form of perturbation theory the first-order wave functions are found and from these the matrix components of electric and magnetic dipole moment correct to the first order are computed.

If we consider that the "molecules" represented by the model are in their lowest quantum state (0 0 0) then in the zeroth approximation only three upper states may be reached by ordinary electric dipole absorption of light, namely (1 0 0), (0 1 0) and (0 0 1). Likewise in the zeroth approximation the magnetic dipole moment has nonvanishing matrix components connecting (0 0 0) only with (0 1 1), (1 0 1), and (1 1 0). As these selection rules are mutually exclusive it follows that there is no jump from the normal state which has simultaneously nonvanishing matrix components of both $\mathbf{p}$ and $\mathbf{m}$, hence the rotatory power vanishes.

In the first approximation, however, one may get nonvanishing rotatory power associated with transitions from the normal state to all six of the excited states just enumerated. It will be convenient to consider explicitly the pair (0 0 0) → (1 0 0) and (0 0 0) → (0 1 1), as the other two pairs behave in exactly the same way and the results for them can be obtained by cyclic permutation of the indices. With regard to (0 0 0) → (1 0 0), here one has a zero-order $\mathbf{p}$ matrix component and is provided with a first-order $\mathbf{m}$ matrix component by the dissymmetric perturbation. In consequence this line is strong in ordinary absorption and also contributes to the rotatory power. On the other hand, $(0\,0\,0) \rightarrow (0\,1\,1)$ has a vanishing zero-order $\mathbf{p}$ component so its ordinary absorption power will be weak as the $\mathbf{p}$ component arises entirely from the dissymmetric perturbation. Its rotatory strength however will be comparable with that of the other absorption line, and in fact comes out actually equal and opposite to it.

This important qualitative distinction between the two lines corresponds exactly to an empirical generalization made by Kuhn.[34] He points out that empirically the strong bands ($f \approx 1$) have very small anisotropy factors ($g \approx 10^{-5}$) whereas the weak absorption bands ($f \approx 10^{-3}$) have much greater anisotropy factors ($g \approx 10^{-2}$) so that in order of magnitude the rotatory strengths ($fg \approx 10^{-5}$) are all about the same.

This result receives a very simple interpretation on the single oscillator model independently of the special form assumed in (76) for the effective force field. The dissymmetry in the effective field in which a particular electron moves is due to the action of the other atoms than the one or two to which the electron essentially belongs and therefore is weak, because the other atoms are farther away and because the dissymmetry is a residual high order effect due to the joint action of several neighbors. Hence there is a tendency for the electron to be governed by the selection rules which would hold rigorously if the dissymmetry were absent. In these the selection rules are mutually exclusive, $\mathbf{p}$'s matrix component vanishing when $\mathbf{m}$'s does not, and *vice versa*. When the selection rule is broken down by the dissymmetric field this will produce two classes of active bands, those which have a rotatory strength arising from a large $\mathbf{p}$ factor and a small $\mathbf{m}$ which will also have a large f value, and those which arise from a large $\mathbf{m}$ combined with a small $\mathbf{p}$ which will have a small f value.

Continuing the discussion of the particular single oscillator model which uses the field (76) we find that the value of β for a particle of charge e and mass μ when it is in the normal state is

$$\beta_{000} = \frac{A\hbar e^2}{12(2\pi)^5\mu^3} \left\{ \left(\frac{1}{v_2} - \frac{1}{v_3}\right) \frac{1}{(v_2 + v_3)^2 - v_1{}^2} \left[-\frac{1}{v_1{}^2 - v^2} + \frac{1}{(v_2 + v_3)^2 - v^2} \right] \right.$$

$$+ \left(\frac{1}{v_3} - \frac{1}{v_1}\right) \frac{1}{(v_3 + v_1)^2 - v_2{}^2} \left[-\frac{1}{v_2{}^2 - v^2} + \frac{1}{(v_3 + v_1)^2 - v^2} \right]$$

$$\left. + \left(\frac{1}{v_1} - \frac{1}{v_2}\right) \frac{1}{(v_1 + v_2)^2 - v_3{}^2} \left[-\frac{1}{v_3{}^2 - v^2} + \frac{1}{(v_1 + v_2)^2 - v^2} \right] \right\}. \quad (78)$$

A similar formula may be obtained for molecules in an arbitrary state

---

[34] Kuhn, Trans. Faraday Soc. "Discussion on Optical Rotatory Power," **300** (1930). Also Kuhn and Freudenberg, *Hand und Jahrbuch der chemischen Physik*, Vol. 8, part 3, p. 84.

(n_1, n_2, n_3). The general formula is given by Condon, Altar and Eyring.[29] It is interesting here only in this connection: it can be written in such a way that Planck's constant cancels out of the expression for the rotatory power from which it follows, with the aid of the correspondence principle, that this same model would show optical activity even if it were treated by classical mechanics. In other words, the nonvanishing rotatory power of the model is not a specific effect of quantum mechanics but is an essential property of the model from either the classical or the quantum-theoretic point of view.

Turning now to the question of how this model is to be applied to actual molecules, we see that it is first necessary to make a definite postulate concerning the part of the molecule which is responsible for a particular absorption band. This part is generally called the *chromophoric group* and there exists a fairly good body of empirical material about absorption spectra of polyatomic molecules which enables a decision to be made. The strongest part of the field in which the chromophoric electron moves is that due to the atom to which it belongs or the two atoms which it is bonding. This part of the field has to be estimated by the methods which are being developed in other connections for dealing with molecular orbitals.

Superimposed on this local field is the field due to the other atoms and it is this which produces the dissymmetry. One can calculate an approximation to this field by assuming it to be due to point charges located at the center of each atom, the magnitudes of these charges being chosen in such a way as to represent the observed values of the static dipole moments arising from each bond. Fairly definite information about this can be obtained from the literature on dipole moments.[35] However, at present one is balked to some extent by lack of information about the extent to which partially hindered free rotations are hindered and such questions. The relation of optical rotatory power to these questions however probably means that in future developments study of rotatory power can be made to throw light on these and related problems of molecular structure.

Having made the best assumption possible about the system of effective charges arising in this way one may next calculate the effect of this field on a particular electron by developing the field in powers of the displacement of that electron about its mean position. In this way quadratic and cubic terms of the type introduced in (76) arise which result in a nonvanishing rotatory strength for that electron's transitions. This method of handling the field due to neighboring atoms is, of course, very much the same as that used so successfully by Van Vleck and others[36] in dealing with the effects of crystalline fields on magnetic susceptibility.

Some detailed calculations of this type may be found in the paper of Condon, Altar and Eyring.[29]

---

[35] See for example, C.P. Smyth, *Dipole Moments and Molecular Structure*.
[36] Bethe, Ann. d. Physik **3**, 133 (1929); Kramers, Proc. Amsterdam Acad. **32**, 1176; **33**, 959 (1929-30); Penney and Schlapp, Phys. Rev. **41**, 194 (1932); Van Vleck, *Electric and Magnetic Susceptibilities*, Chap. XI.

7a. Rotatory Strength and Polarizability of Groups

Just as this review uas being finished there appeared an important paper by Kirkwood[a] who shows how to relate the quantum-mechanical theory of rotatory power to the polarizability of the groups and their mutual coupling. This section provides a brief review and commentary of the paper, written as an insert after completion of the rest of the report so the equations here will be numbered in a separate sequence denoted by (1a), (2a), etc. The notation of Kirkwood's paper has been changed where necessary to conform to that used elsewhere in this report.

We suppose that the electrons in the molecule may be unambiguously assigned to N different groups attached to a central group. Then for the total electric dipole moment we may write

$$\mathbf{p} = \sum_{i=1}^{N+1} \mathbf{p}^{(i)} \tag{1a}$$

where $\mathbf{p}^{(i)}$ is the electric dipole moment of the ith group defined as

$$\mathbf{p}^{(i)} = \sum_s e\mathbf{r}_s$$

in which $\mathbf{r}_s$ is the position vector of an electron in the ith group referred to an origin at the center of mass of that group and not at the center of mass of the molecule as formerly. Let R_k be the position vector of the center of mass of the kth group relative to that of the entire molecule then for the magnetic moment we have

$$\mathbf{m} = \frac{e}{2mc} \sum_k \mathbf{R}_k \times \mathbf{P}_k + \sum_k \mathbf{m}^{(k)} \tag{2a}$$

in which $\mathbf{P}_k$ stands for the total electronic momentum of the electrons in the kth group and $m^{(k)}$ stands for the magnetic dipole moment of the kth group computed with the position vectors of the electrons measured from the center of mass of that group. Now if we consider the electric and magnetic dipole moments written as sums of terms from different groups as in (1a) and (2a) then the rotatory strength R_{ba} associated uith any one transition $a \to b$ can be written, from (32)

$$R_{ba} = \operatorname{Im}\left\{\left(a\left|\sum_i \mathbf{p}^{(i)}\right|b\right) \cdot \left(b\left|(e/2mc)\sum_k \mathbf{R}_k \times \mathbf{P}_k + \sum_k \mathbf{m}^{(k)}\right|a\right)\right\}$$

$$= \operatorname{Im}\left\{\sum_i (a|\mathbf{p}^{(i)}|b) \cdot (b|\mathbf{m}^{(i)}|a) + \sum_{i \neq k} (a|\mathbf{p}^{(i)}|b) \cdot \mathbf{R}_k \times (b|\mathbf{P}_k|a)(e/2mc)\right.$$

$$\left. + \sum_{i \neq k} (a|\mathbf{p}^{(i)}|b) \cdot (b|\mathbf{m}^{(k)}|a)\right\}. \tag{3a}$$

---

[a] Kirkwood, J. Chem. Phys. **5**, 479 (1937).

This equation corresponds to (24) of Kirkwood's paper. On the first line we have the sum of the contributions from the separate groups. These have a nonvanishing value because of the action of the neighboring groups in producing a dissymmetric field as discussed in the preceding section. The second line corresponds to the part which is of most importance in the coupled oscillator theories and is here in a form which lends itself to a discussion which is the main point of Kirkwood's paper. The third line is dismissed by Kirkwood as unimportant but without a thorough investigation of its magnitude. This point deserves to be studied more carefully.

The terms of the second line will now be discussed more fully. From the commutation rules we have for any one electron $H\mathbf{r}_i - \mathbf{r}_i H = \hbar \mathbf{p}_i/im$ and hence summing over the electrons in any, say the kth group

$$H\mathbf{p}^{(k)} - \mathbf{p}^{(k)}H = (e\hbar/im)\mathbf{P}^{(k)} \tag{4a}$$

which enables us to express the electronic momentum matrix components in terms of those of the group dipole moments:

$$\frac{e}{2mc}(b|\mathbf{P}^{(k)}|a) = (\pi i v_{ba}/c)(b|\mathbf{p}^{(k)}|a)$$

and therefore the particular contribution to R_{ba} on the second line of (3a) which we denote with $R_{ba}{}^\circ$ is, since $\text{Im}\{iz\} = R\{z\}$

$$R_{ba}{}^\circ = (\pi v_{ba}/c)R\left\{\sum_{i \neq k}(a|\mathbf{p}^{(i)}|b)\cdot R_k \times (b|\mathbf{p}^{(k)}|a)\right\}.$$

Using standard properties of the mixed triple product and the fact that $R\{\bar{z}\} = R\{z\}$ this can be written

$$R_{ba}{}^\circ = (\pi v_{ba}/2c)R\left\{\sum_{i \neq k}(\mathbf{R}_k - \mathbf{R}_i)\cdot(a|\mathbf{p}^{(k)}|b) \times (b|\mathbf{p}^{(i)}|a)\right\} \tag{5a}$$

In the sum each pair (i, k) appears twice, once as (i, k) and again as (k, i). This result exhibits clearly the features that were emphasized in §6 in the more classical discussion of the coupled oscillators: $R_{ba}{}^\circ$ has the same dimensions as the ordinary strength S_{ba} namely square of dipole moment, but is smaller for two reasons. Firstly $R_{ba}{}^\circ$ is small because of the fact that $(\mathbf{R}_k - \mathbf{R}_i)$, the vector distance from group i to group k, is divided by c/v_{ba} the wave-length of the quantum jump frequency of the active band in question which is large compared to the size of the molecule. Secondly, $R_{ba}{}^\circ$ is small because it owes its existence entirely to the weak dynamic coupling between different groups.

To see this, consider the case in which the dynamic coupling between groups could be ignored. This means that the Hamiltonian can be represented adequately as a sum of separate Hamiltonians, one for each group and that the ensemble of quantum numbers symbolized by a and b would break up into sets referring separately to each group. This is, a would be $a_1 a_2 \ldots a_{n+1}$ where a_k refers only to the kth group. If this were the case then the only nonvanishing matrix components $(a|\mathbf{p}^{(i)}|b)$ would be those in which b is the same as a with regard to all other quantum numbers than those of the ith group and similarly for $(a|\mathbf{p}^{(k)}|b)$. Therefore, in the absence of dynamic coupling the selection rules for $\mathbf{p}^{(i)}$ and $\mathbf{p}^{(k)}$ are mutually exclusive and $R_{ba}{}^\circ$ vanishes. Therefore, to get a nonvanishing rotatory power it is really essential to consider the coupling of the electronic groups in the molecule.

Various assumptions about the nature of this coupling might be made but if one assumes ordinary Coulomb interaction between the electrons in the groups then the first term in a development in inverse powers of the distance is the dipole-dipole interaction energy. This can be written

$$V = \sum_{l>j=1}^{N} \mathbf{p}^{(l)} \cdot \mathbf{T}_{lj} \cdot \mathbf{p}^{(j)} \tag{6a}$$

where
$$\mathbf{T}_{lj} = R_{jl}^{-3}[1 - 3\mathbf{R}_{jl}\mathbf{R}_{jl}/R_{jl}{}^2].$$

The effect of this on the wave functions and hence on the matrix components can be taken into account by ordinary perturbation theory. The details of this will be found in Kirkwood's paper, the end result being a formula for the rotatory strengths which depends on the polarizabilities of the interacting groups.

8. Solvent Effect and the Effective Field Correction

According to the arguments of §2 which led up to (22), the parameter β is a property of individual active molecules. If the rotatory power, φ, is measured and the molecular density N_1 and the refractive index n are known, then by means of (22) one may calculate an empirical value of β. This will be the effective value of β for the active molecules in the average conditions in which they exist in the particular medium for which the measurements were made.

The question arises: Is the parameter β a constant property of the individual molecule quite independent of the molecular environment? In this section we wish to review the evidence which shows that the answer to this question is in the negative. It turns out that β is generally quite sensitive to the molecular environment. This is called the solvent effect. In this respect β is quite different from the ordinary polarizability α which generally shows a constant value leading to the well-known additivity laws for ordinary molecular refractivity.

First let us consider what happens on change of state from liquid to vapor. In the vapor state the molecules are far apart while in the liquid they are densely packed together. The simplest thing to assume would be that in spite of the change in average environment the value of β is the same in both states. If this is the case it is easily seen from (22) that

$$[\varphi]/\tfrac{1}{3}(n^2 + 2)$$

should be continuous at the change of state since N_1 is proportional to the

TABLE II. Rotatory power in vapor and liquid states.
(Guye and Amaral.)

Substance	[φ] Vapor		[φ] Liquid
Valeric aldehyde	7.1 to	6.4	14.6
Amyl acetate	2.6	3.2	2.8
Methyl valerate	14.3	14.5	16.4
Amyl chloracetate	1.9	1.6	3.1
Diamyl	10.7	10.9	11.1
Amyl amine	2.1	2.2	1.8
Amyl bromide	1.9		2.8
Amyl iodide	3.9 to	4.1	5.6
Amyl alcohol	5.8	6.5	5.1
Valeric acid	10.7	10.9	13.5

density, ρ. For all vapors the index of refraction is so close to unity that one may set $\frac{1}{3}(n^2 + 2) = 1$. For liquids, however, the factor is usually between 1.30 and 1.50. Hence if β does not change, the specific rotatory power [φ] should show the same discontinuity at the change of state as the factor $\frac{1}{3}(n^2 + 2)$ namely a 30 to 50 percent decrease in going from liquid to vapor.

This is in conflict with the usual statement[37] that [φ] itself is continuous with change of state. But the available data are quite meager. The most detailed investigation is that of Guye and Amaral.[38] Their results, however, are not accurate enough for a definite conclusion. Their final table of results (reference 38, p. 527) is given in Table II. From the table we see that there is no great discontinuity in [φ] at the change of state, that in most cases [φ]; for liquid is greater than for the vapor as would be required by constant β but this is not true in every case and in no case are the data good enough for a definite check. It appears therefore that one can only say that β does not change much on change of state, if at all.

A modern investigation of the rotatory dispersion of camphor in the vapor phase has been made by Lowry and Gore.[39] They measured both the vapor at 180°C and a solution of camphor in cyclohexane at 20°C. These data are not comparable because of the difference in temperature involved.

Similarly in case of inactive solvents where there is very small disturbance of the solute molecules by those of the solvent we should expect to find the same value of β on applying (22) to the observed rotations independently of the choice of the solvent. This question was first studied from this point of view by Wolf and Volkmann[40] and has recently been the subject of a long

---

[37] For example Lowry, *Optical Rotatory Power*, p. 102 or Bruhat *Traité de Polarimetrie*, p. 194.

[38] Guye and Amaral, Arch. Sci. Phys. Nat. Geneva **33**, 409, 513 (1895).

[39] Lowry and Gore, Proc. Roy. Soc. **A135**, 13 (1932).

[40] Wolf and Volkmann, Zeits. f. physik. Chemie **B3**, 139 (1929); Volkmann, Zeits. f. physik. Chemie **B10**, 161 (1930).

series of experimental investigations by Rule[41] and his associates. In general the results indicate that for a nonpolar active substance in a nonpolar solvent the quantity $[\varphi]/\frac{1}{3}(n^2 + 2)$ is more nearly constant than $[\varphi]$. But in case of polar solvents and polar active substances there are great differences which indicate large changes in the effective value of β produced by solvation. Thus Pickard and Kenyon[42] found that the sign of the rotatory power is different for β-hexyl stearate in two different solvents:

$$[\varphi] = +20.21 \quad \text{in alcohol,}$$

$$[\varphi] = -8.93 \quad \text{in carbon disulphide.}$$

A great deal of information of this kind is to be found in Lowry's book.

What is very important and what has been very much neglected hitherto is the study of rotatory dispersion in relation to solvent effect. Clearly one may expect the rotatory contributions of different chromophoric groups to be differently affected by association or loose compound formation with the solvent. The data ought to be complete enough to show how the individual rotatory strengths, R_{ba}, of the various bands, v_{ba}, change on solution.

On the theoretical side a somewhat phenomenological treatment of solvent action has been given recently by Beckmann and Cohen.[43] In this they attempt to connect the solvent action principally with a deformation of the molecular frame.

9. Circular Dichroism

As mentioned in the introduction, circular dichroism consists in a difference in the absorption of right and left circularly polarized light by the medium. It is observed by determining the ellipticity of the elliptically polarized light transmitted by the active medium from a beam that is initially linearly polarized. It is the property that is related to optical rotatory power in the same way that ordinary dispersion is related to ordinary absorption.

Experimental technique for observing circular dichroism is described in the books by Lowry and by Bruhat already cited, also in the Cornell lectures of Jaeger.[44]

In the older electronic theories of dispersion the connection between refraction and dispersion is always obtained by formal introduction of a *damping* term in the equation of motion of the electron. The damping term is in the form of a force proportional to the velocity and opposite to it in direction so

---

[41] Rule, various papers in the J. Chem. Soc. London from 1931 to 1937, under the general heading, "Studies in Solvent Action."
[42] Pickard and Kenyon, J. Chem. Soc. **105**, 830 (1914).
[43] Beckmann and Cohen, J. Chem. Phys. **4**, 784 (1936).
[44] Jaeger, *Optical Actiuity and High Temperature Measurements* (McGraw-Hill, 1930), Appendix, page 215.

that no matter how the electron moves energy must always be given up
to do work against this force. As is well known this acts to cut down the
sharpness of resonance and leads to a considerable absorption of energy
at frequencies near the resonant frequency where the forced oscillation is
relatively great. Various proposals have been made to give a physical account
of the origin of the damping term, the two most important being the *radiation
damping* of Planck[45] and the *collision damping* of Lorentz.[46] On the radiation
damping picture the loss of energy from the original beam is due to the
scattering of radiation in all directions by the radiation due to the forced
oscillations of the electron. The collision damping is the expression for the
average energy loss associated with the interruption of the forced oscillations
by collisions of other molecules with the resonator.[47]

In dealing with the absorption of light by matter a number of different
measures of absorptive power are in common use. For theoretical purposes
the most convenient mode of description is by means of the complex index of
refraction, usually uritten $\mathbf{n} = n(1 - i\kappa)$. With this form of the index of refraction in (3) the equation for the propagation of the electric induction for
example is

$$\mathbf{D} = R\{D_0 e^{i\psi}\} = \exp(-2\pi\nu n\kappa \mathbf{k} \cdot \mathbf{r}/c)R\{\mathbf{D}_0 \exp(2\pi i\nu(t - n\mathbf{k}\cdot\mathbf{r}/c))\}, \quad (79)$$

so the amplitude of $\mathbf{D}$ diminishes with an exponential factor which depends
on the imaginary part of the index. Since the intensity of the light wave is
proportional to the mean value of $\mathbf{D}^2$, the intensity falls off exponentially
according to the formula,

$$I = I_0 e^{-4\pi\nu n\kappa z/c}, \tag{80}$$

where z is the distance traveled in the medium. The coefficient of z in the
exponential here is called the *extinction coefficient* of the medium and is
usually (as in §1) denoted by ε so

$$\varepsilon = \frac{4\pi\nu n\kappa}{c} = \frac{4\pi n\kappa}{\lambda}, \tag{81}$$

where, as always in this report, λ means the *vacuum* wave-length. (In some
accounts λ means the wave-length in the medium, i.e., our λ/n so the n does
not appear explicitly in the numerator.)

The reader will have no difficulty in showing that (6) and (10) can be neatly
unified with the aid of the complex index of refraction. Let us write φ' for Ψ/d,
which is the ellipticity per unit length and consider φ and φ' to be united into

[45] Planck, Ann. d. Physik **60**, 577 (1897).
[46] Lorentz, Proc. Amsterdam Acad. **14**, 518, 577 (1906).
[47] For modern accounts of these theories see Margenau and Watson, Rev. Mod. Phys.
8, 22 (1936); also Born, *Optik* (Berlin, 1933), Chapter 8.

the single complex quantity $(\varphi - i\varphi')$ which we shall call the *complex rotatory power*.

Then (6) and (10) can be combined into the single equation

$$(\varphi - i\varphi') = (\pi/\lambda)(\mathbf{n}_l - \mathbf{n}_r), \tag{82}$$

in which $\mathbf{n}_l$ and $\mathbf{n}_r$ are the *complex* indices of refraction for left and right circularly polarized light respectively. In other words, *the complex rotatory power is related to the complex indices in the same way that ordinary rotatory power is related to the real indices*. This compact result is extremely useful in discussing the connections of rotatory power and circular dichroism.

Before considering the theory of circular tichroism further it will be convenient to define another measure of absorptive power that is often found in the experimental literature. This is called the molecular or molar absorption and is also usually denoted by κ but will here be denoted by κ' to distinguish it from the imaginary part of the complex index of refraction. It is defined by the equation

$$I = I_0 10^{-\kappa' C z}, \tag{83}$$

in which z is in cm and C is the concentration of the absorbing material in mole/liter. The connection between κ' and ε is evidently

$$\varepsilon = 2.303\kappa' C. \tag{84}$$

One can also introduce an easily visualized quantity which is closely related to the molecular absorption coefficient, namely the effective cross section of the molecule for absorption of a light quantum of the type under consideration. Let A be this effective cross section, then if there are N_1 absorbing molecules per cm^3 the probability of a light quantum getting through a thickness z without being absorbed is, by familiar arguments, $e^{-N_1 A z}$. On the other hand if the concentration is C mole/liter then $N_1 = NC/1000$ where N is Avogadro's number and therefore

$$A = (2.303/N)\kappa' = \kappa' \cdot 3.81 \cdot 10^{-21} \text{ cm}^2. \tag{85}$$

As κ' in ordinary absorption bands is of the order 10 to 10^3 we see that the effective cross-sectional area of the molecules is in general small compared to the actual cross-sectional area of the molecule (10^{-16} cm^2).

In the classical electron theory the absorption is treated by introducing the damping term into the equation of motion of the electronic oscillator. If the electron is elastically bound to the origin so its natural frequency is ν_0 then its equation of motion is assumed to be

$$m\ddot{r} + 2\pi m \Gamma_0 \dot{r} + (2\pi\nu_0)^2 m r = e\mathbf{F}, \tag{86}$$

where Γ_0 measures the strength of the damping term and in what follows we assume that $\Gamma_0 \ll \nu_0$. Here $\mathbf{F}$ stands for the effective field acting on the electron and is the same as the $\mathbf{E}'$ of (17). If $\mathbf{F} = \mathbf{F}_0 e^{2\pi i \nu t}$ so the frequency of the light responsible for the forced vibrations is v then the steady state solution of (86) becomes, in the familiar way,

$$er = er_0 e^{2\pi i v t} = \frac{e^2/m}{4\pi^2[(v_0{}^2 - v^2) + iv\Gamma_0]} \mathbf{F}. \tag{87}$$

This gives us the dipole moment due to the coherent forced oscillation and thus is the classical analog of the quantum-theoretic dispersion formula (56α). In establishing the correspondence the coefficient of $\mathbf{F}$ in (87) is multiplied by the oscillator strength f_{ba} as defined in (26) and equated to the coefficient of $\mathbf{F}$ in (87). If this be done we obtain for the polarizability a modified form of (27) in which the effect of damping is included in the denominator,

$$\alpha_a = \frac{2}{3h} \sum_b \frac{v_{ba} S_{ba}}{(v_{ba}{}^2 - v^2) + iv\Gamma_{ba}}. \tag{88}$$

This result, obtained by formal alteration of the denominators is correct for quantum theory, but the full proof of it is rather elaborate and leads into the Wigner-Weisskopf theory of natural line widths.[48] The corresponding generalization of the theory of rotatory power to include damping has not been worked out but from general considerations it is fairly clear that the end result would be a similar alteration of (56β) or (23) with the same change in the resonance denominators. Thus it is probable that (23) would become

$$\beta_a = \frac{c}{3\pi h} \sum_b \frac{R_{ba}}{v_{ba}{}^2 - v^2 + 2\pi i v\Gamma_{ba}}, \tag{89}$$

in which the damping constants Γ_{ba} are the same as in (88). With this complex value for β_a we obtain for the complex rotatory power an obvious generalization of (22)

$$(\varphi - i\varphi') = \frac{16\pi^3 N_1}{\lambda^2} \cdot \frac{n^2 + 2}{3} \cdot \beta. \tag{90}$$

Separating the real and imaginary parts we obtain formulas for the rotation per unit length and the ellipticity per unit length respectively,

$$\varphi = \frac{16\pi^2 N_1}{3hc} \sum_b \frac{v^2(v_{ba}{}^2 - v^2)R_{ba}}{[(v_{ba}{}^2 - v^2)^2 + v^2\Gamma_{ba}{}^2]},$$

$$\varphi' = \frac{16\pi^2 N_1}{3hc} \sum_b \frac{v^3\Gamma_{ba}R_{ba}}{[(v_{ba}{}^2 - v^2)^2 + v^2\Gamma_{ba}{}^2]}. \tag{91}$$

These formulas are of the same form as those obtained originally by Drude[49]

---

[48] See Breit, Rev. Mod. Phys. **5**, 91 (1933).
[49] Drude, Ann. d. Physik **48**, 536 (1896).

except that they have been brought into correspondence with the quantum mechanically defined rotatory strengths, R_{ba}.

In applying these formulas to the experimental data the fact of the extreme smallness of the radiation damping coefficient must be borne in mind. Hence when experiment gives a broad band several hundred angstrom units in width it will not do to choose a Γ_{ba} large enough to make the absorption band extend over this great a region according to (91). (It is easy to see that Γ_{ba} has the meaning of being the full width from half-maximum absorption on the low frequency to half-maximum absorption on the high frequency side of the band.) To do this would call for impossibly large values of Γ_{ba} and would give the wrong shape to the absorption band in that weak absorption would persist to too great a frequency range around the maximum.

Instead the actual absorption bands due to molecules in the gas phase or especially when in solution have to be regarded as due to an enormous number of quite sharp lines corresponding to a vast number of possible rotation and vibration transitions. It is even convenient to think of the broadening of individual lines by collision in this same way. A broadened line may be regarded as the resultant effect of a large number of sharp lines all close together produced by different individual molecules which have been perturbed by the surrounding molecules to various degrees.

Equations of the form (91) were derived also by Natanson[50] in an important classical treatment of circular dichroism. In this paper was enunciated a generalization known as Natanson's rule: *The wave that is most strongly absorbed also travels more slowly, for frequencies less than the absorption frequency.* The equations were studied from an experimental point of view by Bruhat[51] who found that they represented his observations quite accurately, these being largely confined to studies on tartrates which involve metal ions with visible absorption like chromium and copper.

The modern work on circular dichroism is largely due to Kuhn and his associates.[52] Bielicki and Henri[53] showed that in many absorption bands the absorption coefficient for complicated molecular bands follows the law,

$$\varepsilon(v) \sim e^{-(v - v_0)^2/\theta^2}.$$

This simply means that if the logarithm of the absorption coefficient is plotted against the frequency the resulting curve is a parabola with vertex upward. Kuhn and his associates have adopted a similar representation for the spectral distribution of rotatory power in an absorption band for the empirical representation of their data. This has proved to be much more satisfactory than

[50] Natanson, J. de phys. **8**, 321 (1909).

[51] Bruhat, Ann. d. Physik **3**, 232, 417 (1915).

[52] See especially Kuhn and Braun, Zeits. f. physik. Chemie **B8**, 281 (1930) and numerous other papers in the same journal subsequently.

[53] Bielicki and Henri, Physik. Zeits. **14**, 516 (1913).

the way which accounts for the width of the band by simply choosing a very large value of Γ for the band in question.

In the paper by Condon, Altar and Eyring[29] explicit calculations are carried out quantum-mechanically to show that the aborption transition probability for right and left circularly polarized light are different when the calculations are carried to the same degree of approximation as is necessary to get rotatory power as in §4.

One-Electron Rotatory Power*

EDWARD U. CONDON, WILLIAM ALTAR AND HENRY EYRING

Princeton University
Princeton, New Jersey

It is shown that a single electron moving in a field of suitable dissymmetry can give rise to optical rotatory power in a medium containing molecules of this type. This effect is called *one-electron rotatory power* and is in striking contrast to the models developed by Born, Oseen, Gray and others in which a dynamic coupling between several electronic oscillators is responsible for the rotatory power. The detailed calculations are carried out for the potential function,

$$V = \tfrac{1}{2}(k_1 x^2 + k_2 y^2 + k_3 z^2) + Axyz,$$

which shows rotatory power both quantum mechanically and classically. Next it is shown how fields of this type may be adapted to an approximate description of the field in which the chromophoric electrons of a molecule move, the constants k_1, k_2, k_3 and A being largely determined by the average charge on the different atoms of the molecule as found from additivity of dipole moments due to the bonds in the molecule. As illustrations of the theory absolute calculations are made for the contribution of the nitrite group to the rotatory power of methyl phenyl carbinol nitrite which agrees satisfactorily with experimental data, for the phenyl group contribution of the same molecule and for the hydroxyl group contribution in secondary butyl alcohol. In the latter two instances the theoretical values are much too small but as there is a great deal of freedom in the choice of the configuration because of free rotation these results merely indicate need for more detailed calculations. The general quantum-mechanical theory of circular dichroism is developed and a quantum-mechanical derivation of Natanson's rule is given. The quantum-mechanical definition of Kuhn's "anisotropy factor" is given and an alternative measure of the contribution of an absorption band to rotatory power called the *rotatory strength* is defined and discussed. It is shown that spin magnetic moments may generally be neglected in calculation of rotatory power owing to the weakness of spin-orbit interaction. Interpretation of

---

* A preliminary account of the main results of this paper was presented at the Denver Meeting of the American Physical Society, June 26, 1937.

experiments of Schwab, Rost and Rudolph on the catalytic dehydration of butyl alcohol on active quartz is discussed and kinetic arguments advanced to show how the relative configuration of quartz and butyl alcohol may be inferred from such data. The relation of one-electron rotatory poner to known dipole moment and solvent effects on rotatory power is briefly discussed.

1. Introduction

The property of some liquids to rotate the plane of polarization of linearly polarized light traversing them has played a great role in the development of stereochemistry. In spite of the fact that an enormous amount of work has been devoted to the subject it cannot be said that the theory of the molecular actions responsible for this behavior is in a very satisfactory state. In this paper we shall attempt to carry the quantum-mechanical treatment of the subject further in terms of a simple model which exhibits the main features of the phenomenon.

By way of review let it be recalled that the rotation is produced as an indirect effect of circular double refraction, that is, because of the medium's having slightly different indices of refraction for right and left circularly polarized light. The rotation is called positive if the plane is turned in a clockwise direction as viewed by an observer into whose eyes the light enters and the classical analysis due to Fresnel[1] shows that φ, the rotation per unit length of path, is given by

$$\varphi = (\pi/\lambda)(n_l - n_r) \tag{1}$$

in which λ is the vacuum wave-length and n_l, n_r are the refractive indices for left and right circularly polarized light, respectively. On the electromagnetic theory of light circular double refraction can be described phenomenologically by assuming that the medium is characterized by a parameter g defined by the equation

$$\mathbf{D} = \varepsilon \mathbf{E} - g\dot{\mathbf{H}},$$
$$\mathbf{B} = \mathbf{H} + g\dot{\mathbf{E}}, \tag{2}$$

in which $\mathbf{D}$ and $\mathbf{B}$ are the electric and magnetic inductions and $\mathbf{E}$ and $\mathbf{H}$ the field strengths, ε the effective dielectric constant for the frequencies involved, so that ε equals the square of the mean refractive index. Then a solution of Maxwell's equations by standard methods gives for the rotatory power

$$\varphi = (2\pi/\lambda)^2 cg. \tag{3}$$

The parameter g is the macroscopic representative of a corresponding microscopic parameter representing the response of individual molecules to

---

[1] Fresnel, Ann. chim. phys. **28**, 147 (1825).

the applied fields of the light wave. Let p_1 be the induced electric moment of the molecule due to the fields of the light wave and m_1 the corresponding induced magnetic moment. Then we assume that these are of the form

$$\mathbf{p}_1 = \alpha \mathbf{E}' - (\beta/c)\dot{H},$$
$$\mathbf{m}_1 = \qquad + (\beta/c)\dot{E}', \tag{4}$$

where α is the quantity commonly called the polarizability of the molecule, β is a new parameter that is responsible for the optical activity, and $\mathbf{E}'$ is the effective field acting on a molecule. Accepting the Lorentz' result that

$$\mathbf{E}' = \mathbf{E} + 4\pi \mathbf{P}/3, \tag{5}$$

where $\mathbf{P} = N_1 \mathbf{p}_1$ is the electric moment in unit volume, N_1 being the number of molecules in unit volume, then the standard definitions $\mathbf{D} = \mathbf{E} + 4\pi \mathbf{P}$ and $\mathbf{B} = \mathbf{H} + 4\pi \mathbf{M}$ lead to the following connections between α and ε, and β and g, respectively,

$$\frac{4\pi N_1 \alpha}{3} = \frac{\varepsilon - 1}{\varepsilon + 2}; \qquad \frac{4\pi N_1 (\beta/c)}{3} = \frac{g}{\varepsilon + 2}. \tag{6}$$

The first of these is the familiar result from the ordinary theory of dispersion, the second is its analog for the theory of optical activity. Using the well-known result $n^2 = \varepsilon$ where n is the mean index of refraction, a combination of (6) and (3) permits the expression of the rotatory power φ in terms of the molecular parameter β and other defined properties of the medium,

$$\varphi = (16\pi^3/\lambda^2) N_1 \beta \cdot \tfrac{1}{3}(n^2 + 2). \tag{7}$$

With this equation the task is completed of expressing the observed rotation per unit length φ, in terms of the molecular parameter β, the number of active molecules in unit volume N_1, the vacuum wave-length λ and the mean index of refraction, n. In this way the theory of optical activity is referred back to a molecular theory of β.

Instead of dealing with the rotatory power itself it is often convenient to deal with the molecular rotatory power which is defined as $\varphi M/\rho$ where M is the molecular weight and ρ is the density. Denoting the molecular rotatory power with Φ one has

$$\Phi = (16\pi^3/\lambda^2) N\beta(n^2 + 2)/3, \tag{8}$$

where N is Avogadro's number. Either (7) or (8) may be used to compute value of β from experimental data on the rotatory power. Unfortunately values of the refractive index are often not given in the literature so the factor $(n^2 + 2)/3$ has to be estimated.

The appropriate extension of the quantum-mechanical theory of dispersion needed to give the theory of the parameter β has been given by Rosenfeld.[2]

---

[2] Rosenfeld, Zeits. f. Physik **52**, 161 (1928); see also Condon, "Theories of Optical Rotatory Power," Rev. Mod. Phys. (in press).

Suppose the quantum states of the molecule are labeled by a, b, ... where a single letter stands for the totality of quantum numbers needed to specify the state. Then the matrix components of the electric moment of the molecule are denoted by $(a|\mathbf{p}|b)$ and those of magnetic moment by $(a|\mathbf{m}|b)$ where

$$\mathbf{p} = e \sum_i \mathbf{r}_i \qquad \mathbf{m} = \frac{e}{2mc} \sum_i (\mathbf{r}_i \times \mathbf{p}_i + 2\mathbf{S}_i), \qquad (9)$$

the sum being extended over all the electrons in the molecule, $\mathbf{r}_i$ being position, $\mathbf{p}_i$ momentum and $\mathbf{S}_i$ spin angular momentum of the ith electron. The necessary generalization of the ordinary dispersion theory is obtained by taking into account to the first order the finite ratio of molecular diameter to wave-length of light. The end result of Rosenfeld's calculations is that

$$\beta_a/c = \frac{1}{3\pi h} \sum_b \frac{\mathrm{Im}\{(a|\mathbf{p}|b)\cdot(b|\mathbf{m}|a)\}}{v_{ba}^2 - v^2}, \qquad (10)$$

where the expression has been simplified to take account of the fact that the molecules are oriented at random. In this expression v is the light frequency and v_{ba} is the frequency associated with a transition from state a to state b. The symbol $\mathrm{Im}\{\ \}$ means "imaginary part of" in the sense

$$\mathrm{Im}\{u + iv\} = v \qquad (11)$$

if u and v are real. The subscript a on β means that the formula gives the appropriate value of β for molecules in the state a. If in the actual material the molecules are distributed over various initial states such that $N_1(a)$ is the number of molecules in unit volume in the state a then the effective value of β for use in (4) is the weighted average

$$\beta = (1/N_1) \sum_a N_1(a)\beta_a. \qquad (12)$$

Equation (10) for (β_a/c) is entirely analogous to the more familiar equation for the ordinary polarizability, α_a,

$$\alpha_a = \frac{2}{3h} \sum_b \frac{v_{ba}|(a|\mathbf{p}|b)|^2}{v_{ba}^2 - v^2}. \qquad (13)$$

The work of Rosenfeld thus refers the parameter β and hence the optical activity back to the matrix components of electric and magnetic moment as contained in (10). It appears that in all the years since Rosenfeld's paper no attempts have been made to discuss the quantum-mechanical formula (10) in relation to an actual molecular model. Instead there has been a tendency to revert to the original coupled oscillator model of Born,[3] Oseen[4] and Gray[5] in

---

[3] Born, Physik. Zeits. **16**, 251, 437 (1915); Ann. d. Physik **55**, 177 (1917).
[4] Oseen, Ann. d. Physik **48**, 1 (1915).
[5] Gray, Phys. Rev. **7**, 472 (1916).

what attempts have been made recently[6-8] to discuss the theory of optical activity. The opinion prevalent in the literature[9] seems to be that the only models which need be considered to explain optical activity are those involving coupled oscillators. The model originally proposed by Drude[9] involved only one oscillating particle, and its rejection by Born and Kuhn[9] has done much to strengthen the common belief, that single oscillators are neither adequate nor important in providing a mechanism. In the following we propose to show that electric fields known to be present in molecules actually endow single dispersion electrons with optical activity of the magnitude observed. Consequently we feel that more is to be gained by attempting to discuss the actual application of (10) to molecular models. That is what is done in this paper.*

2. A Simple Model for Optical Activity

A characteristic feature of the modern theory of atomic and molecular spectra is the use of what may be called the Hartree approximation, in which each electron in the system moves in an average field due to the average action on it of all the other atoms and the nuclei. The detailed coupling of the electrons may then be taken into account in higher approximations. It seems, therefore, that the best approach to optical activity is gained by studying some simple potential field, in which a single electron moves, of such a character as to give activity. Such a calculation will be made in this section and later sections will be devoted to the correlation of the simple model with the structure of actual molecules.

The force field, $V(x, y, z)$, must be such that it has no planes of symmetry and no center of symmetry. This is true because if there were such an element of symmetry then the states could be classified as even or odd with regard to this symmetry element, and $\mathbf{p}$ and $\mathbf{m}$ would have selection rules of opposite kind with regard to this character so there would be no states a and b for which $(a|\mathbf{p}|b)$ and $(b|\mathbf{m}|a)$ were both different from zero.

A simple field satisfying this requirement is given by

$$V(x_1 x_2 x_3) = (1/2)(k_1 x_1^2 + k_2 x_2^2 + k_3 x_3^2) + A x_1 x_2 x_3, \qquad (14)$$

where the Cartesian coordinates of the particle are x_1, x_2, x_3. The motion of

[6] Kuhn, Zeits. f. physik. Chemie **B4**, 14 (1929); Trans. Faraday Soc. **26**, 293 (1930).

[7] Born, Proc. Roy. Soc. **A150**, 84 (1935).

[8] Oke, Proc. Roy. Soc. **A153**, 339 (1935).

[9] See: Lowry, *Optical Rotatory Power*, p. 372 f. Kuhn has actually demonstrated that Drude's result was obtained by unjustifiably neglecting certain terms.

* *Added in proof:* These remarks, of course, do not apply to the recent important paper of Kirkwood, J. Chem. Phys. **5**, 479 (1937) who has taken steps toward the use of (10) in definite molecular models.

a particle of mass μ in such a field may be treated by the standard form of perturbation theory, treating the $Ax_1x_2x_3$ term as a perturbation.

In the zeroth approximation the energy levels are given by

$$W(n_1, n_2, n_3) = (n_1 + \tfrac{1}{2})hv_1 + (n_2 + \tfrac{1}{2})hv_2 + (n_3 + \tfrac{1}{2})hv_3 \qquad (15)$$

and the zero-order wave functions are

$$\psi(n_1, n_2, n_3) = \varphi_{n1}(x_1/a_1)\varphi_{n2}(x_2/a_2)\varphi_{n3}(x_3/a_3), \qquad (16)$$

in which

$$2\pi v_i = (k_i/\mu)^{1/2}, \qquad (2\pi a_i)^2 = h/\mu v_i; \qquad (17)$$

and the φ functions are the normalized orthogonal system involving the Hermitian polynomials

$$\varphi_n(\xi) = (2^n n! \pi^{1/2})^{-1/2} H_n(\xi) e^{-\xi^2/2}. \qquad (18)$$

The calculation of the matrix components of the perturbation energy $Ax_1x_2x_3$ with respect to the wave functions (16) is very readily made since the perturbation energy and the wave functions each factor into simple functions of an individual coordinate. Since the recursion formula for the functions (18) can be written

$$\xi\varphi_n(\xi) = [(n + 1)/2]^{1/2}\varphi_{n+1}(\xi) + (n/2)^{1/2}\varphi_{n-1}(\xi), \qquad (19)$$

the nonvanishing matrix components of ξ are in a one-dimensional scheme

$$(n|\xi|n') = [(n)/2]^{1/2} \qquad \text{where} \qquad n' = n \pm 1 \qquad (20)$$

and (n) means the larger of n or n'. Hence the nonvanishing matrix components of the perturbation energy are given by

$$(n_1 n_2 n_3|Ax_1 x_2 x_3|n_1'n_2'n_3') = Aa_1 a_2 a_3 8^{-1/2}[(n_1)(n_2)(n_3)]^{1/2} \qquad (21)$$

where $n_1' = n_1 \pm 1$, $n_2' = n_2 \pm 1$ and $n_3' = n_3 \pm 1$ and (n_i) means the larger of n_i and n_i'. It will be convenient to introduce an abbreviation

$$\lambda = Aa_1 a_2 a_3/8^{1/2}h. \qquad (22)$$

Then by the standard perturbation theory we see that to the first order the energy levels remain the same as (15) since the diagonal elements of the perturbation vanish. The first-order perturbed wave function is

$$\Psi(n_1 n_2 n_3) = \psi(n_1 n_2 n_3) - \sum_{ni'} \frac{\psi(n_1'n_2'n_3')(n_1'n_2'n_3'|Ax_1 x_2 x_3|n_1 n_2 n_3)}{W(n_1'n_2'n_3') - W(n_1 n_2 n_3)}, \qquad (23)$$

where the summation contains at most eight terms because of the simplicity of (21).

Now for the optical activity one needs the matrix components of $\mathbf{r}$ and $\mathbf{L}$, the orbital angular momentum calculated with respect to the perturbed wave functions (23). Denoting these with square brackets and matrix components with respect to unperturbed functions uith round brackets one has, to the first approximation for any quantity F,

$$[n_1 n_2 n_3 | F | n_1' n_2' n_3'] = (n_1 n_2 n_3 | F | n_1' n_2' n_3') - \lambda \{ \bar{\psi}_1(n_1 n_2 n_3) F \psi(n_1' n_2' n_3')$$
$$+ \bar{\psi}(n_1 n_2 n_3) F \psi_1(n_1' n_2' n_3') \}, \tag{24}$$

in which $\lambda \psi_1(n_1 n_2 n_3)$ is written for the second term on the right side of (23) and where $\bar{\psi}_1 F \psi$ means that the expression is to be integrated over the configuration space.

The matrix components of $\mathbf{r}$ with regard to the unperturbed functions follow at once from (20) the result being

$$(n_1 n_2 n_3 | \mathbf{r} | n_1' n_2' n_3') = 2^{-1/2} \{ \mathbf{i}(n_1)^{1/2} \delta(n_2, n_2') \delta(n_3, n_3') a_1$$
$$+ \mathbf{j}(n_2)^{1/2} \delta(n_1, n_1') \delta(n_3, n_3') a_2$$
$$+ \mathbf{k}(n_3)^{1/2} \delta(n_1, n_1') \delta(n_2, n_2') a_3 \}. \tag{25}$$

Here $\mathbf{i}, \mathbf{j}$ and $\mathbf{k}$ are the usual unit vectors and $\delta(x, y) = 0$ for $y \neq x$ and 1 for $y = x$. Hence to get nonvanishing matrix components two of the n's must be unchanged and the third must be given by $n_i' = n_i \pm 1$.

The matrix components of $\mathbf{L}$ with regard to the unperturbed functions may be obtained as follows: from the theory of the Hermite polynomials in one dimension one has for the momentum operator p,

$$p\varphi_n = \frac{i\hbar}{a} [\tfrac{1}{2}(n + 1)]^{1/2} \varphi_{n+1} - \frac{i\hbar}{a} [\tfrac{1}{2} n]^{1/2} \varphi_{n-1}. \tag{26}$$

From the definition of orbital angular momentum $\mathbf{L}$,

$$L_1 = x_2 p_3 - x_3 p_2,$$

so using (19) and (26)

$$L_1 \psi(n_1 n_2 n_3) = \frac{i\hbar a_2 a_3}{2} \sum [\pm a_3^{-2} \mp a_2^{-2}][(n_2)(n_3)]^{1/2} \psi(n_1, n_2 \pm 1, n_3 \pm 1), \tag{27}$$

in which the $\sum$ extends over the four possible signs for $n_2 \pm 1$ and $n_3 \pm 1$ taken independently and where the sign choice on $\pm a_i^2$ is the same as that on $n_i \pm 1$. As before (n_2) means the larger of n_2 or $n_2 \pm 1$. The expressions for $L_2 \psi(n_1 n_2 n_3)$ and $L_3 \psi(n_1 n_2 n_3)$ may be written down by cyclic permutation of the indices 1, 2, 3. In other words the nonvanishing unperturbed matrix components of $\mathbf{L}$ are therefore

$$(n_1, n_2 \pm 1, n_3 \pm 1 | L_1 | n_1 n_2 n_3) = \frac{i\hbar a_2 a_3}{2} \left(\pm \frac{1}{a_3^2} - \pm \frac{1}{a_2^2} \right) [(n_2)(n_3)]^{1/2},$$

$$(n_1 \pm 1, n_2, n_3 \pm 1 | L_2 | n_1 n_2 n_3) = \frac{i\hbar a_3 a_1}{2} \left(\pm \frac{1}{a_1^2} - \pm \frac{1}{a_3^2} \right) [(n_3)(n_1)]^{1/2}, \quad (28)$$

$$(n_1 \pm 1, n_2 \pm 1, n_3 | L_3 | n_1 n_2 n_3) = \frac{i\hbar a_1 a_2}{2} \left(\pm \frac{1}{a_1^2} - \pm \frac{1}{a_1^2} \right) [(n_1)(n_2)]^{1/2}.$$

From (28) and (25) it is obvious why an ordinary anisotropic harmonic oscillator is not optically active, for the nonvanishing matrix components in (25) require that two n's be unchanged whereas in (28) only one n is unchanged; hence there are no pairs of levels $(n_1 n_2 n_3)$ and $(n_1' n_2' n_3')$ for which both the r and the L matrix components have a nonvanishing value.

The optical activity is produced by the perturbing effect of the $Ax_1 x_2 x_3$ term by means of the terms proportional to λ in (24). For the sake of clarity and in view of the applications to be made, it will be sufficient to work out in detail the special case of activity due to molecules in the lowest state (000). Consideration of (23) combined with (21) shows that the perturbed wave function for (000) involves only (000) and (111) wave functions,

$$\Psi(000) = \psi(000) - \frac{\lambda}{v_1 + v_2 + v_3} \psi(111).$$

Hence by (25) the only states for which $[000|r|n_1 n_2 n_3]$ has a nonvanishing zero-order value are (100), (010) and (001). These are the ordinary one-quantum jumps of the anisotropic oscillator. For these transitions L has a vanishing zero-order matrix component but a nonvanishing first-order component owing to the fact that by (23) and (21), $\Psi(100)$ contains $\psi(001)$. More explicitly

$$\Psi(100) = \psi(100) - \frac{\lambda 2^{1/2}}{v_1 + v_2 + v_3} \psi(211) + \frac{\lambda}{-v_1 + v_2 + v_3} \psi(011)$$

so using (24) and (28)

$$[100|L|000] = -\frac{i\lambda \hbar (a_2{}^2 - a_3{}^2)}{2 a_2 a_3} \left[\frac{2 v_1}{(v_2 + v_3)^2 - v_1{}^2} \right] i, \qquad (29)$$

and by cyclic substitution the formulas for $[010|L|000]$ and $[001|L|000]$ can be written down. Hence the three absorption lines $(000 \to 100)$ $(000 \to 010)$ and $(000 \to 001)$ give rise to optical activity in the first order, the appropriate strengths as occurring in (10) being given by

$$[000|r|000] \cdot [100|L|000] = -i \frac{Ah^2}{128\pi^5\mu^2} \left(\frac{1}{v_2} - \frac{1}{v_3} \right) \frac{1}{(v_2 + v_3)^2 - v_1{}^2}; \qquad (30)$$

the other two of similar type being obtainable by cyclic substitution. As the right-hand side of (30) is pure imaginary, the imaginary part needed for (10) is just equal to the right-hand side of (30) with the i omitted.

These three lines therefore give rise to optical activity of the first order in virtue of zero-order matrix components of r combined with the first-order contributions to the matrix components of L. Similarly there are three more lines which will give first-order optical activity in virtue of the first-order change in the matrix components of r combined with the zero-order matrix components of L. These two sets of lines will exhaust the possibilities for optical activity of the ground state to the first order in the asymmetry coefficient A.

There will be an interesting contrast in type of the two sets of lines. For the set given by (30) the ordinary optical absorption will be strong since $[a|\mathbf{r}|b]$ has a nonvanishing zero-order value. But for the other type, whose optical activity is of the same order of magnitude, the ordinary optical absorption will be weak. Thus the model gives a division of the critical frequencies into two types—both of comparable contribution to the rotatory power but one strong in absorption, the other weak. This corresponds to a general experimental fact as recognized by Kuhn and interpreted by him in terms of the crossed oscillator model.

The states for which $[n_1 n_2 n_3|\mathbf{L}|000]$ has a nonvanishing zero-order value are, by (28), (110), (101) and (011). For these we need to calculate the first-order matrix components of $\mathbf{r}$ by a combination of (24), (23) and (25). The result is

$$[000|\mathbf{r}|011] = 2^{1/2} \lambda a_1 \frac{v_1}{(v_2 + v_3)^2 - v_1^2} i, \tag{31}$$

the other two being obtained by cyclic substitution. Hence for this line the contribution to the optical activity is given by

$$[000|\mathbf{r}|011] \cdot [011|\mathbf{L}|000] = +i \frac{Ah^2}{128\pi^5 \mu^2} \left(\frac{1}{v_2} - \frac{1}{v_3} \right) \frac{1}{(v_2 + v_3)^2 - v_1^2}, \tag{32}$$

the other two being obtained by cyclic substitution. It is interesting to note that (32) is exactly equal and opposite to (30). Hence the sum of the rotatory strengths associated with the jumps $(000) \rightarrow (100)$ and $(000) \rightarrow (011)$ is zero, verifying the general sum rule for the numerators in the rotatory dispersion formula.

The general sum rule is that the sum of the numerators in (10) is zero. This may be proved very simply. One has for this sum

$$\text{Im} \left\{ \sum_b (a|\mathbf{p}|b) \cdot (b|\mathbf{m}|a) \right\} = \text{Im} \{ (a|\mathbf{p} \cdot \mathbf{m}|a) \}$$

by the matrix law of multiplication. The quantity on the right vanishes since $(a|\mathbf{p} \cdot \mathbf{m}|a)$ is a diagonal matrix element of a real observable and hence is real.

In view of the fact that $\mathbf{p} = e\mathbf{r}$ and $\mathbf{m} = (e/2\mu c)\mathbf{L}$ for a single charged particle one can write down the complete formula for the parameter β by combining (10), (30) and (32). The result is

$$\beta_{000} = \frac{Ahe^2}{12(2\pi)^5 \mu^3} \left\{ \left(\frac{1}{v_2} - \frac{1}{v_3} \right) \frac{1}{(v_2 + v_3)^2 - v_1^2} \left[-\frac{1}{v_1^2 - v^2} + \frac{1}{(v_2 + v_3)^2 - v^2} \right] \right.$$
$$+ \left(\frac{1}{v_3} - \frac{1}{v_1} \right) \frac{1}{(v_3 + v_1)^2 - v_2^2} \left[-\frac{1}{v_2^2 - v^2} + \frac{1}{(v_3 + v_1)^2 - v^2} \right]$$
$$+ \left(\frac{1}{v_1} - \frac{1}{v_2} \right) \frac{1}{(v_1 + v_2)^2 - v_3^2} \left[-\frac{1}{v_3^2 - v^2} + \frac{1}{(v_1 + v_2)^2 - v^2} \right] \right\}.$$
$$\tag{33}$$

It will be noticed that if any two of the frequencies v_1, v_2, v_3 are equal the whole expression vanishes. This is in accord with the symmetry properties, for if, say, v_2 and v_3 are equal this implies k_2 and k_3 are equal so the harmonic oscillator potential has the x_1 axis as an axis of symmetry and therefore any plane through this axis is a plane of symmetry for the harmonic oscillator potential. Now $x_1 x_2 x_3$ has the plane $x_2 = x_3$ as a plane of symmetry so if $v_2 = v_3$ the plane $x_2 = x_3$ is a plane of symmetry of the model so the optical activity should vanish.

By combining (33) with (7) or (8) one has an explicit formula for the rotatory power of a molecule which contains electrons moving in a field of the type postulated in (14).

3. Optical Activity of the Model in Higher Quantum States. Classical Mechanics as a Limiting Case

The extension of the preceding calculation to states with three arbitrary quantum numbers $(n_1 n_2 n_3)$ instead of (000) is of considerable interest not only because it applies to media with part of their molecules in excited states but also because the generalized formula taking the place of (33) leads for high quantum numbers asymptotically, in the sense of the correspondence principle, to an expression for the rotatory power of the classical model. We shall see that in going to this limit the system retains a finite rotatory pouer, and a simple expression in terms of the amplitudes will be given.

The procedure is of course closely analogous to the derivation previously given for the ground state. Again we start from Eqs (10), (23), (25) and (28). When applied to our model, Eq. (23) gives:

$$\Psi(n_1 n_2 n_3) = \psi(n_1 n_2 n_3) - \lambda \sum \psi(n_1 \pm 1, n_2 \pm 1, n_3 \pm 1) \frac{[(n_1)(n_2)(n_3)]^{1/2}}{\pm v_1 \pm v_2 \pm v_3},$$
$$(34)$$

where the summation is over all possible sign combinations and where equal signs are taken for the change of quantum number in any of the three axes, n_i, and the corresponding v_i in the denominator. Consequently one obtains

$$[n_1 n_2 n_3 |\mathbf{r}| n_1' n_2' n_3'] \cdot [n_1' n_2' n_3' |L| n_1 n_2 n_3]$$

$$= -\lambda \sum \frac{[(n_1)(n_2)(n_3)]^{1/2}}{\pm v_1 \pm v_2 \pm v_3} \{(n_1 n_2 n_3 |\mathbf{r}| n_1' n_2' n_3')$$

$$\times (n_1' n_2' n_3' |L| n_1 \pm 1\, n_2 \pm 1\, n_3 \pm 1)$$

$$+ (n_1 \pm 1\, n_2 \pm 1\, n_3 \pm 1 |\mathbf{r}| n_1' n_2' n_3')(n_1' n_2' n_3' |L| n_1 n_2 n_3)\}$$

$$- \lambda \sum \frac{[(n_1')(n_2')(n_3')]^{1/2}}{\pm v_1 \pm v_2 \pm v_3} \{(n_1 n_2 n_3 |\mathbf{r}| n_1' n_2' n_3')$$

$$\times (n_1' \pm 1\, n_2' \pm 1\, n_3' \pm 1 |L| n_1 n_2 n_3)$$

$$+ (n_1 n_2 n_3 |\mathbf{r}| n_1' \pm 1\, n_2' \pm 1\, n_3' \pm 1)(n_1' n_2' n_3' |L| n_1 n_2 n_3)\}. \quad (35)$$

The zero-order term $(n|\mathbf{r}|n') \cdot (n'|\mathbf{L}|n)$ is absent in (35) because the selection rules of $\mathbf{r}$ and $\mathbf{L}$ mutually exclude each other in the unperturbed case which has three planes of symmetry. The following transitions are found to contribute to $(n|x|n')(n'|L_x|n)$:

(a) $(n_1 n_2 n_3) \rightarrow (n_1 + 1, n_2, n_3)$,

(b) $\quad\quad\quad \rightarrow (n_1 - 1, n_2, n_3)$,

(c) $\quad\quad\quad \rightarrow (n_1, n_2 + 1, n_3 + 1)$,

(d) $\rightarrow (n_1, n_2 - 1, n_3 + 1)$,

(e) $\rightarrow (n_1, n_2 + 1, n_3 - 1)$,

(f) $\rightarrow (n_1, n_2 - 1, n_3 - 1)$.

If we denote these six transitions for brevity by letters a to f in the order listed, it is easily seen that a and b, c and f, d and e have equal and opposite transition frequencies, namely $\pm v_1$, $\pm(v_2 + v_3)$ and $\pm(v_2 - v_3)$ respectively. Since transition frequencies $v_{nn'}$ enter into (10) for β, in the second power, each pair of transitions will jointly give rise to one dispersion term of the general type that was previously obtained in Eq. (30). Actually it is found that only the first and third term in (35) will contribute by virtue of a transition a or b, and similarly only the second and the fourth term of (35) contribute in transitions c, d, e, f. There are, in all, 32 terms to be considered for the matrix product $(n|x|n') \times (n'|L|n)$ and 64 more terms are obtained by cyclic substitution to give contributions touard the y and the z matrix components. A simple though lengthy calculation leads after much simplification and canceling to the final formula for β:

$$\beta_{n_1 n_2 n_3} = \frac{A\hbar e^2}{12(2\pi)^5 \mu^3} \left\{ \left(\frac{1}{v_2} - \frac{1}{v_3}\right) \frac{(n_2 + n_3 + 1)}{(v_2 + v_3)^2 - v_1^2} \left[-\frac{1}{v_1^2 - v^2} + \frac{1}{(v_2 + v_3)^2 - v^2} \right] \right.$$

$$+ \left(\frac{1}{v_2} + \frac{1}{v_3}\right) \frac{(n_2 - n_3)}{(v_2 - v_3)^2 - v_1^2} \left[-\frac{1}{v_1^2 - v^2} + \frac{1}{(v_2 - v_3)^2 - v^2} \right]$$

$$+ \left(\frac{1}{v_3} - \frac{1}{v_1}\right) \frac{(n_3 + n_1 + 1)}{(v_3 + v_1)^2 - v_2^2} \left[-\frac{1}{v_2^2 - v^2} + \frac{1}{(v_3 + v_1)^2 - v^2} \right]$$

$$+ \left(\frac{1}{v_3} + \frac{1}{v_1}\right) \frac{(n_3 - n_1)}{(v_3 - v_1)^2 - v_2^2} \left[-\frac{1}{v_2^2 - v^2} + \frac{1}{(v_3 - v_1)^2 - v^2} \right]$$

$$+ \left(\frac{1}{v_1} - \frac{1}{v_2}\right) \frac{(n_1 + n_2 + 1)}{(v_1 + v_2)^2 - v_3^2} \left[-\frac{1}{v_3^2 - v^2} + \frac{1}{(v_1 + v_2)^2 - v^2} \right]$$

$$\left. + \left(\frac{1}{v_1} + \frac{1}{v_2}\right) \frac{(n_1 - n_2)}{(v_1 - v_2)^2 - v_3^2} \left[-\frac{1}{v_3^2 - v^2} + \frac{1}{(v_1 - v_2)^2 - v^2} \right] \right\}.$$

(36)

This clearly contains the previous result, (33), as a special case, but also exhibits some new features. First of all we have now three resonance frequencies for the rotatory dispersion, instead of the two previously obtained. Of these, v_1, represents a frequency which is also strongly pronounced in absorption, but there are now two weak absorption bands $(v_2 + v_3)$ and $(v_2 - v_3)$ which may exhibit strong rotatory pouer. Again the sum rule is obviously fulfilled as the numerator going with the strong absorption frequency is equal and opposite to the sum of the other two. Eq. (36) may also be written in a slightly modified form

$$\beta_{n_1 n_2 n_3} = -\frac{A\hbar e^2}{12(2\pi)^5 \mu^3} \sum_{\text{cycl.}} \left\{ \left(\frac{1}{v_k} - \frac{1}{v_l}\right) \frac{(n_k + n_l + 1)}{[(v_k + v_l)^2 - v^2][v_i^2 - v^2]} \right.$$
$$\left. + \left(\frac{1}{v_k} + \frac{1}{v_l}\right) \frac{(n_k - n_l)}{[(v_k - v_l)^2 - v^2][v_i^2 - v^2]} \right\}. \tag{36a}$$

Here $\sum_{\text{cycl.}}$ means sum over $(kli) = (2, 3, 1), (3, 1, 2)$ and $(1, 2, 3)$.

In order to derive a limiting expression for β in terms of the amplitudes $A_1 A_2 A_3$ of the particle's motion in the three axial directions we express the quantum numbers in terms of the amplitudes by means of the energy relation valid for a harmonic oscillator:

$$n_1 h v_1 = (1/2)k_1 A_1^2 = 2\pi^2 v_1^2 A_1^2 \mu,$$
$$h v_1 = 2\pi^2 \mu A_1 v_1^2, \tag{37}$$

and two similar equations for the other two quantum numbers. Substituting from (37) into (36) we have at once

$$\beta(A_1 A_2 A_3)$$
$$= \frac{e^2 A}{384\pi^4 \mu^2} \left\{ \left(\frac{1}{v_2} - \frac{1}{v_3}\right) \frac{v_2 A_2^2 + v_3 A_3^2}{(v_2 + v_3)^2 - v_1^2} \left[-\frac{1}{v_1^2 - v^2} + \frac{1}{(v_2 + v_3)^2 - v^2} \right] \right.$$
$$+ \left(\frac{1}{v_2} + \frac{1}{v_3}\right) \frac{v_2 A_2^2 - v_3 A_3^2}{(v_2 - v_3)^2 - v_1^2} \left[-\frac{1}{v_1^2 - v^2} + \frac{1}{(v_2 - v_3)^2 - v^2} \right]$$

$$\left. + \text{ four more lines obtained from these by cyclic permutation of the indices} \right\}. \tag{38}$$

This result does not contain h or other quantum-mechanical concepts and represents in the limit, according to the correspondence principle, the response of our model if it were a system obeying classical mechanics.

4. Qualitative Properties of the Field

Before discussing the application of the results of Section 2 to actual molecules it is worth while to form a mental picture of the kind of potential field represented by (14). This is best done by considering the shape of the equipotential surfaces. Suppose for definiteness that $k_1 > k_2 > k_3$ and at first that $A = 0$. Then the equipotential surfaces will be ellipsoids whose principal axes are the coordinate axes and whose axes increase in the order $x_1 < x_2 < x_3$. The sections of any such ellipsoid by a plane $x_3 = $ constant will then be ellipses all of whose principal axes are parallel to the x_1 and x_2 axes. Now suppose that the $Ax_1x_2x_3$ term is present, say with $A > 0$. Then the equation of the intersection of an equipotential with the plane $x_3 = Z$ is

$$\frac{k_1}{2}x_1^2 + \frac{k_2}{2}x_2^2 + AZx_1x_2 = (V - \tfrac{1}{2}k_3Z^2).$$

This is the equation of an ellipse whose axes have been twisted relative to parallelism with the x_1 and x_2 axes in the counter-clockwise sense as viewed from the positive x_3 axis, the angle of twist being θ where

$$\tan 2\theta = \frac{AZ}{\tfrac{1}{2}(k_1 - k_2)}. \tag{39}$$

The angle of twist therefore increases with increasing Z, in other words the equipotential surface is a kind of twisted ellipsoid something like what one might get if an ordinary ellipsoid of three unequal axes were subjected to a tortional stress twisting it around the x_3 axis in the sense of a right-handed screw. The angle through which an axis is twisted can never exceed 45°.

Similarly the sections of the equipotentials by the planes $x_1 = X$ are also ellipses which are skewed around in the same screw sense for the angle of twist is given by (39) with the cyclic substitutions $Z \to X, k_1 \to k_2, k_2 \to k_3$ and the sense remains the same because we have assumed $k_2 > k_3$. The sections of the equipotentials by the plane $x_2 = Y$ are also a set of ellipses, but they are skewed around in the opposite sense for the form of (39) applicable to this case involves $(k_3 - k_1)$ in the denominator which is negative in the case we are discussing.

It should be clear that no matter what are the relative magnitudes of k_1, k_2 and k_3 the situation will always be the same in that the sections of the equipotential surface by planes parallel to any of the coordinate planes will be sets of ellipses, two of which screw around in the same sense while the third has the opposite screw sense.

Comparing these qualitative properties of the field with the formula (33) for β_{000} we see that if $A > 0$ and $k_1 > k_2 > k_3$ then the v_1 and v_3 absorption lines contribute positive quantities to β while the v_2 line contributes a negative quantity. There is, therefore, an absolute correlation of the screw sense in the potential field and the sign of the contribution to β. The rule is: *the contribution to the rotation of the absorption line v_i is dextro-rotatory if the screw sense of the equipotential surfaces with regard to the x_i axis is positive.* This rule affords a basis for discussion of the age-old problem of absolute configuration in optically active molecules.

The rule is exactly what one would expect from a simple consideration of the relation of the field to formula (4). The absorption line v_3 corresponds to the excitation of a vibratory motion that is essentially parallel to the x_3 axis. Owing to the spread of the wave function in the x_1 and x_2 directions the motion is not confined entirely to motion precisely along the x_3 axis. Owing to the screw twist in the field there is a corresponding screw twist in the wave functions so the electron tends to screw around in a somewhat helical way, following the places where the potential energy is least. Suppose now the motion is in phase with the electric vector, then the velocity will be in phase with $\dot{E}$ and so owing to the particle's following a somewhat helical path there will be a circular current around the x_3 axis, in phase with which will produce the effective induced magnetic moment $\mathbf{m}_1$, directed along the x_3 axis and so vectorially proportional to the $\dot{E}$ which produced it.

The induction of an electric moment vectorially proportional to $\dot{H}$ but of opposite sign as in (4) may also be easily visualized. Suppose that $\dot{H}$ is parallel to the x_3 axis. Then it corresponds to a rate of change of flux through the $x_1 x_2$ plane which therefore is accompanied by induced electromotive forces tending to make the electron move around in the $x_1 x_2$ plane in the clockwise, or negative sense, owing to Lenz's law for induced currents. Owing to the helical constraints, when the electron does this it is caused to move toward the $-x_3$ direction and thus to produce an electric moment in the direction $-\dot{H}$ as given by (4). (The foregoing statements are made for a positive electron—for a negative electron the results are the same because change in sign not only reverses the direction of the forced motion but also reverses the sense of the moments resulting from the forced motion, corresponding to the occurrence of e^2 in (33).)

5. Adaptation of the Model to Actual Molecules

The simplest kind of optically active molecule conceivable is that in which four different single atoms are attached to a central carbon atom. Such compounds have never been prepared, at least not in active form. The next simplest type is that in which the four attachments to a central carbon atom may be organic radicals, as in the case of secondary butyl alcohol,

$$
\begin{array}{c}
\text{H} \\
| \\
\text{CH}_3\text{—C—OH} \\
| \\
\text{C}_2\text{H}_5
\end{array}
$$

Such a molecule will be characterized by a very complicated electronic absorption spectrum, lying entirely in the ultraviolet in the case of colorless liquids. The absorption spectrum is complicated not only because of the large number of electronic transitions possible but also because of the complexity of structure arising from the great variety of vibrational and rotational jumps which can be associated with a given electronic jump. In the face of such a complex situation some simplifying assumptions must be made.

Study of polyatomic absorption in the ultra-violet has shown that it is possible in most cases to associate a given set of bands with an electron transition occurring in a fairly definite part of the molecule. The group of atoms or radical in which the transition occurs is then said to be the chromophoric group or the chromophore for that system of bands.

The optical activity associated with a given electronic jump in the molecule arises essentially from the cooperation of two factors: (l) the electronic jump occurs in the molecule in a fairly well-localized place in some definite group in the molecule, (2) the electron which makes a transition is moving in a force field principally due to the immediate atoms to which it belongs, but secondarily due to the fields set up by the effective charge distributions of the other atoms in the molecules. It is the contribution of the neighboring atoms to the force field in which the chromophoric electron moves which is responsible for the optical activity.

This action of neighboring groups on the chromophoric group has generally been called *vicinal action* in a general way. In what follows we give a more precise picture of the nature of vicinal action.

Of course any electron in the molecule is the chromophoric electron for those transitions in which its state changes. But the important electron transitions will be those lying in the near ultraviolet which will mean that those electrons contained in groups that absorb in the near ultraviolet, or in the visible when present, will dominate the situation. This statement is much nearer the truth here than in the case of ordinary refraction for here the contributions to β of different bands in the deep ultraviolet are of different sign and so tend to cancel whereas in ordinary refraction they are all of the same sign so the cumulative effect of many of them may be considerable.

Suppose, for example, that the chromophoric electron in the near ultraviolet is a bonding electron between C and the OH in secondary butyl alcohol. Then in the absence of the other groups, of if the three other groups were all alike, the effective field in which that electron would move has the C—O direction as an axis of symmetry and would be more extended along the bond direction than transverse to it. The equipotential surfaces would be shaped like rather prolate ellipsoids of revolution. The actual space dependence of the potential energy would, of course, go asymptotically to a constant value like the Coulomb law for large distances.

Now the behavior of the field at large distances does not matter much if we confine ourselves to the two lowest states for in these the wave functions are essentially confined to a small region near the potential minimum. Therefore for discussion of the two lowest states it will be fairly accurate to represent the effective field due to the C and O atoms by a harmonic oscillator potential, having the C—O line as an axis of revolution. Choosing the C—O direction as the Z axis the potential will then be approximately

$$V_0 = \tfrac{1}{2}k_1(x_1{}^2 + x_2{}^2) + \tfrac{1}{2}k_3 z^2, \tag{40}$$

where the origin of coordinates will be somewhere between the carbon and oxygen nuclei at a place that can only be made definite on the basis of more exact information about the actual molecular orbitals in this bond.

Next we have to consider the way in which (40) is affected by the vicinal action of the neighboring groups. As a preliminary, consider the potential

$$U = [(X_1 - x_1)^2 + (X_2 - x_2)^2 + (X_3 - x_3)^2]^{-1/2} \tag{41}$$

at the position (x_1, x_2, x_3) due to unit charge located at $X_1 X_2 X_3$. If $x_1 x_2 x_3$ are all small compared to $X_1 X_2 X_3$ as will be the case if $x_1 x_2 x_3$ are the coordinates of the chromophoric electron and $X_1 X_2 X_3$ are those of the distant vicinal group, then the potential can be developed as

$$U = \frac{1}{R} - x_i \frac{\partial}{\partial X_i}\left(\frac{1}{R}\right) + \tfrac{1}{2} x_i x_j \frac{\partial^2}{\partial X_i \partial X_j}\left(\frac{1}{R}\right)$$

$$- \frac{1}{6} x_i x_j x_k \frac{\partial^3}{\partial X_i \partial X_j \partial X_k}\left(\frac{1}{R}\right), \tag{42}$$

in which summations over repeated indices are implied and $R = [X_1{}^2 + X_2{}^2 + X_3{}^2]^{1/2}$. The constant term is of no interest, since the zero of energy has to be arbitrarily fixed anyway. The linear terms represent a certain general average field acting on the electron which will produce a slight displacement of the equilibrium position.

The important terms in (41) for application of the results of Section 2 are the quadratic and the cubic. The total action of the charge distribution in the vicinal groups will be obtained by summing (37) over the charges in the vicinal groups. The quadratic and cubic terms coming from the summed expression may be written

$$\tfrac{1}{2} a_{ij} x_i x_j + \tfrac{1}{6} b_{ijk} x_i x_j x_k. \tag{43}$$

These have to be added to (40) to get our approximation to the full field in which the chromophoric electron moves. The terms in (43) are supposed to be small compared with those in (41).

Considering first the effect of adding the quadratic part of (43) to (40) we see that the resulting harmonic potential will in general be one in which the principal axes are tilted with respect to the original coordinate axes and in which the axis of symmetry has, in general, been removed. The tilting of one principal axis (x_3) away from the bond axis will in general be negligibly small if the original field V_0 is sufficiently anisotropic, i.e., if the cross terms a_{13} and a_{23} (which cause the tilting) are small compared with $|k_1 - k_3|$ in (40). Neglecting this slight tilting amounts to neglecting completely the a_{13}, and a_{23} terms. The terms involving a_{11} a_{12} and a_{22} however are important for it is these which determine the previously indeterminate principal axes by removing the symmetry of (40) about the x_3 axis.

After adding the important quadratic part of (40) the dependence of the potential on $x_1 x_2$ becomes

$$\tfrac{1}{2}(k_1 + a_{11})x_1{}^2 + a_{12} x_1 x_2 + \tfrac{1}{2}(k_1 + a_{22})x_2{}^2$$

A rotation of the axes through an angle γ is described by

$$x_1 = x_1' \cos \gamma - x_2' \sin \gamma$$
$$x_2 = x_1' \sin \gamma + x_2' \cos \gamma \tag{44}$$

and the potential energy in terms of the x_1' and x_2' is

$$\tfrac{1}{2}(k_1 + a_{11} \cos^2 \gamma + 2a_{12} \sin \gamma \cos \gamma + a_{22} \sin^2 \gamma)x_1'^2,$$
$$+\tfrac{1}{2}(k_1 + a_{11} \sin^2 \gamma - 2a_{12} \sin \gamma \cos \gamma + a_{22} \cos^2 \gamma)x_2'^2, \tag{45}$$

provided that γ has been chosen to make the coefficient of $x_1'x_2'$ vanish. The value of γ which does this is given by

$$\tan 2\gamma = \frac{2a_{12}}{a_{11} - a_{22}}. \tag{46}$$

Hence the important effect of the quadratic terms due to the vicinal groups is the removal of the axis of symmetry of the effective field of the chromophoric electron.

If now one transforms the cubic terms of (43) to the coordinate system (x_1', x_2', x_3) defined by (44) and (46) then the particular term which involves the product of the coordinates $x_1'x_2'x_3$ is the important one and is the term to be identified with the $Ax_1x_2x_3$ term of Section 2.

At first sight it might seem that all of the cubic terms in the potential would contribute to the optical activity and it is true that they do in higher approximations. But in the first approximation of the perturbation theory as used in Section 2 the contribution due to each cubic term would be linear in the coefficient of that term so the first approximation cannot take account of the joint action of several such terms. On the other hand, of the various kinds of cubic terms $Ax_1x_2x_3$ is the only one which removes all the planes of symmetry possessed by the harmonic oscillator potential, and therefore it is the only one which produces optical activity in the first approximation. For instance the potential x_1^3 leaves the x_1x_2 and x_1x_3 planes as planes of symmetry. A combined action of $(x_1^3 + x_2^3 + x_3^3)$ in the cubic potential would remove all the symmetry planes in the field but its contribution to the rotatory power would not appear in the first approximation as it depends on the joint presence of three cubic terms.

To summarize: we suppose the effective field in which the chromophoric electron would have moved if the vicinal action were neglected to be given by (40). As this is simply a formal representation to produce wave functions of approximately the correct spatial extent, one has to pick k_1 and k_3 in this to correspond to the frequencies of the lowest electronic jumps for the electron, the k's being related to the v's by (17).

In general the bonding electron in this approximation will be a Σ electron and the first excited state corresponding to motion along the figure axis will also be a Σ state for the electron so k_3 is the oscillator force constant which gives the frequency through (17) of the lowest $\Sigma \rightarrow \Sigma$ electronic transition. Similarly the excited states corresponding to excitation of transverse motion will be Π states so k_1 will be chosen to make v_1 the frequency of the lowest $\Sigma \to \Pi$ electron transition in that chromophore.

The next step consists in analyzing the probable charge distribution of the vicinal groups. If one had exact knowledge of the molecular orbitals this would be given at once by the density of the orbitals of the rest of the molecule. In the absence of such knowledge one has to try to piece together the main essentials by use of the results of dipole moment studies. These assign measured dipole moments to various bonds in the molecule which may be represented by net charges on the two atoms at opposite ends of a bond chosen so the product of net charge by bond distances is equal to bond dipole moment. This will be discussed later in connection with a specific example.

Finally with a known charge distribution, say charges e_s located at points X_{1s}, X_{2s}, X_{3s}, one has to calculate the quadratic terms in the potential energy due to the charges of the vicinal group as in (41) and (42). This will determine the way in which the axis of symmetry of the chromophoric field is removed through (44) and (46). Finally with axes chosen along the principal axes of the quadratic part of the chromophoric field of all the cubic potential terms one has to calculate only $Ax_1x_2x_3$. In this way we see that k_1 and k_3 are principally determined by the local field to which the chromophore electron belongs, while $k_1 - k_2$ and A are determined by the fields of the vicinal groups.

The discussion may be made more precise by writing the explicit formulas for the important quantities. Calculating the derivations in (42) one finds for the potential energy of an electron of charge e in the vicinal field of charges $e_1 \ldots e_s \ldots$

$$a_{11} = e \sum_s e_s (3X_{1s}{}^2 - R_s{}^2) \frac{1}{R_s{}^3}, \tag{48}$$

$$a_{12} = e \sum_s e_n \frac{3X_{1s}X_{2s}}{R_s{}^5}. \tag{49}$$

and a corresponding formula for a_{22}, obtained by substituting 2 for 1 in (18). Then the transformation to principal axes is given by (45) and (46). Finally the coefficient of the cubic term in the potential energy $Ax_1'x_2'x_3$ is given by

$$A = +(5/2)e \sum_s e_s \frac{X_{1s}'X_{2s}'X_{3s}}{R_s{}^7}, \tag{50}$$

in which X_{1s}' and X_{2s}' are coordinates of vicinal charges referred to the principal axis coordinate system defined by (44) and (46).

6. Application to Methyl Phenyl Carbinol Nitrite

We next proceed to test our thesis that coulombic force action from vicinal groups is responsible for the term $Ax_1x_2x_3$ in the potential function (14). To do this we compute the rotatory power of the nitrite band and the phenyl band in methyl phenyl carbinol nitrite.[10] In molecules containing the nitrite group —O—N=O the existence of an absorption band well removed from the neighborhood of other bands, and in the experimentally accessible region around 3650A, provides a convenient means for such a study because here one can identify with reasonable assurance what part of the rotatory power measured at any given wave-length is contributed by that particular band. The nitrite contribution appears surprisingly rich in structural detail, superimposed on the much stronger and slowly varying contribution from the phenyl band. The assignment to phenyl follows from a comparison with methyl cyclohexyl carbinol nitrite, and with methyl n-hexyl carbinol nitrite. In both of these cases where the phenyl has been replaced by other substituents we find that the contribution allegedly coming from the phenyl band is suppressed as one would expect from this assignment.

In what follows we adopt with minor modifications the general outline given in Section 5. In one essential point, however, we allow our conception of the nitrite band to differ from an assumption made in that section. We regard the chromophoric electron which is instrumental in the nitrite absorption at 3650A as being isotropically bound to its equilibrium position; isotropically, that is, except for the perturbation effect which the electric dipole field from the other three groups exert upon it. In view of the presence of big charges in the nitrite group itself this assumption can hardly be more than an approximation. At the same time it is dictated by the very uncertain way in which these charges are distributed with respect to the rest of the molecule. Since the distribution varies with the free rotation of the ONO group around the single C—O bond the actual orientation of the "unperturbed" polarization ellipsoid is extremely elusive. If, then, we simply consider the unperturbed problem as threefold degenerate it is quite clear that in the perturbed case the orientation of all three principal axes will be sensitive to the perturbing field, and not merely two as was assumed in the last section. The remark should be added that, since the rotatory power of our model is proportional to the three differences between v_1, v_2 and v_3, one must expect the resulting β to be small of higher order in the case where the frequencies are different from each other merely as a consequence of the perturbing field than if they had been different in the unperturbed oscillator.

In Section 5 it was explained that the numerical problem is essentially a determination of the asymmetry coefficient A of (50), when the potential has been referred to three axes $x_1'x_2'x_3'$ in which the quadratic terms of V reduce to a sum of squares. An isotropic oscillator which will absorb at 3650A, the absorption wave-length of the nitrite band, has three equal force constants $k_1 = k_2 = k_3 = 24,000$, in absolute units. Thus, to second-order terms the potential can be represented by

---

[10] Kuhn and Biller, Zeits. f. physik. Chemie **B29**, 1 (1935).

TABLE I

Atom:	H_0	H_1	H_2	H_3	C_1	C_2	O_1
Charge:	0.28	0.28	0.28	0.28	−0.28	−0.84	0.00
X_1	0.93	−0.93	0.93	0.00	0.00	0.00	0.00
X_2	−.53	1.67	1.67	2.16	0.00	0.00	0.00
X_3	−1.80	−2.54	−2.54	−1.03	−1.43	−1.92	0.00

$$U_2 = \sum_{ik} a_{ik} x_i x_k = 12{,}000(x_1{}^2 + x_2{}^2 + x_3{}^2)$$

$$+ \sum x_i x_k \sum_s \frac{ee_s}{2R_s{}^5} (3X_{is} X_{ks} - \delta_{ik} R_s{}^2),$$

where δ_{ik} is 1 or 0 depending on whether i and k are equal or not. This follows from (38) and (49). The second term of the above equation can be obtained from a list of coordinates of the individual vicinal charges e_s if these charges are referred to a right-handed coordinate system with the supposed chromophoric electron at the origin. We have tentatively chosen this position to coincide with the oxygen nearest to the asymmetric carbon atom. A list of coordinates and charges is given in Table I. The vicinal charges have been located at the position of the atoms so as to be compatible with the known bond moments and interatomic distances. The dipole moments used were: +0.35 for the CH bonds, −0.85 for the C—O, +0.30 for O—N, −2.15 for N=O. The last two moments were determined so as to give a resultant dipole moment of −1.90 for the nitrite group. All moments are given in Debye units. Zero bond moment has been assumed for the phenyl group.

Principal axes of U_2 as is well known are most conveniently determined by solving the homogeneous set of three linear equations

$$(a_{11} - k)u + a_{12}v + a_{13}w = 0,$$

$$a_{21}u + (a_{22} - k)v + a_{23}w = 0,$$

$$a_{31}u + a_{32}v + (a_{33} - k)w = 0,$$

where k must be determined so that this set has a solution different from zero. In other words k must be a solution of the secular equation $|a_{ik} - \delta_{ik}k| = 0$. For each of the three roots k_1', k_2', k_3' there exists one solution of the linear set which can be so normalized as to represent the direction cosines α_{i1}, α_{i2}, α_{i3} ($i = 1, 2, 3$), of the three principal axes. These α_{ik} represent the matrix of the linear orthogonal transformation which provides us with the coordinates X_1', X_2', X_3' of the vicinal charges in the new coordinate system. Using (50) one thus determines the numerical value of A for the nitrite absorption band in methyl phenyl carbinol nitrite to be

$$A = 1.46 \cdot 10^{10}.$$

FIGURE 1. Dotted curve,
experimental [M] values
(Fig. 1a of Kuhn and Biller,
Zeits. f. physik. Chemie
B29, 1 (1935)) translated
into B values. Broken line,
estimated contribution of
bands in far ultraviolet.
Full lines, calculated.

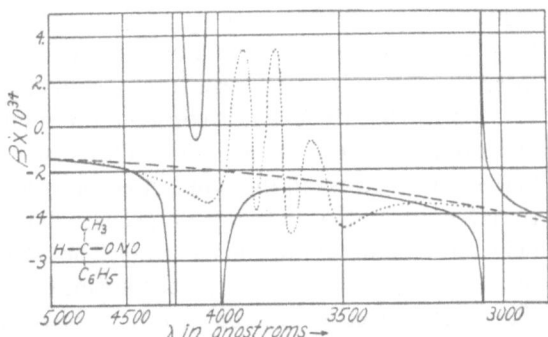

The new wave-lengths into which the threefold degenerate band at 3650 is split are the following: $\lambda 4215$, $\lambda 4026$ and $\lambda 3054$. With the numerical values of A and v_1, v_2, v_3 given one is in a position to apply (33) to calculate β as a function of the frequency of the incident light, and to compare it with the experimental dispersion curve as reported by Kuhn.[11] Fig. 1 gives the experimental values together with a theoretical curve which was obtained by adding the calculated contribution from the nitrite band to the part presumably coming from the phenyl band which can be fairly well estimated from the empirical curve by smooth interpolation (dotted curve). The agreement between theory and experiment is satisfactory particularly in the long wave range of the diagram. As a matter of fact one should keep in mind that our oscillator model was primarily designed to represent the low energy quantum states approximately. But actual transitions which involve higher energies than hv of the harmonic oscillator are probably not well represented by matrices derived from oscillator eigenfunctions, and the calculated energy differences are larger than the transition energies in the actual physical system.

An interesting consequence of the computation just given is that it implies an *absolute configuration* for the compound discussed. The absolute configuration for the methyl phenyl carbinol nitrite which we used in the above computation of A is

$$
\begin{array}{c}
CH_3 \\
| \\
H\text{———}ONO \\
C \\
| \\
C_6H_5
\end{array}
$$

Our computation shows that this should be identified with the laevorotatory isomer. When chemists relate this compound to other d- or l-compounds their absolute configuration will automatically follow.

---

[11] W. Kuhn and H. Biller, Zeits. f. physik. Chemie **B29**, p. 1 (1935). See especially Fig. 1.

If one tries to compute the phenyl contribution using the assumption that the phenyl chromophor is bound isotropically to its equilibrium position, with force constants giving the absorption band of phenyl at 2600A, one obtains a value $A = 0.83 \cdot 10^{10}$, that is considerably smaller than was obtained for the contribution of the nitrite band. This calculation takes into account the perturbing effect of all the other three charges, including the nitrite group as the latter no doubt is the controlling factor in the production of an asymmetric field at the center of the phenyl group. The configuration most likely to occur on statistical grounds, and at the same time the one most effective in producing the desired effect is that configuration in which the nitrite group lies in the plane through the two bonds C_1—O and C_1—phenyl with the double-bonded oxygen in a position closest to the center of the phenyl group. In view of the large dipole moments of the nitrite group and the high polarizability of the phenyl group this configuration (or a very similar one) must be quite stable compared with others. The result of the calculation is instructive in the following detail. It is found that only about two percent of the total value calculated for A is caused by the highly charged oxygen nearest to the phenyl; the reason being that its charge determines the principal axes of the perturbed polarization ellipsoid to such an extent that, lying almost exactly in one of the new coordinate planes it contributes practically nothing to the sum $\Sigma ee_s R_s{}^{-7} X_{1s}' X_{2s}' X_{3s}'$. On the other hand, our assumption that the phenyl chromophor is isotropically bound is far more objectionable according to all that is known about the electronic structure of the benzene ring, than was the analogous assumption in the case of the nitrite chromophor. As the phenyl electron is actually much more stiffly bound in the direction at right angles to the ring than in the other two directions, it will offer a certain amount of resistance to tilting of its principal axes, with the result that the oxygen will lie well away from all three coordinate planes and thus will contribute generously to the asymmetry factor A. In this connection it is interesting to see in a simple way how much a given charge e_s at a given distance R_s can at best contribute to A. Since the optimal position for e_s is to make equal angles with all three principal axes of U_2, it is easily seen from (50) that with

$$X_1' = X_2' = X_3' = R/(3)^{1/2},$$

$$A_{max} = (5/2)\frac{ee_s}{R^4(27)^{1/2}} = \frac{229.5e_s}{R^4}10^{10}, \qquad e_s \text{ in } 10^{-10} \text{ e.s.u., } R \text{ in A.}$$

Since the double-bonded nitrite oxygen represents a charge of $-1.76 \cdot 10^{-10}$ e.s.u. at a distance $R = 2.05$A, we find that its contribution to A might under favorable circumstances reach a value as big as 23×10^{10}. Rapidly diminishing optimal contributions are obtained for the other charges as their distance from the phenyl grows bigger ($A_{max} = 4.9 \times 10^{10}$ for N, and $= 1.13 \cdot 10^{10}$ for O_1). On the whole one sees from this over-all estimate that a potential reservoir of rotatory power is available in our model of this compound which is easily sufficient to account for the observed big rotations associated with the phenyl group.

7. Quantum Mechanics of Circular Dichroism

Closely associated with optical activity is the phenomenon of circular dichro-ism which is observed experimentally as the Cotton effect.[12] The quantum-mechanical formulation of this effect is a fairly obvious extension of the ordinary theory of absorption transition probabilities but, as it appears not to have been worked out explicitly before, it is perhaps useful to work it out here.

The effect in question is that the absorption coefficients for right and left circularly polarized waves are different for the absorption bands which con-tribute to the optical activity. It may be worked out either by the original theory,[13] in which the fields of the light wave are regarded as causing transi-tions, or by the quantized field method due to Dirac.[14] We shall use the former method as this involves somewhat simpler equations. The end result is the same by both methods, of course.

Suppose then the light wave is represented by a vector potential

$$\mathbf{A} = \tfrac{1}{2}[\mathbf{A}\exp[iE(t - \mathbf{k}\cdot\mathbf{r}/c)/\hbar] + \bar{A}\exp[-iE(t - \mathbf{k}\cdot\mathbf{r}/c)/\hbar]], \qquad (51)$$

in which $E = h\nu$ and ν is the frequency of the light. The interaction energy of the light wave and the charges in the molecule is represented by

$$H = -(e/mc)\sum_i \mathbf{A}_i\cdot\mathbf{p}_i - (e/mc)\sum_i \mathbf{S}_i\cdot(\text{curl }\mathbf{A})_i, \qquad (52)$$

the sum being over the various electrons in the molecule. The second term represents the direct interaction of the spin magnetic moment with the magne-tic vector of the light wave.

Let the molecule be initially in quantum state a and the light wave begin to act at $t = 0$. Then for $t > 0$ we can write

$$\Psi = \psi(a)\exp[-iW_a t/\hbar] + \psi_1(a) \qquad (53)$$

and the dynamical equation of quantum mechanics for $\psi_1(a)$ is

$$\left(H_0 - i\hbar\frac{\partial}{\partial t}\right)\psi_1(a) = -H\psi(a)\exp(-iW_a t/\hbar), \qquad (54)$$

where H_0 is the Hamiltonian of the unperturbed molecule. This has to be solved with the initial condition, $\psi_1(0)$ for $t = 0$. The solution is found in the usual way. Expand the right side of (54) in a series of unperturbed wave functions,

---

[12] Cotton, Ann. chim. phys. **8**, 347 (1896).
[13] Schrödinger, Ann. d. Physik **81**, 109 (1926); Slater, Proc. Nat. Acad. **13**, 7 (1927); Pauling and Wilson, *Quantum Mechanics*, p. 302.
[14] Dirac, Proc. Roy. Soc. **A112**, 661 (1926); **A114**, 243 (1927); or *Quantum Mechanics*, second edition, Chap. XI.

$$-H\psi(a)\exp(-iW_a t/\hbar)$$

$$= (1/2)\sum_b \psi(b)\exp(-iW_a t/\hbar)[(b|H+|a)\exp i(W_{ba}+E)t/\hbar$$

$$+ (b|H-|a)\exp i(W_{ba}-E)t/\hbar] \tag{55}$$

and assume for $\psi_1(a)$ a similar expansion. The coefficients are then determined to satisfy (54) and the initial condition. The result is

$$\psi_1(a) = (1/2)\sum_b \psi(b)\exp(-iW_a t/\hbar)$$

$$\times \left[\frac{(b|H+|a)(\exp[i(W_{ba}+E)t/\hbar]-1)}{W_{ba}+E}\right.$$

$$\left. + \frac{(b|H-|a)(\exp[i(W_{ba}-E)t/\hbar]-1)}{(W_{ba}-E)}\right]. \tag{56}$$

Here $W_{ba} = W_b - W_a$ and the coefficients in (55) are given explicitly by

$$(b|H\pm|a) = \int \psi^-(b)\left(\sum_i (e/mc)\mathbf{p}_i \exp(\pm i\mathbf{k}\cdot\mathbf{r}_i E/\hbar c)\psi(a)\cdot\mathbf{A}\right.$$

$$\mp \frac{iE}{\hbar c}\int \bar\psi(b)\left(\sum_i (e/mc)\mathbf{S}_i\right)\psi(a)\cdot(\mathbf{k}\times\mathbf{A}), \tag{57}$$

in which $\bar A$ is to be written for $\mathbf{A}$ when the lower sign is used, the retardation of the wave has been neglected in the smaller spin term of the third line, and $\int$ means the usual integration over the entire configuration space involved in calculating matrix components.

In (56) if $W_{ba} > 0$ as is the case in absorption only the term in $(b|H-|a)$ is important owing to the resonance denominator. Assuming E almost equal to W_{ba} so $W_{ba} - E = \Delta$, a small quantity one may write

$$\psi_1(a) = (1/2)\psi(b)\exp[-iW_b t/\hbar]\frac{(b|H-|a)\sin \Delta t/2\hbar}{\Delta}\cdot 2i\exp(i\Delta t/2\hbar).$$

The probability that the molecule is in state b at time t is given by the absolute square of the coefficient of $\psi(b)$ in $\psi_1(a)$. The transition probability therefore is

$$P(a\to b) = |(b|H-|a)|^2 \frac{\sin^2 \Delta t/2\hbar}{(\Delta t/2\hbar)^2}\cdot\left(\frac{t}{2\hbar}\right)^2. \tag{58}$$

This is exactly the same as in the usual theory where the retardation of the light wave is neglected. Hence there is no need to repeat the standard treatment of (58) here. The next step is to consider $(b|H-|a)$ more closely. By expanding the exponential function in (57) and saving only the first two terms this can be written

$$(b|H-|a) = (i/\hbar c)W_{ba}(b|\mathbf{p}|a)\cdot\bar A - \frac{EW_{ba}}{2\hbar^2 c^2}\mathbf{k}$$

$$\times (b|\mathbf{N}|a)\cdot\bar A + (iE/\hbar c)(b|\mathbf{m}|a)\cdot(\mathbf{k}\times\bar A), \tag{59}$$

in which **p** and **m** are electric and magnetic dipole moments as in section 1 and **N** is the electric quadrupole moment. The quadrupole moment term may be neglected for present purposes as it contributes nothing to circular dichroism. Neglecting the quadrupole moment term we see from (59) that the reduction of (58) will proceed exactly as in the usual theory except that $(b|\mathbf{p}|a) \cdot \bar{A}$ will be replaced by

$$(b|\mathbf{p}|a) \cdot \bar{A} + (E/W_{ba})(b|\mathbf{m}|a) \cdot (\mathbf{k} \times \bar{A}). \tag{60}$$

As the square of the magnitude of $(b|H-|a)$ appears in (58) this means that in place of $|(b|\mathbf{p}|a) \cdot \bar{A}|^2$ in the usual discussion we have here

$$|(b|\mathbf{p}|a) \cdot \bar{A}|^2 + 2(E/W_{ba})R\{\mathbf{A} \cdot (a|\mathbf{p}|b)(b|\mathbf{m}|a) \cdot (\mathbf{k} \times \bar{A})\}. \tag{61}$$

We have now to consider the average of this over all orientations of the molecules. For brevity write $\mathbf{P} = (a|\mathbf{p}|b)$ and $\mathbf{M} = (b|\mathbf{m}|a)$ in the following calculation. The first term is

$$\begin{aligned}
|\mathbf{P} \cdot \mathbf{A}|^2 &= |(\mathbf{P}_1 + i\mathbf{P}_2) \cdot (\mathbf{A}_1 + i\mathbf{A}_2)|^2 \\
&= (\mathbf{P}_1 \cdot \mathbf{A}_1)^2 + (\mathbf{P}_1 \cdot \mathbf{A}_2)^2 + (\mathbf{P}_2 \cdot \mathbf{A}_1)^2 \\
&\quad + (\mathbf{P}_2 \cdot \mathbf{A}_2)^2 + (\mathbf{P}_1 \cdot \mathbf{A}_1 \mathbf{P}_2 \cdot \mathbf{A}_2 + \mathbf{P}_2 \cdot \mathbf{A}_1 \mathbf{P}_1 \cdot \mathbf{A}_2).
\end{aligned} \tag{62}$$

The average of $(\mathbf{P}_1 \cdot \mathbf{A}_1)^2$ over all orientations of P_1 is $(1/3)P_1{}^2 A_1{}^2$ and similarly for the other three terms of this form. For the last two terms in (62) we use the identity

$$\mathbf{P}_1 \cdot \mathbf{A}_1 \mathbf{P}_2 \cdot \mathbf{A}_2 + \mathbf{P}_2 \cdot \mathbf{A}_1 \mathbf{P}_1 \cdot \mathbf{A}_2 = (\mathbf{P}_1 \times \mathbf{P}_2) \cdot (\mathbf{A}_1 \times \mathbf{A}_2).$$

Averaging over all directions for $\mathbf{P}_1$ and $\mathbf{P}_2$ with a fixed angle between them amounts to averaging over all directions of $(\mathbf{P}_1 \times \mathbf{P}_2)$ so the term on the right vanishes. Hence on averaging the last two terms of (62) cancel so the first term of (61) can be written

$$(1/3)|(b|\mathbf{p}|a)|^2|\mathbf{A}|^2.$$

The second term of (61) can be averaged as follows: the dyad **PM** multiplied into $(\mathbf{k} \times \bar{A})$ gives a vector along **P** which when averaged over all values of **P** making the fixed angle with **M** gives a vector along **M** of amount $\mathbf{P} \cdot \mathbf{M}_0 \mathbf{M} \cdot (\mathbf{k} \times \bar{A})$ where $\mathbf{M}_0$ is a unit vector in the direction of **M**. When this is averaged over all values of **M** which make a fixed angle with $(\mathbf{k} \times \bar{A})$ the result is a vector along $(\mathbf{k} \times \bar{A})$ of magnitude $\mathbf{P} \cdot \mathbf{M} \cos^2 \theta$ where θ is the angle between **M** and $(\mathbf{k} \times \bar{A})$. Averaging over all values of θ one has Avge$[\cos^2 \theta] = \frac{1}{3}$ so the average of $\mathbf{PM} \cdot (\mathbf{k} \times \bar{A})$ is a vector of magnitude $\frac{1}{3}\mathbf{P} \cdot \mathbf{M}\mathbf{k} \times \bar{A}$ hence the average obtained is

$$\text{Avge}\{\mathbf{A} \cdot \mathbf{PM} \cdot (\mathbf{k} \times \bar{A})\} = (1/3)\mathbf{P} \cdot \mathbf{MA} \cdot (\mathbf{k} \times \bar{A}).$$

Now writing $\mathbf{A} = \mathbf{A}_1 + i\mathbf{A}_2$ one has

$$\mathbf{A} \cdot (\mathbf{k} \times \bar{A}) = i[\mathbf{A}_2 \cdot (\mathbf{k} \times \mathbf{A}_1) - \mathbf{A}_1 \cdot (\mathbf{k} \times \mathbf{A}_2)],$$

so this factor is pure imaginary and therefore on taking the real part as needed for the second term of (61) only the imaginary part of $\mathbf{P} \cdot \mathbf{M}$ will come in. In particular if the wave traveling in the $\mathbf{k}$ direction is circularly polarized then

$$\mathbf{A} = (1/\sqrt{2})A(\mathbf{i} \pm i\mathbf{j}), \tag{63}$$

where A is a real amplitude and where the upper sign represents right, the lower left circularly polarized light. For this one has

$$\mathbf{A} = (\mathbf{k} \times \bar{A}) = \pm iA^2.$$

Hence finally the averaged form of (61) can be written

$$(1/3)A^2[|(b|\mathbf{p}|a)|^2 \mp 2(E/W_{ba}) \operatorname{Im}\{(a|\mathbf{p}|b) \cdot (b|\mathbf{m}|a)\}], \tag{64}$$

where the upper sign applies for right, the lower for left circularly polarized light.

Hence the mean absorption transition probability ε will be the same as when retardation is neglected. But the difference between the two kinds will be in the following ratio to the mean

$$\frac{\varepsilon_l - \varepsilon_r}{\varepsilon} = 4 \cdot \frac{v}{v_{ba}} \cdot \frac{\operatorname{Im}\{(a|\mathbf{p}|b) \cdot (b|\mathbf{m}|a)\}}{|(b|\mathbf{p}|a)|^2}, \tag{65}$$

in which ε_l and ε_r are written for the absorption transition probability for right and left circularly polarized light respectively.

These calculations provide the quantum-mechanical proof of Natanson's rule[15] previously obtained by classical methods. The rule is: if an absorption band absorbs left-circularly polarized light more strongly than right, then it contributes dextro-rotation to the medium for frequencies lower than that of the absorption band.

The results follow by comparing (65) with a combination of (7) and (10) resulting from elimination of β.

8. Rotatory Strength and Anisotropy Factor

In numerical discussions of the rotatory dispersion formula it is convenient to have a handy measure of the contribution of a given band to the rotatory power. For the absorption line $a \to b$ this is measured by

$$R(b, a) = \operatorname{Im}\{(a|\mathbf{p}|b) \cdot (b|\mathbf{m}|a)\} \tag{66}$$

where we shall call $R(b, a)$ the rotational strength of the line $a \to b$. It has the same dimensions as the ordinary line strength,[16] $(a|\mathbf{p}|b)^2$. It will be convenient to measure $R(b, a)$ in atomic units, that is, to measure $\mathbf{p}$ in units of ea where a

[15] Natanson, J. de phys. **8**, 321 (1909).
[16] Condon and Shortley, *The Theory of Atomic Spectra* (Cambridge, 1935), p. 98.

is the radius of the first Bohr orbit, and **m** in Bohr magnetons. Both of these units are rather large for the quantities to be measured so we shall expect $R \ll 1$ in general.

Conventionally φ is given in degrees/decimeter and what is conventionally called the molecular rotatory power M is

$$[M] = (\varphi/\rho)M/100, \tag{67}$$

where φ is in degrees per decimeter, ρ is in gram/cm^3 and M is the molecular weight. Hence the connection between $[M]$ and Φ is

$$[M] = (18/\pi)\Phi \tag{68}$$

so, combining (8) and (10) one has

$$\frac{[M]}{(1/3)(n^2 + 2)} = 24N(\hbar/\mu c)^2 \sum_b \frac{R_{ba}\nu^2}{\nu_{ba}^2 - \nu^2}. \tag{69}$$

Here R_{ba} is expressed in atomic units. The coefficient in (69) has the numerical value

$$24N(\hbar/\mu c)^2 = 21,700. \tag{70}$$

It is easy to calculate the value of the rotatory strength of an absorption band from the dispersion data as usually given by using (69) and (70). As an example consider Hunter's data[17] on d-sec-octyl alcohol. This extends from $\lambda6708$ to $\lambda3650$ and is represented by

$$[M] = \frac{4.08}{\lambda^2 - 0.0283},$$

where λ is in microns. Hunter has measured the refractive index n and from his data we find that $(1/3)(n^2 + 2)$ is 1.34 at $\lambda6438$ and 1.36 at $\lambda3650$ so this factor is essentially equal to 1.35 throughout the range. One has therefore

$$\frac{[M]}{(1/3)(n^2 + 2)} = \frac{3.02}{\lambda^2 - 0.0283} = \frac{106.9\sigma^2}{35.4 - \sigma^2},$$

where σ is the wave number in reciprocal microns. Hence comparing with (69) one has that the main absorption band is at $\lambda1680$ and the strength of this is

$$R_{ba} = 106.9/21,700 = 4.92 \times 10^{-3}.$$

In this way it is easy to calculate the rotatory strengths of the absorption lines when a rotatory dispersion formula is given.

Kuhn[18] has introduced another way of describing the rotatory strength of an absorption band by means of what he calls the anisotropy factor. Although the definitions he gives are in terms of his coupled oscillator model it is readily

[17] Hunter, J. Chem. Soc. **123**, 1671 (1923).
[18] Kuhn, Trans. Faraday Soc. **26**, 300 (1930).

seen that they amount to the following: the oscillator strength of the absorption line v_{ba} is defined as the dimensionless number

$$f(b, a) = (8\pi^2\mu/3e^2h)v_{ba}|(a|\mathbf{p}|b)|^2. \tag{71}$$

Then Kuhn defines the anisotropy factor $g(b, a)$ for this line as

$$g(b, a) = \text{Im}\{(a|\mathbf{p}|b)\cdot(b|\mathbf{m}|a)\}/|(a|\mathbf{p}|b)|^2. \tag{72}$$

The relation of Kuhn's anisotropy factor to our rotatory strength is then

$$f(b, a)g(b, a) = (2\pi/3)(a/\lambda_{ba})R(b, a), \tag{73}$$

in which $R(b, a)$ is expressed in atomic units, a is the radius of the first Bohr orbit and λ_{ba} is the wave-length of the absorption line $a \rightarrow b$. Thus the product fg is generally smaller by a factor of one to three thousand than the rotatory strength $R(b, a)$.

9. The Contribution of Spins to Optical Rotation

If a molecule contains an even number, $2n$, of valence electrons they can be paired in $(2n)!/n!2^n$ ways to give this number of distinguishable singlet bond eigenfunctions corresponding to each particular electronic configuration. To the very good approximation that the various operators for spin commute with the energy operator (i.e., neglecting magnetic interactions which are responsible in atoms for the fine structure) we can write down a bond function

$$B_i = \frac{1}{((2n)!)^{1/2}} \sum \pm P[\varphi_i(x_1y_1z_1\ldots z_n) \prod_{\substack{\text{all bond} \\ \text{pairs}}} (\alpha_i\beta_j - \alpha_j\beta_i)]$$

for each one of the $(2n)!/n!2^n$ ways of choosing the n bonds. Here φ_i is a function of space coordinates only and

$$\prod_{\substack{\text{all bond} \\ \text{pairs}}} (\alpha_i\beta_j - \alpha_j\beta_i) \equiv g_i$$

of spins alone. The antisymmetrizing operator $\sum \pm P$ indicates summation over all permutations of the electrons in the product $\varphi_i g_i$ taking the positive sign for even and the minus sign for odd permutations. This antisymmetric B_i satisfies the Pauli principle. Of these singlet bond functions only $(2n)!/n!(n + 1)!$ are linearly independent. Now the lowest energy state of molecules having an even number of electrons is always a singlet so that the expression for optical activity will involve the term

$$\text{Im} \sum_{n'} (n|\mathbf{r}|n')\cdot(n'|\mathbf{L} + 2\mathbf{S}|n),$$

where n indicates the singlet bond function B_n which therefore satisfies the equation

$$\mathbf{S}^2B_n = 0B_n = \bar{S}_xS_xB_n + \bar{S}_yS_yB_n + \bar{S}_zS_zB_n.$$

Since each term after the last equality is zero or positive, each must be zero, so that $2(iS_x + jS_y + kS_z)B_n = 0$. Thus spin will not contribute to the optical activity of unexcited molecules containing an even number of electrons. Since it is never possible to obtain excited molecules in sufficient amounts to contribute to optical activity, even if the spin were effective, the spin from such molecules never need be considered.

Next consider the case where there are $2n - 1$ electrons, n still being an integer. Actual examples are provided by free radicals. States of lowest energy as before correspond to maximum electron pairing. Corresponding to each of the B_i functions for $2n$ electrons we now have a bond function

$$B_i' = (1/[(n-1)!]^{1/2}) \sum \pm P[\varphi_i'(x_1 \ldots z_{n-1}) \cdot \alpha_k \prod_{\substack{\text{over} \\ n-1 \\ \text{bond pairs}}} (\alpha_i \beta_j - \alpha_j \beta_i)]$$

which contains one unpaired electron and therefore satisfies the equation $S_z B_i' = (h/4\pi)B_i'$. Still another set B'' which satisfy the equation $S_z B'' = -(h/4\pi)B_i''$ is obtained by replacing α_k by β_k in each B_i'. Again consider the quantity $\mathrm{Im}\{\sum (n|\mathbf{r}|n') \cdot (n'|\mathbf{L} + 2\mathbf{S}|n)\}$. Now

$$2\mathbf{S}B_n' = 2(iS_x + jS_y + kS_z)B_n'$$
$$= (1/[(n-1)]^{1/2}!) \sum \pm P[\varphi_n'(x_1 \ldots z_{n-1})$$
$$\cdot 2(i\beta_k + j(i\beta_k) + k\alpha_k) \prod_{\substack{\text{over} \\ n-1 \\ \text{bond pairs}}} (\alpha_i \beta_j - \alpha_j \beta_i)].$$

Thus if both $B'_{n'}$ and

$$B_n' \text{ are real } (n'|2\mathbf{S}|n) \equiv \int \sum_{\substack{\text{spin} \\ \text{space}}} B'_{n'} 2\mathbf{S}B_n' d\tau$$

contains an imaginary term only if in B_n' and $B'_{n'}$ the odd electron is assigned to an α state in one eigenfunction and a β state in the other; but in this case $(n|\mathbf{r}|n')$ is zero because of this difference in spin of the odd electron in the initial and final states.

The only way then that

$$\mathrm{Im}\{\sum (n|\mathbf{r}|n') \cdot (n'|2\mathbf{S}|n)\}$$

can be different from zero (for real eigenfunctions) would be for $\mathrm{Im}(n|\mathbf{r}|n')$ to be different from zero for the transition but this is impossible. However, suppose B_n' or $B'_{n'}$ contains an imaginary part. Since the operator H is real both the real and imaginary parts of any complex B_j' is separately an eigenfunction of H as one sees immediately by equating both the real and imaginary part of the corresponding Schrödinger equation to zero. There is therefore no loss in generality in assuming all B's to be either purely real or purely imaginary functions. Suppose now one of the B's is imaginary then the preceding argument about $\mathrm{Im}\{(n|\mathbf{r}|n') \cdot (n'|2\mathbf{S}|n)\}$ being zero can be taken over completely since the only effect is to multiply the result for real eigenfunctions by -1. Similarly if both eigenfunctions are imaginary the only effect is to multiply them by $+1$. Thus we have shown that to the approximation in which magnetic terms in the energy can be neglected the spins do not contribute to optical activity.

It follows from the above formulations and from the fact that r and L are each the sums of terms involving only one electron that exactly the same value is obtained for the optical activity when the φ's are used to calculate the matrices for r and L as when the antisymmetric B's are used. This is an important simplification. It is useful to make one further simplification which is justified by the observation that most transitions are satisfactorily described as single rather than multiple electron jumps. Thus we write $\varphi(x_1 \dots z_n) = \psi_1(x_1 y_1 z_1)\psi_2(x_2 \dots z_n)$ where $\psi_2(x_2 \dots z_n)$ is unchanged in the transition and so can be integrated out before calculating the matrix components for r and L. It is this approximation which forms the basis of our present treatment.

10. The Absolute Configuration of Secondary Butyl Alcohol

The procedure which we have applied to methyl phenyl carbinol nitrite leads to a definite statement for the absolute configuration of any molecule where the charge distribution about the chromophoric electron is known. Where the total rotatory power can be analyzed into the contribution from the different chromophoric electrons the theory provides a way of checking the assigned structure since a single assignment of charge distribution must give correctly the optical activity of all the different chromophoric electrons of the molecules. Such crucial tests will help to establish the range of validity of the present theory.

At least three determinations of the absolute configuration of butyl alcohol have been published. We indicate the configuration of butyl alcohol by the structural formula

$$CH_3$$
$$|$$
$$HO{-}C{-}H$$
$$|$$
$$C_2H_5$$

Configuration (*I*)

where the full lines indicate that H and OH lie in the plane of the paper and the dotted lines joining CH_3 and C_2H_5 to C indicate these groups lie below the plane of the paper. Boys'[19] rule applied to this configuration leads to the

---

[19] Boys, Proc. Rov. Soc. **A144**, 655 (1934).

dextro-rotatory assignment. Kuhn[20] on the other hand using the theory of coupled oscillators comes to the conclusion that this isomer is laevo-rotatory. Finally Kirkwood,[21] concludes this isomer is dextro-rotatory in agreement with Boys.

While we shall use our theory to calculate the absolute configuration of secondary butyl alcohol it is of interest to consider another method of obtaining this same information. Schwab, Rost and Rudolph[22] report that dextro-rotatory quartz partially covered with copper or nickel provides a catalyst which dehydrates laevo-rotatory secondary butyl alcohol faster than the dextro-rotatory form. Now quartz is known to crystallize in the form of spirals in which silicon and oxygen atoms alternate. Each spiral is connected with neighboring spirals since each silicon is bound to 4 oxygens and each oxygen to two silicons. Now there are right- and left-hand quartz spirals and they rotate plane polarized light in opposite directions. If some means can be devised to determine the absolute configuration of quartz, then as we shall see, a proper interpretation of experiments such as those of Schwab, Rost and Rudolph will fix the absolute configuration of secondary butyl alcohol.

First suppose that when water is removed from secondary butyl alcohol[23] the reaction may be represented as follows:

$$\text{(74)}$$

---

[20] Kuhn, Zeits. f. physik. Chemie **B31**, 18 (1935).
[21] Kirkwood, J. Chem. Phys. **5**, 479 (1937).
[22] Schwab, Rost and Rudolph, Kolloid Zeits. **63**, 157 (1934).
[23] Schwab, Rost and Rudolph imply that the butylene formed from the dehydration of secondary butyl alcohol has the double bond in the 1,2 position but do not say they have investigated this point.

Suppose further that the butyl alcohol reacting belongs to configuration (I). Then as viewed normal to a plane passing simultaneously through the reacting H and OH and through the two carbon atoms to which these atoms are attached the molecule has the properties of a left-hand screw, since the ethyl group occupies more space than the hydrogen attached to the same carbon atom. If as we now explicitly assume there is an increased activation energy due to the steric hindrance of nonreacting groups then such a left-hand screw alcohol should react faster with right-hand quartz since screws of unlike sense fit along side of each other better than like screws. Such a consideration relates the configuration of the alcohol to that of the quartz. Two different methods for arriving at the absolute configuration of macroscopic hemihedral crystals such as quartz can be suggested. (1) As Pasteur demonstrated long ago it is frequently possible to pick out the two kinds of optically active crystals by the fact that different faces are developed on the two kind of crystals. The identification is then completed if crystal faces can be related to the structure of the crystal. Thus this problem resolves itself into an investigation of the factors responsible for the growth of crystal faces and avoids optical theory entirely in establishing absolute configuration. Such a method could of course be applied to any optically active crystal such as the tartrates o thus avoiding uncertainties in deducing relative configurations. We now consider the second method.

(2) Optical theory can be applied directly to the crystal to deduce its absolute configuration. We leave to a future time the application of the present theory to quartz and accept provisionally Hylleraas'[24] application of the Born coupled oscillator theory in which he finds that right spiraling β-quartz is dextro-rotatory. Assuming this same relation holds for the slightly different low temperature α-quartz we are led by using our other considerations to the decidedly provisional result that laevo-rotatory butyl alcohol has configuration (I). This would be in agreement with Kuhn. While this result must be regarded as purely tentative it seems of very considerable interest to develop further both theoretically and experimentally this approach to absolute configuration because of the light it throws on surface reactions and optically active synthesis.

We now examine in more detail some of the steps in the preceding argument. Suppose that the dehydration of configuration (I) actually results in the formation of a double bond between the second and third carbon atoms instead of 1–2 butylene as we first supposed. Examination of the model shows that this can occur by the OH giving off with either of the two hydrogens on the third carbon atom. In the mechanism we will call (1) the four carbon atoms will lie in a plane so that such a molecule would react equally well with either right or left quartz. Thus whatever part of the reaction goes this way cannot be responsible for the observed differential rate and is therefore not

---

[24] Hylleraas, Zeits. f. Physik **44**, 871 (1927).

interesting for optical activity. It is very likely that steric repulsions between the hydrogens on the end carbon atoms are responsible for reducing the importance of this mechanism. The product of such a mechanism would be 2–3 *cis* butylene.

By mechanism (2) 2–3 *trans* butylene will be the product and configuration (*I*) can be seen, by examining a model to behave like a right screw so that by our former assumption, of fastest reaction between unlike screws, it should go fastest on left spiraling quartz. Thus if 2–3 *trans* butylene (and not 1–2 butylene) is the chief decomposition product in the Schwab, Rost and Rudolph experiment then keeping our tentative assignment of the absolute configuration of quartz we would conclude that configuration (*I*) is the dextro form.

The preceding discussion then suggests that experiments designed to answer the following questions will be particularly valuable.

(1) Is the product of dehydration of secondary butyl alcohol 1–2 or 2–3 butylene and, if the latter, is it *cis* or *trans*?

(2) What is the function of the metal in the dehydration?

(3) Is the rate of dehydration of the secondary alcohol slowed down by replacing nonreacting radicals in the molecule by larger substituents?

From the point of view of the catalytic theory it is particularly significant that the spiral character of quartz in bulk still persists in the surface layer.

We have made a provisional calculation of the rotatory power of butyl alcohol considering only the contribution from the hydroxyl as chromophoric group. The configuration adopted is that of (*I*), the carbon atoms are numbered in order along the chain with the OH group on C_2. Hydrogen atoms 0 and 1 are on C_1, 3 is on C_2, 4 and 5 are on C_3 and 6, 7, 8 on C_4 while H_9 is the hydrogen of the hydroxyl group. The assumed coordinates of the atoms are given in Table II. Here the coordinates are given in angstrom units. The charges assumed for the atoms are (in 10^{-10} e.s.u.): given in the last line of the table. This set of charges and coordinates determines the principal axes of the quadratic potential terms in accordance with the following linear transformation

$$u = -0.9797x + 0.510y + 0.1939z,$$

$$v = +0.0015x - 0.9691y + 0.2465z,$$

$$w = +0.1999x + 0.2398y + 0.9498z,$$

in terms of which it is found that the dissymmetry coefficient has the value, $A = 2.8153 \cdot 10^{10}$. If the unperturbed force constant is chosen so that the longest wave-length absorption band is at 1620A (this is the critical frequency in the empirical one-term dispersion formula) the three calculated frequencies associated with the three axes become,

$$v_u = 2.3873 \cdot 10^{15}, \qquad v_v = 1.8518 \cdot 10^{15}, \qquad v_w = 2.2568 \cdot 10^{15}.$$

These lead to a theoretical value of the parameter β which is

$$\beta = 3.644 \cdot 10^{-38}$$

Table II

	O	H_1	H_2	H_3	H_4	H_5	H_6	H_7	H_8	H_9	H_0	C_1	C_2	C_3	C_4
x	0	+1.85	+0.96	+1.85	−0.92	+0.92	−0.92	+0.92	0	0	−0.92	+1.24	0	0	0
y	0	−1.08	−1.60	−0.04	+1.60	+1.60	+3.05	+3.05	+1.85	−0.94	−0.53	−0.77	0	+1.43	+2.40
z	0	−1.08	−2.55	−2.55	−2.55	−2.55	−0.78	−0.78	+0.18	+0.24	−1.80	−1.94	−1.42	−1.94	−0.76
e	−2.16				all hydrogens +0.28							−0.84	+0.32	−0.56	−0.84

as the contribution of the hydroxyl group to rotatory power at the NaD line wave-length. The experimental value of β calculated from the observed rotatory power by means of (6) is $\beta = 1.40 \cdot 10^{-35}$. This value of course includes the contribution from all electrons and is the average effect allowing for all configurations assumed under conditions of more or less free rotation about the bond axes. The large discrepancy may of course mean that the one-electron model is of secondary importance in this case. On the other hand other possible configurations and assumptions concerning the unperturbed state would give a much more favorable result. Thus a moderate change in the difference between v_u and v_w would introduce a power of ten and a more favorable relation of the principal axes would help very much. We shall consider these possibilities more fully at a later time.

11. Other Consequences of the Theory

The known dipoles of various bonds lead us to locate charges on the atoms from which we can calculate the contribution to the potential coming from these coulomb forces. Other contributions toward rotatory power will arise from the coupling of the chromophoric electron with other electrons in the molecule. It is this latter effect which in the past has been considered to the complete exclusion of other mechanisms. Wherever charges are present both mechanisms undoubtedly contribute to the total effect.

Empirically it has been known[25] for some time that there is a relationship between total dipole moment and optical activity. Furthermore Rule[26] has attempted to relate optical activity to the constituent moments particularly with respect to their size and propinquity to the optically active carbon atom. From the theory presented here we are compelled to postulate the existence of such relationships in a definite quantitative way. Particularly large effects are to be expected wherever the chromophoric electron is closely and asymmetrically surrounded by ions such as are found in ionic crystals (quartz, cobaltioxalate crystals, etc.) Another effect the quantitative treatment of which falls naturally within the scope of this approach is the large change caused by polar and polarizable solvents on rotatory power. This effect interpreted in accord with the present formulation makes it possible to apply optical activity in a quantitative way to the study of liquid structure. As Wolf and Volkmann,[27] Beckmann and Cohen[28] and others have pointed out it is due to the electric forces originating from solvent dipoles surrounding the active molecules. We now readily understand that such forces will change the potential in

[25] Betti, Trans. Faraday Soc. **26**, 227 (1930).
[26] See Lowry, *Optical Rotatory Power*, pp. 326 *et seq.*
[27] Zeits. f. physik. Chemie **B3**, 139 (1929).
[28] Beckmann and Cohen, J. Chem. Phys. **4**, 784 (1936).

which the chromophoric electron is moving without having to postulate, as Beckmann and Cohen do, an actual distortion of the molecular frame.

A polar solvent tends to form a compound with highly charged parts of the optically active molecule; in so doing it partially neutralizes these charges thereby reducing the rotatory power. Thus the general effect of polar solvents should be to lower optical activity except in special cases where enhancement results from neutralization of charges which were themselves previously acting to lower the optical activity. These expectations are born out by experiment.[29]

Comparatively small changes in charge distribution may in fact lead to large changes or even reversal of sign of the asymmetry coefficient A. Thus when the shape of the molecule is not rigidly fixed the over-all rotatory power of the compound will show great sensitivity to all factors which modify the statistical weight of particular configurations. The classical example for this is furnished by tartaric acid and its salts and esters where free rotation about single bonds is restricted by steric effects and compound formation. This results in the well-known temperature and concentration anomaly.

Our thesis that any dispersion electron is made optically active whenever it is brought into a dissymmetric field provides a clear explanation of "induced optical activity" in the sense discussed by Lowry in cases where the dispersion electron is not directly attached to an asymmetric center. The carbonyl group in sugars is a striking example of this possibility. In such a case as this the asymmetric influence cannot be transmitted by way of bonds so that the conclusion seems inescapable that the effect is a coulombic one transmitted through space.

---

[29] Lowry, *Optical Rotatory Power* (Longmans Green 1936), p. 350.

The Theory of Nuclear Structure[*]

E.U. CONDON

Associate Director,
Westinghouse Research Laboratories
Pittsburgh, Pennsylvania

It seems appropriate to open this meeting with an attempt at a broad survey of what we think we know about nuclear structure, and of what are the most serious gaps in our knowledge. We may start by listing what are the main points in the theoretical views that are held today:[1]

(1) We suppose that the nucleus which has a charge Z and mass number A consists of Z protons and $(A - Z)$ neutrons, the associated neutral atom containing Z electrons in its extra-nuclear structure.

(2) We suppose that these particles obey the principles of quantum mechanics as set forth, for example, in Dirac's book. Also, we accept the relativistic principle of equivalence of mass and energy.

(3) We suppose that electrons, protons, and neutrons obey Fermi-Dirac statistics, i.e., that all wave functions involving more than one of any of these kinds of particles is anti-symmetric in the coördinates of like particles. Also, that these particles each have a spin angular momentum of $\frac{1}{2}\hbar$.

(4) We suppose further that charged particles interact according to the Coulomb law, and interact with the electromagnetic field in the sense implied by Maxwell's equations, both through their charges and through their magnetic moments.

(5) We suppose that protons and protons, and protons and neutrons interact with each other by means of additional specifically nuclear interactions which are negligible at greater distances than 10^{-12} cm. and at distances of the order 10^{-13} cm. are much greater than the Coulomb interactions.

(6) We suppose that a proton can be converted into a neutron plus a positron or a neutron into a proton plus an electron by laws which are very imperfectly understood, and which apparently involve violations of the conservation laws of energy, momentum, spin, and statistics. We suppose, also, that it is possible to relate the phenomenologically introduced magnetic

---

[*] Based upon a paper presented at the University of Chicago Symposium on Nuclear Physics, June, 1938.
[1] The outstanding general summary of the literature up to two years ago is Bethe and Bacher, *Rev. Mod. Phys.*, **8**, 82 (1936).

moments, and the specifically nuclear interactions to relations of the heavy particles to an electron-neutrino field or to some other type of quantized wave field suited to the purpose.

The principles have been listed in an order of decreasing certainty or definiteness of our knowledge about them. While the subject seems to fall quite naturally into the parts as listed, we must be prepared from past experience in other branches of theoretical physics to have the resolution of later difficulties come by radical revision of the parts of the subject which are currently taken for granted.[2]

Let us now turn to some of the important results and problems which are related to the principles just enumerated.

(1) This view of proton-neutron composition instead of the previously held hypothesis of proton-electron composition was introduced speculatively by Harkins long before the discovery of the neutron, and was developed by Heisenberg as a quantum mechanical theory soon after the discovery of the neutron. It avoids difficulties characteristic of the proton-electron hypothesis arising from spin and statistics of nuclei and also connected with the fact that the nucleus is too small to hold electrons without introduction of extraordinary forces.

(2) As to the use of quantum mechanics, this is done simply because there is no reason not to extend its use from the well-tested extra-nuclear domain into nuclear problems. The principle of equivalence of mass and energy has received abundant and exact confirmation on a large number of nuclear reactions, especially with mass numbers below 20 where the mass measurements are quite exact.

(3) The validity of F. D. statistics for electrons underlies the whole theory of atomic spectral, of chemical valence, and of metals, and so is established beyond doubt by the many successes of these theories.

For protons we have as to spin and statistics the direct evidence of the alternating intensities in hydrogen molecular bands, and as to spin the additional evidence of the hyperfine structure of atomic hydrogen as revealed in the atomic beam experiments of Rabi, Kellogg, and Zacharias, using the Breit-Rabi method.[3]

For neutrons, however, it is well to emphasize that the evidence is less direct. The customary argument runs as follows: Assuming neutrons to obey F. D. statistics and to have a half-integral spin, then on the proton-neutron composition hypothesis a composite nucleus will have

Einstein-Bose statistics and integral spin for A even,

Fermi-Dirac statistics and half-integral spin for A odd.

Let us emphasize that this work does not show that the neutron spin is $\frac{1}{2}$, but only that it is half-integral. The usual assumption that it is $\frac{1}{2}$ has only the status of being the simplest possibility that is in accord with the data.

---

[2] Heisenberg, *Ann. der Phys.*, **32**, 20 (1938).

[3] Rabi, Kellogg, and Zacharias, *Phys. Rev.*, **45**, 769 (1934); **50**, 472 (1936).

(4) The general validity of the Coulomb Law down to nuclear dimensions is shown for proton-electron interaction by the success of theories of extra-nuclear structure. Similarly for proton-proton interaction by the experiments on angular scattering of protons, deuterons, and alpha-particles at energies sufficiently low that the specifically nuclear forces are inoperative. This is also indicated by the success of the barrier leakage theory for alpha-particle disintegration.

The general validity of the Maxwell equation method of calculating interactions with the electromagnetic field is fairly well indicated by the success of calculations of effective cross sections for dissociation of the deuteron by gamma rays and the converse effect of radiative capture of neutrons by protons. The same is indicated by the general success of calculations on internal conversion of gamma rays.

Interactions with the electromagnetic field are also satisfactorily calculated in the underlying theory of the atomic beam experiments of Rabi and his associates and also in the molecular beam experiments of Stern's laboratory.[4] Out of this work has come the surprising result that the magnetic moment of the proton is 2.5 (Stern, et al.) or 2.85 (Rabi, et al.) times $e\hbar/2Mc$ where M is the proton mass, i.e., about two and one-half times "what it ought to be"! The sign of the proton's magnetic moment is positive, i.e., related to the angular momentum as one would expect from the simple picture of the magnetic moment's being produced by rotating charge.

The analogous experiments for the deuteron give a magnetic moment of $+0.8$ in the same unit.

From evidence on proton-neutron scattering, we conclude that the normal state of the deuteron corresponds to zero orbital angular momentum, in which case the spins are parallel if the neutron spin is $\frac{1}{2}$ giving -2 for the magnetic moment of the neutron. If the neutron spin is 2 then the spins are antiparallel giving $+3.6$ for the neutron magnetic moment. That the magnetic moment of the neutron is antiparallel to the angular momentum is definitely shown by the recent experiments in Columbia and in Copenhagen, on the precession of neutrons in a magnetic field, so the fact of neutron spin of $\frac{1}{2}$ is now definitely established.[5]

(5) The information we have about the specifically nuclear forces is of two types, sharply distinguished by the accuracy with which the associated mathematical problems can be handled. The first type is that which is afforded by study of the two-body problems.

For proton-neutron interaction we have as essential data:

(a) The binding energy of the deuteron

$$8.97 + 8.13 - 14.73 = 2.37 \text{ millimass units.}$$

[4] Estermann, Simpson and Stern, *Phys. Rev.*, **52**, 535 (1937).
[5] Powers, Carroll, Beyer, and Dunning, *Phys. Rev.*, **52**, 38 (1937); Frisch, von Halban, and Koch, *Phys. Rev.*, **53**, 719 (1938).

(b) The experimental fact of the isotropic angular distribution of scattering of neutrons by protons for neutrons up to about 10 mmu neutron energy.

(c) The variation with energy of the total effective cross-section of scattering of neutrons by protons, varying from $1 \cdot 10^{-24}$ cm.^{2} for energies of about 10 mmu up to a cross-section of about $30 \cdot 10^{-24}$ for neutrons of thermal energy.

(d) The results on the scattering of slow neutrons by ortho and para hydrogen obtained by Stern, et al., and by Dunning, et al., and interpreted in accordance with the theoretical ideas of Schwinger and Teller.[6]

Before considering what can be learned from these experiments, let us first review briefly the formal possibilities for interaction laws which present themselves most simply. In classical mechanics the only possibility was that of a simple potential energy described by a general potential energy function $V(r)$. Quantum mechanics introduces us to wider possibilities in the way of exchange and spin dependent forces.

The wave equation for a two-body problem for two particles having position coordinates r_1 and r_2 and spin coördinates σ_1 and σ_2 will involve the kinetic energy terms

$$\frac{p_1{}^2}{2M_1} + \frac{p_2{}^2}{2M_2},$$

where M_1 and M_2 are the masses of the two particles. The wave function will be $\psi(r_1, \sigma_1; r_2, \sigma_2)$ dependent on spin and position of the two particles. Formally we may introduce a set of exchange operators 1, H, M, B, defined by the equations:

$$1\psi(r_1, \sigma_1; r_2 \sigma_2) = \psi(r_1, \sigma_1; r_2, \sigma_2),$$

$$H\psi(r_1, \sigma_1; r_2, \sigma_2) = \psi(r_2, \sigma_2; r_1, \sigma_1),$$

$$M\psi(r_1, \sigma_1; r_2, \sigma_2) = \psi(r_2, \sigma_1; r_1, \sigma_2),$$

$$B\psi(r_1, \sigma_1; r_2, \sigma_2) = \psi(r_1, \sigma_2; r_2, \sigma_1)$$

and may then introduce a more general interaction law

$$V(r) = V_1(r) + V_H(r)H + V_M(r)M + V_B(r)B,$$

where the four $V(r)$ functions are arbitrary and independent. Introducing coördinates of the center of mass and relative coördinates

$$R = \frac{M_1 r_1 + M_2 r_2}{M_1 + M_2}, \qquad r = r_2 - r_1$$

in the usual way, the wave equation then becomes

[6] Schwinger and Teller, *Phys. Rev.*, **51**, 775 (1937); Halpern, Estermann, Simpson, and Stern, *Phys. Rev.*, **52**, 142 (1937); Dunning, Manley, Hoge, and Brickwedde, *Phys. Rev.*, **52**, 1076 (1937).

$$\left[\frac{\mathbf{P}^2}{2(M_1 + M_2)} + \frac{\mathbf{p}^2}{2M} + V(r)\right]\psi = W\psi,$$

in which $M = M_1 M_2/(M_1 + M_2)$, the usual reduced mass.

This formalism makes sense only if the particles are of the same mass, for otherwise, we are unable to separate the motion of the center of mass from the internal motion, and also if the particles have the same spin in order that the field of values of the two spin coordinates shall be the same. It is customary to neglect the small difference of proton and neutron masses (1.38 mmu) which though small compared to the total mass is not small compared to the energy values of the problem.

If this is done, we may separate the motion of the center of mass in the usual way and deal simply with the factor of the wave function which refers to the internal motions and depends on r, σ_1 and σ_2. The states of the system may be labelled by the values of orbital angular momentum and by resultant spin using the familiar spectroscopic notation in which orbital values are denoted by S, P, D, F for $L = 0, 1, 2, 3$ and denoting resultant spin values by superscript 1 (singlet) for $S = 0$ or superscript 3 (triplet) for $S = 1$. Since

$$1\psi_{LS} = \psi_{LS},$$

$$H\psi_{LS} = -(-1)^{L+S}\psi_{LS},$$

$$M\psi_{LS} = (-1)^L\psi_{LS},$$

$$B\psi_{LS} = -(-1)^S\psi_{LS},$$

the effective law $V_{LS}(r)$ for a state of orbital angular momentum L and spin S is

$$V_{LS}(r) = V_1(r) - (-1)^{L+S}V_H(r) + (-1)^L V_M(r) - (-1)^S V_B(r).$$

This gives four types of force law in the two-body problem:

$$^3S: V_{01}(r) = V_1 + V_H + V_M + V_B,$$

$$^1S: V_{00}(r) = V_1 - V_H + V_M - V_B,$$

$$^3P: V_{11}(r) = V_1 - V_H - V_M + V_B,$$

$$^1P: V_{10}(r) = V_1 + V_H - V_M - V_B.$$

Since we know the resultant spin of the deuteron is 1, we know that the normal state of the deuteron must be one of the four possibilities $^3S_1$, $^1P_1$, $^3P_1$, or $^3D_1$. From the observed isotropy of the neutron scattering up to 5 mmu internal energy, we infer that the force laws are such that the radial wave function phase shifts δ_{LS} which determine the scattering according to the formula

$$\sigma(\theta)\,d\omega = \frac{\lambda^2}{16\pi^2}\,d\omega\left[\frac{1}{4}\left|\sum_L (2L+1)(e^{2i\delta L_0} - 1)P_L(\theta)\right|^2\right.$$

$$\left. + \frac{3}{4}\left|\sum_L (2L+1)(e^{2i\delta L_1} - 1)P_L(\theta)\right|^2\right]$$

are all essentially zero, except for $L = 0$, that is, δ_{00} for ^1S scattering and δ_{01} for ^3S scattering. In this case the total scattering cross-section reduces to

$$\sigma = \frac{\lambda^2}{\pi}\left[\tfrac{1}{4}\sin^2 \delta_{00} + \tfrac{3}{4}\sin^2 \delta_{01}\right].$$

Here λ is the de Broglie wave-length $h/\sqrt{2MW}$. If the law of interaction is such that the phase shifts for scattering for $L \neq 0$ are essentially zero, we can regard this as proof that the normal state of the deuteron must be a state for which $L = 0$ and hence must be $^3S_1$.

From general considerations about the binding energy of nuclei in general, we can conclude that the range of action of the specifically nuclear forces is of the order of 10^{-13} cm. and from the known binding energy of the deuteron can get an idea of the general character of the interaction law. The data are insufficient to distinguish between different similarly shaped analytical forms as Gauss error curves or square wells.

Simple calculations of Bethe and Peierls show that the scattering cross section is related to the binding energy by

$$\frac{\lambda^2}{\pi}\sin^2 \delta_0 = \frac{(\lambda')^2}{\pi} \qquad \text{where} \qquad \lambda' = \frac{h}{\sqrt{2M(\varepsilon + W)}}.$$

Numerically this is

$$\sigma = 2.35 \cdot 10^{-24} \frac{\varepsilon}{\varepsilon + W}\, \text{cm}^2.,$$

where ε is the binding energy of the deuteron and W is the internal energy of the scattering problem, half the incident neutron energy. For high energy neutrons this is an agreement with the facts, but does not agree with experiment for slow neutrons where $\sigma = 35 \cdot 10^{-24}$ cm². The answer is that this calculation tacitly supposes the law of force in the ^1S and ^3S states to be the same. Wigner pointed out that one can get agreement by supposing that the binding energy ε' of the singlet state is much smaller than ε for the triplet state. This gives for the scattering cross-section

$$\sigma = \left[\frac{3}{4}\frac{\varepsilon}{\varepsilon + W} + \frac{1}{4}\frac{\varepsilon'}{\varepsilon' + W}\left(\frac{\varepsilon}{\varepsilon'}\right)\right]2.35 \cdot 10^{-24}\ \text{cm}^2.$$

This agrees quite well with experiment if we suppose that

$$\varepsilon' = 40000 \text{ ev.},$$

which implies, of course, a considerable difference between the singlet and triplet interaction laws.

An important and peculiar feature of quantum mechanics is that an attractive potential may not be strong enough to give a stable level—the zero-point energy may be greater than the depth of the well. We can see that this might be so from the formula

$$W = \int \left[\frac{\hbar^2}{2\mu} \left(\frac{\partial \psi}{\partial x} \right)^2 + V\psi^2 \right] dx.$$

The second integral will be small if V represents a short range force unless ψ^2 is bunched up mostly into the region where V is large. But if this region is very small such a large degree of bunching up implies that the first integral is very large, making the first integral, which is essentially positive, be numerically larger than the second.

When this point is taken into consideration, it can be shown that the Wigner explanation of the large scattering cross-section for slow neutrons does not necessarily imply that the ^1S state of the deuteron is a stable state. If the well is not deep enough to give a stable state, the theory nevertheless, gives a formula of the same type for the variation with energy W of the scattering cross-section containing a parameter characteristic of the force law. A mode of speech has grown up whereby this parameter is referred to as a "virtual level" if the force law is such that there is no stable level. This is apt to be confusing in that in case ε' corresponds to a so-called virtual level there is nothing very sharply characteristic about this energy value in the positive continuum.

It becomes an important question to determine whether the parameter ε' of the ^1S proton-neutron interaction law refers to a stable level or to a virtual level in the sense just explained. One cannot tell from the scattering data. The distinction is given by consideration of the phenomenon of photo-disintegration of the deuteron and its converse effect the radiative capture of neutrons by protons to form deuterons. The difficulty of the measurements has thus far prevented any very definite results from being obtained.

The question is involved in the theory of the difference between the scattering of slow neutrons in ortho and in para hydrogen as worked out by Schwinger and Teller. The experiments of Stern and associates, and also those made in Dunning's laboratory have given quite definite evidence of the virtual character of the ^1S level.

There is thus far no definite evidence concerning the ^1P and ^3P interaction laws for proton-neutron from direct study of two-body problems. These are involved in the theory of scattering of neutrons of much higher energy than any at present available and also in the theory of photo disintegration of the deuteron by high energy x-rays.

For proton-proton interaction, we have as essential data

(a) the absence of any evidence for a helium isotope of mass 2,
(b) the proton-proton scattering data obtained by Tuve, Hafstad, and Heydenburg up to incident proton energies of 900 kv., recently extended up to 2400 kv. by Herb and associates in Madison.[7]

---

[7] Tuve, Hafstad, and Heydenburg, *Phys. Rev.*, **49**, 402 (1936); Hafstad, Heydenberg, and Tuve, *Phys. Rev.*, **53**, 239 (1933); Breit, Condon, and Present, *Phys. Rev.*, **50**, 825 (1936); Cassen and Condon, *Phys. Rev.*, **50**, 846 (1936); Breit and Feenberg, *Phys. Rev.*, **50**, 850 (1936); Breit, *Rev. Sci. Inst.*, **9**, 63 (1938); Wheeler, *Phys. Rev.*, **50**, 643 (1936).

Here we can use that same kind of formalism as was introduced for the interaction of protons and neutrons with this difference that there is no doubt connected with inequality of the masses, and secondly, that owing to the fact that the particles obey Fermi-Dirac statistics, the ^3S and ^1P laws are excluded (here we use S and P to symbolize all states of even and odd orbital angular momentum). This means that it is meaningless to express the interaction in terms of the four types introduced before. The two laws are

$$^1S: V_{00} = V_1 - V_H + V_M - V_B,$$
$$^3P: V_{11} = V_1 - V_H - V_M + V_B,$$

so we see that for proton-proton interaction the Heisenberg type is in principle indistinguishable from the ordinary type and the Bartlett spin exchange type indistinguishable from the Majorana type.

At present, the only information we have concerns the ^1S law of interaction. The proton-proton scattering data are fairly well represented by assuming that the departure from Coulomb law scattering is given entirely by the effects of the ^1S specific interaction. From this work, the most striking thing that emerges is the fact that the ^1S interaction of protons with protons is quite closely the same as the corresponding law for protons and neutrons. When one considers the work of Herb at higher energies, it appears that there are appreciable departures from the scattering to be expected solely from the effects of ^1S interaction. In this connection, it should be mentioned that there are other possibilities for interaction laws than those we have been considering, particularly forms involving the scalar product of each spin with the vector separation of the two particles which may be necessary to account for the high energy proton-proton scattering. All in all it seems that the careful extension of these scattering experiments to increasingly higher voltages with as great accuracy as possible is the most important fundamental experiment for which we can clearly see the need at present.

Let us turn now to a brief consideration of the information about interaction laws obtained from calculations on mass defects of the many-body nuclei. This is a very complicated subject in which approximation methods of doubtful validity mar the conclusions obtainable from laborious computations.

Work on the calculation of the energy levels of complex nuclei has been greatly stimulated in the last year by the conclusion drawn from the proton-proton scattering experiments that proton-proton and proton-neutron interactions are essentially alike in the ^1S states.[8] This makes it possible to use, in a fruitful way, a coordinate having two values $+1$ if the particle is a proton

---

[8] Wigner, *Phys. Rev.*, **51**, 106 (1937); **51**, 947 (1937); Feenberg and Wigner, *Phys. Rev.*, **51**, 95 (1937); Feenberg and Phillips, *Phys. Rev.*, **51**, 597 (1937); Hund, *Zeits für Phys.*, **105**, 202 (1937); Wheeler, *Phys. Rev.*, **52**, 1083, 1107 (1937); Euler, *Zeits für Phys.*, **105**, 553 (1937); Volz, *Zeits für Phys.*, **105**, 537 (1937); Flügge, *Zeits für Phys.*, **105**, 522 (1937); Kemmer, *Nature*, **146**, 192 (1937); Wefelmeier, *Zeits für Phys.*, **107**, 332 (1937).

and -1 if it is a neutron. When this extra coördinate is introduced, the state of being a proton or a neutron is regarded as two quantum states of the same particle and the Fermi-Dirac statistics are applied by requiring the wave function to be antisymmetric in regard to proton-neutron exchange as well as in regard to proton-proton and neutron-neutron exchanges separately.

Introduction of this coördinate, which has been called isotopic spin by Wigner and the character coördinate by Cassen and Condon gives a convenient formalism for application of group-theoretic methods to the description of the possible low energy levels in isobaric nuclei. This point of view has been developed by Wigner and by Hund independently.

Breit and Feenberg first investigated the relation of the symmetrical Hamiltonian (in which the specifically nuclear forces are assumed to be independent of the character of the particle) to the statistical calculation of binding energies of heavy nuclei. This same point of view has been followed up in the Leipzig institute by Flügge, Volz, and Euler. In these calculations, the dependence of $V(r)$ on distance is assumed to be that of the Gauss error function

$$e^{-r2/a2},$$

where a is of the order of $e^2/mc = 2.80 \cdot 10^{-13}$ cm. Volz' conclusion concerning the proportions of the four basic types of exchange force were later revised by Kemmer and at present the best supposition seems to be

$$V_M = \frac{10}{12}V_{01}, \qquad V_B = \frac{5}{12}V_{01},$$

$$V_H = -\frac{2}{12}V_{01}, \qquad V_1 -\frac{1}{12}V_{01};$$

where V_{01} is the ^3S law which is the sum of the four types. This means that the binding forces are contributed by Majorana and Bartlett types, while the Heisenberg and ordinary types are repulsive and weaker.

It is interesting to consider the form assumed by the force law when the exchange operators are written in terms of the spin vectors σ_a and σ_b and the formally analogous isotopic spin vectors τ_a and τ_b. We have

$$H_{ab} = -\tfrac{1}{2}(1 + \tau_a \cdot \tau_b), \qquad B_{ab} = +\tfrac{1}{2}(1 + \sigma_a \cdot \sigma_b),$$
$$M_{ab} = -\tfrac{1}{4}(1 + \sigma_a \cdot \sigma_b)(1 + \tau_a \cdot \tau_b).$$

On substituting these expressions and combining them in the proportions given above, one finds that the constant term and the term in $\sigma_a \cdot \sigma_b$ have zero coefficients, so the force law assumes the relatively simple form

$$V = Ae^{-r2/a2}(\tau_a \cdot \tau_b)[1 + \tfrac{5}{3}\sigma_a \cdot \sigma_b].$$

It appears from the simplicity of this result that the expression of exchange forces in terms of τ and σ is more fundamental than in terms of the previously introduced H, M, and B operators. The data are insufficient to determine the value of a except that it must be less than $2(e^2/mc^2)$ but for each assumed range, a, the value of A is quite accurately determined. The values $A = 4.75$ mmu and $a = 0.8\,(e^2/mc^2)$ give a satisfactory representation of data as known at present. The factor containing σ and τ has the values

Type	$(\tau_a \cdot \tau_b)(I + \tfrac{5}{3}\sigma_a \cdot \sigma_b)$
^3S............	$-8/\ +8$
^1S............	-4
^3P............	$+8/3$
^1P............	$+12$

It will be interesting to see whether future experiments on scattering with high energy particles are in accord with the conclusion that the forces are so strongly repulsive for odd values of orbital angular momentum.

(6) Finally, let us consider briefly the situation in regard to β disintegration and the field-theoretic interpretation of magnetic moments and specifically nuclear forces.[9]

As is well known, it has been necessary to invent the neutrino in order to balance the books in regard to energy, momentum spin and statistics in the β disintegration process. This way out was first suggested by Pauli and was made the subject of specific calculations by Fermi in 1934.

On this view, a nucleus emits a β particle as a result of a basic process in which a neutron is converted into a proton with simultaneous emission of an electron and a neutrino. Similarly, if a positron is emitted, the basic process is that in which a proton is converted into a neutron with simultaneous emission of an electron and a neutrino. The continuous spectrum of energies of the β particles emitted then arises from a statistical division of the constant total energy of the process between the observed electron and the unobserved neutrino.

In setting up such a theory, one has to make a definite postulate concerning the form of the term to be put in the Hamiltonian to bring about the transitions. To some extent the choices are restricted by requirements of relativistic invariance, but this still leaves a rather wide field of possible interaction forms available. Fermi chose one of the simpler forms in his first calculations; Konopinski and Uhlenbeck examined other possibilities, and especially emphasized the introduction of another form which they thought would give a theoretical distribution of energy in the β ray spectrum in better accord with experiment. This ambiguity as to possible interaction forms has caused a good deal of dissatisfaction with the electron-neutrino theory, but in this respect the

[9] Wentzel, *Naturwiss*, **26**, 273 (1938); Yukawa, *Proc. Phys. Math. Soc. Japan*, **17**, 48 (1935); Yukawa and Sakata, *Proc. Phys. Math. Soc. Japan*, **19**, 1084 (1937); Critchfield and Teller, *Phys. Rev.*, **53**, 812 (1938); Kemmer, *Proc. Roy. Soc.*, **166**, 127 (1938); Frohlich, Heitler, and Kemmer, *Proc. Roy. Soc.*, **166**, 154 (1938).

theory is no worse than the phenomenological theory of specifically nuclear interactions. To its credit must be acknowledged a good measure of success in accounting for the shape of the β distribution function and for the relation between disintegration constant and upper limit of the energies which is exhibited in the Sargent curves. An important feature of this work is the recognition that disintegration constants of different orders of magnitude for the same upper limit can be satisfactorily correlated with an approximate selection rule on the angular momentum whereby large changes in spin of the nucleus during a β disintegration are improbable.

Soon after the electron-neutrino theory was first presented, it was hoped that it would provide a field mechanism for the phenomenologically introduced nuclear interaction forces. This involves theory closely analogous to that whereby the Coulomb interaction of two charged particles may be regarded as taking place as a result of the coupling separately of each particle to the electromagnetic field. In the case of protons and neutrons, as discussed by Tamm and Iwanenko on the basis of the Fermi theory, the interaction arises from emission of a β particle and neutrino by the neutron which are absorbed by the proton. The net result is that the neutron becomes a proton and the proton a neutron, so the interaction provided by this mechanism is of exchange type. The calculations show, however, that the mechanism is inadequate to account for the observed forces by orders of magnitude—if the coupling of each heavy particle with the electron-neutrino field is made small enough to account for the observed β disintegration constants then it is much too small to provide the observed specifically nuclear interactions.

Likewise, it was thought that the electron-neutrino field might provide a mechanism for explaining the "anomalous" magnetic moments of proton and neutron. If a heavy particle like the proton spends a small part of its time as a neutron plus a positron and neutrino which are emitted and reabsorbed, then one can expect extra magnetic moment to arise in the time average because of the large magnetic moment of the positron. But these calculations are divergent just as in the case of the calculation of the self-energy of a charged particle which arises from its being coupled to the electromagnetic field. The proper handling of those divergent calculations in quantum electrodynamics and in the theory of the electron neutrino fields is a complete mystery which has baffled all attempts at resolution up to now. The solution of the mystery will probably call for an entirely new approach to these questions—at present, theory is up against a stone wall.

In the last couple of years, a number of suggestions for ways to provide a field theory for the nuclear forces have been actively studied. At present a good deal of interest attaches to a suggestion of Yukawa that the interaction results through a coupling of the heavy particles to a field of particles of intermediate mass—variously called heavy electrons, dynatons, barytrons, and Yukons. Yukawa postulated the existence of particles of intermediate mass (100 to 300 times electronic mass) of integral spin and obeying Einstein-Bose statistics to provide a field by means of which protons and neutrons could interact. He showed that in the non-relativistic limit for the heavy particles this would provide an effective interaction potential whose range of action was of the order $\hbar/\mu c$ where μ is the mass of the new particle. Since $\hbar/\mu c$ for the electron is $380 \cdot 10^{-13}$ cm. it is seen that we must postulate the new particle mass to be 100 to 300 times the electronic to get agreement with the observed range of nuclear forces.

Since the whole field theory is being devised solely to account for the nuclear interaction forces, one is at liberty to take the coupling parameter large enough to get the observed magnitude as well as the observed range. Of course, this theory implies the existence of a new kind of radio-activity whereby a proton in the nucleus emits a positive Yukon and becomes a neutron (or conversely a neutron in the nucleus emits a negative Yukon and becomes a proton). But this would require that the initial nucleus have a mass greater than the resulting isobar by an amount in excess of the mass of the emitted Yukon—as no such cases are known, the absence of this type of radio-activity is explained. Perhaps if we can produce nuclei that are far enough away from the minimum energy on an isobaric line, this kind of radio-activity will be observed.

As long as the Yukawa calculations served merely to describe nuclear forces by *ad hoc* adjustment of its parameters, it was not regarded as anything but an interesting possibility. But since the recognition last year of the existence in cosmic radiation of particles of intermediate mass[10] it has attracted a good deal of attention for the possibility of correlating data from different experiments is thereby introduced.

It should be pointed out that the relativistic quantum mechanics always brings in automatically a particle of opposite charge—as in the theory of the positron and electron, so there is no need to postulate the existence of both kinds of Yukon as regards charge—if one is postulated you get the other one at no extra expense! But if one is to account for nuclear interaction of like particles by such a field, it is necessary to postulate also the existence of uncharged Yukons. The charged Yukons can give interaction between heavy particles of the same kind, but does so in a higher order calculation. This makes the apparent equality of proton-proton and proton-neutron interactions hard to understand. It seems better to assume in addition the existence of neutral Yukons whose coupling with the heavy particles is essentially the same as for the charged Yukons.

Finally, it should be mentioned that theories of this kind involve a careful reconsideration of our views of the interaction of the heavy particles with the electromagnetic field.[11] This comes about because the proton, at least, is not only directly coupled to the electromagnetic field, but both proton and neutron are indirectly coupled to it by means of the coupling of the charged Yukons to the electromagnetic field.

---

[10] Neddermeyer and Anderson, *Phys. Rev.*, **51**, 884 (1937); Street and Stevenson, *Phys. Rev.*, **52**, 1003 (1937); Corson and Brode, Phys. Rev., **53**, 215 (1938).
[11] Lamb and Schiff, *Phys. Rev.*, **53**, 651 (1938).

Note on the External Photoelectric Effect of Semi-Conductors

Westinghouse Reseach Laboratories
Pittsburgh, Pennsylvania

An analysis of the effect of contact potentials on photoelectric measurements with semi-conductors is made, which indicates a new method of determining the width of the forbidden energy interval for electrons in a semi-conductor. A possible application to precision determination of h/e by the photoelectric effect is indicated.

The usual textbook and lecture account of Einstein's photoelectric equation does not state the case very carefully in regard to the effect of contact potentials on the actual experimental arrangements. The situation is correctly presented in Hughes and DuBridge, *Photoelectric Phenomena*, p. 22, where attention is called to some anomalous results obtained with copper oxide in Millikan's early work, which Millikan described as "spurious" contact potential differences. The point of this note is to show that these "spurious" contact potential differences have a natural explanation in terms of current theories of semi-conductors and that accurate measurement of them can give information of importance for our knowledge of semi-conductors.

For the reader's convenience, the argument concerning the stopping potentials for photoelectrons will be repeated here in its usual form. Fig. 1 shows the experimental setup with light of frequency v falling on the photoelectrically sensitive plate A, and potential V adjustable until none of the electrons emitted from A are able to reach the collector C. Then V is said to be the stopping potential for this combination of electrodes and this frequency of light. Of course, as is well known from the recent careful researches of DuBridge, Houston, and others,[1] there is no sharply defined value of V at which the current abruptly cuts off, but rather it tails off as an exponential function of Ve/kT. This is as good as a sharp cut-off if we ignore quantities of the order kT (1/40 electron volt at room temperature) which will be done in the first part of the discussion.

Figure 2 is a schematic energy level diagram showing the potential energy

---

[1] DuBridge, *New Theories of the Photo-electric Effect*, Actualités scientifiques et industrielles, 268 (Herman et Cie., Paris, 1935): Houston, Phys. Rev. **52**, 1047 (1937).

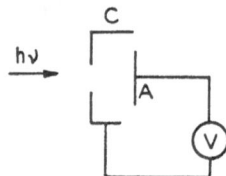

FIGURE 1. Experimental arrangement.

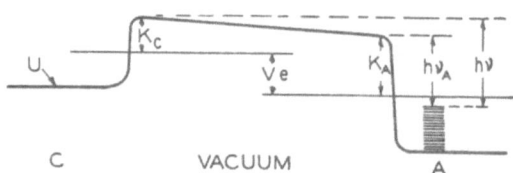

FIGURE 2. Schematic energy level diagram.

of an electron when in A, in C, and in the space behveen. The curve, potential energy, labeled U, rises abruptly by 10 or 20 electron volts in a distance of $10^{-\varepsilon}$ cm or so at the surface of each conductor and is approximately constant over the macroscopic distance of separation of the two conductors. For each conductor there is drawn in Fig. 2 a horizontal line at the level corresponding to the thermodynamic potential per electron, $u - Ts + pv$, of the electrons in that conductor. If the two conductors are allowed to come into thermo-dynamic equilibrium by mutual exchange of charges, they will take up total charges that will set up a field such that the thermodynamic potential of the electrons in each metal is the same. However, if they only exchange charges rapidly through the agency of an external source of electromotive force located between them as at V in Fig. 1, then a quasi-equilibrium state is set up in which the thermodynamic potentials of the electrons in the two conductors are related as in Fig. 2.

In the theory of thermionic emission, it is shown that the thermionic work function is equal to the difference between potential energy of an electron just outside the emitter and the thermodynamic potential per electron of an electron inside the emitter, so K_A and K_C are the respective work functions in Fig. 2. In considering the photoelectric emission of electrons from A, the usual Einstein relation is used, which says that in the direct act of absorption of a quantum of light, all the energy hv of that quantum is directly given to the single electron, which makes a transition in the absorption act. This means that the most energetic electrons emitted have an energy hv greater than that of the most energetic electrons in the emitter that are able to absorb light. Ignoring quantities of the order kT we may say that the most energetic electrons in A are those whose energy is hv_A less than the potential energy of a free electron at rest just outside of A. Then if V has been so adjusted that it just stops the fastest electrons from reaching the collector C, we shall have a situation in which the level that is $h\nu$ above the limiting energy of electrons in A will be equal to the energy of an electron at rest just outside of C, that is, calling this stopping potential V_A,

$$eV_A + K_C = (K_A - h\nu_A) + h\nu. \tag{1}$$

If arrangements are made to substitute an emitter B in place of A without disturbing the properties of C, then for its stopping potential V_B, the analogous equation holds

$$eV_B + K_C = (K_B - h\nu_B) + h\nu. \tag{2}$$

Suppose we measure V_A and V_B for the same light frequency. Subtracting the two equations, we have

$$e(V_A - V_B) = (K_A - h\nu_A) - (K_B - h\nu_B), \tag{3}$$

which gives, on the right, the theoretical interpretation of the experimentally observed quantity on the left.

The experiments[2] of Millikan and also of Kadesch and Hennings show that when various metals are used for the different emitters A, B, the quantity $V_A - V_B$ is zero. From this we conclude that the difference $(K_A - h\nu_A)$ is the same for all metals tested which is quite in accord with modern theory of metals. In the modern theory[3] the thermodynamic potential appears in the Fermi distribution function factor

$$(e^{(E-K_A)/kT} + 1)^{-1},$$

which gives the probability of occupation of an allowed state by an electron if the allowed state is of energy E. This factor changes rapidly from unity down to zero over a region of the width kT at the place where $E = K_A$. Hence the theory predicts that $(K_A - h\nu_A)$ is the same for all metals and is actually equal to zero. The stopping potential measurements verify the first part of this statement and are consistent with the second.

However, the experiments of Millikan showed that when one of the pair of emitters was a metal, say B, and the other, say A, is copper oxide, a semi-conductor, then $V_A - V_B$ is not zero, but is of the order of one volt. This effect was described by Millikan by saying that copper oxide showed a *spurious* contact potential difference as contrasted with the *true* contact potential difference shown by metals. Attempts have been made to find an origin of the spurious contact potential difference in terms of grease films, etc.

It is clear that the origin of the so-called spurious potential difference is an intrinsic property of the semi-conductor itself. If B is a metal, then $K_B - h\nu_B$

[2] Millikan, Phys. Rev. 7, 18, 355 (1916); **18**, 236 (1921); Kadesch and Hennings, Phys. Rev. **8**, 221 (1916).
[3] See for example, Sommerfeld and Bethe, *Handbuch der Physik*, Vol. 24/2 (Springer, Berlin, 1933), p. 342, or Fowler, *Statistical Mechanics* (Cambridge University Press, 1936), Chap. XI.

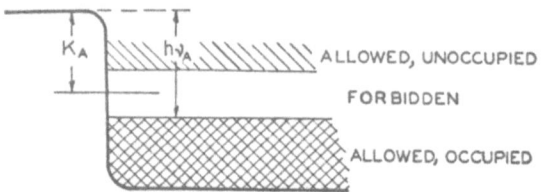

FIGURE 3. Energy level
diagram for semi-conductor.

is actually zero and the experimentally measured quantity $(V_A - V_B)$ is a direct measure of the quantity $(K_A - h\nu_A)$ for the semi-conductor. This interpretation is also in accord with the modern theories of semi-conductors as developed principally by A.H. Wilson[4] and presented in Fowler's *Statistical Mechanics*. In a semi-conductor, it will be recalled, we have a band of allowed energy states which at absolute zero is completely filled. Above it is a forbidden range of energy, then one or more bands of allowed levels which are normally not occupied. In the case of an impurity semi-conductor, there are some extra levels associated with wave functions localized around the impurities or lattice imperfections in a manner that has often been discussed in the literature of semi-conduction. At any finite temperature there will be a very few electrons thermally excited into the upper band with a corresponding small deficiency of electrons from the otherwise filled band. The thermodynamic potential K_A has to take up such a value that the number of electrons thermally excited to the "empty" band is equal to the number missing from the "filled" band. Under simple, plausible assumptions, this makes K_A have a value at the center of the interval between the "filled" and "empty" bands. (See Fig. 3.) This neglects impurity levels and supposes the effective mass of electrons in each band to be the same. If these conditions are not fulfilled, then K_A will not be exactly in the middle of the forbidden interval, but in any case its location can be calculated in terms of basic features of the band structure and the location and number of impurity levels.

As to photoemission from the semi-conductor, the electrons of highest energy will result from the absorption of a quantum $h\nu$ by the very few electrons in the "empty" band. But these will be so few in number that it may be expected that they will have escaped detection in experiments where no special effort was made to find them. Then the highest energy electrons which appear in abundance will be those which come from the top of the "filled" band. Hence $(K_A - h\nu_A)$ will theoretically be not zero, and will in fact, be equal to one-half the width of the forbidden energy range in the simpler cases. This is the interpretation which is offered for Millikan's observations on copper oxide.

The width of the forbidden energy range also plays a role in the temperature dependence of the bulk conductivity of the semi-conductor on Wilson's

---

[4] A.H. Wilson, Proc. Roy. Soc. **A133**, 458 (1931); **A134**, 277 (1932).

theory. While the existing data do not permit of any precision test of whether the photoelectric effect and the temperature dependence of conductivity give the same results, still both methods are in agreement for copper oxide in giving about two volts for the width of the forbidden range.[5]

Another interesting consequence of these ideas was pointed out to me by Dr. W.H. Wells, in a conversation at these Laboratories. He notes that the sharpness of cut-off of the photoelectron current with gradual increase of the retarding potential V ought to be much sharper in the case of a semi-conductor than in the case of a metal. This fact is of considerable interest in itself in view of the large amount of work that has gone into studying the details of the cut-off for metals.[1] It also seems to have an important bearing on experiments such as those of Houston and his associates[6] who are making precision determinations of h/e by measuring the change of stopping potential with change of light frequency. Difficulty of exact definition of what is meant by stopping potential is one of the most troublesome features of such experiments, and it seems that the difficulty would be obviated by using a semi-conductor instead of a metal as the photoemitter.

It appears, therefore, that the ideas here presented should provide a useful new approach to the study of semi-conductors and afford a means of unraveling the hitherto chaotic field of photoemission by compounds. They also suggest an improved way of determining h/e from the photoelectric effect. Experiments from the point of view of application to semi-conductors are being planned in these Laboratories.

---

[5] Gudden, Ergebn. d. exakt. Naturwiss. **13**, 223 (1934).
[6] Overhage, Phys. Rev. **52**, 1039 (1937). Also unpublished work reported by Houston at a colloquium during the Physics of Metals summer session of the University of Pittsburgh, 1938.

Electronic Generation of Electromagnetic Oscillations

EDWARD U. CONDON

Westinghouse Reseach Laboratories
Pittsburgh, Pennsylvania

A general formulation of the problem of transfer of power from the electron beam to an oscillating circuit by moving charged particles is given. The method is applied specifically to calculation of the behavior of the klystron oscillator in making approximations analogous to those made by Webster in his treatment of the same problem.

The general problem of the electronic generation of electromagnetic oscillations is that of the construction of a system of conductors with or without auxiliary constant magnetic fields such that if electrons are emitted from one conductor, called the cathode, they are given energy by a source of steady electromotive force and in the course of their motion to one or more collecting conductors called anodes, they deliver some of this energy to oscillations of the electromagnetic field.

In the following discussion it will be supposed that the density of charge everywhere outside the conductors is small enough that space charge is negligible. Also, that whenever an electron strikes a conductor it sticks to it without emission of secondary electrons, its energy of motion at the moment of impact being dissipated as heat at the anode which it strikes. The method used here is familiar to students of quantum theory[1] but its use in this connection is believed to be new.

Neglecting the effect of the presence of the electrons, the electromagnetic field in the region where the electrons move can be represented as the super-position of the fields associated with normal modes of oscillation. The field is governed by Maxwell's equations

$$\operatorname{div} E = 0, \qquad \operatorname{div} H = 0,$$

$$\operatorname{curl} E = -\frac{1}{c}\frac{\partial H}{\partial t}, \qquad \operatorname{curl} H = \frac{1}{c}\frac{\partial E}{\partial t},$$

(1)

where E is in electrostatic units (300 volt/cm) and H in gauss.

---

[1] See e.g., W. Heitler, *The Quantum Theory of Radiation* (Oxford, 1936).

These fields can be derived from a vector potential A by the relations

$$H = \text{curl } A, \qquad E = -\frac{1}{c}\frac{\partial A}{\partial t}. \tag{2}$$

The equation div $H = 0$ is then satisfied automatically, the equation div $E = 0$ requires that we impose on A the condition

$$\text{div } A = 0.$$

Substituting in the first curl equation

$$-\frac{1}{c}\frac{\partial}{\partial t}\text{curl } A = -\frac{1}{c}\frac{\partial}{\partial t}\text{curl } A$$

and see that it is satisfied identically. The second curl equation gives

$$\text{curl curl } A = -\frac{1}{c^2}\frac{\partial^2 A}{\partial t^2}$$

or

$$\nabla^2 A = \frac{1}{c^2}\frac{\partial^2 A}{\partial t^2}.$$

Neglecting the finiteness of the conductivity of the conductors we must have E normal to the bounding surfaces which implies A normal to boundary surfaces. Hence the conditions on A are

$$\text{div } A = 0, \qquad \begin{array}{c} A \text{ normal at} \\ \text{bounding} \\ \text{surfaces,} \end{array} \qquad \nabla^2 A = \frac{1}{c^2}\frac{\partial^2 A}{\partial t^2}. \tag{3}$$

These equations determine a set of proper frequencies and normal modes for the field. If v_λ is a proper frequency, then $A_\lambda(r,t) = A_\lambda(r)e^{2\pi i v_\lambda t}$ and the spatial dependence of A is determined by

$$\text{div } A_\lambda = 0, \qquad \begin{array}{c} A_\lambda \text{ normal at} \\ \text{boundaries,} \end{array} \qquad \nabla^2 A_\lambda + \left(\frac{2\pi v_\lambda}{c}\right)^2 A_\lambda = 0.$$

The normal fields $A_\lambda(r)$ have the orthogonal property

$$\int A_\lambda(r) \cdot A_\mu(r)\, d\tau = \delta_{\lambda\mu} = \begin{cases} 0 & \lambda \neq \mu \\ 1 & \lambda = \mu \end{cases} \tag{4}$$

if normalized to unit amplitude.[2] Here it is assumed that v_λ and v_μ are unequal. For the degenerate case in which v_λ and v_μ are equal, it is always possible to choose normal fields that are orthogonal, so it will be supposed in what follows that the normal fields are so chosen.

---

[2] Frenkel, *Elektrodynamik*, Vol. 2, p. 320.

The general oscillating field determined by the conductors is then a super-position of these normal fields

$$A = \sum_\lambda q_\lambda(t) A_\lambda(r) \tag{5}$$

and the state of the field is completely specified by giving the $q_\lambda(t)$ which will be called the coordinates of the field.

The equation of motion of an electron moving in an electrostatic field described by the static potential φ and the static vector potential A_0 is

$$m\frac{d^2r}{dt^2} = c\left(E + \frac{1}{c}v \times H\right), \tag{6}$$

where the E and H include the field due to the coordinates q_λ and also the field due to the static potentials φ and A_0 and are to be evaluated at the position of the electron. The coupling of the electron's motion to the field coordinates thus comes from the fact that E and H here depend both on the field coordinates and on the electron's coordinates.

The vector potential must satisfy an altered equation of motion when an electron is present. For then we have

$$\operatorname{curl} H = \frac{4\pi\rho v}{c} + \frac{1}{c}\frac{\partial E}{\partial t}, \tag{7}$$

which becomes

$$\operatorname{curl} \operatorname{curl} A + \frac{1}{c^2}\frac{\partial^2 A}{\partial t^2} = \frac{4\pi\rho v}{c}.$$

Inserting

$$A = \sum_\lambda q_\lambda A_\lambda(r),$$

we get

$$\sum_\lambda \left[-\nabla^2 A_\lambda q_\lambda + \frac{\ddot{q}_\lambda}{c^2}A_\lambda\right] = \frac{4\pi\rho v}{c}.$$

Multiplying by A_μ and integrating over all space outside conductors, using the equation

$$\nabla^2 A_\lambda = -\left(\frac{2\pi\nu_\lambda}{c}\right)^2 A_\lambda$$

and the normal-orthogonal property (4) of the A_λ, we get

$$\frac{1}{c^2}\ddot{q}_\mu + \left(\frac{2\pi\nu_\mu}{c}\right)^2 q_\mu = \frac{4\pi}{c}\int \rho v \cdot A_\mu \, d\tau. \tag{8}$$

In the applications we can neglect the variation of A_μ with position over the small volume of the electron and write

$$\int \rho v \cdot A_\mu \, d\tau = ev \cdot A_\mu(r),$$

where e is the charge on the electron and $A_\mu(r)$ is to be evaluated at the electron's position. Thus the equation of motion for any field coordinate becomes

$$\ddot{q}_\lambda + (2\pi v_\lambda)^2 q_\lambda = 4\pi ecv \cdot A_\lambda(r). \tag{9}$$

We will now see that the rate of change of the electron's kinetic plus potential energy due to its coupling with the field is just equal to the rate of change of the field's energy. Multiply the electron's equation of motion through by v

$$mv \cdot \frac{dv}{dt} = -ev \cdot \operatorname{grad} \varphi - \frac{e}{c} \sum_\lambda \dot{q}_\lambda(t) v \cdot A_\lambda(r)$$

or

$$\frac{d}{dt}(\tfrac{1}{2}mv^2 + e\varphi) = -\frac{e}{c} \sum_\lambda \dot{q}_\lambda(t) v \cdot A_\lambda(r).$$

The terms coming from the magnetic field vanish in the energy equation since $v \cdot (v \times H) = 0$. To complete the proof, we need the expression for the field energy in terms of the field coordinates. It is

$$U = \frac{1}{8\pi}(E^2 + H^2)\,d\tau = \frac{1}{8\pi} \int \frac{1}{c^2}(\dot{q}_\lambda A_\lambda) \cdot (\dot{q}_\mu A_\mu)\,d\tau$$

$$+ \frac{1}{8\pi} \int (q_\lambda \operatorname{curl} A_\lambda) \cdot (q_\mu \operatorname{curl} A_\mu)\,d\tau.$$

The electric terms reduce to

$$\frac{1}{8\pi c^2} \sum_\lambda \dot{q}_\lambda{}^2$$

by direct use of the normal orthogonal property of the A's. The magnetic terms can be reduced by using the formula

$$\int (\operatorname{curl} a \cdot \operatorname{curl} b + \operatorname{div} a \operatorname{div} b + a\nabla^2 b)\,d\tau = \int [(a \times \operatorname{curl} b)_n + a_n \operatorname{div} b]\,d\sigma,$$

where the second integral is over the bounding surfaces. If $a = A_\lambda, b = A_\mu$, we have

$$\int \left[\operatorname{curl} A_\lambda \cdot \operatorname{curl} A_\mu - \left(\frac{2\pi v_\mu}{c}\right)^2 A_\lambda \cdot A_\mu \right] d\tau = 0,$$

since the surface integral vanishes. Hence

$$\int \operatorname{curl} A_\lambda \cdot \operatorname{curl} A_\mu \, d\tau = \left(\frac{2\pi v_\mu}{c}\right)^2 \delta_{\lambda\mu}$$

and the magnetic energy terms reduce to

$$\frac{1}{8\pi} \sum_\lambda \left(\frac{2\pi v_\lambda}{c}\right)^2 q_\lambda{}^2,$$

so the whole field energy becomes

$$U = \sum_\lambda U_\lambda$$

where

$$U_\lambda = \frac{1}{8\pi c^2}[\dot{q}_\lambda{}^2 + (2\pi v_\lambda)^2 q_\lambda{}^2]. \tag{10}$$

We can get an equation for dU_λ/dt from the equation of motion of the field coordinate q_λ:

$$\frac{d}{dt}\frac{1}{2}[\dot{q}_\lambda{}^2 + (2\pi v_\lambda)^2 q_\lambda{}^2] = 4\pi e c \dot{q}_\lambda v \cdot A_\lambda(r).$$

So, dividing by $4\pi c^2$

$$\frac{d}{dt}U_\lambda = \frac{e}{c}\dot{q}_\lambda v \cdot A_\lambda(r) = -ev \cdot E_\lambda(r). \tag{11}$$

This is exactly the negative of the rate of increase of the electron's kinetic plus potential energy, when summed over all the field coordinates.

In the steady operation of an oscillation generator converting d.c. power into electromagnetic oscillations, we will suppose that electrons are emitted at a steady rate, and that there is a load applied to the oscillating power such that the value of the amplitude of q_λ remains constant. Our problem is, then, to calculate the mean value of dU_λ/dt, the power flow from electron to field, averaged over a complete cycle of phase of motion of the electron and over a complete family of trajectories. This gives the useful power delivered by the generator. Likewise, if we calculate the energy with which the electron reaches an anode and average this over all phases of the motion, we will find the power wasted as heat at the anodes. Another source of power loss is that in the actual device, other field modes than the one to which the load is applied will be excited and, since they are all dissipative to some extent, this will represent an absorption of energy.

In what follows, we will assume these parasitic fields to be negligible in calculating the electron's trajectory.

As an illustration of the use of the formulation here presented, let us consider the operation of the "klystron"* recently developed by the group at Stanford University.[3] (During a recent visit to Stanford University I learned from

---

* Trademark registered by Sperry Gyroscope Company.
[3] R.H. Varian and S.F. Varian, J. App. Phys. **10**, 321 (1939); W.W. Hansen, J. App. Phys. **9**, 654 (1938).

FIGURE 1. Klystron recently
developed by the group at
Stanford University.

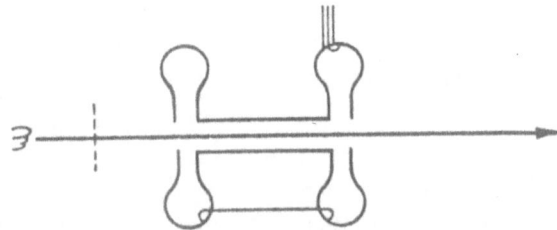

Professor Hansen that he also also been using the method presented here for
some time past.) Here the electronic motion is entirely one-dimensional.
Electrons are emitted at a uniform rate from a cathode-ray gun entering the
oscillating field at velocity v_0. In the work done at Stanford, the oscillating
field is that of two "rhumbatrons" electromagnetically coupled and of the form
shown in Fig. 1. Let us suppose that these or any other electrode structure are
provided, which makes the electric force on an electron be entirely in the x
direction and equal to

$$E_x = \frac{m}{e} F(x) \sin \omega t, \tag{12}$$

so the equation of motion of the electron is

$$\frac{d^2 x}{dt^2} = F(x) \sin \omega t \tag{13}$$

and the initial condition defining the motion $x(t, \delta)$ of a particular electron
entering at phase δ is

$$x\left(\frac{\delta}{\omega}, \delta\right) = 0,$$

$$\left[\frac{dx(t, \delta)}{dt}\right]_{i=\delta/\omega} = v_0. \tag{14}$$

Let it be supposed that $F(x)$ is zero beyond $x = a$ and that the field is weak
enough that every electron gets through the apparatus, i.e., that $dx/dt > 0$ for
all $t > \delta/\omega$ and for all δ. Then the velocity of the electron entering at phase δ
after it has gone through the apparatus will be $dx(t, \delta)/dt$ evaluated at a
sufficiently large value of t that it is out of the field and so moving with constant
velocity. Let us write $u(\delta)$ for this speed

$$u(\delta) = \frac{dx(t, \delta)}{dt} \qquad \text{for large } t. \tag{15}$$

Then the mean energy of motion of the electrons after going through the field is

$$\frac{m}{2} \cdot \frac{1}{2\pi} \int_0^{2\pi} [u(\delta)]^2 \, d\delta. \tag{16}$$

If this is less than $mv_0^2/2$ it means that the electrons on the average lose energy to the oscillating field and the efficiency of the device as an oscillation generator, neglecting losses in the conductors which are used to produce the field, is

$$\mathscr{E} = 1 - \frac{1}{2\pi} \int_0^{2\pi} \left(\frac{u(\delta)}{v_0}\right)^2 d\delta. \tag{17}$$

If the quantity (16) comes out greater than $\frac{1}{2}mv_0^2$ it means that the average electron has gained energy from the oscillating field. Instead of operating as an oscillation generator, the device then operates as a means of accelerating electrons with a high frequency field.

This formulation of the problem brings out clearly that the descriptive language which has been used about "bunching" the electrons is not an essential part of the operation of such a device. "Bunching" in fact does occur with practical beam currents, but the device would operate as an oscillation generator even if operated at such a low electron beam current that only one electron were actually in the oscillating field at a time.

More specifically, we may illustrate this method of handling the problem of the klystron using the same approximations as those used by Webster[4] in his method.

Suppose the electron path in the first rhumbatron is h_1 and the field there is $E_1 \sin \omega t$, counting E_1 as positive when it is in the sense to accelerate the electron. Let e be the magnitude of the electronic charge. Neglecting the phase change in the oscillation which occurs in the electron's flight through the rhumbatron, the energy of an electron after going through the first rhumbatron is

$$e(V + h_1 E_1 \sin \delta),$$

if V is the voltage used to give it its initial energy and δ the phase at which it enters. If d is the distance between the rhumbatrons, then the phase at which it arrives at the other rhumbatron is

$$\delta + \frac{\varphi}{(1 + \alpha \sin \delta)^{1/2}} \quad \text{where} \quad \alpha = \frac{h_1 E_1}{V} \quad \text{and} \quad \varphi = \frac{\omega d}{v_0}.$$

If the length of path in the second rhumbatron is h_2 and the field there is $E_2 \sin \omega t$, again counted positively in a direction to accelerate the electron, then the energy received by the electron in going through the second rhumbatron is

$$eh_2 E_2 \sin\left(\delta + \frac{\varphi}{(1 + \alpha \sin \delta)^{1/2}}\right).$$

The energy gained by an electron entering at phase δ is the sum of what it gains in each rhumbatron and the average energy is obtained by averaging

---

[4] D.L. Webster, J. App. Phys. **10**, 501 (1939).

over all values of δ. The average of $eh_1 E_1 \sin \delta$ is zero, indicating that in this approximation no power is consumed by the action of the first rhumbatron on the electron beam. The whole effect arises from the average action of the second rhumbatron. This calls for calculation of

$$I(\varphi, \alpha) = \frac{1}{2\pi} \int_0^{2\pi} \sin\left(\delta + \frac{\varphi}{(1 + \alpha \sin \delta)^{1/2}}\right) d\delta.$$

Now $\alpha < 1$ is required to keep the first rhumbatron from throwing back some of the electrons that enter it. We may make an approximate calculation by supposing $\alpha \ll 1$ so that the radical may be approximated by $1 - \alpha/2 \sin \delta$ giving

$$I(\varphi, \alpha) = \frac{1}{2\pi} \int_0^{2\pi} \sin(\delta + \varphi - \tfrac{1}{2}\alpha\varphi \sin \delta)\, d\delta$$

$$= \sin \varphi \cdot J_1(\tfrac{1}{2}\alpha\varphi),$$

where $J_1(x)$ is the Bessel function of order one usually denoted in this way.

In this approximation, therefore, the mean energy gained by an electron going through the apparatus is

$$eh_2 E_2 \sin \varphi \cdot J_1(\tfrac{1}{2}\alpha\varphi).$$

For generation of oscillations, this should come out negative either by making E_2 negative or by making $\sin \varphi$ negative. By choosing $\varphi = \pi/2$, $3\pi/2 \ldots$, etc. the $\sin \varphi$ factor can be made unity so that the efficiency of power conversion becomes

$$\mathscr{E} = \frac{h_2 |E_2|}{V} J_1(\tfrac{1}{2}\alpha\varphi).$$

The maximum value of the Bessel function is 0.58, and this value is attained when its argument is equal to 1.84. The factor $h_2 |E_2|/V$ can be made quite close to unity but if it is made too large the formula will not be valid because some of the electrons would not go through the second rhumbatron. It also has to be remembered that the formula is not valid except for $\alpha \ll 1$. Some relevant values of the Bessel function are those given in Table I. This indicates that efficiencies of the order of 50 percent might be attainable, perhaps even higher if the errors introduced by the approximations are in the direction to underestimate the efficiency.

TABLE I

α	$\varphi = \pi/2$	$\varphi = 3\pi/2$
0.1	0.040	0.117
.2	.077	.230
.4	.155	.421
.6	.230	.544

Forced Oscillations in Cavity Resonators

Edward U. Condon

Westinghouse Reseach Laboratories
Pittsburgh, Pennsylvania

Formulas are developed for calculation of the impedance of a cavity resonator when excited by a coupling loop or by a capacitative coupling.

There is a rapidly growing appreciation of the importance of cavity resonators in ultra-high frequency work especially where the frequencies are of the order of 300 megacycles/sec. or greater. The theory for the resonant modes for simple shapes is well known from the classical work[1] and has recently been restated in convenient form for practical use.[2] This paper discusses the coupling of a transmission line to a cavity resonator.[3]

General Ideas

The cavity resonator is assumed to be bounded by highly conducting walls, so the Q values of the various resonant modes are large. The resonator will have a number of resonant frequencies, v_a, associated with each of which there is a standing wave field derivable from the vector potential, $\mathbf{A}_a$ by the formulas

$$\mathbf{E}_a = -\frac{1}{c}\frac{\partial \mathbf{A}_a}{\partial t}, \qquad \mathbf{H}_a = \operatorname{curl} \mathbf{A}_a.$$

It is convenient to normalize the $\mathbf{A}$'s in such a way that they are dimensionless,

$$\int \mathbf{A}_a(\mathbf{r}) \cdot \mathbf{A}_b(\mathbf{r})\, dV = \begin{cases} 0 & a \neq b \\ V & a = b. \end{cases} \tag{1}$$

---

[1] H. Bateman, *Electrical and Optical Wave Motion* (Cambridge University Press, 1915).
[2] W.W. Hansen, J. App. Phys. **9**, 654 (1938); W.W. Hansen and R.D. Richtmyer, J. App. Phys. **10**, 189 (1939).
[3] Compare as to method, E.U. Condon, J. App. Phys. **11**, 502 (1940).

An arbitrary electromagnetic field inside the resonator which satisfies the boundary conditions can be expanded in terms of the resonant wave patterns, $A_a(r)$,

$$A(r, t) = \sum_a p_a(t) A_a(r). \qquad (2)$$

Suppose we are given the time and space distribution of current density $i(r, t)$ inside the resonator and wish to calculate what resonant modes are excited. First we assume that i can be expanded in terms of the $A_a(r)$ functions

$$i(r, t) = \sum_a I_a(t) A_a(r). \qquad (3)$$

The coefficients $I_a(t)$ are easily found formally in view of (1)

$$I_a(t) = \frac{1}{V} \int i(r, t) \cdot A_a(r) \, dV. \qquad (4)$$

The equation for the $p_a(t)$ is found by substituting in the Maxwell equation

$$\text{curl } H - \frac{1}{c} \dot{E} = 4\pi i, \qquad (5)$$

which leads at once to the following equations, one for each resonant mode,

$$\ddot{p}_a + (2\pi v_a)^2 p_a = 4\pi c^2 I_a(t). \qquad (6)$$

The theory leading to this equation is approximate in that it neglects damping. If damping is small and is measured by the quantity Q_a in the usual way, then (6) can be replaced by

$$\ddot{p}_a + \frac{2\pi v_a}{Q_a} \dot{p}_a + (2\pi v_a)^2 p_a = 4\pi c^2 I_a(t). \qquad (7)$$

This equation permits the calculation of p_a, the amplitude of excitation of the ath mode. We see that $I_a(t)$ given by (4) measures the effectiveness of the current distribution in exciting the ath mode.

If, as is usual in practice, the current i is confined to a wire of small cross section with $I(t)$ being the total current in the wire, then

$$I_a(t) = \frac{1}{V} I(t) \int A_a \cdot ds. \qquad (8)$$

By Stokes' theorem

$$\int A_a \cdot ds = \iint \text{curl } A_a \cdot dS = M_a. \qquad (9)$$

The quantity M_a is the flux through the current-carrying loop caused by unit amplitude of excitation of the ath mode. It is appropriate, therefore, to call it the mutual inductance of the loop with this mode.

Impedance of Coupling Loop

If the wire loop in which the current flows has resistance R, then the electromotive force needed when current I flows is given by

$$\mathscr{E} = RI + \sum_a M_a \dot{p}_a, \tag{10}$$

where p_a is given in terms of I by

$$\ddot{p}_a + \frac{2\pi v_a}{Q_a} \dot{p}_a + (2\pi v_a)^2 p_a = \frac{4\pi c^2}{V} M_a I(t). \tag{11}$$

Suppose the currents and voltages have a time dependence of frequency v. Writing $\mathscr{E}$, I, p_a, etc. for the amplitudes of $e^{2\pi i v t}$, the equations for the amplitudes are

$$\mathscr{E} = RI + \sum_a 2\pi i v M_a p_a,$$

$$\frac{\pi V}{c^2}\left(v_a{}^2 - v^2 + i\frac{v v_a}{Q_a}\right) p_a = M_a I$$

and therefore, eliminating the p_a between these equations one finds for the effective impedance of the coupling loop

$$Z = R + \sum_a \frac{2\pi i v M_a{}^2}{\dfrac{\pi V}{c^2}\left(v_a{}^2 - v^2 + i\dfrac{v v_a}{Q_a}\right)}. \tag{12}$$

The important thing to notice about (10) and hence about (12) is that there is no self-inductance term. The effects of self-inductance are contained in (12) since the entire field inside the resonator is represented in terms of the p_a. At low frequencies such that $v \ll v_a$, the second term in (12) simplifies into

$$2\pi i v \sum_a \frac{\lambda_a{}^2 M_a{}^2}{\pi V},$$

which is the inductive reactance due to an effective self-inductance of the loop as modified by the presence of the walls.

If the driving frequency is exactly equal to the resonant frequency of one particular mode, $v = v_a$ then that particular term in the sum in (12) will be large compared to all others. The resonant amplitude of the ath mode is

$$(p_a)_{\text{resonant}} = -iQ_a \frac{\lambda_a{}^2 M_a}{\pi V} I, \tag{13}$$

where p_a lags $\pi/2$ behind I, as shown by the factor $-i$.

Example

As a specific example, consider the simplest mode of oscillation of a copper cube whose edge is $A = 20$ cm. The frequency is $v_a = 1.06 \cdot 10^9$ cycle/sec. and $Q_a = 33,000$. The normalized vector potential is

$$A_{ax} = 4 \sin(\pi y/A) \sin(\pi z/A),$$

$$A_{ay} = 0, \qquad A_{az} = 0,$$

giving for the magnetic field at unit amplitude

$$\operatorname{curl} A_a = \frac{4\pi}{A}\left[j \sin\frac{\pi y}{A}\cos\frac{\pi z}{A} - k\cos\frac{\pi y}{A}\sin\frac{\pi z}{A} \right].$$

Suppose the coupling loop is a single turn of wire whose area is 1 cm², located near the point on the wall whose (x, y, z) coordinates are $(A/2, 0, A/2)$. Near this point the magnetic field is in the k direction, and has a magnitude $4\pi/A$. Hence if the plane of the loop is normal to the z axis, the mutual inductance M_a is

$$M_a = 4\pi/A = 0.628 \text{ cm}.$$

From (12) we see that at resonance the coupling of the ath mode contributes a resistive term to the loop's impedance, whose magnitude is

$$R_a = 2v_a Q_a \frac{\lambda_a{}^2 M_a{}^2}{V}.$$

In (12), $\mathscr{E}$ and I are in electromagnetic units, so to get R_a in ohms, divide by 10^9. Substituting for this example, we get

$$R_a = 2760 \text{ ohms}.$$

The resonant amplitude p_a of this mode is

$$\frac{p_a}{I} = \frac{R_a}{2\pi v_a M_a} = 660,$$

where both p_a and I are in the same units.

The electric vector along the line, $y = z = A/2$ is

$$E = \frac{2\pi}{\lambda} \cdot 4p_a \text{ stat. volt/cm},$$

so the line integral of E from $x = 0$ to $x = A$ is

$$V_B = EA = \frac{8\pi}{\sqrt{2}} p_a \text{ stat. volts}.$$

The voltage drop in the loop is $R_a I$, so the voltage multiplication is

$$\frac{V_B}{V_L} = \frac{8\pi}{\sqrt{2}} \cdot 300 \cdot \frac{p}{R_a I} = 127.5.$$

For example, if an r.m.s. voltage of 1000 is applied to the loop, the loop current will be 0.36 ampere and the power consumed is 360 watts. The resonant amplitude $p_a = 23.8$ abamperes, so the magnetic field amplitude at the loop is 14.9 gauss. The electric vector along the center line is 6.33 kv/cm, giving a total voltage amplitude along the center line of 127 kv.

Capacitative Coupling

In the preceding, we have assumed the charge to be zero everywhere inside the resonator. The more general case is that in which there is a charge density ρ as well as a current i in the cavity. In practice this might be done by inserting a short wire through a hole in the wall; the driving electromotive force is then applied between the wire and the wall.

With charge in the cavity it is necessary to introduce a scalar potential φ, as well as the vector potential. The fields are given by

$$\mathbf{H} = \text{curl}\,\mathbf{A}, \qquad \mathbf{E} = -\frac{1}{c}\frac{\partial \mathbf{A}}{\partial t} - \text{grad}\,\varphi. \tag{14}$$

Substitution in Maxwell's equations gives

$$-\frac{1}{c}\frac{\partial}{\partial t}\,\text{div}\,\mathbf{A} - \nabla^2\varphi = 4\pi\rho,$$

$$\text{curl curl}\,\mathbf{A} + \frac{1}{c^2}\frac{\partial^2 \mathbf{A}}{\partial t^2} = 4\pi\left(\mathbf{i} - \frac{1}{4\pi c}\frac{\partial}{\partial t}\,\text{grad}\,\varphi\right). \tag{15}$$

One may still use the same vector potential solutions as before. For these div $\mathbf{A} = 0$, hence the first equation of (15) becomes

$$-\nabla^2\varphi = 4\pi\rho.$$

Therefore, at each instant φ is to be determined from the charge distribution exactly as if ρ were independent of the time. The boundary condition is $\varphi = 0$ on the walls.

The equation for $\mathbf{A}$ in (15) is of the same form as before, except that the displacement current associated with φ has to be included in the complete expression for i. When one calculates $I_a(t)$ by (4),

$$I_a(t) = \frac{1}{V}\int\left(\mathbf{i} - \frac{1}{4\pi c}\frac{\partial}{\partial t}\,\text{grad}\,\varphi\right)\cdot \mathbf{A}_a\,dV,$$

it turns out that the term arising from the displacement current vanishes, so the expression for $I_a(t)$ is actually the same as before.

If the current is really confined to a wire which ends in an internal electrode, and I is the total current, then the total rate at which work is done on the field by the current is

$$-c \int \mathbf{i} \cdot \mathbf{E} dV = I \sum_a M_a \dot{p}_a + c I \varphi_0 \tag{16}$$

where φ_0 is the potential of the internal electrode. If C is the capacity of this electrode relative to the walls, reckoned electrostatically, then $\varphi_0 = q/C$ where $I = \dot{q}$ so the electromotive force which must be applied between the wire and the wall is

$$\mathscr{E} = c\frac{q}{C} + \sum_a M_a \dot{p}_a. \tag{17}$$

Here $\mathscr{E}$ is in electromagnetic units, while q and C are in electrostatic units. Applying (17) to a steady state with frequency v, we find for the impedance of a capacitative coupling

$$Z = \frac{c^2}{2\pi i v C} + \sum_a \frac{2\pi i v M_a^2}{\frac{\pi V}{c^2}\left(v_a^2 - v^2 + i\frac{v v_a}{Q_a}\right)}. \tag{18}$$

Here C is the capacity calculated electrostatically and M_a is to be calculated from (9) on the lead-in wire. The excitation of the resonant modes is entirely caused by coupling with the lead-in wire, the capacity of the internal electrode merely makes it possible to put current into the lead-in wire.

To give a specific example, suppose a fine wire enters the center of the base of the same copper box that was considered in the previous section. On its end 1 cm above the base, suppose there is a ball 0.1 cm in radius. The electrostatic capacity to a good approximation is

$$C = 0.105 \text{ cm.}$$

The coupling of the wire to the mode considered before is

$$M_a = \int \mathbf{A}_a \cdot d\mathbf{s} = 4 \text{ cm.}$$

Hence the input impedance in ohms at resonance, $v = v_a$,

$$Z = 10^{-9}\left[\frac{c^2}{2\pi i v C} + \frac{2 v_a Q_a \lambda_a^2 M_a^2}{V}\right]$$

$$= 112{,}000 - 1280i.$$

The resonant frequency with capacitative coupling is not exactly $v = v_a$; to get vanishing reactance it is necessary to choose v enough different from v_a to balance out the capacitative reactance of the condenser formed by inner electrode and resonator walls.

Principles of Micro-Wave Radio[*]

Edward U. Condon

*Westinghouse Research Laboratories
Pittsburgh, Pennsylvania*

Introduction

By micro-wave radio is meant the science of electromagnetic radiations in, roughly, the range of wave-lengths from one meter down to one millimeter, that is, of frequencies in the approximate range 3×10^8 to 3×10^{11} cycle/sec. This region of the spectrum is marked off at its high frequency end by the fact that at higher frequencies the techniques become more "optical" than "electrical." At the lower end it is marked off by the fact that for frequencies below 300 megacycles/sec., the conventional methods of radio engineering based on lumped constant circuit analysis are quite adequate for understanding the phenomena.

The micro-wave field is thus principally characterized by these three features:

(1) Its techniques are essentially electrical rather than optical, particularly in the sense that the sources are man-made oscillators built on a macroscopic scale, rather than the non-coherent superposition of radiations from a large number of atoms or molecules.

(2) The apparatus employed is always at least comparable in size with the wave-length and usually large compared with the wave-length. This fact invalidates, or at least greatly complicates, any attempt to understand the phenomena with the aid of conventional circuit analysis or even usual distributed parameter transmission line theory. It looks as if the engineers will

---

[*] *Prefatory Note*—The following material is presented, in some respects, particularly in the inclusion of exercises for the reader, more in the style of a textbook than of a review article. That is because it was originally intended for publication as a textbook. Decision to publish it as a paper in the *Reviews of Modern Physics* was based on the fact that pressure of war work is likely greatly to delay completion of the manuscript, and on the fact that it appears desirable to give the completed portion of the manuscript wide circulation now. These circumstances also account for the fact that the bibliographic notes are not as complete as they should be. Nevertheless, it is felt that they afford a reasonably adequate guide to the literature. Plans are made for concluding chapters in a later issue of this journal.

at last really have to learn electrical field theory! At any rate, so far there exists no *technique of evasion* of the use of field theory that corresponds to the use of complex number algebra to avoid the consideration of differential equations in analysis of steady state alternating current circuit problems. No doubt something of this sort will be worked out to correspond with the growing practical needs, but at this stage it seems desirable to consider the subject from the viewpoint of electrical field theory. The pedagogical tricks will come in due course.

(3) The micro-wave electronic apparatus is characterized by the fact that the time of flight of individual electrons is not negligible compared with the time of one cycle. Electrons in ordinary apparatus go with speeds from 0.01 to 0.1 of the velocity of light. Therefore they travel from 0.01 to 0.1 of a wave-length in one cycle. Usually it is impracticable to design tubes for such short paths when the wave-length itself is of the order of centimeters. This fact brought about a breakdown in the usual modes of thinking about electronic tubes. Recent progress in the field is largely due to the discovery of ways to put finite transit time to good use. In other words, finite transit time is not a limitation on the electronics, but a limitation on the trasitional thinking about the subject.

Historically, the earliest work of Hertz, by which electric waves were first intentionally produced by electrical means, was done in what we here call the micro-wave region. But it was characterized by very low radiated power and by the fact that the oscillations were a succession of highly-damped wave trains instead of the much more useful continuous waves which modern technique provides.

From the point of view of application to communication, the principal importance of micro-waves derives from two things: (1) new frequency channels are made available in an already crowded medium, and (2) owing to the fact that for production of very sharply directed beams the antenna must be large compared to the wave-length, this requirement may be satisfied with structures of more convenient size than in the case of longer wave-lengths. The subject is so new that very little work has been done so far on the propagation of these waves over land or sea, or in relation to the ionosphere. Much work needs to be done in this direction.

With the current development of experimental techniques and equipment in the micro-wave field, physicists will have in their hands a tool for investigation of properties of matter, opening up a field that is at present essentially unknown. Yet we do know already that some molecules (e.g., ammonia) have characteristic frequencies in this range that are of the utmost significance for the understanding of molecular structure. Probably much will be learned through the molecular micro-wave spectroscopy of the future. When ferromagnetic conductors are placed in a micro-wave radiation field, the skin depth to which the waves effectively penetrate is of the same order as the size of the ferromagnetic domains. There is, therefore, no question but that the study of ferromagnetics at micro-wave frequencies will contribute to a better understanding of ferromagnetism. Similarly, many dielectric substances show maximum dielectric absorption in the micro-wave frequency range so study of their properties in this range will be essential to a better understanding of the properties of such substances, especially of the modern synthetic resins and rubbers.

All such contributions to a better understanding of dielectric and ferromagnetic materials are, in a larger sense, topics in "applied physics." However, if we look for future fields of application of the micro-wave equipment outside of the research laboratory, it is at once evident that most of the developments now being made will be directly applicable as aids to marine and aerial navigation. Moreover we must not overlook the fact that applicability of micro-waves to medical diathermy is thus far completely unexplored and that they may well prove to have specific therapeutic effects not possessed by the lower frequencies in use at present.

So there is plenty to be done for a long time to come. It is sincerely hoped that the exposition which follows of some parts of the subject will contribute usefully to a vigorous future development of the subject.

Chapter I. Cavity Resonators

Instead of the conventional coil and condenser as the basic resonant circuit element, in microwave radio the cavity resonator is used. By a cavity resonator is meant a region of space essentially totally enclosed by walls made of good conductors which is used as an oscillating circuit element. It is therefore desirable to begin the study of micro-wave radio by getting a thorough familiarity with the properties of cavity resonators.

1^1. Maxwell's Equations

All the electromagnetic field problems are governed by the basic equations of Maxwell which we shall write in the following form:

$$\text{div}\,\mathbf{D} = 4\pi\rho, \qquad \text{div}\,\mathbf{B} = 0,$$

$$\text{curl}\,\mathbf{E} = -(1/c)\dot{\mathbf{B}}, \qquad \text{curl}\,\mathbf{H} = 4\pi\mathbf{i} + (1/c)\dot{\mathbf{D}}, \tag{$1^1$1}$$

in which

  $\mathbf{E}$ is the electric field in statvolt/cm;
  $\mathbf{D}$ is the electric induction in statvolt/cm;
  ρ is the electric charge density in electrostatic units of charge per cm^3;
  $\mathbf{H}$ is the magnetic field in gauss;
  $\mathbf{B}$ is the magnetic induction in gauss;
  $\mathbf{i}$ is the conductive current density in abamp./cm^2.

This is only of many of the systems of units now competing for public favor, but it is a system which will be found convenient and useful in practical work.

Anyway we shall not fall into the common error of becoming slave to a particular unit system and shall not hesitate to change the units whenever it is advantageous to do so. In any ordinary medium we have,

$$\mathbf{B} = \mu\mathbf{H} \quad \text{and} \quad \mathbf{D} = \varepsilon\mathbf{E}, \tag{1^12}$$

where μ is the magnetic permeability and ε is the dielectric constant of the medium. The coefficients μ and ε are here pure numbers and equal to unity for a vacuum. They are thus equal to the values always listed in the tables in the reference books for these quantities.

In a conducting medium, the electric field needed to produce a current density is given by

$$\mathbf{E} = \rho\mathbf{i}, \tag{1^13}$$

in which ρ is the resistivity of the material. A simple check of the dimensions will show that this resistivity is measured in cm. The usual reference tables give ρ in ohm-cm which is the relation between $\mathbf{E}$ in volt/cm and $\mathbf{i}$ in amp./cm^2. If ρ' is the resistivity in ohm-cm and ρ is the equivalent quantity in the unit defined here, which is the statvolt-cm/abamp. = cm, the relation is

$$\rho = \rho'/30.$$

Thus for copper at room temperature for which $\rho' = 1.7 \times 10^{-6}$ ohm-cm the resistivity in cm is $\rho = 5.7 \times 10^{-8}$ cm.

At the boundary between two different nonconducting media the conditons are:

> Normal components of **D** and **B** continuous,
> Tangential components of **E** and **H** continuous. $\qquad(1^14)$

Continuity of the normal component of **D** implies that there is no surface charge density on the interface. If there is a charge of σ e.s.u./cm^2 on the interface then there is a discontinuity of $4\pi\sigma$ in the normal component of **D** at the interface.

The charge and current density are connected by the relation

$$\operatorname{div}\mathbf{i} + (1/c)\dot{\rho} = 0, \tag{1^15}$$

which expresses the fact that a new flow of electric charge out of a region is always accompanied by a corresponding diminution of the charge density there.

In applications **E** is usually expressed in volt/cm and the connection is 1 statvolt = 300 volts. Likewise current is usually measured in amperes, 1 abampere = 10 amperes, and charge in coulombs, 1 coulomb = 3×10^9 e.s.u. Likewise power is expressed in watts or volt-amperes, whereas the unit of our system would be the statvolt-abampere, 1 statvolt-abampere = 3 kilowatts. Although **H** is usually expressed in gauss also in practical work, some people like to express it in amp./cm which is the field in an infinitely long solenoid excited with 1 ampereturn/cm. The connection is 1 amp./cm = 0.4π gauss.

From Maxwell's equations we may derive the general relation,

$$\operatorname{div}\mathbf{S} + \frac{1}{4\pi}\left(\mathbf{H}\cdot\frac{\partial\mathbf{B}}{\partial t} + \mathbf{E}\cdot\frac{\partial\mathbf{D}}{\partial t}\right) = -c\mathbf{i}\cdot\mathbf{E}, \tag{1^16}$$

in which

$$\mathbf{S} = (c/4\pi)\mathbf{E}\times\mathbf{H}\ \mathrm{erg/cm^2\ sec.} \tag{1^17}$$

The vector $\mathbf{S}$ is called Poynting's vector and is interpreted as giving the flow of electromagnetic energy in the field. The exact flow of the field energy is really not known from this or any other consideration, since to $\mathbf{S}$ could be added any other vector field $\mathbf{S}'$ whose divergence vanishes everywhere without affecting the validity of (1^16). However, since electromagnetic energy is only observed by the effects it produces when converted into mechanical or thermal forms, this ambiguity in the flow pattern does not affect any observable results of the calculations.

In ordinary media having constant ε and μ the second term in (1^16) is the time derivative of the quantity

$$W = \frac{1}{8\pi}(\mu\mathbf{H}^2 + \varepsilon\mathbf{E}^2)\mathrm{erg/cm^3}. \tag{1^18}$$

This is interpreted as the local density of electrical energy in the field, the first term being the magnetic energy density and the second term being the electric energy density.

One advantage of the system of units we are using is that in a plane electromagnetic wave, $\mathbf{E}$ in statvolt/cm is equal to $\mathbf{H}$ in gauss. However for practical work it is handly to have the formula for Poynting's vector expressed in practical units, thus,

$$\mathbf{S} = (1/0.4\pi)\mathbf{E}\times\mathbf{H}\ \mathrm{watt/cm^2}, \tag{1^19}$$

where $\mathbf{E}$ is in volt/cm and $\mathbf{H}$ in gauss. In a vacuum in these units H in magnitude is $(1/300)E$ so the magnitude of the Poynting vector is $S = (1/120\pi)E^2$. The numeric $120\pi = 377$ is expressed in ohms and in the literature is often dignified by giving it the imposing name, *impedance of free space*.

Referring again to (1^16) we see that in a region where there is no current,

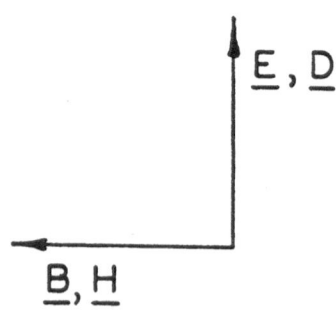

FIGURE 1^1. The relation of vectors in plane wave propagated toward the reader. σ is out from the paper.

so the right side is zero, the equation expresses the conservation of electromagnetic field energy. It also shows that changes in the field energy in any totally enclosed region where there is no outward flow across the boundaries occur only by virtue of flow of electric currents in a direction having a component along the electric field direction. If the current flows with the field the electromagnetic energy diminishes, if the current flows against the field the electromagnetic energy increases.

The whole art and science of micro-wave radio consists in dealing with the generation, transmission, and reception of electromagnetic energy at frequencies so great that the wave-length of the associated waves is not large compared to the apparatus involved. For this reason we have to deal with the distributed fields in accordance with the field equations. In other phases of electrical work, the wave-lengths involved are large compared to the size of the equipment. It is this fact that has made it possible to avoid the use of the field equations in developing the usual lumped constant circuit theory which is the basis of nearly all electrical engineering.

2^1. Plane Waves

Before taking up the problem of the fields in a cavity resonator it is instructive to get some familiarity with the simpler solutions of the field equations which correspond to progessive and to standing plane waves.

We assume each field vector to be the real part of a constant vector multiplying the factor,

$$e^{2\pi i(vt - \sigma \cdot \mathbf{r})}. \tag{$2^1$1}$$

Here σ is the vector whose magnitude indicates the number of waves per unit length and whose direction is normal to the plane wave fronts in the direction of propagation of the phase of the wave, v is the frequency in cycle/sec. The results which follow could equally well be derived by choosing the opposite sign for the exponent and, in fact, the other choice is more common in the literature of electric waves. But this choice is made to get a positive time factor because that is the custom in other parts of electrical engineering where the vectors in a vector diagram in alternating circuit theory are always regarded as rotating in the counterclockwise sense.

The two equations, div $\mathbf{D} = 0$ and div $\mathbf{B} = 0$ give

$$\mathbf{D} \cdot \sigma = 0 \quad \text{and} \quad \mathbf{B} \cdot \sigma = 0,$$

showing that the wave amplitudes must be transverse to the direction of propagation. We shall refer to the direction of $\mathbf{D}$ or $\mathbf{E}$ as the direction of polarization of the wave.

The two curl equations in $(1^1 1)$ give

$$\sigma \times \mathbf{E} = -(v/c)\mathbf{B}, \qquad \sigma \times \mathbf{H} = +(v/c)\mathbf{D}, \tag{$2^1$2}$$

from which we readily find that

$$\boldsymbol{\sigma} \times (\boldsymbol{\sigma} \times \mathbf{E}) = -(v/c)^2 \varepsilon \mu \mathbf{E}, \tag{2^{1}3}$$

and hence, using $\boldsymbol{\sigma} \cdot \mathbf{E} = 0$, that

$$|\boldsymbol{\sigma}| = (v/c)(\varepsilon\mu)^{1/2}. \tag{2^{1}4}$$

Therefore, the phase velocity of the wave is $c/(\varepsilon\mu)^{1/2}$, that is, the refractive index of the medium is $n = (\varepsilon\mu)^{1/2}$. From (2^{1}2) it is easy to see the vector directions are related as in Fig. 1[1]. Also $\sqrt{\varepsilon}\mathbf{E} = \sqrt{\mu}\mathbf{H}$. In empty space all four vectors, $\mathbf{D}, \mathbf{E}, \mathbf{B}, \mathbf{H}$, are numerically equal. For a plane wave the mean energy transport in terms of the amplitude of $\mathbf{E}$ in volt/cm becomes

$$\left(\frac{\text{watt}}{\text{cm}^2}\right)\mathbf{S} = \frac{1}{2}\frac{1}{120\pi(\mu/\varepsilon)^{1/2}}E^2. \tag{2^{1}5}$$

As already remarked in the previous section the coefficient in the denominator is expressed in ohms. Hence we shall say that a medium is characterized by an impedance for plane waves of

$$120\pi(\mu/\varepsilon)^{1/2} \text{ ohms.}$$

The impedance of the plane wave in ohms can also be defined as the ratio of the electric vector (volt/cm) to the magnetic vector (ampere-turns/cm). This definition leads to the same numerical value.

Standing waves arise from the superposition of two progressive plane waves of equal amplitude travelling in opposite directions. Suppose, for example, one has a wave travelling in the $+z$ direction, polarized in the x direction. Then the electric and magnetic vectors are given by

$$E_x = E_1 \cos 2\pi(vt - \sigma z), \qquad E_y = E_2 = 0,$$
$$H_y = (\varepsilon/\mu)^{1/2}E_1 \cos 2\pi(vt - \sigma z), \qquad H_x = H_z = 0. \tag{2^{1}6}$$

Similarly a wave polarized the same way but travelling in the opposite direction has fields given by,

$$E_x = E_2 \cos 2\pi(vt + \sigma z),$$
$$E_y = E_z = 0,$$
$$H_y = -(\varepsilon/\mu)^{1/2}E_2 \cos 2\pi(vt + \sigma z), \tag{2^{1}7}$$
$$H_x = H_z = 0.$$

Suppose now the plane $z = 0$ is a perfect conductor. At its surface the tangential component of $\mathbf{E}$ must vanish, and therefore the two waves must be related in such a way that $E_2 = -E_1$. The combined fields of the incident and reflected waves are then represented by,

$$E_x = 2E_1 \sin 2\pi\sigma z \sin 2\pi vt,$$
$$H_y = 2(\varepsilon/\mu)^{1/2}E_1 \cos 2\pi\sigma z \cos 2\pi vt. \tag{2^{1}8}$$

We observe that in the progressive waves $\mathbf{E}$ and $\mathbf{H}$ are in time phase at each place, but that in the standing wave they are in time and space quadrature.

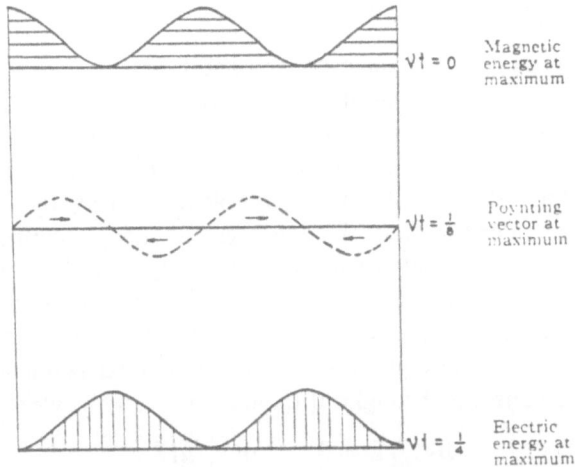

FIGURE 2^1. The pulsation of energy in a standing plane wave.

With the phases as expressed in (2^18), the energy is all magnetic at $t = 0$, and a quarter cycle later it is all electric. The energy in a standing wave does, therefore, but stand entirely still, but pulses back and forth a little as set forth in Fig. 2^1.

The reflection of the plane wave by the perfect conductor comes about by virtue of the flow of induced currents in the conductor. In Chapter IV to be published in a later issue of this journal we shall show how to calculate the radiation from a given current distribution. Here we may anticipate by saying that the induced current sheet flowing in the surface radiates a wave which just cancels the incident wave on the far side of the surface and also radiates the reflected wave on the near side of the surface.

To see what is the magnitude of the induced currents in the reflecting surface one may proceed as follows: Looking at the y^z plane near the surface, $z = 0$, one has,

$$H_y = 2E_1(\varepsilon/\mu)^{1/2} \cos 2\pi vt \quad \text{for } z > 0,$$

$$H_y = 0 \qquad\qquad\qquad \text{for } z < 0.$$

Hence the line integral around a path extending for unit length in the y direction just outside the metal and returning just inside the metal does not vanish. By Maxwell's equations it gives the conduction current flowing across the area enclosed, namely, in the surface of the metal, since the displacement current contribution is zero because tangential E vanishes at the surface. Therefore the surface current density is

$$i_x = -\frac{2E_1}{4\pi}(\varepsilon/\mu)^{1/2} \cos 2\pi vt, \tag{2^19}$$

where i_x is in abamp./cm^2 if E_1 is in statvolt/cm.

3^1. Rectangular Cavity Resonators[1]

Any region of space totally enclosed by a good metallic conductor may serve as a cavity resonator or "rhumbatron." Any such resonator has an infinite number of resonant frequencies and associated wave fields. First we develop the theory for walls of zero resistivity and later consider the effect of the resistivity of the walls. Also it is more suitable to illustrate by working out the case of a rectangular box since this involves only trigonometric functions.

   We have to solve the field equations (1^11) subject to the boundary conditions that **E** must be normal and **H** tangential to the perfectly conducting boundaries. Assume each vector to have a time dependence represented by the factor $e^{2\pi i v t}$. Then the equations for the positional dependence, when ε and μ are assumed constant throughout the medium, may be written

$$\text{div}\sqrt{\varepsilon}\mathbf{E} = 0, \qquad \text{div}\sqrt{\mu}\mathbf{H} = 0,$$

$$\text{curl}\sqrt{\varepsilon}\mathbf{E} = -i(2\pi n v/c)\sqrt{\mu}\mathbf{H}, \qquad (3^11)$$

$$\text{curl}\sqrt{\mu}\mathbf{H} = +i(2\pi n v/c)\sqrt{\varepsilon}\mathbf{E}.$$

Here $n = (\varepsilon\mu)^{1/2}$, the refractive index as defined in Section 2^1 and the combination $2\pi n v/c$ will be denoted by k. From the form of the equations it is evident that $\sqrt{\varepsilon}\mathbf{E}$ and $\sqrt{\mu}\mathbf{H}$ satisfy the same equations, with c replaced by c/n, as do **E** and **H** in free space. From this it follows that the theory for a resonator filled with any ordinary medium can be easily derived from the theory for the corresponding shape of empty resonator. For this reason and especially because nearly all the resonators used in practice so far are empty, we shall henceforth suppose ε and μ equal to unity.

   From (3^11) we readily find by taking the curl of each curl equation and doing a little reducing that **E** has to satisfy.

$$\nabla^2\mathbf{E} + k^2\mathbf{E} = 0. \qquad (3^12)$$

If an appropriate solution for **E** has been found it is not necessary to solve the corresponding equation for **H** separately, since the associated magnetic field can be calculated from the **E** field by means of the curl **E** equation of (3^11) in the form

$$\mathbf{H} = (i/k)\,\text{curl } \mathbf{E}. \qquad (3^13)$$

The magnetic field so calculated will automatically satisfy the correct boundary condition. If we take any path lying in the boundary then the line integral

$$\int \mathbf{E} \cdot \mathbf{ds} = 0$$

---

[1] Earliest development of this topic in physics was made in connection with the theory of blackbody radiation. Compare Jeans, *Dynamical Theory of Gases* (Cambridge University Press, London, 1921), third edition, chapter 16; or Fowler, *Statistical Mechanics* (Cambridge University Press, London, 1936), second edition, chapter 4.

since **E** is everywhere normal to the bounding surface. Therefore

$$\iint \text{curl } \mathbf{E} \cdot \mathbf{sS} = 0,$$

where the surface integral extends over any portion of the bounding surface. Hence it follows that the normal component of curl **E** vanishes everywhere on the boundary which is therefore true of the magnetic field calculated from (3^13).

There is no general way of solving (3^12) for cavities of arbitrary shape, and in fact solutions are only known for a very few special shapes. The problem has many points in common with the corresponding acoustical problem of finding the resonant sound waves in a closed cavity. However the electromagnetic problem is more complicated because the wave amplitude is a vector, each component of which must satisfy (3^12) and also satisfy div $\mathbf{E} = 0$; whereas in the acoustic problem there is only a single scalar wave amplitude, for example, the pressure in the wave.

We now consider the solution of the special problem of the rectangular cavity resonator whose walls are at the ends of the ranges,

$$0 < x < A, \qquad 0 < y < B, \qquad 0 < z < C.$$

If we write

$$E_x = E_1 \, {\textstyle{\cos \atop \sin}} \, k_1 x \, {\textstyle{\cos \atop \sin}} \, k_2 y \, {\textstyle{\cos \atop \sin}} \, k_3 z,$$

$$E_y = E_2 \, {\textstyle{\cos \atop \sin}} \, k_1 x \, {\textstyle{\cos \atop \sin}} \, k_2 y \, {\textstyle{\cos \atop \sin}} \, k_3 z,$$

$$E_z = E_3 \, {\textstyle{\cos \atop \sin}} \, k_1 x \, {\textstyle{\cos \atop \sin}} \, k_2 y \, {\textstyle{\cos \atop \sin}} \, k_3 z,$$

then (3^12) is satisfied for any combination of cos or sin provided the three k's are such that

$$k_1{}^2 + k_2{}^2 + k_3{}^2 = k^2.$$

To make **E** normal to all walls we have to specialize the cos or sin alternative and restrict the k's to the following discrete set of values:

$$k_1 = l\pi/A, \qquad k_2 = m\pi/B, \qquad k_3 = n\pi/C \qquad (3^14)$$

in which (l, m, n) are integers. The solution for **E** is therefore

$$E_x = E_1 \cos(l\pi x/A) \times \sin(m\pi y/B) \sin(\eta\pi z/C),$$

$$E_y = E_2 \sin(l\pi x/A \times \cos(m\pi y/B) \sin(n\pi z/C), \qquad (3^15)$$

$$E_z = E_3 \sin(l\pi x/A) \times \sin(m\pi y/B) \cos(n\pi z/C).$$

The three constant amplitudes, E_1, E_2, and E_3, cannot be chosen independently, but the condition div $\mathbf{E} = 0$, imposes the restriction,

$$(l\pi/A)E_1 + (m\pi/B)E_2 + (n\pi/C)E_3 = 0,$$
$$\vec{k} \cdot \vec{E} = 0.$$
(3¹6)

Therefore, for each set of integers there are two linearly independent modes of oscillation: if we think of (k_1, k_2, k_3) as the components of a vector and (E_1, E_2, E_3) as those of another vector, then any vector (E_1, E_2, E_3) perpendicular to k is permissible.

The possible resonant frequencies are given by

$$(v/c)^2 = (l/2A)^2 + (m/2B)^2 + (n/2C)^2$$
(3¹7)

where (l, m, n) are integers, at least two of which are not zero.

For the modes in which one of the integers is zero, the electric vector is everywhere parallel to the axis whose integer is zero, and the resonant frequency is independent of the dimension along that axis. For each such set of integers there is only one vector satisfying (3¹6), and so only a single solution for such a set, although as already remarked there are in general two linearly independent solutions associated with each set (l, m, n).

The least resonant frequency for the box is that which corresponds to putting the integers associated with the two larger dimensions each equal to unity, and the third one equal to zero. Thus if A and B are the two larger dimensions, the lowest mode will be polarized with the electric vector along the shortest dimension and the wave-length λ will be

$$\lambda = \frac{2}{(A^{-2} + B^{-2})^{1/2}}.$$

In particular for a cubical box the lowest mode has a wave-length equal to the face diagonal of the box, $\lambda = \sqrt{2}A$.

The number of different resonant modes mounts very rapidly as one goes up the frequency scale. For example consider a shallow square box ($B = A$, $C \ll A$) for which the lower frequency modes will all correspond to $n = 0$. The values of $2A\sigma$ are given by

$$2A\sigma = [l^2 + m^2 + n^2(A/C)^2]^{1/2}.$$

One can easily count up and find that there are 33 different sets of the integers giving rise to frequencies less than or equal to five times the lowest frequency. It should also be observed that the frequencies can be arranged in series: the fundamental is accompanied by all its integral multiples forming the series (110), (220), (330), etc.; another series begins with (120) and (210) and includes their integral multiples as (240) and (420), (360) and (630), etc. This occurrence of the integral multiples among the allowed frequencies is, however, a special property of the rectangular box which other sharpes do not possess.

It is important to define some terms which will be used in discussing cavity resonators. Each frequency for which there exists a solution of the field equations satisfying the boundary conditions will be called an *allowed frequency* or a *proper frequency*. The least allowed frequency is called the *fundamental*. The higher allowed frequencies are only called *harmonics* if they are integral multiples of the fundamental.

A particular solution for **E** and **H** will be referred to as a *mode* of oscillation: Any frequency for which there is more than one mode is referred to as a *degenerate* frequency.[2] The *order* of degeneracy is the number of linearly independent *modes* associated with the degenerate frequency. Thus in the example just considered the fundamental is not degenerate, but the next higher frequency is, because $(1, 2, 0)$ and $(2, 1, 0)$ are linearly independent solutions each having this same frequency. As we have seen all of the modes for which no one of the integers is zero are twofold degenerate because of the two linearly independent solutions of (3^16) that are possible. This type of degeneracy we shall call *polarization degeneracy*. Degeneracy also arises from symmetry in the shape of the resonator, thus $(1, 2, 0)$ and $(2, 1, 0)$ have the same frequency only because we assumed a square cross section $A = B$. Degeneracy arising in this way we shall can *symmetry degeneracy*.

A slight departure from the condition $A = B$, intentionally or due to some imperfection of manufacture or an unsymmetric location of the coupling device by which the resonator is placed in a circuit, will cause the degenerate frequencies to become slightly separated. We say that such changes remove the degeneracy.

It is important to recognize a lack of uniqueness in the wave fields associated with a degenerate frequency. For example, in the case of polarization degeneracy, one may choose any two (perferably mutually perpendicular) vectors satisfying (3^16) as the basic modes. Any linear combination of them is a possible mode of oscillation associated with that frequency. Similarly, though our particular analysis may present us with certain particular forms for the degenerate modes in the case of symmetry degeneracy, these have really no special standing in the physics of the problem and the actual mode of oscillation may be any linear combination of these.

For example, consider the modes $(1, 2, 0)$ and $(2, 1, 0)$. From (3^15) these have only a z component of **E** which in the two cases is

$$E_{120} = C \sin(\pi x/A) \sin(2\pi y/A),$$

$$E_{210} = D \sin(2\pi x/A) \sin(\pi y/A),$$

where C and D are arbitrary relative amplitudes. At this frequency quite a variety of different modes are possible according to the relative magnitudes of C and D and the phase relation existing between them. Some of the field distributions which can arise from different relative excitations of these two modes are sketched in Fig. 3^1. Since the combination with $C = D$ has a node

---

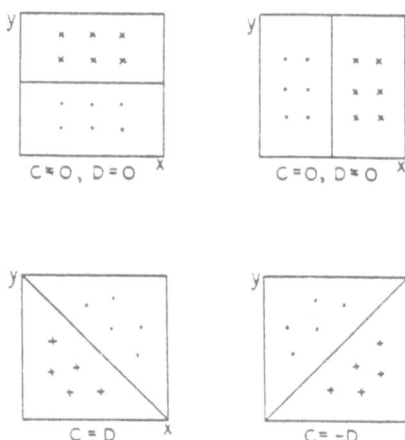

FIGURE 3[1]. The different types of field distribution resulting from co-existence of the degenerate modes (1, 2, 0) and (2, 1, 0).

on the line $y = A - x$ as well as $y = 0$ and $x = 0$, it obviously satisfies all the conditions to be the fundamental mode for a right triangular prism whose cross section is formed by these three lines. By this means one can often find particular solutions for special shapes which would not otherwise be easy to find. The trick does not work for a right triangular prism of unequal sides for in that case the two modes of the corresponding rectangular prism would not belong to the same frequency and so could not be superposed in this way.

Exercise: Discuss the modes corresponding to $D = \pm iC$, that is, where the two degenerate modes are in time quadrature.

4[1]. *Resonator Coordinates*[3]

By working out in detail the solution for a rectangular cavity resonator in the preceding section, we have learned most of the general properties which are applicable to resonators of any shape. These are, that the fields in the resonator can be made to satisfy the boundary conditions only for certain discrete allowed frequencies and associated with each frequency there may be one or more wave patterns. Evidently the most general state of excitation of a resonator would be for all of the possible modes to be simultaneously present, just as many of the different possible modes of vibration of a drum-head a simultaneously present when the drum is struck. To deal mathematically with this situation calls for introduction of a convenient means of describing such

---

[3] Most of this section can be skipped at a first reading, but it should be scanned to see the main results concerning orthogonality of the wave functions (4[1]5) and the dynamical equation for a mode amplitude (4[1]10). The results are an application of the formalism used in quantum electrodynamics. Compare E. Fermi, *Rev. Mod. Phys.* **4**, 87 (1932), or W. Heitler, *Quantum Theory of Radiation* (Oxford University Press, London, 1936), p. 40.

general states of oscillation. This we can do by means of resonator coordinates, which are simply the amplitudes of each of the basic wave fields in the actual state of motion.

Vector Potential

Instead of dealing directly with **E** and **H** it is convenient to derive the electro-magnetic fields from the usual scalar and vector potentials, **A** and φ, according to the relations

$$\mathbf{E} = -(1/c)\dot{\mathbf{A}} - \text{grad } \varphi, \qquad \mathbf{H} = \text{curl } \mathbf{A}. \qquad (4^1 1)$$

Here the vector potential is measured in the same units as current, namely abamperes, and the scalar potential is in statvolts. This mode of representation of the field satisfies the field equation for curl **E** automatically, as also the equation div **H** = 0. Substituting from $(4^1 1)$ into the other two field equations we find,

$$-\nabla^2 \mathbf{A} + (1/c^2)\ddot{\mathbf{A}} + \text{grad}\left(\text{div } \mathbf{A} + \frac{1}{c}\dot{\varphi}\right) = 4\pi\mathbf{i},$$

$$-\nabla^2 \varphi + (1/c^2)\ddot{\varphi} - \frac{1}{c}\frac{\partial}{\partial t}\left(\text{div } \mathbf{A} + \frac{1}{c}\dot{\varphi}\right) = 4\pi\rho. \qquad (4^1 2)$$

We are at liberty to assume some further relation involving div **A** to simplify the equations, and evidently they will be simplified very considerably if we put div $\mathbf{A} + (1/c)\dot{\varphi} = 0$, which gives us the following set of equations for the potentials:

$$\frac{1}{c^2}\ddot{\mathbf{A}} - \nabla^2 \mathbf{A} = 4\pi\mathbf{i},$$

$$\frac{1}{c^2}\ddot{\varphi} - \nabla^2 \varphi = 4\pi\rho, \qquad (4^1 3)$$

$$\text{div } \mathbf{A} + (1/c)\dot{\varphi} = 0.$$

These same equations will play a basic role in Chapter IV when we develop the theory of radiation from a system of moving charges and conductive currents. If we take the divergence of the first equation and take $(1/c)$ times the time derivative of the second and add, we find an equation for the time dependence of $[\text{div } \mathbf{A} + (1/c)\dot{\varphi}]$, namely,

$$\left(\frac{1}{c^2}\frac{\partial^2}{\partial t^2} - \nabla^2\right)\left(\text{div } \mathbf{A} + \frac{1}{c}\dot{\varphi}\right) = 0.$$

The right side is zero since the charge and current satisfy the conservation equation $(1^1 5)$. This shows that if we admit only solutions which at $t = 0$ satisfy the third equation of $(4^1 3)$ together with the time derivative of that equation, then the third equation will be satisfied at all times.

Suppose now that the mathematical problem of finding the allowed frequencies and associated wave patterns for the cavity has been solved by some such procedure as that of the preceding section. This means that we know the set of values k_1, k_2, k_3, etc., and associated solutions A_1, A_2, A_3, etc., with satisfy the equations

$$\nabla^2 A + k^2 A = 0, \qquad \text{div } A = 0, \tag{4¹4}$$

A normal to walls or zero.

As we have seen, it is possible for an allowed value k_a to be degenerate, that is to have more than one linearly independent solution A_a associated with it. Thus a complete enumeration of the A's requires another "degeneracy" index, to distinguish the different A_a belonging to the same k_a. Ordinarily it will not be necessary to write this explicitly: in the formal mathematics we can regard the index a as labelling all of the independent wave functions, then it will happen that the associated k_a are equal for several different values of the index a.

Orthogonality of Wave Functions

The A_a wave patterns have an important orthogonality property which makes it convenient to use them to represent other functions in a manner similar to Fourier series. Write down the curl curl equation satisfied by A_a and by A_b, multiply the former by A_b, the latter by A_a and subtract:

$$A_b \cdot \text{curl curl } A_a - A_a \cdot \text{curl curl } A_b = (k_a^2 - k_b^2) A_a \cdot A_b.$$

Now use one of the basic identities of vector analysis,

$$\text{div}(u \times v) = v \cdot \text{curl } u - u \cdot \text{curl } v$$

to transform the left side of this equation into

$$\text{div}(A_a \times \text{curl } A_b + (\text{curl } A_a) \times A_b).$$

Next integrate both sides over the volume of the cavity. The integral of the left side vanishes because it can be transformed to a surface integral over the surface which vanishes because A_a and A_b are normal to the walls. Hence,

$$\iiint A_a \cdot A_b \, dV = 0, \qquad \text{if} \qquad k_a \neq k_b. \tag{4¹5}$$

In the case of a degenerate value of k, it is possible to choose the several A's belonging to it so they are mutually orthogonal by taking appropriate linear combinations of the original ones if those found in the original solution do not already have this property.

Since a particular solution is still a solution when multiplied by a constant, the constant multiplier of each A_a may be chosen to suit one's convenience. It turns out that for the work which follows it is convenient to *normalize* the functions in such a way that

$$\iiint \mathbf{A}_a \cdot \mathbf{A}_a^* dV = V \qquad (4^16)$$

in which $\mathbf{A}_a^*$ is the conjugate complex function to $\mathbf{A}_a$. With this choice of normalization, the $\mathbf{A}_a$ functions are physically dimensionless; V is the volume of the cavity.

Let us first consider the case in which the currents in the cavity are distributed in such a way that the charge density is everywhere zero at all times, then $\varphi = 0$ and we may try to find a solution of the first of (4^13) by writing

$$\mathbf{A} = \sum_a J_a(t)\mathbf{A}_a(x, y, z). \qquad (4^17)$$

There is one time-dependent coefficient for each wave pattern. Since the $\mathbf{A}_a$ as normalized by (4^16) are dimensionless, the $J_a(t)$ are measured in the same units as $\mathbf{A}$, namely abamperes. Each J_a gives the amplitude of excitation of its particular mode at a particular instant and is for that reason called a *resonator coordinate*. There is one for each mode and hence an infinite number of them for a particular cavity resonator.

Exciting Current

Similarly we may expand the given current distribution inside the resonator $\mathbf{i}(x, y, z, t)$ in terms of the $\mathbf{A}_a$ functions, denoting the time-dependent coefficients by $I_a(t)$,

$$\mathbf{i}(x, y, z, t) = \sum_a I_a(t)\mathbf{A}_a(x, y, z). \qquad (4^18)$$

The determination of the coefficients in this expansion is particularly easy formally. Because of the orthogonality property of the $\mathbf{A}$ functions it is just like the method used in Fourier series,

$$I_a(t) = (1/V) \iiint \mathbf{i} \cdot \mathbf{A}_a^* \, dV. \qquad (4^19)$$

Thus each $I_a(t)$, like $\mathbf{i}$ is a current density, abampere/cm^2. We shall call $I_a(t)$ the *exciting current* of the ath mode. It may be remarked in passing that a current distribution $\mathbf{i}$ is most effective in exciting the ath mode if its spatial distribution is like that of the mode it is to excite.

Dynamical Equation

Now substituting (4^18) and (4^17) in the first of (4^13) we may equate coefficients of each $\mathbf{A}_a$ and thus obtain a simple differential equation for the time dependence of each resonator coordinate,

$$\ddot{J}_a(t) + (ck_a)^2 J_a(t) = 4\pi c^2 I_a(t). \qquad (4^110)$$

This equation is just like that for a simple harmonic oscillator of natural frequency $(ck_a/2\pi)$ driven in forced oscillations by the exciting term on the right side. If the exciting term is zero then the corresponding J_a executes harmonic time variation at its natural frequency with constant amplitude. The free oscillations are undamped because we have supposed the walls to be of zero resistivity: the effect of finite resistivity is considered in Section 8[1].

It will make the rest of the discussion easier to follow if we suppose that all of the A_a are real functions: this is no restriction as it is always possible to choose them in this way.

EXPRESSIONS FOR ENERGY

The electric field energy in the cavity is (in ergs)

$$W_e = \iiint (E^2/8\pi)\, dV = (V/8\pi c^2) \sum_a \dot{J}_a^2 \qquad (4^1 11)$$

as can be found by substituting the expression for E in terms of A and using $(4^15), (4^16)$, and (4^17). Similarly the magnetic field energy in the cavity (in ergs) is

$$W_m = \iiint (H^2/8\pi)\, dV = \sum_{a,b} \iiint J_a \cdot J_b (\text{curl } A_a \cdot \text{curl } A_b)\, dV.$$

To simplify this we need to evaluate,

$$\iiint (\text{curl } A_a \cdot \text{curl } A_b)\, dV = \iiint \text{div}(A_a \times \text{curl } A_b)\, dV + k_b^2 \iiint A_a \cdot A_b\, dV.$$

The integral of the divergence vanishes because it can be transformed to a surface integral of $A_a \times \text{curl } A_b$ which has a vanishing normal component at the boundary. Therefore,

$$\iiint (\text{curl } A_a \cdot \text{curl } A_b)\, dV = \begin{cases} 0 & b \neq a \\ k_a^2 V & b = a, \end{cases}$$

so the magnetic energy is

$$W_m = (V/8\pi) \sum_a k_a^2 J_a^2. \qquad (4^1 12)$$

We shall write W_a for the energy associated with the ath mode of oscillation. There is no mutual energy between different modes in consequence of the orthogonality of the A functions. For W_a we have

$$W_a = (V/8\pi c^2)[\dot{J}_a^2 + (ck_a)^2 J_a^2]. \qquad (4^1 13)$$

We can derive an expression for dW_a/dt from $(4^1 10)$ in just the same way as is done in obtaining the energy integral in particle dynamics. Multiply through by $(V/4\pi c^2)$ to obtain,

$$(dW_a/dt) = V I_a(t) \dot{J}_a(t). \qquad (4^1 14)$$

In words: the instantaneous rate of increase of the field energy in the ath mode is equal to the volume times the ath exciting current times the rate of increase of the ath resonator coordinate. This is fully analogous to the expression for power as a force (in this case I_a) multiplied by a velocity, in this case proportional to the rate of increase of J_a.

EFFECTIVE INDUCTANCE AND CAPACITY

Those who are acustomed to thinking in terms of resonant circuits in terms of their inductances and capacities will grope for a definition of some sort of effective inductance and capacity which is applicable to the cavity resonator. This can be done as soon as one has fixed on a proper current coordinate by means of which to measure the amplitude of excitation. In an ordinary inductance the magnetic energy is $W_m = Li^2/2$ where W is in ergs if L is in cm and i in abamperes. Here we measure the amplitude of the ath mode by giving the value of its resonator coordinate J_a which is a current, hence we may properly identify the coefficient of $J_a{}^2$ in the expression for the magnetic energy as half the *effective inductance* L_a of the ath mode. This gives, for L_a in cm

$$L_a = Vk_a{}^2/4\pi = \pi V/\lambda_a{}^2. \qquad (4^115)$$

For conversion to practical units, note that 1 cm of inductance is equal to 10^{-9} henry. To get a correspondingly appropriate definition of the capacity of the ath mode we must choose C_a in such a way that the product $L_a C_a$ gives the correct resonant frequency in accordance with the equation.

$$\lambda_a = 2\pi(L_a C_a)^{1/2}.$$

In this way it is easily found that the electrostatic capacity in cm to be associated with the ath mode is

$$C_a = 4\pi/Vk_a{}^4. \qquad (4^116)$$

Exercise: Show that the normalized A_a for the (110) mode of a cubical resonator of edge A is

$$A_{110} = 2k\sin(\pi x/A)\sin(\pi y/A),$$

where k is a unit vector parallel to the z axis. Also show that the maximum value of the electric vector occurs at points along the line $x = A/2$, $y = A/2$ and that when the resonator coordinate $J_{110} = 1$ abampere, this maximum electric vector equals $(4\pi/\lambda)$ statvolt/cm where $\lambda = \sqrt{2}A$ as worked out in the preceding section.

EFFECT OF CHARGE IN CAVITY

Let us now turn to the more general case in which there is charge density as well as current in the resonator. The expansion (4^17) is no longer adequate,

since it gives div $\mathbf{A} = 0$ which is no longer true. A similar remark holds for
(4¹8). The necessary generalization runs as follows:

Suppose the scalar boundary value problem,

$$\nabla^2 \varphi_b + k_b{}^2 \varphi_b = 0,$$

$$\varphi_b = 0 \quad \text{on walls,}$$
(4¹17)

has been solved so its allowed functions and allowed values are known. The
functions φ_b may be proved to be orthogonal:

$$\varphi_c \nabla^2 \varphi_b - \varphi_b \nabla^2 \varphi_0 + (k_b{}^2 - k_c{}^2)\varphi_b \varphi_c = 0,$$

$$\varphi_c \nabla^2 \varphi_b - \varphi_b \nabla^2 \varphi_c = \text{div}(\varphi_c \, \text{grad} \, \varphi_b - \varphi_b \, \text{grad} \, \varphi_c).$$

Integrate over the volume. The volume integral of the divergence may be seen
to vanish by transforming to a surface integral. Therefore $\iiint_{\varphi_b \varphi_c} dV = 0$ if
$k_b \neq k_c$. In the case of degeneracy we may choose orthogonal linear combinations of the degenerate wave functions so all φ's are orthogonal. We shall
consider the φ's to be normalized as the $\mathbf{A}$'s [compare (4¹6)]

$$\iiint \varphi_b{}^2 \, dV = V.$$
(4¹18)

Next we assume that $\rho(x, y, z, t)$ is expanded in terms of the φ_b, so, analogous
to (4¹8),

$$\rho = \sum_b R_b(t)\varphi_b(x, y, z),$$
(4¹19)

and likewise that the scalar potential can be so expanded,

$$\varphi = \sum_b \Phi_b(t)\varphi_b(x, y, z).$$
(4¹20)

Substitution of these in the equation for φ in (4¹3) leads to equations of motion
for the coefficients in (4¹20), analogous to (4¹10),

$$\ddot{\Phi}_b + (ck_b)^2 \Phi_b = 4\pi c^2 R_b.$$
(4¹21)

In addition the expansions for $\mathbf{A}$ and $\mathbf{i}$ have to be extended by bringing in
additional terms for which the divergence does not vanish. The functions

$$\mathbf{B}_b = (1/k_b) \, \text{grad} \, \varphi_b, \qquad \text{curl} \, \mathbf{B}_b = 0$$
(4¹22)

are appropriate for this purpose. They are all orthogonal to each other and
to the $\mathbf{A}_a$ functions. To prove $\mathbf{B}_b$ orthogonal to A_a apply the general formula

$$\iiint (\text{curl} \, \mathbf{u} \cdot \text{curl} \, \mathbf{v} + \text{div} \, \mathbf{u} \, \text{div} \, \mathbf{v} + \mathbf{u} \cdot \Delta \mathbf{v}) \, dV$$

$$= \iint \mathbf{u} \times \text{curl} \, \mathbf{v} \cdot \mathbf{dS} + \iint \text{div} \, \mathbf{v} \mathbf{u} \cdot \mathbf{dS}.$$

Identify **v** with $\mathbf{A}_a$ and **u** with $\mathbf{B}_b$. Then on the left the first integral vanishes since curl $\mathbf{B}_b = 0$, the second since div $\mathbf{A}_a = 0$, and the third reduces to $-k_a^2 \iiint \mathbf{A}_a \cdot \mathbf{B}_b \, dV$. On the right the first integral vanishes because $\mathbf{B}_b$ is normal to the surface and the second because div $\mathbf{A}_a = 0$. Hence the functions are orthogonal.

The factor $(1/k_b)$ is inserted in $(4^1 22)$ to take care that the B_b are normalized like the A's:

$$\iiint \mathbf{B}_b \cdot \mathbf{B}_c \, dV = \begin{cases} 0 & b \neq c \\ V & b = c. \end{cases} \tag{$4^1 23$}$$

This follows from the relation,

$$\iiint \operatorname{grad} \varphi_b \cdot \operatorname{grad} \varphi_c \, dV = \iiint \operatorname{div}(\varphi_b \operatorname{grad} \varphi_c) \, dV - \iiint \varphi_b \Delta \varphi_c \, dV,$$

both as regards orthogonality and normalization.

We may now assume $(4^1 7)$ and $(4^1 8)$ extended as follows:

$$\begin{aligned} \mathbf{A} &= \sum_a J_a(t) \mathbf{A}_a + \sum_b K_b(t) \mathbf{B}_b, \\ \mathbf{i} &= \sum_a I_a(t) \mathbf{A}_a + \sum_b H_b(t) \mathbf{B}_b. \end{aligned} \tag{$4^1 24$}$$

Substituting these into the equation of motion for **A** in $(4^1 3)$ we find

$$\ddot{K}_b + (ck_b)^2 K_b = 4\pi c^2 H_b \tag{$4^1 25$}$$

which, together with $(4^1 21)$ and $(4^1 10)$, gives the equations of motion of all of the resonator coordinates. A considerable complication has resulted from the introduction of charge into the cavity: not only was it necessary to introduce the scalar potential, but to extend the expansion for **A** as well. The unnumbered equation following $(4^1 3)$ gives a proof that if

$$\dot{\Phi}_b - ck_b K_b = 0$$

and

$$\frac{d}{dt}(\dot{\Phi}_b - ck_b K_b) = 0$$

initially, then they will remain zero at all times. Hence the solutions of $(4^1 10)$, $(4^1 21)$, and $(4^1 25)$ have to be chosen so as to satisfy these conditions as part of the initial conditions.

If now we compute the electric and magnetic energy expressions, we find no change in the magnetic energy since the curl of the additional terms in **A** is zero. But there are added terms in the electric energy and the complete expression to replace $(4^1 11)$ is

$$W_e = (V/8\pi c^2) \sum_a \dot{J}_a^2 + (V/8\pi c^2) \sum_b \dot{K}_b^2 + (V/8\pi) \sum_b k_b^2 \Phi_b^2. \tag{$4^1 26$}$$

5^1. *Cylindrical Resonators*[4]

By a cylindrical resonator is meant one which is bounded by the planes $z = 0$ and $z = C$ and whose cross section in any plane $z = $ constant is the same curve. The three-dimensional problem can be reduced quite generally to a two-dimensional problem for such resonators. We start with Eqs. ($3^1$1) in which $\varepsilon = \mu = 1$,

$$\text{div}\,\mathbf{E} = 0 \qquad \text{div}\,\mathbf{H} = 0$$
$$\text{curl}\,\mathbf{E} = -ik\mathbf{H} \qquad \text{curl}\,\mathbf{H} = +ik\mathbf{E}. \tag{$5^1$1}$$

It is natural to expect the solutions to depend on z through a $\cos k_3 z$ or $\sin k_3 z$ factor as in the case of the rectangular box (which is a special case of the class of cylindrical resonators).

The modes can be classified into types as follows:

$$E \text{ type, for which } E_2 \neq 0, \text{ but } H_z = 0,$$
$$H \text{ type, for which } H_z \neq 0, \text{ but } E_z = 0. \tag{$5^1$2}$$

E (ELECTRICS) MODES

Let us consider first the modes of E type. Since $H_z = 0$ we have, from the curl $\mathbf{H}$ equation,

$$ikE_x = -(\partial H_y/\partial z),$$
$$ikE_y = +(\partial H_x/\partial z),$$
$$ikE_z = (\partial H_y/\partial x) - (\partial H_x/\partial y), \tag{$5^1$3}$$

and, similarly, the curl $\mathbf{E}$ equation gives

$$-ikH_x = (\partial E_z/\partial y) - (\partial E_y/\partial z),$$
$$-ikH_y = (\partial E_x/\partial z) - (\partial E_z/\partial x),$$
$$-ikH_z = (\partial E_y/\partial x) - (\partial E_x/\partial y). \tag{$5^1$4}$$

Using the x and y components of these equations we can express E_x and E_y in terms of derivatives of E_z:

$$\frac{\partial^2 E_x}{\partial z^2} + k^2 E_x = \frac{\partial}{\partial x}\left(\frac{\partial E_z}{\partial z}\right),$$

$$\frac{\partial^2 E_y}{\partial z^2} + k^2 E_y = \frac{\partial}{\partial y}\left(\frac{\partial E_z}{\partial z}\right).$$

---

[4] The literature dealing with special properties of resonators of particular shapes is becoming quite extensive. Some useful general references are: Stratton, *Electromagnetic Theory* (McGraw-Hill, New York, 1941), chapters 6 and 7; Bateman, *Electrical and Optical Wave Motion* (Cambridge University Press, London, 1915); Borgnis, Ann. d. Physik **35**, 359 (1939).

The left-hand side of these becomes $(k^2 - k_3^2)$ times E_x or E_y no matter whether the z dependence contains a cos or sin factor, so we can write for the component of $\mathbf{E}$ in the cross section

$$\mathbf{E}_s = E_x\mathbf{i} + E_y\mathbf{j},$$
$$(k^2 - k_3^2)\mathbf{E}_s = \text{grad}_s(\partial E_z/\partial z), \tag{5^{1}5}$$

where grad$_s$ means gradient in the section and is the usual gradient with the z component omitted. Next use (5^{1}4) to eliminate the H components from the z component of (5^{1}3) to obtain the basic differential equation which governs the variation of E_z over the cross section

$$\nabla_s^2 E_z + (k^2 - k_3^2)E_z = 0, \tag{5^{1}6}$$

where ∇_s^2 is the sectional Laplacian, obtained by omitting the z component from the three-dimensional Laplacian.

Finally we may use (5^{1}4) to express $\mathbf{H}$ in terms of $\mathbf{E}$:

$$H_x = \frac{ik}{k^2 - k_3^2}\frac{\partial E_z}{\partial y}, \qquad H_y = \frac{-ik}{k^2 - k_3^2}\frac{\partial E_z}{\partial x}, \tag{5^{1}7}$$

or, in vector form,

$$\mathbf{H} = \frac{-ik}{k^2 - k_3^2}\mathbf{k} \times \text{grad}_s E_z, \tag{5^{1}8}$$

in which $\mathbf{k}$ is the unit vector in the z direction.

The boundary conditions on $\mathbf{E}$ require that $\mathbf{E}$ be normal to all bounding surfaces. This calls for the choice of the factor $\cos k_3 z$ instead of $\sin k_3 z$ in the expression for E_z in order that the sectional component $\mathbf{E}_s$ shall vanish at the ends. The condition on the cylindrical walls requires that only solutions of (5^{1}6) which vanish at the boundary be admitted.

Suppose we denote by $\psi_a(x, y)$ and k_a the associated functions and proper values which satisfy the two-dimensional boundary value problem

$$\nabla_s^2 \psi_a(k, y) + k_a^2\psi_a(x, y) = 0,$$
$$\psi_a(x, y) = 0, \quad \text{on boundary.} \tag{5^{1}9}$$

This two-dimensional problem defines a sequence of proper functions and proper values. They are known in the mathematical literature for a variety of shapes since they occur in other branches of mathematical physics, for instance, the vibrations of a drumhead of the same shape as the resonator section.

Summarizing the results for the E type modes we have,

E type:

$$E_z = A\psi_a(x, y)\cos k_3 z,$$
$$k^2 = k_a^2 + k_3^2,$$

$$\mathbf{E}_s = -(k_3/k_a{}^2)A\left(\frac{\partial\psi_a}{\partial x}\mathbf{i} + \frac{\partial\psi_a}{\partial y}\mathbf{j}\right)\sin k_3 z,$$

$$\mathbf{H}_s = -i(k/k_a{}^2)A\left(-\frac{\partial\psi_a}{\partial y}\mathbf{i} + \frac{\partial\psi_a}{\partial x}\mathbf{j}\right)\cos k_3 z. \tag{5^{1}10}$$

In terms of the normalized vector potentials of Section 4^1 we have for the vector potential $\mathbf{A}$

$$\mathbf{A} = B\{\psi_a(x, y)\cos k_3 z\mathbf{k} - (k_3/k_a{}^2)\operatorname{grad}_s\psi_a\sin k_3 z\}, \tag{5^{1}11}$$

where B is the normalizing factor to be chosen to satisfy $(4^1 6)$. We have

$$\iiint \mathbf{A}^2\, dV = B^2(C/2) \times \left[\iint \psi_a{}^2\, dx\, dy + \frac{k_3{}^2}{k_a{}^4}\iint (\operatorname{grad}\psi_a)^2\, dx\, dy\right].$$

Since $(\operatorname{grad}\psi)^2 = \operatorname{div}(\psi\operatorname{grad}\psi) - \psi\nabla^2\psi$ this reduces to

$$\iiint \mathbf{A}^2\, dV = B^2(C/2) \times [1 + (k_3{}^2/k_a{}^2)]\iint \psi_a{}^2\, dx\, dy,$$

so if V is the volume of the resonator, the normalizing factor is

$$B^2 = \frac{2Vk_a{}^2}{Ck^2\displaystyle\iint \psi_a{}^2\, dx\, dy}. \tag{5^{1}12}$$

H (MAGNETIC) MODES

The theory for the modes of H type is quite similar. In place of $(5^1 3)$ and $(5^1 4)$ we have

$$ikE_x = \frac{\partial H_z}{\partial y} - \frac{\partial H_y}{\partial z}, \qquad ikE_y = \frac{\partial H_x}{\partial z} - \frac{\partial H_z}{\partial x}, \qquad ikE_z = \frac{\partial H_y}{\partial x} - \frac{\partial H_x}{\partial y},$$

$$-ikH_x = -\frac{\partial E_y}{\partial z}, \qquad -ikH_y = \frac{\partial E_x}{\partial z}, \qquad -ikH_z = \frac{\partial E_y}{\partial x} - \frac{\partial E_x}{\partial y}.$$

These permit us to express H_x and H_y in terms of H_z, and yield a relation analogous to $(5^1 5)$

$$(k^2 - k_3{}^2)\mathbf{H}_\delta = \operatorname{grad}_s\left(\frac{\partial H_z}{\partial z}\right). \tag{5^{1}13}$$

Similarly we find in analogy with $(5^1 6)$

$$\nabla_\delta{}^2 H_z + (k^2 - k_3{}^2)H_z = 0, \tag{5^{1}14}$$

and in analogy with $(5^1 8)$

$$\mathbf{E}_s = \frac{ik}{k^2 - k_3{}^2}\mathbf{h} \times \operatorname{grad}_s H_z. \tag{5^{1}15}$$

Since the boundary conditions require that **H** be tangential at the walls, this calls for the $\sin k_3 z$ factor in H_z which will also make $\mathbf{E}_s$ vanish at the two ends. From (5¹13) we see that, in order to make $\mathbf{H}_s$ be tangential at the cylindrical surfaces, the normal gradient of H_z must vanish at the walls.

Suppose we denote by $\varphi_b(x, y)$ and k_b the proper functions and proper values which satisfy the two-dimensional problem

$$\nabla_\delta^2 \varphi_b + k_b^2 \varphi_b = 0,$$
$$\partial \varphi_b / \partial n = 0 \quad \text{on boundary,}$$

(5¹16)

where $\partial/\partial n$ means differentiation in a direction normal to the boundary. The difference in boundary conditions between (5¹9) and (5¹16) gives rise to a different set of proper functions and proper values in the two cases.

Summarizing the results for the H type modes we have,

H type:

$$H_z = A\varphi_b(x, y)\sin k_3 z,$$
$$k^2 = k_b^2 + k_3^2,$$
$$\mathbf{H}_s = (k_3/k_b^2)A\,\mathrm{grad}_\delta\,\varphi_b \cdot \cos k_3 z,$$
$$\mathbf{E}_s = i(k/k_b^2)A\left(-\frac{\partial \varphi_b}{\partial y}\mathbf{i} + \frac{\partial \varphi_b}{\partial x}\mathbf{j}\right)\sin k_3 z.$$

(5¹17)

For the vector potential describing these modes we may take

$$\mathbf{A} = B\left(-\frac{\partial \varphi_b}{\partial y}\mathbf{i} + \frac{\partial \varphi_b}{\partial x}\mathbf{j}\right)\sin k_3 z,$$

where B is the normalizing factor whose value is readily calculated to be,

$$B^2 = \frac{2V}{Ck_b^2 \iint \varphi_b^2\,dx\,dy}.$$

(5¹18)

Note that there are modes of E type for which $k_3 = 0$, and that for these the resonant frequencies are independent of the height of the cylinder, but that the H type modes require $k_3 \neq 0$. The allowed values of k_3 are, of course,

$$k_3 = n\pi/C \quad (n, \text{ an integer}).$$

(5¹19)

We shall need a notation to designate a particular mode in a cylindrical resonator. A convenient notation is $E(n, a)$ and $H(n, b)$ to denote an E type or H type mode, respectively, built on the use of $k_3 = n\pi/C$ and the scalar functions ψ_a or φ_b, respectively. When we deal with cylindrical resonators of particular cross section the general notations a and b are replaced by more specific designations referring to special properties of the functions ψ_a and φ_b.

DOUBLE-WALLED RESONATORS

If the section of the cylindrical resonator consists of the region of space external to curve C_1 and internal to curve C_2 as in Fig. 4^1, then the interior of the cavity resonator is not a singly-connected region (which means simply that an arbitrary closed path in the region cannot be shrunk continuously to zero while staying entirely in the region). This gives rise to some special electromagnetic properties which are important. In practice the curves C_1 and C_2 are usually concentric circles but we shall see that the general properties to be discussed are independent of this particular shape.

The two most important properties to be developed are, first, that there exist zero frequency modes giving rise to an internal magnetic field not associated with an electric field, and secondly, that there exist modes for which *both* E_z and H_z vanish, whose frequency depends only on the length, not on the cross section of the cylinder. These will be called coaxial cable modes.

If E_z as well as H_z vanishes then (5^13) and (5^14) become

$$ikE_x = -\partial H_y/\partial z, \qquad ikE_y = +\partial H_x/\partial z, \qquad 0 = (\partial H_y/\partial x) - (\partial H_x/\partial y),$$

$$-ikH_x = -\partial E_y/\partial z, \qquad -ikH_y = +\partial E_x/\partial z, \qquad 0 = (\partial E_y/\partial x) - (\partial E_x/\partial y).$$

The z component of these shows that $\mathbf{E}$ or $\mathbf{H}$ may be expressed as the gradient of a scalar function $U(x, y)$. Write

then
$$\mathbf{E}_s = -\mathrm{grad}_s\, U(x, y, z);$$
$$ik\mathbf{H}_s = \mathbf{k} \times \mathrm{grad}(\partial U/\partial z). \qquad (5^120)$$

The first two equations require

$$\frac{\partial^2}{\partial z^2}\left(\frac{\partial U}{\partial x}\right) + k^2\left(\frac{\partial U}{\partial x}\right) = 0,$$

$$\frac{\partial^2}{\partial z^2}\left(\frac{\partial U}{\partial y}\right) + k^2\left(\frac{\partial U}{\partial y}\right) = 0,$$

and the third requires that

$$\left(\frac{\partial^2}{\partial x^2} + \frac{\partial^2}{\partial y^2}\right)\left(\frac{\partial U}{\partial z}\right) = 0.$$

The boundary conditions require that $\mathbf{E}_\delta$ vanish at $z = 0$ and $z = C$ so

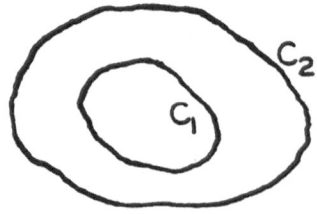

FIGURE 4^1. Sketch of double-walled resonator section.

$U(x, y, z)$ must contain the factor $\sin n\pi z/C$. Write

$$U(x, y, z) = u(x, y) \sin n\pi z/C,$$

where

$$\nabla^2 u(x, y) = 0.$$

To satisfy the boundary conditions on the cylindrical walls we must have $u = $ constant on the boundary curves C_1 and C_2.

From potential theory it is known that if the boundary consists of the single curve C_2 so the region is the entire region interior to C_2, then if $u = $ constant on the boundary it is constant throughout the inside. Such a solution for u gives vanishing electric and magnetic fields inside the resonator which shows that for such a resonator there are no modes with E_z and H_z both zero. But if the region is bounded by two curves C_1 and C_2 we may satisfy the boundary conditions by putting $u = u_1$ on C_1 and $u = u_2$ on C_2 where u_1 and u_2 are two different constants. This gives rise to a non-constant solution $u(x, y)$, which in fact is the same function of position as the electrostatic potential distribution between the two cylinders if C_1 is at potential u_1 and C_2 at potential u_2.

Since k is not involved in the boundary value problem in the section it follows that k is determined entirely by the length, so $k = n\pi/C$. Hence the result: For any shape of the bounding curves C_1 and C_2 the double-walled cylindrical resonator possesses modes whose frequencies are such that $C = n\lambda/2$ where n is an integer.

If we pass to the limit $k \to 0$ in $U = u(x, y) \sin kz$ we get $\mathbf{E}_s = 0$, but

$$\mathbf{H}_\delta = \mathbf{k} \times \operatorname{grad} u.$$

Hence the equations are satisfied by a steady magnetic field, produced by circulation of steady current up the inner cylinder and down the outer cylinder. Such zero-frequency modes always occur if the interior of the resonator is a multiply-connected region.

USE OF FUNCTION THEORY

Since u satisfies Laplace's equation in two dimensions, many results obtainable from the theory of functions of a complex variable are applicable here. For the calculations which follow let $z = x + iy$ (not to be confused with previous use of z as the coordinate along the length of the cylinder). Also let $w = u + w$. Then if $w = f(z)$ is an analytic function of z we have,

$$w = u(x, y) + iv(x, y). \tag{5^{1}21}$$

The Cauchy-Riemann conditions for the existence of a unique derivative $f'(z)$ are:

$$\partial u/\partial x = \partial v/\partial y, \qquad \partial u/\partial y = -\partial v/\partial x, \tag{5^{1}22}$$

from which it follows that u and v each satisfy Laplace's equation.

From $(5^1 19)$ it follows that, except for the fact that $\mathbf{E}$ contains a sin factor and $\mathbf{H}$ a cos factor in its dependence on the coordinate along the length of the cylinder, we have:

$$E_x = -(\partial u/\partial x), \qquad E_y = -(\partial u/\partial y),$$
$$iH_x = -(\partial v/\partial x), \qquad iH_y = -(\partial v/\partial y),$$

which can be summarized in the vector formula

$$\mathbf{E} - i\mathbf{H} = -\operatorname{grad} w. \tag{$5^1 23$}$$

Consider now any function, $w = f(z)$ such that the equation $u(x, y) = $ constant, defines a family of closed curves, successive members of which enclose the preceding members. Any two of these curves may be chosen as the bounding curves C_1 and C_2 of a double-walled cylindrical cavity resonator, and therefore each such function provides the solution for the fields in a whole family of such cavity resonators.

CIRCULAR COAXIAL CABLE

The simplest application of this general method is the solution which applies to the circular coaxial cable. This is given by the function,

$$w = \log z, \tag{$5^1 24$}$$

or

$$e^{u+iv} = z,$$

from which we find,

$$e^u = |z| = r, \quad \text{that is,} \quad u = \log r,$$

and

$$v = \operatorname{arc} \quad z = \varphi,$$

so

$$\log z = \log r + i\varphi.$$

Therefore the lines of constant u are the circles $r = $ constant and the lines of constant v are the radial lines of constant φ. The electric and magnetic fields are given by

$$\mathbf{E} - i\mathbf{H} = -\operatorname{grad} w = -(1/r)\mathbf{r}_0 - i(1/r)\boldsymbol{\varphi}_0. \tag{$5^1 25$}$$

The electric field is directed radially and the magnetic field is directed circumferentially and each varies as the inverse first power of the radius.

If the inner radius is a and the outer radius is b, then the current flowing axially in the inner conductor is $1/4\pi a$ abamp./cm and on the outer conductor is $1/4\pi b$ abamp./cm. The total current flowing on either is the same and is equal to $\frac{1}{2}$ abamp. The line integral of $\mathbf{E}$ from $r = a$ to $r = b$ is $\log(b/a)$ statvolt.

Therefore if the amplitude of excitation of the resonator is such that the maximum current amplitude is 1 ampere at $z = 0$ or any other place where $\cos n\pi z/C$ equals ± 1, the maximum voltage amplitude, which occurs at places where $\sin n\pi z/C$ equals ± 1, is equal to $60 \log(b/a)$ volts. This relation is expressed by saying that the impedance of the circular coaxial cable resonator is $60 \log(b/a)$ ohms.

More general shapes may be treated by remarking that the positional coordinates (x, y) must be periodic functions of v. There is no loss of generality in assuming the period to be 2π, and the most general complex function having this property is the Fourier series,

$$z = \sum_{m=-\infty}^{+\infty} A_m e^{m(u+iv)}. \tag{5^{1}26}$$

No extra generality arises from the inclusion of the A_0 term since this simply provides for a shift of origin in the (x, y) plane. The circular coaxial cable, already discussed, is obtained by putting $A_1 = 1$, and $A_m = 0$ for $m \neq 1$.

ELLIPTIC COAXIAL CABLE

An important simple interesting case is obtained by using (5^{1}26), and by putting

$$A_1 = A_{-1} = f/2, \qquad A_m = 0 \quad \text{for} \quad m \neq \pm 1,$$

which gives,

$$z = f(\cosh u \cos v + i \sinh u \sin v).$$

From this it follows that the curves of constant u are the ellipses

$$\left(\frac{x}{f \cosh u}\right)^2 + \left(\frac{y}{f \sinh u}\right)^2 = 1.$$

that is, confocal ellipses whose foci are at the points $(x, y) = (\pm f, 0)$. Hence this special case is appropriate to the case of a double-walled resonator where the bounding curves are confocal ellipses. Thus if the inner and outer ellipses have semi-major axes a and b, respectively, (both greater than f) then the inner and outer walls are given by u_1 and u_2 where

$$\cosh u_1 = a/f \quad \text{and} \quad \cosh u_2 = b/f.$$

The line integral from inner to outer wall is

$$u_2 - u_1 = \cosh^{-1}(b/f) - \cosh^{-1}(a/f).$$

The magnetic field at the point (u, v) is $-\operatorname{grad} v$, so the axial current per unit length on either wall is $(1/4\pi) \operatorname{grad} v$ abamp./cm. Therefore the total current on either conductor is $\frac{1}{2}$ abampere since the integral of $\operatorname{grad} v$ around either cylinder is 2π. Hence, using the definition of impedance introduced in discussing the circular coaxial cable, we find

$$60[\cosh^{-1}(b/f) - \cosh^{-1}(a/f)] = 60\log\frac{b + (b^2 - f^2)^{1/2}}{a + (a^2 - f^2)^{1/2}} \quad \text{ohms} \quad (5^127)$$

for the impedance of the cylindrical resonator whose walls are confocal elliptic cylinders.

We can give a more general result for the impedance of the resonator formed by two cylinders of arbitrary shape. No matter what the coefficients in (5^126), the total current flowing in inner or outer conductor is $\frac{1}{2}$ abampere, and the line integral of the electric vector from inner to outer conductor is $u_2 - u_1$ statvolt if the conductors are given by $u = u_1$ and $u = u_2$, respectively. Therefore the impedance of such a resonator is given by

$$60(u_2 - u_1) \quad \text{ohms} \tag{5^128}$$

for any shape whatever.

6^1. Circular Cylinder

The general results of the preceding section may be illustrated and useful practical results obtained by specializing to the case of a circular cylinder of radius R. In place of the coordinates (x, y) it is convenient to use polar coordinates (r, φ).

Equations (5^19) and (5^116) are satisfied by

$$J_m(k_a r)e^{im\varphi}, \tag{6^11}$$

where $J_m(x)$ is the Bessel function usually denoted this way, and m is an integer. The boundary conditions for modes of E type are satisfied by choosing k_a such that

$$J_m(k_a R) = 0. \tag{6^12}$$

This leads us naturally to replace the general label "a" by two integers m and p, where m is the order of Bessel function used and p is the ordinal number of the root when they are numbered in order of increasing magnitude.

Some of the roots are given in Table I. Thus we denote a particular E mode by $E(n, m, p)$ and the frequencies are given in terms of the dimensions R and C by

$$k_{Enmp}^2 = X_{mp}^2/R^2 + n^2\pi^2/C^2. \tag{6^13}$$

TABLE I. Values of X_{mp} for which $J_m(X_{mp}) = 0$

	$m = 0$	1	2	3
$p = 1$	2.405	3.832	5.135	6.379
2	5.520	7.016	8.417	9.760
3	8.654	10.173	11.620	13.017
4	11.792	13.323	14.796	16.229

TABLE II. Values of Y_{mp} for
which $J_m'(Y_{mp}) = 0$

	$m = 0$	1
$p = 1$	3.832	1.840
2	7.016	5.335
3	10.173	8.535
4	13.323	11.705

The E mode of lowest frequency is $E(0, 0, 1)$ for which

$$k_{E001} = 2.405/R \quad \text{or} \quad \lambda_{E001} = 2.61R. \tag{6^{1}4}$$

The next higher mode in the symmetric $m = 0$ series is $E(0, 0, 2)$ for which

$$k_{E002} = 5.520/R \quad \text{or} \quad \lambda_{E002} = 1.14R. \tag{6^{1}5}$$

Notice that the frequency of $E(0, 0, 2)$ is considerably greater than twice that of $E(0, 0, 1)$.

Similarly for the modes of H type the boundary conditions require that k_b be such that

$$J_m'(k_b R) = 0. \tag{6^{1}6}$$

Hence for H type waves we need a table of roots of the equation $J_m'(x) = 0$. (See Table II.) The frequencies of the H modes are therefore given by,

$$k_{Hnmp}^2 = Y_{mp}^2/R^2 + n^2\pi^2/C^2. \tag{6^{1}7}$$

From Table II we see that Y_{11} is smaller than any of the X_{mp}. However since $n = 0$ is not allowed with an H wave we find that $E(001)$ has a lower frequency than $H(111)$ for $C < 1.15R$ but the order is reversed for $C > 1.15R$.

The modes for a resonator whose shape is that of a sector of a circular cylinder are obtained by an easy generalization of the foregoing. Suppose the sector is bounded by the planes $\varphi = 0$ and $\varphi = \alpha$ where $\alpha < 2\pi$.

For the E modes we must have $E_z = 0$ on these bounding planes, which will be satisfied by using for $\psi(r, \varphi)$

$$\psi_a(r, \varphi) = (Jm\pi/\alpha)(k_a r)\sin(m\pi\varphi/\alpha) \tag{6^{1}8}$$

in place of (6^{1}1). The boundary condition at $r = R$ will require the Bessel function to vanish, and therefore the determination of the allowed frequencies calls for a knowledge of the roots of Bessel functions of fractional order for which the designation $X_{m\pi/\alpha, p}$ is a natural notation.

Similarly for the H modes we must use,

$$\varphi_b(r, \varphi) = (Jm\pi/\alpha)(k_b r)\cos(m\pi\varphi/\alpha), \tag{6^{1}9}$$

and the allowed values of k_b will be determined by requiring the radial derivative of the Bessel function to vanish at $r = R$. This calls for a knowledge of the roots $Y_{m\pi/\alpha,\,p}$ in an obvious way.

Exercise: If the cross section is a sector of opening α bounded by two circular radii, $A < r < B$, show that the appropriate Bessel junction for the E modes is

$$\psi = [CJ(k_a r) + DN(k_a r)] \sin m\pi\varphi/\alpha,$$

where J and N are two associated Bessel functions of order $m\pi/\alpha$. Discuss the dependence of the fundamental frequency on α and on A/B.

7^1. *Figure of Revolution*

In practice, resonators in the form of figures of revolution are often useful. In discussing them we use cylindrical polar coordinates (r, φ, z) whose axis is the axis of symmetry of the resonator. Such resonators possess symmetrical modes in which $E_\varphi = 0$ and H_φ is independent of φ. This section will develop the theory for modes of this class.

In cylindrical coordinates the curl equations of $(5^1 1)$ become:

$$ikE_r = \frac{1}{r}\frac{\partial H_z}{\partial \varphi} - \frac{\partial H_\varphi}{\partial z},$$

$$ikE_\varphi = \frac{\partial H_r}{\partial z} - \frac{\partial H_z}{\partial r},$$

$$ikE_z = \frac{1}{r}\frac{\partial}{\partial r}(rH_\varphi) - \frac{1}{r}\frac{\partial H_r}{\partial \varphi},$$

$$-ikH_r = \frac{1}{r}\frac{\partial E_z}{\partial \varphi} - \frac{\partial E_\varphi}{\partial z},$$

$$-ikH_\varphi = \frac{\partial E_r}{\partial z} - \frac{\partial E_z}{\partial r},$$

$$-ikH_z = \frac{1}{r}\frac{\partial}{\partial r}(rE_\varphi) - \frac{1}{r}\frac{\partial E_r}{\partial \varphi}.$$

$$(7^1 1)$$

Now assume that $H_r = H_z = 0$ and that H_φ is independent of φ. These reduce to

(a)
$$ikE_r = -\frac{\partial H_\varphi}{\partial z},$$

(b)
$$ikE_\varphi = 0,$$

(c)
$$ikE_z = \frac{1}{r}\frac{\partial}{\partial r}(rH_\varphi),$$

$$(7^1 2)$$

(d)
$$0 = \frac{1}{r}\frac{\partial E_z}{\partial \varphi},$$

(e)
$$-ikH_\varphi = \frac{\partial E_r}{\partial z} - \frac{\partial E_z}{\partial r},$$

(f)
$$0 = -\frac{1}{r}\frac{\partial E_r}{\partial \varphi}.$$

Of these (d) and (f) are satisfied because by (c) and (a), E_z and E_r are expressed in terms of H_φ which is independent of φ. A differential equation for H_φ is obtained from (e), by the use of (a) and (c):

$$\frac{\partial^2 H_\varphi}{\partial r^2} + \frac{1}{r}\frac{\partial H_\varphi}{\partial r} + \frac{\partial^2 H_\varphi}{\partial z^2} + \left(k^2 - \frac{1}{r^2}\right)H_\varphi = 0. \tag{7^13}$$

The boundary condition on **H** is satisfied without restriction on solutions of 7^13, but these must be restricted to fit the boundary conditions on **E**. From (a) and (c) of (7^12) we have

$$\mathbf{E}_s = E_r\mathbf{r}_0 + E_\varphi\varphi_0 = (i/kr)\varphi_0 \times \mathrm{grad}(rH_\varphi). \tag{7^14}$$

Suppose the boundary of the figure of revolution is the curve $f(r,z) = 0$, so the normal to it is given by

$$\mathbf{n} = \mathrm{grad}\, f.$$

The boundary condition on **E** requires that **E** have a vanishing tangential component, that is $\mathbf{n} \times \mathbf{E}_s = 0$ which gives the boundary condition in the form

$$\frac{\partial f}{\partial r}\frac{\partial}{\partial r}(rH_\varphi) + \frac{\partial f}{\partial z}\frac{\partial}{\partial z}(rH_\varphi) = 0 \quad \text{on} \quad f(r,z) = 0. \tag{7^15}$$

The form of this suggests the convenience of introducing the quantity

$$u = rH_\varphi \tag{7^16}$$

as the basic scalar function from which solutions are to be derived. The differential equation for u follows from (7^13):

$$\frac{\partial^2 u}{\partial r^2} - \frac{1}{r}\frac{\partial u}{\partial r} + \frac{\partial^2 u}{\partial z^2} + k^2 u = 0,$$

$$u = 0 \quad \text{at} \quad r = 0, \quad \frac{\partial u}{\partial r} = 0 \quad \text{at} \quad f(r,z) = 0. \tag{7^17}$$

As a simple example let us consider a length $0 < z < C$ of coaxial cable bounded by the radii $r = A$ and $r = B$. Write

$$u(r,z) = v(r)\cdot w(z);$$

then (7^17) is satisfied if

$$v'' - (1/r)v' + k_a^2 v = 0, \qquad w'' + k_3^2 w = 0, \qquad k^2 = k_a^2 + k_3^2. \tag{7^18}$$

The equation for w together with its boundary conditions is satisfied by writing

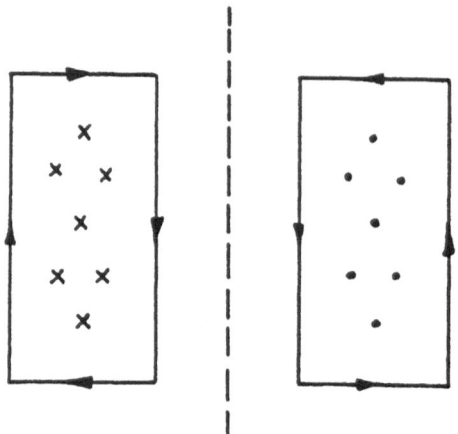

FIGURE 5^1. Sketch of currents and magnetic field in zero frequency mode of coaxial cable resonator.

$$w(z) = \cos k_3 z \quad \text{with} \quad k_3 = n\pi/C.$$

The simplest solution for v is obtained by putting $v = 1$ and $k_a = 0$. The lowest frequency mode is that corresponding to this solution for v, and to $n = 0$. Its frequency is zero and it corresponds to a steady magnetic field, unaccompanied by an electric field due to circulating steady current as shown in Fig. 5^1. The first mode of non-zero frequency corresponds to $n = 1$. Its frequency is independent of the radii A and B and is such that the length C is half a wave-length. The series of higher modes going with $v = 1$ are the harmonic series such that $C = n\lambda/2$.

The equation for $v(r)$ for $k_a \neq 0$ is satisfied by

$$v(r) = rZ_1(k_a r), \tag{$7^1 9$}$$

in which Z_1 is written for the general Bessel function of the first order. We have

$$Z_1(x) = aJ_1(x) + bN_1(x). \tag{$7^1 10$}$$

The radio $a : b$ and the parameter k_a now have to be chosen in such a way that $v'(r) = 0$ at $r = A$ and $r = B$. In this way the frequencies of the higher modes can be calculated if necessary.[5] This boundary value problem defines a sequence of values of k_a each of which can be associated with any value of n to give a mode of symmetric type.

QUARTER-WAVE COAXIAL RESONATOR

A form which has found much practical application is the quarter-wave coaxial resonator. It is generated by revolving the figure shown in Fig. 6^1 about the z axis. It is not susceptible of exact calculation. An approximate

---

[5] This problem is treated in explicit detail by Borgnis, Zeits. f. Hochfrequenztechnik **56**, 47 (1940).

FIGURE 6[1]. Sketch of cross section of quarter-wave coaxial resonator.

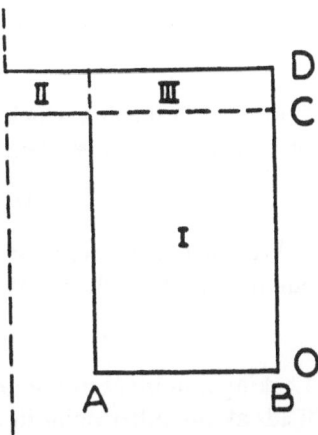

treatment runs as follows.[6] Consider separately the three regions I, II, and III. In I, especially near $z = 0$ we expect the fields to be quite accurately represented by the coaxial cable solution, so we assume,

$$u_I = a \cos kz. \quad (0 < z < C) \tag{7^{1}11}$$

Here $k = k_3$ since $k_a = 0$. By analogy with lumped-constant circuit ideas we might expect a voltage node at the end $z = C$ or D, provided the capacity formed by the region II is great enough to store the charge carried by the current in the walls without developing an appreciable voltage across the region II. That would call for infinite capacity in II: since it is actually finite there will actually be voltage across II.

In the region II we assume

$$u_{II} = brJ_1(kr). \tag{7^{1}12}$$

($k = k_a$ since here $k_3 = 0$.) In most practical cases B is small compared with a quarter wave-length so $kB \ll \pi/2$. In this region $xJ_1(x)$ is practically equal to the first term in its power series $x^2/2$ so to a good approximation

$$u_{II} = bkr^2/2,$$

which represents a uniform axial field

$$E_z = -ib$$

in the region II; hence the line integral from $z = C$ to $z = D$ on $r = A$ is $-ib(D - C)$.

Along the line $z = C$, from $r = A$ to $r = B$, we have approximately

$$E_r = -(ia/r) \sin kC,$$

[6] A more accurate analysis of this problem was given by W. W. Hansen, J. App. Phys. **10**, 38 (1939). Some interesting experimental results are given by Barrow and Mieher, Proc. I. R. E. **38**, 184 (1940).

so the line integral of E on this path is

$$-ia \sin kC \ln(B/A).$$

If we neglect the flux which goes through the region III then the line integral of E around III must vanish and therefore

$$a \sin kC \ln(B/A) = b(D - C). \tag{7^{1}13}$$

We must also have continuity of the magnetic field in the two regions, so equating them at the point $r = A$, $z = C$ we have

$$a \cos kC = bkA^2/2. \tag{7^{1}14}$$

Lacking a more accurate analysis we might just as well equate the magnetic fields at any other point in III. This point is simply preferred because u_I and u_{II} are probably better approximations to the true function there than deeper in region III.

Dividing (7^113) by (7^114) we get

$$\tan kC = \frac{2(D - C)/A^2}{k \ln B/A} \tag{7^{1}15}$$

which determines the value of k in terms of the given dimensions.

The preceding analysis is quite crude, and should leave the reader dissatisfied. Nevertheless, it corresponds to the result obtained by applying the standard engineering form of transmission line theory, as will be shown in Section 8^2. This mode of derivation has the merit that it shows more vividly what approximations have been made. The ratio of a to b can be obtained from either (7^113) or (7^114), after k has been calculated.

$$\frac{a}{b} = \frac{(D - C)}{\sin kC \cdot \ln B/A} = \frac{kA^2}{2 \cos kC}. \tag{7^{1}16}$$

The foregoing theory will now be illustrated by a numerical design example. Suppose the frequency is 150 mc so the wave-length is 200 cm. Suppose we wish to make a resonator with $D - C = 3$ inches, $A = 6$ inches, $A = 6$ inches, $B = 10$ inches, what is the proper value of C? From (7^115), since $k = 1/12.5$ in.$^{-1}$,

$$\tan kC = 4.07 \quad \text{so} \quad kC = 76.2°$$

and therefore the proper value of C is

$$C = (76.2/90)(\lambda/4) = 16.6 \text{ inches.}$$

Therefore the resonator in this case is about 15 percent shorter than a quarter wave-length. We find from (7^116)

$$b/a = 0.0647.$$

Hence, if the excitation is such that the magnetic field amplitude is 1 gauss in the corner, $z = 0$, $r = A$, the amplitude of the electric field on the axis in region

II is *b* statvolts with $a = 15.26$ (since $A = 15.26$ cm); therefore $b = 0.986$ so the line integral of electric vector from $z = C$ to $z = D$ on $r = 0$ is 2260 volts.

Another method of deriving the symmetric modes of a figure of revolution is sometimes useful. If ψ is any function satisfying the scalar wave equation

$$\nabla^2 \psi + k^2 \psi = 0, \tag{7^1 17}$$

then it is easily verified that if **C** is any constant vector, the vector,

$$\mathbf{A} = \mathbf{C} \times \operatorname{grad} \psi \tag{7^1 18}$$

satisfies the vector wave equation

$$\operatorname{curl} \operatorname{curl} \mathbf{A} = k^2 \mathbf{A} \tag{7^1 19}$$

as does also $\mathbf{B} = \operatorname{curl} \mathbf{A}$.

Suppose that we use particular solutions of $(7^1 17)$ that are independent of φ and that we choose for **C** the unit vector **k** in the direction of the axis of revolution. Then **A** is entirely in the direction of φ_0 and so is suitable to represent the magnetic field in the symmetric modes of a figure of revolution. Writing

$$\mathbf{H} = \mathbf{k} \times \operatorname{grad} \psi = \frac{\partial \psi}{\partial r} \varphi_0 \tag{7^1 20}$$

we have for the associated electric field

$$ik\mathbf{E} = \operatorname{curl} \mathbf{H} = -\frac{\partial^2 \psi}{\partial z \partial r} \mathbf{r}_0 + \frac{1}{r}\frac{\partial}{\partial r}\left(r\frac{\partial \psi}{\partial r}\right)\mathbf{k}. \tag{7^1 21}$$

The boundary conditions, when we compare with $(7^1 5)$, are that the normal derivative of $u = r(\partial\psi/\partial r)$ be zero on the boundary and that $u = 0$ at $r = 0$. The solutions of $(7^1 17)$ that are finite on the axis and independent of φ are of the form,

$$\psi = J_0(\alpha r)e^{i\beta z} \qquad \alpha^2 + \beta^2 = k^2;$$

hence a general solution will be

$$\psi(z, r) = \int_{-k}^{+k} g(\beta)J_0(k^2 - \beta^2 r)^{1/2}e^{i\beta z}\, d\beta,$$

where $g(\beta)$ is an arbitrary function. Since $J_0'(x) = -J_1(x)$,

$$u = r\frac{\partial \psi}{\partial r} = r\int_{-k}^{+k} h(\beta)J_1(k^2 - \beta^2 r)^{1/2}e^{i\beta z}\, d\beta,$$

where $h(\beta)$ is an arbitrary function.

8^1. Skin Effect

In preparation for developing the theory of losses in a cavity resonator due to the finite conductivity of the metal walls let us consider the propagation of an electromagnetic wave in a good conductor. From the field equations $(1^1 1)$

for a medium with constants ε, μ, and ρ we can find the wave equations that are satisfied by the space dependence of the field quantities whose time dependence is represented by the factor $e^{i\omega t}$ to be

$$\operatorname{curl} \operatorname{curl} \mathbf{E} = k^2 \mu(\varepsilon - 2i\lambda/\rho)\mathbf{E},$$
$$\operatorname{curl} \operatorname{curl} \mathbf{H} = k^2 \mu(\varepsilon - 2i\lambda/\rho)\mathbf{H}, \tag{8^{1}1}$$

where, as usual, $k = \omega/c$. These equations are of the same form as those governing propagation in a non-conducting medium except that the medium is characterized by a complex index of refraction,

$$n^2 = \mu(\varepsilon - 2i\lambda/\rho). \tag{8^{1}2}$$

Since for metals ρ is of the order of 10^{-8} cm it follows that the pure imaginary component of n^2 is very large compared to the real part. In fact this is true even at optical frequencies. Hence we may neglect ε in comparison with λ/ρ, which amounts physically to neglecting the displacement current relative to the conduction current, which gives as an entirely adequate approximation to the index of refraction,

$$n = (\mu\lambda/\rho)^{1/2}(1 - i). \tag{8^{1}3}$$

The general plane wave solution thus appears in the form

$$\mathbf{E} = \mathbf{E}_0 e^{i\omega t - knx} = \mathbf{E}_0 e^{-x/\delta} \cos(\omega t - x/\delta), \tag{8^{1}4}$$

in which the quantity δ is called the *skin depth* and is given by

$$\delta = \frac{1}{2\pi}(\rho\lambda/\mu)^{1/2}. \tag{8^{1}5}$$

Evidently the length δ in cm, if ρ and λ are in cm, gives a measure of the depth of penetration of the rapidly damped wave in the metal. For copper the values of δ at some representative wave-lengths are given in Table III. At 60 cycle/sec. the value of δ in copper is 0.85 cm.

If we have a wave propagated into a metal in the z direction with its electric vector along the x axis when its magnetic vector is along the y axis so,

$$H_y = H_{y0} e^{-z/\delta} \cos(\omega t - z/\delta), \tag{8^{1}6}$$

TABLE III. Depth of penetration (δ) in copper

λ cm	δ cn in copper
1	0.368×10^{-4}
3	0.670×10^{-4}
10	1.22×10^{-4}
30	2.11×10^{-4}
100	3.86×10^{-4}
1000	12.2×10^{-4}

then the associated set of conduction currents is obtained by calculating $i = (1/4\pi)$ curl H which gives $i_y = i_z = 0$ and

$$i_x = -(H_{y0}/4\pi\delta)e^{-z/\delta}[\cos(\omega t - z/\delta) - \sin(\omega t - z/\delta)], \qquad (8^17)$$

where, of course, i_x is in abamp./cm^2 if H is in gauss.

The power instantaneously converted into heat in unit volume is $\rho c i^2$ erg/cm^3 sec., so the power loss below unit area at all depths and averaged over a cycle is

$$\rho c \frac{H_{y0}^2}{4\pi\delta^2} \int_0^\infty e^{-2z/\delta} \, dz = \frac{\pi\delta}{\lambda} \cdot c \frac{\mu H_{y0}^2}{8\pi}. \qquad (8^18)$$

In this expression, $\mu H_{y0}^2/8\pi$ represents magnetic energy density at the surface (erg/cm^3) which multiplied by c gives power per unit area (erg/cm^2 sec.), a small fraction of which, $\pi\delta/\lambda$, represents the power per unit area that is absorbed in the walls.

In view of the extreme smallness of δ, we may neglect the curvature of the walls in all practical work (except very fine wires) and suppose the actual losses in unit area of a wall to be given by (8^18) even where the walls are curved. If this assumption is made, the whole power loss in the walls of a cavity resonator is given by

$$\frac{c\delta}{8\lambda} \iint \mu \mathbf{H}^2 \, dS, \qquad (8^19)$$

where the integral is extended over the whole bounding surface and $\mathbf{H}$ is the tangential magnetic vector at the surface.

Because of the finite conductivity of the walls, the electric vector is not strictly normal to a metal wall. From the expression for i_x evaluated at $z = 0$, we find

$$E_{x0} = \rho i_x = \frac{\sqrt{2\rho}H_{y0}}{4\pi\delta} \sin\left(\omega t - \frac{\pi}{4}\right). \qquad (8^110)$$

Therefore the actual tangential electric field at the surface is small compared with the tangential magnetic field in the ratio of $\sqrt{2\rho}/4\pi\delta$.

As will be seen in later sections, the losses in one cycle in a cavity resonator are small compared to the energy stored in the resonator. For this reason the approximation procedure by which we first find the fields which would exist in case of infinite conductivity and then calculate the losses gives a very good approximation.

9^1. Resonator Losses

In radio engineering, the losses of an oscillatory system are conventionally measured by giving the Q value, a kind a figure of merit which is an inverse measure of the damping.[7] The quantity Q can be defined by saying that the

---

[7] Compare Terman, *Radio Engineering* (McGraw-Hill, New York, 1937), p. 37 *et seq.* and chapter 3.

damping of a free oscillation is such that the amplitude of free oscillation contains a factor

$$e^{-\omega t/2Q}, \tag{9¹1}$$

so the total field energy in the oscillator during free oscillation is

$$W = W_0 e^{-\omega t/Q}, \tag{9¹2}$$

which amounts to saying that

$$Q = 2\pi \frac{\text{Energy stored in oscillating system}}{\text{Energy dissipated in a cycle}}. \tag{9¹3}$$

Using the formula of the preceding section for the losses in a cavity resonator we have, therefore,

$$Q = (2/\delta\mu) \frac{\iiint H^2 \, dV}{\iint H^2 \, dS}. \tag{9¹4}$$

For rough order of magnitude ideas, we observe that since $\mathbf{H}$ has a loop at the surface, the mean value of $\mathbf{H}^2$ on the surface will be roughly twice the mean value in the volume, and therefore, very roughly,

$$Q = V/\delta\mu S, \tag{9¹5}$$

where V is the volume and S the bounding area of the resonator. Hence for a resonator whose linear dimensions are large compared with δ we may expect that Q will be of the order of a linear dimension divided by the skin depth. Since the ratio of integrals in (9¹4) has the dimensions of length and δ varies as $\sqrt{\lambda}$, it follows that the Q value of geometrically similar resonators varies as the square root of a linear dimension and hence as the square root of the wave-length of any particular resonant mode.

In practical cavity resonators in the microwave region one may expect $Q > 1000$ and therefore the actually existing fields in the cavity are only very slightly different from those calculated on the assumption of perfect conductivity of the walls. In applying (9¹4) one uses therefore the fields as calculated by assuming perfect conductivity. This is the basis of all the calculations of Q that have thus far been made.

As an illustrative example, consider the Q value for the $(0, m, n)$ mode of a rectangular resonator of edges A, B, and C as discussed in Section 3^1. It is easily calculated to be

$$Q = \frac{1}{\delta\mu} \frac{ABC}{BC + 2AC \dfrac{(m/B)^2}{(m/B)^2 + (n/C)^2} + 2AB \dfrac{(n/C)^2}{(m/B)^2 + (n/C)^2}}. \tag{9¹6}$$

If the prism is square, $B = C$, this reduces to

$$Q = (B/\delta)\frac{A/B}{1 + 2(A/B)},$$

and for a cube this reduces to $Q = A/3\delta$. For the mode of lowest frequency $\lambda = \sqrt{2}A$ and, therefore,

$$Q = (\lambda/3\sqrt{2\delta\mu}).$$

The Q value of a copper cube filled with material of unit permeability is, therefore,

$$Q = 5920\sqrt{\lambda} = 7040\sqrt{A},$$

so the Q value for a copper cube resonator designed for $\lambda = 10$ cm is $Q = 18,800$.

Next we consider how the losses affect the equations of motion of the resonator coordinates $J_a(t)$ introduced in Section 4^1. When we take account of the losses it becomes necessary to introduce a damping term in (4^110). It is easily verified that the damping is correctly represented if we replace (4^110) by

$$\ddot{J}_a + (\omega_a/Q_a)\dot{J}_a + \omega^2 J_a = 4\pi c^2 I_a(t). \tag{9^17}$$

Multiplying this through by $(V/4\pi c^2)\dot{J}_a$ we find the equation for $\dot{W}_a$ which replaces (4^114):

$$\dot{W}_a = VI_a(t)\dot{J}_a(t) - (V\omega_a/4\pi c^2 Q_a)\dot{J}_a^2. \tag{9^18}$$

The second term on the right, which represents the losses, is essentially negative, since it contains the square of the speed of the resonator coordinate.

In the steady state in which $J_a(t)$ executes a harmonic time variation, the mean rate of conversion of energy into heat due to losses in the cavity walls is, therefore,

$$P = (V\omega_a/4\pi c^2 Q_a)(\dot{J}_a^2/2) = (2\pi^2 Vc/\lambda_a^3 Q_a)(J_a^2/2),$$

in which J_a stands for the amplitude of the sinusoidal variation of $J_a(t)$. If P is to be expressed in watts and J_a in amperes, then we need to write $c = 30$ ohms. The coefficient of $J_a^2/2$ in this expression will be called the resonator resistance R_a of the ath mode. Hence we have,

$$R_a = 60\pi^2 V/Q_a\lambda_a^3 \quad \text{ohms.} \tag{9^19}$$

Thus, for a cube resonator of edge A with copper walls, we have for the (011) mode, $R_a = 0.0112$ ohms.

SHUNT RESISTANCE

Another measure of the losses which is often more convenient, is called the shunt resistance of the resonator. In some calculations of the theory of electronic oscillator tubes involving cavity resonators we like to specify the amplitude, not by J_a, but by the magnitude of the line integral in volts of the electric vector along some particular path through the resonator. We have

$$\mathbf{E} = -(1/c)\dot{J}_a\mathbf{A}_a,$$

so the amplitude of the electric vector is $k_aJ_a\mathbf{A}_a$ statvolt/cm if J_a is in abamperes or $30k_aJ_a\mathbf{A}_a$ if $\mathbf{E}$ is in volt/cm and J_a is in amperes, so the line integral of $\mathbf{E}$ in volts is

$$V_a = 30k_aJ_a\int^c \mathbf{A}_a \cdot \mathbf{ds}. \tag{9^{1}10}$$

We can now calculate the resistance which when shunted across a voltage of this magnitude will dissipate power at the same rate as the actual dissipation in the resonator. This is called the *shunt resistance* and is dependent not only on the resonator and the particular mode of oscillation in question, but also on the particular path in the resonator along which V_a is calculated. Calling the shunt resistance S_a, we have

$$V_a^2/2S_a = (60\pi^2 V/Q_a\lambda_a^3)(J_a^2/2),$$

and, therefore,

$$S_a = 60Q_a\lambda_a \frac{\left(\int \mathbf{A}_a \cdot \mathbf{ds}\right)^2}{\iiint \mathbf{A}_a^2\, dV}. \tag{9^{1}11}$$

On substituting for Q_a its expression by (9^{1}4) and using the vector formula in the equation just preceding (4^{1}12) we may write the following expression for the shunt resistance:

$$S_a = \frac{480\pi^2}{\lambda_a\delta\mu} \frac{\left(\int \mathbf{A}_a \cdot \mathbf{ds}\right)^2}{\iint \mathbf{H}^2\, dS}. \tag{9^{1}12}$$

By taking the path to be from $z = 0$ to $z = A$ on the line $x = A/2$, $y = A/2$, one readily finds, for the (110) mode of a copper cube of edge A, the shunt resistance to be $S = 105600A^{5/2}$ ohms where A is in cm.

$E(00p)$ Modes of Circular Cylinder

It is, of course, not necessary nor even desirable to refer the calculation back to normalized vector potentials when one is seeking explicit results for a particular mode. To illustrate, let us derive the formulas for the $E(00p)$ modes of a circular cylinder.

From (6^1) we know that

$$E_z = AJ_0(kr) \text{ statvolt/cm},$$

where $kR = X_{0p}$, the pth root of $J_0(x) = 0$. Hence the statvoltage amplitude for a path along the axis from $z = 0$ to $z = C$ is

$$V = AC \text{ statvolt.}$$

From (5^{1}8), the magnetic field, except for the time phase which does not enter this calculation, is

$$\mathbf{H} = AJ_1(kr)\phi_0 \text{ gauss.}$$

From (8^{1}9) the power loss on each end is

$$P_E = A^2(\pi c \delta \mu/4\lambda) \int_0^R J_1^{\,2}(kr) r \, dr = A^2(\pi c \delta \mu/4\lambda)\frac{R^2}{2}J_1^{\,2}(X_{0p}),$$

and the power loss on the cylindrical wall is

$$P_C = A^2(\pi c \delta \mu/4\lambda) RC J_1^{\,2}(X_{0p}),$$

so the total power loss is

$$(P_C + 2P_E) = A^2(\pi c \delta \mu/4\lambda)(R^2 + RC)J_1^{\,2}(X_{0p}).$$

The energy stored is

$$W = \iiint \mathbf{H}^2/8\pi \, dV = (A^2/8)CR^2 J_1^{\,2}(X_{0p});$$

therefore from the definition of Q in (9^{1}3) we have

$$Q_p = (2\pi R/\rho\mu)^{1/2}\frac{C}{R+C}(X_{0p})^{1/2}. \tag{9^{1}13}$$

It is interesting to note the effect of varying C with a fixed value of R. This does not affect the frequencies of the modes, which depend solely on R. For C small compared to R, the Q value is small because the end losses remain as a "fixed charge" although there is very little volume for stored energy. As the height C is increased, Q increases, but approaches a limit as C becomes large compared to R for then the gain in extra field energy stored is offset by the corresponding increase in losses in the extra length of side wall.

The shunt resistance S_p is now obtained by equating $V^2/2S_p$ with V in volts to the total power loss expressed in watts. This gives,

$$S_p = 120(2\pi R/\rho\mu)^{1/2} \times \frac{C^2}{R^2 + RC}\frac{1}{(X_{0p})^{1/2}J_1^{\,2}(X_{0p})} \text{ ohm.} \tag{9^{1}14}$$

Comparing the formula for S_p with the one for Q_p we find that they are simply related,

$$S_p = 120(C/R)[1/X_{0p}J_1^{\,2}(X_{0p})]Q_p \text{ ohm.} \tag{9^{1}15}$$

In making numerical applications of (9^{1}14) Table IV is useful. Here X_{0p} is the pth root of $J_0(x) = 0$ as in Section 6^1.

TABLE IV. Values useful in calculating the
shunt resistance S_p

p	X_{op}	$J_1(X_{op})$	$\sqrt{X_{op}}J_1{}^2(X_{op})$
1	2.4048	+0.5191	0.417
2	5.5207	− .3403	.272
3	8,6537	+ .2705	.216
4	11.7915	− .2325	.186
5	14.9309	+ .2065	.165

COAXIAL CABLE MODES

Another example that is important in practice is the circular coaxial cable of
inner radius $r = a$ and outer radius $r = b$, terminated by $z = 0$ and $z = C$. The
fields have been discussed in (5^1) and are given by $(5^1 25)$. From $(5^1 25)$, inserting
the dependence on z which is not explicitly written there, we have,

$$\mathbf{H} = (A/r)\cos n\pi z/C\phi_0 \text{ gauss.}$$

By applying $(8^1 9)$ we find the losses in either end are

$$P_E = A^2(\pi c\delta\mu/4\lambda)\log b/a,$$

and the losses on the inner and outer walls are, respectively,

$$P_I = A^2(\pi c\delta\mu/4\lambda)C/2a, \qquad P_0 = A^2(\pi\delta\mu/4\lambda)C/2b.$$

Hence the total power loss is

$$(P_1 + P_0 + 2P_E) = A^2(\pi c\delta\mu/4\lambda) \times (2\log b/a + C/2a + C/2b).$$

The energy stored is

$$W = (A^2/8)C\log b/a,$$

and, therefore, from the definition of Q we have

$$Q_n = (2C/\rho\mu)^{1/2}\frac{2\pi\log b/a}{4\log b/a + C/a + C/b}\sqrt{n}. \tag{$9^1 16$}$$

The most natural path with respect to which we may define the shunt
resistance is from $r = a$ to $r = b$ in a plane of constant z for which $\sin n\pi z/C = \pm 1$, that is, at a voltage loop. On any such path

$$V = A\log b/a \text{ statvolt.}$$

Calculating the shunt resistance S_n from this expression and that for the power
loss, we get, in ohms,

$$S_n = 120(2C/\rho\mu)^{1/2}\frac{2(\log b/a)^2}{4\log b/a + C/a + C/b}\frac{1}{\sqrt{n}}. \tag{$9^1 17$}$$

Hence the relation between S_n and Q_n which is the analogue of (9^115) is

$$S_n = (120Q_n/\pi n)\log b/a \text{ ohms.} \tag{9^118}$$

Exercise: Consider a resonator for which $C = 150$ cm, and hence the wave-length of the lowest resonant frequency is 3 meters. Suppose $a = 30$ cm and $b = 45$ cm, what power will be required in a copper resonator to get a voltage amplitude, at a voltage loop, of one million volts? *Answer*: 1720 kilowatts.

10^1. *Spherical Resonators*

The theory of the modes of oscillation of a spherical cavity resonator may be developed as a special case of a method which is capable of more general application. Use an orthogonal curvilinear coordinate system x_1, x_2, x_3 such that the line element is

$$ds^2 = e_1{}^2 dx_1{}^2 + e_2{}^2 dx_2{}^2 + e_3{}^2 dx_3{}^2. \tag{10^11}$$

For example, for spherical polar coordinates we have

$$ds^2 = dr^2 + r^2 d\theta^2 + r^2 \sin^2 \theta \, d\varphi^2. \tag{10^12}$$

and hence

$$x_1 = r, \quad x_2 = \theta, \quad x_3 = \varphi; \quad e_1 = 1, \quad e_2 = r, \quad e_3 = r \sin \theta.$$

In such a general curvilinear coordinate system the curl equations of (5^11) become:

$$\begin{aligned}
ike_2 e_3 E_1 &= (\partial/\partial x_2)(e_3 H_3) - (\partial/\partial x_3)(e_2 H_2), \\
ike_3 e_1 E_2 &= (\partial/\partial x_3)(e_1 H_1) - (\partial/\partial x_1)(e_3 H_3), \\
ike_1 e_2 E_3 &= (\partial/\partial x_1)(e_2 H_2) - (\partial/\partial x_2)(e_1 H_1), \\
-ike_2 e_3 H_1 &= (\partial/\partial x_2)(e_3 E_3) - (\partial/\partial x_3)(e_2 E_2), \\
-ike_3 e_1 H_2 &= (\partial/\partial x_3)(e_1 E_1) - (\partial/\partial x_1)(e_3 E_3), \\
-ike_1 e_2 H_3 &= (\partial/\partial x_1)(e_2 E_2) - (\partial/\partial x_2)(e_1 E_1).
\end{aligned} \tag{10^13}$$

Suppose further that the choice of x_1 is such that $e_1 = 1$, as is the case with ordinary spherical polar coordinates, and, moreover, that the coordinate system is such that e_2/e_3 is independent of x_1. We find now that the modes fall into two types, each derivable from a scalar wave function. The two modes will be called E type and H type, corresponding to the classification already introduced in (5^1).

For the E type modes $H_1 = 0$. The fourth of (10^13) is satisfied if we write,

$$e_2 E_2 = (\partial P/\partial x_2), \qquad e_3 E_3 = (\partial P/\partial x_3).$$

If we write $P = (\partial U/\partial x_1)$ then the second and third of (10^13) give

$$H_2 = (ik/e_3)(\partial U/\partial x_3), \qquad H_3 = -(ik/e_2)(\partial U/\partial x_2).$$

The first and fifth of (10^13) give two different expressions for E_1 in terms of U. The condition that these be consistent leads to an equation for U:

$$\frac{\partial^2 U}{\partial x_1{}^2} + \frac{1}{e_2 e_3}\left(\frac{\partial}{\partial x_2}\frac{e_3}{e_2}\frac{\partial U}{\partial x_2} + \frac{\partial}{\partial x_3}\frac{e_2}{e_3}\frac{\partial U}{\partial x_3}\right) + k^2 U = 0. \qquad (10^14)$$

the most convenient expression for E_1 in terms of U being

$$E_1 = k^2 U + (\partial^2 U/\partial x_1{}^2).$$

With these choices the sixth of (10^13) is satisfied identically.

In a similar way the H type modes are given by putting $E_1 = 0$. This leads to a similar scheme of equations for deriving field components from a scalar function U which satisfies the same wave equation (10^14). To summarize, the equations for the field components in terms of U are:

E type:

$$E_1 = k^2 U + \frac{\partial^2 U}{\partial x_1{}^2}, \qquad H_1 = 0,$$

$$E_2 = \frac{1}{e_2}\frac{\partial^2 U}{\partial x_1 \partial x_2}, \qquad H_2 = \frac{ik}{e_3}\frac{\partial U}{\partial x_3}, \qquad (10^15)$$

$$E_3 = \frac{1}{e_3}\frac{\partial^2 U}{\partial x_1 \partial x_3}, \qquad H_3 = -\frac{ik}{e_2}\frac{\partial U}{\partial x_2}.$$

H type:

$$E_1 = 0, \qquad H_1 = k^2 U + \frac{\partial^2 U}{\partial x_1{}^2},$$

$$E_2 = -\frac{ik}{e_3}\frac{\partial U}{\partial x_3}, \qquad H_2 = \frac{1}{e_2}\frac{\partial^2 U}{\partial x_1 \partial x_2}, \qquad (10^16)$$

$$E_3 = +\frac{ik}{e_2}\frac{\partial U}{\partial x_2}, \qquad H_3 = \frac{1}{e_3}\frac{\partial^2 U}{\partial x_1 \partial x_3}.$$

Now we may specialize this general method to the case of a sphere, using spherical polar coordinates as in (10^12). Equation (10^14) for U becomes,

$$\frac{\partial^2 U}{\partial r^2} + \frac{1}{r^2}\left(\frac{1}{\sin\theta}\frac{\partial}{\partial\theta}\left(\sin\theta\frac{\partial U}{\partial\theta}\right) + \frac{1}{\sin^2\theta}\frac{\partial^2 U}{\partial\varphi^2}\right) + k^2 U = 0. \qquad (10^17)$$

The general solution of this is of the form

$$U = R(r)\Theta(\theta, \varphi),$$

where

$$\frac{1}{\sin\theta}\frac{\partial}{\partial\theta}\left(\sin\theta\frac{\partial\Theta}{\partial\theta}\right) + \frac{1}{\sin^2\theta}\frac{\partial^2\Theta}{\partial\varphi^2} + l(l+1)\Theta = 0 \qquad (10^18)$$

and

$$\frac{d^2R}{dr^2} + \left(k^2 - \frac{l(l+1)}{r^2}\right)R = 0. \tag{10^{1}9}$$

The quantity l assumes integral values for solutions of (10^{1}8) that are finite and single-valued in all directions. Any solution of (10^{1}8) is known as a spherical harmonic and a great deal of information can be found about them in books on harmonic analysis. The solutions of (10^{1}9) will be called spherical Bessel functions, from a terminology introduced by Morse.

$$R(r) = (kr)z_l(kr), \tag{10^{1}10}$$

where

$$z_l(x) = (\pi/2x)^{1/2}Z_{l+1/2}(x).$$

For any spherical Bessel function we have,

$$z_{n-1} + z_{n+1} = (2n + 1/x)z_n,$$

$$(d/dx)z_n(x) = [nz_{n-1} - (n+1)z_{n+1}]/(2n+1),$$

$$(d/dx)x^{n+1}z_n = x^{n+1}z_{n-1}, \tag{10^{1}11}$$

$$(d/dx)x^{-n}z_n = -x^{-n}z_{n+1}.$$

It is necessary to specialize to the particular. Bessel function which is finite at $r = 0$. These are denoted by $j_l(x)$. The functions j_lx are given in Table V.

TABLE V. Values of $j_l(x)$

l	$j_l(x)$
0	$\sin x/x$
1	$\sin x/x^2 - \cos x/x$
2	$(3/x^3 - 1/x)\sin x - (3/x^2)\cos x$
3	$(15/x^4 - 6/x^2)\sin x - (15/x^3 - 1/x)\cos x$

The spherical harmonics will be used in the following notation:

$$\Theta(\theta, \varphi) = \Theta(l, m)e^{im\varphi}, \tag{10^{1}12}$$

where

$$m \geq 0 \begin{cases} \Theta(l, m) = (-1)^m\left[(2l+1)\dfrac{(l-m)!}{(l+m)!}\right]^{1/2} \times \sin^m\theta\,\dfrac{d^m}{d(\cos\theta)^m}P_l(\cos\theta) \\ \Theta(l, -m) = +\text{same expression} \end{cases}$$

in which $P_l(\cos\theta)$ is the lth Legendre polynomial.

A list of explicit expressions for some of the spherical harmonics is given below:

$$\Theta(0,0) = 1,$$

$$\Theta(1,0) = \sqrt{3}\cos\theta,$$

$$\Theta(2,0) = \sqrt{5/2}(3\cos^2\theta - 1),$$

$$\Theta(3,0) = \sqrt{7/2}(2\cos^3\theta - 3\cos\theta\sin^2\theta).$$

The coefficients appearing here are chosen to normalize in such a way that

$$\iint |\Theta|^2 \sin\theta\, d\theta\, d\varphi = 4\pi,$$

the integral extending over all directions in space.

In calculations involving the spherical harmonics the following properties of them are useful,

$$\frac{\partial}{\partial\theta}\Theta(l,m) = \tfrac{1}{2}[(l-m)(l+m+1)]^{1/2}\Theta(l,m+1)$$

$$- \tfrac{1}{2}[(l+m)(l-m+1)]^{1/2}\Theta(l,m-1),$$

$$\cos\theta\,\Theta(l,m) = \Theta(l+1,m)\left[\frac{(l+1-m)(l+1+m)}{(2l+1)(2l+3)}\right]^{1/2}$$

$$+ \Theta(l-1,m)\left[\frac{(l-m)(l+m)}{(2l-1)(2l+1)}\right]^{1/2}, \qquad (10^1 13)$$

$$\sin\theta\,\Theta(l,m) = -\Theta(l+1,m+1)\left[\frac{(l+m+1)(l+m+2)}{(2l+1)(2l+3)}\right]^{1/2}$$

$$+ \Theta(l-1,m+1)\left[\frac{(l-m)(l-m-1)}{(2l-1)(2l+1)}\right]^{1/2}$$

$$= \Theta(l+1,m-1) - \Theta(l-1,m-1).$$

The final result is that the E and H modes of the sphere may be derived from the following expression for U:

$$U = kr \cdot j_l(kr)\Theta(l,m)e^{im\varphi}, \qquad (10^1 14)$$

with $|m| \leqslant l$, and $l = 0, 1, 2, 3\ldots$.

The modes of *electric type* are given by applying $10^1 5$:

$$E_r = (k^2/r^2)l(l+1)U,$$

$$E_\theta = \frac{k}{r}\frac{\partial}{\partial(kr)}[krj_l(kr)] \cdot \frac{\partial}{\partial\theta}\Theta(l,m) \cdot e^{im\varphi},$$

$$E_\varphi = \frac{ikm}{r\sin\theta}\frac{\partial}{\partial(kr)}[krj_l(kr)]\Theta(l,m)e^{im\varphi}, \qquad (10^1 15)$$

$$H_r = 0,$$

$$H_\theta = \frac{-km}{r\sin\theta}U,$$

$$H_\varphi = -\frac{ik}{r}[krj_l(kr)]\frac{\partial}{\partial\theta}\Theta(l,m)e^{im\varphi}.$$

(10¹15)

The boundary conditions require that $E_\theta = E_\varphi = 0$ at $r = R$, if R is the radius of the sphere. Hence for the E type modes we must have

$$kR = S_{nl},$$

(10¹16)

where S_{nl} is a root of the equation

$$(d/dx)(xj_l(x)) = 0.$$

Similarly for modes of *magnetic type*:

$$E_r = 0,$$

$$E_\theta = (km/r\sin\theta)U,$$

$$E_\varphi = \frac{ik}{r}krj_l(kr)\frac{\partial}{\partial\theta}\Theta(l,m)e^{im\varphi},$$

$$H_r = \frac{k^2}{r^2}l(l+1)U,$$

$$H_\theta = \frac{k}{r}\frac{\partial}{\partial(kr)}[krj_l(kr)]\frac{\partial}{\partial\theta}\Theta(l,m)e^{im\varphi},$$

$$H_\varphi = \frac{ikm}{r\sin\theta}\frac{\partial}{\partial(kr)}[krj_l(kr)]\Theta(l,m)e^{im\varphi}.$$

(10¹17)

For the modes of magnetic type the boundary conditions require that

$$kR = T_{nl},$$

(10¹18)

where T_{nl} is a root of the equation $j_l(x) = 0$.

The modes will be designated by the notation $E(n,l,m)$ and $H(n,l,m)$. Inspection of (10¹15) and (10¹17) shows that there is no solution corresponding to $l = 0$, $m = 0$, hence the least value of l is unity. Since the roots S_{nl} and T_{nl} are independent of m, it follows that there are $(2l+1)$ modes of either E or H type going with a particular (n,l) all of which have the same frequency. This degeneracy arises from the spherical symmetry.

The values of the roots for the fundamental modes of each type are

$$S_{11} = 2{,}74, \qquad T_{11} = 4.49.$$

(10¹19)

Therefore, the resonant wave-length for the $E(11)$ modes is $2.29R$.

The spherical harmonics introduced in (10¹12) are appropriate for problems involving the complete sphere for they are finite at $\theta = 0$ and $\theta = \pi$, the singular points of (10¹8). Spherical coordinates may also be used to discuss the fields in a conical resonator bounded by $\theta = \theta_0$ as well as $r = R$, or in a region consisting of a sphere with two conical dimples cut out of it, that is the region $0 < r < R$ and $\theta_0 < \theta < \theta_1$. To deal with such cases it is necessary to introduce more general solutions of (10^18) which have singularities at the excluded poles 0 or π for θ.

SPHERE WITH CONIAL DIMPLES

As a specific example consider the fundamental electric mode for the sphere with conical dimples.[8] For this the appropriate U function is

$$U = \log \tan(\theta/2) \cdot (\sin kr/kr) \tag{10^120}$$

which would not be admissible for a complete sphere because of its logarithmic singularities. Since this corresponds to an $l = 0$ solution, it follows that $E_r = 0$, as well as $H_r = 0$. Applying (10^115) we find that the only non-vanishing field components are

$$E_\theta = k^2(1/\sin \theta)(\cos kr)/kr,$$
$$H_\varphi = -ik^2(1/\sin \theta)(\sin kr)/kr. \tag{10^121}$$

Since E_r and E_φ vanish everywhere the only boundary condition to be imposed is that $E_\theta = 0$ at $r = R$, which requires that

$$kR = (n + \tfrac{1}{2})\pi$$

independently of the location of the angular boundaries. The lowest mode is that for which $n = 0$ and for this the wave-length is exactly equal to four times the radius of the sphere.

For this solution E_θ becomes infinite as r approaches zero, but in such a way that the line integral of E along a path of constant r from one dimple to the other is finite. The solution is therefore appropriate for representation of the fields which exist when the apices of the two conical dimples are not quite in electrical contact.

Exercises: 1. Show that the Q value for this mode of the dimpled spherical resonator is

$$Q = \left(\frac{4R}{\rho\mu}\right)^{1/2} \frac{\log \tan \theta_1/2 - \log \tan \theta_0/2}{\log \tan \theta_1/2 - \log \tan \theta_0/2} + I(\csc \theta_1 + \csc \theta_0)$$

in which

$$I = \int_0^{\pi/2} \frac{\sin^2 x}{x} dx = 0.825.$$

Show that if $\theta_1 = \pi - \theta_0$, the Q value as a function of θ_0 has a maximum at about $\theta_0 = 34°$.

---

[8] W.W. Hansen and R.D. Richtmyer, J. App. Phys. **10**, 189 (1939).

2. Show that the shunt resistance for this mode, with voltage measured from apex to apex is

$$S = 120 \left(\frac{4R}{\rho\mu}\right)^{1/2} \frac{\log^2(\tan\theta_1/2/\tan\theta_0/2)}{\log\dfrac{\tan\theta_1/2}{\tan\theta_0/2} + I(\csc\theta_1 + \csc\theta_0)}.$$

Show that for $\theta_1 = \pi - \theta_0$, the shunt resistance as a function of θ_0 has a maximum at about 9°.

Chapter II. Transmission Lines[1]

In low frequency radio work, power is transmitted from one place to another by transmission lines consisting of two conductors, such as parallel wire lines, or coaxial cable. Such lines also play a great role in micro-wave radio. In addition it becomes practical at the shorter wave-lengths to transmit power through hollow pipes. In the literature it has been customary to call two-conductor lines transmission lines and to call hollow pipes wave guides. This chapter deals with two-conductor lines while the properties of hollow wave guides will be developed in Chapter III.

1^2. Two-Conductor Transmission Lines

The commonest form of two-conductor transmission line is the coaxial cable, consisting of an inner circular conductor of radius $r = a$, and an outer circular conductor of radius $r = b$. The theory is very closely related to that of co-axial cavity resonators discussed in (5^1) under the subhead, double-walled resonators.

Let z be the coordinate along the length of the line and suppose any section by a plane $z = $ constant, gives a region bounded by two curves C_1 and C_2, the latter enclosing the former, as in Fig. 4^1.

We seek solutions of (5^11) in which the dependence on z is given by a factor $\exp(-ik_3 z)$, which when combined with the time factor $e^{i\omega t}$, represents a progressive wave moving in the $+z$ direction. The phase velocity v_p is given by

$$v_p = w/k_3. \tag{$1^2$1}$$

If one writes $k = \omega/c$ and assumes $E_z = H_z = 0$, then Eqs. (5^11) are found to give the following for the factors which represent the dependence of the field components on x and y,

---

[1] The general literature on electrical transmission lines is very extensive since this topic is important for long power transmission lines as well as in telegraphy and telephony. In this chapter a brief account of the subject is given from the point of view of micro-wave applications.

$$kE_x = k_3 H_y, \qquad kE_y = -k_3 H_x, \qquad 0 = (\partial H_y/\partial x) - (\partial H_x/\partial y),$$
$$-kH_x = k_3 E_y, \qquad -kH_y = -k_3 E_x, \qquad 0 = (\partial E_y/\partial x) - (\partial E_x/\partial y).$$

The z component of these shows that $\mathbf{E}$ or $\mathbf{H}$ may be expressed as the gradient of a scalar function, $u(x, y)$. Write

$$\mathbf{E}_s = -\operatorname{grad} u(x, y),$$
$$-k\mathbf{H}_s = k_3 \mathbf{k} \times \operatorname{grad} u(x, y), \tag{1²2}$$

which are the transmission line analogues of (5¹19). These equations imply that $k_3 = \pm k$, hence, the phase velocity of the wave is $\pm c$. The z component of the equation for curl $\mathbf{H}$ requires that $u(x, y)$ satisfy Laplace's equation in the cross section

$$\nabla^2 u(x, y) = 0. \tag{1²3}$$

Since this is the same as (5¹20) with the same boundary conditions, namely $u = $ constant on C_1 and C_2 it follows that the discussion following (5¹20) is applicable here.

Suppose $u(x, y)$ is a solution of the boundary value problem such that the coordinates (x, y) are periodic functions with period 2π in the conjugate harmonic function $v(x, y)$, as in (5¹26). Let the values of u corresponding to the inner and outer conductors be u_1 and u_2, respectively. Then for a wave propagated in the $+z$ direction we have

$$\mathbf{E}_s = -\operatorname{grad} u(x, y) \cos(\omega t - kz),$$
$$\mathbf{H}_s = -\mathbf{k} \times \operatorname{grad} u(x, y) \cos(\omega t - kz), \tag{1²4}$$

and for a wave propagated in the $-z$ direction,

$$\mathbf{E}_s = -\operatorname{grad} u(x, y) \cos(\omega t + kz),$$
$$\mathbf{H}_s = +\mathbf{k} \times \operatorname{grad} u(x, y) \cos(\omega t + kz). \tag{1²5}$$

CHARACTERISTIC IMPEDANCE

As remarked just before Eq. (5¹28), the solution in which (x, y) have the period 2π in v, corresponds to a current amplitude of $\frac{1}{2}$ abampere in the inner and outer conductors, and to a statvoltage amplitude $(u_2 - u_1)$ in the line integral of $\mathbf{E}$ from one conductor to the other in a plane of constant z. Hence the ratio of voltage amplitude to current amplitude in amperes is

$$Z = 60(u_2 - u_1) \text{ ohm.} \tag{1²6}$$

This quantity is called the characteristic impedance, or surge impedance of the transmission line. The surge impedance of the line, as thus defined, is therefore the same as the impedance of the double-walled cylindrical cavity resonator as introduced in Section 5¹.

Another way of looking at the surge impedance may help to bring out more clearly its physical significance. Let C_1 be the capacity per unit length of the condenser formed by the two conductors of the line. Suppose one of them is at potential O and the other at V statvolt. Then the charge per unit length is $C_1 V$ e.s.u./cm. If now the line is to be fed in such a way as to set up on it a wave travelling from left to right with speed c, then at the input end one must supply current which will keep the charge on each conductor at its requisite amount. This is a current $cC_1 V$ e.s.u./sec. or $C_1 V$ abamp. Hence the input impedance in ohms, the ratio of voltage to current in amperes, is

$$Z = (300\, V/10C_1 V) = 30/C_1 \text{ ohms.} \qquad (1^2 7)$$

If the space between the conductors is filled with a medium whose constants are (ε, μ) it is easy to see that the input impedance of the line is

$$Z = (\mu/\varepsilon)^{1/2} \cdot 30/C_1 \text{ ohm,} \qquad (1^2 8)$$

in which C_1 is the geometrical capacity per unit length in the absence of the medium.

For circular coaxial cable, the capacity per unit length is

$$C_1 = 1/(2 \log b/a) \qquad (1^2 9)$$

and, therefore, the surge impedance of such a transmission line is

$$Z = (\mu/\varepsilon)^{1/2} 60 \log b/a. \qquad (1^2 10)$$

Likewise for coaxial confocal elliptic cylinders of focal length f and inner and outer semi-major axes a and b, as in $(5^1 27)$, the surge impedance is

$$Z = (\mu/\varepsilon)^{1/2} 60[\cosh^{-1}(b/f) - \cosh^{-1}(a/f)]. \qquad (1^2 11)$$

Transmission lines in which one conductor completely surrounds the other are to be preferred to "open" lines like a pair of parallel wires, because they are self-shielding and do not interact with nearby conductors. The theory developed in this section is, however, equally applicable to open lines in which the curves C_1 and C_2, which bound the conductors, lie external to each other. The result in $(1^2 8)$ is applicable in this case as well, it being supposed that the line is "balanced to ground," that is, that the potential of C_2 is as much negative with respect to distant points as C_1 is positive.

An important special case is the pair of round wires each of radius a, whose center-to-center distance is d, for which

$$C_1 = 1/(4 \cosh^{-1} d/2a)$$

so

$$Z = (\mu/\varepsilon)^{1/2} 120 \cosh^{-1} d/2a. \qquad (1^2 12)$$

Another example is the parallel plate transmission line, made of two plates whose width is b, separated by a distance d which is small compared with the width. For such a line

$$C_1 = d/4\pi b \quad \text{so} \quad Z = (\mu/\varepsilon)^{1/2} 120\pi d/b. \qquad (1^2 13)$$

Exercise: What separation between the surfaces of the wires of a parallel wire transmission line is required to make the surge impedance of the line equal to 73 ohms? *Answer*: About 19 percent of the diameter of a wire.

What should be the ratio of the radii of a coaxial cable in order to give a surge impedance of 73 ohms? *Answer*: $b/a = 3.36$.

(The point of these questions is that the radiation resistance of a half-wave dipole is approximately 73 ohms.)

TRANSMISSION LINE EQUATIONS

We have developed the theory of transmission lines from the point of view of the field theory. In the engineering literature[2] the subject is usually approached as an extension of the theory of networks having lumped constants. This method will now be briefly presented in order to compare it with what has gone before. (See Fig. 1[2].)

The line is regarded as equivalent to the limiting case of a circuit of the type shown, in which the meshes are assigned smaller parameters and more meshes are put in per unit length in such a way that, for example, the inductance per mesh multiplied by the number of meshes per unit length approaches a definite limit L, the inductance per unit length. A similar situation exists for the resistance R, the conductance G and the capacitance C per unit length.

If $V(z, t)$ is the potential difference (volts) of the upper line with respect to a point on the lower at the same z, and if $I(z, t)$ is the current (amperes) flowing toward the right in the upper line and toward the left in the lower line, then we must have

$$(\partial V/\partial z) = -RI - L(\partial I/\partial t), \qquad (1^2 14)$$

where R and L are resistance (ohms) and inductance (henries) per unit length. Similarly if G and C are conductance (mhos) and capacitance (farads) per unit length then

$$(\partial I/\partial z) = -GV - C(\partial V/\partial t). \qquad (1^2 15)$$

These two equations form the basis of the circuit theory approach to trans-

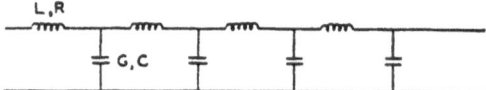

FIGURE 1[2]. Equivalent lumped-constant circuit of a transmission line.

[2] For a good elementary introduction see Everitt, *Communication Engineering* (McGraw-Hill, New York, 1937), chapters 4 and 5. Also Guillemin, *Communication Networks* (John Wiley, New York, 1935), Vol. 2. Some important recent papers are: Nergaard, RCA Rev. **3**, 156 (1938); Nergaard and Salzberg, Proc. I. R. E. **27**, 579 (1939); Reukema, Elec. Eng. **56**, 1002 (1937); King, Proc. I. R. E. **23**, 885 (1935); Mason and Sykes, Bell Sys. Tech. J. **16**, 275 (1938).

mission line theory. The field theory treatment given in the first part of this section corresponds to the ideal case in which R and G are negligible.

Before going on to discuss solutions of (1^214) and (1^215) it is desirable to connect their derivation with the field theory. In the first place we speak of "potential difference" between the two lines. Yet we know that a rapidly-varying electric field is not derivable from a scalar potential. We can remove this ambiguity by agreeing that $V(z, t)$ means the line integral of $\mathbf{E}(x, y, t)$ on a path from one line to the other, in a plane of constant z. Since we have seen that in such a plane the $\mathbf{E}_s$ is derivable from a scalar potential (1^22) there is no need further to specify the path in the plane of constant z.

The connection of "current" with the field quantities is to be understood as follows. We take the line integral of $\mathbf{H}$ around a closed path surrounding either line in a plane of constant z and very close to one of the conductors. From the equation curl $\mathbf{H} = 4\pi\mathbf{i} + (1/c)\mathbf{D}$ we have

$$\int \mathbf{H} \cdot d\mathbf{s} = \iint \operatorname{curl} \mathbf{H} \cdot d\mathbf{S} = 4\pi \iint \mathbf{i} \cdot d\mathbf{S}.$$

The line integral is over the path just described; the surface integral is on a plane of constant z bounded by this path. The displacement current makes no contribution to this line integral because it is everywhere normal to the conductor. Hence the total current $I(z, t)$ in a conductor is

$$I(z, t) = (1/4\pi) \int \mathbf{H} \cdot d\mathbf{s}, \tag{1^216}$$

where I is in abamperes if H is in gauss.

As to inductance per unit length, that is to be understood as follows. The spatial distribution of magnetic field in the space between the conductors is the same for the high frequency case as it is for the direct current. The magnetic field energy stored between z and $z + dz$ can be regarded as $\iiint (\mathbf{H}^2/8\pi)\, dv$ in the space between these planes. Equating this to $(L dz)I^2/2$ we obtain a suitable precise definition of L, the inductance per unit length. If I is in abamperes and the energy is in ergs then L is a pure number.

In the same way the capacitance per unit length is related to the electric field energy stored between the planes z and $z + dz$ by the relation,

$$(C\, dz)V^2/2 = \iiint (E^2/8\pi)\, dv.$$

If I is in statvolts and the energy is in ergs, then C is a pure number. It can be shown that $LC = 1$ at frequencies such that the magnetic flux in the conductors is negligible. The resistance per unit length has to be defined with due regard to the skin effect and is the sum of the resistance per unit length in each of the two lines. The conductance per unit length arises from the dissipative characteristic of the dielectric as discussed further in Section 5^2.

VOLTAGE AND CURRENT DISTRIBUTION

We look now for a solution of $(1^2 14)$ and $(1^2 15)$ in which the dependence of V and I on position is that associated with progressive simple harmonic waves, hence,

$$V = V e^{i(\omega t - kz)}, \qquad I = I e^{i(\omega t - kz)}. \tag{$1^2 17$}$$

Substituted in $(1^2 14)$ and $(1^2 15)$ this gives for the voltage and current amplitudes

$$ikV = (R + i\omega L)I, \qquad ikI = (G + i\omega C)V. \tag{$1^2 18$}$$

This pair of equations leads to non-vanishing values of V and I only if the propagation constant k have the value

$$k^2 = \omega^2 (L - iR/\omega)(C - iG/\omega) = -(R + i\omega L)(G + i\omega C). \tag{$1^2 19$}$$

In case the line is without loss, so that $R = 0$ and $G = 0$, this reduces to

$$k = \omega (LC)^{1/2}, \tag{$1^2 20$}$$

and the waves are propagated without attenuation in either direction and with the phase velocity $1/(LC)^{1/2}$. From the definitions of L and C it follows that this is equal to c, the velocity of light.

In the general case of a line with loss, Eq. $(1^2 19)$ leads to a complex value of k with means simply that the wave is attenuated in being propagated along the line. In general the magnitude of the attenuation (measured by the imaginary part of k) depends on the frequency, and the line introduces distortion in transmitting a signal which is not a monochromatic wave. However, in the special case that $LG = RC$ it is easily seen that the attenuation is independent of ω and the real part of k is proportional to ω, hence, the phase velocity is the same for all frequencies. Such a line is called distortionless.

Further developments of the theory along these lines are of the greatest importance in power engineering for long-distance transmission of electric power, and in telephony at audio- or carrier frequencies. For that reason the theory has had a very thorough practical development which can be found in standard textbooks and will not be fully developed here.

2^2. Transmission Line with Load

Consider a transmission line terminated at $z = L$ by a load of arbitrary impedance Z_L. In general, waves will exist on the line which are travelling both to and from the load. These interfere with each other, producing a standing wave system superposed on a progressive wave. In consequence, the ratio of voltage to current is different at different points on the line. Suppose the line characteristic impedance is Z_0.

If the voltage amplitudes of the waves travelling toward $+z$ and $-z$ are V_1 and V_2, respectively, then the voltage at any point z is the real part of

$$V = (V_1 e^{-kz} + V_2 e^{+ikz})e^{i\omega t}, \tag{$2^2$1}$$

and the total current flowing in the line at z is

$$I = (1/Z_0)(V_1 e^{-ikz} - V_2 e^{+ikz})e^{i\omega t}. \tag{$2^2$2}$$

Note that V_1 and V_2 are in general complex numbers.

At $z = L$, where the line is terminated in the impedance Z_L we must have $V = Z_L I$ so

$$Z_0 \frac{V_1 e^{-ikL} + V_2 e^{+ikL}}{V_1 e^{-ikL} - V_2 e^{+ikL}} = Z_L. \tag{$2^2$3}$$

At the other end of the line $z = 0$, the input impedance Z is the ratio of V to I so

$$Z_0(V_1 + V_2)/(V_1 - V_2) = Z. \tag{$2^2$4}$$

Equation ($2^2$3) determines the ratio V_2/V_1 of reflected to incident waves. Solving for this ratio we have,

$$(V_2 e^{ikL}/V_1 e^{-ikL}) = (Z_L - Z_0)/(Z_L + Z_0). \tag{$2^2$5}$$

Here $V_2 e^{ikL}$ is the amplitude of the reflected wave at $z = L$ and $V_1 e^{-ikL}$ is the amplitude of the incident wave at the load. From ($2^2$5) we see that $V_2 = 0$ if $Z_L = Z_0$, that is, the reflected wave vanishes if the load impedance matches that of the line.

It is convenient to introduce an auxiliary quantity ψ by the defining relation

$$V_2/V_1 = -e^{-2\psi}, \tag{$2^2$6}$$

in terms of which we note that ($2^2$3) and ($2^2$4) can be written

$$Z_L = Z_0 \tanh(\psi - ikL), \tag{$2^2$7}$$

$$Z = Z_0 \tanh \psi. \tag{$2^2$8}$$

If we write

$$Z = Z_0 \tanh(u + iv),$$

$$Z_L = Z_0 \tanh(u_L + iv_L),$$

we see that

$$u + iv = \psi \quad \text{and} \quad u_L + iv_L = \psi - ikL,$$

and, therefore,

$$u = u_L \quad \text{and} \quad V = v_L + kL. \tag{$2^2$9}$$

Suppose now that on an impedance plane, $Z = R + iX$, we plot the two mutually orthogonal families of curves corresponding to $u = $ constant and $v = $ constant. The load impedance Z_L will correspond to a pair of values u_L, v_L. From ($2^2$9) we see that the impedance transformation produced by putting in an electrical length kL of the transmission line corresponds to a displacement along the curve $u = j_L$ from the point $v = v_L$ to the point $v = v_L + kL$. It is therefore of great importance to learn more about the curves defined by the transformation

$$Z = Z_0 \tanh(u + iv) = Z_0 \frac{\tanh u + i \tan v}{1 + i \tanh u \tan v}.$$

The curve $u = 0$, gives $Z = iZ_0 \tan v$, hence, Z sweeps out the imaginary axis as v increases from 0 to π. For u infinite we have $Z = Z_0$ for all values of v, and the "curve" has shrunk to a point. For $v = 0$, we have $Z = Z_0 \tanh u$ which sweeps over the part of the real axis between 0 and Z_0 as u increases from 0 to infinity. For $v = \pi/2$, we have $Z = Z_0 \coth u$ which sweeps over the real axis from infinity to Z_0 as u increases from 0 to infinity. Therefore the curve $u = $ constant intersects the real axis at two points, namely $(Z_0 \tanh u, 0)$ and $(Z_0 \coth u, 0)$. The curve $u = $ constant is, in fact, a circle whose center is at

$$(Z_0 \coth 2u, 0)$$

and whose radius is $Z_0/\sinh 2u$.

Similarly the curves $v = $ constant form an orthogonal family of circles with center at

$$(0, -Z_0 \cot 2v)$$

and with radius equal to $Z_0/\sin 2v$.

Suppose we are given Z_L and wish to determine what kind of line, as regards the value of Z_0, and how much, given by kL, should be introduced in order to transform to a given input impedance Z. In Fig. 2^2A, we draw the perpendicular bisector of the line $Z_L Z$. Its intersection with the real axis will be the center of the circle $u = u_L$ along which the transformation proceeds as various lengths of line of the as yet unknown Z_0 are introduced. This circle will intersect the real axis in two points, the product of whose abscissas is equal to the square of Z_0. This enables the calculation of Z_0 after which it can be plotted on the diagram. With Z_0 known we can now carry out the construction indicated in detail in the upper part of Fig. 2^2A which permits us to locate $C(v_L)$ and $C(v_L + kL)$, the centers of the circles $v = v_L$ and $v = v_L + kL$. Finally $2kL$ is the angle at Z_0 subtended between the lines drawn out to the centers of the two circles.

It is evident that the frequency dependence of Z arises jointly from any inherent frequency dependence there may be in Z_L and the variation of electrical length of the line due to the variation in k. If $Z_L(k)$ is given one may construct a series of points giving the corresponding values of $Z(k)$ as a means of determining the frequency dependence of line and load. In this connection it is instructive to note that if the line is many wave-lengths long then a small fractional change in k will cause kL to change by several times 2π. This in itself produces a variation of several revolutions around the Z_0 point on the plane when connected with what is usually a rather slow variation with k of Z_L, except in case Z_L exhibits a sharp resonance in the range of frequencies involved.

FIGURE 2^2A. Geometrical construction for determining Z_0 and kL, given Z_0 and Z_L. Detailed construction for one case is shown in the upper figure.

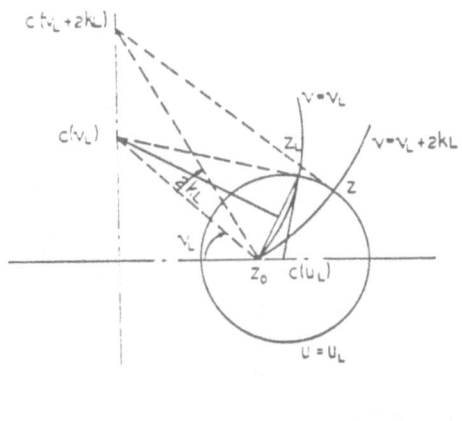

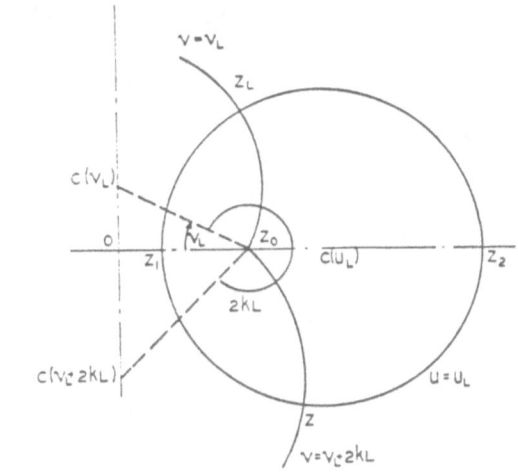

If very many calculations of this kind are to be made it is convenient to prepare special plotting paper, as in Fig. 2^2B, on which the circles u = constant and v = constant are superposed on ordinary Cartesian scales for R and X, for some particular value of Z_0. The values of u and v corresponding to given R and X then can be read at a glance with sufficient accuracy for most purposes.

These geometrical constructions suffer from the complication that the motion of Z along the circle $u = u_L$ is non-uniform as kL is increased at a constant rate. Another disadvantage is that an infinite half-plane is required on which to carry out the calculations for all possible impedances. This suggests seeking a diagram in which the circles $u = u_L$ are all concentric and the curves v = constant become straight lines running out from the common center, as in ordinary polar coordinates.

We can see that this is possible, and how to construct the new kind of diagram, by making use of the properties of the stereographic projection of a plane on a sphere. Suppose in Fig. 2^2C a sphere of diameter Z_0 is tangent to

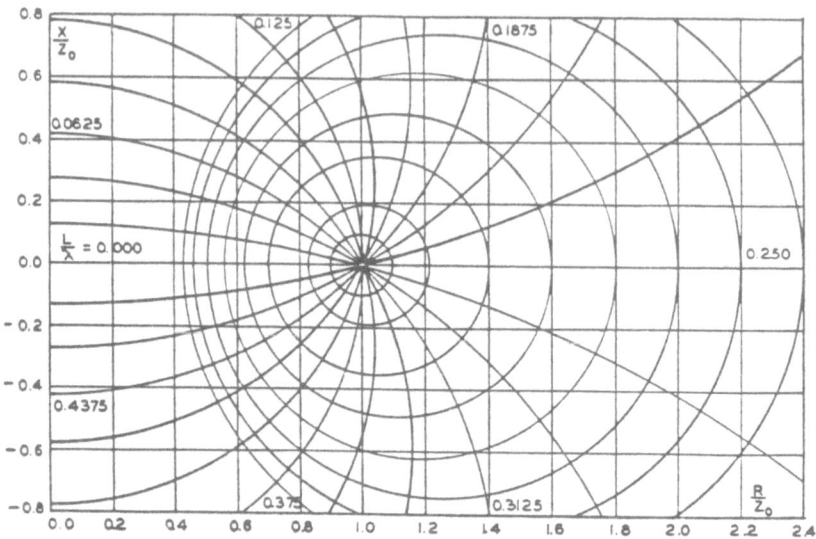

FIGURE 2^2B. Special plotting paper for impedance calculations with the circles, $u =$ constant, $v =$ constant superposed on Cartesian scales for R and X.

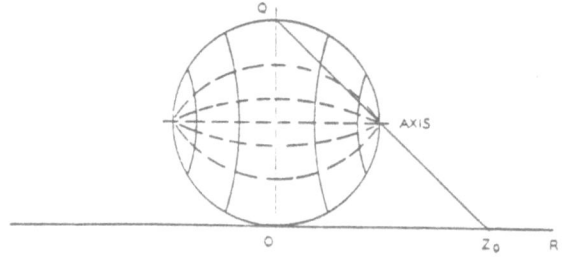

FIGURE 2^2C. Stereographic projection of the RX plane on a sphere with diameter Z_0.

the RX plane at the origin and let Q be the end of the diameter through O that is opposite O. Any point Z in the RX plane is associated with a point Z' on the sphere which is the intersection of the line QZ with the sphere. It is a property of this projection that any circle on the plane transforms into a circle on the sphere, and vice versa.

If now we think of the system of small circles and meridian great circles laid out on the sphere, having as its axis a diameter parallel to the R axis, then we can readily see that the system of circles $u =$ constant on the Z plane corresponds to the circles of constant "latitude" and that the circles $v =$ constant on the Z plane correspond to great circles on constant "longitude" on the sphere.

The situation for $u = 0$ is shown in Fig. 2^2D, a section through the imaginary axis and Q. Since for $u = 0$ we have $Z = iZ_0 \tan v$, it follows that the

FIGURE 2^2D. A section through the
imaginary axis and $Q \cdot u = 0$.

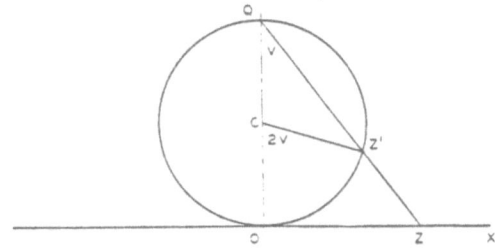

FIGURE 2^2E. A section
through the real axis and
$Q \cdot v = 0$.

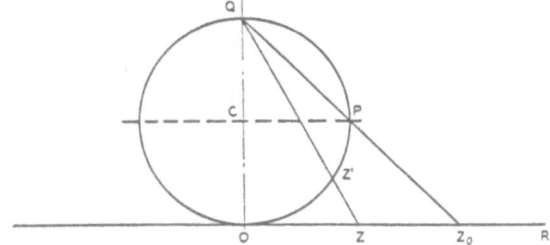

angle OQZ is v and therefore the angle OCZ' is $2v$. Hence an increase of v by π corresponds to a variation of $2v$ through its entire period 2π. The lines of constant v are therefore the meridian circles for which the longitude is $2v$.

Figure 2^2E shows a section through the real axis and Q. The quadrantal arc OP is the locus $v = 0$, and the quadrantal arc PQ is the locus $v = \pi/2$. The value of u corresponding to any particular small circle is that for which $u = \tanh^{-1}(R/Z_0)$, the small circle being drawn through the point Z' on the sphere associated with the point $(R, 0)$ on the plane. In particular, the equator is the circle $u = 0$, and the pole is the limiting circle $u = \infty$.

We may also project on the sphere the Cartesian coordinate lines for $R = $ constant and $X = $ constant. Evidently the locus of points belonging to $R = $ constant will be the small circle which is the intersection with the sphere of the plane through Q and the line $R = $ constant. This family of circles will have a common tangent at Q. Similarly the lines $X = $ constant project into a similar set of circles orthogonal to the first set and also having a common tangent at Q. Thus the appearance of the sphere in the neighborhood of Q with lines of constant R and X drawn on it will be as in Fig. 2^2F.

With this system of $R = $ constant and $X = $ constant circles mapped out on the sphere, one can now dispense with the impedance plane altogether. On the sphere we have two systems of mutually orthogonal circles, one giving the (R, X) coordinates of a point, the other its (u, v) coordinates. If we are given R_L, X_L we locate it on the sphere by using the (R, X) nets. Then the change in impedance due to a length of line kL is obtained by moving along the small circle $u = u_L$ until the longitude has been increased by an amount $2kL$.

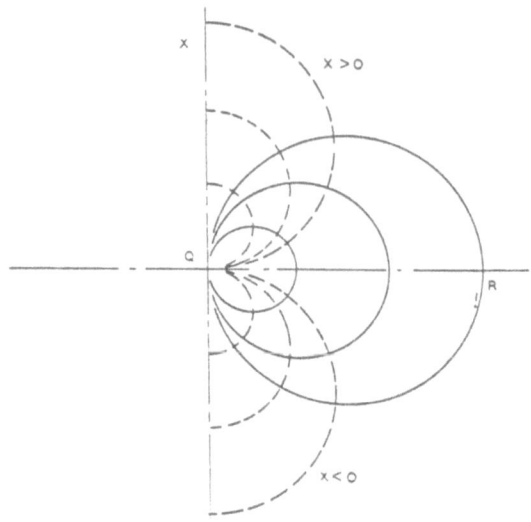

FIGURE 2²F. Projection
onto the sphere of the
Cartesian coordinate lines,
R = constant, X = constant.

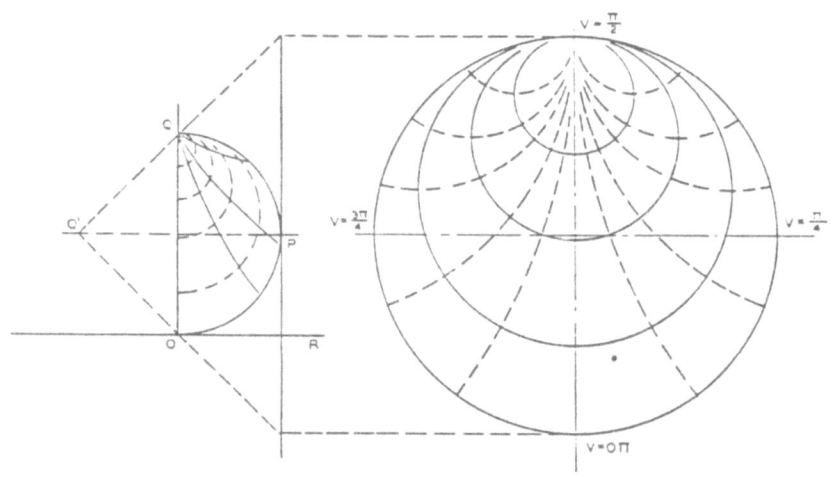

FIGURE 2²G. Reprojection of the sphere from any point Q' on it to a plane tangent at
the opposite end of the diameter through Q'.

It is quite instructive to think this all through but it is not very practical to
work out the impedance transformations by reference to curves drawn on a
sphere. But we can now go back to a wide variety of diagrams on a plane by
reprojecting the sphere from any point Q' on it to a plane tangent at the
opposite end of the diameter through Q'. Of all the plane diagrams which
might be made in this way one is particularly valuable, namely, that in which
Q' is chosen to be the pole opposite the point corresponding to $Z = Z_0$ on the
sphere. It is evident in Fig. 2²G that the hemisphere corresponding to positive

FIGURE 2^2H. Nets for impedance calculations, in which the
circles u = constant are concentric and the circles v = constant
are equally spaced radii.

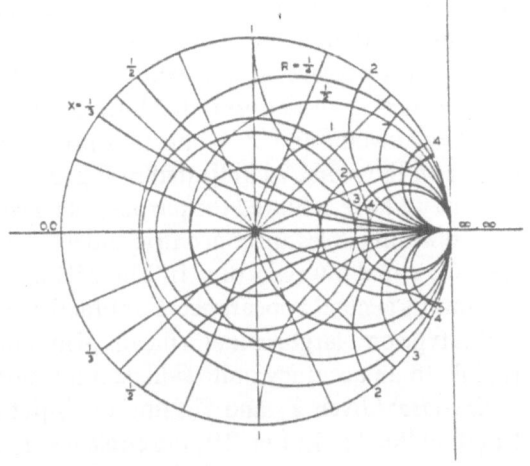

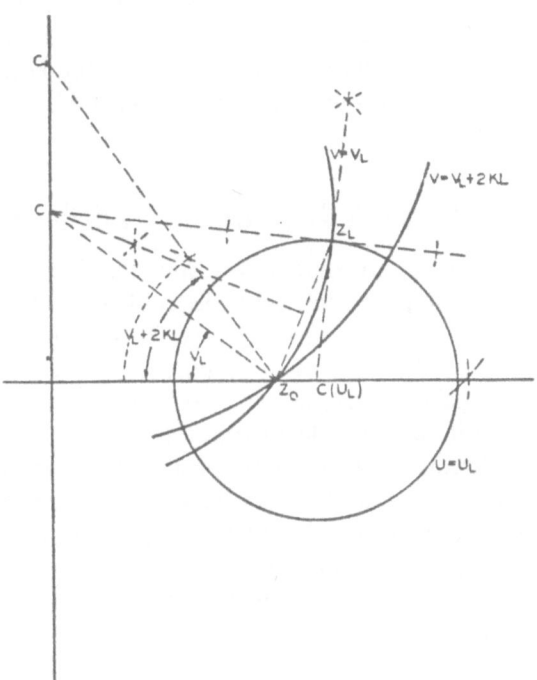

FIGURE 2^2I. Construction for
determining Z for any KL,
given Z_0 and Z_L.

resistances projects into a circle of radius Z_0, which is the projection of the equator $u = 0$, and the center corresponds to the pole $u = $ infinity. Other values of u are represented by concentric circles. Likewise the meridian circles, $v = $ constant, project into radial lines on the plane. The circles on the sphere for $R = $ constant and $X = $ constant project into a similarly disposed set of circles on the plane, as indicated in Fig. 2^2G.

Thus we have achieved the purpose of constructing a diagram on which the $u = $ constant circles are concentric and $v = $ constant circles are equally spaced radii for equal intervals of v. In Fig. 2^2H a diagram of this type is presented to show its general appearance. For practical work one may prepare diagrams of this type on a large scale as a means of making transmission line calculations rapidly to an accuracy quite sufficient for most purposes.

Exercise: Given Z_L and Z_0, find the input impedance graphically for any length of line kL. In Fig. 2^2I, the circle $v = v_L$ has its center on the imaginary axis and passes through both Z_L and Z_0. Construct the perpendicular bisector of the line $Z_0 Z_L$, its intersection of the imaginary axis is the center of the circle $v = v_L$. To find the center of the circle $u = u_L$ draw a perpendicular to CZ_L at Z_L; the center is at the intersection of this perpendicular with the real axis. Next draw the circle $u = u_L$. The impedance transformation introduced by the line length kL is obtained by adding kl to v_L, finding the new center C', and drawing the new circle $v = v_L + kL$ to its intersection with the circle $u = u_L$.

3^2. *Variable Impedance Transformers*

Since there are losses in transmission lines as well as possibility of insulation failure in power lines, it is desirable to lead power into a load in such a way that there is no reflected wave. This requires that the load impedance Z_L be "matched" to the line impedance Z_0, which then raises the question of design of adjustable transformers to be inserted between the line and the load to permit matching the load to the line.

First let us see what can be done by connecting a shorted line of adjustable length L_1 in parallel with the load. Assume the surge impedance of the parallel unit to be the same as that of the line. By Eq. ($2^2$9) the impedance of the unit is $iZ_0 \tan kL_1$. Since it is connected in parallel with the load it is more convenient to carry out the calculations with reciprocal impedances, that is admittances.

Let $Y_L = G_L - iB_L$ be the admittance of the load and $-iY_0 \cot kL_1$ be that of the parallel shorted line, where $Y_0 = 1/Z_0$ is the surge admit tance of the transmission line. Then the combined admittance of the two in parallel is

$$Y_1 = Y_L - iY_0 \cot kL_1. \tag{$3^2$1}$$

By varying L_1 over the range of one-half wave-length the second term can be made to take on any numerical value, hence, the resultant admittance can be made to assume any value on i vertical line through Y_L on the complex admittance plane. Therefore if the real part of Y happened to be equal to the characteristic admittance of the line it would be possible to get a perfect match by an appropriate choice of L_1.

Since a complete match involves equating two complex numbers, it is evident that a transformed suitable for all cases must involve at least two adjustable elements. Let us see what can be done by inserting another parallel shorted line of adjustable length L_2 into the line at a distance L_3 away from the load.

The admittance given by (3^21) is transformed by the length L_3 of line to

$$Y = Y_0 \frac{1 - w}{1 + w} \quad \text{where} \quad w = \frac{Y_0 - Y_1}{Y_0 + Y_1} e^{-2i\theta},$$

and

$$\theta = kL_3,$$

and the effect of the second shorted line in parallel with this will be to add $-iY_0 \cot kL_2$. The second shorted line or piston can thus be used to balance out any reactive component there is in Y. The problem thus reduces to a study of the range of values which the real part of Y may be made to assume for various choices of L_1 and L_3. Writing $Y_1/Y_0 = g - ib$ we find

$$\frac{Y}{Y_0} = \frac{g}{(\cos\theta - b\sin\theta)^2 + g^2 \sin^2\theta}$$
$$- i \frac{\sin\theta\cos\theta(1 - g^2 - b^2) + b(\cos^2\theta - \sin^2\theta)}{(\cos\theta - b\sin\theta)^2 + g^2 \sin^2\theta}. \tag{3^22}$$

Hence, by varying b we can make the real part of Y/Y_0 take on all values from 0 (for b infinite) to $1/g \sin^2\theta$ [for $(\cos\theta - b\sin\theta) = 0$]. Therefore it will be possible to match any load to the line for which $g \sin^2\theta$ is less than unity. Since the real part of Y_1 is the same as that of Y_L it follows that with the two-piston transformer it will be possible to match any load admittance such that

$$G_L \sin^2\theta < Y_0. \tag{3^23}$$

At first sight it might appear that this restriction could be removed simply by choosing L_3 such that $\theta = n\pi$, so $\sin\theta = 0$. However if this were done the first piston loses control, since its position appears in the combination $b \sin\theta$ in the real part of (3^22). Therefore one is confronted by the need to compromise as follows: In order to make (3^23) as little restrictive as possible one should design for a small value of $\sin\theta$, but in doing this it becomes necessary to be able to make very accurate adjustments of position of the piston L_1.

A reasonable choice of θ is to make $L_3 = \lambda/8$ or $3\lambda/8$ so the sines and cosines are each equal to $1/\sqrt{2}$ in magnitude. This permits matching of all impedances for which $G_L < 2Y_0$ without a very great sacrifice in control by the first piston. If $Z_L = Ae^{ia}$ then $G_L = A^{-1}\cos a$, hence, (3^23) requires that $A^{-1}\cos a$ be less

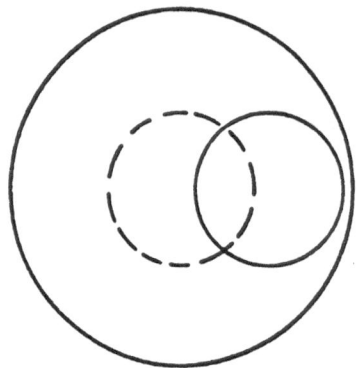

FIGURE 3^2. A quarter wave-length coaxial section with variable surge impedance.

than $2/Z_0$. On the impedance plane for Z_L this means that Z_L must lie outside a circle of radius $Z_0/4$ whose center is at $(Z_0/4, 0)$.

Another useful type of variable element consists of a section one quarter wave-length long whose surge impedance can be continuously varied from a maximum to a minimum value. A suitable construction is indicated in Fig. 3^2. An inner conductor is eccentrically mounted on an eccentric shaft so on turning it through 180° it varies from the coaxial position (dotted) to one in which it comes very close to the outer conductor. In the coaxial position the surge impedance of the section is a maximum while in the position where the center element is closest to the wall it is a minimum.

If Z_L is the load impedance connected to such a unit then the input impedance from $(2^2 6)$ is,

$$Z = Z_a^{\ 2}/Z_L.$$

Thus one can use a single piston in parallel with Z_L to cancel out the reactive component of Z_L followed by a quarter-wave unit of the type just described to transform the magnitude of R_L so as to make it match the line impedance Z_0. Alternatively one can use the quarter-wave section first to effect a reciprocal transformation on Z_L followed by a piston in parallel to cancel out the reactive component remaining after the reciprocal transformation.

4^2. Losses in Transmission Lines

Losses in transmission lines arise from the lack of perfect conductivity of the conductors and from the imperfection of the dielectric which is between them. In coaxial cable, for instance, it is necessary to use dielectric to give mechanical support to the center conductor. If the cable is to be flexible it is almost necessary to use solid (plastic) dielectric filling the whole cable to keep the center conductor in place when the cable is bent.

The losses in the conductors have to be handled as in Section 8^1. From $(8^1 9)$ the power loss in unit length of the line is

$$\frac{c\delta\mu}{8\lambda}\iint H^2\,dS, \tag{$4^2$1}$$

where the integration extends over unit length of both line conductors. For coaxial cable, if I is the current amplitude in abamperes in either conductor $[I(t, z) = I\cos(\omega t - kz)]$, then the field at the inner conductor is $(2I/a)\cos(\omega t - kz)$ and at the outer conductor it is $(2I/b)\cos(\omega t - kz)$. Hence the time average of the power loss in unit length is

$$(\pi c\delta\mu/2\lambda)(1/a + 1/b)I^2$$

which means that the line has an effective resistance per unit length of

$$R_1 = 15(\rho\mu/\lambda)^{1/2}(1/a + 1/b)\ \text{ohm/cm}. \tag{$4^2$2}$$

The mean power flow down the line at a place where the current amplitude is I is $Z_0 I^2/2$ watts if I is in amperes and Z_0 is the surge impedance in ohms, while the mean power loss per unit length is $R_1 I^2/2$ in watts/cm if I is in amperes and R_1 in ohm/cm. Hence the power level P is attenuated according to the law.

$$(dP/dx) = -(R_1/Z_0)P,$$

and

$$P(z) = P(0)\exp[-(R_1/Z_0)z]. \tag{$4^2$3}$$

Therefore, the quantity Z_0/R_1 gives the distance along the line in which the power level drops by a factor e^{-1} because of the losses in the conductors. This quantity is large if the losses are low and so is qualitatively analogous to the Q value for a cavity resonator. We shall denote it by L. For coaxial cable we have

$$L = 4b(\lambda/\rho\mu)^{1/2}\frac{\log b/a}{1 + b/a}\ \text{cm for } e^{-1}\ \text{loss}. \tag{$4^2$4}$$

This factor is quite closely analogous to ($9^1$16) for the Q value of a coaxial cable resonator of finite length. The principal difference arises from the fact that here there are no end losses to be considered.

In radio engineering power ratios are usually expressed in decibels (db) where 1 db corresponds to a power ratio of $10^{0.1} = 1.258$. Since $\log_{10} e = 0.434$, a factor e^{-1} corresponds to a loss of 4.34 db in the power level. Since the losses are not great in copper coaxial cable it is convenient to express L in meter/db loss. Using $\rho = 5.7 \cdot 10^{-8}$ cm we have the practical formula

$$L = 10.72b\sqrt{\lambda}\frac{\log b/a}{0.279(1 + b/a)}\ \frac{\text{meter}}{\text{db}}, \tag{$4^2$5}$$

where b, a, and λ are in cm. The factor depending on b/a has a rather flat maximum at $b/a = 3.58$, the maximum value being equal to 1. Since the maximum is so flat it is not necessary to design close to the optimum value

TABLE I[2]. List of values for Eq. (4[2]5)

x	$\dfrac{\log x}{0.279(1+x)}$
1.5	0.58
2.0	.83
2.5	.94
3.0	.98
3.5	1.00
4.0	.99
4.5	.98
5.0	.96

to get a good line, as the list of values in Table I[2] shows. As a specific design example, suppose the outer diameter is $\frac{5}{8}$ inch and the line is used for 15-cm waves, then the maximum value of L is obtained if the inner diameter is 175 mils. For such a line $L = 33$ meter/db loss.

Exercise: Show that if the cable is filled with perfect dielectric of dielectric constant ε, and that if the inner and outer conductors are made of different metals having resistivity and permeability, $\rho_a \mu_a$ and $\rho_b \mu_b$, respectively, then the appropriate generalization of (4[2]4) is

$$L = 4b \left(\frac{\lambda/\varepsilon}{\rho_b \mu_b} \right)^{1/2} \frac{\log b/a}{1 + (\rho_a \mu_a / \rho_b \mu_b)^{1/2} (b/a)}. \qquad (4^2 4a)$$

In Section 5[1] we learned how functions of a complex variable of the form (5[1]26) can be used to work out the fields in two conductor lines of more general shape. Let us now consider the losses in such lines. If $z = f(w)$ and the inverse function is $w = g(z)$, then

$$\text{grad}^2 v = (\partial v/\partial x)^2 + (\partial v/\partial y)^2 = |g'(z)|^2.$$

On a curve of constant u,

$$ds = [(\partial x/\partial v)^2 + (\partial y/\partial v)^2]^{1/2} = |f'(w)| \, dv.$$

Since $|g'(z)| = 1/|f'(w)|$, the integral appearing in (4[2]1) is

$$\int H^2 \, ds = \int_0^{2\pi} \frac{dv}{|f'(w)|},$$

where the integral is to be evaluated with $u = u_1$ for the inner conductor and $u = u_2$ for the outer conductor.

This gives the losses associated with a current of $\frac{1}{2}$ abampere in either conductor. Therefore, by steps analogous to those used in deriving (4[2]2), the effective resistance per unit length in ohm/cm is

$$R_1 = 15 \cdot (\rho \mu / \lambda)^{1/2} \cdot \frac{1}{2\pi} \left[\int_0^{2\pi} \frac{dv}{|f'(w)|_{u_1}} + \int_0^{2\pi} \frac{dv}{|f'(w)|_{u_2}} \right], \qquad (4^2 6)$$

and the quantity L is easily obtained from the relation $L = Z_0/R_1$ on using the formula ($5^1 28$) for Z_0.

As a specific example consider the calculation of R_1 for the line consisting of confocal elliptic cylinders whose surge impedance was calculated in ($5^1 27$). We have $z = f \cosh w$, hence, $f'(w) = f \sinh w$ and the integral to be calculated is

$$\frac{1}{2\pi} \int_0^{2\pi} \frac{dv}{f'(w)} = \frac{1}{2\pi f \cosh u} \int_0^{2\pi} \frac{dv}{(1 - k^2 \sin^2 v)^{1/2}},$$

where $k^2 = 1/\cosh^2 u$. This is a complete elliptic integral (see Peirce's Tables, No. 524 for definition and p. 121 for tables). Since $\cosh u_1 = a/f$ and $\cosh u_2 = b/f$ the final result for R_1 is

$$R_1 = 15(\rho\mu/\lambda)^{1/2} \left[\frac{1}{a} \frac{2}{\pi} K\left(\frac{f}{a}\right) + \frac{1}{b} \frac{2}{\pi} K\left(\frac{f}{b}\right) \right]. \tag{$4^2 7$}$$

If f/a and f/b are small compared with unity, the two elliptic integrals approach $\pi/2$ and this result reduces to the formula for the effective resistance of circular coaxial cable as it should. The first-order correction to R_1 for small values of f is obtained by using power series expansions for the elliptic integrals to give the result

$$R_1 = 15(\rho\mu/\lambda)^{1/2} \left[\frac{1}{a} + \frac{1}{b} + \frac{f^2}{4}\left(\frac{1}{a^3} + \frac{1}{b^3}\right) + \cdots \right]. \tag{$4^2 8$}$$

Since the first-order correction depends only on f^2 it is evident that the losses are not changed much by moderate flattening of the conductors.

5^2. Dielectric Losses

Suppose the space between the conductors is filled with dielectric of dielectric constant ε. Then, according to ($1^2 8$) the surge impedance is changed from its vacuum value by the factor $\varepsilon^{-1/2}$. If the dielectric shows loss then it can be described by a complex dielectric constant.

It is worth while to go back to ($1^1 1$) to the equation for curl $\mathbf{H}$. By assuming a time dependence by the factor $e^{i\omega t}$ it becomes

$$\text{curl } \mathbf{H} = 4\pi\mathbf{i} + ik\mathbf{D}.$$

If the material has a resistivity ρ and a dielectric constant ε then the right side of this can be written

$$ik\mathbf{E}(\varepsilon - 2i\lambda/\rho),$$

as was already remarked in dealing with skin effect in metals in Section 8^1. In metals ρ is so small that the second term is very much greater than the first. For dielectrics the reverse is true.

The actual phenomena which occur in real dielectrics are much more complicated than is usually admitted in discussions of the formal mathematical field theory.[3] The actual molecular processes involve dissipative energy losses by other mechanisms than those represented by ohmic conduction. Among these, for example, is the dissipation represented by turning of molecules with permanent dipole moments against viscous dragging forces. But all such dissipative processes have this in common, that they give rise to current density in the dielectric, that is, in time phase with $\mathbf{E}$ and which can be taken into account formally by means of an imaginary term in the dielectric constant. In addition there may be ohmic conduction of a sort which would be represented by a resistivity. These losses are dependent upon the frequency and there is no unambiguous way in which dipole turning losses can be separated experimentally from ohmic conduction losses.

It is therefore more satisfactory to discard any attempt at distinction between "true" ohmic conduction and other dissipative mechanisms. Phenomenologically the imperfect dielectric is to be described by means of a complex dielectric constant

$$\varepsilon = \varepsilon' - i\varepsilon'' \tag{5^{2}1}$$

in which the quantities ε' and ε'' are frequency dependent quantities characteristic of the material. Sometimes the losses are measured by giving the magnitude and phase angle of the complex dielectric constant

$$\varepsilon = \varepsilon_0 e^{-i\delta}. \tag{5^{2}2}$$

Before considering losses in lines due to imperfect dielectric it will pay to reconsider the work of Section 3^1 to see how a cavity resonator is affected by being filled with leaky dielectric. Referring to Eqs. $(3^1 1)$, let us agree to work with $\sqrt{\varepsilon}\mathbf{E}$ and $\sqrt{\mu}\mathbf{H}$ as the basic field vectors. In $(3^1 1)$ introduction of a complex dielectric constant results in a complex index of refraction $n = (\varepsilon\mu)^{1/2}$. The index of refraction appears in the combination $k = n\omega/c$. Most of the theory of Chapter I consisted in devising ways to find allowed values of k which would give fields which fit the boundary conditions. Since now n is complex and the allowed values of k are real, this gives rise to complex values for the frequency v.

Suppose that for a particular mode we have found that k_a is an allowed value. Then in vacuum the resonator fields can execute undamped free oscillations of frequency $ck_a/2\pi$. But when the resonator is filled with leaky dielectric the frequency becomes,

$$v_a = v_a' + iv_a'' = k_a c/2\pi n = (k_a c/2\pi)(\varepsilon' - i\varepsilon'')^{-1/2}. \tag{5^{2}3}$$

The physical meaning of the imaginary part v_a'' is that the time factor now is

$$\exp(2\pi i v_a' t) \cdot \exp(-2\pi v_a'' t),$$

and the free oscillations are damped by the losses in the dielectric.

---

[3] Manning and Bell, Rev. Mod. Phys. **12**, 215 (1940); W. Kauzman, Rev. Mod. Phys. **14**, 12 (1942).

The situation thus closely resembles that in Section 9^1 where damping due to finite conductivity of the walls was considered. We can define a Q' factor which measures the dielectric damping in analogy with the definition of Q in (9^11). For a resonator with walls of perfect conductivity the damping factor will be $\exp(-\omega t/2Q')$ and

$$Q' = v_a'/2v_a'' = (1/2)\cot \delta/2, \tag{5^24}$$

where δ is the phase angle of the dielectric constant.

In a resonator where there are additional losses due to the finite conductivity of the walls the factor (9^11) with Q defined as in (9^14) will also affect the decay of the free oscillations and therefore the complete damping Q_c will be given by

$$1/Q_c = 1/Q + 1/Q'. \tag{5^25}$$

From the form of this result it is evident that if the power factor of a dielectric is 1 percent (which is another way of saying that $\tan \delta = 0.01$), then the Q value of a resonator filled with this material cannot exceed 100. Moreover in such a case the dielectric losses will be large compared to those in the walls, under ordinary circumstances.

We consider now the effect of leaky dielectric on a transmission line. A glance over the equations of Section 1^2 shows that they are satisfied for a dielectric on writing $\sqrt{\varepsilon}\mathbf{E}$ in place of $\mathbf{E}$ and by writing

$$k = n\omega/c$$

and using the complex index of refraction in place of its previous real value, $n = 1$. The complex index of refraction gives rise to a complex k which can be written

$$k = k' - ik'' = (\omega/c)(\varepsilon' - i\varepsilon'')^{1/2}, \tag{5^26}$$

which gives rise to damped propagation along the line. The current, for example, is now given by

$$I = I_0 e^{-k''z} \cos(\omega t - k'z), \tag{5^27}$$

and, therefore, the power level in the line dies off like $e^{-2k''z}$. Therefore the loss length L' for a line with imperfect dielectric is

$$L' = 1/2k'' = (\lambda/4\pi\sqrt{\varepsilon})\csc(\delta/2), \tag{5^28}$$

where L' is expressed in cm per e^{-1} power loss if λ is in cm. It should be noted that λ is the vacuum wave-length and $\lambda/\sqrt{\varepsilon}$ is the wave-length in the medium.

Since this loss is additional to the loss arising from finite conductivity of the walls, the total effective L_c when both losses are present is given by

$$1/L_c = 1/L + 1/L', \tag{5^29}$$

where L represents loss in the walls as in (4^24) and L' arises from dielectric loss.

6^2. Reflection at Supports

Thin buttons of dielectric may be used to hold the center conductor in place in coaxial cable. Such buttons necessarily introduce wave reflections at each surface. However, by choosing the button spacing properly one may reduce the reflection to zero. Also, with a proper understanding of the effects of such buttons one may design micro-wave filters which are analogous to the recurrent-section lumped-constant wave filters in use at lower frequencies.

Let $n = \sqrt{\varepsilon}$ be the refractive index of the dielectric material. At any place z along the cable there will be an advancing wave (propagated from left to right, toward $+z$) and a returning wave. Let A be equal to nV_a where V_a is the voltage amplitude of the advancing wave, and let B equal nV_b be a corresponding measure of the amplitude of the returning wave. Then the electrical condition at a given point is described by the two-component quantity $\begin{pmatrix} A \\ B \end{pmatrix}$. This will be handled as a one-column two-row matrix in the calculations which follow.

We assume, as always, a time factor $e^{+i\omega t}$. The dependence of A on position is given by a factor e^{-ikz} where $k = n\omega/c$. Similarly, the dependence of B on position is given by a factor e^{+ikz}. Therefore the amplitudes $\begin{pmatrix} A \\ B \end{pmatrix}$ at any point can be expressed in terms of those a distance z to the right, in the same medium, denoted by $\begin{pmatrix} A_1 \\ B_1 \end{pmatrix}$ by means of the matrix equation

$$\begin{pmatrix} A \\ B \end{pmatrix} = \begin{pmatrix} e^{ikz} & 0 \\ 0 & e^{-ikz} \end{pmatrix} \begin{pmatrix} A_1 \\ B_1 \end{pmatrix}. \tag{$6^2$1}$$

For the reader who is not familiar with matrix algebra it may be remarked that the matrix equation

$$\begin{pmatrix} a \\ b \end{pmatrix} = \begin{pmatrix} c & d \\ e & f \end{pmatrix} \begin{pmatrix} g \\ h \end{pmatrix}$$

is simply a concise way of writing the two linear equations

$$a = cg + dh, \qquad b = eg + fh.$$

In particular ($6^2$1) is a particular notation for the pair of equations

$$A = e^{ikz} A_1, \qquad B = e^{-ikz} B_1.$$

The occurrence of the zeros in ($6^2$1) expresses the fact that A depends only on A_1 and not on B_1 which is the mathematical expression of the fact that there is no reflection produced along a uniform cable.

Now consider what occurs to $\begin{pmatrix} A \\ B \end{pmatrix}$ in going from left to right across an interface where the refractive index changes from 1 to n. The conditions to be fulfilled are that the radial electric vector must be continuous, and the circular magnetic field must be continuous. Letting $\begin{pmatrix} A_1 \\ B_1 \end{pmatrix}$ and $\begin{pmatrix} A_n \\ B_n \end{pmatrix}$ be the amplitudes on the two sides of the interface, we find these conditions are expressed by

$$A_1 + B_1 = n^{-1}(A_n + B_n), \qquad A_1 - B_1 = A_n - B_n.$$

Solving these for A_1 and B_1 in terms of A_n and B_n we find the result can be written in matrix notation as

$$\begin{pmatrix} A_1 \\ B_1 \end{pmatrix} = (1/2)\begin{pmatrix} n^{-1}+1 & n^{-1}-1 \\ n^{-1}-1 & n^{-1}+1 \end{pmatrix}\begin{pmatrix} A_n \\ B_n \end{pmatrix}. \tag{6^{2}2}$$

To bring out the physical significance of this result suppose the dielectric fills the cable to the right of the interface and that the cable is completely empty to the left of the interface. Suppose the cable is properly terminated so that $B_n = 0$. Then we have

$$A_1 = (1/2)(n^{-1}+1)A_n, \qquad B_1 = (1/2)(n^{-1}-1)B_n.$$

The energy flow in the incident wave is proportional to A_1^2 or $(1/4)(n^{-1}+1)^2 A_n^2$, and in the reflected wave to $B_1^2 = (1/4)(n^{-1}-1)^2 A_n^2$. We assume for simplicity that n is real. The fraction of the incident energy that is reflected is

$$R = \frac{(n^{-1}-1)^2}{(n^{-1}+1)^2} = \frac{(n-1)^2}{(n+1)^2}. \tag{6^{2}3}$$

The equivalence of the two forms shows that the reflecting power at a single interface is the same whether the refractive index goes from 1 to n or from n to 1.

As a numerical example, if the dielectric is polystyrene for which $\varepsilon = 2.7$, we have

$$n = 1.65, \qquad R = 5.8 \text{ percent.}$$

With such a large amount of reflection at a single interface it is obviously important to take steps to produce destructive interference between waves reflected from the different interfaces in a cable.

At an interface where the index changes from n to 1 we find, analogous to (6^{2}2),

$$\begin{pmatrix} A_n \\ B_n \end{pmatrix} = (1/2)\begin{pmatrix} n+1 & n-1 \\ n-1 & n+1 \end{pmatrix}\begin{pmatrix} A_1 \\ B_1 \end{pmatrix}. \tag{6^{2}4}$$

This matrix is the reciprocal of the one occurring in (6^{2}2) as it should be. The rule for multiplying matrices will be needed in verifying this statement and in the following calculations. It is this: If

$$\begin{pmatrix} a & b \\ c & d \end{pmatrix} = \begin{pmatrix} e & f \\ g & h \end{pmatrix}\begin{pmatrix} i & j \\ k & l \end{pmatrix},$$

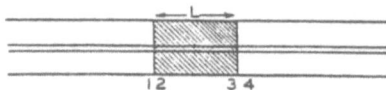

FIGURE 4^2. Single-button support in coaxial cable.

then

$$a = ei + fk, \qquad b = ej + fl, \qquad c = gi + hk, \qquad d = gj + hl.$$

Consider now the over-all effect of a single button of thickness L. (See Fig. 4^2.) By means of (6^24) we can express $\begin{pmatrix} A_3 \\ B_3 \end{pmatrix}$ in terms of $\begin{pmatrix} A_4 \\ B_4 \end{pmatrix}$. Then (6^21) gives $\begin{pmatrix} A_2 \\ B_2 \end{pmatrix}$ in terms of $\begin{pmatrix} A_3 \\ B_3 \end{pmatrix}$ and finally $\begin{pmatrix} A_1 \\ B_1 \end{pmatrix}$ in terms of $\begin{pmatrix} A_2 \\ B_2 \end{pmatrix}$ is given by (6^22). Hence the over-all expression for $\begin{pmatrix} A_1 \\ B_1 \end{pmatrix}$ in terms of $\begin{pmatrix} A_4 \\ B_4 \end{pmatrix}$ is given by

$$\begin{pmatrix} A_1 \\ B_1 \end{pmatrix} = \frac{1}{2}\begin{pmatrix} n^{-1} + 1 & n^{-1} - 1 \\ n^{-1} - 1 & n^{-1} + 1 \end{pmatrix}\begin{pmatrix} e^{ia} & 0 \\ 0 & e^{-ia} \end{pmatrix} \times \frac{1}{2}\begin{pmatrix} n + 1 & n - 1 \\ n - 1 & n + 1 \end{pmatrix}\begin{pmatrix} A_4 \\ B_4 \end{pmatrix}$$

where $a = \omega n L/c$. Multiplying together the three matrices (remembering that the order of the factors is important) we find a single matrix representing the effect of a single button,

$$\frac{1}{4n}\begin{pmatrix} [(n + 1)^2 e^{ia} - (n - 1)^2 e^{-ia}] & -2i(n^2 - 1)\sin a \\ 2i(n^2 - 1)\sin a & [(n + 1)^2 e^{-ia} - (n - 1)^2 e^{ia}] \end{pmatrix}. \qquad (6^25)$$

This will be called the one-button matrix. For most purposes it is more convenient to write (6^25) in the form

$$\begin{pmatrix} A_1 \\ B_1 \end{pmatrix} = \begin{pmatrix} P_1 & Q_1{}^* \\ Q_1 & P_1{}^* \end{pmatrix}\begin{pmatrix} A_4 \\ B_4 \end{pmatrix},$$

with

$$P_1 = \cos a + i\frac{n^2 + 1}{2n}\sin a, \qquad (6^26)$$

$$Q_1 = i\frac{n^2 - 1}{2n}\sin a.$$

From this we find that the reflecting power of a single button is

$$R_1 = \frac{(n^2 - 1)^2 \sin^2 a}{(n^2 + 1)^2 \sin^2 a + 4n^2 \cos^2 a}. \qquad (6^27)$$

For polystyrene, $n = 1.65$, this is

$$R_1 = \frac{2.89 \sin^2 a}{13.7 \sin^2 a + 10.8 \cos^2 a}.$$

From these results we see that the reflecting power vanishes if $a = m\pi$, that is for L such as to give an integral number of half wave-lengths in the material. Maximum reflection occurs for an odd integral number of quarter wave-lengths in the material. For polystyrene the maximum is 21 percent.

From a mechanical point of view $L = \frac{1}{8}''$ is a good thickness for ordinary coaxial cable. With polystyrene and a 15-cm vacuum wave-length, this makes $a = 120°$ and so R_1 is about 1.1 percent. It is worth noting that these same beads would give an extremely low reflecting power if the cable is used at considerably longer wave-lengths. That is why the problem of reflection from the beads is not such an important one in ultra-high frequency work as it is in the micro-wave region.

7^2. Chokes and By-Pass Condensers

Suppose we wish to continue a transmission line as a circuit for low frequency currents while having the high frequency power not go beyond a certain point. A suitable element for this is called a choke. In Fig. 5^2 suppose that a cupshaped member is attached to the inner conductor of a coaxial cable as shown. Suppose the load impedance as regarded from the closed end of the cup is Z_L. If the length of the cup is L, the impedance presented at the open end of the cup is, from (2^25),

$$Z' = Z_2 \frac{Z_L \cos kL + iZ_2 \sin kL}{iZ_L \sin kL + Z_2 \cos kL}, \qquad (7^21)$$

where Z_2 is the characteristic impedance of the element of concentric line formed by the outer conductor and the outside wall of the cup.

Likewise the input impedance presented at the open end of the cup is

$$Z'' = iZ_1 \tan kL,$$

where Z_1 is the characteristic impedance of the line formed by the inner conductor and the inside wall of the cup. Regarded from the cross section at the open end of the cup, these two impedances are in series, for the currents flow as marked in the sketch and the total voltage drop is the sum of that over the two elements. Hence the total input impedance is

$$Z = Z' + Z''. \qquad (7^22)$$

If the length of the cup is a quarter wave-length then

$$Z = iZ_1 \infty + Z_2{}^2/Z_L = \infty. \qquad (7^23)$$

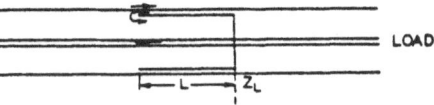

FIGURE 5^2. High frequency choke with voltage node at far end of cup.

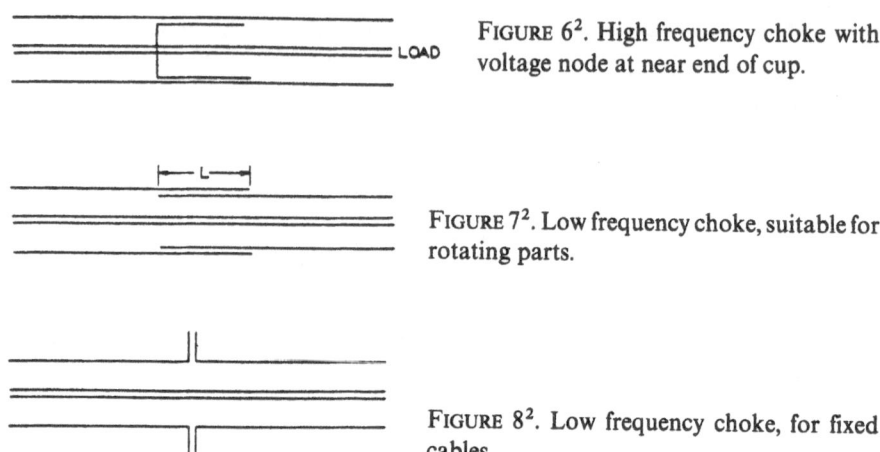

FIGURE 6[2]. High frequency choke with voltage node at near end of cup.

FIGURE 7[2]. Low frequency choke, suitable for rotating parts.

FIGURE 8[2]. Low frequency choke, for fixed cables.

Thus the impedance at the open end of the cup is infinite. Hence there will be total reflection at the cup of a radiofrequency wave in such a way that there is a voltage loop and a current node at the mouth of the cup, exactly as if the line terminated there in an open circuit.

Next let us consider the same structure with the cup turned the other way. (See Fig. 6[2].) In this case the cup impedance is $iZ_1 \tan kL$ and this is in series with the load impedance Z_L. Therefore the resultant of the two is $(Z_L + iZ_1 \tan kL)$. If the length of the cup is a quarter wave-length, this is infinite. Viewed from the bottom end of the cup this infinite impedance becomes a zero impedance. Therefore, a wave coming from the left is totally reflected with a voltage node at the outside of the bottom of the cup, just as if the cup constituted a complete short circuit of the end of the line.

In other circumstances one may need to have a break in the line for low frequency currents while not interfering with the flow of the high frequency power. This can be done as in Fig. 7[2]. If the overlapping portion of the separated outer conductors is equal to a quarter wave-length then the infinite impedance at the open end between the two outer conductors transforms to zero impedance between the two outer conductors at the left end of the overlap. Hence the current flow in the large outer conductor is carried on in the inner one without a voltage drop, so the wave goes on, although low frequency currents are blocked by the lack of contact between the two outer conductors.

Another way of doing the same thing is to put circular flanges on the ends of the two portions of outer conductor of the same size, as in Fig. 8[2]. In this case the correct radius of the flanged ends has to be calculated as follows. Suppose $r = a$ is the radius of the outer conductor; then we must have $E_z = 0$ at $r = a$, so there will be no potential drop across the gap between the flanges, just as if the outer conductor were continuous. We have, quite generally,

$$E_z = AJ_0(kr) + BN_0(kr),$$

where J_0 and N_0 are the two Bessel functions of zero order. The requirement $E_z = 0$ at $r = a$ gives the equation

$$AJ_0(ka) + BN_0(ka) = 0,$$

which determines the ratio of B to A. At the outer radius of the flange the radial current must sink to zero, giving $H_\varphi = 0$ which requires that $\partial E_z/\partial r = 0$. This gives

$$AJ_0'(kb) + BN_0'(kb) = 0,$$

which is the equation to determine b, the outer flange radius.

As a specific example, suppose $a = 0.5$ cm and we are dealing with $\lambda = 3$ cm, giving $k = 2.08$ and $ka = 1.04$. We have $J_0(ka) = 0.7473$ and $N_0(ka) = 0.1188$ and, therefore, if we write

$$E_z = 0.1188J_0(kr) - 0.7473N_0(kr),$$

we have a suitable expression which vanishes at $r = a$. We have now to find the value of kb such that $\partial E_z/\partial r = 0$. This is best done by making a graph of E_z against r from standard tables of Bessel functions. In this way we find the function has a maximum at $kb = 2.4$ or $b = 1.15$ cm, as the proper outer radius of the flanges.

8^2. Transmission Line Resonators

Any finite section of transmission line may be used either by itself or in connection with lumped inductance and capacity to make resonant circuits. First let us consider a length z of transmission line which is closed at one end and open at the other. The impedance at the open end is, by (2^27),

$$Z = iZ_0 \tan kz.$$

The free oscillations must be such that there is zero current flowing even though there is a finite voltage amplitude. Hence, the natural resonant frequencies will be such as to make the impedance at the open end be infinite. Therefore, the resonant values of k are

$$kz = (n + \tfrac{1}{2})\pi$$

which can be written

$$z = (n/2 + \tfrac{1}{4})\lambda, \tag{$8^2$1}$$

where n is an integer. The resonance of lowest frequency is such that the length is a quarter wave-length.

If the resonator is closed at both ends then the frequency has to be such as to give zero impedance at either end. This means one must have $kz = n\pi$, and the length must be an integral number of half wave-lengths. This in agreement with the field theory treatment given in Section 5^1.

Suppose now that there is a condenser of capacity C (farad) across the otherwise open end of a line that is closed at the other end as in the sketch. The shorted line is in series with the condenser so the input impedance at the terminals 1, 2 is the sum of the separate impedances. It is therefore,

$$Z = iZ_0 \tan kz + 1/i\omega C, \tag{8²2}$$

and the resonances are frequencies for which $Z = 0$, since in the actual resonator in which the terminals 1, 2 are joined together current must flow there without a potential drop. This condition gives the equation,

$$kz \tan kz = z/cCZ_0. \tag{8²3}$$

If C is small the roots of (8²3) are close to those given by (8²1). If L and C and Z_0 are given the possible values of k can be conveniently found by graphing $x \tan x$ against x from which the allowed values of kz are readily found. From such a graph it can be immediately seen that an increase of the capacity has the general effect of reducing all of the resonant frequencies. This approximate treatment based on transmission line theory should be compared with the field theory discussion given in Section 7¹.

A resonator can be made as in Fig. 9² by joining together a length z_1 of shorted line of characteristic impedance Z_1 and a shorted length z_2 of characteristic impedance Z_2. The input impedance presented by this combination to terminals mounted on the disconnected outer conductors is then

$$i(Z_1 \tan kz_1 + Z_2 \tan kz_2)$$

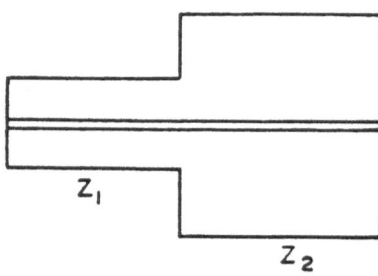

FIGURE 9². Resonator made from two shorted sections of line.

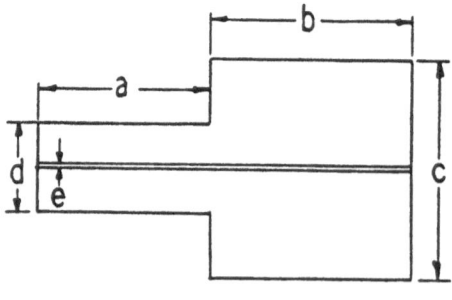

FIGURE 10². Resonator with dimensions for exercise.

which must vanish at the resonant frequencies. These can easily be located graphically by plotting $Z_1 \tan kz_1$ and $-Z_2 \tan kz_2$ against k and noting the values of k at which the two curves intersect.

Exercise: Calculate the lowest resonant frequency of the resonator shown in Fig. 10^2 (figure of revolution about the horizontal center line), where the dimensions are, in inches, $a = 2$, $b = 3$, $c = 1$, $d = \frac{1}{2}$, and $e = \frac{1}{4}$. *Answer*: 468 megacycle/sec.

9^2. *Tapered Lines*

By a tapered line is meant one in which the proportions change along the length of the line—for example, a coaxial cable with a variable ratio of inner to outer diameter.[4]

It will be supposed that the dimensions in a section are all small compared with the wavelength. The theory may be developed by using $(1^2 14)$ and $(1^2 15)$ as a starting point. Since in practice tapered lines will be used only in short transition sections we shall neglect losses, that is, assume $R = 0$ and $G = 0$.

The generalization now being considered is that L and C are here functions of z. The basic line equations are:

$$(\partial V/\partial z) = -L(\partial I/\partial t),$$
$$(\partial I/\partial z) = -C(\partial V/\partial t),$$

$(9^2 1)$

where the units are: V, volt; z, cm; L, henry/cm; I, ampere; t, sec.; and C, farad/cm. Assuming harmonic time dependence through the factor $e^{i\omega t}$ we find that V and I satisfy the following differential equations:

$$V'' - \frac{d \log L}{dz} V' + \omega^2 LCV = 0,$$

$$I'' - \frac{d \log C}{dz} I' + \omega^2 LCI = 0.$$

$(9^2 2)$

If $\varepsilon = \mu = 1$ we have $LC = 1/c^2$ and the characteristic impedance of the line Z in ohms is related to L and C by the expressions, $Z = cL = 1/cC$. Therefore the two logarithmic derivatives appearing in $(9^2 2)$ can be expressed in terms of the logarithmic derivative of Z. With $k = \omega/c$ we have

$$V'' - \frac{d \log Z}{dz} V' + k^2 V = 0,$$

$$I'' + \frac{d \log Z}{dz} I' + k^2 I = 0.$$

$(9^2 3)$

---

[4] Eckart, Zeits. fur Hochfrequenztechnik **55**, 173 (1940) gives a very general treatment of the theory. Other important references are: Ballantine, J. Frank. Inst. **203**, 561 (1927); Wheeler and Murnaghan, Phil. Mag. **6**, 146 (1928); Starr, Proc. I. R. E. **20**, 1052 (1932); Burrows, Bell Sys. Tech. J. **17**, 555 (1938); Wheeler, Proc. I. R. E. **27**, 65 (1939).

It is only necessary to discuss one of these, since, if the solution is known for $V(z)$, that for $I(z)$ may be obtained from the first of (9²1) in the form

$$I = (i/\omega L)(\partial V/\partial z) = (i/kZ)(\partial V/\partial z). \qquad (9^24)$$

We have now to discuss the properties of the first of (9²3) which determines the variation along the line of the potential difference between conductors in the line. For a line of uniform properties $d \log Z/dz = 0$ and the equation reduces to one which is satisfied by e^{+ikz} or e^{-ikz} giving the usual propagation of undistored harmonic waves at the velocity of light. The term in the derivative of V can be transformed away by writing

$$V = \sqrt{Z}U, \qquad (9^25)$$

in which case the differential equation for U is

$$U'' + [k^2 + (Z''/2Z) - (3Z'^2/4Z^2)]U = 0. \qquad (9^26)$$

There are two special cases in which the equation for V can be solved in terms of known functions.

EXPONENTIAL LINE

The simplest special case is that in which the line is tapered in such a way that the characteristic impedance varies exponentially along the line. Suppose

$$Z(z) = Z_0 \exp(2k_0 z), \qquad (9^27)$$

then the differential equation for U becomes

$$U'' + (k^2 - k_0{}^2)U = 0. \qquad (9^28)$$

and, therefore, the solutions for U depend on the sign of

$$k'^2 = k^2 - k_0{}^2.$$

If k'^2 is positive the solution for U is undamped and oscillatory and there is real wave propagation along the line, with the voltage amplitude building up exponentially as one goes in the direction in which the characteristic impedance increases. But if k'^2 is negative the solution for U is a real exponential function and the wave is attenuated in going along the line. Such an exponentially tapered line therefore behaves like a high-pass filter. It passes only those waves for which k is greater than k_0. Therefore the cut-off frequency is greater for more rapid rates of taper.

Suppose we have a wave traveling toward $+z$. The voltage is represented by,

$$V = V_0 \exp(k_0 z) \cdot \exp[i(\omega t - k'z)],$$

and, therefore, by (9²4), the current is represented by

$$I = (V_0/Z_0)\frac{(k' + ik_0)}{k} \times \exp(-k_0 z) \cdot \exp[i(\omega t - k'z)].$$

The ratio of voltage to current at any place gives the impedance of a load which could terminate the line at that place without producing a reflected wave. This terminating impedance is

$$Z_i = \frac{k}{(k' + ik_0)} Z_0 \exp(2k_0 z).$$
(9²9)

This terminating impedance must therefore be somewhat reactive although its phase angle tends to zero if the frequency used is large compared with the cut-off frequency so k' is large compared to k_0.

Let us consider a particular example. Suppose it is desired to design a transition section of coaxial cable to pass from a characteristic impedance of 50 ohms to a characteristic impedance of 100 ohms in a meter of line length. If the inner conductor is the same throughout and is 125 mils in diameter, then the diameter of the outer conductor at the two ends must be 288 mils and 660 mils, respectively. Since the transition takes place in one meter we have $200k_0 = x \ln 2$ or $k_0 = 3.47 \cdot 10^{-3}$ cm^{-1}. Therefore the cut-off wave-length is $2\pi/k_0 = 1810$ cm.

If this transition section is used for radiation of 15-cm wave-length, or $k = 0.418$ cm^{-1} then it can be calculated that the phase-angle of the terminating impedance is less than one degree.

LINE WITH Z VARYING A POWER OF z

Another case which can be treated in terms of known functions is that in which $Z(z)$ is a simple power of z measured from some origin. Suppose

$$Z(z) = Z_1 z^n,$$
(9²10)

where Z_1 is the characteristic impedance at a point at unit distance from the place where Z would vanish if this law were valid everywhere. In practice, one will be dealing with finite sections of tapered line for which $z \neq 0$, say the portion extending from $z = +a$ to $z = +b$; hence, no difficulty arises from the vanishing or negative values of Z seemingly implied by (9²10).

For this case $d \log Z/dz = n/z$ and (9²3) becomes

$$V'' - (n/z)V' + k^2 V = 0,$$
(9²11)

an equation which can be solved in terms of Bessel functions. The solution is

$$V(z) = z^m Z_m(kz) \quad \text{with} \quad m = (1 - n)/2,$$
(9²12)

where $Z_m(kz)$ stands for the general Bessel function of order m. By making use of known properties of Bessel functions it is therefore possible to make a detailed study of tapered lines of this kind.

Detection of Metastable Ions With the Mass Spectrometer

J.A. Hipple and E.U. Condon

Westinghouse Research Laboratories
Pittsburgh, Pennsylvania

It has usually been customary to operate mass spectrometers with the filament and electron gun at a positive potential with respect to the analyzer section which is grounded. Under these conditions the ions pass through the exit slit and strike the ion-collector plate with the full energy acquired in the ion gun since this plate is at ground potential. Recently with an instrument of the sector type employing a magnetic deflection of 90°, the first slit in the ion gun was connected to ground while the analyzer section was made negative with respect to ground. Since the ions are formed in a region one or two volts above ground, they reach the collector plate in this arrangement with an energy of only a few electron volts.

This change has had a very interesting effect on the spectra of hydrocarbons. In Fig. 1 a portion of the spectrum of normal butane is shown. The lower curve shows the baseline with the conventional arrangement, and the upper curve shows the improved baseline when the ion source is grounded. These curves were obtained with an automatic recorder employing a non-linear scale with provision for recording large peaks (such as mass 43 which was allowed to go off scale since it is of no concern at the moment). The "hump" on the side of the peak at mass 39 has not previously been investigated carefully but was usually attributed to the formation of $C_3H_3^+$ with kinetic energy. However, it should still be present in the upper curve of Fig. 1 if this explanation is correct. Its disappearance means it must be associated with a loss in energy. Similarly, the disappearance of the diffuse peaks around 30.5 and 32 requires a loss in energy.

The experimental results may be explained by supposing that some of the ions dissociate after acquiring energy in the ion gun, i.e., if metastable ions are formed in the ion source. If this dissociation occurs in the region of the instrument between the ion source and the magnet, the ion will have an apparent mass m^* which has the following relation to m_0, the mass of metastable ion, and m_1, the mass of the ion to which the metastable dissociates,

$$m^* = m_1{}^2/m_0.$$

This simple relation assumes that the metastable ion dissociates into two

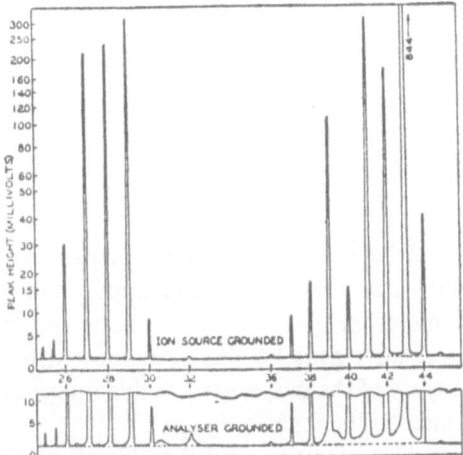

FIGURE 1. Portion of the spectrum of normal butane obtained on an automatic recorder employing a non-linear scale. The two curves show the effect on the spectrum of operation of the mass spectrometer with and without the ion source grounded. The disappearance of the diffuse peak at mass 32, for instance, is attributed to the presence of metastable ions in the mass spectrometer. The small residual peak at mass 32 is caused by O_2-impurity in the tube. The original charts have been retouched for reproduction.

fragments with negligible release of internal kinetic energy. In general, of course, there may be some energy release. Also the dissociation processes in general will follow an exponential decay distribution in time, and so some will occur in the magnetic analyzer and other parts of the instrument giving rise to the observed unsharpness of the peaks. On dissociation the ion has therefore only that fraction of the kinetic energy represented by its fraction of the original mass and is therefore unable to reach the collector when there is a full retarding field between the main analyzer tube and the collector.

The peak around mass 32 can be explained by the dissociation

$$C_4H_{10^+} \rightarrow C_3H_{7^+} + CH_3,$$

giving

$$m^* = (43)^2/58 = 31.9.$$

In this case the parent ion dissociates, and one fragment is the most abundant ion in the n-butane spectrum.

Similarly the ionization around mass 30.5 in the spectrum can be explained by the dissociation

$$C_4H_{10^+} \rightarrow C_3H_{6^+} + CH_4,$$

giving

$$m^* = (42)^2/58 = 30.4.$$

The hump on the side of mass 39 might be explained by

$$m^* = (40)^2/41 = 39,$$

or

$$m^* = (41)^2/43 = 39.2.$$

Similar peaks have been noticed in the spectra of a great many hydro-carbons, and in all cases the explanation appears to fit into the pattern outlined above. Some other observations on mass spectra of hydrocarbons, such as variation of relative peak heights from one instrument to another, and with initial temperature of the gas, find plausible interpretation as effects due to dissociation of metastable ions.

We wish to thank H.A. Thomas and R.E. Fox for their assistance in this experiment. A more complete report on metastable ions in the mass spectro-meter is in preparation.

Metastable Ions Formed by Electron Impact in Hydrocarbon Gases

J.A. HIPPLE, R.E. FOX, AND E.U. CONDON*

Westinghouse Research Laboratories
Pittsburgh, Pennsylvania

It has been recently reported that the non-integral masses appearing in the mass spectra of various hydrocarbons may be explained by the spontaneous dissociation of some of the ions into fragments of lighter mass after they have been accelerated and emerge from the ion gun. By means of an energy filter, an energy analysis of the non-integral masses in *n*-butane, butadiene, and ethane has been made and the values obtained agree with those predicted on the basis that they arise from metastable ions. Variation of the pressure and electrode potentials confirms that the dissociation is spontaneous. The formation of metastable ions appears to be a general occurrence in the ionization and dissociation of hydrocarbons and is shown here to occur in ethane, propane, 1,3-butadiene, butene-1, *cis*-butene-2, isobutylene, normal butane, iso-butane, pentene-2, normal pentane, iso-pentane, and methyl-cyclo-pentane.

Introduction

A preliminary report[1] has been made on evidence obtained with the mass spectrometer indicating that some kinds of primary ions formed by electron impact in several hydrocarbon gases are metastable. They are apparently stable enough to hold together for a time of the order of 10^{-6} sec. during which they are drawn out of the ion source region and accelerated, but dissociate before completing their trip through the analyzing magnetic field. The observed facts are in accord uith the interpretation that the metastable ions simply fall apart so that the fragments have a small energy of relative motion. This "small" energy may amount to some volts but is small compared to the 500 to 1000 electron volts of energy which the ions are given for analysis in the mass spectrometer.

This paper reports further data obtained by means of the mass spectrometer with energy analysis of the ions which fully confirms and considerably extends

---

* At present Director of National Bureau of Standards, Washington, D.C.
[1] J.A. Hipple and E.U. Condon, Phys. Rev. **68**, 54 (1945).

the results given earlier. Broadly the phenomenon here discussed bears some resemblance to pre-dissociation, first discovered by Henri[2] in optical band spectra, but it should be observed that when pre-dissociation results in broadening of the lines in the band spectrum, the mean life of the molecule for pre-dissociation must be short compared to the usual radiative life of 10^{-8} sec. Hence optical pre-dissociation studies deal with cases in which the mean life is much smaller than in the cases described in this paper. The range of values of the mean life which may be studied in a particular mass spectrometer is of course dependent on potentials of the electrodes and the dimensions of the tube in the ion source and ion accelerating regions.

These studies grew out of a desire to understand the origin of several broad peaks, some of which occur at positions corresponding to non-integral masses, which were observed in the mass spectrometer. In the first experiments, reported earlier, an experiment was performed on the effect on the mass spectrum of operating with the ion source and the collector both at ground while the main accelerating potential was applied to the analyzer section of the tube. Since the ions are formed in a region that is one or two volts above ground, only those ions reach the collector which have essentially as much energy on leaving the analyzer as they acquired on entering it.

Figure 1 shows the mass spectrum of normal butane taken in the two ways, the ion source grounded curve being superposed above the curve taken in the more familiar way in which the analyzer is grounded. Comparing the two curves, the following differences are at once evident:

---

[2] V. Henri, Comptes rendus **177**, 1037 (1923).

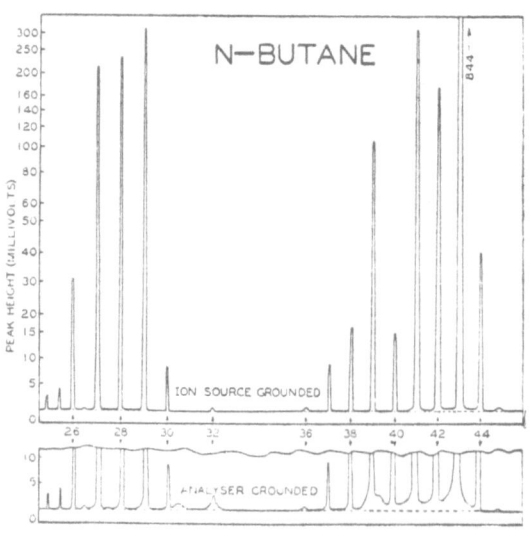

FIGURE 1. Showing effect on observed mass spectrum of *n*-butane of operating with ion source grounded instead of with analyzer grounded.

(a) The broad background from mass 38 to 44 is removed.
(b) The hump on the side of the mass 39 peak is removed.
(c) The mass 32 peak is considerably reduced and sharpened indicating that in the usual spectrum it is really compounded of a broad peak, which is removed when the ion source is grounded, together with a small sharp peak which is unaffected by the change. In later experiments with a tube that was vacuum-tight this mass 32 peak (O_2) was entirely eliminated when the ion source was grounded.
(d) The broad peak at mass 30.5 is removed.

The material which is not collected with the ion source grounded must consist of ions which lose energy in some way in going through the tube. Of course, it is possible that some energy losses occur by impact with the residual gas in the tube but it was thought that this could only, account for a general raising of the background around regions of intense ionization.

The hypothesis was considered that these effects might be due to unstable ions which hold together long enough to be accelerated through all or most of the accelerating potential drop but which dissociate, perhaps with a small release of internal energy, somewhere in the tube before they have entered the analyzer, or at least before they have gone through the field. Let the potential be reckoned as zero at the place where the ion is first formed, and suppose the ion has mass m_0 which it retains until it has moved to a place where the potential is V_1. At this place suppose it dissociates with negligible release of internal energy into an ion of mass m and a neutral fragment of mass $(m_0 - m)$. The new ion will continue on with little immediate change of velocity, but will in turn be accelerated by the electric field in going from the region of potential V_1 to that of the full voltage V of the accelerating field.

Its kinetic energy just before dissociation is eV_1 but after dissociation the ion fragment has only the kinetic energy, $(m/m_0)eV_1$, the other part being carried off by the neutral fragment. After traversing the whole accelerating field the kinetic energy of the ion fragment is

$$T = (m/m_0)eV_1 + e(V - V_1).$$

Now the radius in which it moves in the analyzer is determined by the usual formula,

$$R = \frac{c}{eH}(2mT)^{1/2}.$$

Therefore the ion fragment will appear at the collector for the same combination of V and H as does a normal ion of mass m^* where

$$m^* = \frac{m^2}{m_0}\frac{V_1}{V} + m\left(1 - \frac{V_1}{V}\right).$$

If dissociation occurs before the ion is accelerated, then its effective mass is m. However, if dissociation occurs after full acceleration but before entering the analyzer then ($V_1 = V$)

$$m^* = m^2/m_0.$$

It has been found that the observations correlate with this expression. The peaks have width which is partly due to the fact that some dissociations occur elsewhere and partly to the effect of the neglected release of internal energy. An additional reason for the diffuseness of the metastable peaks is the focusing property of the instrument. The instrument has been designed to focus ions that maintain a constant m/e in passing through the analyzer. Thus it is probable that the observed diffuse peaks are caused by transitions occurring very close to the exit slit of the ion gun and before the ion has traversed very far into the analyzer. This is confirmed by the spectrum of n-butane obtained by Washburn, Wiley, and Rock[3] on an instrument employing a deflection in the magnetic field of 180°. Although there is no field-free region between the ion source and the analyzer, the diffuse peaks at mass 32 and 39.2 are quite prominent, and there is some indication that those at 25.2 and 30.4 are also present. The ions dissociating elsewhere in the tube will contribute to the background with no pronounced peaks appearing.

Of course, an observation of m^* alone does not suffice to determine both m_0 and m and hence to fix the particular dissociation process responsible for a peak. This was done by arranging to measure the kinetic energy of the ions after they have gone through the analyzer by adjusting the retarding field between the analyzer region and the ion collector, as explained in a later section.

Peaks arising from reactions occurring during the transit of the ions through the mass spectrometer have previously been reported. Smyth reported two peaks in hydrogen[4] having an apparent mass less than one (throughout this paper singly charged ions are being discussed unless explicitly stated). Smyth[5] also reported a peak at mass 7 in N_2 which occurred at low electron voltages. The size of these peaks relative to the parent ion was very sensitive to the pressure in the instrument and they were attributed to collisions with other molecules during the transit through the instrument. Discussing the case of N_2, Smyth[6] concludes "that there are N_2^+ ions, probably those formed at the first ionization potential which dissociate on collision after acquiring large kinetic energies." Hogness and Lunn[7] reported a similar effect in NO—"... some of the NO$^+$ ions are unstable toward collision with gas molecules, and on collision dissociate into either N^+ or O^+."

Dissociation of molecules during transit has been observed in several mass

[3] H.W. Washburn, H.F. Wiley, and S.M. Rock, Ind. Eng. Chem. Anal. Ed. **15**, 541 (1943).
[4] H.D. Smyth, Phys. Rev. **25**, 452 (1925).
[5] H.D. Smyth, Proc. Roy. Soc. **A104**, 121 (1923).
[6] H.D. Smyth, Rev. Mod. Phys. **3**, 373 (1931).
[7] T.R. Hogness and E.G. Lunn, Phys. Rev. **30**, 26 (1927).

spectrographs by the appearance of diffuse traces at non-integral mass numbers on the photographic plate. These are greatly enhanced by increasing the pressure in the field-free region between the electrostatic energy filter and the magnetic analyzer. A very careful study of these "bands" has been made by Mattauch and Lichtblau,[8] who have described 28 different processes of dissociation of this type induced by collision. All of these processes differ from those reported in the present work which it is believed are not induced by collision.

With the improved techniques now available, mass spectrometers are now operated at a pressure 100 to 1000 times lower than in the early experiments and the possibility of dissociation induced by collision is remote. For instance, Smyth found a peak at mass 7 with 30-volt electrons amounting to 3 percent of mass 28 in N_2 in one experiment.[5] Hagstrum and Tate[9] reported the same peak having an appearance potential of 64 ± 2 volts and an abundance relative to $N_2{}^+$ of 0.006 percent. This relative abundance was confirmed approximately in the course of the present investigation using 100-volt electrons and no ionization was evident at 50 volts. This work was done with an automatic recorder without pushing the sensitivity beyond 1 part in 20,000, but the ionization below 50 volts at mass 7 must be less than the above figure in our equipment. The conclusion is that mass 7 ($m/e = 7$) in N_2 must be attributed to N^{++} in present instruments at the usual operating pressures and the large peak at this mass observed in the early experiments arises from dissociation induced by ion-molecule collisions.

In order to show that the ions at non-integral masses occurring in the spectra of hydrocarbons are caused by spontaneous dissociation and are not to be attributed to ion-molecule collisions, the spectrum of n-butane was studied as the pressure in the tube was varied. This tube did not have differential pumping[6] and the analyzer section was not pumped separately; therefore, the pressure in the analyzer varied in the same manner as that in the ion source and any peaks conditioned by collisions in the analyzer should not vary linearly with the pressure. The pressure in back of the leak bleeding the gas into the mass spectrometer tube was used as a measure of the pressure in the tube—the leak had previously been found to have a linear flow characteristic in the range used. The pressure in the large gas reservoir used for stabilizing the flow through the leak was determined by measuring the pressure in a smaller pre-expansion volume with a manometer before the sample was expanded into the larger reservoir. In this way the pressure could be measured accurately. In Fig. 2 the pressure in the pre-expansion reservoir is taken as the abscissa since it is a linear function of the pressure in the tube. This is evident by the linearity of the ion current corresponding to the

---

[8] J. Mattauch and H. Lichtblau, Physik. Zeits. **40**, 16 (1939); see also F.W. Aston, *Mass Spectra and Isotopes* (Edward Arnold and Company, London, 1942), page 61.
[9] H.D. Hagstrum and J.T. Tate, Phys. Rev. **59**, 354 (1941).

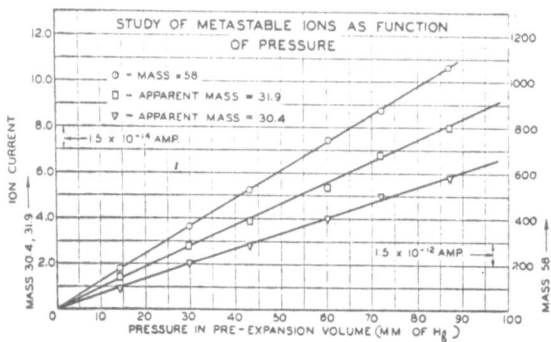

FIGURE 2. Variation of ion current with pressure. The linearity of the curves shows that the peaks at 30.4 and 31.9 are not induced by ion-molecule collisions.

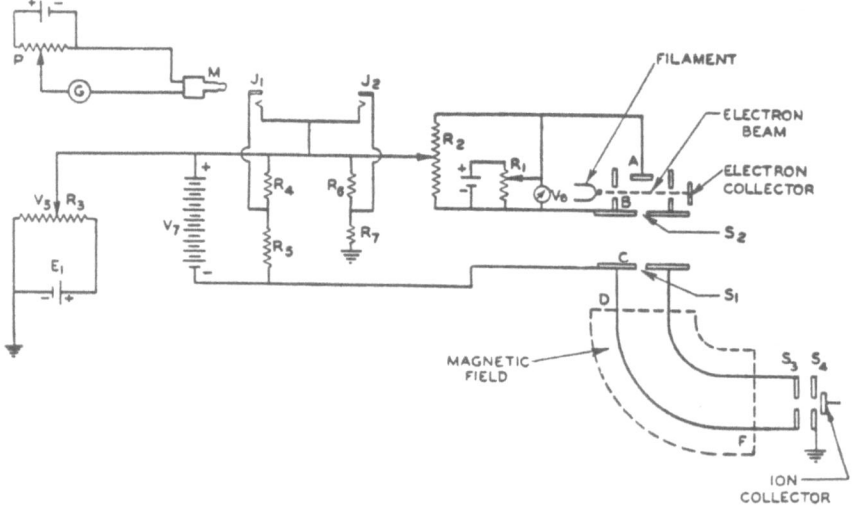

FIGURE 3. Arrangement for studying energies of ion fragments from metastable ions.

parent ion $C_4H_{10}^+$ in normal butane. The other curves show that the peaks arising from the metastable ions

$$C_4H_{10}^+ \rightarrow C_3H_7^+ + CH_3, \quad m^* = 31.9,$$
$$C_4H_{10}^+ \rightarrow C_3H_6^+ + CH_4, \quad m^* = 30.4,$$

also vary linearly with the pressure and are, therefore, to be attributed to a primary process. Other strong evidence for this is the effect of the ion-draw-out voltage (V_6 in Fig. 3) on the intensity of the metastable ions relative to the rest of the spectrum which is described in detail later. When less time is spent in the ion source, the abundance of the metastable ions rises rapidly as compared with the rest of the spectrum.

Apparatus and Experimental Procedure

The mass spectrometer employed in these experiments was of the type with a sectored magnetic field in which the ions were deflected through 90 degrees on a five-inch radius. An ion accelerating voltage variable from 300–1300 volts was available. The electron energies could be varied by changing the accelerating voltage on the electrons from 0–100 volts.

The ions were measured with a pen recording system whose sensitivity enabled signals of 0.2 millivolts to be recorded. This represented an ion current of about 5×10^{-15} ampere. This recorder, along with certain features of the mass spectrometer has been previously described.[10]

In order to obtain the retarding potential which formed the energy filter, the electrical circuit was arranged as shown in Fig. 3.

An electron beam creates the ions in the region AB. By means of a small potential V_6 these ions are drawn through slit S_2 into the accelerating field BC. The ions emerging from slit S_1 pass through a field-free region CD into the magnetic field DF. After leaving the magnetic field, they pass through a slit S_3 into the retarding field which is applied between S_3 and S_4. Those ions which are able to penetrate this retarding field are caught by the ion collector and measured with the recording system.

An ion accelerating voltage V_7 was applied between slits S_1 and S_2 in which S_1 was negative with respect to S_2. The ion source was connected to ground through a variable voltage supply V_5. S_4 was maintained at ground potential. Since the potential of the ion source and analyzer could be varied with respect to ground potential by means of R_3, the retarding potential between S_3 and S_4 could be adjusted so as to allow the metastable ions to be completely filtered. A measure of the voltage V_5 which would just enable these ions to get through to the ion collector would then be a measure of the energy lost in the dissociation. The potentials V_5 and V_7 were obtained with banks of batteries.

Since the ions are drawn out of AB with about 3 volts, it was necessary to take this voltage into consideration as the ions would then not be formed at ground potential. This was done by means of the resistor R_2. This resistor was placed across plates A and B and in parallel with the source of potential used to draw out the ions from that region. R_3 was adjusted so that V_5 was zero. R_2 was then adjusted until the primary ions were just beginning to be cut off. Thus the voltage V_5 which just cut off the ions in question was then a good measure of the energy lost.

The voltages V_5 and V_7 were measured with a potentiometer P. This potentiometer measured the voltage drop across the bleeder resistors R_4 and R_6. The potential drop across the resistor R_4 gave a value of the voltage

[10] J.A. Hipple, D.J. Grove, and W.M. Hickam, Rev. Sci. Inst. **16**, 69 (1945).

V_7 while that across R_6 gave the voltage V_5. R_4 and R_6 were 1500 ohms each while R_5 and R_7 were 3×10^6 ohms each. The resistors in each bleeder were matched to better than 0.1 percent. The potentiometer circuit measured each of the voltages by means of the two jacks J_1 and J_2, the accuracy being about 0.2 volt.

With this arrangement the spectra of n-butane, 1,3-butadiene and ethane were studied to determine the transitions giving rise to the metastable ions. The spectra were scanned by varying the magnetic field.

Discussion of some Hydrocarbon Spectra

1. Normal Butane

In addition to the normal sharply defined peaks in the mass spectrum, there are broad peaks at effective masses about 32 and 30.5 and a large hump on the side of the 39 peak which appears to be superposed on the normal 39 peak. There is also a slight hump at 25.2.

The peak at 32 is associated with the transition

$$C_4H_{10}{}^+(58) \rightarrow C_3H_7{}^+(43) + CH_3(15),$$

for which $m^* = (43)^2/58 = 31.9$. This assumption has in its favor the fact that mass 43 is a very prominent peak in the normal spectrum. If the assumption is correct, then the ions responsible have only (43/58) of the kinetic energy corresponding to the total applied potential of 361 volts. This means that in the energy filter this peak should disappear if the ions are caused to go against a retarding field of more than

$$(43/58) \times 361 = 268 \text{ volts.}$$

They, therefore, will have lost 93 electron volts of kinetic energy, which actually went to the unobserved neutral methyl radical.

Figure 4 shows the effect of the energy filter on the mass 32 region as V_5, the potential of the ion source with respect to ground is varied. The diffuse peak is present for $V_5 = 95$ volts but begins to go out as V_5 is further diminished and is quite gone at $V_5 = 90$ volts. In curve F of Fig. 4 the remaining small sharp peak at mass 32 is due to $O_2{}^+$ from a small trace of that gas in the tube.

Similarly the diffuse peak at $m^* = 30.5$ also shown in Fig. 4 may be associated with

$$C_4H_{10}{}^+(58) \rightarrow C_3H_6{}^+(42) + CH_4(16),$$

for which $m^* = (42)^2/58 = 30.4$. In this transition 16/58 of the original energy is lost to the ion fragment on dissociation. This is 99.5 volts and in Fig. 4 it is clearly seen that the 30.5 peak starts to disappear at 99 volts and is cleaned out at 95 volts—a higher value of V_5 than the diffuse 32 peak, in accord with this interpretation.

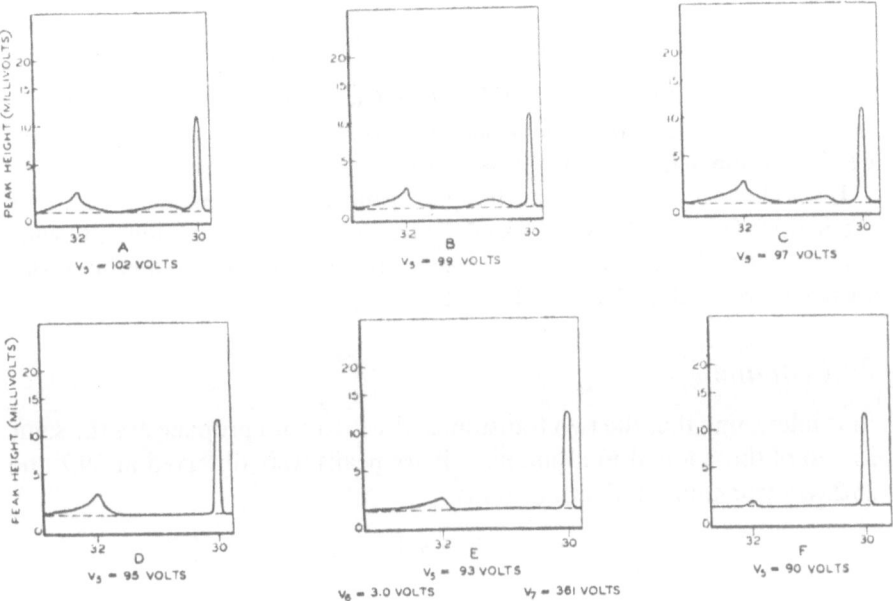

FIGURE 4. Effect of energy filter on mass spectrum of *n*-butane in the mass 30 region.

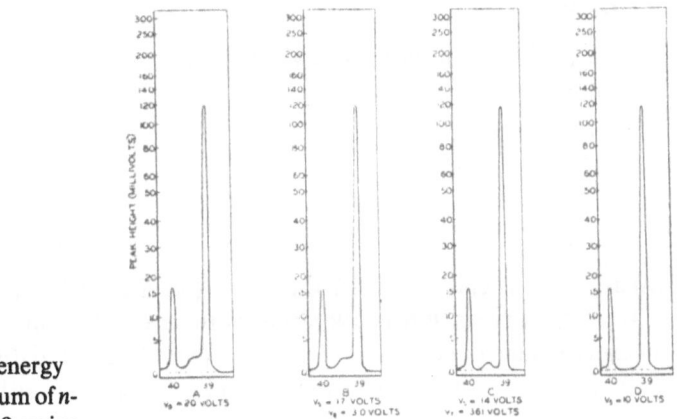

FIGURE 5. Effect of energy filter on mass spectrum of *n*-butane in the mass 39 region.

The hump at $m^* = 39$ is associated with the transition

$$C_3H_7^+(43) \rightarrow C_3H_5^+(41) + H_2(2)$$

giving a calculated $m^* = 39.2$. The calculated energy loss of the ion fragment is 2/43 of 361 or 16.3 volts. Figure 5 shows the effect of the energy filter on this region of the mass spectrum. Again the expected behavior is obtained, a considerable portion of the hump being removed in going from 17 to 14 volts.

The weak ion current appearing at 25^+ may be explained by the transition

$$C_2H_5{}^+(29) \rightarrow C_2H_3{}^+(27) + H_2(2)$$

from which $m^* = (27)^2/29 = 25.1$. It has been found that this diffuse peak may be erased from the spectrum when $V_5 = 24.0$ which agrees well with the deduced value of $(2/29) \times 361 = 24.8$ volts.

It should be emphasized that these experiments do not reveal the nature of the neutral fragment that is formed. Thus, in the preceding equations what we write as CH_4 may possibly represent production of this material in dissociated form such as $CH_3 + H$ or $CH_2 + H_2$, etc.

2. Propane

It is interesting that the two transitions observed with propane are the same as two of those found in n-butane. Diffuse peaks were observed at 39.2 and 25.2 corresponding to the transitions

$$C_3H_7{}^+(43) \rightarrow C_3H_5{}^+(41) + H_2(2), \qquad m^* = 39.2,$$

and

$$C_2H_5{}^+(29) \rightarrow C_2H_7{}^+(27) + H_2(2), \qquad m^* = 25.2.$$

These transitions were not studied with the energy filter.

3. Ethane

Diffuse peaks are observed at 24^+, 25^+, and 26^+. These may be explained by the transitions

$$C_2H_6{}^+(30) \rightarrow C_2H_4{}^+(28) + H_2(2), \qquad m^* = 26.2, \qquad V_L = 34.4,$$

$$C_2H_5{}^+(29) \rightarrow C_2H_3{}^+(27) + H_2(2), \qquad m^* = 25.2, \qquad V_L = 35.4,$$

$$C_2H_4{}^+(28) \rightarrow C_2H_2{}^+(26) + H_2(2), \qquad m^* = 24.2, \qquad V_L = 36.8,$$

where V_L is the energy lost by the ion to the neutral fragment.

In this study an ion accelerating voltage of 516 volts was used and the measured value of voltages at which the peaks started to disappear were 35.0, 36.0, and 37.0 volts, respectively—in nice agreement with the values of V_L given above.

4. 1,3-Butadiene

In the mass spectrum of 1,3-butadiene,

FIGURE 6. Effect of energy filter on mass spectrum of 1,3-butadiene in the mass 28 region.

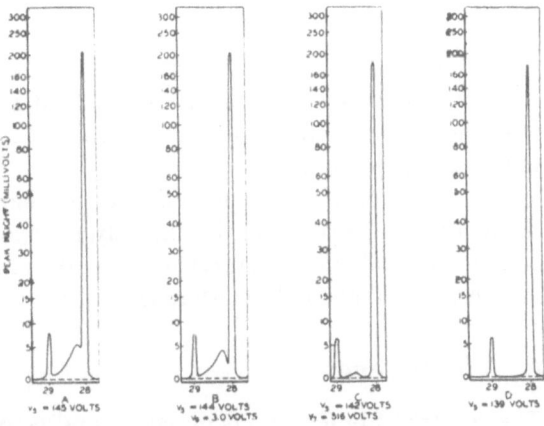

FIGURE 7. Effect of energy filter on mass spectrum of 1,3-butadiene in the mass 14 region.

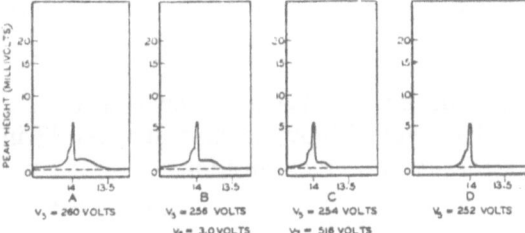

there is a large diffuse peak at about mass 28.2 and a smaller one at about mass 13.8. The parent mass is 54 and the large diffuse peak at 28.2 can be associated with the transition $54^+ \rightarrow 39^+ + 15$ for which the calculated m^* is $(39)^2/54 = 28.2$ although not immediately evident from the structure. However, 39^+ and 54^+ are the largest peaks in 1,3-butadiene.

Figure 6 shows the effect of the energy filter on the mass 28 region which shows that the peak begins to disappear for $V_5 = 144$ volts and is removed when $V_5 = 139$ volts. The expected loss is $(15/54) \times 516 = 143$ volts.

The diffuse peak at mass 13.5 is shown in Fig. 7 for various values of V_5. This peak is associated with a simple splitting in half of the parent ion, $54^+ \rightarrow 27^+ + 27$, which leads to a calculated m^* of 13.5. On this view the ion fragment has only one-half of the original energy and so can be stopped by a retarding potential of 258 volts in good agreement with the experimental value ($V_5 = 256^+$ volts) shown in the figure. The remaining asymmetry on mass 14 is caused by the production of CH_2 with kinetic energy.

These interpretations indicate that there are two different ways in which the parent ion can break up, the dropping off of a methyl group being considerably more probable than the process of splitting in half.

5. Iso-butane

Iso-butane shows the same broad peaks as normal butane except for some differences in relative intensity. The peaks at 31.9 and 30.5 are much weaker than those from n-butane, and the "hump" at 39.2 is about twice as strong as in the case of n-butane.

These differences correlate with the fact that 58^+, which is interpreted as giving rise to 31.9 and 30.5, is relatively stronger in n-butane than isobutane, whereas 43^+, which is interpreted as the source of the 39.2 hump, is relatively stronger in isobutane than n-butane.

6. The butenes

The spectra of butene-1, *cis* butene-2, and isobutylene were studied and no important differences were noted with respect to peaks due to metastable ions.

A broad diffuse peak appears at mass 30 similar in appearance to the 31.9 peak in the butanes. This is interpreted as being due to decomposition of the parent ion by the dropping of a methyl group,

$$56^+ \rightarrow 41^+ + 15,$$

the calculated m^* being $(41)^2/56 = 30.0$.

The "hump" now appears at mass 37 instead of mass 39 and may be associated with the transition,

$$41^+ \rightarrow 39^+ + 2.$$

7. Normal pentane

With normal pentane there is a very diffuse peak at mass 24.5. This peak can be associated with the transition

$$72^+ \rightarrow 42^+ + 30, \qquad m^* = 24.5.$$

There is also a very broad peak between 29 and 30 which can be regarded as due to

$$57^+ \rightarrow 41^+ + 16, \qquad m^* = 29.5.$$

A "hump" is also present on the 39 peak as in the butanes.

This spectrum has not been studied with the energy filter to give detailed confirmation of the suggested transitions. It might be expected that the parent 72^+ ion could drop a methyl group. This would lead to a predicted diffuse peak at $m^* = 45.3$ but there is no indication of such a process in our observations.

8. Iso-pentane

The same diffuse peaks are observed here as with n-pentane, presumably arising from the same transitions. In iso-pentane the 57 peak is about twice that for n-pentane and correspondingly the diffuse 29.5 peak is about twice as intense as in n-pentane. On the other hand the situation is reversed with regard to the parention 72^+ and its associated diffuse peak at 24.5 where the n-pentane intensity is about four times that in iso-pentane.

9. Pentene-2

Pentene-2 shows diffuse peaks at masses 24, 15.3, 27.8, and 51. These peaks can be interpreted by the following transitions, with the corresponding calculated m^*,

$$70^+ \rightarrow 41^+ + 29, \qquad m^* = 24.0,$$
$$55^+ \rightarrow 39^+ + 16, \qquad m^* = 27.7,$$
$$55^+ \rightarrow 29^+ + 26, \qquad m^* = 15.3,$$
$$55^+ \rightarrow 53^+ + 2, \qquad m^* = 51.1.$$

10. Methyl Cyclo-Pentane

Diffuse peaks are observed in methyl cyclopentane at 56.8 and 37.4. The following transitions could account for these:

$$C_6H_{12}^+(84) \rightarrow C_5H_7^+(69) + CH_3(15), \qquad m^* = 56.7,$$
$$C_6H_{12}^+(84) \rightarrow C_4H_8^+(56) + C_2H_4(28), \qquad m^* = 37.4.$$

Effect of Ion-Draw-Out Potentials

Since the ions spend a considerable portion of their lifetime in the ion source, it should be possible to change the magnitude of those peaks originating from the metastable ions by changing the time spent in this region. This can be accomplished by changing the potential with which the ions are drawn from the ion source into the main accelerating field. This potential is represented by V_6 in Fig. 3. With an increase in this voltage the ions are drawn out faster and hence they spend a shorter time in this region. Thus more ions have a chance to dissociate in the regions where the effect can be detected.

A complication arises, however, in that an increase in V_6 not only increases these metastable peaks, but also affects all of the peaks in general because of the change in the focusing conditions. In order to obtain some idea as to the effect of V_6 on the dissociation phenomenon, it is necessary to look at the peak ratios.

Table I presents the peak heights in arbitrary units of some masses in the normal butane spectrum taken at several values of V_6.

Table II is derived from Table 1. The data are normalized to show more clearly the relative change in the peak heights as V_6 is increased. From this table it is noted that from 2.0 through 5.0 volts, peaks 58, 57, 43, 42, 41, 40, and 39 remained fairly constant, and that they all changed by approximately

TABLE I. Peak heights in normal butane spectrum

Mass	$V_6 = 1.0$	2.0	3.0	4.0	5.0
58	246.0	425.6	437.9	431.3	430.2
57	41.08	68.67	71.22	70.25	69.55
43	1243	2016	2129	2119	2129
42	174.6	278.2	291.6	288.2	287.2
41	349.0	541.1	563.4	553.5	553.5
40	15.38	24.16	25.66	25.93	26.00
39	114.0	178.9	189.5	187.6	189.7
39.2	1.28	2.84	3.36	3.99	4.26
31.9	0.88	1.94	2.39	2.61	2.89
30.5	0.27	1.03	1.37	1.49	1.64

TABLE II. Relative peak heights at various voltages

Ion-draw-out voltage	Mass									
	58	57	43	42	41	40	39	39.2	31.9	30.5
1.0	1.00	1.00	1.00	1.00	1.00	1.00	1.00	1.00	1.00	1.00
2.0	1.73	1.67	1.62	1.59	1.55	1.57	1.57	2.22	2.21	3.82
3.0	1.78	1.73	1.71	1.67	1.61	1.67	1.66	2.62	2.72	5.26
4.0	1.75	1.71	1.70	1.65	1.59	1.69	1.65	3.11	2.97	5.52
5.0	1.75	1.70	1.71	1.65	1.59	1.69	1.66	3.33	3.29	6.07

the same percentage between 1.0 and 2.0 volts. However, those peaks which are attributed to the metastable ions changed by a greater percentage between 1.0 and 2.0 volts, and they continued to increase as V_6 was increased. In fact it is noted that the 30.5 peak is six times as large at 5.0 volts as at 1.0 volt, while the 31.9 peak is over three times as large.

Effect of Electron Energy

A study of the metastable ions in n-butane[11] as a function of the electron voltage V_8 is shown in Fig. 8. The spectrum in the 28–44 mass region is shown with $V_8 = 12, 17, 22$, and 27 volts. With 12-volt electrons the metastable ions at 30.5 and 31.9 are just appearing whereas the "hump" at 39.2 has not appeared. At 17 volts the 39.2 peak has appeared, although the true mass 39 peak is barely evident. Since the metastable peaks are weak, it is difficult to measure their appearance potentials accurately. However, it is interesting that mass 43 and the apparent mass 31.9 appear at the same voltage within

---

[11] D.P. Stevenson and J.A. Hipple, J. Am. Chem. Soc. **64**, 1588 (1942).

FIGURE 8. Mass spectrum of *n*-butane in the mass 28–44 region showing the effect of variation of the electron voltage. The discontinuities indicate a change to $\frac{1}{10}$ sensitivity to record the tops of the large peaks.

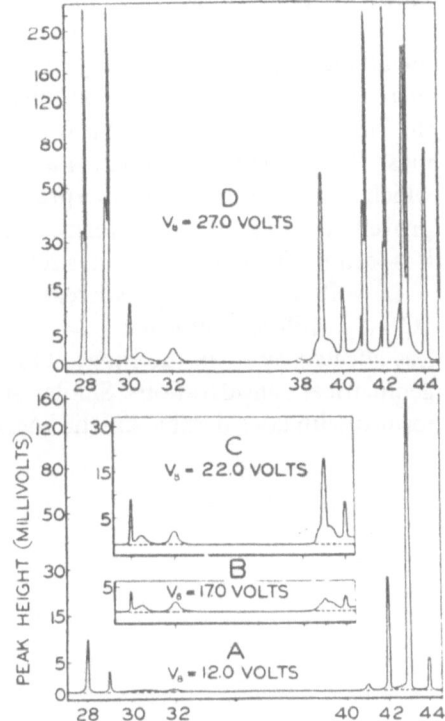

the accuracy of the measurements (about 0.5 volts); in the former case the reaction

$$C_4H_{10}{}^+ \rightarrow C_3H_7{}^+ + CH_3$$

occurs before the acceleration in the ion gun (region AB in Fig. 3), and in the latter case after this acceleration (after BC in Fig. 3). The appearance potential of 39.2 is about 2.5 volts higher than that for 31.9.

Ions Formed with Kinetic Energy

Since the mass spectrometer of the type used in this work will not focus ions of the same m/e which differ in energy, those ion fragments which are formed from the parent molecule with appreciable kinetic energy will give rise to asymmetrical peaks. A measurement of the kinetic energy released in the ionization and dissociation of some less complicated molecules has provided a greater insight into the dissociation processes involved in these molecules.[12,9] Part of the asymmetry in Fig. 7 was ascribed to the formation of $CH_2{}^+$ with kinetic energy in butadiene. Similarly, in *n*-butane there appears to be a weak

---

[12] W. Bleakney, Phys. Rev. **35**, 1180 (1930).

kinetic energy peak at mass 14 and a pronounced one at mass 15. These ions still reached the ion collector when electrode A (Fig. 3) was grounded, eliminating all the other peaks in the spectrum between mass 15 and mass 60. Under this condition only those ions formed initially with kinetic energy can reach the ion collector. Although this is not a metastable effect, the investigation of mass 15 is described in this paper as there is a non-integral mass involved here which might be mistakenly interpreted as arising from a metastable ion unless proven otherwise. If the ion formed with kinetic energy is designated $*CH_3^+$, it is seen in Fig. 9 that this ion is relatively unaffected when V_5 is varied from -1.0 volts to -2.8 volts whereas CH_3^+ disappears at -2.8 volts. In fact, $*CH_3^+$ is still present when $V_5 = -5.0$ volts. Figure 10 shows that $*CH_3^+$ is less affected by V_6 in the range studied than CH_3^+ which is reasonable on geometrical considerations. Similar studies have shown the presence of ions formed with considerable kinetic energy in the C_2 and C_3 regions.

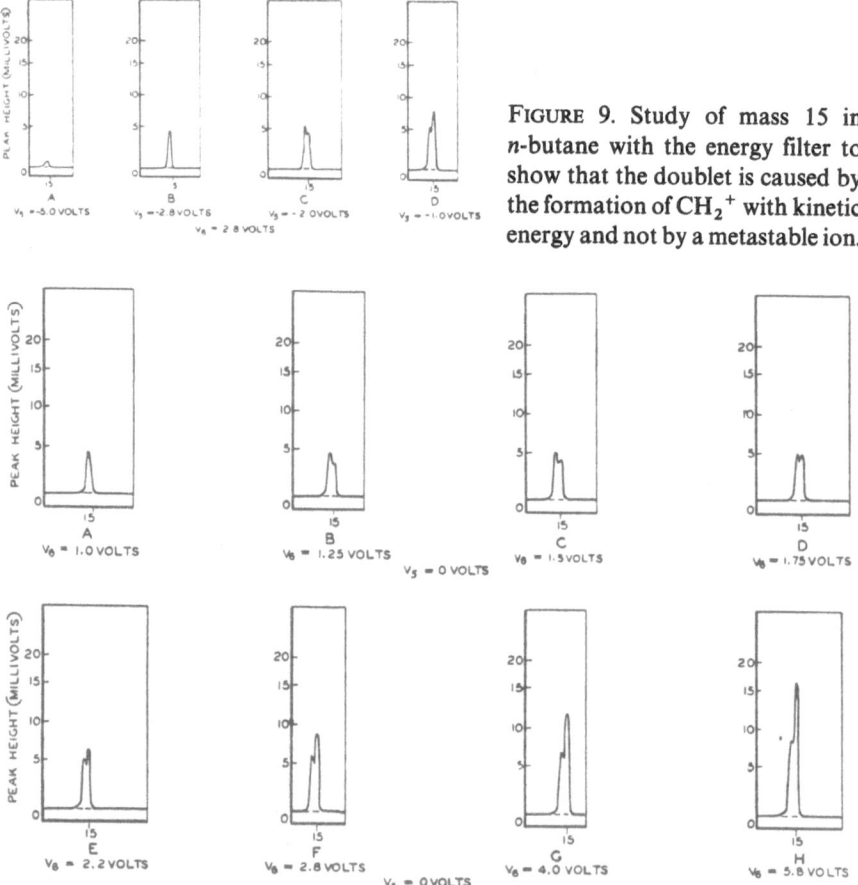

FIGURE 9. Study of mass 15 in n-butane with the energy filter to show that the doublet is caused by the formation of CH_2^+ with kinetic energy and not by a metastable ion.

FIGURE 10. Showing the difference in the effect of the ion-draw-out voltage (V_6) on the CH_3^+ in n-butane formed with considerable kinetic energy as compared with the fragment formed with little or no kinetic energy.

The Franck-Condon Principle and Related Topics*

E.U. CONDON

National Bureau of Standards
Washington, D.C.

A review of the historical development and present status of certain topics in molecular physics may serve as a reminder of the progress made in the past two decades. Let us recall that in 1925 there was no subject of nuclear physics, and high voltage equipment was almost non-existent except in the x-ray laboratories. About all that was known about cosmic rays was that the ionization increases in high balloon flights in the atmosphere and that its cause penetrates deeply into lakes. Nobody had heard of closed electron bands in a solid, and the word semi-conductor was essentially unknown.

"What did physicists find to be interested in?," young physicists may wonder. The answer is, of course, that they worried about a lot of things that are nowadays taught to and learned by beginning graduate students with such glibness that it is hard to realize there ever was a time when these things were not known.

In 1925 there was no quantum mechanies. Though de Broglie's thesis was published in 1924, I have never met anyone who read it seriously until later. We worked with quantization by the $\int p \, dq = nh$ method of Bohr and Sommerfeld and tried to get at radiative transition probabilities by approximate and unclearly formulated procedure based on the Bohr correspondence principle, according to which quantum radiative jumps were associated with, or set in correspondence with certain terms in the Fourier analysis of the quasi-periodic motions in the mechanical system.[1]

The Bohr theory had given a beautiful account of the spectrum of atomic hydrogen and of the arc spectra of the alkalis. However, why the *D* lines of

---

* Address of the Retiring President, American Physical Society, New York, January 31, 1947. My apology for including my own name in the title of this paper is that it conforms to a well-established usage. In the text I shall refer simply to the Principle. Professor Rabi once said to me that the Principle was a great boon to lecturers in courses on atomic physics, for it is so easy to understand that they are always assured of one or two lectures which they do not have to prepare.

[1] W. Lenz, *Zeits. f. Physik* **25**, 299 (1924). This paper represents an attempt to deal with the problem of nuclear transitions associated with electronic jumps by means of the correspondence principle.

sodium were double was a great mystery that was only cleared up by the
electron-spin hypothesis of Goudsmit and Uhlenbeck. Similarly, the old quantum theory had thrown a good deal of light on the infra-red and electronic
spectra of diatomic molecules. Pure rotation, rotation-vibration, and electronic band systems were recognized, analyzed and utilized to get interesting
quantitative data on important molecules. Nevertheless, here too there were
puzzling things; for example, the isotope shift between HCl^{35} and HCl^{37}, and
in other cases, required that the vibrational levels be assigned half-integral
quantum numbers.

First Ideas on the Principle

At that time I was a graduate student in the University of California in
Berkeley. Ernest Lawrence was a graduate student at Yale, and there were no
cyclotrons in the world—not one, not even in Berkeley. Professors R.T. Birge
and L.B. Loeb were the stimulating research leaders of the department, the
former on molecular spectra, the latter on every phase of electric conduction
in gases.

During the year 1925–26, Birge conducted a seminar on molecular spectra
which it is impossible for me to praise too highly. Then as always his work
was distinguished by the most exact and painstaking scrutiny of the data in
their relation to the theory. I still remember how exciting it was when he first
showed, with precision, that the swelling of molecules by rotation, as inferred
from their pure rotation spectrum agreed with the values found from the
vibration-rotation spectrum and from electronic systems. It was he who had
carefully compiled all the existing data on analyzed electronic band systems—
then about a dozen in number—and who recognized the basic empirical facts
of the intensity distribution, later to be explained by the Principle. I mention
this point so explicitly because without his stimulating guidance I would never
have been aware of the problem of intensity distribution in band systems.

In Göttingen, Professor James Franck was very much interested in photochemical reactions and, in particular, the dissociation of iodine vapor by
absorption of light. He gave a paper before the Faraday Society in London[2]
in which the basic idea of the Principle was first presented in connection with
this problem.

Proof sheets of this paper were sent by Franck to his student, Dr. Hertha
Sponer, who was in Berkeley that year on an International Education Board
Fellowship. She let me read them, and, because I had just learned the empirical
problem about intensity distribution in band systems from Birge's seminar, it
was immediately evident how to generalize Franck's ideas a little more in

[2] J. Franck, *Trans. Faraday Soc.* **21**, 536 (1925). The paper in which the Principle was
first advanced to explain the photochemical dissociation of iodine vapor.

FIGURE 1. Potential-energy
curves from Franck's Faraday
Society paper (reference 2).

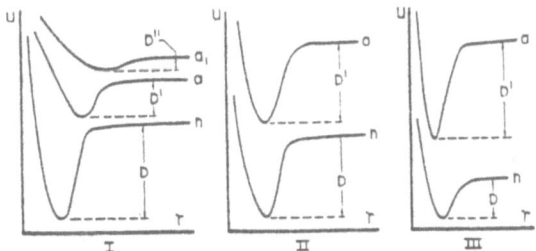

order to get the full story. What is more from Birge's seminar I had at hand
a good critical compilation of the existing data which made possible a quick
quantitative test of the ideas.

This work was all done in a few days. Doctor Sponer showed me Franck's
paper one afternoon, and a week later all the quantitative work for my
1926 paper[3] was done. But let us see just what was the situation.

Figure 1 is from Franck's paper.[2] He pointed out that, owing to the large
masses of the nucle in a molecule, their relative momentum cannot be directly
affected by an electronic transition, so that those transitions will be most likely
that conform most closely to the Principle. Therefore, *if* in the iodine molecule
the curves are related as in set *I* of Fig. 1, a molecule initially not vibrating
will most readily absorb light that carries it into states of high vibration, or
to states higher than the energy needed for dissociation, resulting in photochemical dissociation of the molecule.

It was, of course, not a very difficult step to recognize that if the molecule
is vibrating initially, then the Principle asserts that transitions are favored to
those states which require least instantaneous adjustment of the relative
position *and momentum*. Moreover, it was intuitively felt that the electronic
transition was sufficiently independent of the nuclear vibration that it was
equally likely to occur at any phase of the nuclear vibratory motion. This is
not self-evident and perhaps not exactly true. It might be, for example, that
the electron jump is stimulated by vibratory motion in the molecule, perhaps
in such a way that it is more likely to occur at a phase of maximum relative
velocity. However, this is not the case. With all times of electron jump equally
likely, the most favored vibrational transitions will be those associated with
the turning points of the nuclear vibration; for, as the nuclei move slowly here,
a larger fraction of the time is spent in such regions.

This idea led to Fig. 2, taken from my 1926 paper,[3] which indicates that
there are two most favored vibrational quantum-number changes associated
with nonvanishing values of the initial vibrational quantum number.

The proof of the pudding lay in the fact that a number of band systems were
well analyzed, and therefore it was possible to put down the potential-energy

---

[3] E.U. Condon, *Physical Rev.* **28**, 1182 (1926). First application of the Principle to
intensity distribution in band systems.

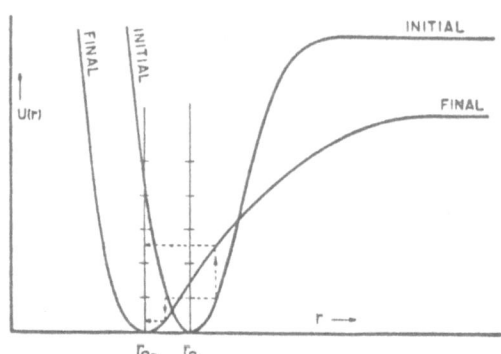

FIGURE 2. Graphical construction of favored transitions from my 1926 paper (reference 3).

curves in their approximately correct form and relative location. Near the minimum a curve is parabolic:

$$V(r) = \tfrac{1}{2}k(r - r_0)^2 + \cdots .$$

One can get the value of k from the vibration frequency v, since, if μ is the reduced mass, $2\pi v = (k/\mu)^{1/2}$; and one gets r_0 from the moment of inertia, which in turn is given quantitatively from the observed rotational energy levels. Evidently in Fig. 2, when there is little change in v or r_0, the curves are similar and lie directly over each other, so the Principle requires small or zero change in the vibrational quantum number. But if there is a big change in r_0, the curves are widely displaced and the Principle requires large changes in the vibrational quantum number. Both cases were found in the data available in 1926. The Principle was triumphant in that those with large changes in vibrational quantum number were correctly correlated with large changes in the equilibrium internuclear distance, and *vice versa*.

It must be remembered that the intensity data available were simply rough estimates of plate blackening, uncorrected for variations in plate sensitivity over the rather great range of wave-length involved in some band systems. Nevertheless, there was no doubt of the essential correctness of the Principle.

Curiously enough, my calculations indicated a bad disagreement with the facts for iodine, the very molecule which led Franck to his qualitative discovery of the idea. This caused me much worry, and I searched for an error in calculation a long time before sending in the 1926 paper with such a bad discrepancy for iodine. About a year later the error was discovered by Professor Wheeler Loomis, then of New York University. I had used the value of the moment of inertia for the 26th vibrational state in the electronically excited level, which is involved in the fluorescence spectrum of iodine, thinking it was the moment of inertia of the nonvibrating molecules. When this error was corrected, iodine agreed with the Principle as well as the others.[4]

---

[4] E.U. Condon, *Physical Rev.* **32**, 858 (1928). Development of quantum-mechanical formulation of the principle, including basic idea of internal diffraction.

The faculty of the University of California was broad-minded enough to accept the paper as a doctor's thesis. However, it is interesting to note that orthodox theoretical physics was so tied to Bohr's correspondence principle at the time that the referee for the *Physical Review* was reluctant to recommend the paper for publication. He felt it could not be correct because it did not go at the problem in terms of Fourier amplitudes of the classical motion.

But the whole story was not given by the simple classical picture of 1926. The main puzzle might be stated thus: How exact is the Principle? or, what determines the extent of its inexactness?

The potential-energy curves lie in a definite location, so the Principle leads to a rather definite indication of most favored transition. Actually, although the predictions of the most favored transition agreed well with the facts, there was no indication of why other transitions could occur or how to calculate their intensities. This was a serious shortcoming of the theory, which Professor Birge did not hesitate to point out to me at the time.

A year later, in 1927, we all became familiar with Heisenberg's uncertainty principle as one of the broad implications of quantum mechanics. But in 1926 it did not occur to anyone that there was anything basically wrong with talking about definite values of both nuclear position and momentum being carried over without alteration in an electronic transition. Looking back on it later one can see that it might have been thought of, because statistical mechanics was dealing in terms of elementary cells of finite extension in phase space and the uncertainty principle is closely related to that. However, as elsewhere, hindsight is often better than foresight in theoretical physics.

Quantum-Mechanical Formulation

All that I just described happened in the spring of 1926. Matrix mechanics had been developed, but the matrix calculus was so difficult mathematically that it was hard to put any physical insight into it.* Schrödinger's famous series of papers on wave mechanics was just appearing; but it was not until the fall of 1926 that Born first recognized the probability interpretation of $|\psi|^2$ which collision problems forces on us, as contrasted with the more hydrodynamic views which are possible alternatives for closed systems.

I had the good fortune to be sent by the International Education Board from Berkeley to Göttingen in the fall of 1926 and thus was able to plunge into the study of the new quantum mechanics at one of the few points of high concentration of original thinking in this field. Great ideas were coming out so fast that period (1926–27) that one got an altogether wrong impression of the normal rate of progress in theoretical physics. One had intellectual indigestion most of the time that year, and it was most discouraging.

---

* In fact, I remember very well, in the fall of 1926 at Göttingen, that Professor David Hilbert said to his class in this connection, "die Physik wird zu schwer für die Physiker!"

Besides studying the current papers I tried to solve the basic problem in quantum mechanics that underlies all molecular dynamics—the basic justification for the method of first working out the electronic states for fixed positions of the nuclei, so that the electronic energy levels depend parametrically on the nuclear coordinates and serve as the potential-energy functions for determination of the nuclear motions.

The justification is, of course, connected with the smallness of electron mass relative to nuclear mass, but I never could see how to work it out. Later the problem was handled in a basic paper by Born and Oppenheimer[5] which, however, I have never felt that I properly understood. But Born is such a great master at seeing all physics in terms of "Entwicklungen nach einem kleinen Parameter Kappa" that I suppose it must be all right. The paper of Born and Oppenheimer is among those difficult ones that are more often cited than read.

These studies did, however, serve to make clear the place of the Principle in relation to the general ideas of quantum mechanics. Although this topic was treated in a short paper written from Göttingen, it was not properly handled until the fall of 1928 in a paper[4] written in Princeton. The essence of the argument is that the wave function for a diatomic molecule is approximately of the form of a product of an electronic wave function, in which the nuclear coordinates appear as parameters, and of a wave function for the nuclear motion, say

$$\Psi = u(x_e, x_n) \cdot v(x_n),$$

where x_e stands symbolically for the electron coordinates and x_n for the nuclear coordinates. Therefore, the matrix element for a quantity like the dipole moment, which determines the radiation transition probabilities, is of the form

$$M_{12} = \int_e \int_n \Psi_1 M(x_e, x_n) \Psi_2 \, dx_e \, dx_n$$

$$= \int_n \bar{v}_1(x_n) M_{12}(x_n) v_2(x_n) \, dx_n,$$

where

$$M_{12}(x_n) = \int_e \bar{u}_1(x_e, x_n) M(x_e, x_n) u_2(x_e, x_n) \, dx_e.$$

The quantity $M_{12}(x_n)$ is the matrix element of the dipole moment calculated by regarding the nuclear coordinates as parameters of the electronic problem rather than as dynamical coordinates. It is characteristic of the electronic jump in question and so is the same for all the vibrational transitions of a band system.

---

[5] M. Born and J.R. Oppenheimer, *Ann. Physik* **84**, 457 (1927). Basic quantum-mechanical justification of use of electronic potential-energy curves to determine nuclear motion in molecules.

The nuclear wave functions $v(x_n)$ may be quite complicated for a polyatomic molecule, but for a diatomic molecule they are of the form of a function of the radial coordinate $R(r)$ multiplied by a spherical harmonic for the simple motion under conservation of angular momentum.

It is hard to say much in general about $M_{12}(x_n)$. In fact, no explicit calculation of an example has been made even yet. But it is natural to suppose that it will be a slowly varying or smooth function of r over the small range of r in which the radial wave functions have appreciable value.

The radial functions $R(r)$ are rather closely related to the corresponding classical vibratory motion, according to the general correlation provided by the Wentzel-Brillouin-Kramers approximation. According to this, the wave function is of the form

$$R(r) = \frac{1}{4(p)^{1/2}} \cos\left[(2\pi/h) \int^r p\, dr + \alpha \right]$$

within the range of the classical motion, and falls off rapidly to zero outside the range of the classical motion, being dominated by a factor of the form

$$\exp \pm \left[(2\pi/h) \int^r |p|\, dr \right],$$

where p is given by

$$(1/2\mu)p^2 + V(r) = W,$$

where $V(r)$ is the effective potential energy of the radial motion, including the effects of the rotational energy;

$$V(r) = V_0(r) + \frac{\hbar^2}{2\mu} \frac{J(J+1)}{r^2},$$

when J is the rotational quantum number and $V_0(r)$ is the potential-energy function applicable in the case of a nonrotating molecule.

It is easy to see qualitatively that the value of an integral of the form

$$\int R_1(r)R_2(r)\, dr,$$

with wave functions given by the Wentzel-Brillouin-Kramers approximation, is in the main determined by the conditions set up in the classical formulation of the Principle as given in my 1926 paper:

(1) The integral will be small unless the wave functions overlap, that is, unless there is little sudden change in the internuclear distance called for in the transition;

(2) The integral will be small if the wave functions are related in such a way that a nonoscillatory part of one wave function overlaps a rapidly oscillatory part of the other, for this means that the electron jump would have to be accompanied by a considerable change in the radial momentum.

This formulation goes considerably further than that of the 1926 paper in three main respects:

(i) It follows in a definitely deductive way from a well-founded theory whose other successes have given it a definite standing as an adequate basis for atomic mechanics;
(ii) It gives, in principle, definite results for the relative strength of each possible quantum transition; it therefore goes beyond the simple classical picture based on the potential-energy curves and provides a basis for calculating the extent to which the less favored transitions actually do occur;
(iii) It leads to the prediction of some specific effects arising from the wave nature of matter for which I want to introduce the term "internal diffraction;" these will be dealt with at length in the next section.

When it comes to setting up explicit formulas for the integrals $\int R_1(r)R_2(r)\,dr$ it is not possible to get results of very general applicability. The natural thing to do is to start with the approximation that the initial and final potential-energy functions are Hooke's law parabolas, differing as to force constant and as to equilibrium separation. The formulas on this supposition were set up and evaluated by Hutchisson.[6] However, the results are rather complicated.

But a more severe limitation is that the cases which are most interesting physically are those in which there is a fairly large change in the equilibrium separation occasioned by the electronic transition, so that large changes in the vibration quantum number occur. In such cases the wave functions have to be known with fair accuracy at some distance away from the equilibrium separation. The actual force law is not parabolic, so that the harmonic-oscillator wave functions are no longer good approximations over an important part of the range of the coordinates.

Since the actual departure from the harmonic law is different for each molecule, general formulas are of small value. The only proper test of the theory is therefore explicit calculation based on the specific facts in particular cases.

Internal Diffraction

I particularly want to stress the specifically nonclassical, or wave-mechanical, features of the theory, for it seems to me they should be more widely known. These features seem to me to be more than mere quantum-mechanical refinements of detail, as they have sometimes been presented in books. On the contrary, internal diffraction is to me just as real and forceful a proof of the wave nature of nuclear motion as any of the basic external diffraction experi-

---

[6] E. Hutchisson, *Physical Rev.* **36**, 410 (1930); **37**, 45 (1931). Explicit calculation of integrals for transitions using harmonic-oscillator wave functions.

FIGURE 3. Wave functions as related for internal diffraction
in continuous spectra, from my 1928 paper (reference 4).

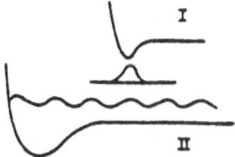

ments, such as those in which a beam of electrons or neutral hydrogen atoms
is diffractively scattered by a crystal.

Figure 3, from my 1928 paper,[4] presents one particular situation in which
internal diffraction might rise. It does not matter whether curve I is above or
below curve II: if I is above, the phenomenon will appear in emission, while
if I is below, then it will appear in absorption. The wave function of the lowest
vibrational level in state I will be approximately a Gauss error function as
shown. A radial wave function for a typical energy value in the continuum
above the dissociation limit of II will look like the one sketched in the figure.
As the energy of the final state is increased the radial wave function will vary
in this way: its first loop will have a fairly constant relation to the turning
point for the classical motion; but, as the de Broglie wave-length slowly
decreases, the nodes become more closely spaced, so that the phase of the
quasisine wave located under the center of the Gauss error wave function
gradually changes. Clearly, when a loop of the sinoidal wave function is under
the center of the Gaussian wave function we shall get a large value of the
integral which governs radiative transition. But when a node is under the
Gaussian wave function, the integral is very small.

In this way the initial-state wave function is able to "see" the wave nature
of the final-state wave function. Transitions can occur to the energies of the
continuum for which a loop is under the initial wave function and are much
weaker when they occur to the energies having a node there. The result
is rippling variation in the intensity of the continuous spectrum. Such a
manifestation of the wave detail of the ψ function I call *internal diffraction*, in
analogy with external diffraction, which is also determined by a phase relation
between initial and final wave functions.

The term internal diffraction is a convenient one to use more generally to
describe specific variations of the transition probabilities away from those
indicated by the Principle after it is fuzzed out by uncertainty principle
requirements. In the more general case we may suppose that two wave
functions for initial and final states are related approximately as in Fig. 4.
Evidently, the exact value of the integral of the product of two such functions
is quite sensitive to the exact relative positions of the nodes and loops. The
integrand is itself a roughly oscillatory function, being positive where the
factors are of the same sign and negative where they are of opposite signs.
Hence the value of the integral is quite sensitive to the average relative phase
of the oscillations in the wave functions of initial and final states. This is a
specifically quantum-mechanical effect, and definite results calculated from it

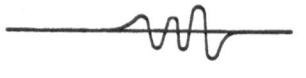

FIGURE 4. Rough sketch of wave functions showing sensitivity of integral to relative location of nodes.

afford a sensitive test of the reality of de Broglie waves associated with the nuclear motions in the molecules.

Although it did not seem feasible to make sufficiently accurate calculations to check the point at the time, this view was put forward in 1928 to explain the intensity fluctuations observed by Wood in the fluorescence spectrum of iodine vapor. On excitation by the green line of mercury, diatomic iodine molecules are raised to the 26th vibrational level of an excited electronic state from which they omit a long series of fluorescence doublets on jumping to various vibrational levels of the normal electronic state. These doublets form a series the intensity of which varies in quite an irregular way, namely: 10, 9, 1, 9, 3, 8, 8, 2, 9, 0, 8, 3, 2, 7, 0, 7, 0, 2, 1, 0, ... All of these transitions are permitted by the approximate form of the Principle. Undoubtedly the fluctuations are due to the particular phase relations of the nodes in the initial and final states, although because of the high quantum numbers involved an explicit calculation to test this statement would be very laborious and has not been made.

But the lack has been pretty well supplied by two explicit calculations which will now be mentioned. A similar irregular variation of intensity occurs in the fluorescence bands of Na_2. W.G. Brown[7] in 1933 calculated the approximate integrals and obtained the comparison of observed and calculated intensities that is shown in Fig. 5.

In 1939 similar calculations were made by Gaydon and Pearse[8] the band system of rubidium hydride. Figure 6, from their paper, shows the potential-energy functions and wave functions that they used. The dashed parabolas show the Hooke law approximation to the potential energy, and the full curves show the more accurate potential-energy functions inferred from the levels. The wave functions used were obtained by a reasonable approximate transformation from harmonic-oscillator wave functions rather than by numerical integration of the wave equation.

The result of their calculations is shown in Fig. 7. The table on the left gives observed values (estimates) and that on the right the results of calculation. The main parabolic locus of strong bands is the familiar result of the classical Principle. The secondary weaker loci are the result of special phase relations between initial and final wave functions, that is, internal diffraction.

[7] W.G. Brown, *Zeits. f. Physik* **82**, 768 (1933). Explicit calculation of internal diffraction intensity variations in fluorescence bands of diatomic sodium molecule.

[8] A.G. Gaydon and R.W.B. Pearse, *Proc. Roy. Soc.* **A173**, 37 (1939). Careful calculation of transition probabilities in band spectrum of rubidium hydride, showing internal diffraction effects quantitatively.

FIGURE 5. Brown's calculations on sodium fluorescence bands (reference 7).

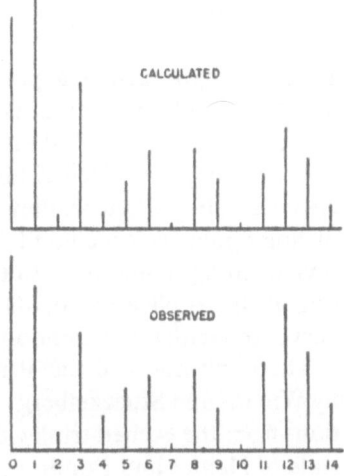

FIGURE 6. Potential curves and wave
functions used by Gaydon and Pearse
in calculation on rubidium hydride
(reference 8).

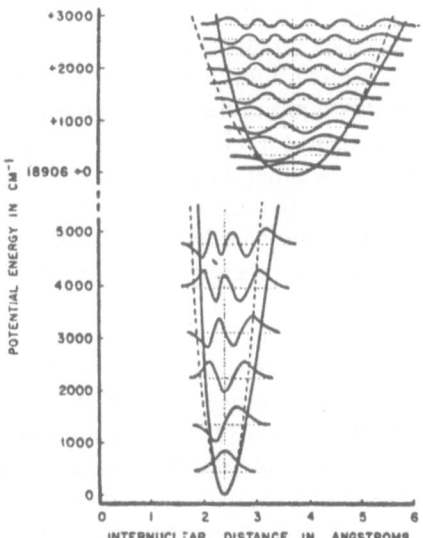

FIGURE 7. Intensities for rubidium
hydride as given by Gaydon and
Pearse: (a) observed visual estimates
on a scale of 10; (b) calculated,
reduced to a scale of 25.

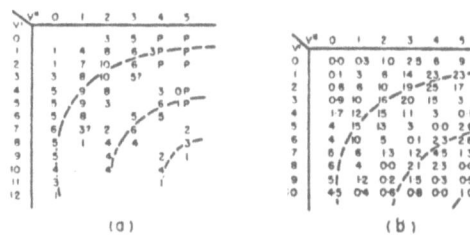

Molecular Hydrogen

Diatomic hydrogen is a particularly interesting molecule since it has the simplest structure, one that is simple enough to permit of making some fairly accurate theoretical calculations. These calculations, initially made by Heitler and London,[9] provided a host of new viewpoints applicable in general to chemistry. In particular, they cleared up the basic nature of the electron-pair or homopolar valence bond. There has to be some limit to the scope of this review, so my remarks will be confined to some interesting points that grew out of the application of the Principle to the theoretical potential-energy curves of Heitler and London, and of Burrau[10] for H_2^+.

The application of the Principle to molecular hydrogen was first made by Winans and Stueckelberg[11] in 1928. They recognized that radiative transitions from the excited triplet levels on to the repulsive-force triplet sigma ($^3\Sigma$) state of Heitler and London could account for the very extensive ultraviolet continuous spectrum of molecular hydrogen. At about the same time a paper written with Smyth[12] gave some interpretations of observed critical potentials in terms of the repulsive curves. These two papers may be said to have provided the first evidence of the physical reality of the repulsive-force curves given by quantum mechanics.

A much more convincing proof of the physical reality of such electronic states in molecular hydrogen was given in 1930 by Bleakney[13] when we were together at the University of Minnesota.

In the older data on critical potentials in hydrogen a value close to 31.5 v had usually been reported. As this agrees quite well with the minimum value of the energy needed to strip the two electrons off a hydrogen molecule, it was assumed to correspond to the process of total dissociation of the hydrogen molecule into two protons and two electrons.

But how could this be? According to the Principle, the colliding electron would have to strip off the two electrons, leaving the two protons at essentially the same distance apart as they are in the normal molecule. At this distance, the two protons have a Coulomb interaction energy of about 20 ev; therefore the process of total dissociation would require about 51.5 v instead of 31.5 v. Subsequently, of course, the two protons would fly apart under their mutual repulsion, each of them getting 10 ev of energy.

The situation is shown in Fig. 8, taken from Bleakney's 1930 paper.[13]

[9] W. Heitler and F. London, *Zeits. f. Physik* **44**, 455 (1927). First quantum-mechanical calculation of potential-energy curves of molecular hydrogen.

[10] O. Burrau, *Kgl. Danske Videnskab. Selskab. Mathfysiske Medd.* **7**, 14 (1927). First calculation of potential-energy curves for normal state of ionized hydrogen molecule.

[11] J.G. Winans and E.C.G. Stueckelberg, *Proc. Nat. Acad. Sci.* **14**, 867 (1928). Interpretation of the continuous ultraviolet emission spectrum of molecular hydrogen.

[12] E.U. Condon and H.D. Smyth, *Proc. Nat. Acad. Sci.* **14**, 871 (1928). Interpretation of the critical potentials of molecular hydrogen.

[13] W. Bleakney, *Physical Rev.* **35**, 1180 (1930). Experimental proof that some ions formed on electronic impact with molecular hydrogen have kinetic energy.

FIGURE 8. Potential-energy curves for molecular hydrogen, from Bleakney's 1930 paper (reference 13).

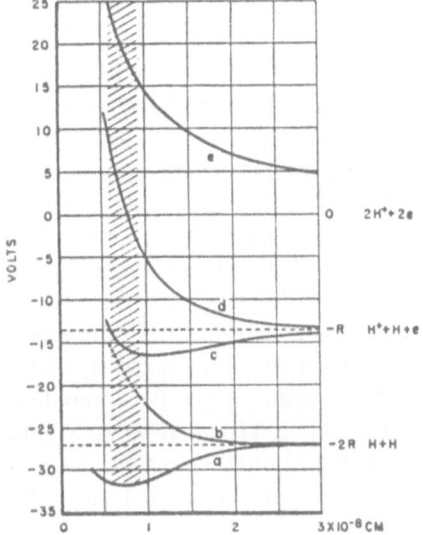

Curves (a) and (b) are derived from the basic Heitler and London solution for the mutual interaction of two normal hydrogen atoms. Curve (c) is from Burrau's calculation for the normal state of H_2^+. Curve (d) is the analogous repulsive-force state in H_2^+ as calculated by Morse and Stueckelberg.[14] Finally, curve (e) is simply e^2/r, the potential-energy curve of the molecule H_2^{++}.

As soon as these curves were drawn some interesting conclusions were evident. In the first place it is clear that the Principle does not allow us to dissociate the molecule simply by striking it with an electron of the minimum necessary energy, in this case about 4.4 v, for that would require a large change in position or momentum of the nuclei to accompany the electronic transition.

However, at about 11 v it ought to be possible to dissociate the molecule by electron impact by inducing transitions from state (a) to state (b). This transition is forbidden for light absorption but allowed for electron impact. The excited H_2 molecules would immediately fly apart, giving two normal hydrogen atoms each with about 3.5 ev of kinetic energy. As both products are neutral they could not be observed in a mass spectrometer. Observations have been made that indicate a rapid clean-up—that is, adsorption on the walls—of hydrogen, which could be due to formation of atomic hydrogen at about this energy, but not at lower voltages.

Considering state (c), the change in equilibrium separation indicates that the transition from H_2 to normal H_2^+ not vibrating is quite unlikely, and that it is likely that there will be a good yield of $H + H^+$ by direct transition from

---

[14] P.M. Morse and E.C.G. Stueckelberg, *Physical Rev.* **33**, 932 (1929). First calculation of repulsive-force potential-energy curve for ionized hydrogen molecule.

the normal state of the part of (c) lying about the dissociation limit. This was observed to the case.

Coming now to curve (d), we notice the explanation of the 31.5-v critical potential. It has nothing to do with total dissociation; that would require transitions to curve (e). The 31.5-v potential is in reality the transition, indicated by the Principle, to curve (d). If this is the correct explanation, then the H^+ ions formed in this process should have about 6.5 ev of kinetic energy. By an appropriate use of retarding fields in his mass spectrometer, Bleakney could show that this was so. It constituted the first instance of observation of molecular-ion fragments formed with kinetic energy and the most direct and unambiguous proof yet found of the physical reality of the repulsive-force molecular states predicted by quantum mechanics.

Later more careful data were taken by Tate and Lozier,[15] establishing that molecular-ion fragments can be formed with kinetic energy in other molecules as well, although here the quantum-mechanical calculations are too difficult to permit detailed predictions to be made.

There is another point in this connection, which was studied briefly by Hipple in 1936, that provides another interesting example of a quantum-mechanical phase of the Principle. Consider again the transitions from curve (a) to curve (c), induced by electron impact. The minimum of curve (a) is so related to (c) that the most favored transitions are those leading to H_2^+ molecules in a rather high vibrational state. If we bombard H_2 with, say, 18-v electrons, we will get some H^+ but transitions to H_2^+ are more favored.

What happens if we use deuterium instead of hydrogen? Theory tells us that the potential-energy curves are quite accurately the same in the two isotopic molecules. Theory also tells us that the wave function of the normal zero-vibrational state in D_2 will be narrower than in H_2; because of its greater mass it behaves more classically. Hence we can predict a lower yield of D^+ relative to D_2^+ at 18 v in D_2 than of H^+ relative to H_2^+ in H_2. Hipple tried this and found a considerable effect. It would be interesting to get accurate data on this and to attempt a precise calculation.

In 1931 Finkelnberg and Weizel tried to get definite information on the shape of the basic Heitler-London repulsive-force curve by an interpretation of data on the potentials needed to excite different parts of the continuous spectrum of molecular hydrogen. Such a procedure involved using the Principle in a more precise way than was ever intended, as was pointed out in 1936 by Coolidge, James and Present.[16] They did a very careful job, first on an improved calculation from theory of the repulsive-force curve, then on numerical integrations of the radial wave-function products occurring in the theory, in order to get a proper treatment instead of the usual rough graphical construction from the potential-energy curves.

[15] J.T. Tate and W. Lozier, *Physical Rev.* **39**, 254 (1932).
[16] A.S. Coolidge, H.M. James and R.D. Present, *J. Chem. Physics* **4**, 193 (1936). Careful discussion of application of the Principle to continuous spectrum of molecular hydrogen.

The work of Coolidge, James and Present brought out some fairly good evidence that in this case one cannot treat the electronic dipole-moment matrix component as constant. This meant that the rough assumption I had made earlier in order to get the main idea straightened out needed to be improved. Because of this point they were led to say in the abstract that "It is concluded that the Franck-Condon principle leads to results definitely incompatible with observations." I cannot let that pass, even ten years later, without a word of protest. They can say that I misused the Principle if they like, but not that the Principle is incompatible; for the Principle is properly to be judged by its correct quantum-mechanical formulation, which they so beautifully worked out, rather than by the rough criterions I gave as approximate rules.

The whole situation with regard to these special properties of molecular hydrogen seems to me to be highly satisfactory.

Conclusion

In the past decade, despite both the competitive attraction of nuclear physics and the wasteful interruptions of the war, many more band systems have been analyzed so that today there are a large number of instances of electronic transitions in a wide variety of molecules, all of which conform nicely to the semiquantitative intensity relations given by the approximate form of the Principle. Nevertheless, the fact remains that the Principle has never been subjected to a really severe test in which intensities are carefully measured by good photographic photometry and compared with those calculated from highly accurate wave functions.

I will conclude this review by merely mentioning one other application. The Principle has also contributed a good deal to the elucidation of many points concerned with predissociation of molecules.[17] Here, except for a few cases in which feebly bound molecules can dissociate by rotational instability aided by potential barrier leakage, we are concerned with radiationless transition between two electronic states of equal energy. The Principle acts restrictively here also, in that such transitions cannot occur if they would require too much readjustment of the nuclear motions.

The Principle also applies "in principle" to polyatomic molecules,[18] although here the situation is much more complicated than in the atomic case. Not only are the band systems much more complicated, so that thus far little progress has been made in analyzing them, but also the potential-energy curves have to be replaced by potential-energy surfaces, which are not well known either empirically or theoretically.

---

[17] L. A. Turner, *Zeits. f. Physik* **68**, 178 (1931). Application of the Principle to predissociation.

[18] J. Franck, H. Sponer and E. Teller, *Zeits. f. physik. Chemie* **18**, 88 (1932). Application to predissociation in polyatomic molecules.

Investigation of the Attractive Forces Between the Persistent Currents in a Superconductor and the Lattice*

E.U. Condon and E. Maxwell

National Bureau of Standards
Washington, D.C.

We have investigated the behavior of a superconducting sphere in a magnetic field in order to answer the question of whether the persistent surface currents can move with respect to the metal or are rigidly bound to it. An early experiment of Tuyn's[1] has been interpreted as indicating that the supercurrents are in fact rigidly fixed in the superconductor. Since Tuyn employed a hollow sphere, one would expect the intermediate state to set in and produce a strong frozen-in moment. Therefore his experiment is not conclusive.

It also appears that the hypothesis of currents rigidly bound to the metal is incompatible with the Meissner effect. Since

$$\mathbf{B} = \mathbf{H} + 4\pi\mathbf{I}, \tag{1}$$

we may regard the condition $\mathbf{B} = 0$, the Meissner effect, as resulting from the building up of a $4\pi\mathbf{I}$ in the metal which is everywhere equal to $\mathbf{H}$ and oppositely directed to it, inasmuch as (1) is a vector equation and $\mathbf{I} = -\mathbf{H}/4\pi$ vectorially. If we were now to regard the surface currents, or what amounts to the same thing, the magnetization $\mathbf{I}$, as fixed to the metal sphere, then on turning the sphere in a magnetic field, the vector $\mathbf{I}$ would rotate with respect to $\mathbf{H}$, and therefore the condition $\mathbf{H} + 4\pi\mathbf{I} = 0$ could not be fulfilled.

We have investigated this matter experimentally by suspending a tin spheroid of radius 1.8 cm in a magnetic field and observing the period of oscillation of the resulting torsion pendulum as a function of the field strength. Observations have been made at temperatures down to 1.74°K and at field strengths sufficiently below the critical field to avoid the appearance of the intermediate state. The earth's field was compensated to within one or two milligauss before cooling so as to avoid the possibility of frozen-in moments.

If the currents were rigidly fixed to the sphere, then we should have, in addition to the restoring torque of the suspension, a torque of $\mathbf{H}^2/4\pi$ dyne-

---

* Supported by the ONR.
[1] W. Tuyn, *Quelques essais sur les courants persistants*, Leiden Comm. No. 198.

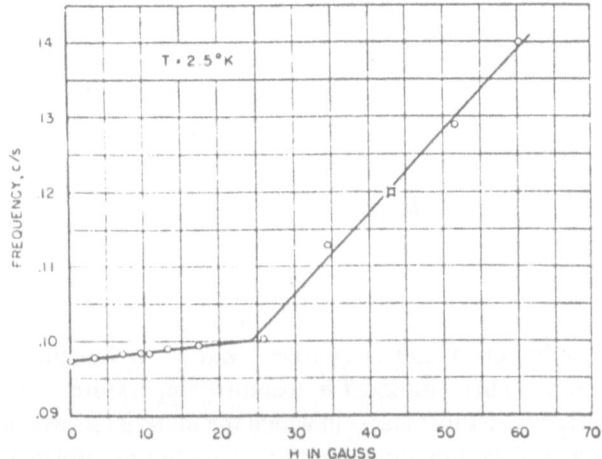

FIGURE 1. Relation between frequency of oscillation and magnetic field for tin spheriod. ⊙ Points taken with ascending field. × Point taken with descending field.

cm/rad. cm³ due to the interaction between the currents and the applied field. We have observed no torques greater than about 1 percent of this quantity and therefore conclude that except for effects of the order of 1 percent or less, the currents do not remain bound to the metal.

We have, however, observed a restoring torque which is 1 percent or less of the value $H^2/4\pi$ and which is an interesting function of the applied field strength. Since we used a rather soft suspension (approximately 130 dyne-cm/rad.) these effects were well within the precision of measurement. A typical result is plotted in Fig. 1 which gives the frequency of oscillation *vs.* the applied field for a temperature of 2.5°K. Note the curious discontinuity in the slope at 24 gauss. The same general behavior has been observed at other temperatures. No hysteresis is apparent as long as we do not exceed the critical field.

The experiments are still in a preliminary stage so that it is entirely possible that the effect we observe is due to small departures from true cylindrical symmetry or other secondary features. We have some indication of dependence on the relative orientation of field and spheroid and are investigating this point further. It is also planned to measure the torque under static conditions, to extend the measurements to other superconductors, and possibly to study the effect using a long cylinder with circumferential persistent currents.

After we observed these phenomena it was brought to our attention that effects of this sort have been observed and briefly reported on by Mendelssohn.[2]

[2] K. Mendelssohn, *Report of an International Conference on Fundamental Particles and Low Temperatures* (Physical Society, Cambridge, 1946), p. 128.

Superconductivity and the Bohr Magneton

EDWARD U. CONDON

National Bureau of Standards
Washington, D.C.

Each superconducting metal is characterized by a function $H_c(T)$ where H_c is a magnetic field in gauss and T is absolute temperature, which determines the boundary between those states in which the material shows superconducting properties and the region where the properties are normal. That is, for each T, the magnetic field H must be less than $H_c(T)$ for the metal to be in the superconducting state.

This functional relation plays an important rôle in the thermodynamics of superconductivity. If G_n is the Gibbs free energy of unit volume of the metal in its normal state and G_s is that in the superconducting state, but in zero magnetic field, then the Gibbs free energy of the superconductor in a magnetic field H is $G_s + H^2/8\,\pi$. At the field $H = H_c$ the normal and superconducting phases are in equilibrium so $G_n = G_s + H_c{}^2/8\,\pi$. This relation is the starting point of the Keesom-Rutgers-Gorter analysis of thermodynamic properties of superconductors, which has been so extraordinarily fruitful in the last decade.

The general form of $H_c(T)$ is usually that given by the formula

$$H_c(T) = a(T_0{}^2 - T^2)$$

which forms a basis for extrapolation of the observational data to $T = 0$. There is a nice graph of the data in figure 23 of Schoenberg's book on Superconductivity.[1] Aside from the question of the exact validity of this particular formula, the boundary curve has two particularly interesting parameters, namely, the critical field at absolute zero, $H_c(0)$, and the temperature T_0 for which H_c is zero.

If a long cylinder of superconducting material is magnetized by a field parallel to its axis it is known to behave as a perfect diamagnetic so that $B = 0$ inside it except for a skin layer of the order of 10^{-6} cm. thickness in which B drops from the external H down to zero in virtue of the building up of a $4\,\pi\,I$ oppositely directed to H and equal to it in magnitude when

---

[1] Schoenberg, D., *Superconductivity*, Cambridge University Press, 1938.

TABLE 1

Element	$I_0 = N\beta$	$I_c(0) = H_c(0)/4\pi$	$K^{-1} = I_0/I_c$
Pb	306	63	4.8
Ta	513	92	5.6
Sn	271	26	10.4
Hg	376	35.8	10.5
In	356	23.1	15.5
Tl	324	15.9	20.4
Cb	506	207	2.5
Zr	393	17.9	21.9
Cd*	430	4	117
Al*	559	14	40

* The data for the last two, Cd and Al, are so uncertain that it is doubtful whether they should be included.

one gets below the skin depth. From this point of view, $I_c = H_c/4\pi$ is the maximum amount of magnetic moment per unit volume which can be set up in the superconductor without destruction of the superconducting state.

When one focusses attention on $I_c = H_c/4\pi$ as a limiting magnetic moment per unit volume it is natural to inquire whether empirically there is any relation between this number and the product $N\beta$ where N is the number of atoms per unit volume of the metal and β is the Bohr magneton. No reason at all is offered for doing this except that these are of like dimensions and are both related to fundamental magnetic properties.

The results are shown in the table, the values of $H_c(0)$ having been read by rough extrapolation of the curves in Schoenberg's book and the values of N being calculated from standard density and atomic weight data.

It seems remarkable that in all cases the quantities, I_0 and I_c, are of the same general order of magnitude. It is also striking that the values of I_0/I_c are roughly close to being an integral multiple of 5 (except columbium which requires a half quantum number!) so that the data here presented points roughly to a possible empirical relation,

$$I_c(0) = KN\beta,\tag{1}$$

in which, approximately, $K = \frac{1}{5}n$, where n is an integer (except that $n = \frac{1}{2}$ for columbium) characteristic of the metal. Whether this integer rule will stand up to a more critical study of the data remains to be seen.

It is interesting to compare the relation of I_c to $N\beta$ with the relation between mass transport of electrons by the superconducting electrons and Planck's angular momentum unit which was discovered by F. London.[2] London remarks that the electromagnetic measure of current per unit length of the

---

[2] London, F., *Rev. Mod. Phys.*, **17**, 310 (1945).

cylinder wall, integrated through the skin depth is equal to $H/4\pi$ abamp./cm. where H is in gauss. To find the maximum mass transport in gram/cm. sec., $H_c/4\pi$ must therefore be multiplied by mc/e. London next remarks that this has the same dimensions as $n_e\hbar$ where $\hbar$ is the Planck unit of angular momentum and n_e is the number of superconducting electrons in unit volume, and then proceeds to present some evidence that these are of the same order of magnitude.

London's relation is thus

$$mcH/4\pi e = K'n_e\hbar$$

where K' is an empirical coefficient which he finds to be of the order of unity on the basis of some scanty data for lead and tin.

In view of the fact that the Bohr magneton, β, is related to the fundamental constants by $\beta = e\hbar/2mc$, London's relation can be written

$$I_c = H_c/4\pi = 2K'n_e\beta \tag{2}$$

so that it is very closely similar to the relation given in equation (1). The two relations can be put together in such a way as to give an expression for the atomic concentration of superconducting electrons as follows:

$$n_e/N = K/2K'. \tag{3}$$

Of course it is not really necessary to go through these relations to get a value of n_e/N. Basically, one finds n_e, in terms of London's theory, by making some kind of experimental determination of the skin-depth parameter, λ, from which n_e is calculated by the relation, $n_e = (mc^2/e^2)/4\pi\lambda^2$. The real difficulty lies in experimental measurement of λ. For this the available data are sparse and uncertain.

In London's paper[2] he only presents rough data for mercury and tin which lead him to say that $K' \sim 1$ for Hg and $K' \sim \frac{1}{6}$ for Sn. Taking these values in combination with the values of K given in table 1, one gets

$$n_e = \tfrac{1}{20} \quad \text{for } Hg \text{ and } n_e/N = \tfrac{1}{2} \quad \text{for Sn.}$$

These are rather larger values than are usually supposed to be applicable, which may indicate that the experimental values for λ used by London are rather too small.

These considerations strongly point to the desirability of obtaining more accurate measurements of λ and of $H_c(0)$ for as many superconductors as possible.

Physics of the Glassy State I: Constitution and Structure[*]

EDWARD U. CONDON

Corning Glass Works
Corning, New York

Modern views of the constitution and structure of inorganic glasses are outlined. Materials used in glass making are classified according to their role as network formers and network modifiers. Evidence of structure from x-ray scattering is reviewed. Examples of linear and nonlinear dependence of specific volume on composition are presented, and the classical factors for composition dependence of heat capacity are related to quantum theory of specific heat.

1. Glassy State

The glassy or vitreous state of matter is that which arises when a liquid is cooled sufficiently rapidly that it solidifies (becomes rigid) without crystallizing. It is an intermediate state between that of the usual solid and the liquid, possessing the rigidity characteristic of solids and the amorphous disordered structure characteristic of liquids.

Favorable to the production of the glassy state is for the liquid to be relatively viscous at temperatures near the *liquidus*, the temperature at which a solid crystalline phase is in equilibrium with the liquid. The crystalline or devitrified condition is the state of thermodynamic equilibrium so all ordinary glass exists in the vitreous condition because the rates of crystallization have become negligible. However these rates become quite appreciable for glass in the molten condition and are one of the important limiting factors as to the time available for glass forming operations.

Strictly speaking, the viscosity as ordinarily defined—shear stress needed to maintain unit gradient of shear flow—is probably not the rate determining property for the more microscopic flows required for crystal nucleation and growth. Nevertheless there is a close qualitative correlation with ordinary viscosity. The factors governing the kinetics of the devitrification process have not been studied much thus far.

---

[*] Based on the first ot four lectures delivered at the Fifteenth Annual Colloquium of College Physicists at the State University of Iowa, June 19 and 20, 1953.

Possibly all liquids could be brought into the glassy state if cooled rapidly enough. Pryde and Jones[1] have reported some observations on water in the vitreous state. Much more readily brought into the glassy state is the more viscous glycerol ($CH_2OH \cdot CHOH \cdot CH_2OH$) whose properties have recently been studied by Davies and Jones[2] and shown to have transformation-range properties like the silica-rich glasses, as discussed in the second lecture.

2. Glass Forming Oxides

Nearly all glasses made industrially are based on silica SiO_2, the material which exists in the crystalline forms known as quartz, tridymite, and cristobalite. Most of our knowledge of the glassy state is derived from study of such glasses so for that reason primary emphasis is placed on them in these lectures.

When one starts the study of the subject he is likely to be overwhelmed by the great variety of compositions of glass described in the literature, which are hard to remember since they usually involve many components. The mind gropes for a way of bringing this situation into some sort of order which is, of course, to be found in connection with the periodic system of the elements and the modern quantum theory of atomic structure.

The glasses which we shall consider will all he the product of fusing together various oxides. The composition is usually reported by giving the weight-fractions of the various oxide constituents. Some glasses may contain about one percent of fluorine which is usually reported as such with the understanding that in the actual glass two atoms of fluorine replace a single atom of oxygen in the oxide composition as reported. Although the compositions are reported in terms of oxides it needs to be clearly understood at the outset that this fact does not imply that the oxides exist in definite molecular form in the glass.

Although a very large number of oxides find some application in glass-making, only a very few of these play an important role in the subject. For this reason it is possible to start with an extremely abbreviated glassmakers periodic table (Table I). The table is arranged in a way which conforms to the modern theory of atomic structure, identifying the various groups by the kinds of angular momenta of the electronic orbits involved in the outer-electronic shells. The oxides of really major importance are shown in black type. We see that there are four involving incomplete p shells, which the chemists call *acidic oxides*:

$$B_2O_3$$
$$(Al_2O_3) \qquad SiO_2 \qquad P_2O_5.$$

Of these SiO_2 is most important and P_2O_5 least. Al_2O_3 is listed in parentheses to indicate its amphoteric character in acid-base relationships.

---

[1] J.A. Pryde and G.O. Jones, Nature **170**, 685 (1952).
[2] R.O. Davies and G.O. Jones, Proc. Roy. Soc. (London) **A217**, 26 (1953).

TABLE I. Glass-makers' periodic table

	s		d			p			
1	H_2O								
2	Li_2O	BeO				B_2O_2			F^-
3	Na_2O	MgO				Al_2O_2	SiO_2	P_2O_5	
4	K_2O	CaO		(V, Cr, Mn, Fe, Co, Ni)Cu, Zn			GeO_2	As_2O_2	
								As_2O_5	
5	Rb_2O	SrO		Ag	Cd	SnO_2		Sb_2O_2	
								Sb_2O_5	
6	Cs_2O	BaO	La, Ce	Au		PbO			
7			UO_2						

These oxides, especially SiO_2 and B_2O_3, are known as *network formers*, a term which will become clearer when we consider the modern views of the arrangement of atoms in glasses. Alumina has an intermediate role, partly functioning as a network former.

Other important oxides are known as *network modifiers*. These are:

$$Li_2O$$
$$Na_2O \quad MgO$$
$$K_2O \quad CaO$$
$$BaO,$$

as well as ZnO and PbO. The other alkalies and alkaline earths are also usable but are not used much in practice because of their comparative rarity. Lead oxide is important in making glasses of high refractive index. The oxides of arsenic and antimony are introduced in small quantities as *fining agents*; they play a role in the oxygen equilibrium of the glass because of the ease with which their valence state can be changed. ZnO is an important constituent of some optical glasses. The oxides of the iron-group metals are important in colored glasses, as are also various materials like cadmium sulphide and cadmium selenide which as colloidal precipitates give rise to the ruby red glasses used to stop auto traffic. Finally Cu, Ag, and Au are important as imparting photosensitive properties to glass as discussed in the fourth lecture.

In spite of the complexity of detail the subject is thus in the main reducible to a consideration of the general properties of three main classes of oxides: the network formers which are principally silica and boric oxide, the network modifiers which are principally of two kinds, the alkali oxides, and the alkaline earth oxides.

3. Some Specific Glass Compositions

Fused silica or vitreous silica itself is an excellent glass, but it requires such high temperatures for fusion that its practical usefulness is for this reason severely limited. First and foremost the importance of the modifiers is then as

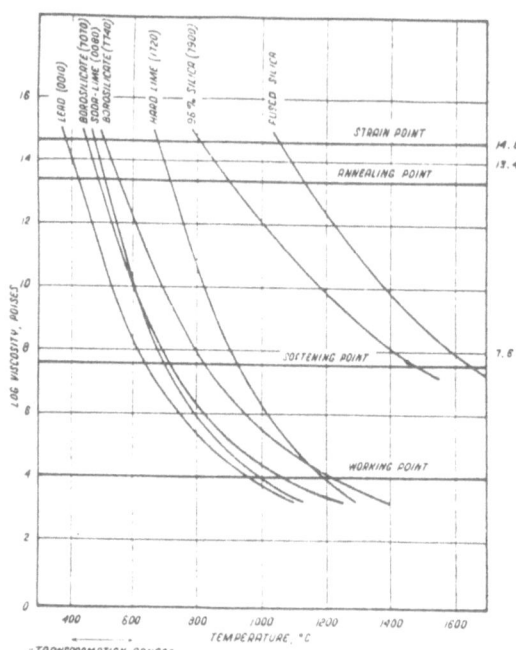

FIGURE 1. Viscosity-temperature
curves for several kinds of glass.

fluxes, that is, to reduce the temperature needed to make the glass fluid enough for the various forming operations.

We are not dealing here with sharp melting points. The viscosity varies over many powers of ten in the different conditions of interest. It is customary to describe the viscosity of the glass by giving the temperatures at which the viscosity assumes various arbitrarily chosen standard values. This is illustrated in Fig. 1 which will serve to illustrate the terms and presents actual data on several glasses of considerable importance.

Starting at the highest temperatures, the viscosity of molten glass in tanks is of the order of 10^2 poises, which may be compared with glycerol at 0°C which is 120 poises. When being pressed into molds or drawn into tubes, it is used in the range of temperatures at which its viscosity is 10^3 to 10^6 poises, so arbitrarily the temperature at which $\eta = 10^4$ poises, is called the *working point*. The *softening point* is defined as the temperature at which $\eta = 10^{7.6}$ which corresponds to a condition in which a moderate-sized piece deforms under its own weight at an easily noticeable rate. The *annealing point* is defined as the temperature at which $\eta = 10^{13.4}$, for this is the temperature at which internal strains are quickly relieved. The *strain point* is the temperature for which $\eta = 10^{14.6}$ and is the highest temperature from which the glass can be rapidly cooled without developing permanent internal strain. The fractional exponents occurring in these definitions came about because the reference temperatures were at first defined in terms of certain practical laboratory test

TABLE II. Approximate compositions of the glasses of Fig. 1

Glass	Network formers			Alkalies		Alkaline earths		Lead
	SiO_2	B_2O_3	(Al_2O_3)	Na_2O	K_2O	MgO	CaO	PbO
Fused silica	0.998+						(0.002)	
"Vycor" 7900	0.96+	0.03+	(0.01−)					
1720	0.57	0.04	0.20	0.01		0.12	0.06	
"Pyrex" 7740	0.80	0.14	0.02	0.036	0.004			
0080								
Soda-lime	0.72		0.01	0.20		0.03	0.04	
7070	0.71	0.27		(0.02)				
0010	0.63		0.01	0.08	0.06		0.01	0.21
Lead								

procedures which only later were calibrated in terms of absolute values of viscosity.

The approximate compositions of the glasses shown in Fig. 1 expressed as weight fractions of the constituent oxides are shown in Table II. In Fig. 1, fused silica and 7900 stand out as being verv much harder than the others. It is interesting to note nevertheless how much softer the 7900 glass is than fused silica.

1720 is a hard-lime glass of high alumina content of a type used for cooking utensils.

7740 is a borosilicate glass that is widely used for all forms of chemical glassware.

7070 is a borosilicate glass of high electrical resistance specified for certain electrical applications.

0080 is typical of a soda-lime glass as used for electric lamp bulbs, window glass, and bottles. Glasses close to this in composition account for 90 percent of all glass melted.

0010 is a potash-soda-lead glass of high electrical resistance as used for electric lamp bulb stems and sign tubing.

In the first approximation the softness of the glass is mainly dependent upon the SiO2 content expressed as weight percent as is indicated in Fig. 2, in which the various characteristic temperatures for these glasses are plotted against the weight fraction of SiO_2.

The differences in the electrical volume resistivity of these glasses are brought out in Fig. 3 which shows the log resistivity against temperature for the same glasses.

From the foregoing it is evident that the properties are strongly dependent on the composition which therefore has to be carefully controlled to produce

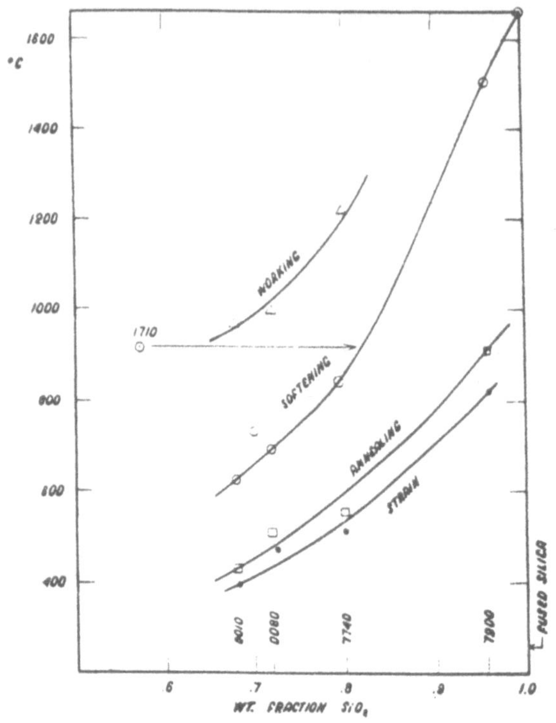

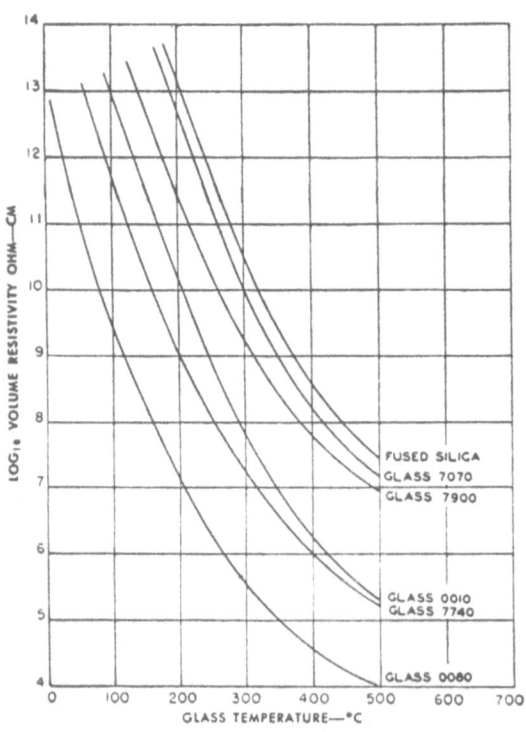

FIGURE 2. Dependence of hardness on silica content.

FIGURE 3. Volume resistivity *vs* temperature.

product with properties within narrowly specified limits. The strong dependence on composition makes possible the development of glasses having widely different properties adapted to special uses.

4. The Glass Network

We turn now to the modern ideas of the network constitution of glass.[3] These are an out growth of the study of silicate crystalline materials by means of x-ray diffraction analysis, the general ideas on the subject laid down in a classic paper by Zachariasen,[4] and their fuller working out in a series of experimental researches by Warren and his associates[5] which were a natural outgrowth of the pioneering researches on x-ray diffraction in liquids made at the State University of Iowa by Stewart[6] and his associates about a quarter century ago.

Let us first recall briefly what one can learn from x-ray diffraction studies. Each electron in the material scatters[7] coherently with a cross section $\sigma = (e^2/mc^2)^2$ according to the theory of J.J. Thomson, and the total observed scattering intensity in any direction is obtained by the addition of *amplitudes* of scattered waves—it is this fact which makes the resultant be dependent on the structure in the scatterer. Let the wave vector for the incident wave be **k** and for the scattered wave be **k**', where both have the magnitude $2\pi/\lambda$, and suppose that $\rho(r)$ is the volume density of electrons at the vector position **r**. Then the *amplitude* of the scattered wave due to coherent scattering by the electrons will be proportional to

[3] The following review papers are recommended: H.G. Vogt, Encyclopedia of Chemical Technology (Interscience Publishers; Inc., New York), Vol. 7, pp. 195–206; B.E. Warren, J. Am. Ceram. Soc. **24**, 256 (1941); B.E. Warren, J. Appl. Phys. **13**, 602 (1942); K.H. Sun, J. Am. Ceram. Soc. **30**, 277 (1947); as well as the following book references: W.A. Weyl, *Coloured Glasses* (Society of Glass Technology Sheffield, 1951), Chap. 2; J.E. Stanworth, *Physical Properties of Glass* (Clarendon Press, Oxford, 1950), Chap. 2; J.M. Stevels, *Progress in the Theory of the Physical Properties of Glass*, (Elsevier Press, New York, 1948), Chap. 1.

[4] W.H. Zachariasen, J. Am. Chem. Soc. **54**, 3841 (1932).

[5] B.E. Warren, J. Appl. Phys. **8**, 645 (1937) (Vitreous Silica); B.E. Warren and C.F. Hill, Z. Krist. **89**, 481 (1934) (GeO₂ Glass); B.E. Warren, H. Kruther and O. Morningstar, J. Am. Ceram. Soc. **21**, 259 (1938) (Soda-silica Glass): B.E. Warren and J. Biscoe, J. Am. Ceram. Soc., **21**, 287 (1938) Soda boric oxide Glass); Biscoe, Pincus, Smith, and Warren, J. Am. Ceram. Soc. **24**, 116 (1941) (Phosphate Glasses), J. Biscoe, J. Am. Ceram. Soc. **21**, 262 (1941) (Soda-lime-silica Glasses); R.L. Green, J. Am. Ceram. Soc. **25**, 83 (1942) (Potash-boric oxide Glass); G.J. Bair, J. Am. Ceram. Soc. **19**, 339 (1936) (PbO-silica Glasses).

[6] G.W. Stewart, Revs. Modern Phys. **2**, 116 (1930); Phys. Rev. **29**, 232 (1927); **31**, 174 (1928); **35**, 726 (1930); **37**, 9 (1931), and others.

[7] Compare A.H. Compton and S.K. Allison, *X-Rays in Theory and Experiment* (D. Van Nostrand Company, Inc., New York, 1935), Chap. III.

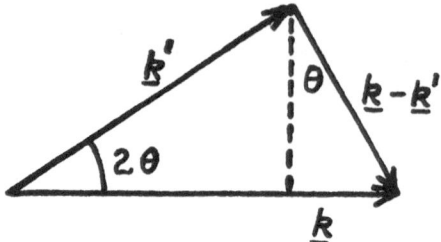

FIGURE 4. Scattering of x-rays by electrons.

$$\iiint \rho(\mathbf{r})e^{i(\mathbf{k}-\mathbf{k}')\cdot\mathbf{r}}\,dV,$$

the volume integral extending over the whole of the scatterer that is illuminated by the incident beam.

Thus the amplitude of the scattered wave is proportional to the $(\mathbf{k}-\mathbf{k}')$ component of the Fourier transform of $\rho(\mathbf{r})$. For scattering through the angle 2θ (Fig. 4), this is equivalent to reflection from the planes normal to $(\mathbf{k}-\mathbf{k}')$ in accordance with the well-known Bragg law. Since $|\mathbf{k}-\mathbf{k}'| = (4\pi\sin\theta)/\lambda$ the scattering through angle 2θ arises from the component of the Fourier transform of the electron density which has a wave-length in the material equal to $\lambda/2\sin\theta$. In other words, small-angle scattering is related to the part of the structure involving periodicities over larger distances.

Looked at in this way there is an essential unity between scattering by amorphous liquids, by polycrystalline solids, and by single crystals. In the latter case the material is quite accurately periodic in its structure so that scattering is almost completely absent except in the allowed directions for which $(\mathbf{k}-\mathbf{k}')$ in magnitude and direction agrees with one of the actual periodicities in the lattice. In the powder method of Debye and Scherrer, which averages over many randomly-oriented tiny crystals, the conditions are such that the condition on $(\mathbf{k}-\mathbf{k}')$ as to direction is always fulfilled by the fact that some of the scatterer is properly oriented, but the condition on $|\mathbf{k}-\mathbf{k}'|$ is determined by the definite lattice spacing of the tiny crystals thus giving rise to sharp rings. In amorphous materials the scattering density function $\rho(\mathbf{r})$, being aperiodic, is represented by a Fourier integral rather than a Fourier series so that scattering occurs for a continuum of values of $|\mathbf{k}-\mathbf{k}'|$ giving rise to broad fuzzy rings.

By measuring the scattered intensity at various angles and performing a Fourier transform one can find the distribution of scattering power in the amorphous material. Figure 5 shows the microphotometer record of the intensity of scattering as a function of $\sin\theta/\lambda$ as measured by Warren for vitreous silica, cristobalite, and silica gel. It is remarkable that the broad peak in the two amorphous forms came at nearly the same value of $\sin\theta/\lambda$ as the main sharp peak in the crystalline cristobalite. Also the presence of a large amount of small angle scattering in silica gel in the range for $\sin\theta/\lambda$ from 0.02

FIGURE 5. Microphotometer records of x-ray diffraction patterns of vitreous silica, cristobalite, and dried commercial silica gel.

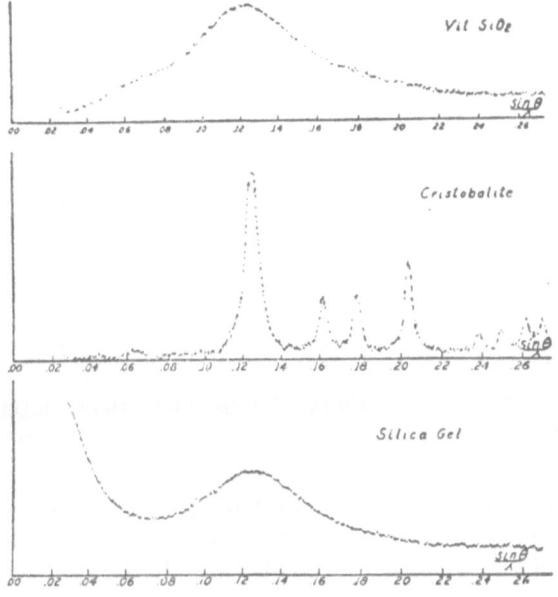

to 0.06 points to the existence of structure on a scale of periodicity from 8 to 25A that is absent in the vitreous silica and in the crystal.

From more detailed analysis of the vitreous silica curve along the lines indicated, it is inferred that

(a) the average Si—O distance is 1.62A in close agreement with 1.60A found in crystalline silicates, and
(b) the average number of oxygens surrounding each Si is 4, closely resembling crystalline silicates.

These two facts give rise to a picture of vitreous silica as being built up of a random distribution of tetrahedral (SiO_4) groups interlocking in the sense that each O belongs to two adjacent tetrahedra. The tetrahedra are thus linked together to form one giant molecule, a fact which correlates with the strength and hardness of silica in all its forms.

It is customary to discuss the structure as if the individual atoms are ions held together by ionic bonds. On this view (SiO_4) is an ionic tetrahedral structure with a central Si^{++++} surrounded by four O^{--} ions each of which obtained one of its electrons from each of the two silicons to which it is bonded. Throughout inorganic crystal chemistry the ions tend to keep rather constant ionic radii, tables of which have been made by Pauling,[8] based on an earlier tabulation by Goldschmidt, which is exhibited graphically in periodic table

---

[8] L. Pauling, J. Am. Chem. Soc. 49, 765 (1927).

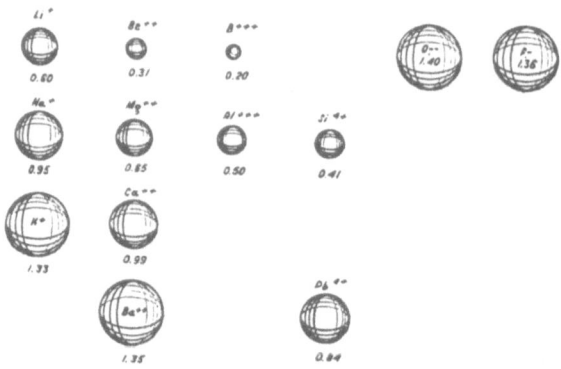

FIGURE 6. Ions of importance in glass structures.

form for the ions of interest in Fig. 6. Of interest here is that B^{3+}, Si^{4+}, and Al^{3+} are all small compared with O^{--}, in contrast with the larger cations of lower charge number which enter into the modifier oxides. In Fig. 7 a plan view of a number of interlocking SiO_4 tetrahedral units is drawn. We regard vitreous silica then as an extended random network of such tetrahedra joined together by shared oxygen atoms or ions.

For brevity, we use the ionic radii and the terminology of ionic bonds as is customary in the literature of glass. The more modern viewpoint regards the bonds as partly ionic and partly covalent in character.[9] Thus if ψ_1 represents a wave function for the ionic structure of SiO_4 and ψ_2 that for a covalent bond structure, a better approximation is an appropriate linear combination $\psi = (a\psi_1 + b\psi_2)$ determined by resonance energy calculations. Here $|a|^2$ is regarded as the fractional ionic character. Pauling has attempted to estimate the fractional ionic character for various links with the following results of interest in glass-making shown in Fig. 8. According to this, the basic SiO bond is only 50 percent ionic, so that the usual ionic descriptions are not to be taken literally.

Corresponding work on the Na_2O—SiO_2 glasses, up to 0.35 Na_2O (wt.) showed the same SiO_4 tetrahedral structure in the main. However, as there are now more oxygens than in SiO_2, it is not possible for all to be bonded between two Si as in silica; some must be bonded to only one Si, and it is to this fact that the softening action of Na_2O is due. The diffraction work gives further indication of 2.35A for the average Na—O distance as would be expected from the ionic radii of Fig. 6 and also gives an average coordination of 6 ± 1 oxygens around each sodium atom.

The soda-boric acid series introduces a new feature. Diffraction results

---

[9] For an excellent up-to-date survey see the book by C.A. Coulson, *Valence* (Oxford University Press, New York, 1952).

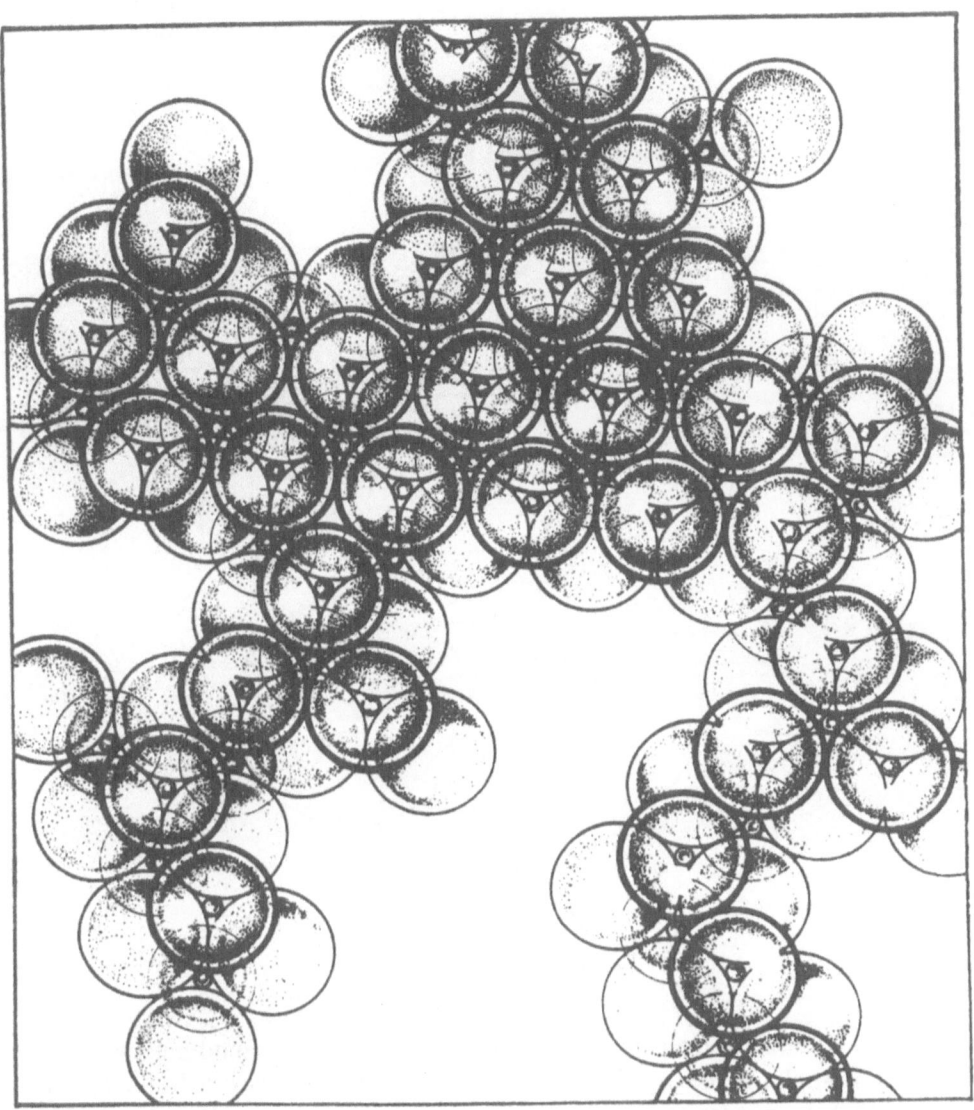

FIGURE 7. Groups of SiO$_4$ units.

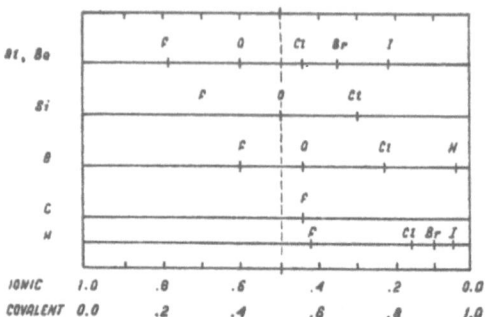

FIGURE 8. Ionic-covalent character of chemical bonds.

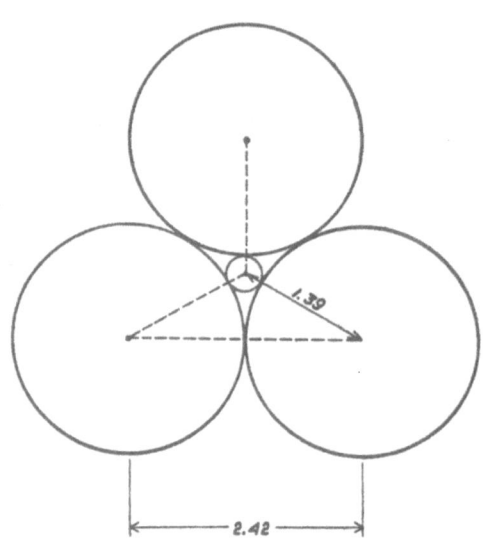

FIGURE 9. BO₃ triangular unit.

indicate[10] that vitreous boric oxide itself is made up of a random network of BO_3 triangles linked together by shared oxygens, analogously to the linkage of the SiO_4 tetrahedra, the B—O distance being 1.39A, rather less than the 1.60A expected from the Pauling ionic radii. A typical such triangular unit is shown in Fig. 9.

With the addition of Na_2O, it is found that the B—O distance gradually increases up to 1.48A and the coordination number definitely increases from 3 to 4 indicating a *gradual shift in the structure* from one of interlocking BO_3 *triangles* over to one of partially interlocking BO_4 *tetrahedra*. A similar behavior was observed in an x-ray diffraction analysis of several glasses in the

---

[10] See, however, K. Fajans and S.W. Barber, J. Am. Chem. Soc. **74**, 2761 (1952), who conclude that B_2O_3 is made up of molecular dimers B_4O_6 relatively loosely held together, accounting for its low melting point relative to that of fused silica.

K_2O—B_2O_3 series. This behavior is the basis for many peculiar properties of borosilicate glasses, known in the literature broadly as the *boric oxide anomaly*.

5. Density of Glass

The older glass literature abounds with papers which attempt to express various properties of glass as linear functions of the composition. These attempts are useful for concise presentation of empirical data and also in serving to direct attention to unusual cases involving structural peculiarities. Let us consider the density of glass from this point of view.

In physical chemistry we learn that the molal volume of many compounds and solutions is a sum of contributions from the constituents. Hence if x_r is the weight fraction of the rth oxide in a glass whose specific volume is v ($= 1/\rho$, where ρ is the density), it is natural to suppose that, in a first approximation,

$$v = \sum_r v_r x_r,$$

where v_r is the effective specific volume of the rth oxide in glass. This kind of formula is discussed in Morey,[11] Chap. X.

We have seen that the largest ion present in glass is the oxygen atom. The network forming cations are much smaller. It is therefore interesting to consider the density data on a uniform basis with regard to the volume contributed by one gram molecular weight of oxygen. For an oxide whose formula is written A_aO_b let M_r be $1/b$ times the usual molecular weight, so M_r grams of the rth oxide contain 1 gram mole of oxygen atoms. If v_r is the specific volume of the free oxide, then $v_r M_r$ is *the volume in cm^3 of an amount containing one mole of oxygen atoms*. Alternatively, if v_r is the effective specific volume of the oxide in a glass, $v_r M_r$ is the volume per mole of oxygen for that oxide constituent. Various sets of effective factors are presented in Morey's Table X.1 from the literature. These were used to prepare the values of Table III.

These factors are the values to be used in the linear formula expressed as

$$v = \sum_r (v_r M_r)(x_r/M_r),$$

where (x_r/M_r) is the number of moles of oxygen in unit weight of glass coming from the rth oxide.

Aside from the fact that the different authors give rather different values, we note that

(a) the effective values for network formers are close to the value for the free oxide, and
(b) for the alkalies the effective value is decidedly less than the free oxide value.

---

[11] G.W. Morey, *The Properties of Glass* (Reinhold Publishing Corporation, New York, 1938), p. 224.

TABLE III. Volume (cm^3) per gram mole of oxygen atoms

Oxide	Free oxide	Winkelmann and Schott	Baillie	English and Turner	Tillotson
			Effective glass factors		
SiO_2	13.65 (vitr.)	13.05	13.40	13.65	13.05
B_2O_3	12.60	12.25			
Al_2O_3	8.53	8.34	12.40	12.40	12.40
P_2O_5	11.90	11.15			
Li_2O	14.9				8.06
Na_2O	27.3	23.9	19.4	17.9	
K_2O	40.6	33.6			
MgO	10.9	10.6	12.4	11.9	10.1
CaO	16.5	17.	13.05	11.2	13.7
BaO	26.8	21.9			
ZnO	14.85	13.8			
PbO	23.4	23.2			

Both these observations are qualitatively consistent with the view that the alkali ions fill in the interstices in the network and so do not add much to the space occupied by the glass. Note how much smaller is the value for the network formers than it is for water (18 cm^3), serving to emphasize the known fact of the open structure of liquid water compared even with the open structure of the glass networks.

In recent years a good deal of attention[12] has been given to departures from the linear formula. We shall merely consider three binary systems and one ternary system to provide simple examples of how it applies and how it breaks down. The first example is of the binary glasses made of two network formers SiO_2 and B_2O_3, illustrated in Fig. 10. Ordinates are the volume occupied by enough glass to contain one mole of oxygen atoms and the scale of abscissas shows the fraction u of all the oxygen atoms contributed by the silica

$$u = (x_1/M_1)/[(x_1/M_1) + (x_2/M_2)],$$
$$V = (1/\rho)/[(x_1/M_1) + (x_2/M_2)],$$

where the subscript 1 refers to silica and 2 to boric oxide. The experimental points lie quite nicely on the straight line joining the values derived from the pure vitreous oxides at either end. [The experimental points are calculated from a series given in Morey,[11] (p. 231).]

The second example provides a striking contrast. It is the series SiO_2 and Na_2O consisting of a network former with increasing amounts of network modifier. The points in Fig. 11 are calculated in the same way as before from a series given in Morey,[11] (p. 232), where now subscript 1 refers to soda and 2

---

[12] J.M. Stevels, see reference 3, Chap. 2. M.L. Huggins and K.H. Sun, J. Am. Ceram. Soc. **26**, 4 (1943); **27**, 10 (1944).

FIGURE 10. Molar volumes in
the silica-boric oxide system.

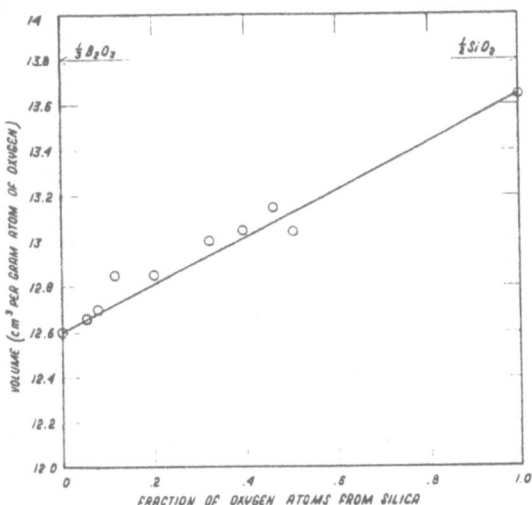

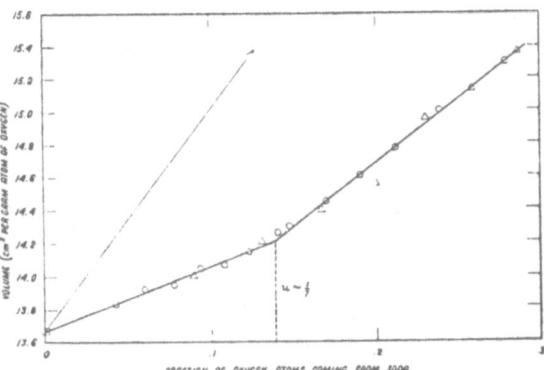

FIGURE 11. Molar volumes
in the silica-soda system.

to silica. The line with an arrowhead is the one joining to the volume 27.3 cm³
of pure Na_2O. The experimental points in the range $0 < u < 0.14$ lie on a
straight line of much lower slope, appropriate to an effective molal volume of
only 17.65 cm³ for Na_2O dissolved in the SiO_2 vitreous network. For $u > 0.14$
the curve is again straight but with a considerably greater slope indicating a
swelling of the network after about one in seven oxygens are contributed by
the soda—otherwise expressed, the abrupt change comes about where there
are two sodium atoms for every three (SiO_4) tetrahedra, so one would expect
most of the interstices to be filled.

The second example exemplifies well (a) the departure of the effective
volumes from the free oxide values and (b) the nonlinearity of behavior
of a system containing a network modifier in a network former.

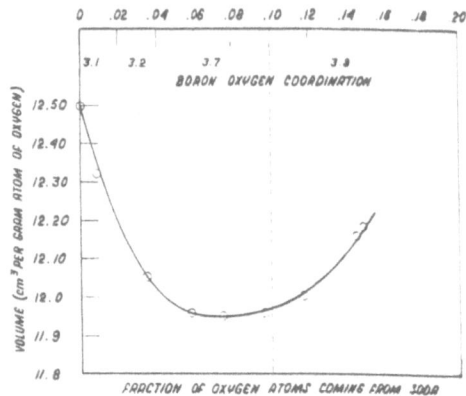

FIGURE 12. Molar volumes in the boric oxide-soda system.

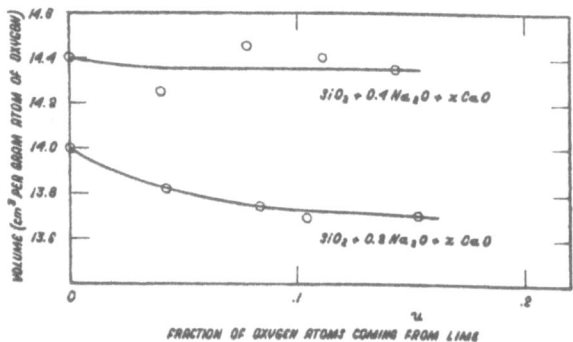

FIGURE 13. Molar volumes in some soda-silica glasses.

The third example is the binary system $(B_2O_3 + Na_2O)$, calculated from the new data of Shartsis, Capps, and Spinner.[13] In Fig. 12, V is plotted against u, as before.

Addition of Na_2O causes the total volume per oxygen atom to *contract* up to about $u \sim 0.07$ at least, the total shrinkage being some 5.5 percent of the initial value. The minimum volume occurs for compositions near to $Na_2O \cdot 4B_2O_3$ and $Na_2O \cdot 5B_2O_3$.

The fourth example is based on data by Peddle (see Morey,[11] p. 376) for the two series (expressed in moles):

$$SiO_2 + 0.20 \, Na_2O + xCaO,$$

$$SiO_2 + 0.40 \, Na_2O + xCaO \qquad \text{(see Fig. 13)}.$$

The fraction of all oxygen atoms coming from the lime (in the first series) is

$$u = x/(2.2 + x),$$

[13] Shartsis, Capps, and Spinner, J. Am. Ceram. Soc. **36**, 35 (1935).

and the associated volume per mole of oxygen atoms is

$$V = (72.4 + 56.1x)/(2.2 + x),$$

and similarly for the second series. Incidentally, on p. 376 is also given[11] another set of data for $(SiO_2 + xNa_2O)$ which checks perfectly the other series in Fig. 11, the points from this series being plotted as triangles with center dots on Fig. 11. Here there is definite contraction in the first series with increasing proportion of lime and an almost exact constancy in the second series. This is associated with the effect of the double valence of Ca so it can act to pull the network together, and the fact that there is only one such ion to be accommodated for each lime-oxygen.

In conclusion, it may be said that the traditional linear specific volume factors are too inaccurate to be very useful, and they fail to draw attention to important facts related to changes in the glass network with changes in composition.

6. Heat Capacity of Glass

Data on heat capacity of glass are understandable in terms of the network structural ideas. According to Morey[11] (Chap. VIII), the specific heat may be more accurately represented as a linear function of the composition than most other properties.

According to the quantum theory of heat capacity, we expect that at sufficiently high temperatures, but such that the glass is still quite rigid, each atom will behave like a three-dimensional harmonic oscillator. Classically such an oscillator has a heat capacity of $3R$ or 6 cal/(mole deg) in the limit of high temperatures. According to quantum theory the actual specific heat will be less than this:

$$C = 3Rf(u),$$

where

$$f(u) = \frac{u^2 e^{-u}}{(1 - e^{-u})^2}, \qquad u = hc\omega/kT,$$

if ω is the frequency of oscillation in cm^{-1}. Also $hc/k = 1.438$. In Fig. 14 is a graph of $f(u)$ against u.

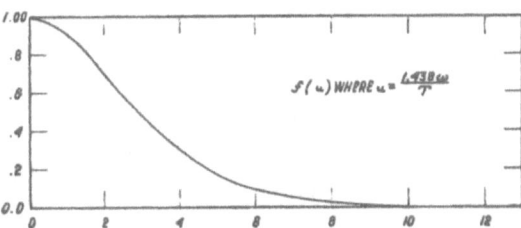

FIGURE 14. Quantum specific heat function.

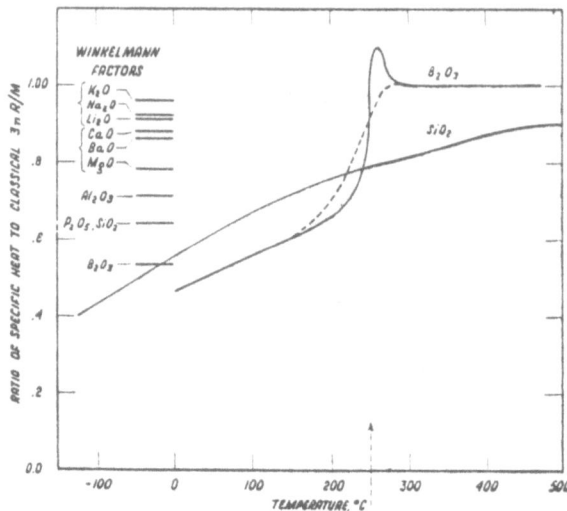

FIGURE 15. Heat capacity of SiO_2 and B_2O_3 and specific heat factors.

In an actual solid different atoms will be diffferently bound so that the actual specific heat will be given by a sum of functions $f(u_1) + f(u_2) + \cdots$, where the u's correspond to their different frequencies $\omega_1, \omega_2, \ldots$. In the modern theory of crystal lattices, much attention has been given to calculations of the distribution of frequencies to be used.

Measurements of the heat capacity of vitreous silica and vitreous boric oxide over quite a range of temperatures are reported by Morey and are plotted in Fig. 15. Here the ordinate is the value of the specific heat, not in cal/g deg, but in the ratio of the actual value to the classical theoretical limit, that is, to $3nR/M$, where $n = 3$ for SiO_2 and 5 for B_2O_3 and M is molecular weight.

The curve for SiO_2 is smooth and of the type usually shown by crystalline solids and to be expected from the quantum theory of specific heat. On the other hand, that for B_2O_3, while showing a similar behavior below 200°C, shows a relatively sharp peak at about 260°C and then levels off at the full classical value for temperatures above 300°C. The melting point is 577°C. The exact shape of the curve depends on the previous thermal history of the sample as indicated by the two different representative curves.

These curves indicate that in SiO_2 and B_2O_3 such strong interatomic forces prevail that the atomic oscillation frequencies are high enough that the heat capacities at room temperature are only about 0.6 and 0.5, respectively, of the classical limit.

It is now of interest to see how the traditional Winkelmann factors for specific heat compare with the classical values, $3nR/M$, for each oxide. The factors c_r (cal/g deg) for an additive representation of the heat capacity,

TABLE IV. Heat capacity factors of oxides in glass

Oxide	Winkelmann Observed factor	Ratio to $3nR/M$
SiO_2	0.1913	0.64
B_2O_3	0.2272	0.53
Al_2O_3	0.2074	0.71
P_2O_5	0.1902	0.64
Li_2O	0.5497	0.91
Na_2O	0.2684	0.92
K_2O	0.1860	0.96
MgO	0.2439	0.78
CaO	0.1903	0.88
BaO	0.067	0.86
ZnO	0.1248	0.84
PbO	0.0512	0.95

$$c = \sum_r c_r x_r,$$

in terms of weight fractions x_r of the oxides in a glass are taken from Table VIII-1 in Morey[11] and put here in Table IV along with the ratio of the factor to the $3nR/M$ classical limit.

These results correlate clearly with what would be expected from the network-structure theory. The strongly bonded atoms in the network are considerably affected by the quantization of the oscillator energy levels, to the extent that their effective heat capacity is some 50 percent to 70 percent of the classical value, the alkaline earths contributed much more nearly the classical value and the alkalies contribute almost completely what would be expected, corresponding to the fact that they are much more loosely bound in the network. The observed ratios are shown on the left side of Fig. 15.

Physics of the Glassy State
II: The Transformation Range[*]

EDWARD U. CONDON

Corning Glass Works
Corning, New York

The concept of fictive temperature as a means of describing nonequilibrium states of glass is described. Activation energies and the modern views of chemical rate processes are discussed in relation to the kinetics of the approach of these nonequilibrium states to equilibrium.

1. Fictive Temperature

The various physical properties of most crystalline solids change with time at a sufficiently rapid rate that they keep up with the rate of change of temperature occurring in usual situations. Thus if we heat up some zinc in a crucible and use a thermocouple to measure the temperature, as the temperature passes through the melting point in either direction, the metal will change quite promptly from the crystalline to the liquid condition. This behavior is the basis of a familiar technique in metallurgical studies, in which transformation points are located by irregularities on heating and cooling curves occasioned by abrupt absorption or evolution of latent heats of transformation occurring at these transformations.

However, the situation is quite different in silicate systems (crystalline as well as vitreous) where the time rates of transformation may be so sluggish as to require very long periods of time to bring about thermodynamic equilibrium.[1] In consequence, it frequently happens, in the laboratory and in industrial processes, that one is dealing with material that is not in internal structural equilibrium appropriate to its temperature. A proper approach to the description of the phenomena involved requires careful consideration of the liinetics of these slow approaches to equilibrium; that is, to the time and temperature relations.

---

[*] The second of a series of lectures delivered at the Fifteenth Annual Colloquium of College Physicists, State University of Iowa, Iowa City, June 19, 1953.
[1] F.C. Kracek, *Phase Transformations in Solids* (John Wiley and Sons, Inc., New York, 1951), Chap. 9. See also M.J. Buerger's Chap. 6, of the same book.

Most glasses show interesting behavior of this sort in the general neighborhood of 400° to 600°C, which has come to be known as the *transformation range*.[2] Above the transformation range the rates of internal structural change are so great that, for the rates of change of temperature ordinarily encountered, internal changes are always essentially in equilibrium with the instantaneous temperature. Below the transformation range, these rates are so slow that they are negligible in periods of time like weeks, months, or even years. Therefore, at temperatures below the transformation range, the internal structure remains sensibly unaltered with time, even when the material is warmed or cooled or put under mechanical or electrical stress.

The particular internal structure thus frozen into the material will depend on its thermal history; that is, on the time-temperature cycle involved in bringing it down from high temperatures through the transformation range. Strictly speaking, the resulting structure probably depends on every detail of the time-temperature cycle, but it has proved useful to introduce a simplifying hypothesis, which is probably not strictly true, but which gives a good first approximation. The hypothesis is that the internal structure below the transformation range may be characterized by a single parameter τ, introduced by Tool and called by him the *fictive temperature*, (fictive for short) independently of the details of the time-temperature cycle by which the final state was produced.

On this view the condition of a glass at temperatures below about 600°C is characterized by *two temperatures*, the usual T, which is what a thermometer in contact with the material reads, and τ, the fictive which characterizes the frozen-in internal structure as determined by the previous thermal cycle of the material.

Operationally the fictive τ is defined as follows: for temperatures T, below the transformation range, the material is said to have a fictive τ if it is in the *state* produced by rapid quenching from complete equilibrium at temperature $T = \tau$. The statement should be made more precise by saying how the *state* is to be characterized; that is, what measurable physical property is to he used to determine the fictive temperature scale. Density or linear expansion of the glass are commonly used as will appear in detail later.

Since rapid quenching will necessarily occur at a finite rate, the resulting

---

[2] Phenomena associated with the transformation range in glass already have a large literature. See W.A. Weyl, Chap. 11 of the book referred to in footnote 1, also J.E. Stanworth, *Physical Properties of Glass* (Clarendon Press, Oxford, 1950), Chap. 8, for general reviews and specific journal references and G.O. Jones, J. Soc. Glass Technol. **32**, 382 (1948). The basis for the modern approach was laid by A.Q. Tool of the National Bureau of Standards whose many papers should be consulted. A few of the most important of these are: (a) A.Q. Tool and C.G. Eichlin, J. Opt. Soc. Am. **4**, 340 (1920); A.Q. Tool and C.G. Eichlin, J. Research NBS **6**, 523 (1931). (b) A.Q. Tool and E.E. Hill, J. Soc. Glass Technol. **9**, 185 (1925). (c) A.Q. Tool and J. Valasek, Sci. Papers Bur. Standards. No. 358. (d) Tool, Lloyd, and Merritt, J. Am. Ceram. Soc. **13**, 632 (1930). (e) A.Q. Tool, J. Am. Ceram. Soc. **29**, 240 (1946).

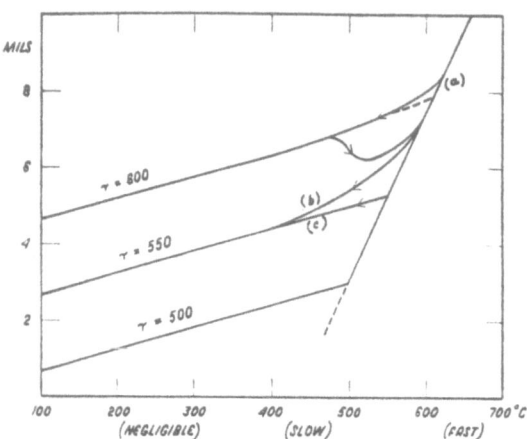

FIGURE 1_2. Idealized linear expansion plot in the transformation range.

structure will always correspond to a somewhat lower fictive temperature than that from which the quenching took place. This effect becomes more pronounced in quenching from higher temperatures where transformation rates are highest. Because quenching necessarily occurs at a finite rate, there will be a practical *upper* limit to the values of the fictive τ, which can be attained. Also because complete equilibrium at a temperature T requires longer and longer times for its establishment as T becomes lower, there will be a practical *lower* limit to the attainable values of the fictive τ. The transformation range is the more or less inexactly defined range between these practical upper and lower limits.

These ideas may he made more clear by reference to Fig. 1_2, an idealized plot of the total fractional expansion in length (thousandths or mils) from 0°C to the temperature shown as abscissas. The steeply sloping line (a) in the range $T > 500°C$ is the expansion of the glass when held long enough at each temperature to be in complete equilibrium. Above 600°C the rates of internal structural change are so fast that we would have to quench a small specimen from somewhat more than 600°C to put it in a condition such that for $T < 400°C$, its expansion varies along the line $\tau = 600°C$, which corresponds in slope to the low temperature coeffcient of expansion and in location is the line which intersects (a) at $T = 600°C$.

Suppose the cold glass has a fictive of 600°C and that it is warmed up *very slowly*. It expands along the $\tau = 600$ line until, somewhere near 500°C, the rates of internal transformation become appreciable, permitting the fictive τ to decrease. This may produce a greater contraction than the expansion due to increasing T, with the result that the linear dimensions go through a maximum followed by a minimum, as indicated.

For an appropriate rate of cooling from equilibrium at $T = 575°$ the curve might be like the one marked (b) joining asymptotically with the line $\tau = 550$, which corresponds to a rapid quench from equilibrium at $T = 550°C$, in-

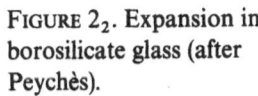

FIGURE 2_2. Expansion in borosilicate glass (after Peychès).

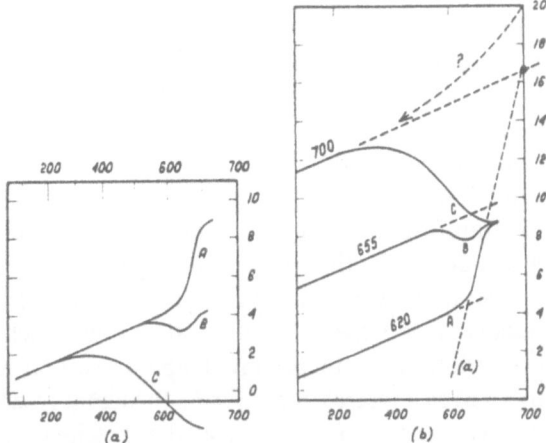

dicated in curve (c). The basic hypothesis is that for $T < 400°C$ the properties of the glass will be the same, whether the actual thermal history was along (b) or along (c). Thus different thermal histories may lead to the same τ and, if they do, in this approximation the properties will be the same.

The working value of the hypothesis, that the thermal history is adequately described by the single parameter τ will depend on the extent to which values of τ are the same when different measurable properties are used to establish the fictive temperature scale. This question has not been studied as critically up to now as one would like, so we shall proceed to make the underlying hypothesis and go on, bearing in mind that it may need elaboration when more careful studies are made.

An excellent example of the qualitative behavior indicated in Fig. 1_2 is shown in Fig. 2_2, part (a) of which is taken from Peychès.[3] Ordinates are the total fractional increase in length in thousandths of three pieces of the same glass, a borosilicate, on being warmed up from room temperature. A refers to a carefully annealed sample, B refers to a rapidly chilled massive sample, and C refers to a much more rapidly chilled fiber, 6μ in diameter.

According to the point of view presented in the foregoing discussion, it is more natural to regard the glasses as being in the same structural condition when *above* the transformation range, and so to displace the three curves vertically with respect to each other until they coincide at, say, $650°C$, producing the result shown in part (b) of Fig. 2_2. The dotted lines have been sketched in to correspond with Fig. 1_2. The steep dotted line (a) corresponds to (a) in Fig. 1_2 and represents the expansion when complete equilibrium is established. As A was carefully annealed the glass presumably followed along (a) when it

[3] I. Peychès, J. Soc. Glass Technol. **36**, 164 (1952).

was prepared, to $T = 620°$, and then broke away to contract on the line of gentler slope, the same as the line on which expansion was observed. Had the annealing been still more "careful," that is, slower in the range 600–650°C, it would have been possible to follow (a) farther down, producing a still lower fictive in the glass. But the transformation rates become so slow at this end of the range that the cooling rate has to be reduced enormously to produce a slight lowering of fictive. Thus 620°C is essentially the practical lower limit of the transformation range for this glass. For B the fictive is 655°. This glass on warming up in the expansion apparatus expanded because of increasing temperature, went through a maximum at 580°C and then began to contract because the fictive is cooling at a rate which more than counteracts the expansion due to increasing temperature. Curve C shows the same behavior in an even more extreme form. Presumably when it was originally formed it traversed some such curve as the sketched one marked "?" and ended with a fictive of approximately 700°C which is so high that it started to contract, on being warmed up, at a temperature somewhat below 400°C.

Evidently one of the most important questions we can ask is: what are the laws governing the temperature dependence of the rates of these structural changes in the transformation range?

2. Activation Energies and Time Rates

Throughout the whole of the modern theory of physico-chemical rate processes, no matter how or by whom the details are formulated,[4] the dominant factor in the temperature dependence is the Arrhenius function

$$e^{-E/RT} = 10^{-A/T},$$

where T is the absolute temperature, R the gas constant in cal/mole degree, and E the *activation energy* in cal/mole. The constant A is in degrees and what is derived directly from the slope of a plot of $\log_{10}$(rate) against $1/T$. It is related to E by

$$E = 4.57A \text{ cal/mole.}$$

The Arrhenius function varies over enormous ranges in various physico-chemical phenomena such as the vaporization of helium and the vaporization of tungsten (vaporization can be considered as a rate process or as an equilibrium—in fact, vapor "pressure" is often measured by the rate of loss of weight of an evaporating solid). It is therefore important to become familiar with the values of E for various processes and the associated behavior of the Arrhenius function. Figure 3_2 shows the log of the function plotted against

---

[4] Glasstone, Laidler, and Eyring, *Theory of Rate Proccsses* (McGraw-Hill Book Company, Inc., New York, 1941), is a good general presentation of modern application of statistical mechanics to rate processes.

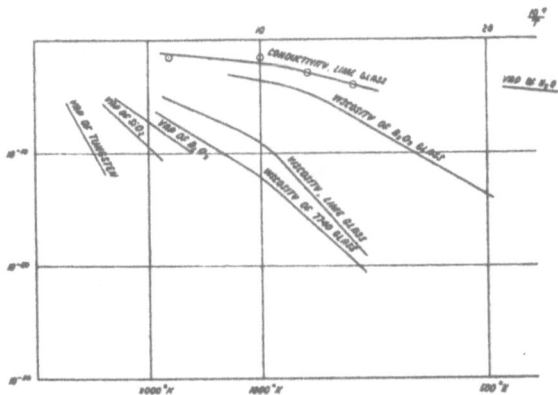

FIGURE 3$_2$. Arrhenius function.

the reciprocal of the absolute temperature. On it are plotted the Arrhenius functions governing the vaporization of several substances, in the range of temperature in which they are usually considered, also the curves for the function governing the viscosity of a lime glass and of B_2O_3, and also a curve for the temperature dependence of the electrical conductivity of the same lime glass whose viscosity variation is plotted.

These curves have been displaced vertically parallel to themselves in such a way that their high-temperature portions extended would pass through the origin. The interesting special feature about the curves for glass is that they are not straight, bending in such a way as to have considerably greater slope in their low-temperature portions, that is, they show higher activation energies in the low-temperature range.

Every student of physico-chemical rate processes will find it highly instructive to collect his own private file of experimental values of activation energy. A start toward this is afforded by Fig. 4$_2$, in which the values[5] of E from 1 to 200 kcal/mole are laid out vertically on a logarithmic scale for a variety of processes. The range of values for the glasses is indicated, high values referring to the low-temperature regime, and low values to high temperatures.

Let us now have a very general look at physico-chemical rate process theory. The observed macroscopic rate, whether it be of diffusion, of shear

---

[5] The diffusion values are from F. Seitz, Chap. 4 of the book cited in footnote 1, as are the viscosity and electrical conductivity data used here and in Fig. 3. The activation energy for 7740 glass is based on viscosity data from the Corning Glass Works bulletin, "Properties of Selected Commercial Glasses," Corning, 1949. The viscosity and heat of vaporization data for B_2O_3 are from M.P. Volarovich and D.M. Tolstoi, J. Soc. Glass Technol. **18**, 209 (1934) and G.S. Parks and M.E. Spaght, Physics **6**, 69 (1935). The value $E = 20$ labeled "internal friction" is for the temperature dependence of internal friction of a commercial plate glass measured by J.V. Fitzgerald, J. Am. Ceram. Soc. **34**, 388 (1951).

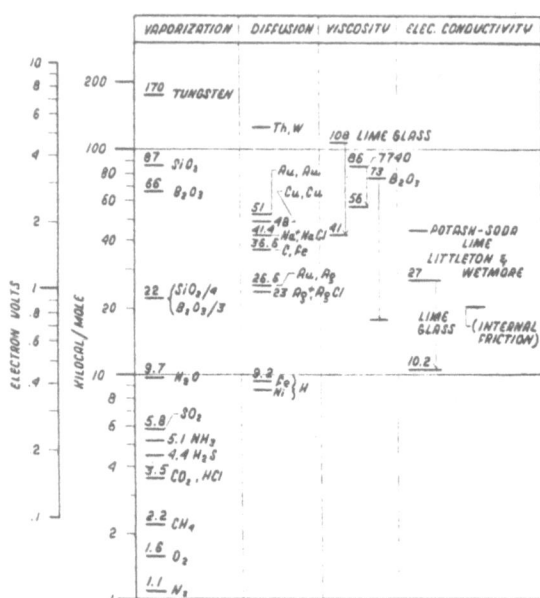

FIGURE 4_2. Some activation energies.

flow, of transport of electricity, of evaporation, of gas phase chemical reaction, of oxidation of metallic surface, is the *net* statistical effect of many elementary atomic or molecular processes. A given process will be called a *mechanism* and will be called *forward* if its direction contributes to the observed macroscopic rate, and *backward* if it works against this rate. Both forward and backward processes are going on all the time. In physico-chemical equilibrium they go on at equal rates in both directions (principle of detailed balancing). Macroscopically observed rate phenomena are the resultant of usually an extremely slight unbalance of the two rates, expressed as a fraction of either of the two nearly equal rates.

The first step in building a rate theory consists in recognizing the possible mechanisms which might be contributing. The next step involves application of statistical mechanics to the forward and backward rates. This involves calculation of the relation between the macroscopic rate parameter and the net frequency of occurrence of forward processes. It involves calculation of the way in which an external influence like applied stress or applied electric field affects the unbalance of these rates. It involves a calculation of the rates of forward and backward processes by application of statistical mechanics, making use of the fact that the unbalance is small, so the departure from thermodynamic equilibrium will be small enough to validate equilibrium distributions of energy among atomic coordinates.

In processes governed by an activation energy factor the common element is that some atomic or molecular configuration must pass from a state A to a state B in the forward process, and from B to A in the backward process, and that transition between these states is only possible when the energy in the coordinate which describes motion from A to B exceeds some minimum activation energy E. Thus the factor $e^{-E/RT}$ enters as the fraction of all possible motions in this coordinate which meet this condition.

The unstable configuration between A and B through which the system must pass is called the *activated complex*. Only a certain fraction of the events, even if the energy is sufficient, may be successful in bringing about a passage, $A \rightarrow B$. For example, suppose we are considering the kinetics of the reaction

$$H_2 + D_2 \rightarrow 2HD.$$

First we may adopt effective sizes for the molecules and calculate the total number of collisions in unit time in unit volume. Then the effective number leading to forward reaction will be reduced, not only by the Arrhenius factor, but also perhaps in that only collisions which meet some other conditions of proper mutual orientation at the instant of collision or the like, can lead to reaction.

It has recently become the practice to write this additional factor (usually unknown in any detail) in the form $e^{S/R}$ and speak of S as the entropy (per mole) of the activated complexes, relative to the initial reactants in the free state. Thus if N is the number of collisions of all sorts in unit time per mole of reactants, the forward rate in terms of number of forward processes in unit time per mole is

$$f = Ne^{S/R}e^{-E/RT}$$

or

$$f = Ne^{-F/RT},$$

where

$$F = E - TS$$

and is called the *free energy of the activated complex*, relative to the initial reactants.

In the foregoing highly simplified summary it was implied that only a certain *fraction* of the collisions having requisite energy might be successful implying $e^{S/R} < 1$, or a negative entropy of activation. In cases in which the reactants are large molecules with many internal degrees of freedom, it may happen that the activated complex is less restricted by quantization conditions than the initial reactants which means that S is positive and $e^{S/R} \gg 1$. This is especially true in the kinetics of important biochemical reactions.

This summary is a totally inadequate account of the subject, but it is hoped that it can serve to make a study of the literature easier by high-lighting the structure of the argument that is common to all theories of this kind.

Suppose now that there are several distinct alternative mechanisms $A_1 \rightarrow B_1, A_2 \rightarrow B_2$, etc., which can produce the observed macroscopic result. If these are on an *either-or* basis then the total rate will be the sum of the rates by each

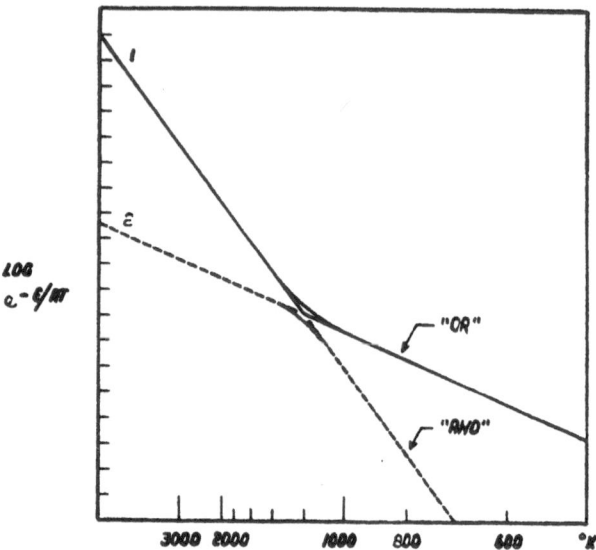

FIGURE 5_2. Change-over between mechanisms.

of the mechanisms. Because these rates vary so enormously with changes in the free energy of activation F, in the factor $e^{-F/RT}$, it will usually happen that the mechanism of the fastest rate will be overwhelmingly fast compared to the others so that the whole rate is essentially that of the fastest mechanism.

If the free energy of activation of two mechanisms was equal for a particular temperature. (as indicated in Fig. 5_2 where the two straight lines represent $\log(e^{-F/RT})$ for the mechanisms 1 and 2, respectively, shown intersecting at 1200°K). At higher temperatures 1 is the faster mechanism and will dominate the rate, while at lower temperatures 2 is the faster and dominates. The two mechanisms may be said to be "inequilibrium" or contribute equally to the rate when each has the same free energy of activiation for its activated complex.

Such a result gives an understanding of a situation in which the log(Rate) against $1/T$ curves have lower activation energy at the *low*-temperature end. But we have seen in Fig. 3_2 that the bend in such curves for glass viscosity is in the opposite sense!

If the bend in the curves for glasses is to be interpreted as a change-over betnecn different mechanisms one has to devise a way of coupling the two in series, so that the over-all rate is dominated bv the slower of the two whose joint operation is required for the total effect. The logical "and" must connect them instead of the "or," as indicated on Fig. 5. Various possibilities are open. Thus suppose transitions go by way of a stable intermediate C

$$A \to C \overset{\nearrow B}{\underset{\searrow D}{}}$$

and each process has its own free energy of activation as indicated by the subscripts. Here it is supposed that $C \rightarrow D$ represents a loss or "quenching" of C's so they can no longer contribute to the process which is only aided by processes ending at B. Write f for F/RT. The net rate for $A \rightarrow C$ is Ne^{-f_3} and for $C \rightarrow B$ and $C \rightarrow D$ it is $N'e^{-f_2}$ and $N'e^{-f_1}$ with N' determined by $Ne^{-f_3} = N'[e^{-f_1} + e^{-f_2}]$ giving as the over-all rate of arrival to B

$$ r = N \frac{e^{-(f_2+f_3)}}{e^{-f_1} + e^{-f_2}}. $$

At high temperatures $e^{-f_1} \gg e^{-f_2}$ as in Fig. 5_2, so the rate has an apparent activation energy $(E_2 + E_3 - E_1)$, while at low temperatures the inequality is in the opposite sense so the rate has an apparent activation energy $(E_2 + E_3 - E_2)$. Since by hypothesis (from the way the curves are drawn in Fig. 5_2) $E_1 > E_2$, it follows that on this model there is a lower effective activation energy at high temperatures than at low, the behavior shown by glass.

What is lacking is a more specific picture, in terms of the random network model, of exactly how such a two-stage mechanism operates.

3. Dependence of Other Properties on the Fictive

Below the transformation range the glass remains with a frozen-in structure characterized by its fictive τ. On being cooled down rapidly it contracts but not nearly as much as if cooled down slowly, as Fig. 2_2 clearly illustrates. With further drop in temperature it does contract somewhat, this being the expansion ordinarily measured as the usual reversible thermal expansion at temperatures below 400°C.

Adopting linear approximations, the glass has an equation of state in which its specific volume is a function both of its temperature and its fictive

$$ V(\tau, T) = V_0[1 + 3\alpha'(\tau - \tau_0) + 3\alpha T]. $$

Here α is the usual coefficient of linear expansion for the glass, V_0 is the volume at zero degrees temperature and at least practicable fictive τ_0 while $V(\tau, T)$ is the specific volume at temperature T and fictive τ. Evidently α' is the analog of α, the fractional change in length per unit change in fictive. Generally α' is about three times α, but the total amount of data capable of being treated in this way is not extensive. The density is given by

$$ \rho(\tau, T) = \rho_0[1 - 3\alpha'(\tau - \tau_0) - 3\alpha T]. $$

If the constants are known, a measurement of density serves as a fictive thermometer.

In the foregoing it is tacitly supposed that the specimens are small enough, relative to the rate of temperature cycling to which they are exposed, that all parts of the specimen have the same thermal history. Otherwise the material will not be in the same condition throughout. It will have fictive gradients in it. This situation is an important source of frozen-in stress distributions occurring in glass articles which have not been fully annealed.

Refractive index is extremely closely related to density and so may be calibrated instead of density as a fictive thermometer. Some recent unpublished measurements of H.N. Ritland made in the Research Laboratory of Corning Glass Works on the relation of density and refractive index for Glass 8370, an optical glass, showed the relation to be linear to high accuracy with $dn/d\rho = 0.167$ cm^3/g, with the experimental uncertainty in each density value being about $\pm 10^{-6}$ g/cm^3 and each refractive index value being good to $\pm 20 \times 10^{-6}$. The observed $dn/d\rho$ is definitely less than that to be expected if the atomic refractivities remained constant, that is if ρ were proportional to $(n^2 - 1)/(n^2 + 2)$. The departure is in the sense of a slight drop in atomic refractivities with increasing density.

Tool[2a] made qualitative observations by the method of differential thermal analysis of the *heat effects* associated with the transformation range. Apparently no accurate calorimetric work has been done. His method was to heat the glass in a furnace whose temperature was increased at a steady rate of 6 deg/min, having one junction of a thermocouple in the glass and the other in a neutral body. If there is no change with temperature of the heat capacity of either then the temperature difference will remain constant as the two bodies heat up together, although the actual temperature of one may lag behind the other by a constant amount. If now the glass gives off some heat in a certain range, its temperature will run ahead of that of the neutral body, while if it absorbs heat it will run behind. Effects of only a few degrees are easily noted but not evaluated in terms of heat units.

His results are consistent with the general picture that the heat content depends on the fictive as well as the temperature,

$$H(\tau, T) = H_0 + c'(\tau - \tau_0) + cT,$$

in which c' is a fictive heat capacity and c the usual heat capacity. Below the transformation range τ is constant so the effective heat capacity is c, but above the range $\tau = T$ so the effective heat capacity is $(c' + c)$. At present it is not known whether such a linear expression for $H(\tau, T)$ is adequate. The results of Thomas and Parks on the heat capacity of boric oxide alluded to in Part I indicate that additional heat changes occur in the transformation range.

One would certainly expect various other properties to depend on τ as well as on T. The available information on elastic properties at high temperatures is rather meager and the literature is confused by the lack of definite specifications of heat treatments used. The most careful work is that of Stong[6] who measured Young's modulus E on a soda-lime glass. He found $E = 7600$ kg/mm^2 at 0°, which decreased to 7000 at 400°C and thereafter decreased even

---

[6] G.E. Stong, J. Am. Ceram. Soc. **20**, 16 (1931).

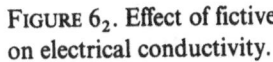

FIGURE 6_2. Effect of fictive
on electrical conductivity.

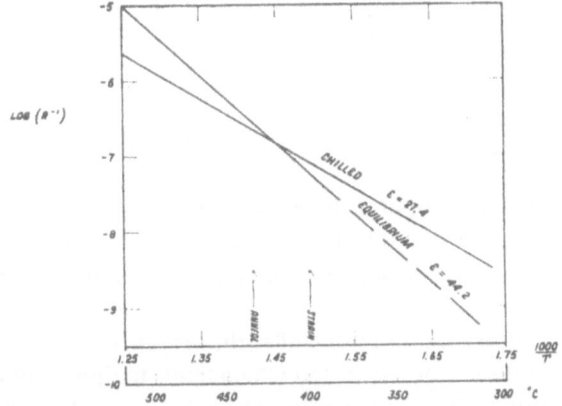

more at a rapidly increasing rate. He found that the curves were displaced
more or less parallel to themselves by amounts up to 300 kg/mm² for different
thermal histories, indicating a dependence on the fictive.

That the electrical resistance of a soda-potash-lead glass depends on the
fictive as well as on the temperature is clearly shown in results of Littleton
and Wetmore.[7] The main results are shown in Fig. 6_2 adapted from their
paper. Ordinates are log of conductivity and abscissas are $1/T$. The long
straight line marked "equilibrium" is presumably what one would obtain
if it were possible to anneal long enough to bring τ down to T, although their
experimental points do not go below $T \sim 370°$ along this curve. The broken
curve marked "chilled" represents the course of the conductivity at lower
temperatures for samples rapidly cooled, whose fictive therefore remains high,
perhaps $\tau \sim 430°$, as T is dropped on down to 300°.

The frozen-in structure distends the network, making it easier for the ions,
presumably mostly Na^+, to move about, hence the effect of higher fictive is to
increase the conductivity. The bend is in the opposite sense of those discussed
in the preceding section for viscosity, at higher temperatures, and of a type
compatible with *additive participation* of alternative mechanisms. The view
then is this: in the open structure of high fictive some of the Na^+ ions can get
through with an average activation energy of 27.4 kcal/mole. A good many
more can get through a chain of higher hurdles requiring an average activation
of 44.2 kcal/mole. At high temperatures this larger number dominates the
conductivity, while at low temperatures, in equilibrium, the open structure
closes up so the conductivity varies along the $E = 44.2$ line. But in the chilled
specimens the smaller number of ions which have a path of low hurdles open to
them dominate, their fewer number being more than made up for by the extra
ease with which they can move. This interpretation leads however to an

---

[7] J.T. Littleton and W.L. Wetmore, J. Am. Ceram. Soc. **19**, 23 (1936).

extended and gradual transition between the intersecting straight lines, instead of to the rather sharp break reported by Littleton and Wetmore.

4. Rate of Change of Fictive

At any temperature the vitreous equilibrium condition is that in which the fictive is equal to the temperature. Tool[8] has set up a semiempirical law for the rate of change of the fictive with time and tested it against experimental data.

The first requirement is that $d\tau/dt$ be equal to a function of τ and T which vanishes for $\tau = T$ and which tends to diminish $|\tau - T|$. The simplest assumption is that $d\tau/dt$ is proportional to $(T - \tau)$, a kind of Newton's law of cooling for the fictive. However, the proportionality factor, whose reciprocal is a time constant measuring the time lag of τ behind T, is itself expected to be strongly temperature dependent.

One expects this to depend both on T and on τ and probably mainly through an Arrhenius function $e^{-E/RT}$, but Tool uses a different form, one suggested by Twyman[9] which is equivalent to it for the fairly small ranges of temperature involved. It consists of approximating to $Ae^{-E/RT}$ with the form $Ke^{T/k}$. Near to T_0 this is equivalent to an effective activation energy of $E = RT_0^2/k$. Tool also assumes a similar variation of the rate with τ, with a different effective activation energy. He thus postulates that

$$\frac{d\tau}{dt} = K \exp\left(\frac{T}{k} + \frac{\tau}{h}\right)(T - \tau),$$

which may be called Tool's law of fictive cooling. He shows that it gives a good account of his observations although some discrepancies are found.

Consider the change of τ with time when T is held constant. This would give the familiar exponential decay of $(\tau - T)$ if the coefficient of $(T - \tau)$ were constant. A few transformations reduce the Tool equation to standard form:

$$\theta = K \exp\left[T\left(\frac{1}{k} + \frac{1}{h}\right)\right]t, \qquad u = \frac{\tau - T}{h} \frac{du}{d\theta} + e^u u = 0.$$

The solution involves different functional forms according as u approaches zero through positive values, or through negative values, because of the temperature dependence of the rate constant. In terms of the exponential integral functions,[10] the two forms are:

$$\text{for } u > 0 \qquad \theta = -Ei(-u) + Ei(-u_0);$$

$$\text{for } u < 0 \qquad \theta = -\overline{Ei}(-u) + \overline{Ei}(-u_0).$$

[8] A.Q. Tool, J. Research Natl. Bur. Standards **37**, 73 (1946).
[9] F. Twyman, J. Soc. Glass. Technol. **1**, 61 (1917).
[10] Jahnke-Emde, *Tables of Functions* (Dover Publications, New York, 1943), p. 6.

FIGURE 7_2. Tool's law of fictive cooling (ordinate is u on log scale; abscissas are $-Eil(-u)$ and $-\bar{E}i(u)$.

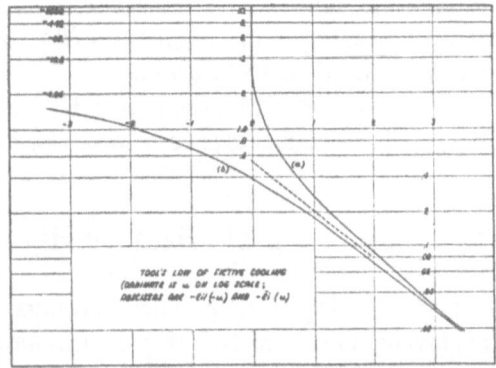

It is natural to plot θ as abscissas and u as ordinates on a logarithmic scale because this gives a straight line plot in the case of exponential decay, appropriate to a constant rate. Tool's law also gives a straight line on this plot for $u \gg 1$, as then the variation of the rate is unimportant.

The theoretical curves are shown on Fig. 7_2. Curve (a) is for $u > 0$, for fictive cooling from above. Large values of u decay with great rapidity relative to the natural time unit of the problem. Curve (b) is for $u < 0$, fictive warming up from below. In the upper left are tabulated the abscissa values at which the curve crosses successive ordinates up to $u = 10$, indicating the extreme slowness with which the fictive warms up if its initial value is much below the temperature. The extreme rapidity for cooling and the extreme slowness for warming up is what defines practical limits to the transformation range, as remarked earlier.

A careful study of the Tool equation in relation to fictive change at constant temperature in 8370 glass was made by Ritland[11] in the Research Laboratory of Corning Glass Works during the past year. He finds his data definitely require a correction term, so he proposes a modified Tool equation

$$\frac{d\tau}{dt} = Ae^{T/K}e^{\tau/h}(T - \tau)[1 + B(T - \tau)],$$

with

$$A = 2.12 \times 10^{-3} \qquad B = 0.244$$

$$k = 9.3°C \qquad\qquad h = 36.4°C$$

This gives an excellent fit for the approach to equilibrium as observed both from above and from below three different constant soaking temperatures, $T = 520$, 526, and 533°C at which times of the order of 10 hours (at 520°C) down to 1 hr (at 533°C) were required to bring the glass essentially to complete

---

[11] H.N. Ritland, J. Am. Ceram. Soc. (to be published).

equilibrium, that is, to $|\tau - T| < 0.5°C$. The values of k and h expressed as activation energies in exponents E/RT and $E'/R\tau$ give rather low values, $E = 15.5$ and $E' = 3.5$ kcal/mole.

As already stated all of this work is on a semi-empirical basis. No attempt at a detailed kinetic rate process model for change of the fictive has been made.

5. Compaction of Glass by High Pressure

Some interesting new experimental results on a quasi-permanent compaction of glass under extremely high pressures have just been reported by Bridgman and Šimon.[12] When fused silica and vitreous B_2O_3 are put under pressures of the order of 10^5 to $2 \cdot 10^5$ atmospheres, the material is compacted so that it shows a density increase of some 7 or 8 percent even after the pressure is released.

In order to achieve such great stresses the specimens were squeezed between two Carboloy dies so the pressures were not exactly hydrostatic and were of somewhat uncertain amounts.

Their main results are shown in Fig. 8_2. The shapes of the curves are quite different which may well be correlated with the very great cllferences in the transformation temperatures, melting points, and so on of the two materials.

One might be inclined to think that the applied pressure simply acts to speed up the rate of transformation to the higher density form so that this proceeds even at room temperature to an extent that enables the effect to reach saturation in about a minute, the time the pressure was applied in this experiment. But the effect must be a more profound change than that. Extra-

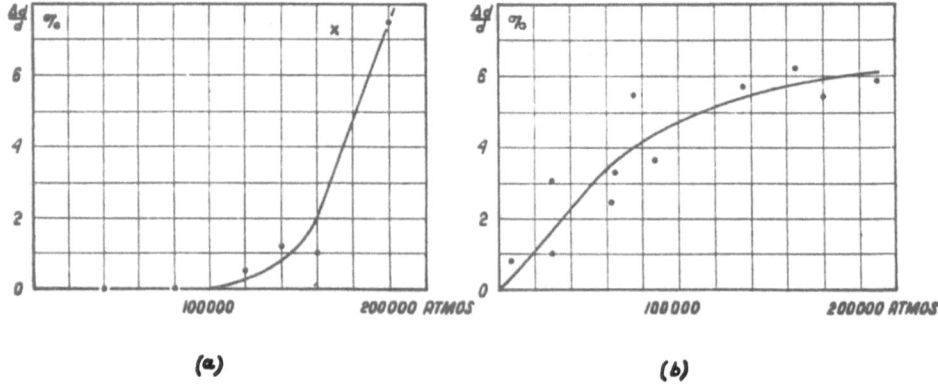

FIGURE 8_2. Quasi-permanent density increase in (a) vitreous SiO_2 and (b) vitreous B_2O_3 as a function of applied pressure.

[12] P.W. Bridgman and I. Šimon, J. Appl. Phys. **24**, 405 (1953).

polation of known high- and low-temperature expansion coefficients for B_2O_3 indicate that there is only about 3- or 4-percent excess volume frozen-in in the glassy B_2O_3 structure at ordinary temperatures, whereas the compaction produced by 200 000 atmospheres is about twice this. There do not seem to be data available on extreme high-temperature expansion coefficients on fused silica itself which would permit estimation of the amount of the frozen-in excess volume in room temperature SiO_2 but it seems likely that this is not much more than about 4 percent either (using $\alpha = 5 \cdot 10^{-7}/°C$, below the transformation temperature, assuming a fictive of 1500°C and that the expansion coefficient above the transformation range is about $15 \cdot 10^{-7}/°C$). On the other hand the relative compaction involved in known crystalline forms of SiO_2 is adequate:

Form	$cm^3/mole$	Ratio
Vitreous	27.2	1.000
Tridymite	26.6	0.975
Cristobalite	25.9	0.953
Quartz	23.3	0.857

Thus spatial arrangements of the Si and O atoms whose volume is even smaller than the compaction due to $2 \cdot 10^5$ atmospheres are possible.

One may speculate that there is a molal volume V_p which is the equilibrium value for each pressure and that when the pressure p is applied and the volume is V the effect is to change the activation energy by an amount $-p(V - V_p)$ for forward compacting processes and by an amount $+p(V - V_p)$ for backward or expanding processes, therefore tentatively one may expect an equation of the form

$$\frac{dV}{dt} = Ae^{-E/RT}\left[e^{p(V-V_p)/RT} - e^{-p(V-V_p)/RT}\right]$$

as governing the rate of approach to equilibrium.

At room temperature $RT \sim 2.5 \times 10^4\ cm^3 \cdot atmos$ and the equilibrium volume *change* for $2 \cdot 10^5$ atmospheres is about $3\ cm^3/mole$ so initially $p(V - V_p)/RT$ is of the order 13 giving an enhancement of the rate by only e^{13} or $10^{5\ or\ 6}$, which is not nearly enough to account for the effects observed relative to the rates of the ordinary transformation process. On the other hand an expression of the form of that suggested above can account for the enormous difference in rate observed on compacting and decompacting. Applying 40 000 atmospheres to vitreous B_2O_3 produced a density increase of 3.3 percent within about a minute. On release of the pressure the material starts to expand, that is, its density decreases but much more slowly as their Fig. 14 shows, the increase in density from its original value only having decreased to 2.2 percent after some 50 days, whereas the original compaction of 3.3 percent occurred in about one minute. The normal molar volume of B_2O_3 is $38\ cm^3/mole$, so the decrease produced in 1 minute by 40 000 atmospheres is about 1.2 $cm^3/mole$. It is not clear whether the value reached in one minute is the final equilibrium value, but it is probablv within 10 percent of it, so the rate of compaction under high pressure is such that the change in the logarithm (to base 10) of $(V - V_p)$ is about 1 in 1 minute. With the pressure off (that is, reduced to 1 atmosphere) one sees roughlv from their Fig. 14 that a corresponding change requires of the order of 300 days or 4×10^5 minutes which is very crudely about the order of ratio of the rates to be expected under the two different pressures.

Much more experimental and theoretical study needs to be devoted to this interesting new phenomenon.

6. A General Transformation Rate Equation

As already remarked, the rate factor for which Tool uses the form $e^{T/k}$ is more probably an approximation to $e^{-E/RT}$ valid only in a small range of temperature. But τ is itself proportional to the discrepancy between the actual volume and the equilibrium volume, and so probably belongs in the numerator of the exponential. The Bridgman-Šimon experiments indicate that the rate of transformation, as well as the equilibrium condition, is pressure sensitive. All these things together suggest a general transformation rate equation.

The transformation may be characterized as going from configuration 1 to 2 and $E_1 S_1 V_1 \ldots$ and $E_2 S_2 V_2 \ldots$ are the molal thermodynamic quantities for these quasi-equilibrium configurations. To pass from 1 to 2 one must go through an activated configuration whose molal quantities will be written E, S, $V \ldots$ without subscript.

The measure of the extent of the transformation is the change in volume $(V_1 - V_2)$ related to χ by

$$V = V_1 \chi + V_2 (1 - \chi),$$

and the transformation proceeds with change of χ from 1 to 0.

The rates in each direction involve the *Gibbs* free energy of activation $(E - TS + pV)$ relative to the initial and final states, hence the general kinetic equation will be assumed to be

$$\frac{d\chi}{dt} = -a\{e^{-[G-G_1\chi]/RT} - e^{-[G-G_2(1-\chi)]/RT}\}.$$

The rates balance, so equilibrium occurs, when $G_1 = G_2$, the familiar criterion. Writing

$$G_1, G_2 = \bar{G} \pm \tfrac{1}{2}g$$

the rate equation becomes

$$\frac{d\chi}{dt} = -2a \exp\left[-\left(\frac{G - \bar{G}}{RT}\right)\right] \sinh \frac{g\chi}{RT}.$$

At constant temperature and pressure this becomes

$$du/dt + A \sinh u = 0$$

with $u = g\chi/RT$ and $A = (ga/RT)\exp[-(G - \bar{G})/RT]$. This integrates to

$$At + \log\frac{\text{tahn}(u/2)}{\tanh(u_0/2)} = 0.$$

The observed change in volume is related to u by the equation

$$v = V + (RTv/g)u.$$

This is qualitatively at least of the form observed for the dependence, starting off at a rapid logarithmic rate of transformation, and slowing down to the usual constant logarithmic decrement as equilibrium is approached, as shown in Fig. 9_2. Abscissas are values of log tanh $u/2$ on a linear scale, and ordinates are values of u on a logarithmic scale. For $u < 1$ the logarithmic decrement of u is essentially constant. The quantities $E_1 E_2$, $V_1 V_2$, $S_1 S_2$, etc., depend on the variables p and T, which affects the kinetics of the process as well as the equilibrium.

At present the available data are insufficient for an adequate test of this general transformation rate equation.

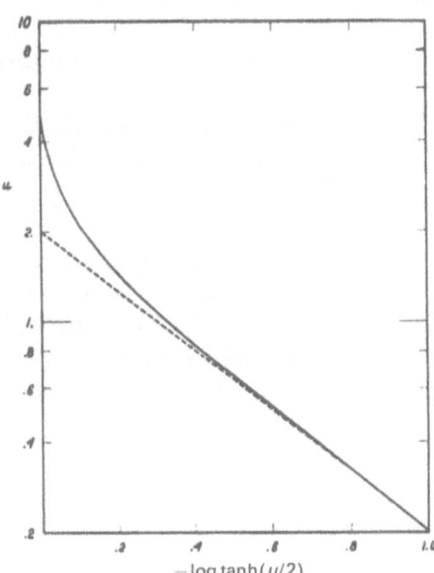

FIGURE 9_2. Approach to equilibrium with generalized kinetic equation.

Physics of the Glassy State
III: Strength of Glass*

EDWARD U. CONDON

Corning Glass Works
Corning, New York

The main empirical facts about the strength of glass are reviewed, including statistical theories effects of adsorbed moisture and of acid treatments.

1. Statistical Variability

When solid bodies are subjected to increasing mechanical stress ultimately they *fail* in the sense that their shape undergoes a permanent *plastic deformation* or the body undergoes *fracture* into several parts. Before failure they undergo elastic deformations, more or less accurately described by Hooke's law, that strain is proportional to stress, sometimes with time-delay effects, and intermingled with various irreversible flow phenomena. These have been the object of study for over a century but still much remains to be done to clarify the field.

More than a century ago (1848) James Thomson[1] found it necessary to caution against oversimplification in studying strength of materials:

The introduction of new, though necessary, elements into the consideration of the strength of materials may, on the one hand, seem annoying from rendering the investigations more complicated. On the other hand, their introduction will really have the effect of obviating difficulties by removing erroneous modes of viewing the subject, and preventing contradictory or incongruous results from being obtained by theory and experiment. In all investigations, in fact, in which we desire to attain or approach nearly to truth, we must take facts as they actually are, not as we might be tempted to wish them to be for enabling us to dispense with examining processes which are somewhat concealed and intricate but are not the less influential from their hidden character.

Reading the complicated and fragmentary literature on strength of glass

---

* The third of a series of lectures delivered at the Fifteenth Annual Colloquium of College Physicists, State University of Iowa, Iowa City, June 20, 1953.
[1] Quoted in the article "Elasticity" in the ninth edition of the Encyclopedia Britannica.

one cannot but feel that our subject would have profited greatly if many of the investigators had been more adequately guided by these words of wisdom.

Glass, like most nonmetallic solids, is a *brittle* material at ordinary temperatures, which means that fracture occurs without being preceded by any appreciable plastic deformation. The problem before us thus involves consideration of (a) basic reasons for the absence of plasticity, (b) conditions determining *initiation* of a fracture, (c) conditions determining *propagation* of a fracture.

As to the absence of plasticity it will merely be remarked that in recent years great advances have been made in understanding the plasticity of ductile materials in terms of motion of dislocations in almost perfect crystals,[2] which makes it evident that such mechanisms cannot operate in amorphous materials.

Consider now glass objects of simple shape such as circular rods or strips of rectangular cross section or plates, to which tension or compression or bending or torsion is applied with the aid of a mechanical testing machine. Under the best circumstances there is a wide statistical variation in the force needed to break a number of specimens that are nominally alike and nominally tested in the same way, and also a wide statistical spread in the nature of the fractures produced.

This indicates the need for careful *statistical* design[3] of the excperimental work. Unfortunately the statistical treatment of most of the work reported in the literature is inadequate. If χ is the force needed to produce fracture then the distribution of values of χ in a population of specimens tested in a certain way will be given by a distribution function $f(\chi)\,d\chi$ giving the fraction of all specimens for which χ lies between χ and $\chi + d\chi$. Usually this may be assumed to be a normal Gaussian distribution,

$$f(\chi) = \frac{1}{\sqrt{(2\pi)}\sigma}\,e^{-(x-a)^2/2\sigma^2}$$

characterized by two parameters, the mean a and the standard deviation σ whose square σ^2 is called the variance. Often the test results show a marked skewness in distribution.

The ratio σ/a is called the *coefficient of variation*. In work on glass strength it usually is between 0.10 and 0.25 but on occasion may be much greater. The standard deviation σ is just as much a fact of observation as is the mean a, and a complete understanding of strength will interpret observations of σ as well as those of a.

The integral form of the normal distribution is

---

[2] National Research Council. Committee on Solids. *Imperfections in Nearly Perfect Crystals* (John Wiley and Sons, Inc., New York, 1952).

[3] Recommended for statistical theory are: (a) C.E. Weatherburn, *A First Course in Mathematical Statistics* (The Cambridge University Press, New York, 1946); (b) M.J. Moroney, *Facts from Figures* (Penguin Ltd., 1951); (c) C. Eisenhart *et al.*, *Selected Techniques of Statistical Analysis* (McGraw-Hill Book Company, Inc., New York, 1947).

TABLE I$_3$. Normal distribution function

$F(u)$	u	$F(u)$	u
0.001	-3.08	0.999	3.08
0.01	-2.32	0.99	2.32
0.1	-1.282	0.9	1.282
0.2	-0.842	0.8	0.842
0.3	-0.525	0.7	0.525
0.4	-0.253	0.6	0.253
0.5	0.000		

$$F(u) = \frac{1}{\sqrt{(2\pi)}} \int_{-\infty}^{u} e^{-u^2/2} \, du.$$

Here $u = (\chi - a)/\sigma$ and $F(u)$ gives the fraction of the population for which $(\chi - a)/\sigma$ has values in the range $-\infty$ to u, likewise $[1 - F(u)]$ is the fraction for which $(\chi - a)/\sigma$ lies in the range u to $+\infty$. For nearly all practical purposes Table I$_3$ is sufficient.

Tests on the loading x to break rods in transverse bending are more usual than loading them in tension. This is because of the difficulty of arranging for accurate axial loading without introduction of uncontrolled bending moments. The tests are not easily comparable in that all of the specimen's surface is in tension in a straight tensile test, whereas only a small part of the surface area has the full tensile stress on it in a bending test.

2. Surface Failure in Tension

The next important generalization is that fractures nearly always initiate at a place on the surface that is in tension.[4] With a circular rod in pure tension, assuming homogeneity, the stress is the same over the cross section. The statement has meaning since we can determine, by inspection of the fractured parts, the *break point* or point of origin of the fracture.

Internal break points seem to occur only when there is an obvious internal inhomogeneity, such as a *stone*, as inclusions of undissolved crystalline material are called. They might also occur when by special heat treatment stress distributions have been frozen-in, in such a way that the internal tensile stress is much greater than that on the surface.

---

[4] For general surveys, (a) G.W. Morey, *The Properties of Glass* (Reinhold Publishing Corporation, New York, 1938), Chap. 13; (b) J.E. Stanworth, *Physical Properties of Glass* (Oxford University Press, New York, 1950), Chap. 4; (c) R.N. Haward, *The Strength of Plastics and Glass* (Cleaver-Hume, 1949), Chap. 3; (d) E. Orowan, Repts. Progr. in Phys. **12** (1948–49).

FIGURE 1_3. Breaking stress in glass fibers (Griffith).

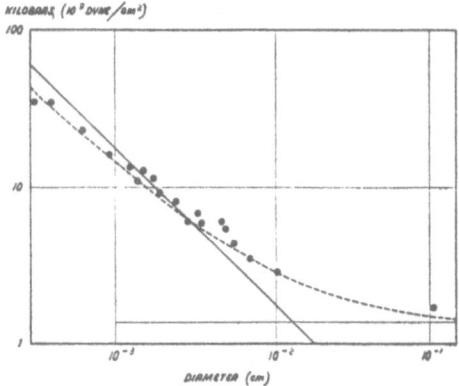

Associated with the fact that fractures start at the surface are: (a) The fact that the breaking *force* for a circular rod in tension is not proportional to its cross-sectional area, but for small fibers more nearly to the first power of the radius. (b) Strength is greatly affected by surface treatments which do not directly affect bulk properties of the glass, these effects involving both the action of easily observable macroscopic scratches and more subtle microscopic alterations of the surface by chemical action.

The first point is illustrated in Fig. 1_3, showing data, due to Griffith,[5] on the breaking *stress* of glass fibers as a function of the radius, in the range 10^{-4} to 10^{-1} cm in diameter.

The points are roughly represented by the solid line which is $S = 18/d$, where S is in bars (10^6 dyne/cm^2) and d is cm, but are better represented by the dotted line which corresponds to

$$S = 1360 + 14/d.$$

This is equivalent to saying that the total load F need to break the fiber is the sum of an ordinary stress part, proportional to the cross section, $A = \pi d^2/4$, and an unusual extra part, proportional to the circumference $C = \pi d$,

$$F = (1360A + 3.5C) \cdot 10^6 \text{ dynes},$$

where A is in cm^2 and C is in cm. The strength of fibers greater than 0.1 cm in diameter is mostly due to the usual volume stress term but for fibers small compared to 0.1 cm in diameter it is mostly due to the surface stress term, although the surface breaking stress, $3.5 \cdot 10^6$ dyne/cm is vastly greater than usual surface tensions. The notion of surface stress is due to Preston.[6] Its physical meaning is not clear.

[5] A. A. Griffith, Trans. Roy. Soc. (London) **A221**, 163 (1920).
[6] F.W. Preston, J. Soc. Glass Technol. **17**, 4 (1933).

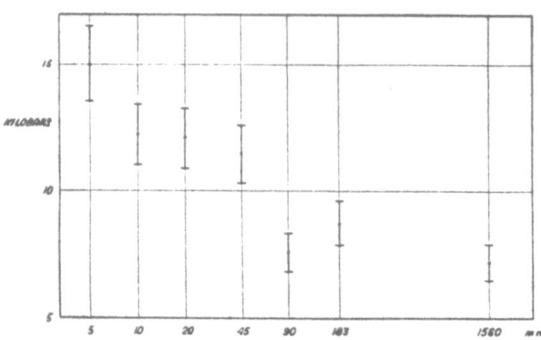

FIGURE 2_3. Variation of breaking stress with length (Anderegg).

Much more information of this sort is in the literature, indicating in a rough way the dependence of the tensile stress to start fracture on the size of the specimen. For example Holland and Turner[7] measured the load p to break rectangular specimens 10 cm long supported between knife edges 7.6 cm apart and with the load p applied in the middle, using specimens of various widths in the range 0.4 to 1.2 cm. With the edges ground and polished they found the mean breaking stress (calculated from $S = 3pl/2bh^2$, where l is the distance between supports, b the width and h the thickness) to decrease with width,

$$S = (470 + 66/b) \cdot 10^6 \text{ dyne/cm}^2.$$

This is of the same form as that used to represent Griffith's fiber data, but here the term varying inversely as a linear dimension is relatively much more important.

Dependence on *length* indicates a weakening with increasing length. On soda-lime glass fibers 13 microns in diameter, Anderegg[8] found the values shown in Fig. 2_3. In the absence of information about the observed spread, the points are plotted with an indicated coefficient of variation of 0.10 although it is probably greater than this. The results indicate a decrease in strength by roughly a factor of 2 on increasing the length by a factor of 300.

Attempts have been made to develop[9] an interpretation of such size effects in terms of statistical distributions of weak spots or flaws distributed either through the volume or on the surface of the glass. Without making special hypotheses about the nature of the flaws, let us suppose they are distributed on the surface as potential break points so that there are n of them on the surface of the specimen and let $F_1(\chi)$ be the fraction of them that will fail

[7] A.J. Holland and W.E.S. Turner, J. Soc. Glass Technol. **20**, 72 (1936).
[8] F.O. Anderegg, Ind. Eng. Chem. **31**, 290 (1939).
[9] J. Bailey, Glass Ind. **20**, 21, 59, 95, 143 (1939); W. Weibull, Ingeniorvetenskapsakad. Handl. **151** (1939).

when the tensile stress to which they are subjected is χ (supposing it were possible to make individual tensile tests on the glass surrounding each flaw).

A specimen having a random sample of n flaws will fail at that stress at which the weakest one fails. We are thus led to this statistical problem: Random samples of specimens containing n flaws are drawn, from a population in which the integral distribution of χ values is $F_1(\chi)$, what is the distribution of values of χ associated with the weakest one in each of the samples?

The solution[10] is simple enough in principle if $F_1(\chi)$ is known. If $F_n(\chi)$ is the integral distribution function for the weakest one in samples of n then

$$dF_n = n(1 - F_1)^{n-1} dF_1,$$

or if $F_n(\chi)$ is the fraction of the samples of n for which the weakest one has a value in the range $-\infty$ to χ,

$$[1 - F_n(\chi)] = [1 - F_1(\chi)]^n.$$

This result is easily stated in words: the fraction of the samples of n in which the *weakest* flaw is *stronger* than χ is equal to the nth power of the fraction of the individual flaws which are stronger than χ. The relation is shown in Fig. 3_3, in which ordinates represent $[1 - F_n(\chi)]$ and abscissas represent values of χ. Evidently $F_n(\chi)$ takes on appreciable values in the neighborhood of the values of χ for which $F_1(\chi) \sim 1/n$. Therefore for large numbers of flaws the shape of the $F_n(\chi)$ curve is mostly determined by the exact shape of the $F_1(\chi)$ curve in this vicinity. Figure 3_3 was drawn for the normal distribution, with the abscissas being $u = (\chi - a)/\sigma$ and for various values of n as labeled. The qualitative behavior is similar for other forms of $F_1(\chi)$.

The most careful attempt to apply this statistical view to glass strength made thus far is that of Fisher and Hollomon.[11] They conclude that the statistical theory indicates the presence of about 10^3 flaws/cm^2 as being active in order to explain the variation of strength with fracture stress with size.

---

[10] M.G. Kendall, *The Advanced Theory of Statistics* (Griffin, 1948), Vol. 1, p. 218.
[11] J.C. Fisher and J.H. Hollomon, Metals Technol. **14**, technical paper 2218 (1947).

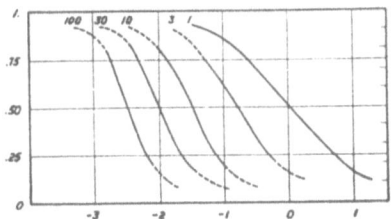

FIGURE 3_3. Distribution of weakest links.

3. Effects of Surface Treatment

Surface treatments have a pronounced effect and, as the glass is sensitive to factors that are hard to control, this probably accounts for much of the scatter in some of the experimental results reported in the literature. This topic is reviewed by Morey[4(a)] and by Stanworth.[4(b)]

Maximum strength is obtained when the surface is freshly drawn from hot glass. Microscopic scratches produced by very slight contact with other solids may reduce strength by a factor of 2. Milligan[12] found a marked effect of moisture on the strength of glass bars with regard to bending after being scratched with a diamond:

	Mean load (g)	No. of samples	Relative load
Dry	786	8	1.00
Wet	622	7	0.79

The reduction in strength occurred immediately after the moistening. The effect of moisture is also shown in Fig. 4_3, based on data of Yurkov[13] on strength of fused silica fibers. Baker and Preston[14] studied the increase in strength which they could get in glass rods 0.22 inch in diameter after baking out in vacuum at 350°C for 10 minutes and being tested at 80°C. Their principal results are shown in Table II_3. The effect of baking out is presumed to be mainly the result of driving off adsorbed moisture.

---

[12] L.H. Milligan, J. Soc. Glass Technol. **13** 351 (1929).
[13] (a) S. Yurkov, Physik. Z. Sowjetunion **l**, 123 (1932); (b) Berdinnikow, Physik. Z. Sowjetunion **2**, 397 (1933).
[14] (a) T.C. Baker and F.W. Preston, J. Appl. Phys. **17**, 179 (1916); (b) F.W. Preston, J. Appl. Phys. **13**, 623 (1942); (c) C. Gurney and Pearson, Proc. Phys. Soc. (London) **B62**, 469 (1949).

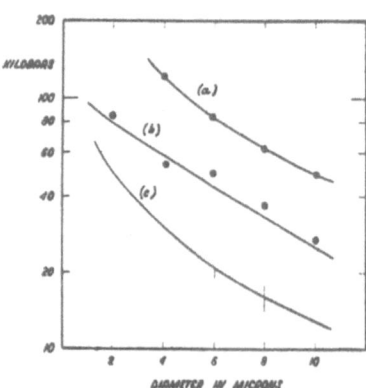

FIGURE 4_3. Strength of fused silica fibers (Yurkov). (a) Vacuum bake at 600–700°C (b) air dried with $CaCl_2$ (c) moist.

TABLE II₃. Effect of baking out glass
in vacuum

| Glass | Strength in 1000 psi | | |
	Before	After bake-out	Ratio
Pyrex brand	17.7	44.3	2.5
Soda lime	12.1	24.7	2.0
Fused silica	16.7	27.9	1.7

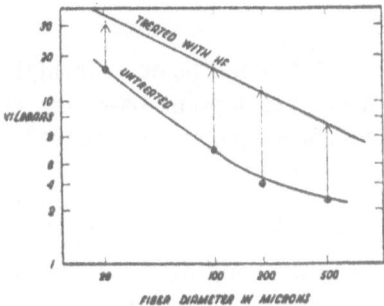

FIGURE 5₃. Effect of HF on strength of
borosilicate glass fibers (Yurkov).

The effect of hydrofluoric acid is opposite to that of moisture. Figure 5₃
shows some data obtained by Yurkov[15] on the effect of HF treatment on fibers
of a borosilicate glass: a general increase in strength of about threefold was
observed.

Another way of producing an increase in strength is by introducing some
SO_2 into the atmosphere of the annealing lehr.[16] The mode of action involved
here is believed to be conversion of some of the Na_2O into Na_2SO_4 leaving a
silica-rich surface layer of lower expansion coefficient which is put into compression by differential contraction relative to the bulk of the glass.

If a sheet of glass is quickly cooled from a high temperature, the surface
layers solidify while the inner layers are still hot and soft. Later when the inner
layers cool and contract, internal stresses are set up which put the inner
layers in tension and the outer layers in compression.[17] Such treatments are
used to give extra strength to glass articles. The product is sometimes called
toughened glass. Figure 6₃ shows data of Haward[18] on bending strength of
annealed and toughened sheet glass for various times of loading in seconds.
The thickness of the toughened sheets was 0.46 cm, while that of the annealed

[15] S. Yurkov, J. Tech. Phys. (U.S.S.R.) 1, 386 (1935).
[16] J. Boow and W.E.S. Turner, J. Soc. Glass Technol. 28, 406 (1944).
[17] (a) J.T. Littleton, Phys. Rev. 22, 510 (1923); (b) J.T. Littleton and F.W. Preston, J.
Soc. Glass Technol. 19, 221 (1935).
[18] R.N. Haward, Nature 157, 21 (1946).

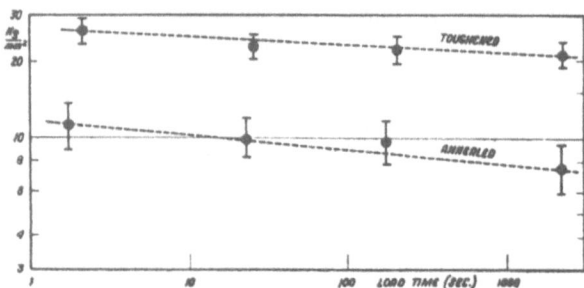

FIGURE 6₃. Effect of toughening on strength of sheet glass in bending (Haward).

was 0.41 cm. The experimental values of the standard deviation are shown around each point, based on samples of 30 to 40 pieces for each point. The toughened pieces show about half as great a coefficient of variation as do the annealed pieces.

The whole toughening process depends on being able to set up a temperature gradient across the thickness and for this reason the process is practically applicable only for thicknesses greater than $\frac{3}{16}$ inch. With greater thickness it is possible to produce greater surface compression and so greater strength—opposite to the behavior of annealed glass. Von Reis[19] was able to get a bending strength of 20 kg/mm² for a thickness of 5 mm increasing to 30 kg/mm² for a thickness of 12 mm.

4. Static Fatigue in Glass

In Fig. 6₃ some data were presented indicating that glass will fail after a considerable time of being loaded to a stress which will not cause it to fail when it is first applied. This effect is sometimes called *fatigue,* but in so doing it is essential not to confuse it with the effect called fatigue in metal testing, which is the reduction in load carrying capacity after many cycles of cyclic variation of the load. As to order of magnitude, one may say that the long time load (~1 day) may be about ⅓ that which can be sustained for a few seconds. This effect makes it necessary to specify the times used in strength measurements, a fact not always appreciated by workers in this field. Often specimens in a test machine have the load increased at a steady rate until they fail: the strength so observed will depend on the time rate of increase of the load as indicated in Fig. 7₃ for example in some data due to Black[20] in which there is about a two-to-one strength change for a several hundred-fold change in the rate of loading. Another way of looking at the time delay effect is provided

---

[19] L. Von Reis, Glastech. Ber. **13**, 239 (1935)
[20] L.V. Black, Bull. Am. Ceram. Soc. **15**, 274 (1936).

FIGURE 7₃. Effect on strength of rate of loading (Black).

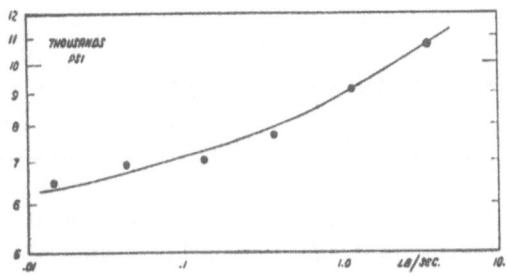

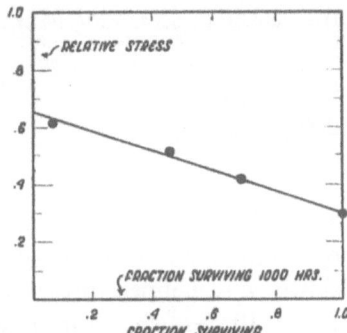

FIGURE 8₃. Stress-mortality relation at 1000 hours (Holland and Turner).

in Fig. 8₃ based on data of Holland and Turner.[21] This is a mortality curve showing the fraction of the specimens able to survive 1000 hours for various stresses expressed as fractions of the stress which would produce rupture quickly.

These effects are believed to be due to slow change in the action of moisture on the glass produced when the glass is under stress because Baker and Preston[14] found the fatigue effects essentially gone if the glass is baked out and kept in vacuum while under load. Likewise these effects are small below 0°C (where the vapor pressure of moisture is small) and are larger at room temperature and become small again above 300°C when adsorbed moisture is driven off.

5. Theoretical Strength

Various ways of calculating theoretical strengths for glass in relation to intermolecular forces have been proposed. These are usually based on order of magnitude arguments, but give values considerably greater (by a factor of 10 to 100) than the observed strengths.

---

[21] A.J. Holland and W.E.S. Turner, J. Soc. Glass Technol. **24**, 46 (1940).

One such estimate was given by Polanyi.[22] Suppose a rod of cross-sectional area A cm^2 having Young's modulus E dyne/cm^2 is extended with a tensile stress S dyne/cm^2. The work done in applying the stress is stored as elastic energy of deformation $(S^2/2E)$ erg/cm^3. When fracture occurs a new surface area $2A$ cm^2 is formed. The surface has extra energy γ erg/cm^2 which in liquids manifests itself as a surface tension giving rise to the phenomena of capillarity. After the fracture therefore the fragments contain additional energy $2A\gamma$ erg because of the increase in surface area.

The assumption is then made that for fracture to occur this amount of energy must be stored elastically in a layer whose thickness is that of one intermolecular distance d. This assumption is reasonable but not obviously true as will be seen later. Making it one gets, when S equals the breaking stress,

$$AdS^2/2E = 2A\gamma$$

giving

$$S = \sqrt{(4\gamma E/d)}.$$

The difficulty now arises that γ is not known for solids. For molten glass it is observed to be in the range 200–500 erg/cm^3, and as it decreases rather slowly with temperature, we may assume $\gamma \sim 10^3$ erg/cm^2. For fused silica $E = 7 \cdot 10^{11}$ erg/cm^3. For d one may take $(V/N)^{1/2}$, where V is the molar volume and N is Avogadro's number which works out to be $d = 3.6 \times 10^{-8}$ cm. These values give a theoretical strength of $S = 2.8 \times 10^{11}$ dyne/cm^2 = 280 kilobars or 4.07×10^6 psi. This assumes validity of Hooke's law up to such a stress value at which the fractional increase in length would be 40 percent. The figure is reasonably close to the value toward which Yurkov's data on fused silica fibers, baked out in vacuum, seems to be trending [Fig. 4$_3$(a)] for fiber diameters of about 1μ. Better fit than is indicated would be meaningless because of a possible several-fold uncertainty in the proper value of γ and also in that for d.

The uncertainty in d lies not so much in doubt that $(V/N)^{1/2}$ is a reasonable value for thickness of the surface layer, as that concerning the validity of the assumption that only that elastic energy which is stored in a layer of this thickness is available to provide the surface energy of the new surface being formed. With $d = 3.6 \times 10^{-8}$ cm there is $d^{-1} \sim 30$ million times as much stored elastic energy per cm of length of the test specimen as is assumed to be effective.

Another approach to the calculation is to remark that as the two new surfaces produced by the fracture pull away from each other, the forces of attraction between them drop from the full value S per unit area to zero in a distance of the order of d. If the fall is linear then the work done is $\frac{1}{2}Sd$ per unit area by the forces acting to pull back on it. It is this energy which

---

[22] M. Polanyi, Z. Physik 7, 323 (1921); E. Orowan, Z. Krist. 89, 327 (1934).

must be equated to the surface energy of each new surface formed because it is in the very breaking of the bonds of the atoms in each surface to those in the other fragment that the surface atoms are given the higher energy per atom which is the origin of what we call surface energy. On this view $\frac{1}{2}S'd = 2\gamma$ giving

$$S' = 4\gamma/d$$

instead of

$$S = [(2\gamma/d)(2E)]^{1/2}.$$

Using the values previously adopted this gives $S = 0.56 \times 10^{11}$ dyne/cm^2 = 56 kilobars or $\frac{1}{3}$ of the other estimate, and more in line with observation on small fused silica fibers, although the uncertainty in γ, already mentioned, must still be borne in mind.

Now we can ask: if the breaking stress is given by S', then what thickness of layer in each fragment must give up its elastic energy to provide the energy for the new surface of that fragment? It is D, where

$$ADS'^2/2E = A\gamma,$$

giving

$$D = \tfrac{1}{2}(E/S')d = 6.2d,$$

from which it would appear that about ten layers are needed to store in each fragment the energy elastically that is used in forming the new surface.

Lacking any measurements of γ for the solid surface one can proceed to use a relation, originally suggested by Stefan,[23] between surface energy and heat of vaporization. If L is the heat of vaporization per unit volume expressed as erg/cm^3, this is the energy which must be supplied to the molecules to spread them out in the gas phase against intermolecular attractions. The molecules in the surface layer of thickness d occupy a volume d cm^3/cm^2 and so the energy needed to vaporize the amount of one monomolecular layer is Ld. But the surface molecules are already partly out from under the influence of their neighbors; that is, have more energy on the average, which Stefan roughly supposed was $\frac{1}{2}Ld$ giving for the surface energy

$$\gamma = kLd,$$

where k is some factor which Stefan supposed to be $\frac{1}{2}$. For H_2O, we have $L = 22.6 \times 10^9$ erg/cm^3, $d = 3.1 \times 10^{-8}$ and $\gamma = 70$ erg/cm^2, indicating $k = 0.1$, that is, the surface molecules only have 10 percent of the energy needed for vaporization. On the other hand for liquid mercury $k = 0.44$. Harkins and Roberts find low values of k for polar liquids and higher values for nonpolar liquids. Using this expression for γ in the S' formula for breaking

[23] Harkins and Roberts, J. Am. Chem. Soc. **44**, 653 (1922).

strength one gets rid of the uncertain d, but ends up with the uncertain k, that is,

$$S' = 4kL.$$

On this view, taking the simple value $k = \frac{1}{2}$ the breaking strength should be about twice the heat of vaporization of unit volume.

Using 88.5 kcal/mole for the heat of vaporization *of fused silica*[24] this gives $L = 136$ kilobars, which leads to the same estimate for S' as before if k for fused silica is about 0.1 which is about that of water and $\frac{1}{4}$th that for liquid mercury, and therefore of a reasonable order of magnitude.

On page 322 of his book, Morey says that usually "calculation of the theoretical strength of glass ... suffers from the fault that the necessary underlying assumptions are usually so obscured by mathematics as to lend to the calculation an appearance of finality wholly unjustified by the facts." I hope that the preceding discussion will be exempted by him from this charge for the analysis given is mathematically simple and by no means definite or final.

Nevertheless the end result is that the estimate

$$S' = 4kL$$

with $k = 0.1$ to 0.3 is not far from the maximum values achieved with small fibers when baked out. Under ideal conditions strengths of this magnitude are attainable and understandable in terms of a simple model. There remains to discuss the reasons why observed strength is less than this by factors as great as 10 to 20, under certain circumstances such as those discussed in the preceding section.

6. Propagation of Fracture

The theoretical analysis of the preceding section is based on an idealized picture, in which the two fragments resulting from a fracture move apart all at once, moving normally to the plane of fracture. Actually the process is different. The fracture begins at some flaw on the surface and spreads, initially in a plane normal to the surface. The process is shown schematically in Fig. 9_3. The fracture starts at A. It spreads outward from A in all directions at the same velocity, producing the smooth surface (a) in the cross section, which may, under some circumstances, occupy the whole cross section. But when it does not it ends in a circular boundary of roughened surface, marked (b). Then the fracture bifurcates producing a crescent-shaped wedge marked (c) which may have radial rib markings on it as well as on the main fragments. Another possible variation is for (a) to have frozen in wave markings in circles with A as center.

---

[24] K.K. Kelley, Bulletin 383, U.S. Bureau or Mines (1935).

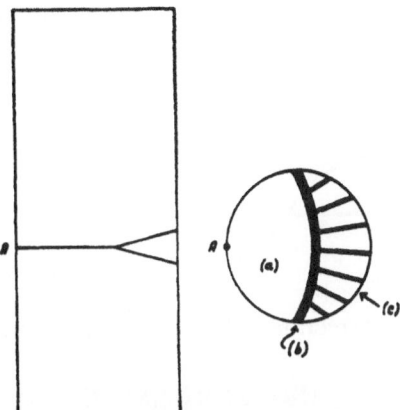

FIGURE 9_3. Mode of fracture of glass rod.

As remarked earlier there is stored elastically in the specimen, at the instant of fracture, millions of times more energy than is needed to provide for the energy of the new surfaces formed. So there is no lack of energy available to the process, once it starts. Our basic view is that *initiation* of fracture is dependent on surface conditions on the old prefracture surface, especially to minute cracks and flaws around which stress concentrations are formed so that local stresses at their bottom may be several times the average applied stress. Once fracture is started, the problem of how it propagates is a separate question involving the dynamics of elastic wave motions in the material.

In the literature a good deal of attention is given to the theory of Griffith[5] which uses calculations of stress concentration factors around sharp-edged cracks to estimate the reduction in strength caused by them. For a flat plate having an elliptical hole of high eccentricity and subject to tensile stress S in a direction normal to the major axis, the maximum tensile stress will occur at the end of a major axis[25] and will have the value S' where

$$S' = S\sqrt{(4c/r)},$$

where $2c$ is the major axis and r is the radius of curvature at the end of the major axis. The same result holds for a semi-elliptical hole at the surface whose major axis is normal to the surface.

The decrease in elastic energy stored in the medium because of the crack is

$$We = \tfrac{3}{2}\pi c^2 S^2/2E$$

for unit thickness of plate (for a thin plate compared to $2c$). If the thickness is large compared to $2c$ this expression has to be multiplied by $(1 - v^2)$, where

[25] S. Timoshenko and J.N. Goodier, *Theory of Elasticity* (McGraw-Hill Book Company, Inc. New York, 1951), p. 197.

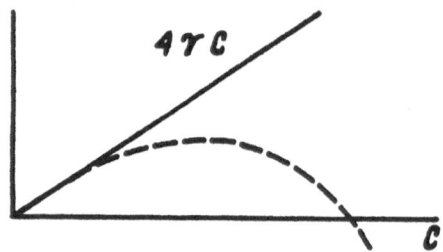

FIGURE 10₃. Total energy of crack formation.

v is Poisson's ratio, a correction of less than 10 percent. This decrease can be written

$$We = \pi(c)(\tfrac{3}{2}c)(S^2/2E),$$

from which we see that it is equal to the elastic energy normally stored ($S^2/2E$ in unit volume) in a much larger volume, namely, that of an ellipse whose semiaxes are c and $\tfrac{3}{2}c$. This decrease comes about because of the large region near the flat sides in which the stress is greatly reduced in spite of there being more energy stored in unit volume in the small region near the ends of the major axis where the stress is raised. The energy of formation of new surface in the crack is $4\gamma c$ per unit thickness and therefore the total energy of the crack as a function of c is

$$V = 4\gamma c - \tfrac{3}{4}\pi c^2 S^2/E.$$

As indicated diagrammatically in Fig. 10₃, $V(c)$ has a maximum, which occurs at

$$S = \left[\left(\frac{4\gamma}{d}\right)(E)\left(\frac{2d}{3\pi c}\right)\right]^{1/2}.$$

If for a given stress S the crack is deeper than the c satisfying this equation, then the energy of the system will decrease for an increase in c, so c will spontaneously increase leading to propagation of the fracture. The expression is written in this particular way to emphasize that the critical stress for crack propagation on this view is much smaller than the previously given theoretical estimates of strength, and can easily become 10^{-2} as great if $c \sim 10^4 d$, that is, if the crack depth is of the order of 10^{-4} cm.

Other shapes of crack give similar results but with somewhat different numerical factors.

If in the relation for S' we substitute the value just found for the S needed for crack propagation, we find

$$S' = \left(\frac{8}{3\pi}\right)^{1/2}\left(\frac{4\gamma}{r}E\right)^{1/2} = 0.92\left(\frac{4\gamma}{r}E\right)^{1/2},$$

so if r is of the order of atomic dimensions, the maximum stress at the root of the crack is just what was estimated from one point of view in the previous section.

For lack of time we shall omit any review of interesting applications of high-speed photography to measure the rate of crack propagation when a glass plate is fractured by impact. A wide variety of speeds is found, with a maximum ranging up to about a third the velocity of longitudinal elastic waves in the glass. Theoretical analysis is almost entirely lacking although interesting first steps have been made by Poncelet.[26]

Much could be learned about the fracture process if piezoelectric transducer techniques were to be applied to study the shock waves set up in a piece of glass during the time that the fracture is taking place.

Some interesting work in this direction has just been published by Miklowitz.[27] We close our brief discussion with some additional references to measurements of the speed of fracture propagation.[28]

---

[26] C.F. Poncelet, Am. Inst. Mining Met. Engrs., Tech. Publs., 1684 (1944). See brief review on p. 58 of Haward's *The Strength of Plastics and Glass*, reference 4(b). Also F.W. Preston, Glass Ind. **32**, 284 (1951).

[27] J. Miklowitz, J. Appl. Mech. **20**, 122 (1953).

[28] H. Schardin, Glastech. Ber. **23**, 325 (1950); H.J. Krug, Glastech. Ber. **18**, 70 (1940); E. Rexer, Glastech. Ber. **17**, 207 (1939); H.M. Dimmick, J. Soc. Glass Technol. **35**, 318 (1951).

Physics of the Glassy State
IV: Radiation-Sensitive Glasses*

EDWARD U. CONDON

Corning Glass Works
Corning, New York

Effects of radiation in producing color changes, often called solarization, in glass are described, in close analogy with modern work on F centers especially in alkali-halide crystals. Photosensitive glasses in which the action of radiation is to produce centers of nucleation around which colloidal or crystalline precipitates can form are also discussed.

I. Light Quanta

For most purposes the relation of light to glass as an optical medium is fully described by giving its index of refraction as a function of wavelength. If the light is absorbed this is described by giving a complex index of refraction, the imaginary part being related to the absorption of the light by the glass. Usually the radiant energy that is absorbed in an optically absorbing medium goes into heat and produces, for sources of moderate intensity, only a slight warming.

But sometimes more profound changes of a photochemical kind are produced by absorption of light. This is what happens when dyed materials fade or when a photographic emulsion is affected by light. Such effects are also produced in certain kinds of glass, giving rise to a new way of studying glass and to a new range of important possible applications. This relatively new study of the *photochemistry of glass* is the subject to be reviewed in the last lecture.

Let us start by recalling that, although light is propagated as a wave motion, it behaves in a corpuscular way when emitted or absorbed by matter. Light of frequency v cycle/sec behaves as if it were a stream of individual energy units called *light quanta, or photons*, containing an amount of energy hv (ergs), where h is Planck's constant, $h = 6.624 \times 10^{-27}$ erg sec. This quantum aspect of electromagnetic radiation manifests itself over a range of many powers of ten

---

* The fourth of a series of lectures delivered at the Fifteenth Annual Colloquium of College Physicists, State University of Iowa, Iowa City, June 20, 1953.

in frequency, from microwaves to the hardest γ rays. Here we shall be principally concerned with visible and ultraviolet light.

Commonly the wavelength $\lambda = c/v$ is given, instead of the frequency v, and for light is expressed either in microns (10^{-4} cm) or angstroms (10^{-8} cm). Visible light ranges approximately from 0.7 micron or 7000 angstroms (red) to 0.36 micron or 3600 angstroms (violet). If λ is in microns the size of a quantum is $(1.98/\lambda) \cdot 10^{-12}$ erg. It is convenient to introduce two other units for quanta. One is the electron volt (ev) which is the energy gained by an electron in falling freely across a potential drop of one volt,

$$e/300 = 1.602 \times 10^{-12} \text{ erg.}$$

This is the energy of a quantum whose wavelength is 1.2395 μ or 12395A, so the energy of a light quantum in electron volts is

$$1.2395/\lambda \qquad \text{or} \qquad 12395/\lambda$$

according as λ is expressed in μ or in A.

The other unit involves expression of a mole of quanta in terms of calories/mole. Since $N = 6.02 \times 10^{23}$ mole^{-1} (Avogadro's number) and $J = 4.185 \times 10^7$ erg/calorie, one mole of quanta of wavelength λ is

$$28.4/\lambda \qquad \text{or} \qquad 2.84 \times 10^5/\lambda \text{ kcal/mole}$$

according as λ is in μ or in A.

Radiations in the visible or ultraviolet thus correspond to quanta whose molal energy equivalents are comparable with chemical bond dissociation energies. For example, the quantum of the prominent 2537 line emitted by the mercury arc corresponds to 4.9 ev or to 112 kcal/mole. Radiations in the infrared correspond to much smaller quanta. The fundamental absorption band associated with the Si—O bond in glasses and crystalline silicates is at about $\lambda = 9\mu$, for which the quantum energy is 0.54 ev or 3.2 kcal/mole. Such small quanta are usually unable to induce electronic transitions—they act directly to change the vibrational motion of the atoms in the absorbing medium.

Ordinary glass is thus transparent to visible light because the ionic vibrations occur at such low frequencies that absorption by them occurs in the infrared, and because the first-excited electronic levels in it lie so high (> 3 ev) above the ground level that absorption by electronic levels occurs first with ultraviolet light. Colored glasses contain materials having electronic levels from 1.8 to 3.4 ev, approximately, above the ground level, which are able to absorb visible light.

The *intensity* of a beam of light ought always to be expressed in units of energy crossing in unit time a unit of area normal to the direction of propagation of the beam. In practice it is more commonly stated in photometric units, in particular the foot-candle. At $\lambda = 0.55\mu$, the wavelength giving maximum visual sensation for unit energy, this corresponds to an energy flux of 17.4 erg/cm^2 sec or 4.87×10^{12} quanta/cm^2 sec. (One foot-candle corresponds to

TABLE 4-I. Wavelength relative-
visibility relation.

$\lambda(\mu)$	$\bar{y}$	$\lambda(\mu)$	$\bar{y}$
0.42	0.0040	0.56	0.9950
0.44	0.0230	0.58	0.8700
0.46	0.0600	0.60	0.6310
0.48	0.1390	0.62	0.3810
0.50	0.3230	0.64	0.1750
0.52	0.7100	0.66	0.0610
0.54	0.9540	0.68	0.0170
		0.70	0.0041

1.08×10^{-3} lumen/cm^2 and the maximum luminous efficiency at $\lambda = 0.55\mu$ is 621 lumens/watt, which is equivalent to 1.61×10^4 erg/sec for one lumen.)

For light of other wavelengths a correspondingly greater energy is needed to give the visual sensation called one foot-candle. The energy flux in erg/cm^2 sec is equal to $17.4/\bar{y}$, where $\bar{y}$ is a factor known as the relative visibility and is listed in Table 4-I.

The flux of quanta/cm^2 sec equivalent to illumination of one foot-candle at wavelength λ in microns is then

$$n = (\lambda/\bar{y})4.87 \cdot 10^{12}.$$

Another comparison of interest is with the dosage required for minimum perceptible erythema, defined as that which produces a slight reddening of the skin of an average untanned individual observable 12 to 24 hours after exposure. The radiation[1] of maximum effectiveness has $\lambda = 0.297\mu$ and for it the dose is $5 \cdot 10^5$ erg/cm^2. This corresponds to $7.52 \cdot 10^{16}$ quanta/cm^2.

Light is absorbed in homogeneous media by the exponential law, $I(x) = I(0)e^{-kx}$, where k is the absorption coefficient in cm^{-1} and x the depth of penetration in cm. This means that a fraction $-(dI/I) = k\,dx$ is absorbed in going a distance dx. If the number of absorbing atoms in unit volume is $n(\text{cm}^{-3})$ and each has an effective cross section for absorption $\sigma(\text{cm}^2)$, then the absorption coefficient is

$$k = n\sigma.$$

The seemingly implied proportionality of k with n is known as Beer's law. It is true to the extent that the structure of the absorbing centers does not change with a change in their concentration. Often there are large departures from Beer's law if the chemical state of the absorbing atoms changes with n. Study of variation of experimental values of k/n with n, is an important tool in the study of photochemical mechanisms.

---

[1] Coblentz, Stair, and Hogue, Proc. Natl. Acad. Sci. U.S. **17**, 401 (1931).

Suppose now a beam of light of intensity I foot-candles, of wavelength λ enters a material whose absorption coefficient is k for a time-interval t seconds. Then the total dosage, expressed as number of quanta absorbed per cm^3 in the exposure is

$$Q = nktI = It\frac{k\lambda}{\bar{y}(\lambda)}4.87\cdot 10^{12}\frac{quanta}{cm^3}.$$

This only applies to the layer in which the intensity is I. As I falls off with depth with a factor e^{-kx}, so also will Q.

The unit for measuring x-ray dosage is the *roentgen*, defined as the amount of radiation which produces 1 esu of charge in 1 cm^3 of air at 0°C and 760 mm. Letting e = charge on electron in esu, 1 esu corresponds to formation of e^{-1} ion pairs, per cm^3; so the number of ion pairs per cm^3 produced by r roentgens is r/e.

If the time of exposure is t sec, I is the intensity of the x-ray beam in erg/cm^2 sec, A is the energy in erg/ion pair needed to produce an ion pair on the average, and $\mu_A(cm^{-1})$ is the linear absorption coefficient of the x-rays in air, then

$$It = rA/\mu_A e \text{ erg/cm}^2.$$

If B is the energy per ion pair in another material, whose linear x-ray absorption coefficient is $\mu_B(cm^{-1})$, and it is exposed to the same dosage, (the same It), then

$$rA/\mu_A e = N_B B/\mu_B,$$

where N_B is the number of ion pair/cm^3 formed in material B, so

$$N_B = (A/B)(\mu_B/\mu_A)(r/e),$$

where $N_A = r/e$, as stated in the foregoing, is the number of ion pairs formed in air for a total dose of r roentgens.

Since with x-rays the mass absorption coefficient is more nearly constant from element to element, it is convenient to write, if ρ is the density in g/cm^3,

$$N_B \doteq C(A/B)\rho_B r(e\rho_A)^{-1} \text{ ion pair/cm}^3,$$

where $C = (\mu/\rho)_B \div (\mu/\rho)_A$.

If E is the energy of an x-ray quantum, the number of ion pairs formed per quantum is (E/B) and therefore the total number of quanta/cm^3 absorbed in material b is

$$Q_B = BN_B/E = C\rho_B r(A/E)(e\rho_A)^{-1} \text{ quanta/cm}^3.$$

For practical applications it is convenient to insert the experimentally determined[2] value $A = 33$ ev or 33/1000 kev and to express E also in kev so

[2] W. V. Mayneard, Rept. Progr. in Phys. **14**, 396 (1951). See also "Report of a discussion on radiation chemistry," The Faraday Society (1952).

TABLE 4-II. Biological effects of x-radiation.

Roetgens (one dose)	Effect
<25	none
25–50	blood changes, no serious injury.
50–100	some injury, no disability.
100–200	possible disability.
200–400	disability certain, death possible.
400	fatal in approximately half the cases.
600	fatal in nearly all cases.

$$Q_B = \frac{C\rho_B r}{E} \frac{33}{e\rho_A} 10^{+3}$$

$$= \frac{C\rho_B r}{E} (5.35 \cdot 10^{10}) \text{ quanta/cm}^3.$$

The ion pairs are not uniformly distributed throughout the volume but occur in clumps, averaging (E/B) ion pairs per clump, each one formed around the site of absorption of an x-ray quantum.

As an orientation on biological effects of x-radiation for a single dose over the whole human body, Table 4-II is useful.[3]

The tolerance dose rate adopted by the U. S. Committee on X-Rays and Radium Protection is 0.3 roentgen/week. On the other hand, highly localized doses up to 1000 roentgens on small parts of the body may be used in cancer therapy.

II. Consequences of Primary Light Absorption

In the study of luminescent materials it is often the case that the ability of the material to absorb the stimulating visible or ultraviolet light is due to the presence of an accidental impurity, or a deliberately added constituent, present in small amount. This active constituent is called an *activator*. For example, pure ZnS shows almost no luminescence, but can be activated by proper incorporation in its lattice of 10^{-4} parts of Au, Ag, Cu, Mn, and some other metals. In such cases the important effects are associated with the electronic properties of what Lenard called *active centers*; that is, the activator atoms and their immediate surroundings. Thus the properties of the active centers depend both on the activator and on the base material which surrounds it. A further complication arises in the fact that some properties depend on the simultaneous presence of two activators. Such materials are said to be *coactivated*; this happens in the solarization of glass as discussed later.

---

[3] Los Alamos Scientific Laboratory *The Effects of Atomic Weapons* (Government Printing Office, Washington, D.C., 1950), Chaps. 8 and 11.

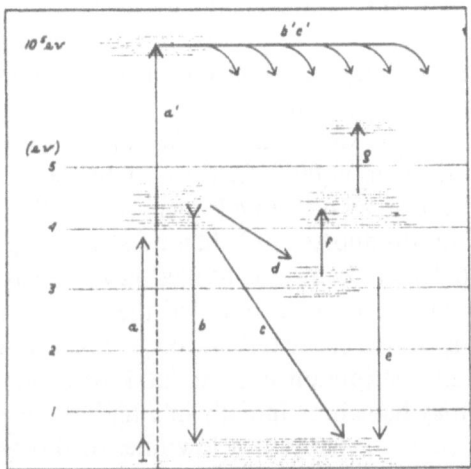

FIGURE 1₄. Scheme of electron transitions in glass.

The primary act of absorption of visible or ultraviolet light raises one of the electrons to a higher-energy level (Fig. 1₄) in a process that is generally accompanied by a change in the vibrational motions of those atoms whose bonding is affected by this change in state of the electron. The nuclear motions thus associated with electron transitions are governed by the Franck-Condon principle,[4] first worked out in 1926 for *diatomic molecules*, and since found to be applicable to polyatomic molecules and to vibrational motions in condensed solids.

For ultraviolet radiation the energy given the electron may not be enough to free it from its local bonds, or if so will not give it more than one or two electron volts in addition to what it takes to free it as in *a* of Fig. 1₄. Therefore he distance it goes before being captured again can hardly be more than a few interatomic distances. It may be captured into some excited levels at *d*, which may be metastable, or it may lose energy by collision in such a way that it is returned essentially to a ground level, as by process *c*, without emitting radiation. In this case the fluorescence is said to have been quenched. With x-rays the primary electron is given some tens-of-thousands of volts of energy as in *a'*. Even so it does not go far, in a macroscopic sense, before it loses this energy and is trapped somewhere. It loses the energy by electron impacts which excite and set free other electrons indicated by *b'* and *c'*. The electrons are trapped so quickly that negligible conductivity results but it seems reasonable to assume the number knocked loose is of the order $E/33$ per quantum, where E is the quantum energy in volts, this being the number of ion pairs produced by x-ray ionization in air where there is no trapping and

---

[4] E.U. Condon, Am. J. Phys. **15**, 365 (1947) reviews its development.

ion recombination is slow enough that the resulting conductivity can be measured.

The excited electron may return at once (within about 10^{-8} sec) to substantially the same electronic level as it came from, but again accompanied by associated changes in the vibrational levels, radiating a suitable light quantum. This is what happens in *fluorescence*. Usually the emitted quantum will be smaller than the incident quantum leaving some vibrational energy in the lattice as heat.[5] Thus the fluorescent light is mostly of longer wavelength than the incident light. This is known as Stokes' law. Probably there are many unreported cases of fluorescence in which the emitted radiation is in the infrared. Minor violations of Stokes' law occur because the initial levels are really a Boltzmann distribution over about 1/40 ev at room temperature and the electron may have been initially in a high thermal state from which it returned by a quantum jump to a low one. In this way the fluorescent quantum can exceed the exciting quantum by about 1/40 ev.

The electron may be *trapped in a metastable excited level*, by making a radiative transition to a high level from which further transition back to the normal state is not allowed as in *d*. By thermal agitation it will have a small probability of returning, by way of allowed radiative transitions to the normal state, (radiative process *e*) and so will give off light for an extended period of time after the incident light is cut off. This is called *phosphorescence*. Since this depends on the electrons passing over an activation energy barrier, the rate at which the stored energy is given off as light will increase rapidly with temperature. Thus it may happen that at room temperature the rate is so low that the material does not seem to give off light and on warming the rate is high enough that it glows brightly for a time. This is *thermoluminescence*.

If the electron is trapped in a metastable level, there is generally another set of higher levels above it to which it may now make allowed transitions by, absorbing light, as in process *f*. Thus a material having a large number of such electrons in metastable states has a different absorption spectrum than the normal material. The normal material may absorb ultraviolet light only, but the metastable electrons may be able to absorb in the visible. This is the basic mechanism by which normally colorless transparent materials may become colored. The metastable trapped electrons (and the surroundings to which they are coupled), which are responsible for the new absorption hands, are called *color centers*.

Absorption of light by the electrons in color centers may put the electrons into still a different class of metastable states giving rise to power to absorb still other new bands, as in process *g*. Thus absorption of light in the color-center band may induce a change which brings in an additional color change in the material; one which could not have been produced by light acting on the original material, because before the color centers were formed it could not absorb visible light.

---

[5] A. Einstein, Ann. Physik **17**, 132 (1905).

Warming the material, in general, will help the electrons trapped at color centers to return to their normal state with or without noticeable thermoluminescence. Thus warming the material will bleach out the color center absorption bands, returning the material to its original colorless condition. Also electrons may be assisted to get out of color centers by direct absorption of infrared light which may therefore act both to increase the brightness of phosphorescence and to bleach the material to its original transparency.

Also the excited electrons in some instances may move comparatively freely through the material giving rise to *photoconductivity*.

The primary act of absorption of ultraviolet light by an electron may thus give rise to a wide variety of interesting and important effects.[6]

Experimental study of these processes has produced an enormous literature of early work, carried out before the basic theoretical picture was clarified. As the effects observed are extremely sensitive to small impurities, and the importance of controlling these was at first not recognized, it is not surprising that in many cases the older literature is almost valueless except for the qualitative indications it gives. Careful studies on crystals of the alkali-halide type with known and controlled added materials and also the modern industrial study of crystalline phosphors recently have done much to remedy what, but a few years ago, was a most unsatisfactory situation.

Even so the quantitative study of radiation sensitivity of glass is still in a very primitive state. Nevertheless some interesting and useful results have been achieved. Stimulation by x-rays and γ rays and also bombardment by high-energy cathode rays or by particles gives rise to phenomena broadly similar to those produced by ultraviolet light, with differences of detail. No attempt will be made here to review this field.[7]

III. Solarization of Glass

By solarization is meant a change in the absorption spectrum of the glass due to prior absorption of light.[8] Usually the light absorption causing the change is in the ultraviolet, and, if the change also occurs in the ultraviolet, the glass may seem to be unaltered so far as visible optical properties are concerned. If

[6] For reviews of this field; (a) N. F. Mott and R. W. Gurney, *Electronic Processes in Ionic Crystals* (Oxford University Press, London, 1948), Chapters 6 and 7; (b) F. Seitz, *The Modern Theory of Solids* (McGraw-Hill Book Company, Inc., New York, 1940), p. 459; (c) F. Seitz, Revs. Modern Phys. **18**, 384 (1946); (d) G. F. J. Garlick, Repts. Prog. in Phys. **12**, 34 (1949); (e) F. Seitz, Revs. Modern Phys. **23**, 328 (1951); (f) Cornell Symposium of American Physical Society, *Preparation and Characteristics of Solid Luminescent Materials* (John Wiley and Sons, Inc., New York, 1948).
[7] K. H. Sun and N. J. Kreidl, Glass Industry **33**, 511, 589, 651 (1952). See also Mayer and Gueron, J. Chem. Phys. **49**, 204 (1952).
[8] For a review, see W. A. Weyl, *Coloured Glasses* (published by Society of Glass Technology, Sheffield, 1951), Chap. 31.

the new absorption band induced by absorption of the ultraviolet is in the visible, then the effect of solarization is to produce a coloration of the glass.

The most common instance of this is in the purple tinting of old windows or of old bottles thrown away on the western deserts, where they lie for years exposed to the ultraviolet rays of the desert sunlight. The effect was first described by Michael Faraday in 1825. In the early days of photography (mid 19th century) when the plates were mainly affected by the "actinic" rays of the near ultraviolet and photographers' studios used natural sunlight, this property of window glass caused difficulty. The explanation of the mechanism of the purple tinting, which in a modernized form is accepted today, goes back to Pelouze in 1867.

Nowadays many such solarizable glasses are known. In every case the basic action is believed to be the absorption of light by an ion in a lower state of oxidation, which frees an electron to go elsewhere in the network; it perhaps being captured in a region of high positive potential which might have held another oxygen ion, or perhaps being captured by another ion of a different kind which is thereby reduced to a lower state of oxidation. Schematically, the light is absorbed by ion A in a reduced state which is thereby oxidized, the electron then being captured by ion B oxidized which is thereby reduced.

The process is shown schematically in Fig. 2_4, where A_r and A_o stand for the relatively reduced and oxidized states of A, respectively, such as Fe^{++} and Fe^{+++} in the case of iron. When the electron is captured by B_o in the more oxidized state there is a binding energy to be given up which may happen by fluorescence or by direct loss to vibrational modes of the network (symbolized by solid and dotted lines, respectively.)

The essential things are (a) the presence in the glass of ions which can be further oxidized by absorption of the light in question and (b) some means of capturing the ejected electron *other* than having it be captured by some A_o ion which would thereby be reduced to the original A_r ion leaving no net change.

From this simple picture several consequences follow:

(a) If the base material can capture electrons readily, then the light can change the state of oxidation of A without the presence of B.

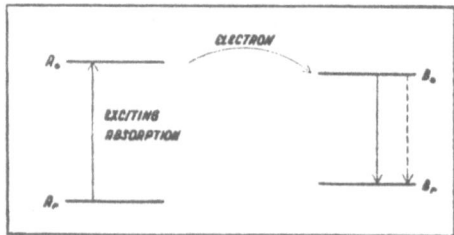

FIGURE 2_4. Solarization mechanism.

(b) The exact optical properties of A_o depend on its coordination to surroundings and so may be different from those obtained from other A_o ions originally present in the melt; i.e., the light generates A_o ions having the network ties appropriate to A_r ions which may be different than the network ties usually appropriate to A_o ions.

(c) If the glass contains both A_o and A_r ions in metastable equilibrium and the total concentration of A is increased enough, then the freed electrons may be captured by other A_o ions present, thus nullifying the over-all effect, or at least diminishing the number of electrons that are captured by B_o. If B capture is responsible for fluorescence, this means that increased concentration of A, whose presence is necessary for fluorescence due to B, will result in a diminished amount of fluorescence.

These principles are illustrated in many combinations and, of course, can be complicated by, the simultaneous presence of several kinds of polyvalent ions, A, B, C ... in the glass.

Three specific examples may be mentioned:

(1) The purple tint of glass containing Mn and Fe, already mentioned, is due to the reaction $Mn^{++} \rightarrow Mn^{+++}(e) Fe^{+++} \rightarrow Fe^{++}$. The resulting increased Mn^{+++} concentration gives the purple tint. At the same time light acting on the Fe^{++} ions produces more Fe^{+++} which is believed to be the cause of decrease in the ultraviolet transmission of such glasses.

(2) Glasses containing Ce and As solarize readily, developing yellow or even dark brown coloration at larger cerium concentrations.[9]

   In Ce containing glasses, both Ce^{3+} and Ce^{4+} ions are present. The first act of light absorption gives $Ce^{3+} \rightarrow Ce^{4+}$ and a free electron. Some of the electrons are either not completely freed or return to the Ce^{++} ion to emit a blue fluorescence which extends into the ultraviolet to about 3000A. If As is also present, electrons can be captured by the As^{5+} ions resulting in yellow-brown coloration.

(3) Glasses containing Ce and V show a striking color change on solarization. The electrons are freed by light absorption in Ce^{3+}, as before. The initially present V^{3+} ions give the glass a green color. The freed electrons reduce these to V^{2+}, first producing a grey and finally a purple color as the reduction becomes essentially complete.

All these solarization effects can be reversed by heat treatment, to some extent by gentle warming, but usually requiring 400 or 500°C for anything like complete restoration of the properties of the original glass. Reversal can in some cases also be caused by red or infrared radiation indicating that the trapped electrons are in such cases not as tightly held as in the original ion from which they were released. This indicates that the actual net effect produced when glass is exposed to all the frequencies in natural sunlight is due to a balance of different tendencies produced by different wavelengths.

---

[9] F. Eckert and K Schmidt, Glastech. Ber. **10**, 80 (1932).

IV. Photosensitive Glasses

Usually the color changes produced by solarization are not very deep, corresponding more to a tinting than a coloring of the glass. In recent years new principles have been discovered in the research laboratories of Corning Glass Works largely by Stookey[10] building on earlier work by R. H. Dalton and W. H. Armistead. These are opening up interesting new research and application possibilities.

The photosensitive glasses are based on the general principle that changes in the state of oxidation of certain metal ions contained in the glass may make it possible for the active centers so formed to act as nuclei for crystal growth. These may, on subsequent heat treatment, coagulate to form colloidal metal particles. These can give colloidal colors of greater density than those produced in ordinary solarized glass, and permit the making of photographic prints of good quality in the glass itself. Because the process involves coagulation into colloidal particles, it is not reversible on heating in the same sense that solarization is reversed by heat, that is, by electron transfer. At sufficiently high temperature, prolonged heating, however, will dissolve the colloidal particles in the glass, destroying the picture, and making it possible to use the same piece of glass for another picture.

A small concentration of cerium oxide is used in a glass which may contain also Ag, Au, or Cu ions. For definiteness consider the case of copper. The glass is melted under mildly reducing conditions. If strongly oxidized the stable blue Cu^{++} ions are formed which cannot be reduced to the metallic state by heat treatment, while, if strongly reduced, metallic copper forms in the melt giving a brown cloudy glass.

The most effective radiation for exposure is in the range 3000–3500A and exposures of the order of 2 milliwatt-minutes/cm^2 are used. This corresponds to an exposure of about $2 \cdot 10^{17}$ quanta/cm^2, roughly of the order of magnitude of the total number of cerium ions contained in an area of 1 cm^2 and a thickness of a few millimeters.

It is believed that the electrons liberated by the oxidation $Ce^{3+} \rightarrow Ce^{4+}$ are captured by the cuprous ions to reduce them to neutral copper atoms, $Cu^+ \rightarrow C$. This constitutes the formation of the latent image.

Development consists of holding the glass at a temperature between its annealing and softening temperatures. The development time decreases logarithmically with increase in temperature, but not as rapidly as does the viscosity of the same glass in this range, indicating a lower activation energy for migration of the atoms toward each other than is involved in viscous flow. A rough calculation from curves in Stookey's paper[6b] indicates an activation energy of 105 kcal/mole for viscous flow and only 40 kcal/mole for the

[10] S. D. Stookey, (a) J. Phot. Soc. Am. **14**, 399 (1948); (b) Ind. and Eng. Chem. **41**, 856 (1949); (c) J. Am. Ceram. Soc. **32**, 246 (1949); (d) Pages 697–706, Vol. 7 of Alexander's *Colloid Chemistry* (D. Van Nostrand Company, Inc., New York, 1950).

development process. This roughly parallels the comparatively low activation energy found for ion mobility in electrolytic condiction in glass as compared with that for viscosity, noted in the second lecture. In practice one carries out the development at as high a temperature as possible without deforming the glass, usually close to 580°C. In passing it may be noted that the glass after exposure but before development shows the printed picture very faintly. This is attributed to the occurrence of solarization by some other mechanism, for this faint image clears up quickly when the glass is warmed to less than 400°C and does not recur if the glass is cooled from that temperature without development.

V. Photosensitive Opal Glasses

Opal glasses are of many kinds, but most of them depend on some constituent of the glass becoming supersaturated and precipitating out in a cloud of minute crystals when the glass is cooled. The various phenomena familiar in other connections of tendencies toward supersaturation and the need for crystallization nuclei to start crystal growth are also to be found here.

Stookey finds that he can take certain opal compositions and sensitize them by the same means as described for photosensitive glasses. They then have the property that the opal phase forms, on suitable heat treatment, in the part that has been exposed to ultraviolet light, but the glass remains clear and transparent in the part that has not been exposed. This material is now in commercial production for various lightdiffusing illuminating fixtures and other uses.

The basic mechanism of formation of the latent image is presumed to be the same as before. Initial heat treatment to nearly 600°C is used to develop the metallic nuclei. It is then necessary to drop the temperature to about 520°C to start the growth of the crystals of the opalizing phase on these metallic nuclei. After this the temperature is again raised to 550°C to speed up crystal growth in the opal phase.

Still another ramification of importance in applications occurred with the discovery[11] of an additional property of some of these photosensitive opal glasses. It is that the opalized phase dissolves much more rapidly in dilute hydrofluoric acid than does the glass in which the opal phase is not developed. This makes possible the precise forming of glass articles by photoengraving techniques, including in fact the making of glass half-tone engravings of good quality for printing. This method of making engravings for printing will have distinct advantages in some fields. The chemically machinable glass is also of importance for other applications such as the aperture masks in one type of color television tube.

---

[11] S. D. Stookey, Ind. Eng. Chem. **45**, 115 (1953).

Conclusion

In these lectures I have tried to indicate to you that there is a great deal of scientific interest in the problems connected with trying to understand the structure and properties of glass. Because of the limited time it has been necessary to deal with only a few of the topics and with these only rather superficially. I hope that some of you will be stimulated by this inadequacy of the discussion to delve more deeply and be richly rewarded by learning for yourselves about the many fascinating and still unsolved problems of the Physics of the Glassy State.

Intensity-Dependent Absorption of Light[*]

EDWARD U. CONDON

Joint Institute for Laboratory Astrophysics
University of Colorado
Boulder, Colorado

Interesting observations on intensity-dependent absorption of light have been reported by Bret and Gires,[1] who have not given a theoretical interpretation of their results. This note provides a simple discussion of the probable mechanism of the effects observed as being due to optical pumping by the absorbed light.

Their work was done with a Q-spoiled ruby laser which provides flashes of $\lambda 6943$ red light in pulses up to an intensity of 10^7 watt/cm^2. With a photocell, they measured the incident intensity, I_0, and transmitted intensity, I, varying I_0 with suitable attenuating absorbers in the path of the beam from the laser. The measurements were made on a 3-mm thick Schott RG10 filter which was found to give $I/I_0 = 0.002$ at $I_0 < 10^3$ watt/cm^2.

For $I_0 > 10^3$ watt/cm^2 the transmittance, I/I_0, was found to increase gradually over several powers of ten on I_0 until it had reached a value of 0.40 at the largest value of I_0, namely, 10^7 watt/cm^2. On another sample a transmittance of 10^{-6} at low intensities increased to 0.35 at $I_0 = 10^8$ watt/cm^2.

In the usual linear theory of light absorption one assumes an absorption cross section σ(cm^2) at each of n absorbing atoms per cm^3 giving an absorption coefficient α(cm^{-1}) $= \sigma n$, or in the heterogeneous case of absorption due to several kinds of absorbing centers $\alpha = \sum_i \sigma_i n_i$. The intensity is reduced by the relation $dI/dz = -\alpha I$ in the material, leading to the integrated relation $I = I_0 e^{-\alpha z}$ for the transmission through a path length z of the absorber.

The usual theory neglects the population change produced by light absorption itself. For simplicity let the absorbers have but two states, 0 and 1, and write n_0 and n_1 for the number per unit volume of absorbers in the lower state and the upper state, respectively. For material in thermal equilibrium at low temperatures and for visible light, $h\nu \gg kT$ so $n_1 \ll n_0$ and is neglected.

* This research was supported in part by the Advanced Research Projects Agency (PROJECT DEFENDER), and was monitored by the U.S. Army Research Office (Durham) (contract DA-31-124-ARO[D]-139).

1 Bret, G., and F. Gires, *Compt. Rend. Acad. Sci.*, **258**, 3469 (1964); Bret, G., and F. Gires, *Appl. Phys. Letters*, **4**, 175 (1964).

When n_1 is not negligible, one ought to write $\alpha = \sigma(n_0 - n_1)$ to take into account the coherent stimulated emission or negative absorption processes discovered by Einstein.[2] Each photon that is absorbed lifts an absorber from state 0 to state 1. Some are returned from 1 to 0 by negative absorption. Others return by spontaneous emission, or by radiationless processes due to interaction with the lattice, or by molecular collisions in the case of a gas. Let $\beta(\text{sec}^{-1})$ be the total rate coefficient for all such $1 \rightarrow 0$ processes other than coherent negative absorption.

In the steady state, under intensity I, the populations n_0 and n_1 are governed by balance of upward and downward transition rates, so

$$\frac{I\sigma}{h\nu}(n_0 - n_1) = \beta_1.$$

It is convenient to introduce a light intensity, K, defined by the absorber parameters as

$$K = \beta h\nu/2\sigma,$$

so we find

$$\sigma(n_0 - n_1) = \frac{\alpha K}{I + K},$$

in which $n = n_0 + n_1$ and α is written for the absorption coefficient which applies for $I \ll K$.

At higher intensities the absorption of light provides its own optical pumping and builds up an appreciable value of n_1 and thus reduces the absorption coefficient by the factor $K/(I + K)$. Thus, K is that light intensity for which the absorption coefficient is half its low intensity values. Ih the heterogeneous case different kinds of absorbers are characterized by $\alpha_i = \sigma_i n_i$ where n_i is the total density of kind i, and these are also characterized by different values β_i and hence K_i. Accordingly, the differential equation for light absorption becomes

$$\frac{dI}{dz} = -\left(\sum_i \frac{\alpha_i K_i}{I + K_i}\right)I.$$

For $I \ll K_i$, this reduces to the usual form, whereas for $I \gg K_i$, the steady state is one in which there is so much optical pumping that the material becomes nearly transparent.

This equation is readily integrated when the α_i and K_i are known. For the homogeneous case we find

$$\alpha z = -\ln(I/I_0) + (I_0/K)[1 - I/I_0]$$

which shows that when I_0/K is not negligible, it is necessary to go to greater thicknesses of absorber to produce a given transmittance, I/I_0.

---

[2] Einstein, A., *Physik Z.*, **18**, 122 (1917).

A quantitative discussion shows that in the homogeneous case the change-over from low transmittance to higher transmittance occurs essentially in one decade of change of I_0. The results of Bret and Gires for the CdSe filter glass show that the change occurred over about four powers of ten. This is a strong indication of a heterogeneous set of absorbers characterized by a spectrum of K_i values. It also indicates that a much more rapid changeover can be obtained with more homogeneous materials.

Although other instances of nonlinear optical effects have been reported,[3] this one seems to have the simplest mechanism. Effects of this kind will need to be considered in studying the photochemical behavior of high-intensity light beams and will probably find useful application in light-switching devices. Some materials may be found having much lower values of K, and for these the intensity-dependent absorption may become appreciable at intensities of a more usual order of magnitude.

---

[3] Franken, P.A., and J.F. Ward, *Rev. Mod. Phys.*, **35**, 23 (1963).

Spin-Orbit Interaction In Self-Consistent Fields

E.U. Condon and H. Odabasi

Joint Institute for Laboratory Astrophysics
University of Colorado
Boulder, Colorado

Although much effort has gone into the calculation of self-consistent fields by the Hartree, Hartree-Fock, and Hartree-Fock-Slater methods (*l*) since the advent of automatic digital computers, little attention has thus far been paid to the use of such solutions for fields and radial functions to calculate spin-orbit interaction parameters. This paper is a report on some calculations we are making on this subject.

In the older accounts (2) the magnetic spin-orbit interaction is represented by a term in the Hamiltonian

$$\mathscr{H}^I = \frac{e^2 \hbar^2}{2m^2c^2} \sum_{i=1}^{N} \xi(r_i) \mathbf{L}_i \cdot \mathbf{S}_i, \tag{1}$$

the sum being over the N electrons in the atom or ion. Here $\mathbf{L}_i$ and $\mathbf{S}_i$ are the orbital and spin-angular momenta of each electron in units of $\hbar$, and $\xi(r_i)$ is commonly taken as

$$\xi(r_i) = \frac{1}{r_i} \frac{\partial U}{\partial r_i}, \tag{2}$$

in which $-e^2 U(r_i)$ is the potential energy at distance r_i from the nucleus of the effective central field in which the ith electron moves.

Measuring energy in Rydbergs, $e^2/2a$, and length in atomic units, $a = \hbar^2/me^2$, the coefficient in front of the sum in Eq. (I) becomes α^2 where $\alpha = e^2/\hbar c$, the fine structure constant. That is,

$$\mathscr{H}^I = \sum_i \alpha^2 \xi(r_i) \mathbf{L}_i \cdot \mathbf{S}_i. \tag{1'}$$

Calculation of matrix elements of $\mathscr{H}^I$ in the SLM_SM_L scheme of zero-order states leads to (2, p. 196)

$$(\gamma SLM_SM_L|\mathscr{H}^I|\gamma SLM_SM_L) = \zeta(\gamma SL)M_SM_L$$

$$= \sum_i \zeta_{nl}^{(i)} m_{si} m_{ii} \tag{3}$$

in which (2, p. 122)

$$\zeta_{nl} = \int_0^\infty \alpha^2 \xi(r) P_{nl}^2(r) \, dr \tag{4}$$

and $P_{nl}(r)$ is written for r times the radial wave function as in Ref. (1).

In Eq. (3) the sum of m_{si} and m_{li} over the states of electrons in a closed shell gives zero, so the sum can be restricted to the electrons not in closed shells.

When hydrogenic radial functions are used in Eq. (4), the integral can be evaluated exactly to give

$$\zeta_{nl} = \frac{\alpha^2 Z^4}{n^3 l(l + \frac{1}{2})(l + 1)} \qquad (l \neq 0). \tag{5}$$

In the pre-quantum-mechanical period much of the empirical analysis of spin-orbit interaction was expressed in terms of a semiempirical formula due to Landé (3) which can be obtained from Eq. (5) by replacing n by n^* in which n^* is the effective total quantum number calculated from the observed term energy by $E_{nl} = -Z_0^2/2n^{*2}$ and by replacing Z^4 by $Z_i^2 Z_0^2$ in which $Z_0 = (Z - N + 1)$ and Z_i is an empirically adjusted effective nuclear charge for the inner part of the orbital.

For Russell-Saunders terms the spin-orbit factors $\zeta(\gamma SL)$ can be expressed in terms of the ζ_{nl} for electrons outside closed shells in the configuration γ. The contribution of spin-orbit interaction energy $\mathcal{H}^I$ to the energy levels is then (2, p. 194)

$$E^I(\gamma SLJ) = \tfrac{1}{2}\zeta(\gamma SL)[J(J + 1) - L(L + 1) - S(S + 1)]. \tag{6}$$

This result gives the theoretical basis of the Landé interval rule according to which inside the same term $E(\gamma, S, L, J) - E(\gamma SL, J - 1) = \zeta(\gamma SL)J$, and so is proportional to the higher of the two J values involved. This result affords the basis for estimation of empirical values of the $\zeta(\gamma SL)$ from observed spectra, from which one obtains empirical values of the ζ_{nl} by using the theoretical connections between the $\zeta(\gamma SL)$ and the individual ζ_{nl}.

In the simplest case of doublet spectra due to a single electron outside of closed shells the doublet interval gives the ζ_{nl} directly,

$$E(\gamma^2 L_{l+1/2}) - E(\gamma^2 L_{l-1/2}) = (l + \tfrac{1}{2})\zeta_{nl}. \tag{7}$$

In other cases the $\zeta(\gamma SL)$ are expressed in terms of the ζ_{nl}. Examples of such relations are given in Chap. VII of (2) and a more complete collection of them is given by Edlén (4).

Before extensive calculations of self-consistent fields became possible, little use could be made of Eq. (4) to compute theoretical values of the ζ_{nl} because of lack of knowledge of the $U(r)$ and of the radial wave functions $P_{nl}(r)$. Now an abundance of such material is available, especially through the work of Herman and Skillman (5). They have calculated the effective central fields and the radial wave functions for the ground-state configurations of all the neutral atoms from $_2$He to $_{103}$Lw. They have also calculated the ζ_{nl} involved in the ground states of the elements of even Z. Dependence on Z is sufficiently smooth that those for odd Z can be found by interpolation.

In the Herman and Skillman work the same radial potential energy function $V(r)$ is used for each electron. This is defined as

$$V(r) = -2(Z - N + 1)/r, \qquad r \geqslant r_0,$$

$$= V_0(r), \qquad r \leqslant r_0, \tag{8}$$

where r_0 is defined as that value of r for which these two expressions are equal, a procedure originally introduced by Latter (6), and

$$V_0(r) = -\frac{2Z}{r} + \frac{2}{r}\left[\sum_{nl} \omega_{nl} \int_0^r P_{nl}^2(s)\,ds\right] + \sum_{nl} \omega_{nl} \int_r^\infty \frac{P_{nl}^2(s)}{s}\,ds - \left[\frac{81}{8\pi}\rho(r)\right]^{1/3}. \tag{9}$$

Here ω_{nl} is the number of electrons occupying the nl orbital in the ground state so that

$$\sum \omega_{nl} = N \tag{10}$$

and $\rho(r)$ is the total charge density

$$\rho(r) = (4\pi r^2)^{-1} \sum \omega_{nl} P_{in}^2(r). \tag{11}$$

The term in Eq. (9) involving $\rho(r)$ is Slater's effective potential energy correction for the exchange terms of the Hartree-Fock equations (1, Vol. 2, pp. 10–14). Wave functions based on the use of Eq. (9) in (8) are customarily referred to as the Hartree-Fock-Slater approximation.

Herman and Skillman use the $V(r)$ so defined in determining the energy parameters E_{nl} and the radial wave function $P_{nl}(r)$ for each type of occupied orbital in the ground states of the neutral elements. Their same computer program can also be used for the ground-state configuration of various stages of ionization for which $N < Z$.

For a given (Z, N) one can also calculate the radial wave functions for excited states. Strictly speaking, this ought to involve a complete recalculation of the $P_{nl}(r)$ for all of the electrons, because excitation of one of them alters $V_0(r)$ which alters each $P_{nl}(r)$. However, experience has shown, as was first pointed out to us by R.N. Zare, that the gain in accuracy is too small to justify the considerable extra computing effort. Therefore, we use the $V(r)$ as found by a self-consistent solution for the ground state also to determine the $P_{nl}(r)$ for excited states. This approximation has the added advantage that the $P_{nl}(r)$ for excited states that are so determined form an orthonormal set with that of the ground state.

Whether fully justifiable or not, what we have done for excited states of the (Z, N) ion is to use the $V(r)$ determined by Eq. (8) using the occupation numbers ω_{nl} that are appropriate to the ground configuration of that ion.

We consider next the question of what $U(r)$ should be used in Eq. (2) for calculation of the ζ_{nl} by (4). Herman and Skillman simply identify $U(r)$ with

$V(r)$ to obtain the ζ_{nl} which they tabulate on pages 2–6 through 2–16 of their book. We followed the same procedure in calculating ζ_{2p}, ζ_{3p} in the $N = 3$(Li) isoelectronic sequence, and ζ_{3p}, ζ_{3p} in the $N = 11$ (Na) sequence, and also for ζ_{3d} for both of these sequences.

But there is considerable question as to whether this procedure represents a good approximation because $V(r)$ includes all N of the electrons and even includes the $\rho^{1/3}$ term which Slater introduced to represent the effective central field of the exchange energy.

Instead of starting with Eq. (1) as is usually done, it seems more appropriate to follow the discussion given by Slater (1, Vol. 2, Chap. 24), which leads to the conclusion (his Eqs. 24–18) that the spin-orbit interaction energy is a sum over the $i = 1, \ldots, N$ electrons of

$$\alpha^2 \mathbf{S}_i \cdot \left[\frac{Z\mathbf{r}_i}{r_i^3} - \sum_j' \frac{\mathbf{r}_i - \mathbf{r}_j}{r_{ij}^3} \right] \times (-i\nabla_i), \tag{12}$$

in which the prime on $\sum'$ calls for the omission of the $i = j$ term. The quantity in brackets is the electric field at the ith electron due to the combined influence of the nucleus and the other $(N - 1)$ electrons.

The contribution of closed shells to this is spherically symmetrical. We may also make the usual spherical average approximation for the other electrons. Accordingly the field at r_i, due to the nucleus and the other electrons, is $Z_{fi}(r_i)/r_i^2$ in which

$$Z_{fi} = Z - \sum_{nl}' \omega_{nl} \int_0^{r_i} P_{nl}^2(s)\, ds, \tag{13}$$

since, from elementary electrostatics, the field at r_i due to a spherical charge distribution is determined by the total charge within a sphere of radius r_i.

This treatment leads to an altered spin-orbit interaction term, in place of Eq. (1),

$$\mathcal{H}^I = \sum_i \frac{\alpha^2 Z_{fi}(r_i)}{r_i^3} \mathbf{L}_i \cdot \mathbf{S}_i. \tag{14}$$

This is close to the usual form but with the ζ_{nl} to be calculated by

$$\zeta_{nl} = \int_0^\infty \frac{\alpha^2 Z_{fi}(r)}{r_i^3} P_{nl}^2(r)\, dr, \tag{15}$$

instead of (4). Eq. (15) has a more secure theoretical basis than (4) when U is replaced by V_0, and so we have also calculated some values of ζ_{nl} this way.

We have calculated a value of ζ_{nl} for several values of (nl) and for several isoelectronic sequences of low values of N, both by Eq. (4) and by Eq. (15). We have also made a systematic collection of the corresponding "observed" values of these ζ_{nl}'s by calculating them from the doublet intervals. In the case of optical spectra the observed intervals are taken from the compilation of Moore (7), and in the case of X-ray spectra from the compilation of Sandström (8).

TABLE 1. ζ_{2p} (in Rydbergs)

$N = 3$

Z		Observed	Calculated (4)	Calculated (15)
3	Li	0.000002066	0.00000710	0.00000562
4	Be	0.000040096	0.00008654	0.00007432
5	B	0.00020716	0.00036336	0.00032438
6	C	0.000650644	0.00101224	0.00092537
7	N	0.001569808	0.00226348	0.00209982
8	O	0.003234995	0.0043939	0.00412358
9	F	0.005935381	0.00774354	0.00732525
10	Ne	—	0.01268848	0.01208730
11	Na	0.016099036	0.01969656	0.01884492
12	Mg	0.02458596	0.02920308	0.02808797
13	Al	0.03596464	0.0418719	0.04035757
14	Si	0.05097016	0.05822746	0.05624969
15	P	0.07004599	0.07888952	0.07641298
16	S	—	0.10453448	0.10154908

$N = 5$

Z		Observed	Calculated (15)
5	B	0.000097202	0.00017107
6	C	0.000388807	0.00060938
7	N	0.001060106	0.00153294
8	O	0.002348029	0.00319950
9	F	0.0045320306	0.00591982
10	Ne	0.007994842	0.01005795
11	Na	0.012994656	0.01603099
12	Mg	0.020072157	0.02430914
13	Al	0.029707278	0.03541674
14	Si	0.0424650	0.04992909
15	P	0.058928548	0.06847842
16	S	0.080592382	0.09174777
17	Cl	—	0.12047313
18	Ar	—	0.15544507

$N = 9$

Z		Observed	Calculated (15)
9	F	0.002454344	0.00337137
10	Ne	0.004750734	0.00619540
11	Na	0.008286447	0.01046786
12	Mg	0.0135231904	0.01661275
13	Al	0.0208983714	0.02510470
14	Si	0.030983051	0.03647289
15	P	0.0441538848	0.05129898
16	S	0.061540844	0.07021835
17	Cl	0.082621468	0.09392031
18	Ar	0.109734674	0.12314656
19	K	—	0.15869125
20	Ca	—	0.20140271
21	Sc	—	0.25217713
22	Ti	—	0.31197428

E.U. Condon and H. Odabasi

The results of this work are shown in Tables 1–5 and also in Figs. 1–7. In Table 1, the observed and calculated values of ζ_{2p} in Rydbergs are given for the isoelectronic sequences $N = 3$, 5, and 9. In the $N = 3$(Li) sequence, the $P_{2p}(r)$ is an excited state in the configuration $1s^2 2p$, the ground state being $1s^2 2s$. So, as already mentioned, the $P_{2p}(r)$ is the one obtained by solving for $P_{2p}(r)$ using the potential function $V(r)$ derived from the self-consistent field for the $1s^2 2s$ configuration.

In the $N = 5$(B) sequence, the $P_{2p}(r)$ refers to the ground configuration, $1s^2 2s^2 2p$ and in the $N = 9$ sequence, it also refers to the ground configuration which is $1s^2 2s^2 2p^5$, giving an inverted ^2P ground term. In Table 1 the values as calculated both by Eq. (4) and by (15) are given, showing that (15) gives somewhat better agreement with observed values than (4), although the differ-

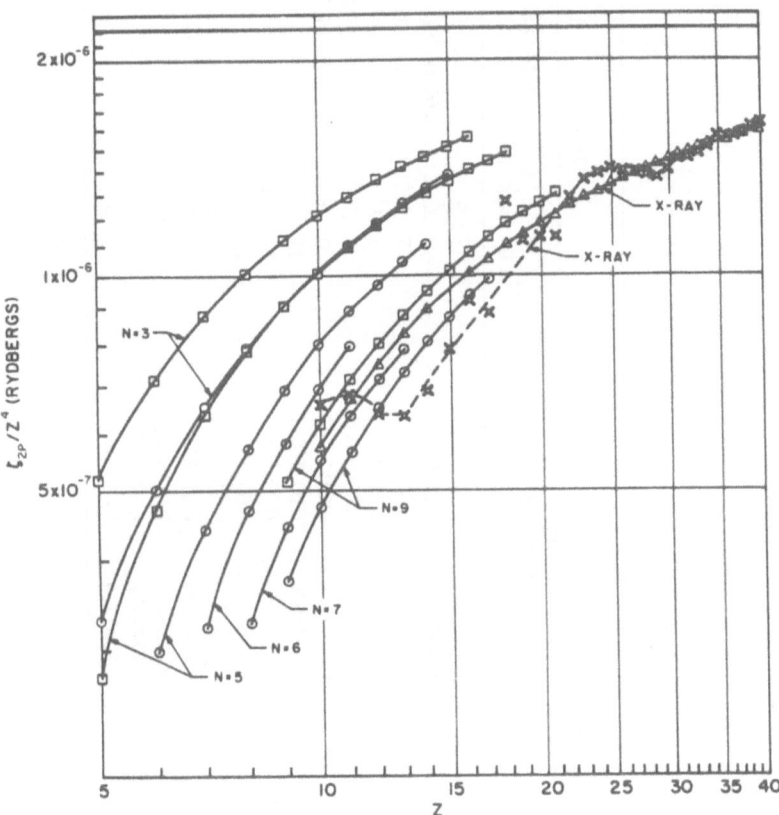

FIGURE 1. ζ_{2p}/Z^4 (Rydbergs) plotted against Z (loganthmic scales). Top of figure is hydrogenic value, $\alpha^2/24$. Squares are calculated, circles observed values for optical spectra, by number of electrons, N, in isoelectronic sequences. Triangles are calculated values and crosses are observed values of the $L^{II} - L^{III}$ doublet interval due to the $(2p)^5$ configuration in X-ray spectra for which $N = Z - 1$.

TABLE 2a. ζ_{3p} (in Rydbergs)

$N = 3$

Z	Observed	Calculated (4)	Calculated (15)
3 Li	—	0.00000230	0.0000018
4 Be	0.000010935	0.00002688	0.00002291
5 B	0.000061966	—	0.00009828
6 C	0.0001913659	—	0.00027797
7 N	0.000450773	0.00067916	0.00062780
8 O	0.000951362	0.00131386	0.00122942
9 F	0.0017131804	0.00231078	0.00218011
10 Ne	—	—	0.00359309
11 Na	0.0046292322	—	0.00559728
12 Mg	0.0079766167	—	0.00833753
13 Al	0.010631439	0.01244394	0.01197438
14 Si	—	0.0172972	0.01668421
15 P	0.03499262	0.02342732	0.02265894
16 S	—	0.03103220	0.03010665

$N = 5$

Z	Observed	Calculated (15)
5 B	—	0.00002017
6 C	0.000067616	0.00009741
7 N	0.0002187039	0.00028567
8 O	0.0005291419	0.00065386
9 F	0.0010752941	0.00128662
10 Ne	0.0017557062	0.00228435
11 Na	—	0.00376329
12 Mg	—	0.00585538
13 Al	—	0.00870839
14 Si	—	0.01248547
15 P	—	0.01736639
16 S	—	0.02354649
17 Cl	—	0.03123594
18 Ar	—	0.04066303

$N = 9$

Z	Calculated (15)
9 F	0.00008458
10 Ne	0.00038563
11 Na	0.000097265
12 Mg	0.00194808
13 Al	0.0034283
14 Si	0.00556395
15 P	0.00849491
16 S	0.01239577
17 Cl	0.01745087
18 Ar	0.02386166
19 K	0.03184432
20 Ca	0.04163288
21 Sc	0.05347656
22 Ti	0.06763368

TABLE 2b. ζ_{3p} (in Rydbergs)

| | | $N = 11$ | |
| | | Calculated | |
Z	Observed	(4)	(15)
11 Na	0.0001044694	0.00011910	0.00011287
12 Mg	0.0005561761	0.00064770	0.00061754
13 Al	0.001412463	—	0.00158117
14 Si	0.0027963722	0.00323256	0.00310858
15 P	0.0048272808	0.00553160	0.00533591
16 S	0.0076728614	0.00870418	0.00841872
17 Cl	0.011481954	0.01292626	0.01253175
18 Ar	0.016512143	—	0.01786591
19 K	0.022878856	—	0.02462716
20 Ca	0.030867624	0.0338976	0.03304006
21 Sc	0.040642472	0.04440932	0.04334373
22 Ti	0.052610435	0.05709084	0.05579154
23 V	0.067312196	0.07222646	0.0706555
24 Cr	0.084565502	0.0900834	0.0882197
25 Mn	0.104491856	0.11099286	0.10878599
26 Fe	0.127395014	0.13524558	0.13267676
27 Co	0.15418624	0.16319978	0.16021989
28 Ni	0.184683282	0.19521012	0.19177413
29 Cu	0.219311398	0.23160832	0.22770043
30 Zn	—	—	0.26837353

ence of the two methods is rather small. For $N = 5$ and 9 we calculated the ζ_{nl} values only on the basis of Eq. (15).

In every case the ζ_{nl} is smaller than the hydrogenic value (5) because of the screening by the other electrons. As the hydrogenic values increase with Z^4, we found it convenient to exhibit the numerical results graphically by plotting ζ_{nl}/Z^4 against Z, choosing the hydrogenic value for the top of the figure in each case. Figure 1 is such a plot for ζ_{2p}/Z^4 against Z on log-log scales, squares showing values inferred from observed intervals. The curves connect points of the same isoelectronic sequence.

The general trend is as expected: for constant N increasing Z gives approach toward the hydrogenic value, and for constant Z increasing N gives a decrease of ζ_{2p} as more electrons produce more screening. Figure 1 includes the observed values for the $N = 6$ and 7 sequences showing clearly how they fit in with the others, but we did not calculate theoretical ζ_{2p} for them.

Table 2 and Fig. 2 are similar presentations of the results for ζ_{3p}. Here the doublet in question for $N = 3$, 5, and 9 corresponds to excited terms. Observations are lacking for $N = 9$ where the ζ_{3p} would have to be inferred from the complex $2p^4 3p$ structure. For the $N = 11$(Na) sequence we calculated ζ_{3p} by both methods, and again we see that Eq. (15) gives better agreement than (4). Also here we see that the agreement of the calculated and

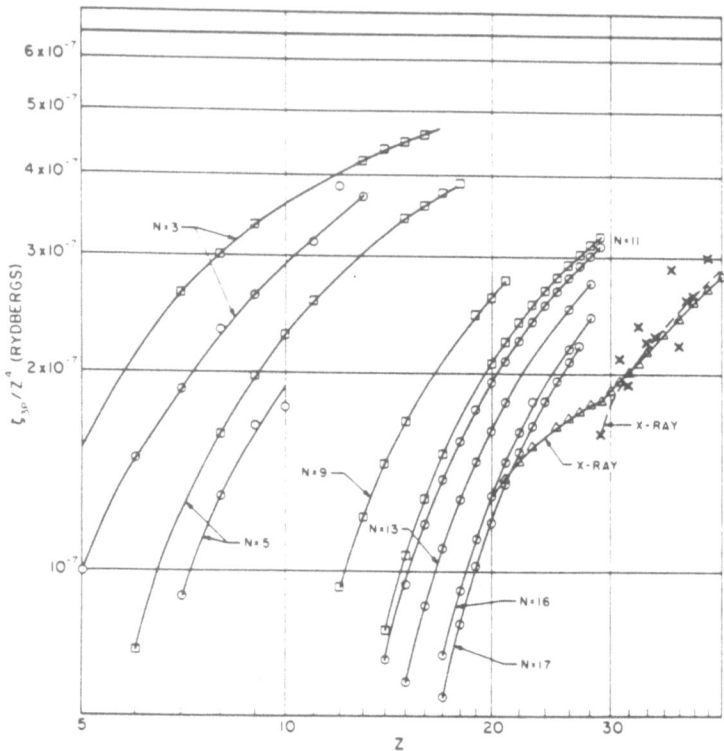

FIGURE 2. ζ_{3p}/Z^4 (Rydbergs, plotted against Z (logarithmic scales). Top of figure is hydrogenic value, $\alpha^2/81$. Points designated as in Fig. 1 except that X-ray values refer to the $M_{II} - M_{III}$ doublet interval of the $(3p)^5$ configuration.

observed values is considerably better than in the other instances covered so far.

Table 3 and Fig. 3 give a similar presentation of the results for ζ_{3d}. Here the ζ's are considerably smaller than for the p-orbitals because of their less penetrating character. Also the "observed" values cannot be directly inferred from the doublet intervals for the low values of Z in the sequences because some of these are actually inverted, due to the additional contribution to the doublet interval arising from higher-order perturbations such as were studied by Phillips (9). Even where the configuration-interaction effect is not large enough to invert the ^2D, its effect may still produce an appreciable effect on the doublet interval which is not considered in these comparisons. Here we see that there is quite close agreement of observed and calculated values in the $N = 3$ and 5 sequences, but less good agreement for the $N = 11$ sequence.

Table 4 and Fig. 4 provide a similar presentation of the results for ζ_{4p}. Here we made calculations for the $N = 11$ sequence, using Eqs. (4) and (15). We did not make any calculations for ζ_{4d}, but Fig. 5 gives a presentation of the

TABLE 3a. ζ_{3d} (in Rydbergs)

| | | N = 3 | | | | | N = 5 | | | | N = 9 |
| | | | Calculated | | | | | Calculated | | | Calculated |
Z		Observed	(4)	(15)	Z		Observed	(15)	Z		(15)
3	Li	0.0000001458	0.00000014	0.00000013	5	B	—	0.00000016	9	F	0.00000015
4	Be	—	0.00000218	0.00000214	6	C	0.000005285	0.00000298	10	Ne	0.00000303
5	B	0.000012393	—	0.00001086	7	N	0.0000215059	0.00001525	11	Na	0.00001737
6	C	0.000039002	—	0.00003430	8	O	0.0000608726	—	12	Mg	0.00005797
7	N	0.0000874816	—	—	9	F	0.0001458026	0.00011266	13	Al	0.00014361
8	O	0.000186263	0.000017540	0.00017333	10	Ne	—	0.00022793	14	Si	0.00029601
9	F	0.000328056	0.000032410	0.000032080	11	Na	0.0004337627	0.00041289	15	P	0.00053973
10	Ne	0.0008164945	—	0.00054673	12	Mg	0.0006196610	0.00069046	16	S	0.00090223
11	Na	0.0017131804	—	0.00087497	13	Al	0.0008748155	0.00108667	17	Cl	0.00141383
12	Mg	0.002041236	0.00196482	0.00133251	14	Si	0.0017131804	0.00163078	18	Ar	0.00210805
13	Al	0.003608614	0.00277946	0.00194949	15	P	0.0024057428	0.00235507	19	K	0.00302145
14	Si	0.004957288	0.00382370	0.00275922	16	S	—	0.00329537	20	Ca	0.00419378
15	P	—	0.00513838	0.00379816	17	Cl	—	0.00448996	21	Sc	0.00566826
16	S	—		0.00510595	18	Ar	—	0.00598100	22	Ti	0.00749033

TABLE 3b. ζ_{3d} (in Rydbergs)

			N = 11	
			Calculated	
Z		Observed	(4)	(15)
11	Na	−0.000000181	0.00000016	0.00000016
12	Mg	0.000003645	—	0.00000352
13	Al	−0.000008311	0.00002548	0.00002207
14	Si	—	—	0.00007680
15	P	+0.000040825	0.00021686	0.00019266
16	S	0.000116642	0.00044114	0.00039622
17	Cl	0.0002660897	0.00078868	0.00071599
18	Ar	0.0005504048	0.00129314	0.00118294
19	K	0.0010643589	—	0.00183076
20	Ca	0.0015090568	—	0.00269633
21	Sc	0.00225994014	—	0.00381951
22	Ti	0.00324410767	0.00559484	0.00524336
23	V	0.0044834297	0.00746142	0.00701386
24	Cr	0.0062330608	0.00972088	0.00918058
25	Mn	0.0079826919	0.01246656	0.01179568
26	Fe	0.010351984	0.01570152	0.01491490
27	Co	0.013231585	0.01953884	0.01859687
28	Ni	0.016767298	0.02401522	0.02290380
29	Cu	0.0211413758	0.02916054	0.02700952
30	Zn	—	—	0.03365597

TABLE 4. ζ_{4p} (in Rydbergs)

			N = 11	
Z	Element	Observed	Calculated (4)	Calculated (15)
11	Na	0.0000342029	0.00003966	0.00003759
12	Mg	0.0001852908	0.00021982	0.00020958
13	Al	0.0004867984	—	0.00055007
14	Si	0.0009829525	0.00114856	0.00110441
15	P	0.0017253307	0.00199928	0.00192845
16	S	0.0027763243	—	0.00308599
17	Cl	0.0041796742	—	0.00464797
18	Ar	0.0060568825	—	0.00669253
19	K	0.0087542305	—	0.00930448
20	Ca	0.011573081	0.01290282	0.01257543
21	Sc	0.015370023	0.01701310	0.01660404
22	Ti	0.020290861	0.02199574	0.02149372
23	V	0.026548222	0.0279600	0.02735821
24	Cr	0.032623330	—	0.03431330
25	Mn	0.040946228	0.043350666	0.04248527
26	Fe	0.048965370	0.05301320	0.05200558
27	Co	0.0612737089	0.06419056	0.06301278
28	Ni	—	0.07701256	0.07565116
29	Cu	—	0.09162308	0.09007627
30	Zn	—	—	0.10643764

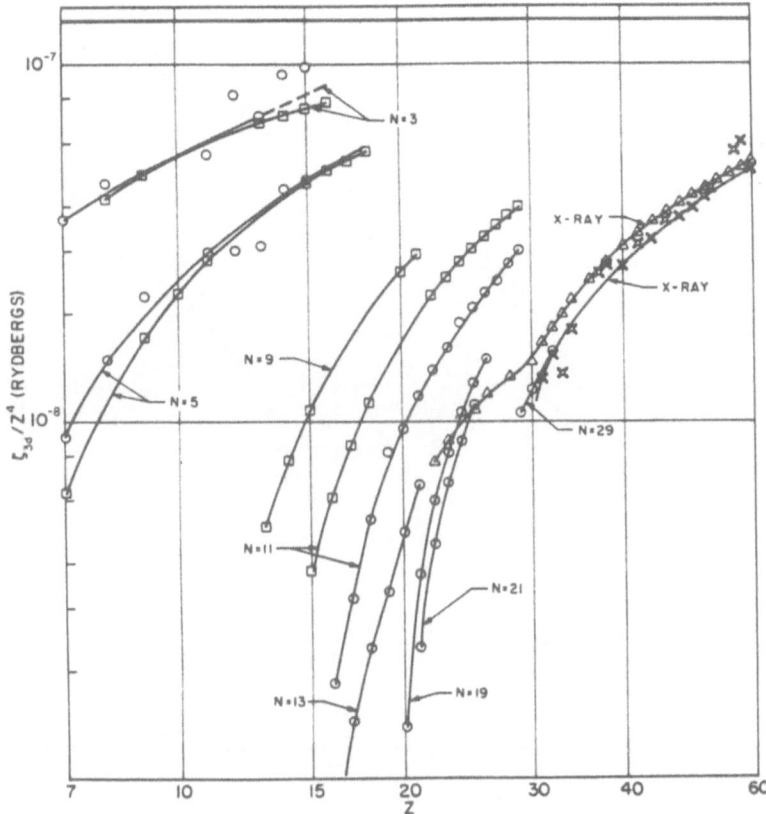

FIGURE 3. ζ_{3d}/Z^4 (Rydbergs) plotted against Z (logarithmic scales). Top of figure is hydrogenic value, $\alpha^2/405$, Points are designated as in Fig. 1 except that X-ray values refer to the $M_{IV} - M_V$ doublet intervals.

observed values indicating that they show general trends similar to the other cases already considered.

Now we turn to a brief discussion of the corresponding doublet intervals in X-ray spectra as also shown in Figs. 1–5 and in Table 5 and Figs. 6 and 7. In X-ray spectra the doublet arises from an electron being removed from a normally closed shell in an atom from which one-electron has been removed, so that $N = (Z - 1)$. Thus the $(L_{II} - L_{III})$ interval arises from the open $(2p)^5$ shell in atoms in which the 2p shell is normally filled. Similarly the $(M_{II} - M_{III})$ interval arises from the open $(3p)^5$ shell in atoms in which the 3p shell is normally filled, and the $(M_{IV} - M_V)$ arises from the open $(3d)^9$ shell, in atoms in which the 3d shell is normally filled.

The range of Z-values in the figures covers that in which the respective X-ray terms first make their appearance through filling of the shells. Here there is a considerable scatter in the X-ray observed values because the doublet intervals

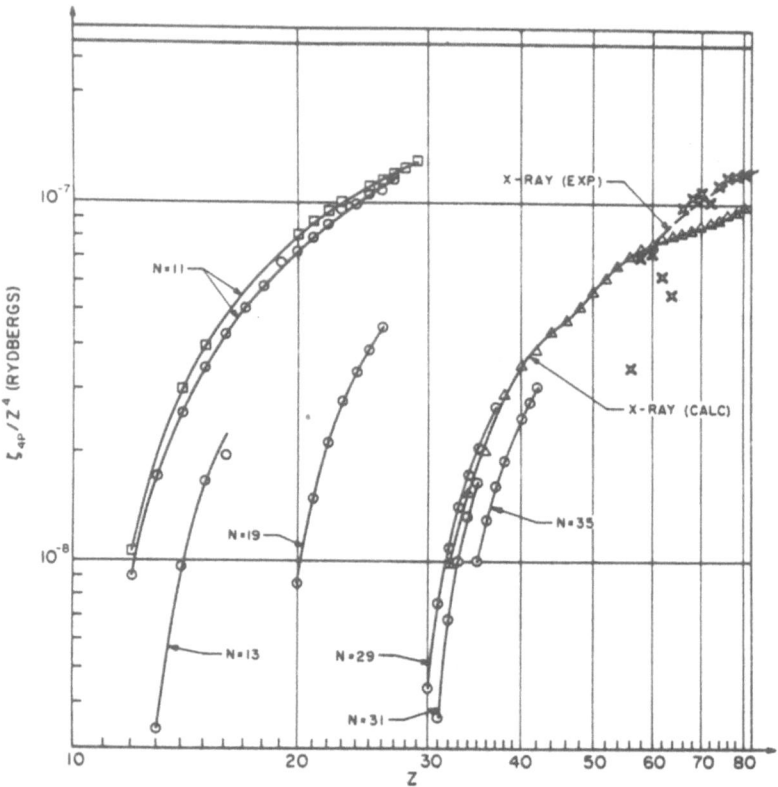

FIGURE 4. ζ_{4p}/Z^4 (Rydbergs) plotted against Z. Top of figure is hydrogenic value, $\alpha^2/192$. Points are designated as in Fig. 1 except that X-ray values refer to the $N_{II} - N_{III}$ doublet interval.

are quite small. The crosses show observed values and the triangles show the calculated values given by Herman and Skillman for even Z, supplemented by our calculations of the values for odd Z, by the same method. Making allowance for the scatter in the observed values, agreement of observation and calculation in all these cases is quite good.

Table 5 and Figs. 6 and 7 bring out a more detailed comparison of the transition between a particular doublet interval in optical spectra for the particular case of ζ_{2p}. The comparison is between the doublet interval due to $(2p)^5\, {}^2P$ in the $N = 9$ optical spectra and the $L_{II} - L_{III}$ interval due to $(2p)^5$ ${}^2P$ in the $N = Z - 1$ X-ray spectra for $Z \geqslant 10$. In the $N = 9$ sequence the configuration remains as $1s^2 2s^2 2p^5$ with increasing Z. In the X-ray spectra for $Z = 10$ we have the same configuration, but for large values of Z we have an increasing number of outer electrons added to this configuration. These are 3s, 3s² for Na and Mg, then a gradual filling of the 3p shell from Al to Ar and the addition of 4s and 4s² for $Z = 19$ and 20.

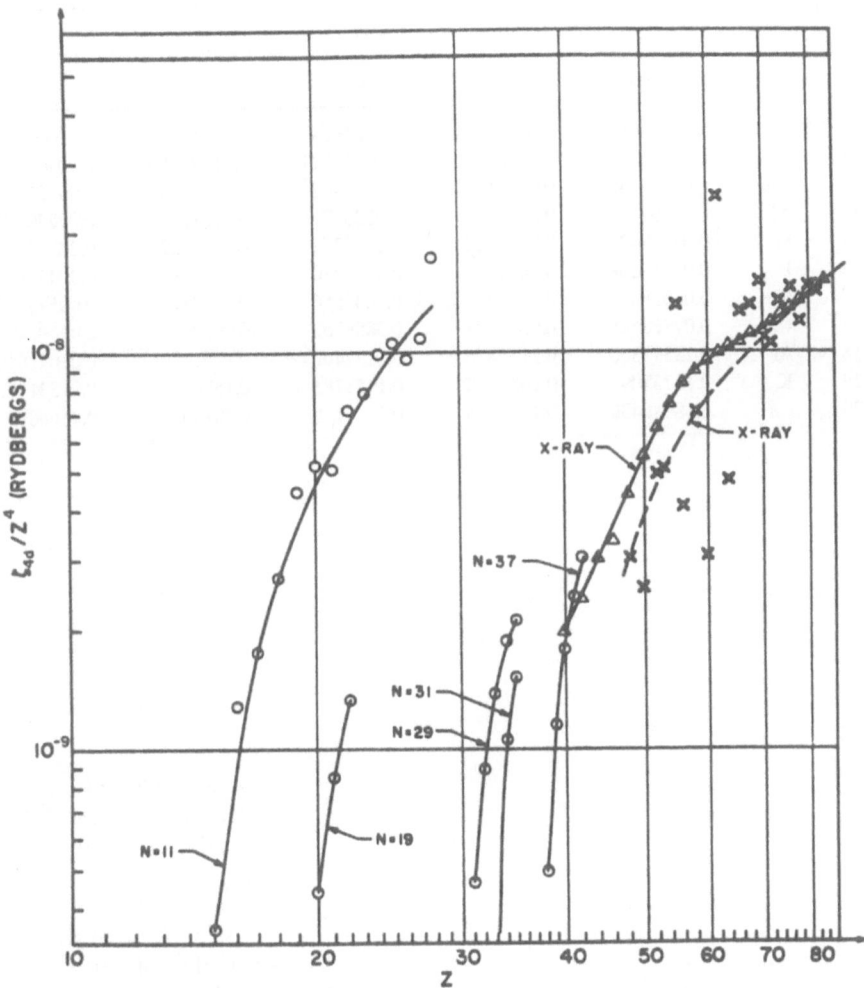

FIGURE 5. Same as preceding figures, but for ζ_{4d}/Z^4. Top of figure is hydrogenic value, $\alpha^2/960$. Points refer to 4^2D intervals in optical spectra and $N_{IV} - N_V$ $(4d)^5$ intervals in X-ray spectra.

Hence the distinction between ζ_{2p} for a given Z is, for the X-ray case, the additional screening produced by these "outer" electrons to the small extent that they penetrate to radial values within the 2p-orbital, as compared with the $N = 9$ sequence for which no outer electrons are added to the $1s^2 2s^2 2p^5$ configuration. This effect means that ζ_{2p} should be smaller for the X-ray sequence than for the $N = 9$ sequence. The calculated values are shown in Table 5, as also a column of differences $[\zeta_{2p}(N = 9) - \zeta_{2p}(N = Z - 1)]$ in the column headed Δ.

TABLE 5. ζ_{2p} (in Rydbergs)

		X-ray			$N = 9$	Δ
Z		Observed	(4)	(15)	(15)	(15)
10	Ne	0.00669868	0.00574064	0.0061954	0.0061954	0.00000000
11	Na	0.01004802	0.0098288	0.00945871	0.01046786	0.00100915
12	Mg	0.01339736	0.0156159	0.0150942	0.0166128	0.00151860
13	Al	0.01808644	0.0236047	0.022897	0.0251047	0.00220770
14	Si	0.02679472	0.0342703	0.0333372	0.0364729	0.00313570
15	P	0.04019208	0.0481425	0.0469410	0.0512990	0.0043580
16	S	0.06028812	0.065802	0.0642856	0.0702184	0.00593280
17	Cl	0.07368548	0.0879598	0.0859962	0.0939203	0.00792410
18	Ar	0.13397360	0.1150476	0.1127460	0.1231470	0.01041010
19	K	0.14737096	0.1481618	0.1453785	0.1586913	0.01331280
20	Ca	0.18086436	0.1880368	0.1847128	0.20140271	0.01668991

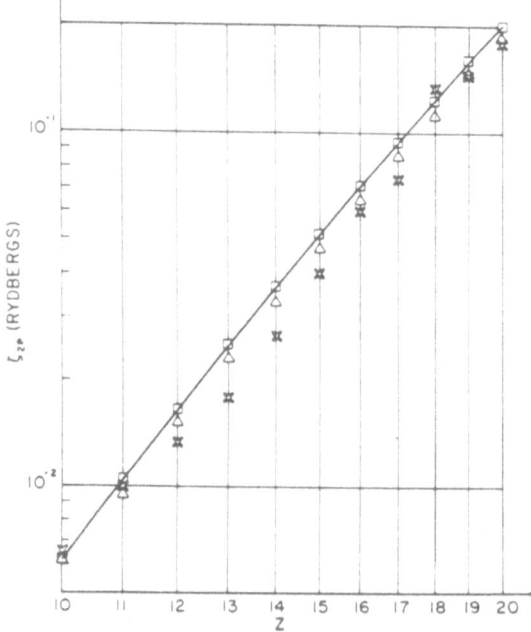

FIGURE 6. Effect of screening on doublet 2^2P interval as shown in value of ζ_{2p}. Squares are calculated values for the $1s^2 2s^2 2p^5$ ($N = 9$) ground states of the fluorine isoelectronic sequence. Triangles are calculated values for the $N = (Z - 1)$ $(2p)^5$ X-ray interval in elements $Z = 10$ to 20. Crosses are observed X-ray doublet intervals.

In Fig. 6 the corresponding ζ_{2p} are shown, the squares referring to calculated values for the $N = 9$ sequence, and the triangles to the smaller calculated values for the ζ_{2p} for the X-ray intervals, while the crosses show the ζ_{2p} inferred from the X-ray levels. These latter despite scatter tend to be systematically smaller than the calculated ζ_{2p}. Figure 7 shows a graph of the difference Δ against Z.

FIGURE 7. Calculated difference between ζ_{2p} in $N = 9$ isoelectronic sequence and in calculated X-ray doublet interval due to $(2p)^5$ with $N = Z - 1$

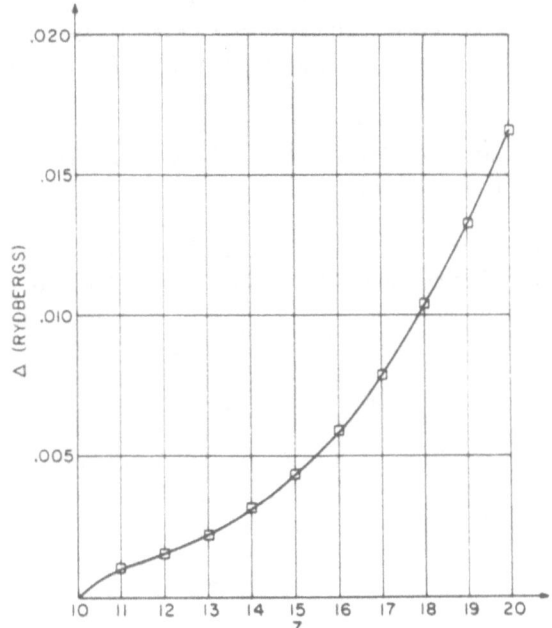

In conclusion it seems fair to say that the calculations show that the self-consistent fields are capable of giving quite good values of the spin-orbit parameters ζ_{nl} both with regard to absolute values and general trends with N and Z. Some discrepancies remain which can be reduced by the use of more elaborate approximations $(10, 11)$, but this work shows that the spin-orbit parameters are quite adequately given for most purposes by calculations of the type considered here.

This research was supported in part by the Advanced Research Projects Agency (Project Defender), monitored by the U:S. Army Research Office, Durham, under Contract DA-31-124-ARO[D]-139 with the University of Colorado.

References

1. C. Slater, "Quantum Theory of Atomic Structure", 2 Vols. McGraw-Hill, New York, 1960. Especially Chap. 17 for exposition of the method and the bibliography, p. 383, Vol. 2.
2. E.U. Condon and G.H. Shortley, "The Theory of Atomic Spectra" pp. 120–125, 144–147, 193–197. Cambridge Univ. Press, London and New York, 1935.
3. A. Landé, *Zeit. Physik*, **25**, 46 (1924).
4. B. Edlén, in "*Encyclopedia of Physics*", Vol. 27, p. 106. Springer-Verlag, Berlin, 1964.
5. F. Herman and S. Skillman, "Atomic Structure Calculations". Prentice-Hall, Englewood Cliffs, New Jersey, 1963.

6. R. Latter, *Phys. Rev.* **99**, 510 (1955).

7. C.E. Moore, "Atomic Energy Levels as Derived from Analysis of Optical Spectra," 3 Vols. National Bureau of Standards Circular 467.

8. A.E. Sandström, Experimental methods of X-ray spectroscopy: ordinary wavelengths, in "*Encyclopedia of Physics*", Vol. 30. Springer-Verlag, Berlin, 1957. See also "*American Institute of Physics Handbook,*" 2nd ed., pp. 77–136–8. McGraw-Hill, New York, 1963.

9. M. Phillips, *Phys. Rev.* **44**, 644 (1933).

10. M. Blume and R.E. Watson, *Proc. Roy. Soc.* (*London*) **A270**, 1340 (1962).

11. M. Blume and R.E. Watson, *Proc. Roy. Soc.* (*London*) **A271**, 1347 (1963).

Self-Consistent Field Calculations of the $1s^2 2s^2 2p\,3p$ Configuration in the Carbon Isoelectronic Sequence

E.U. CONDON AND H. ODABASI

*Joint Institute for Laboratory Astrophysics**

It is a pleasure to dedicate this paper to Professor P.M. Morse, recalling a forty-year association with him of the senior author, which started when we worked together in Palmer Physical Laboratory of Princeton University in the fall of 1928

We have recently completed some self-consistent field calculations of several low configurations in the carbon, nitrogen, and oxygen isoelectronic sequences which are presented in detail in a recent JILA (Joint Institute for Laboratory Astrophysics) report.[1]

In this work, we found an interesting case of strong mixing of two Russell-Saunders levels, because of the magnetic interaction terms, of a type which has not occurred before, so far as we are aware. We will report on this briefly here, referring to the report for fuller details.

The self-consistent field treatment is based on the nonrelativistic Hamiltonian with omission of magnetic terms

$$H = \sum_{i=1} \left[-\frac{\hbar^2}{2m}\nabla_i^2 + \frac{Ze^2}{r_i} \right] + \sum_{\text{pairs}} \frac{e^2}{r_{ij}} \tag{1}$$

We write

$$u_a(i) = r_i^{-1} P(n_a, \ell_a) Y(\ell_a, m_a; \theta_i, \varphi_i)(m_{sa}, \sigma) \tag{2}$$

for the one-electron wave function of the ith electron whose coordinates are $r_i \theta_i \varphi_i$ and σ_i and whose quantum numbers are a which stands for $(n_a, \ell_a, m_{\ell a}, m_{sa})$

In the Hartree method the N-electron wave function is a simple product over N one-electron wave functions

$$\Psi_H = u_1(1)u_2(2)\ldots u_N(N) \tag{3}$$

When a wave function of this type is used, the mean energy of a configuration is readily computed to be

---

* The National Bureau of Standards and the University of Colorado.

$$E_H(\text{config}) = \sum_a [q_a I(a) + \tfrac{1}{2} q_a(q_a - 1) F^0(a,a)] + \sum_{a,b} q_a q_b F^0(a,b) \quad (4)$$

where the sums on a and b are over distinct shells, specified by (n_a, ℓ_a) and q_a is the number of electrons in each shell. The double sum is on each pair of shells counted once. Here

$$I(a) = \int_0^\infty P(a;r) \left[-\frac{\hbar^2}{2m} \frac{d^2}{dr^2} + \frac{\ell_a(\ell_a + 1)}{r^2} - \frac{Ze^2}{r} \right] P(a;r)\, dr \quad (5)$$

$$F^0(a,b) = \int_0^\infty P^2(a;r) \frac{e^2}{r} Y_0(b,b;r)\, dr \quad (6)$$

and

$$Y_0(b,b;r) = \int_0^r P^2(b;s)\, ds + r \int_r^\infty P^2(b;s) \frac{ds}{s} \quad (7)$$

The Hartree equations[2] for the occupied $(q_a \neq 0)$ radial functions are found by varying these functions in the expression for E_H (config) subject to the condition that each one of them is normalized, to find the radial functions which give E_H a stationary value.

In the Hartree-Fock method,[3] the determinantal form of N-electron wave function is used. This gives for the configuration average energy,

$$E(\text{config}) = \sum_a [q_a I(a) + \tfrac{1}{2}(q_a - 1) J(a,a)] + \sum_{a,b} q_a q_b J(a,b) \quad (8)$$

which is formally the same as Equation 4 but with J in place of the F. These are defined as

$$J(a,b) = F^0(a,b) - \frac{1}{2} \sum_k \begin{pmatrix} \ell_b & k & \ell_a \\ 0 & 0 & 0 \end{pmatrix}^2 G^k(a,b) \quad (9)$$

and

$$J(a,a) = \frac{2}{4\ell_a + 1} \left[(2\ell_a + 1) F^0(a,a) - \frac{2\ell_a + 1}{2} \sum_k \begin{pmatrix} \ell_a & k & \ell_a \\ 0 & 0 & 0 \end{pmatrix}^2 F^k(a,a) \right] \quad (10)$$

Here the 3j symbol or Wigner coefficient is related to the reduced element of the Racah[4] tensor $C^{(k)}$ as

$$\begin{pmatrix} \ell & k & \ell' \\ 0 & 0 & 0 \end{pmatrix} = (-1)^{\ell'} \frac{\langle \ell' \| C^{(k)} \| \ell \rangle}{\sqrt{(2\ell + 1)(2\ell' + 1)}} \quad (11)$$

and

$$F^k(a,b) = \int_0^\infty \int_0^\infty P^2(a;r_1) P^2(b;r_2) e^2 \frac{r_<^k}{r_>^{k+1}}\, dr_1\, dr_2 \quad (12)$$

and

$$G^k(a,b) = \int_0^\infty \int_0^\infty P(a;r_1)P(b;r_1)P(b;r_2)P(a;r_2)e^2 \frac{r_<^k}{r_>^{k+1}} dr_1\, dr_2 \qquad (13)$$

where $r_<$ and $r_>$ are respectively the lesser and the greater of r_1 and r_2.

The Hartree-Fock equations for the radial wave functions are those that express the condition that $E(\text{config})$ have a stationary value for variations of the $P(a;r)$ subject to the requirement that the radial functions be normalized. The two methods differ because of the difference between Equations 4 and 8. The presence of the exchange integrals of Equations 13 in 8 greatly complicates the Hartree-Fock equations.

In the Hartree-Fock-Slater[5] method, the effect of the exchange integrals is approximately taken into account by a method which preserves the simpler form of the Hartree equations. In this method, the equations for the radial function are

$$\frac{\hbar^2}{2m}\left[-\frac{d^2}{dr^2} + \frac{\ell_a(\ell_a + 1)}{r^2} \right] P(a;r) + V(r)P(a;r) = E_a P(a;r) \qquad (14)$$

in which

$$V(r) = V_0(r) \qquad\qquad\qquad r < r_0$$
$$= -\frac{(Z - N + 1)e^2}{r} \qquad r > r_0 \qquad (15)$$

in which r_0 is the value of r for which these two expressions are equal and the expression for $V_0(r)$ is

$$V_0(r) = -\frac{Ze^2}{r} + \frac{e^2}{r}\sum_a q_a Y_0(a,a;r) - 3e^2\left[\frac{3}{8\pi}\rho(r)\right]^{1/3} \qquad (16)$$

in which

$$\rho(r) = \frac{1}{4\pi r^2}\sum_a q_a P^2(a;r)$$

The term involving $\rho(r)$ is the Slater approximation to the exchange terms. The device of making the change expressed in Equation 15 to give $V(r)$ the correct behavior was introduced by Latter[6] and adopted by Herman and Skillman.[7]

The electrostatic interaction in Equation 1 has the effect of splitting the configuration energy into *terms*, accurately labelled by S and L, the quantum numbers which give resultant spin and resultant orbital angular momentum. Terms so labelled are said to be in Russell-Saunders coupling.

We write $E(Z, N, \text{config, term})$ for the energy of a term measured from the state of complete N-fold ionization, and $W(Z, N, \text{config, term})$ for the departure of the term energy from the configuration average, that is,

$$E(Z, N, \text{config, term}) = E(Z, N, \text{config}) + W(Z, N, \text{config, term}) \qquad (17)$$

When magnetic interaction within states of the same configuration is taken into account, each term of $(2S + 1)(2L + 1)$ states is split into *levels*, labelled by J values with run from $|L - S|$ to $L + S$ by unit steps. Here $J(J + 1)$ is the value of the squared vector resultant of spin and orbital angular momentum, $(\mathbf{L} + \mathbf{S})^2$.

Nondiagonal matrix elements of the magnetic interaction connect levels of the same J but differing S and L values. In consequence S and L are no longer accurate quantum numbers, the wave function for a particular level becoming a linear combination of those for the interacting Russell-Saunders levels. This situation is called intermediate coupling[8]. Usually one particular Russell-Saunders level is predominant in the linear combination and spectroscopists attach that name to the level, while recognizing that the wave function contains components of other levels as well.

The magnetic terms in the Hamiltonian were first studied by Breit.[9] A modern account is given by Bethe and Salpeter[10] and by Slater[3]. They are all formally proportional to α^2 where $\alpha = e^2/\hbar c$ is the fine structure constant. Written in atomic units, they are conveniently expressed as the sum of three parts,

$$H_a = \frac{1}{2}\alpha^2 Z \sum_i \frac{1}{r_i^3} \mathbf{L}_i \cdot \mathbf{S}_i \tag{18a}$$

$$H_b = -\frac{1}{2}\alpha^2 \sum_{i,j} \frac{1}{r_{ij}^3}(\mathbf{r}_{ij} \times \mathbf{P}_i) \cdot (\mathbf{S}_i + 2\mathbf{S}_j) \tag{18b}$$

$$H_c = \frac{1}{2}\alpha^2 \sum_{\text{pairs}} \frac{3}{r_{ij}^3}\left[\mathbf{S}_i \cdot \mathbf{S}_j - \frac{3}{r_{ij}^2}(\mathbf{S}_i \cdot \mathbf{r}_{ij})(\mathbf{S}_j \cdot \mathbf{r}_{ij})\right] \tag{18c}$$

The H_a represents the interaction of each electron's magnetic moment with the magnetic field arising from the electron's motion in the Coulomb field of the nucleus. In H_b each pair occurs once as (i, j) and again as (j, i) where $i \neq j$. The terms in $\mathbf{S}_i$ represent the interaction of the ith electron's magnetic moment with the magnetic field due to its own motion in the electric field of the other $(N - 1)$ electrons. Its sign is opposite to that of H_a so the effect of these terms is to reduce the effect of H_a in a manner that is analogous to electrostatic screening.

The terms proportional to $\mathbf{S}_j$ in H_b arise from the interaction of the jth electron's magnetic moment with the magnetic field because of the orbital motion of the other $(N - 1)$ electrons.

Finally H_c, commonly called spin-spin interaction, represents the magnetic dipole interaction of all pairs of electrons.

Since the parts H_b and H_c involve sums over all electron pairs of two-electron symmetric operators, their matrix elements with determinantal wave functions will involve exchange integrals, as well as direct integrals, as first pointed out by Marvin.[11] The treatment of these terms was studied later by Horie[12] and Blume and Watson.[13]

The older treatment (Reference 8, p. 195), of magnetic interaction was based on the approximation that the magnetic part could be put in the form

$$H^I = \sum_i \xi(r_i) \mathbf{L}_i \cdot \mathbf{S}_i \tag{19}$$

in which, in atomic units,

$$\xi(r_i) = \frac{1}{2}\alpha^2 \frac{1}{r_i}\frac{\partial U_a(r_i)}{\partial r_i} \tag{20}$$

and $U_a(r_i)$ is the effective central field for the electron with quantum numbers (n_a, ℓ_a), namely,

$$U_a(r) = -\frac{Ze^2}{r} + \frac{e^2}{r}\sum_b (q_b - \delta_{ab}) Y_0(b, b; r) \tag{21}$$

in which b is summed over the occupied shells.

In some presentations $\xi(r)$ is defined as in Equation 20 but using $V(r)$ from Equation 15 rather than $U_a(r)$ from Equation 21. The difference is quite small but Equation 21 is preferable as having a firmer theoretical foundation.

Horie has shown that the major part of H_b can be put in the form of Equation 19. The sum of H_a and this part of H_b is called *spin-orbit* interaction. The residual and the spin-spin interactions are often quite small so that in most cases the magnetic interaction is quite accurately given by the spin-orbit part alone.

The spin-orbit interaction in the Russell-Saunders limit (neglect of non-diagonal matrix elements) gives rise to an energy term (Reference 8, p. 194),

$$\Gamma(\alpha, S, L) = \tfrac{1}{2}\bar{\zeta}(\alpha, S, L)[J(J + 1) - L(L + 1) - S(S + 1)] \tag{22}$$

Here the $\bar{\zeta}$ for levels are related to the one-electron $\zeta(n_a, \ell_a)$ of Reference 8, where

$$\zeta(n_a \ell_a) = \int_0^\infty \xi(r) P^2(n_a \ell_a; r)\, dr \tag{23}$$

The formulas for the $\zeta(n_a, \ell_a)$ are considerably more complicated when the full spin-orbit part of H_b is included with its exchange integrals. For $\zeta(2p)$ the formula is

$$\zeta(2p) = \frac{1}{2}\alpha^2 Z \int_0^\infty \frac{1}{r^3} P^2(2p, r)\, dr - 4M^0(2p, 1s)$$
$$- 4M^0(2p, 2s) + 2V^0(2p, 1s) + 2V^0(2p, 2s) + 2N^1(2p, 1s)$$
$$- 4N^{-1}(2p, 1s) - 4N^{-1}(2p, 2s) + mM^0(2p, 2p) \tag{24}$$

in which $m = 1$ for $N = 6$; $m = 3$ for $N = 7$ and $m = 5$ for $N = 8$ and the various terms in Equation 24 are

$$M^k(a,b) = \frac{1}{4}\alpha^2 \int_0^\infty \int_0^\infty \frac{r_2^{\,k}}{r_1^{k+3}} \varepsilon(r_1 - r_2) P_1(a) P_2^{\,2}(b)\, dr_1\, dr_2$$

$$N^k(a,b) = \frac{1}{4}\alpha^2 \int_0^\infty \int_0^\infty \frac{r_2^{\,k}}{r_i^{k+3}} \varepsilon(r_1 - r_2) P_1(a) P_1(b) P_2(a) P_2(b)\, dr_1\, dr_2$$

$$V^k(a) = \frac{1}{4}\alpha^2 \int_0^\infty \int_0^\infty \frac{r_<^{\,k}}{r_>^{k+3}} r_1 r_2 P_1(a) P_2(b) \left[r_2 \frac{\partial}{\partial r_1} - r_1 \frac{\partial}{\partial r_2} \right]$$

$$\times \frac{P_2(a) P_1(b)}{r_1 r_2}\, dr_1\, dr_2$$

where

TABLE 1. Observed and Theoretical Term Energies Relative to the Configuration Average for the $1s^2 2s^2 2p3p$ Configuration of the $N = 6$ Sequence (Unit: Rydbergs. First entry is observed and second is theoretical)

	^3D	^3P	^3S	^1D	^1P	^1S
$Z = 6$, CI	−0.00862	0.00644	0.00069	0.01771	−0.01649	0.03015
	−0.01103	0.00771	−0.00079	0.02179	−0.01785	0.04292
7, NII	−0.02193	0.01471	−0.00118	0.04730	−0.04020	0.08430
	−0.02826	0.02101	−0.00602·	0.05516	−0.04309	0.10637
8, OIII	−0.03544	0.02194	−0.00372	0.07854	−0.06387	0.14431
	−0.04469	0.03341	−0.01099	0.08724	−0.06716	0.16797
9, FIV	−0.06039	0.04504	−0.01554	0.11808	−0.09030	0.22765
10, NeV	−0.07561	0.05617	−0.01978	0.14810	−0.11283	0.28601
11, NaVI	−0.09051	0.06696	−0.02382	0.17757	−0.13497	0.34349
12, Mg VII	−0.10519	0.07754	−0.02773	0.20666	−0.15683	0.40034
13, AlVIII	−0.11971	0.08796	−0.03153	0.23546	−0.17849	0.45672
14, SiIX	−0.13411	0.09825	−0.3525	0.26406	−0.20001	0.51278
15, PX	−0.14841	0.10846	−0.03891	0.29250	−0.22141	0.56855
16, SXI	−0.16265	0.11859	−0.04252	0.32081	−0.24273	0.62413
17, CXII	−0.17683	0.12867	−0.04610	0.34902	−0.26398	0.67954
18, ArXIII	−0.19097	0.13871	−0.04966	0.37716	−0.28518	0.73483
19, KXIV	−0.20507	0.14872	−0.05319	0.40523	−0.30633	0.79001
20, CaXV	−0.21914	0.15869	−0.05670	0.43325	−0.32744	0.84510

$$\varepsilon(x) = 1 \qquad \text{for } x > 0$$

$$\varepsilon(x) = 0 \qquad \text{for } x < 0$$

These complications over the order theory have a quite small effect in the examples so far calculated. The calculated ζ parameters depend both on the formulas used and the specific choice of radial wave functions. Froese[14] has made the best calculations of any known to us. She used radial functions from a full Hartree-Fock solution and the full theoretical formula as given by Blume and Watson. This gives somewhat smaller values of the $\zeta(n\ell)$ for the cases tested than the ones calculated by us[15] using Equations 20, 21, and 23 with radial wave functions from the Hartree-Fock-Slater solution.

For the study of intermediate coupling, the nondiagonal elements of magnetic interaction must also be included. This gives a matrix to be diagonalized for each J, and results in producing wave functions which are mixed in L and S.

An interesting example of this occurs in $1s^2 2s^2 2p 3p$ of the $N = 6$ carbon isoelectronic sequence. This configuration is known from analysis of spectra for $Z = 6, 7, 8$ but only a few levels are known for $Z > 8$.

For this configuration the matrices of electrostatic interaction are given by Slater[3] (Volume 2, pp. 288–290). The matrices of spin-orbit interaction are given by Condon and Shortley[8] (p. 268). The radial functions were obtained by an adaptation to local computer facilities of the program of Herman and Skillman[7] for the Hartree-Fock-Slater methods. Table 1 presents the observed and theoretical term energies relative to configuration averages. The observed values are calculated from the values given in the compilation by C. Moore.[16] Table 2 gives the theoretical values of the energy levels relative to the configuration average. Figure 1 presents the same material graphically.

Of the four levels having $J = 1$, the one which we denote by "a" is almost purely $^3D_1$ and the one we denote by "b" is almost purely $^1P_1$ for low Z. The coefficients for the expansion of the $\Psi(a)$ and $\Psi(b)$ in terms of Russell-Saunders states are given in Table 3.

With increasing Z, the admixtures of $^3P_1$ and $^3S_1$ in "a" and "b" remain quite small, but the mixing of $^3D_1$ and $^1P_1$ increases rapidly. Figure 2 shows the projections of $\Psi(a)$ and $\Psi(b)$ on the plane of $\Psi(^3D_1)$ and $\Psi(^1P_1)$. Because the $^3P_1$ and $^3S_1$ components remain small, $\Psi(a)$ and $\Psi(b)$ are nearly orthogonal, nearly unit vector as Z increases. Only for $Z > 14$, does $\Psi(a)$ become quite appreciably less than a unit vector in the plane of Figure 2, because of the increase of its components out of this plane.

As figure 2 shows, $\Psi(a)$ and $\Psi(b)$ have a smooth dependence on Z. However, for $Z > 17$, the $\Psi(a)$ contains less of $\Psi(^3D_1)$ than of $\Psi(^1P_1)$. Hence, if the usual rule were to be followed of giving a Russell-Saunders name according to the term of largest coefficient in the state of intermediate coupling, then the names should be interchanged between $Z = 16$ and 17 as indicated by the crossover lines in Figure 1.

The calculated values of $\zeta(2p)$ and $\zeta(3p)$ lead to good agreement with observed term intervals as shown in Table 4. Figure 3 shows a plot of the Z

TABLE 2. Theoretical Energy Levels for $N = 6$ $1s^2 2s^2 2p3p$ Relative to Configuration Average (Unit: Rydberg)

Z	$^3D_3$	$^3D_2$	$^1D_2$	$^3P_2$	a	$^3P_1$	b	$^3S_1$	$^3P_0$	$^1S_0$
6	−0.010773	−0.011155	0.021798	0.007837	−0.011396	0.007594	−0.017860	−0.000793	0.007459	0.042920
7	−0.027556	−0.028627	0.055171	0.021365	−0.029297	0.020680	−0.043133	−0.006027	0.020295	0.106386
8	−0.043118	−0.045510	0.087274	0.034190	−0.046953	0.032679	−0.067302	−0.101999	0.031800	0.168008
9	−0.057349	−0.061988	0.118185	0.046532	−0.064684	0.043654	−0.090699	−0.015551	0.041894	0.227749
10	−0.070273	−0.078444	0.148345	0.058756	−0.082893	0.053809	−0.113855	−0.019800	0.050584	0.286255
11	−0.081796	−0.095212	0.178099	0.071141	−0.101851	0.063272	−0.137355	−0.023835	0.057719	0.344020
12	−0.091725	−0.112598	0.207715	0.083895	−0.121592	0.072123	−0.161973	−0.027691	0.063003	0.401410
13	−0.099791	−0.130916	0.237435	0.097174	−0.141912	0.080439	−0.188776	−0.031349	0.066011	0.458754
14	−0.105679	−0.150529	0.267626	0.111114	−0.162511	0.088352	−0.219078	−0.034731	0.066207	0.516398
15	−0.109014	−0.171833	0.298569	0.125803	−0.183261	0.096035	−0.254190	−0.037654	0.062920	0.574691
16	−0.109394	−0.195317	0.330748	0.141319	−0.204488	0.103760	−0.295288	−0.039806	0.055365	0.634101
17	−0.106367	−0.221538	0.364701	0.157701	−0.226978	0.111868	−0.343381	−0.040678	0.042619	0.695134
18	−0.099443	−0.251151	0.401077	0.174975	−0.251907	0.120786	−0.399450	−0.039568	0.023629	0.758383
19	−0.088096	−0.284906	0.440617	0.193164	−0.280706	0.131006	−0.464511	−0.035605	0.002779	0.824524
20	−0.071769	−0.323652	0.484141	0.212310	−0.314921	0.143081	−0.539643	−0.027860	−0.037882	0.894297

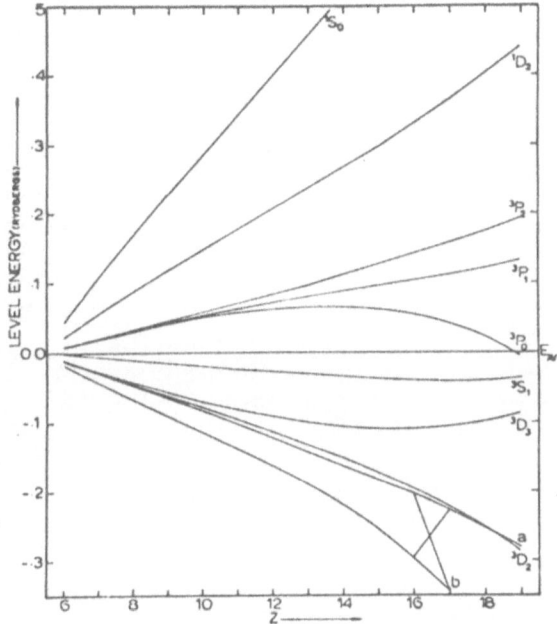

FIGURE 1. Theoretical level energies of the $1s^2 2s^2 2p3p$ configuration in the $N = 6$ sequence.

TABLE 3. Mixing Coefficients for $(^3D_1)$ and $(^1P_1)$ in $N = 6$ $1s^2 2s^2 2p3p$
$(a) = a_1(^3D_1) + a_2(^3P_1) + a_3(^1P_1) + a_4(^3S)$
$(b) = b_1(^3D_1) + b_2(^3P_1) + b_3(^1P_1) + b_4(^3S)$

Z	a_1	a_2	a_3	a_4	b_1	b_2	b_3	b_4
6	0.999	−0.008	−0.035	0.001	0.035	0.006	0.999	−0.012
7	0.999	−0.008	−0.046	0.001	0.047	0.006	0.999	−0.016
8	0.997	−0.010	−0.070	0.003	0.071	0.008	0.997	−0.023
9	0.994	−0.014	−0.106	0.006	0.107	0.011	0.994	−0.033
10	0.987	−0.020	−0.157	0.012	0.158	0.014	0.986	−0.046
11	0.974	−0.029	−0.225	0.023	0.227	0.018	0.972	−0.062
12	0.949	−0.040	−0.311	0.041	0.314	0.021	0.946	−0.079
13	0.909	−0.055	−0.408	0.067	0.415	0.023	0.905	−0.096
14	0.855	−0.075	−0.502	0.104	0.515	0.023	0.850	−0.110
15	0.795	−0.099	−0.580	0.149	0.601	0.022	0.790	−0.121
16	0.737	−0.128	−0.633	0.200	0.667	0.020	0.733	−0.129
17	0.686	−0.162	−0.662	0.255	0.716	0.018	0.685	−0.135
18	0.641	−0.201	−0.672	0.312	0.750	0.015	0.646	−0.141
19	0.602	−0.244	−0.666	0.367	0.775	0.013	0.615	−0.145
20	0.567	−0.288	−0.650	0.417	0.793	0.011	0.591	−0.150

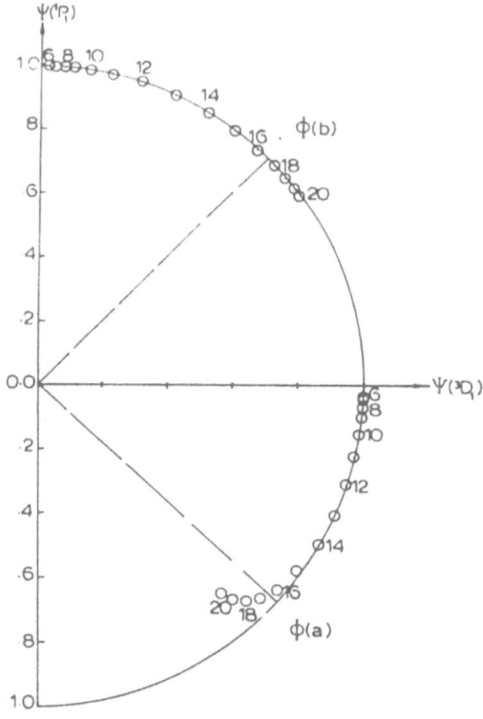

FIGURE 2. Projection of $\Psi(a)$ and $\Psi(b)$ on the $\Psi(^3D_1)$, $\Psi(^1P_1)$ plane.

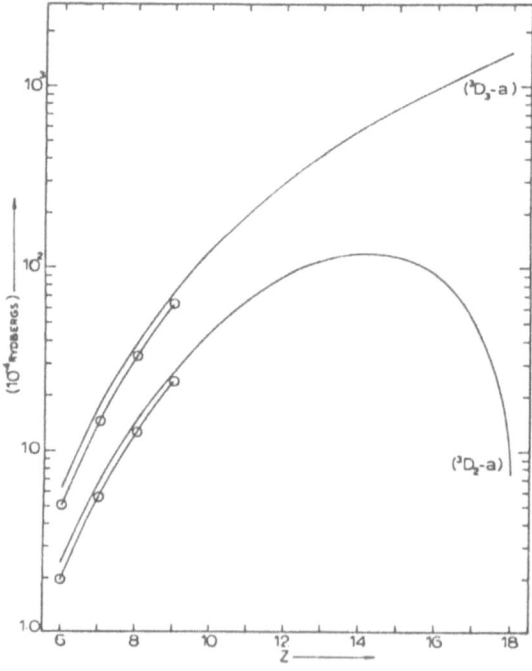

FIGURE 3. Variation with Z of the theoretical and observed $(^3D_3 - a)$ and $(^3D_2 - a)$ intervals in the $N = 6$ sequence $1s^2 2s^2 2p3p$. (Observed points are circled.)

dependence of the $(^3D_3 - a)$ and $(^3D_2 - a)$ intervals, including the few known observed intervals.

The large degree of mixing involved in this example also shows up in the Zeeman effect. (Condon and Shortley,[8] Section 3^{16}). In addition to the strong mixing in $2p3p$ of $^3D_1$ and $^1P_1$ there is also a strong mixing in $2p3d$ of $^1D_2$ and $^3F_2$. Table 5 gives the calculated Landé g-factors of interest in these cases. Figure 4 shows the Z-dependence of the g-factors for $J = 1$ in the $2p3p$, showing the crossover between $Z = 16$ and 17. Figure 5 is a similar graph of the g-factors for the $J = 2$ levels in the $2p3d$ configuration showing the crossover interaction of $^1D_2$ and $^3F_2$ and the lack of it for $^3D_2$ and $^3P_2$ owing

TABLE 4. Intervals of the $^3D_{3,2,1}$ and $^3P_{2,1,0}$ Russell-Saunders Levels for the $1s^2 2s^2 2p3p$ Configuration of the $Z = 6$ Sequence (Unit: 10^{-4} Rydberg. First entry is observed, second is theoretical)

	$^3D_3 - a$	$^3D_2 - a$	$^3P_2 - {}^3P_0$	$^3P_1 - {}^3P_0$
$Z = 6$, CI	4.98	1.93	3.00	1.13
	6.23	2.41	3.78	1.35
7, NII	14.30	5.54	8.53	3.21
	17.41	6.70	10.70	3.85
8, OIII	32.48	12.42	19.38	7.48
	38.35	14.43	23.90	8.79
9, FIV	63.56	23.83	38.77	15.55
	73.35	26.96	46.38	17.60
10, NeV				
	126.20	44.49	81.72	32.25
11, NaVI				
	200.55	66.39	134.22	55.53
12, MgVII				
	298.67	89.94	208.92	91.20
13, A1VIII				
	421.21	109.96	311.63	144.28
14, SiIX				
	568.32	119.82	449.07	221.45
15, PX				
	742.47	114.28	628.83	331.15
16, SXI				
	950.94	91.71	859.54	483.95
17, C1XII				
	1206.11	54.40	1150.82	692.49
18, ArXIII				
	1524.64	7.56	1513.46	971.57
19, KXIV				
	1926.10	−42.00	1903.85	1282.27
20, CaXV				
	2431.52	−87.31	2501.92	1809.63

TABLE 5. Landé g-Factors for the $2p3s$, $2p3p$ and $2p3d$ Excited Configurations of the N = 6 Sequence. Intermediate Coupling

	$L-S$	$Z=6$	7	8	9	10	11	12	13	14	15	16	17	18	19	20
$2p3s\ ^3P_1$	1.5	1.500	1.500	1.499	1.498	1.497	1.494	1.491	1.485	1.479	1.471	1.462	1.452	1.442	1.433	1.423
$2p3s\ ^1P_1$	1.0	1.000	1.000	1.001	1.002	1.003	1.006	1.010	1.015	1.021	1.029	1.038	1.048	1.058	1.067	1.077
$2p3p\ ^3D_2$	1.167	1.167	1.167	1.167	1.167	1.167	1.167	1.167	1.167	1.168	1.168	1.168	1.169	1.170	1.171	1.171
$2p3p\ ^3P_2$	1.5	1.500	1.500	1.500	1.500	1.499	1.498	1.497	1.494	1.491	1.486	1.480	1.471	1.460	1.446	1.431
$2p3p\ ^1D_2$	1.0	1.000	1.000	1.000	1.000	1.001	1.002	1.003	1.005	1.008	1.012	1.019	1.027	1.037	1.050	1.065
$2p3p\ ^3D_1$	0.5	0.501	0.501	0.503	0.506	0.513	0.527	0.553	0.593	0.648	0.711	0.776	0.762	0.739	0.721	0.708
$2p3p\ ^3P_1$	1.5	1.500	1.500	1.500	1.501	1.501	1.502	1.504	1.506	1.510	1.515	1.523	1.534	1.547	1.564	1.585
$2p3p\ ^3S_1$	2.0	2.000	2.000	1.999	1.998	1.996	1.993	1.987	1.978	1.962	1.940	1.907	1.861	1.802	1.731	1.652
$2p3p\ ^1P_1$	1.0	1.000	0.999	0.998	0.996	0.990	0.978	0.957	0.923	0.880	0.834	0.794	0.843	0.912	0.983	1.054
$2p3d\ ^3F_3$	1.083	1.100	1.085	1.084	1.084	1.084	1.085	1.085	1.086	1.087	1.088	1.090	1.092	1.094	1.097	1.100
$2p3d\ ^3D_3$	1.333	1.316	1.331	1.333	1.333	1.332	1.332	1.331	1.330	1.328	1.326	1.323	1.320	1.316	1.312	1.307
$2p3d\ ^1F_3$	1.0	1.001	1.000	1.000	1.000	1.000	1.001	1.001	1.001	1.002	1.002	1.003	1.004	1.006	1.008	1.010
$2p3d\ ^3F_2$	0.667	0.700	0.672	0.675	0.689	0.727	0.789	0.829	0.796	0.777	0.766	0.760	0.756	0.753	0.752	0.752
$2p3d\ ^3D_2$	1.167	1.137	1.165	1.168	1.170	1.175	1.183	1.196	1.210	1.224	1.235	1.239	1.234	1.221	1.200	1.174
$2p3d\ ^3P_2$	1.5	1.498	1.499	1.498	1.495	1.489	1.480	1.465	1.447	1.427	1.409	1.392	1.379	1.367	1.358	1.351
$2p3d\ ^1D_2$	1.0	0.999	0.998	0.994	0.979	0.942	0.882	0.844	0.880	0.905	0.924	0.943	0.965	0.992	1.023	1.057
$2p3d\ ^3D_1$	0.5	0.503	0.502	0.503	0.507	0.514	0.525	0.541	0.560	0.582	0.605	0.628	0.649	0.668	0.685	0.700
$2p3d\ ^3P_1$	1.5	1.496	1.498	1.497	1.483	1.487	1.476	1.460	1.441	1.420	1.398	1.376	1.356	1.339	1.323	1.309
$2p3d\ ^1P_1$	1.0	1.002	1.000	1.000	1.000	1.000	1.000	0.999	0.999	0.998	0.997	0.996	0.995	0.994	0.992	0.990

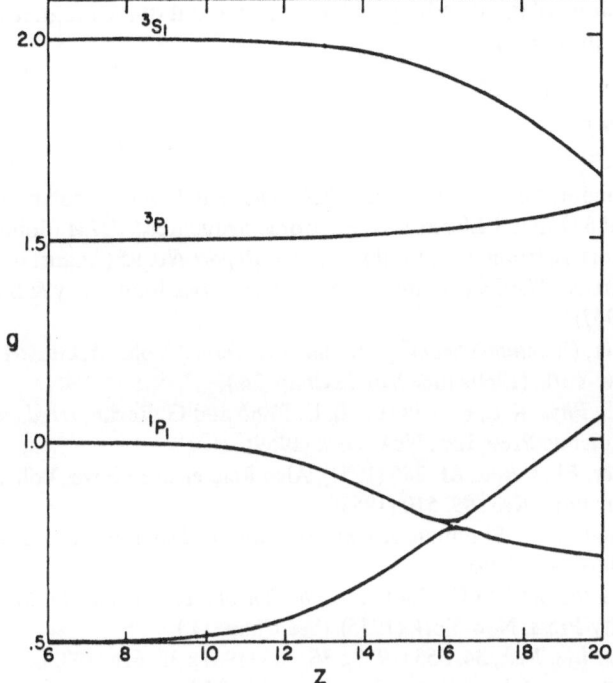

FIGURE 4. Landé g-factors of the $J = 1$ levels of $1s^2 2s^2 2p 3p$.

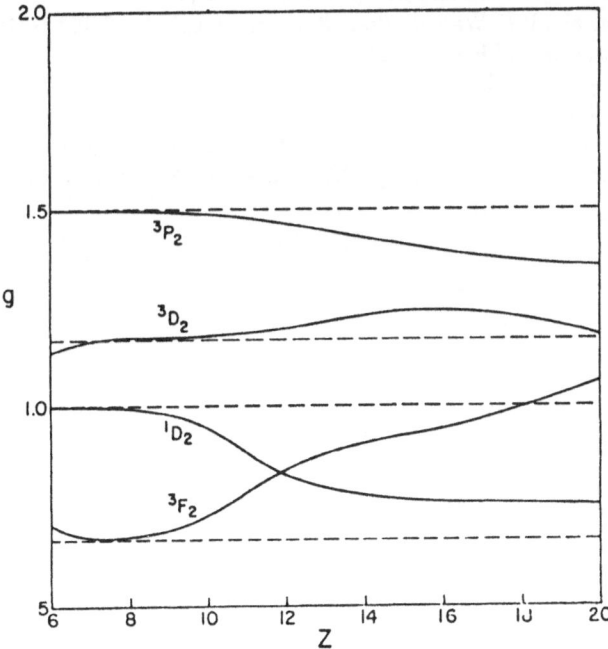

FIGURE 5. Landé g-factors of the $J = 2$ levels of $1s^2 2s^2 2p 3d$.

to the vanishing of the nondiagonal matrix element of the magnetic interaction connecting these two levels.

References

1. E.U. Condon and H. Odabasi, "Self-consistent Field Calculations for Energy Levels of 6, 7 and 8 Electron Isoelectronic Sequences." *JILA* (Joint Institute for Laboratory Astrophysics, Boulder, Colo.) *Report No. 95* (August 6, 1968).
2. D.R. Hartree, *The Calculation of Atomic Structures*. John Wiley & Sons, Inc., New York (1957).
3. J.C. Slater, *Quantum Theory of Atomic Structure*, 2 Vols. McGraw-Hill Book Co. Inc., New York. (1960). (See Vol. 2, Chap. 24.)
4. G. Racah, *Phys. Rev.*, **62**, 438 (1942); U. Fano and G. Racah, *Irreducible Tensorial Sets*, Academic Press Inc., New York (1959).
5. J.C. Slater, *Phys. Rev.*, **81**, 385 (1951). Also Reference 3 above, Vol. 2, Chap. 17.
6. R. Latter, *Phys. Rev.*, **99**, 510 (1955).
7. F. Herman and S. Skillman, *Atomic Structure Calculations*, Prentice-Hall Englewood Cliffs, N.J. (1963).
8. E.U. Condon and G.H. Shortley, *The Theory of Atomic Spectra*, Cambridge University Press, New York (1935). (See Chap. 11.)
9. G. Breit, *Phys. Rev.*, **34**, 553 (1929); **36**, 383 (1931); **39**, 616 (1932).
10. H.A. Bethe and E.E. Salpeter, *Encyclopedia of Physics*, Vol. 35, p. 267. Springer-Verlag, Berlin (1957).
11. H.H. Marvin, *Phys. Rev.*, **71**, 102 (1947).
12. H. Horie, *Progr. Theoret. Phys.* (*Kyoto*), **10**, 296 (1953).
13. M. Blume and R.E. Watson, *Proc. Roy. Soc.* (*London*), **A270**, 127 (1962).
14. C. Froese, *Can. J. Phys.*, **45**, 1501 (1967).
15. E.U. Condon and H. Odabasi, *JILA Report* No. 61, also pp. 185–201 in *Quantum Theory of Atoms, Molecules and the Solid State. A Tribute to John C. Slater*, Per-Olov Löwdin, ed., Academic Press Inc., New York (1966).
16. C. Moore, *Atomic Energy Levels*. Circular 467 of National Bureau of Standards.